S0-AUG-129

DEFORMATION AND FRACTURE OF HIGH POLYMERS

BATTELLE INSTITUTE
MATERIALS SCIENCE COLLOQUIA

Published by Plenum Press

1972: Interatomic Potentials and Simulation of Lattice Defects
Edited by Pierre C. Gehlen, Joe R. Beeler, Jr., and Robert I. Jaffee

1973: Deformation and Fracture of High Polymers
Edited by H. Henning Kausch, John A. Hassell, and Robert I. Jaffee

In preparation:

Defects and Transport in Oxides
Edited by M. S. Seltzer and Robert I. Jaffee

DEFORMATION AND FRACTURE OF HIGH POLYMERS

Edited by

H. Henning Kausch
Materials Sciences Department
Battelle-Institut e. V., Frankfurt/Main, Germany

John A. Hassell
Organic and Polymeric Materials Department
Battelle Memorial Institute, Columbus Laboratories

Robert I. Jaffee
Department of Physics and Metallurgy
Battelle Memorial Institute, Columbus Laboratories

**BATTELLE INSTITUTE
MATERIALS SCIENCE COLLOQUIA**

Kronberg, Germany
September 11–16, 1972

Robert I. Jaffee, Chairman

PLENUM PRESS • New York – London

Library of Congress Cataloging in Publication Data

Battelle Institute Materials Science Colloquia, 7th, Kronberg im Taunus, 1972.
Deformation and fracture of high polymers.

Includes bibliographical references.
1. Polymers and polymerization—Fracture—Congresses. 2. Deformations (Mechanics)—Congresses. I. Kausch, H. H., ed. II. Hassell, John A., ed. III. Jaffee, Robert Isaac, 1917- ed. IV. Title.
TA455.P58B37 1972 620.1′923′23 73-19857
ISBN 0-306-30772-3

A Division of Plenum Publishing Corporation
227 West 17th Street, New York, N.Y. 10011

United Kingdom edition published by Plenum Press, London
A Division of Plenum Publishing Company, Ltd.
Davis House (4th Floor), 8 Scrubs Lane, Harlesden, London, NW10 6SE, England

Printed in the United States of America

To DR. HERMAN F. MARK

whose contributions to high polymer science and technology span the entire time period from the early days of pioneering research to the contemporary scene described in these proceedings

PARTICIPANTS

H. AHLBORN *Battelle-Institut e.V., Frankfurt/Main, Germany*
T. ALFREY, JR. *The Dow Chemical Company, Midland, Michigan, U.S.A.*
E. H. ANDREWS *Queen Mary College, London, England*
K. H. ANTHONY *Universität Stuttgart, Germany*
G. I. BARENBLATT *Moscow University, Moscow, U.S.S.R.*
J. BECHT *Battelle-Institut e.V., Frankfurt/Main, Germany*
G. W. BECKER *Bundesanstalt für Materialprüfung, Berlin, Germany*
H. F. BRINSON *Virginia Polytechnic Institute and State University, Blacksburg, Virginia, U.S.A.*
J. D. FERRY *The University of Wisconsin, Madison, Wisconsin, U.S.A.*
H. GLEITER *Ruhr-Universität, Bochum, Germany*
J. C. HALPIN *Air Force Materials Laboratory, Wright-Patterson Air Force Base, Ohio, U.S.A.*
K. HANSEN *Farbenfabriken Bayer AG, Leverkusen, Germany*
J. A. HASSELL *Battelle-Columbus, Columbus, Ohio, U.S.A.*
L. E. HULBERT *Battelle-Columbus, Columbus, Ohio, U.S.A.*
D. HULL *University of Liverpool, Liverpool, England*
R. I. JAFFEE *Battelle-Columbus, Columbus, Ohio, U.S.A.*
M. F. KANNINEN *Battelle-Columbus, Columbus, Ohio, U.S.A.*
H. H. KAUSCH *Battelle-Institut e.V., Frankfurt/Main, Germany*
W. KNAPPE *Deutsches Kunststoff-Institut, Darmstadt, Germany*
W. G. KNAUSS *California Institute of Technology, Pasadena, California, U.S.A.*

A. S. KOBAYASHI *University of Washington, Seattle, Washington, U.S.A.*
R. F. LANDEL *Jet Propulsion Laboratory, Pasadena, California, U.S.A.*
E. H. LEE *Stanford University, Stanford, California, U.S.A.*
J.C.M. LI *The University of Rochester, Rochester, New York, U.S.A.*
H. MARK *Polytechnic Institute of Brooklyn, Brooklyn, New York, U.S.A.*
G. MENGES *Institut für Kunststoffverarbeitung, Aachen, Germany*
F. H. MÜLLER *Institut für Polymere der Universität Marburg, Marburg, Germany*
T. O'NEILL *Battelle-Geneva, Geneva, Switzerland*
W. PECHHOLD *Universität Ulm, Ulm, Germany*
S. V. RADCLIFFE *Case Western Reserve University, Cleveland, Ohio, U.S.A.*
R. S. RIVLIN *Lehigh University, Bethlehem, Pennsylvania, U.S.A.*
A. R. ROSENFIELD *Battelle-Columbus, Columbus, Ohio, U.S.A.*
E. RYBICKI *Battelle-Columbus, Columbus, Ohio, U.S.A.*
F. R. SCHWARZL *Centraal Laboratorium TNO, Delft, The Netherlands*
M. SHEN *University of California, Berkeley, California, U.S.A.*
M. TAKAYANAGI *Kyushu University, Fukuoka, Japan*
A. G. THOMAS *The Natural Rubber Producers' Research Association, Welwyn Garden City, Hertfordshire, England*
N. W. TSCHOEGL *California Institute of Technology, Pasadena, California, U.S.A.*
P. I. VINCENT *Imperial Chemical Industries Limited, Welwyn Garden City, Hertfordshire, England*
J. G. WILLIAMS *Imperial College of Science and Technology, London, England*
M. L. WILLIAMS *The University of Utah, Salt Lake City, Utah, U.S.A.*
L. ZAPAS *National Bureau of Standards, Washington, D.C., U.S.A.*

PREFACE

DEFORMATION AND FRACTURE OF HIGH POLYMERS is the proceedings of the Seventh Battelle Colloquium in the Materials Sciences. Like its predecessors, it attempts to snapshot an important subfield of materials at a critical point in its evolution, and thereby to make a significant impact, even to change its direction into more fruitful areas. The means for accomplishing this is a gathering together of the leaders in the field in a colloquium format, whereby all participants at some time play a leadership role in the proceedings.

The field of high polymers has had a long and distinguished development essentially from a single point of view – the molecular viewpoint of the polymer chemist, and later the physical chemist. In recent years, the field of high polymers has attracted other disciplines: first, continuum mechanics, later solid state physics and physical metallurgy. Each of the other disciplines has attempted to apply its particular approach to the mechanical behavior of high polymers. Traditionally, the physicists and metallurgists deal with deformation and fracture of crystalline solids where deformation is accomplished by the movement of crystalline defects under the influence of mechanical state variables. The macroscopic viewpoint of continuum mechanics develops constitutive equations without concern for internal structure. One of the major objectives of the Battelle Colloquium was to see whether an accommodation of the three points of view might be accomplished by the outstanding scientists who assembled at the Schlosshotel in Kronberg, Germany, in September, 1972. How well the conciliation of viewpoints was accomplished will be left to the readers of these proceedings to judge. Attention is directed to the five Agenda Discussions. In these structured discussions over 2 to 3-hour periods, the different viewpoints had great opportunity for interaction. Discussions of the individual research papers after the morning sessions also provided opportunities for incisive interaction, and these as well are commended to the reader.

As in previous years, the Colloquium started with a first day of introductory lectures by major contributors, each reflecting a particular point of view. The opening lecture by Herman Mark gave what may be considered the intuitive approach of the traditional school of polymer chemistry, a viewpoint reflecting close interaction with the industrial application of high polymers. Introductory topics based on continuum approaches were given by F. R. Schwarzl and R. S. Rivlin, these being applied to linear and nonlinear viscoelastic solids, respectively. G. I. Barenblatt unfortunately could not attend, but submitted a contribution which applies the mathematical theory of combustion to the mechanical behavior of polymers. Lastly, Turner Alfrey presented a broad discussion of the entire field of mechanical behavior, emphasizing the important role of processing.

The subdivision of the overall field into sessions on phenomenology, molecular descriptions, continuum descriptions, and fracture occasioned some difficulties, because of considerable overlapping among sessions. For example, crazing of polymers cropped up in all four subdivisions of the general field. However, despite the overlap, each session had a topical center of gravity, essentially dictated by the research papers, which dominated the discussion. The Concluding Agenda Discussion led by J. C. Halpin, was initiated by an extensive analysis of the field, which set the stage for a useful discussion of how and where future development of high polymers might proceed profitably.

The Kronberg Colloquium provided an opportunity to honor Dr. Herman Mark, who has contributed importantly to the field of high polymers continuously over a 50-year period. Dr. Mark's autobiographical remarks presented in the next section provide much new historic matter on the development of the field, and the role played by him and his various colleagues.

We wish also to acknowledge our debt to our colleagues at Battelle who have contributed to the Battelle Colloquium. We wish to thank Dr. Sherwood L. Fawcett, President of Battelle Memorial Institute, and Dr. F. J. Milford, Director of Research in the Physical Sciences, under whose support these colloquia are a continuing activity. We wish to thank Dr.

Max Barnick, Director of Battelle-Frankfurt, for superb support provided by members of his staff: Mr. Albert Schwarz who was in charge of local arrangements; the secretariat, consisting of Christel Kohlmann, Karin van Els, and Heike Törber, who provided excellent secretarial assistance; Walter Gunkel and Horst Fischer who efficiently provided sound and lighting; and Dr. Dieter Langbein of the Battelle-Frankfurt Advanced Study Center, who was responsible for slide projection and photography, and who contributed to scientific discussions as well. Lastly we wish to thank Mrs. Robert Jaffee, in charge of the Ladies Program.

The Organizing Committee is grateful to T. Alfrey, E. H. Andrews, J. C. Halpin, R. F. Landel, G. Menges, F. H. Müller, S. V. Radcliffe, and M. L. Williams for their advice during the organization of the technical program.

We trust that this book dealing with the mechanical behavior of high polymers from several viewpoints will be useful to the practitioners in this field and to new workers as well, and will have a beneficial impact on the future course of developments.

The Organizing Committee
R. I. Jaffee, Chairman
H. Ahlborn
J. A. Hassell
M. F. Kanninen
H. H. Kausch
A. R. Rosenfield

AUTOBIOGRAPHICAL REMARKS OF H. F. MARK— THE EARLY DAYS OF POLYMER SCIENCE

H. F. Mark

1 INTRODUCTION

The expression "structure of a molecule" did not become significant and meaningful before the second half of the last century. Up to that date, organic chemists were satisfied to establish for a new molecule which they had synthesized, the chemical *composition* in terms of a stoichiometric *formula,* and to describe its properties – color, specific gravity, refractive index, melting point, boiling point, etc. – as completely as possible.

When, about one hundred years ago, the establishment of a *structure* became an important part of a chemical publication, it was particularly Kekule who became the protagonist of the new approach when – in the 1850's – he had the vision of carbon atoms being bonded together and forming chains to which other atoms such as hydrogen, oxygen, or nitrogen could be attached. Later, Kekule added to the concept of an open chain that of a closed ring, and explained in a global way the essential differences between aliphatic and aromatic chemistry. All formulas of these days referred to the structure and the behavior of ordinary, *small* molecules, but when Kekule in 1877 became Rektor of the University in Bonn and, as usual, delivered an inaugural address of general character and wider scope, he advanced the hypothesis that the natural organic substances which are most directly connected with life – proteins, starch, cellulose – may consist of very *long* chains and derive their special properties from this peculiar structure.

The change in emphasis from *composition* to *structure* led to the demand that any chemist would have to present in his publication the *structural formula* of the material which he was investigating. Since publications are written and printed on paper, it was unavoidable that the two-dimensional character of this commodity created the inclination for simplifications and distortions of what, in the well known sense of Van't Hoff and Le Bel, should have been three-dimensional systems. One of the greatest promoters of structural organic chemistry around the turn of the century was Emil Fischer who ingeniously used two-dimensional formulas to express without any inconsistencies the most complicated three-dimensional structures in the chemistry of sugars and amino acids. As early as 1893 he had already, in general terms, the structure of cellulose as a polysaccharide in mind and expressed the opinion that it might be represented as a chain of glucose units[1], and his later systematic work on polypeptides clearly indicated a long chain structure for natural proteins[2]. For two decades, from the turn of the century to his death (July, 1919), Emil Fischer's Institute in Berlin was the undisputed center of work on natural high polymers and their precursors.

Many of the distinguished scientists who, later, figured prominently in the developments of the 1920's were connected with his school for a shorter or longer period: Abderhalden, Bergmann, Delbrueck, Freudenberg, Gabriel, Harries, Leuchs, and Zemplen (compare also reference 2).

In order to get a correct impression of the general ideas prevailing in the field of such important natural substances as cellulose, starch, proteins, and rubber, early in this century, let us treat each of them separately.

2 POLYSACCHARIDES

French chemists, Braconnot, Payen, Fremy, and Pelouze[3], studied the composition of various vegetable cell walls and arrived at the conclusion that a certain substance, called "cellulose", was their most important component. Analysis showed that cellulose is a carbohydrate, isomeric with starch, and could be degraded hydrolytically into simple sugars. From these early observations it took a long time until H. Ost demonstrated in 1910 that many celluloses can be converted almost quantitatively into d-glucose by acid hydrolysis.[4] In his *Handbuch der Kohlendydrate,* Leipzig, J. A. Barth 1914, B. Tollens presented the concept that cellulose may be a long chain consisting of glucose units; no perimental evidence was offered for this idea.

From 1899 to 1920, several structural proposals were offered for the "formula" of cellulose as shown in Fig. 1; they all are based only on the fact that cellulose consists essentially of glucose but did not have the benefit of any additional and more precise evidence. Some of them favor the chain concept, others that of small cyclic building units.[4] Under these conditions it was clear that the separation of cellulose and other oligosaccharides as intermediate degradation products would be of great importance for the experimental support of any structural concept of cellulose, and considerable efforts were spent on the accumulation of new and reliable experimental data on this point.[5]

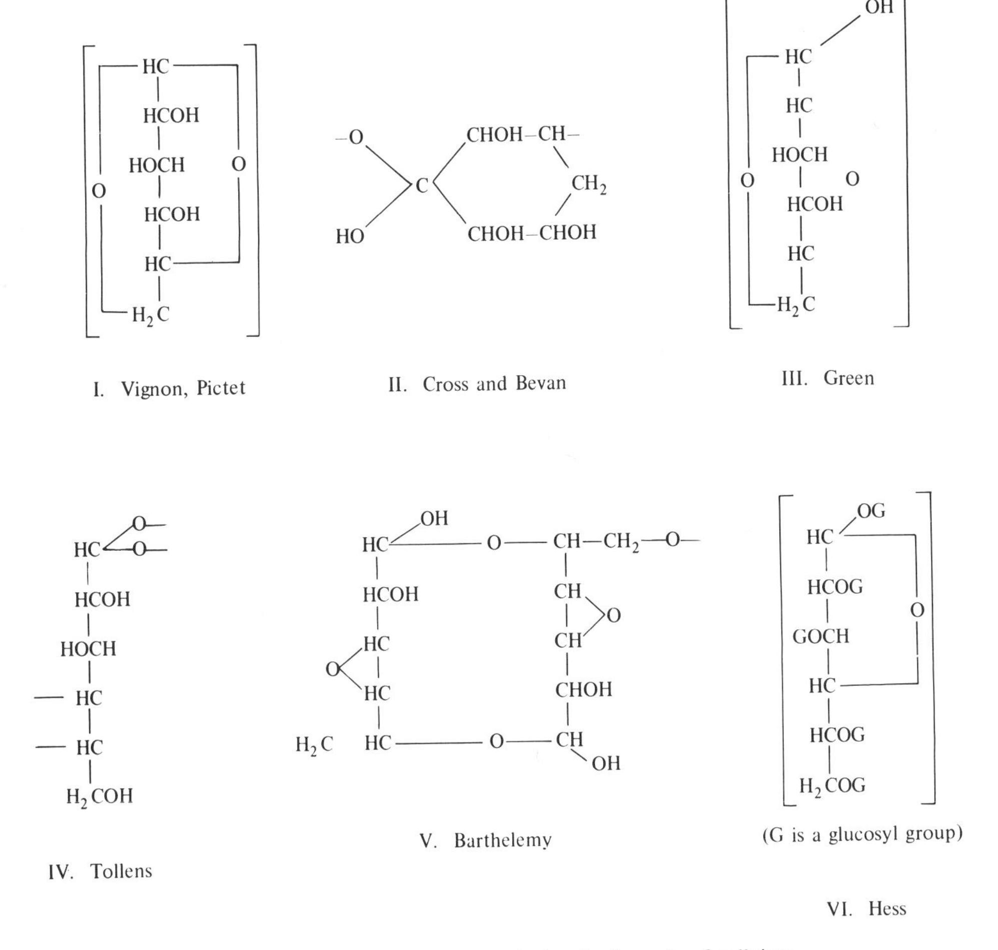

Fig. 1. Several structural proposals for the formula of cellulose

In 1920 and 1921 three important papers appeared which postulated long chain structure for several synthetic and natural compounds on the basis of general considerations and offered specifically for cellulose the long chain character as a preferred alternative in comparison with other structures. The first of these papers was published by Staudinger[6] and proposed for polystyrene, polyoxymethylene, and rubber, formulas which represented linear long chains such as

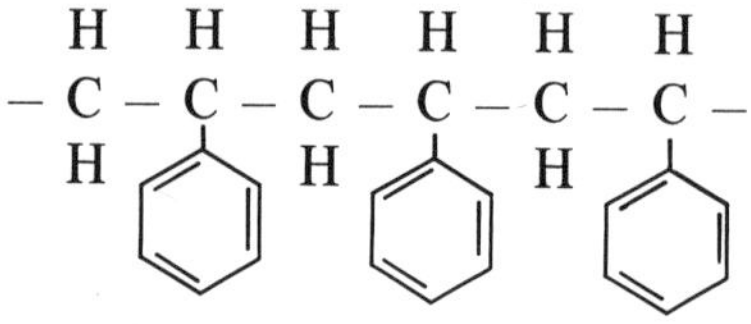

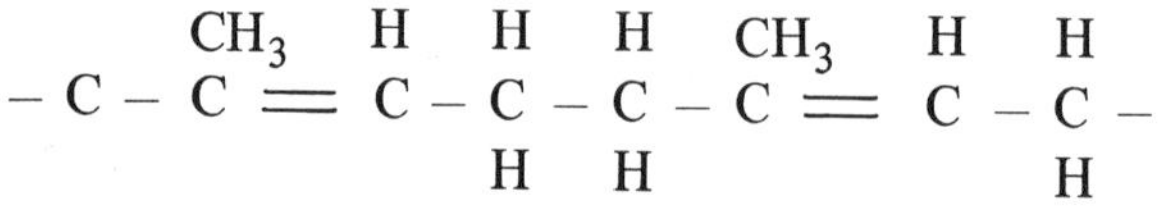

$$\begin{array}{ccccccccccccc}
 & H & & H & & H & & H & & H & & H & \\
- & C & - & C & - & C & - & C & - & C & - & C & - \\
 & H & & \bigcirc & & H & & \bigcirc & & H & & \bigcirc &
\end{array} \qquad \text{polystyrene}$$

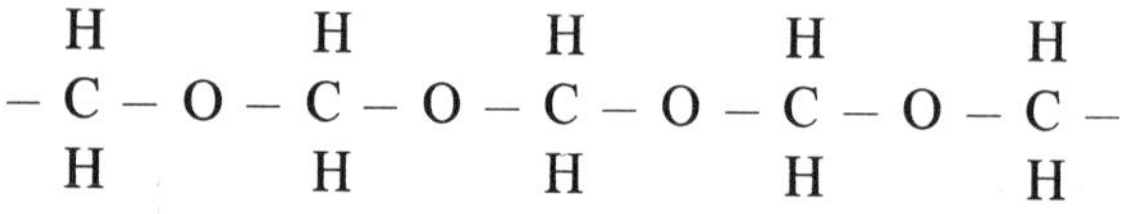

$$\begin{array}{ccccccccccccccccccc}
 & H & & & & H & & & & H & & & & H & & & & H & \\
- & C & - & O & - & C & - & O & - & C & - & O & - & C & - & O & - & C & - \\
 & H & & & & H & & & & H & & & & H & & & & H &
\end{array} \qquad \text{polyoxymethylene}$$

$$\begin{array}{ccccccccccccccccc}
 & & & CH_3 & & H & & H & & H & & CH_3 & & H & & H & \\
- & C & - & C & = & C & - & C & - & C & - & C & = & C & - & C & - \\
 & & & & & & & H & & H & & & & & & H &
\end{array} \qquad \text{rubber}$$

They are actually still accepted today.

The second paper, published by Freudenberg[7], offered new experimental data on the yield of cellobiose during cellulose degradation and stated that the best available data are in conformity with a long chain structure and, certainly, do not contradict this concept. If Freudenberg, at that occasion, would have taken a more aggressive attitude and would have said that his data *proved* the chain structure of cellulose, much confusion would have been spared during the following years; but Freudenberg was much too conservative to go with his statements beyond what he could actually prove. As a consequence, this early paper did not play a very important role in the scramble for a unified and convincing formulation of macromolecular systems.

The third article[8] refers to a lecture which M. Polanyi gave on March 7, 1921, commenting on a paper by Herzog and Jancke[9] which presented X-ray data on various cellulosic samples. Polanyi came to the conclusion that the measured X-ray diffraction spots were in agreement *either* with *long glucosidic chains* or with rings *consisting of two glucose anhydride units.* It was made quite clear at this occasion that on the basis of X-ray data *alone*, one could not distinguish between these two possibilities. The cautious and guarded language of this article gave rise later to the false statement that the small basic unit of the lattice of crystalline cellulose was a proof for a low molecular weight of this material. Although additional attempts were made to correct this wrong position[10], one finds even now the erroneous opinion expressed that the molecular size of a compound cannot be larger than its crystallographic unit cell.

A few years later, Haworth, Hirst, and Irvine[11] demonstrated by a brilliant analytical technique that in cellulose the hydroxyl groups 2, 3, and 6 are still free and, consequently, that the bonding between the individual glucose units must be through the carbon atoms 1 and 4. Based on this new important experimental evidence, Sponsler and Dore[12] in 1926 made another significant step in the interpretation of the existing X-ray data by correlating the Haworth glucose ring with the Bragg atomic radii and arrived at a more detailed structure for the cellulose chain than Polanyi. Unfortunately, they did not take into account the irrevocable chemical evidence of a 1-4 glucosidic bond in cellobiose and arrived, therefore, at a wrong bonding principle along the length of the chain molecules.

If Polanyi, in 1921, would have known of Staudinger's article of 1920 in which chain structures are postulated for several natural and synthetic materials and Freudenberg's article in 1921, he would have referred to them and, probably, placed more emphasis on the chain structure of cellulose, which he offered only as one of two possibilities compatible with the X-ray data alone, with no additional information from chemical sources.

If Sponsler and Dore in 1926 would have seen Polanyi's article of 1921, in which he ruled out a nonpolar sequence of glucose units in the chains, they would not have proposed the incorrect alternating 1-1 and 4-4 glucosidic bonds along the cellulose molecules but would have preferred the correct continuous 1-4 enchainment. Even Staudinger in his book[13] makes no mention of Polanyi's article eleven years after its appearance, although it was the first correct qualitative anticipation of the ultimately accepted macromolecular chain structure of cellulose. These and many similar examples show that at that time there existed only insufficient and slow communications between chemists, even if they worked on the same substances.

K. Hess, for example, in his well-known book on the *Chemistry of Cellulose*[14] presents a detailed account of practically all formulations up to 1928 without arriving at a completely conclusive decision between those proposals which prefer the long chain concept and those which believe that unusually strong association forces between relatively small units are responsible for the "colloidal" character of cellulose and its derivatives.

On the basis of all previously available data, with considerable additional evidence from X-ray diagrams of various cellulose derivatives and from the optical activity of cellulose and its degradation products, Meyer and Mark[15] in 1928 accumulated convincing material for the presently accepted chain structure of cellulose and for the crystalline amorphous character of cellulosic fibers. After the acceptance of this first, approximate draft, many important details had to be settled, such as the exact position of each individual atom in the unit cell, the location and direction of the intermolecular hydrogen bonds, and the relationship between the "crystalline" and "amorphous" domains. Meyer and Mark in 1928[15] spoke of an amorphous "bark substance", Gerngross, Hermann, and Abitz in 1930[16] postulated that gelatin has a "fringe micellar" structure, and more recently Manley[17] has applied the chain folding principle of Keller[18] to arrive at new concepts concerning the true supermolecular structure of cellulose.

All more recent modifications, refinements, and improvements concerning the structure of native cellulose revolve, in essence, around the basic formulas of Freudenberg, Polanyi, Meyer and Mark, but are a classical demonstration for the fact that any new experimental evidence necessitates amendments and correction to earlier formulations.

3 POLYHYDROCARBONS

In 1900, E. Bamberger and F. Tschirner[19] reacted diazomethane with beta-aryl-hydroxylamines in order to methylate the hydroxyl-groups of the substituted hydroxylamine. Instead they obtained a product in which two phenyl hydroxylamines were connected by a methylene bridge and concluded that diazomethane

$$CH_2{=}N{=}N \quad \text{or} \quad \begin{array}{c} CH_2 \\ / \ \backslash \\ N{=}N \end{array},$$

dissociates into N_2 and CH_2 which in turn can either react with the amine or *polymerize* to form *polymethylene.* This material $(CH_2)_n$ was found to be a white, chalklike, fluffy powder, apparently amorphous with a melting point of 128 C.

This is the first correct formulation and description of a polyhydrocarbon – linear polymethylene, which, in its structure and properties, is identical with the linear 1,2 polyethylene. However, at that time, because of lacking interest in "amorphous by-products" of organic chemical syntheses and, evidently, also in view of the difficult availability of diazomethane, the discovery of polymethylene made no impression on the chemists of 1900.

A different attitude, however, was taken at that time toward another polyhydrocarbon (native Hevea rubber), which was a natural product of considerable interest and importance as an essential material in the electrical industry, which was rapidly developing around the end of the last century with special demands for the insulation of cables, wires, and electrical connections. Already in 1886 Tilden had observed the formation of rubberlike products from isoprene, and in 1900 C. O. Weber characterized rubber as a "polymeric isoprene"[20]. At the same time, observations by Ipatiev, Kondratiev, and Euler[21] reinforced the conviction that the rubber molecule consisted essentially of isoprene units.

This left as the remaining important problem the best proposal for the mutual bonding of these units. The solubility characteristics of rubber in aromatic solvents, the high viscosity of the resulting solutions, their adhesivity, and the general soft-elastic properties of rubber resembled the behavior of many other colloids. Since the rubber molecules still contain double bonds, Harries[22] felt compelled to explain their strong association with the aid of Thiele's concept of partial valences.

This theory, an ingenious anticipation of pi-electrons and resonance, was postulated in 1899 in order to explain the unusual reactivity of the aliphatic carbon-carbon double bond.[23] Harries adopted it and assumed that in the rubber molecule the individual isoprene units are combined with each other in the following manner:

```
      CH3                                   CH3
      |                                     |
  ____C  - CH2 - CH2 - CH ____              C  - CH2 - CH2 - CH ___
<----||                 ||    ____          ||                 ||___ --->
  ____CH - CH2 - CH2 - C                    CH - CH2 - CH2 - C
                       |                                     |
                       CH3                                   CH3
```

This formulation provided for Staudinger the background for one of his crucial experiments to disprove the association theory of rubber and to strengthen his own concept of *macromolecules,* i.e., of long chains in which the individual monomers are connected with each other by *normal covalent bonds* just as the carbon atoms in a paraffin chain. He hydrogenated rubber[24] and obtained a saturated paraffinic hydrocarbon of the composition $(CH_2)_x$. If the double bonds are responsible for the polymeric character, then hydrorubber ought to be a normal paraffinic material with a sharp melting point and with the characteristic solubility of such compounds. But it is not. It still does not crystallize, has no sharp melting point, cannot be distilled without decomposition, and gives solutions with very high viscosity. Staudinger, therefore, had now direct experimental evidence for the validity of the rubber formula which he had postulated intuitively in 1920.

Further progress to the final, much more detailed structural formula of rubber was initiated by the discovery of I. R. Katz[25] that stretched rubber gives a fiberlike X-ray diagram and by the quantitative interpretation of this diagram by Hauser and Mark[26]. In 1928 Meyer and Mark[27] proposed for rubber the "cis" and for balata the "trans" structure and correlated all chemical and crystallographic evidence to present a detailed picture of the rubber and balata chains in the crystallized domains of oriented samples of these materials.

As in the case of cellulose, more recent studies have added refinements and improvements to this picture but the essential features are not changed and the macromolecular character of rubber remained established.

4 POLYESTERS AND POLYPEPTIDES

As early as 1881, Michler and Zimmermann reacted phosgen with meta-phenylene-diamine in chloroform and obtained an amorphous powder melting above 300 C which they considered to be a cyclic urea and for which they found the correct elemental analysis. In 1898 Einhorn reacted phosgene with hydroquinone by a two-phase process in toluene and water (interfacial polymerization) and proposed structures such as

$$\left[C_6H_4O_2 \cdot CO\right]_x \quad \text{and} \quad \left[-O-C_6H_4-O-\overset{\displaystyle O}{\overset{\|}{C}}-\right]_x$$

which evidently indicated the polymeric character of the resulting amorphous, high-melting powders.

During many years of intense and systematic studies on aminoacids, polypeptides, and proteins, Emil Fischer never suggested or postulated anything else for the structure of his synthetic products but the character of a linear chain consisting of many amino acid units which are connected with each other by the normal –CO–NH–linkage as it occurs in all amides and peptides.

In a famous lecture which he gave in 1906 at a Meeting of the Deutsche Naturforscher and Aerzte[2], he declared that, in his opinion, there exists an uninterrupted line between the simplest dimeric and trimeric amino acids and the native proteins and illustrated this conviction by the demonstration of a linear synthetic polypeptide which he had synthesized step by step with a complete record of all intermediates; it had a molecular weight higher than 1000.

Hermann Leuchs, one of Fischer's most distinguished associates, concentrated most of his interests and efforts on the study of heterocyclics and alkaloids, made an even bolder step in the direction of true synthetic polyanhydrides[28]; they decompose at elevated temperature and in the presence of traces of moisture under the development of CO_2, into solid bodies which he considered as "polymers of a cyclic monomer".

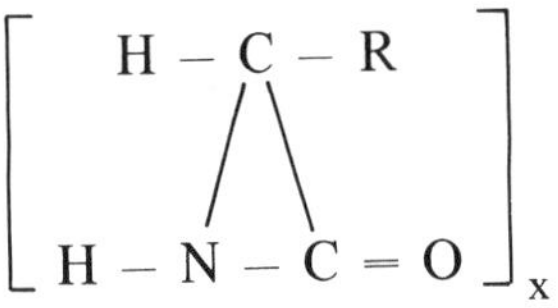

$$\left[\begin{array}{c} \mathrm{H-C-R} \\ / \quad \backslash \\ \mathrm{H-N-C=O} \end{array}\right]_x$$

Leuchs described the synthesis of several representative anhydrides of this class. It is well known that compounds of this type have played and are still playing an important role in the synthesis of numerous linear polypeptides with many important outlooks at the structure and properties of native proteins. It is interesting that, for many years, these substances did not receive much interest because of the general lack of enthusiasm for the polymer field until K. H. Meyer[29] and R. B. Woodward[30] put their existence and significance in the proper perspective.

K. H. Meyer also initiated the first complete X-ray diffraction study of natural silk which was successfully carried out by R. Brill.[31] Later, with the writer of these lines, he postulated for natural silk and for all fibrous proteins a linear chain structure which was capable to explain most chemical and physical properties of these important materials. In this paper[32] one can also find the first experimentally supported proposal for a helicoidal conformation of polypeptide chains.

Adopting Staudinger's general arguments for the existence of long main valence chains and striving for a more quantitative and precise description of their structural details, Meyer and Mark added in 1928 chitin and starch to the list of "truly polymeric" compounds for which, in their opinion at that time, the main valence structure was an established scientific truth.[33]

5 LIVELY OBJECTION AND EVENTUAL CLARIFICATION

While, in this manner, for the most important natural substances (cellulose, rubber, proteins, and starch), the high polymeric or macromolecular character was first postulated, and later more and more reliably established, there were many scientists who were unconvinced and preferred the concept that these substances consist of small building units which, however, are held together by exceptionally strong forces of aggregation or association, which were supposed to be of a new and still unknown character.

The fact that all materials under investigation are the products of *living beings,* plants or animals, was an attractive and probably perfectly legitimate argument in favor of *something new, something which we still have to learn and to clarify* in order to understand the structure and properties of these materials which are so indispensable.

However, as often happens in science and history, this somewhat romantic approach had to fade away gradually under the influence of more and better perimental evidence for the macromolecular theory. This did not happen without contradiction and opposition; on the contrary, your reporter remembers many meetings, symposia, and seminars in the course of which the opposing views were presented and defended with more or less emphasis and success. Strangely enough, even the champions of the long chain or macromolecular aspect – Freudenberg, K. H. Meyer, Staudinger – did not agree with each other, as they easily could have done, because instead of concentrating on the essential principle, they disagreed in specific details and, at certain occasions, they argued with each other more vigorously than with the defenders of the association theory. Of course, at that time, none of them was completely correct in all details of his approach but they all thought and worked in the right direction and, at the end, emerged as the natural leaders for future developments.

There were many factors which, eventually, tipped the scales in favor of the concept of very long, chainlike molecules. One of them was the rapid improvement and refinement of the X-ray diffraction method which, again and again, not only gave answers in favor of long chains but permitted, and still permits, a progressively detailed description of every kink and twist in a macromolecule. As far as proteins are concerned, it is becoming increasingly clear that, in fact, each turn and wiggle represents a significant design and has its far-reaching consequences for the actions of the substance in biochemistry, biology, and medicine.

If I compare the X-ray equipment and the methods of evaluation which Brill, Katz, Polanyi and I used in 1923 – an air-filled X-ray tube, a ruler, and a log table – with the present refined techniques – automatic registration of hundreds of diffraction spots with the aid of a Weissenberg goniometer and direct feed of the output (position and intensity) into a computer, it would be seen that the progress is as large as that from an airplane of the early 1920's to a supersonic jet of the 1970's.

Another important factor was the introduction of Svedberg's ultracentrifuge[34] which played a decisive role because it was the first method which permitted a direct and absolute measurement of the molecular weight in the range between 40,000 and several millions; at the same time, improved osmometers added significance and reliability to these data.[35]

But, probably more than any other single factor did the work of W. H. Carothers and his associates contribute to the ultimate breakthrough in favor of the long chain concept.[36] Their efforts extended from synthesis and characterization to ultimate properties and encompassed with the same emphasis condensation and addition polymers. A careful analysis of all prior art led Carothers early to the conclusion that the macromolecular hypothesis was correct, and all his own experiments strengthened his conviction. For him the controversy – association hypothesis vs. long chain theory – was a matter of the past and he advanced with full scientific and industrial success on the basis of the latter.

Once the basic concepts of the new branch of chemistry were firmly established, the polymer chemists settled to useful and practical work: synthesis of new monomers, quantitative study of the mechanism of polymerization processes in bulk, solution, suspension, and emulsion, characterization of macromolecules in solution on the basis of statistical thermodynamics, and fundamentals of the behavior in the solid state with a resulting basic understanding of the properties of rubbers, plastics, and fibers.

REFERENCES

1. Fischer, E., *Ber. dt. Chem. Ges.*, **26**, 2404 (1893).
2. Hoesch, Kurt, *Emil Fischer, Sein Leben und Sein Werk,* Verlag Chimie, Berlin (1921), p 389 ff; lecture on proteins given on January 6, 1906; also, *Ber. dt. Chem. Ges.*, **54**, special number 1921, 480 pp.
3. Braconnot, H. (1819), Payen, J. (1842), Fremy, E. (1847), Pelouze, T. J. (1859); compare Emil Ott, *Cellulose,* Interscience, New York (1943), pp 29-36; also Heuser, Emil, *Cellulose Chemistry,* Wiley, New York (1944), pp 571 et seq.
4. Compare, e.g., Ost, H., *Chem. Z.*, **34,** 461 (1910); also E. Heuser's book, pp 519 et seq; compare particularly the article of C. B. Purves in Ott's book, p 42.
5. Compare, e.g., Willstätter, R., and Zechmeister, L., *Ber. dt. Chem. Ges.*, **46,** 2403 (1931); Ost, H., *Ann.*, **398**, 323 (1913); Freudenberg, K., *Ber.*, **54**, 771 (1921); Hess, K., and Friese, H., *Ann.*, **456,** 39 (1927); also E. Heuser's book, pp 512-519.
6. Staudinger, H., *Ber. dt. Chem. Ges.*, **53**, 1073 (1920).
7. Freudenberg, K., *Ber. dt. Chem. Ges.*, **54,** 771 (1921).
8. Polanyi, M., *Naturwissenschaften,* **9**, 288, 337 (1921).
9. Herzog, R. O., and Jancke, W., *Ber. dt. Chem. Ges.*, **538**, 2162 (1920).
10. Compare Mark, H., *Ber. dt. Chem. Ges.*, **59**, 2932 (1926).
11. Compare *J. Chem. Soc. London,* **123,** 518, 1352 (1923); also *Nature,* **116,** 430 (1925).
12. Sponsler, O. L., and Dore, W. H., *Colloid Symposium Monograph,* **4** 174 (1926).
13. Staudinger, H., *Die Hochmolekularen Organischen Verbindungen,* Julius Springer, Berlin (1932).
14. Hess, K., *Die Chemie der Zellulose,* Akademische Verlagsgesellschaft m.b.H., Leipzig, 1928.
15. Meyer, K. H., and Mark, H., *Ber. dt. Chem. Ges.*, **61B,** 593 (1928); also the book on

Hochpolymere, Akademishe Verlagsgesellschaft m.b.H., Leipzig (1930).
16. Garngross, O., Hermann, K., and Abitz, W., *Biochemische Zeitschrift,* **228**, 409 (1930).
17. Manley, R. St. J., *J. Polymer Science,* **1A,** 1875 (1963).
18. Keller, A., *J. Polymer Science,* **17**, 291 (1955); compare also particularly Geil, P. H., *Polymer Single Crystals,* Interscience-Wiley, New York (1963).
19. Bamberger, E., and Tschirner, F., *Ber. dt. Chem. Ges.,* **33**, 955 (1900).
20. Weber, C. O., *Ber. dt. Chem. Ges.,* **33**, 784 (1900).
21. Compare Euler, W., *Ber. dt. Chem. Ges.,* **30**, 1989 (1897).
22. Harries, C., *Ber. dt. Chem. Ges.,* **38**, 3985 (1905); **46**, 2590 (1931); **47**, 573 (1941).
23. Thiele, J., *Ann.,* **306,** 92 (1899).
24. Compare Staudinger, H., and Fritsch, J., *Helv. Chim. Acta,* **5**, 785 (1922).
25. Katz, J. R., *Naturwissenschaften*, **13**, 410 (1925).
26. Hauser, E. A., and Mark, H., *Kolloid Chem. Beih.,* **22**, 63 (1926).
27. Meyer, K. H., and Mark, H., *Ber. dt. Chem. Ges.,* **61**, 1939 (1928).
28. Leuchs, H., et al., *Ber. dt. Chem. Ges.,* **39**, 857 (1906); **40**, 3235 (1907); **41**, 1721 (1908).
29. Meyer, K. H., and Go, Y., *Helv. Chim. Acta,* **17**, 1488 (1934).
30. Woodward, R. B., *JACS,* **69**, 1551 (1940); also particularly Bamford, C. H., et al., *Synthetic Polypeptides,* Academic Press, New York (1957).
31. Brill, R., *Ann.,* **434,** 204 (1923).
32. Meyer, K. H., and Mark, H., *Ber. dt. Chem. Ges.,* **61**, 1932 (1928).
33. Meyer, K. H., and Mark, H., *Der Aufbau der Hochmolekularen Organischer. Substanzen,* Akademische Verlagsgesellschaft m.b.H. Leipzig (1930).
34. Compare particularly Svedberg, Th., *The Ultracentrifuge,* Oxford University Press, London (1940).
35. Herzog, R. O., and Spurlin, H. M., *Z. Physik. Chem.,* p 239 (1931); Campen, P. von, *Rec.,* **50,** 915 (1931); Dobry, A., *J. Chim, Phys.,* **32**, 50 (1932); Schulz, G. V., *Z. Physik. Chem.,* **A158,** 237 (1932).
36. Compare *Collected Papers of W. H. Carothers,* Interscience, New York (1940).

CONTENTS

Part Two PHENOMENOLOGY

Part Three MOLECULAR DESCRIPTIONS OF DEFORMATION

Part Four CONTINUUM DESCRIPTIONS OF DEFORMATION

Part Five FRACTURE

Part Six CONCLUDING DISCUSSION: CRITICAL ISSUES

Part One

INTRODUCTORY LECTURES

PROGRESS IN THE SYNTHESIS AND APPLICATION OF POLYMERS

H. F. Mark

Polytechnic Institute of Brooklyn
Brooklyn, N.Y.

1 INTRODUCTION

Before starting to deliver my brief report on the topic of the title, I am anxious to express my sincerest thanks and my deep appreciation to the management of the Battelle Institute and particularly to Dr. Jaffee for giving me the pleasure and honor to introduce this 1972 Battelle Colloquium. Since our meeting deals with the deformation and fracture of high polymers, it was felt that a short review of new methods to prepare polymers and of novel ways to apply them might be of interest to the participants.

Progress in the large field of our common interests is being achieved in many ways and I would like, at this occasion, to select a few particularly lively areas, namely,

(a) Improved control of molecular weight and molecular-weight distribution
(b) New ways to control and influence the molecular structure
(c) Recent advances in polymerization techniques.

2 CONTROL OF MOLECULAR WEIGHT AND MOLECULAR-WEIGHT DISTRIBUTION

Very early in the use of natural and synthetic polymers – starch and glue as adhesives, nitrocellulose, cellulose, and cellulose xanthate or acetate as films and fibers – it was empirically recognized that materials, which were "degraded" during processing would give inferior mechanical properties. When, in the early 1920's, the high-molecular-weight (long chain) concept of these materials was scientifically established by Staudinger and other workers in the field, their "molecular weight" became a key property for basic consideration – and for practical applications. Soon, however, it was realized that these "molecular weights" were, in fact, average values resulting from the existence, in all samples, of a plurality of species with different molecular weights and that a more precise description of a given material would have to proceed to the consideration of the *molecular-weight distribution.* This refinement first became necessary in the production of cellulosic fibers which would not only be sufficient for standard textile uses but also for tire cords and high-wet-modulus rayon fibers, so called polynosic fibers. Instead of specifying only one figure for a commercial wood pulp, namely the alpha cellulose, one was obliged to add numerical values for beta and gamma cellulose which, in fact, amounted to an approximative molecular-weight distribution. Since then, many advances in the technology of polymer conversion – fast spinning and drawing, blow molding, injection molding, and others – have placed increasing demands on molecular-weight uniformity. Analytically, gel-permeation chromatography (GPC) provided a very valuable tool to arrive conveniently at a rather detailed picture of the polymolecularity of at least certain polymers, which permitted the ascertainment of the importance of molecular homogeneity for certain new conversion techniques.[1]

One of these techniques is the preparation of strong, stiff, tough, and opaque two-dimensional systems which range from nonwoven carpet and textile goods to paperlike sheets of superior tear strength and complete insensitivity to moisture. Such structures are formed by the random deposition of very thin and strong fibers on a moving belt and by subsequent compression and heat sealing of the resulting web. This process – known as *spun bonding* – has been carried out with many different polymers, and is the basis of a great variety of products – Tyvek (polyethylene), Typar (polypropylene), and Reemay (polyester) of Du Pont and Cerex (polyamide) of Monsanto. On top of the many useful applications of these webs themselves, one experimenter has also succeeded in building up layer structures with a very thin porous polyethylene web in the middle and one cellulosic tissue layer on each side. The resulting sandwich combines the softness and high water absorptivity of a fine tissue paper with the toughness and wet strength of polyethylene (Crowntex of Crown Zellerbach).[2]

This success in combining the highly hydrophilic cellulose with the highly hydrophobic polyethylene led to the question as to whether it is, indeed, necessary to prepare the polyethylene first in its usual bulk form, as a powder or granulate, and then convert it into fibrous form by an additional melt-spinning process. Would it not be possible to manipulate the Ziegler polymerization of ethylene in such a manner that one would get a pulplike, fibrous polyethylene mass directly from the reactor which could be mixed with the cellulosic pulp and converted on a paper machine in a sheet of composite character, so that its properties would be favorably influenced by the presence of the two components? In fact, experiments which were started in Japan a few years ago and which are now continued also in the U.S.A. and Europe indicate that it is possible to obtain low-pressure polyethylene in the form of short and thin fibers directly from an appropriately designed reactor.

Another method to prepare polymers with a very narrow distribution curve was introduced several years ago by M. Szwarc with the aid of the "living polymer" technique. It relies on the absence of a termination step under certain anionic reaction conditions and is very useful in the production of thermoelastic block copolymers and stereoregulated elastomers.[3] The classical form of the former are molecules of three segments (each having a molecular weight around 20,000): one of them being a plastic with a high T_g, e.g., polystyrene with T_g ~100 C, and the other two being elastomers with a low T_g, e.g., cis polybutadiene with T_g ~-80 C. Optionally, such molecules may also consist of one rubbery block and two hard blocks. The properties of such "molecular composites" are known from the behavior of Kraton and Cariflex (Shell). A certain weakness is the relatively low T_g of the polystyrene segment which causes considerable creep under continued tension at temperatures only slightly above room temperature. In a new thermoelastomer – Hytrel of Du Pont – the amorphous polystyrene is replaced by a crystallizable polyester the T_m of which is substantially above the T_g of polystyrene. The system of fix points produced by the randomly distributed polyester crystallites is firmer and has a higher softening point, and provides for less creep during the use of the material even at somewhat elevated temperatures. Efforts are now being made to go a step further and use for the hard segment a polyamide, e.g., 6-nylon.

Another recent attempt to reduce the polymolecularity of polyethylene and polypropylene on a commercial scale involves the use of a *loop reactor* (Phillips Petroleum) in which, at low temperatures and moderate pressures, the polymerization of the monomer in the presence or absence of an inert diluent is carried out continuously in a recycling system. Only a small percentage (less than 10 percent) of the monomer is converted into polymer in each cycle; this remains suspended in the diluent as a slurry and is continuously removed by filtration or centrifugation. In this manner, polymer which has been formed under the influence of the catalyst is continually and rapidly withdrawn from the reacting system and cannot undergo side reactions which broaden the distribution curve, such as chain transfer, branching, or cross linking. Evidently, a continuous-loop-reactor process also avoids any "hot spots" and any "stagnant polymer accumulation", both of which are largely responsible for the broadening of the molecular-weight distribution of the material.[4]

3 CONTROL OF MOLECULAR STRUCTURE

There exist many ways in which to manipulate the internal structure of macromolecules in order to adapt their properties to specific uses; this section attempts to present to you a few new steps in this direction, which may not yet be too well known but promise already to lead to interesting new products. Two new fibers have been put on the market in Japan by the skillful molecular engineering of known polymeric species. One (Chinon of Toyobo) is made by aqueous-solution spinning of a natural protein and by the grafting of acrylonitrile on the fiber as it is converted from the solution over a gel into its solid form. The final product consists of about 50 weight percent of each component and represents a favorable combination of textile properties, having adequate tensile strength (3 to 4 g/den) and sufficient wet strength and abrasion resistance, and combines excellent (silklike) dying characteristics with equally superior hand drape and crease resistance.

The other new fiber (Cordelan of Kojin and Okamura) consists of a system of fine, highly oriented fibrils of PVC embedded in a matrix of cross-linked polyvinyl alcohol (PVA) and firmly connected with this matrix by a small amount of a PVC-PVA graft. The composition and structure of this fiber provide an attractive combination of textile properties, already known for acrylics, as well as superior moisture metabolism and dying characteristics and a flame resistance similar to that of certain modacrylics. The final fiber, which contains about 50 weight percent PVC, is produced by spinning a PVC emulsion in a PVA solution with subsequent stretching and cross-linking of the PVA matrix. The basic ingredients as well as the fiber-forming technology appear to be capable of producing Cordelan at an attractive price.

Another important manipulation of polymer structure is the synthesis of linear macromolecules having a plurality of bonds with a very low rotational energy barrier (commonly called "flexible chains"). If one combines this structural feature of the individual chains with a very low cohesive energy density of the entire system, one obtains elastomers, which remain soft and supple even at temperatures as low as -100 C. At the same time, by using F instead of H as substituents on the chain, one is able to achieve structures having superior oil resistance and oxidative stability at elevated temperatures. Examples of such structures are the fluorinated siloxanes, the fluorinated polyformaldehydes, and the fluorinated nitrosorubbers.

The opposite effect, namely the increase of modulus, tensile strength, and softening characteristics (T_g and T_m) by the introduction of rigid (mainly cyclic) chain elements, has been known since the introduction of terephthalic acid into polyester technology (Whinfield and Dixon). Since then, many (more than 1000) polyesters, polyamides, polycarbonates, polyureides, and polyurethanes have been prepared in which the mechanical and thermal properties were drastically changed by the introduction of cyclic – mostly aromatic – units. One of the earliest commercial representatives of this new class of film and fiber formers was Nomex of Du Pont. It is an aromatic polyamide (aromide) with the diamine and the dicarboxylic acid in the meta form. Nomex is already a very important structural material; in the form of a spunbonded paper it is an indispensable ingredient in the construction of high-efficiency electronic systems.[5]

This success stimulated research on further improvements, particularly on the use of the para forms instead of the meta forms. Elaborate studies, pioneered particularly by Du Pont and Monsanto demonstrated that, in fact, aromatic, all-para chains would have melting points between 500 and 550 C (about 100 C higher than the corresponding meta chains) and would be greatly superior in modulus and tensile strength. Naturally, these combinations would also be more difficult to prepare, spin, and draw, i.e., more complicated solvents as well as higher temperatures and tensions for drawing would be required, and removal of the last traces of solvent would be more difficult. These difficulties have been overcome, however, and in a recent paper, R. E. Wilfong and J. Zimmerman of Du Pont reported for a para-para aromatic polyamide – currently referred to as Fiber B – the following characteristic data:

	Fiber B	PA 66
Tensile Strength, g/den	Up to 25	Up to 12
Elongation to Rupture, %	Around 4.5	Around 10
Modulus at 25 C, g/den	450	40
Modulus at 150 C, g/den	320	10
Melting Point, C	550	260

These figures show, dramatically, the enormous progress achieved by the reduction of chain flexibility in terms of modulus, ultimate strength, and resistance against elevated temperatures. In fact, the modulus of 450 g/den corresponds to about 9 million psi and is in the rigidity (stiffness) range of the best commercial glass fibers.

As the technology of the preparation and processing of aromatic polyamides, polyesters, and more highly condensed systems such as polyimides, polybenzimidazoles, polyoxadiazoles, and ladder polymers progressed, several additional new approaches were initiated.

One of them is adjustment of the orientation and organization of an aromatic polyamide (a kind of follow up on Fiber B), in such a manner that modulus and tensile strength are substantially increased whereas (understandably) the elongation to break is reduced. Representative values for this fiber, known as PRD-49 of Du Pont, are:

Modulus	1000 g/den (20 million psi)
Tensile Strength	25 g/den (500,000 psi)
Elongation to Rupture	1.8-2.2 percent
Specific Gravity	1.4

Evidently, this material is of considerable interest for the reinforcement of various thermosetting systems – polyesters, polyepoxides, etc. – and has several advantages over the classical glass fibers, namely, at least equivalent mechanical behavior combined with lower specific gravity and better bonding of the fiber to the resinous matrix of the test piece. Systematic studies of PRD-49 fiber-reinforced resins carried out in the Du Pont Laboratories and also by independent investigators have, in fact, demonstrated that it represents a very useful and valuable new material for the design of composites which have superior mechanical properties and retain them for short performance periods up to 400 C and over extended periods up to 250 C. It might be speculated that the outstanding properties of these linear aromatic para polyamides are due to a combination of chain length, chain rigidity, dense packing, and interchain hydrogen bonding.[6]

Fibers made of various polyimides, polybenzimidazoles, and polyoxadiazoles in the laboratories of Monsanto, Celanese, ICI, Bayer, and others possess higher T_m and T_g, but do not (at least at present) have comparable modulus and tensile strength. This might be due to a looser packing, less efficient hydrogen bonding, and the much more difficult processibility of these systems, which do not yet permit the production of fibers or films with a sufficiently small number of flaws or which do not drive the molecular orientation of the systems to sufficiently high values. The importance of high molecular weight and perfect orientation of highly laterally ordered thin domains was demonstrated by Blades and White, and by Porter and Wang, who obtained for linear polyethylene (where there certainly exists no hydrogen bonding) moduli of the order of 400 g/den.

Even somewhat higher thermal resistance is displayed by a series of ladder polymers, but, at present, impressively superior mechanical characteristics could not be obtained. However, all highly aromatic fibers – from Nomex to the most complicated ladder polymers – are very interesting and probably promising precursors for the preparation of carbon fibers. It is well known that great efforts have been expended to arrive at carbonized (and eventually graphitized) fibers, starting with highly oriented filaments of cellulose, polyacrylonitrile, and pitch. The objective is to arrive at the optimum attainable *compromise* among modulus, strength, and elongation to rupture. A desirable (and probably realistic) target was set – a modulus of 100 million psi (7 million kp/cm^2), a tensile strength of 500,000 psi (35000 kp/cm^2), and an elongation to rupture of about 2 percent. Such properties combined with the practical infusibility (T_m about 2000 C) and the low specific gravity (usually around 1.0) would make those materials extremely interesting and valuable for the reinforcement of not only resinous matrices but also metals and ceramics.

At present, the individual values set for modulus, tensile strength, and elongation to rupture have actually been reached, but a simultaneous combination of all three target values has not been attained as yet. For instance, if one prepares a carbon fiber with a modulus of about 80 million psi, its tensile strength will be below 250,000 psi and its elongation to rupture is less than 1.0 percent. Practically, this means that this fiber is rigid but brittle.

The failure to obtain the above-mentioned target, together with the very low speed of the gradual transformation of the precursor into the final filament (about 1 ipm), has recently dampened the enthusiasm of expending more efforts on the preparation of such fibers. However, it must be pointed out that other precursors, particularly the highly aromatic polyamides and, even more so, the higher condensed heterocyclic ring systems offer new opportunities for a more rapid and better controlled process of carbonization; their structure is already rather close to that of carbon or graphite and, according to TGA measurements, some of them lose only 10 percent of their weight when they are converted into a black and infusible carbon fiber. This should, in any event, speed up the process of carbonization and may also, hopefully, lead to a better compromise among modulus, strength, and elongation to rupture.

I don't want to close this discussion without drawing your attention to a phenomenon which has been established experimentally but for which, at present, a completely satisfactory explanation can not be given.

It has been found that certain linear polymers, specifically polypivalolactone, polypropylene, and polyformaldehyde can be obtained as fibers which display a long-range reversible elasticity which deviates from the classical rubberlike elasticity in several ways:

(a) The initial modulus is of the same order as that of a normal drawn fiber, namely, somewhere around 30 g/den, i.e., about 100 times larger than that of the standard elastomeric fibers (Spandex type) which show conventional rubber elasticity.

(b) The elongation to rupture is between 80 and 300 percent, with up to 95 percent instantaneous recovery and almost complete recovery after a period of a few minutes.

(c) The tensile strength at rupture may be high as 5 g/den.

(d) The modulus of elasticity decreases with increasing temperature.

These properties indicate that the difference between the stretched and relaxed state is not a matter of strongly different conformation of the individual chain molecules (entropy effect), but rather the opening of valence angles and/or the stretching of intermolecular forces (van der Waals or hydrogen bonds). The phenomenon of this type of reversible elasticity is reminiscent to a certain extent of the transition of alpha Karatin into the beta form on stretching and to the elastic return into the alpha form on relaxation. From recent observations with the electron microscope, however, it appears more probable that this effect is due to the elastic bending of stacked crystalline lamellae which are prevented from slipping and gliding by a sufficient number of throughgoing (inter-lamallae) chain molecules. The word "springiness" has been proposed for this behavior because the samples act more like a steel spring than does a classical elastomer.[7]

REFERENCES

1. Compare, e.g., Johnson, J. F., and Porter, R. S., *Polymer Symposia* (1968), Vol 21.
2. Battista, O. A., *Synthetic Fibers in Paper Making,* Interscience Publishers, New York (1964); also Kentschel, R.A.A., *TAPPI,* **55**, 1174 (1972).
3. Moacanin, J., Holden, G., and Tschoegl, N. W., *Polymer Symposia* (1969), Vol 26.
4. Compare, e.g., U.S. Patent 3,248,179 (April 26, 1966) D. D. Norwood; assigned to Phillips Petroleum Company.
5. Frazer, A. H., *Polymer Symposia* (1967), Vol 19; also Preston, J., *Applied Polymer Symposia* (1969), Vol 9.
6. Compare, e.g., Moore, J. W., paper presented at the 27th SPI Conference, Washington, D. C. February 10, 1972, and Stone, R. H., paper presented at the 17th SAMPE Conference, Los Angeles, California, April 11, 1972.
7. Quynn, R. G., and Brody, H., *J. Macr. Sci,* **B5**, 721 (1971), and Clark, E.S., and Garber, C. A., *Int. J. Pol. Mat.,* **1**, 31 (1971).

DISCUSSION on Paper by H. F. Mark

HASSELL: Dr. Mark, you mentioned that a narrow molecular weight is needed to obtain the best drawing properties for strength. In your comments on the method of producing a fibrous polymer using high shear during polymerization, how is the molecular weight maintained, since shear will break the polymer down and broaden the distribution?

MARK: The narrow molecular-weight distribution which I mentioned was obtained in a loop reactor, such as described in a patent of Dr. Norwood of the Phillips Petroleum Company. In this reactor, the polymer is formed as a fine slurry at a low conversion (less than 10 percent) per pass and is immediately separated from the reacting mixture by filtration or centrifugation as a fine powder. The design of the reactor completely excludes the formation of hot spots and of stagnant polymer lumps. This appears to be one reason for the reduced polymolecularity. The stirred reactor which I mentioned for the formation of a fibrous type of polyethylene (Offenlegungsschrift 1,951,576, of June 4, 1970) does not claim to form a polymer with narrow molecular-weight distribution.

HASSELL: Since true di-block copolymers will exhibit two glass transitions, how can a three-component block copolymer exhibit only one transition? The definition of a block copolymer usually imples that the blocks are large enough to cause physical properties characteristic of the polymer in the block and not an averaging as observed in the random configuration.

MARK: I must apologize for having expressed myself in a confusing manner. I was talking only about segmented polymers of *two* components which have two or more segments (blocks). The new feature of Hytrel (Du Pont) over Kralon (Shell) is that the hard (plastic) block or blocks in Kralon consist of polystyrene, which is amorphous and is characterized by its T_g whereas in Hytrel, the hard component is a polyester which is crystallizable and has a T_m which is higher than the T_g of polystyrene. Efforts are now being made to go even a step further and use a polyamide as the hard component of a thermoelastic material of this type.

RADCLIFFE: The advances in polymer synthesis based on new processes and control of molecular weight and structure are intriguing scientifically and for their potential applications in providing new or improved products. However, recently recognized social needs point to the desirability of addressing the scientific challenges involved in developing the ability to desynthesize polymers (for recycling) for reasons of aesthetics, health, and finite limitations in the availability of the natural resources involved. How tractable do you see these challenges being?

MARK: Your question brings up a point which is of great social and industrial importance and is, at present, a very vividly discussed subject. From the practical point of view, plastics may be divided in two groups:

(a) Those which are supposed to serve for long periods (from 3 to 30 years), such as pipes and partitions in buildings, wall cover, floor cover, furniture and the like. In these cases, one uses stabilizers which offset the degrading influences of impurities and environment (air, light, acids, etc).

(b) Those which are for short-term use, such as packaging materials, newspapers, cups, napkins, and the like.

The latter can be made "degradable" by the introduction of reactive units which decompose the material under the influence of radiation or heat – essentially that available on a hot, sunny day. For example: if one introduces in a polystyrene molecule 2 to 3 percent p-isopropylstyrene and incorporates into the object (cups or bottles) a small amount of an oxygen transfer agent, the piece will embrittle within a few minutes if exposed to UV light in air. Under normal service conditions, it will behave satisfactorily.

Bio-Degradable Plastics, Inc., in Boise, Idaho, has developed such a self-degrading polystyrene under the name of Sty-Grade and is now carrying out a test with 7.5 million plastic lids for cold-drink cups. Laboratory tests have shown that molded objects of polystyrene, such as cups, egg cartons, bottles, and films, which contain 2 to 3 percent Sty-Grade start to degrade in sunlight within 4 to 5 weeks and are practically reduced to a powder within 3 to 4 months. This powder, in turn, is attacked by several common microorganisms so that the material completely disappears after 10 to 12 months.

This may be good enough for certain cases but probably not for all. Similar tests are now being carried out with polyethylene, polypropylene, and PVC.

One important question is: how much more will the biodegradable plastics cost than their normal nondegradable precursors? Estimates are ranging from a 5 to 15 percent premium.

Summarizing, it can be said: "auto-degradability" of plastics is technically possible; its realization is essentially an economic problem.

FERRY: The properties of the springy polymers you have described are strongly reminiscent of those of the natural protein fibrin, of which blood clots are composed. Fibrin can be prepared in the form of films which can be stretched 100 percent and mimic some aspects of rubber-like elasticity, though they cannot possibly be truly rubber-elastic in view of what is known about the protein structure. When the properties of the springy polymers are understood, the interpretation may also be applicable to proteins such as fibrin, keratin, and collagen.

MARK: As far as I know there are presently two tentative explanations for the behavior of polymers in the "springy" state: One of them assumes a helix-helix transition, similar to that in the protein which you have mentioned. There would be little difference in the entropy of the initial and final state and the restoring force would be the contraction of stretched van der Waals bonds. This would explain the high modulus, its decrease with temperature, and the limit of extensibility around 100 percent. The other explanation is based on electron microscope studies of springy polypropylene and polyformaldehyde and considers the extension to be the result of the bonding of crystalline lamellae which are stacked in the direction of the stress and held together by through-going chain molecules. It seems that one needs more experimental information until this phenomenon can be really understood.

O'NEILL: I have a question relating to the polyphosphoric acid polycondensation technique which is not without relevance to Professor Radcliffe's comments. If one gets complex formation between the condensation polymer and the polyphosphoric acid, then this latter would be expected to remain in the finished material, thus providing what could be an in-built hydrolytic degradation catalyst. Do you know how the hydrolytic stability of polyesters, say, produced by the polyphosphoric acid technique compares with that of conventional polyesters?

MARK: Under normal processing conditions, the polycondensation product of the two tetrafunctional reactants, e.g.,

H_2N NH_2 H_2N NH_2 and HOOC COOH HOOC COOH ,

remains dissolved in the concentrated polyphosphoric acid (PPA) solution. It is then converted into a fiber or film by spinning or casting in a more dilute solution, and the rest of the PPA is ultimately washed out with water. Whatever (firm or loose) complex existed between the polymer and the PPO is, hereby, decomposed. It seems perfectly possible, as you suggested, that some PPO left in the final product (fiber or film) will render it more sensitive to hydrolysis and produce a certain degree of auto-degradability.

MENGES: What are the commercial aspects for synthetic papers and where will they find applications?

MARK: Commercial applications of the spun-bonded type (Tyvek 10) already comprise wallpapers, wall maps, book covers, charts, calling cards, envelopes, tags, labels, bags, signs, posters and road maps.

The blow-molded types Acroart, Celestra, Polyart, Printel, Spiax, etc., are described in detail in R.A.A. Hentschel, *TAPPI,* **55**, 1174 (1972). Present applications and future economic conditions are discussed in this article. At present, the price of a high-grade synthetic printing paper (about 100 g/m^2) is 2.5 to 3 times that of a corresponding cellulosic paper. However, the synthetic paper is stronger and much more resistant to moisture.

SHEN: During the past several decades, there has been an explosive growth in the use of synthetic materials, most of which were made from petroleum products. However, in view of the fact that the petroleum reserve is finite, do you anticipate that there will be a trend to return to regenerable natural products such as wood, cotton, etc.?

MARK: Cellulosic, amylosic, and proteinic polymers have two important advantages:

(1) they are reproduced each year by solar energy in very large quantities and

(2) they are much more biodegradable than the petroleum-based synthetics.

Up to about 1920, all man-made fibers and films were based on natural polymers, such as rayon, cellulose acetate, cellulose nitrate, and respun proteins. Also, several thermoplastic and thermosetting resins had cellulose, starch, gums, and protein (casein) as components. With the advent of PVC, polystyrene, and the other oil-based synthetics, the earlier materials could not compete in cost and performance (stability) and were largely replaced.

Today, about 20 million tons of oil-based plastic is produced and consumed every year compared with about 500,000 tons of cellulose-based thermoplastics.

If, however, oil should become scarcer and the demand for biodegradability should become stronger, then there is a very good chance that derivation of cellulose, starch, gums, and proteins will again receive increased attention and use.

I know that several companies are reinvestigating a number of such derivatives which are well known and which have been abandoned in the past for economic reasons.

WILLIAMS: Please describe the basic mechanics involved in "flash spinning".

MARK: Whenever fibers are made for the purpose of textile application, a very high degree of uniformity of many properties – size and shape of cross section, orientation, crystallinity and fibrillar morphology – is required. This imposes on the processes of spinning, drawing, etc., certain limitations concerning speed, cooling, winding, etc. As soon as one wants the fibers for less delicate operations, such as filling a pillow, making a filter, or depositing a felt or web, one can relax on many restrictions and produce the fibers in a sloppier (and therefore less expensive) way.

Processes of this type are called "flash spinning". Some of them involve the rapid – almost explosive – vaporization of a solvent, while others involve the very fast orientation of the fiber in a semisolid state with the aid of hot air at speeds up to 5000 yd/min (about 150 mph). It is quite possible, as you say, that vacuum and chemical processes could also be used. Good information on the details of flash spinning can be found in several Du Pont patents, such as U.S. Patents 2,988,782, 2,999,788, and 3,068,527 to P. W. Morgan et al., and also in subsequent patents by Blades and White and by Levy and Kinney. The resulting fibers (in the case of polyethylene) have unusual properties in terms of high modulus (up to 400 g/den) and high tensile strength (up to 19 g/den) but are too nonuniform to be used for standard textile operations.

TSCHOEGL: What I find particularly significant in the trends you have outlined is the fact that the tendency is away from the synthesis of ever more new monomers and towards modification of the properties of base polymers through the application of very sophisticated physical chemistry and chemical engineering. The production of spun-bonded polyolefins is an excellent example of this trend. It requires an extremely high degree of technological sophistication which draws on all aspects of physical chemistry.

MARK: I fully agree with Professor Tschoegl that the present trend in polymer science and technology is toward deepening our understanding of the behavior of existing polymers and toward refining and expanding our methods to convert them into valuable materials for new applications. Spun bonding is one good example, the preparation of hollow filaments for reverse osmosis is another, the construction of composite systems for battery separators and fuel cells is a third, and the synthesis of autodegradable plastics is still another. Apparently we can still do a lot of useful things with the polymers which are now within our reach.

MECHANICAL BEHAVIOR OF HIGH POLYMERS—OVERVIEW AND HISTORIC REMARKS

Turner Alfrey, Jr.

The Dow Chemical Company
Midland, Michigan

1 OVERVIEW

The widespread use of organic polymers is based upon the diversity of mechanical behaviors exhibited by these materials under conditions of use and fabrication. The range of mechanical properties, in turn, results from the multiple diversity of possible molecular structures (chemical compositions and molecular architectures).

Some universal molecular features of organic polymers are: (1) covalent chains, (2) rigid bond angles, and (3) restricted bond rotation. At elevated temperatures, the chains exhibit rapid micro-Brownian motion, conferring flexibility upon macroscopic specimens. Upon cooling, polymers can harden by two distinct mechanisms: crystallization and vitrification. Crystalline melting points and glass transition temperatures, which depend on chemical composition, to a large extent govern the mechanical properties.

The topic "Deformation and Fracture of High Polymers" embraces a wide range of material behaviors. Even a single polymer has markedly different properties under different conditions. Consider, for example, a typical crystallizable linear polymer. Depending upon temperature and past history, it may behave in any of the following ways:

1. Viscous fluid, with some elasticity, at very high temperatures.
2. Rubbery solid, with some flow, at lower temperatures.
3. Crystalline, with flexible amorphous regions.
4. Crystalline, with glassy amorphous regions.
5. Glassy amorphous.
6. Metastable, supercooled flexible amorphous.

And, finally, in many cases *phase change* accompanies and governs the mechanical deformation. For example, a polymer may enter a mechanical fabrication operation in a metastable amorphous condition, and undergo crystallization during deformation, to yield an oriented crystalline product.

The *quantitative* representation of mechanical properties includes elastic, viscoelastic, and fracture responses of both crystalline and amorphous polymers. The discipline of *linear viscoelasticity* plays a central role in this endeavor. The rigorous application of linear viscoelasticity is limited to infinitesimal strains in simple materials (such as homogeneous, isotropic, amorphous elastomers) without failure, a scope which excludes many important materials problems. Indirectly, however, its principles are more broadly pertinent, since important features carry over into nonlinear deformation processes, and even fracture. For example, Thor Smith[1] has shown that the fracture of noncrystallizing elastomers exhibits a time-temperature equivalence similar to that encountered in linear viscoelastic response, and Professor Ferry[2] has pointed out that similar effects are encountered in such complicated phenomena as adhesion and abrasion.

The mechanical performance of a polymeric article depends upon both the molecular structure (primarily established during polymerization) and the spatial arrangement of the polymer molecules (governed by fabrication conditions). This latter feature includes morphology of crystalline polymers and molecular orientation in amorphous polymers. Controlled molecular orientation can lead to marked improvements in strength and toughness; uncontrolled orientation can result in weakness and fragility in service.

To illustrate the importance of orientation, let us consider the specific example of polystyrene. At room temperature, unoriented polystyrene is a brittle, glassy, amorphous polymer. Uniaxially oriented polystyrene is highly anisotropic. In the direction of orientation it has a high tensile strength and ductile extensibility; it is resistant to environmental stress-crazing and stress-cracking.[3] On the other hand, in the *transverse* direction its strength is reduced, it is extremely brittle and fragile, and its fracture energy (as measured in a controlled cleavage test) may be reduced by two orders of magnitude.[4]

Molecular orientation occurs inevitably in most mechanical fabrication operations, the pattern of orientation in the part being governed by the process *kinematics* – the pattern of flow and deformation followed by the polymer during fabrication. The *effects* of orientation can be either favorable or unfavorable, depending on the directions of orientation relative to the stresses encountered in service. Planned, controlled molecular orientation can be a valuable aid in achieving optimum properties and performance. Uncontrolled orientation will have beneficial effects only by accident; usually it is a source of weakness and failure.

Controlled molecular orientation in fabricated polymeric articles can be classified as uniaxial, biaxial, and "crossed".

Controlled uniaxial orientation has long been an indispensible tool in the manufacture of synthetic fibers, which are subjected in service mainly to tensile and bending loads.

Biaxially oriented polystyrene sheets (e.g., stretched 3:1 in both directions at 110 C) are strong and tough in all directions in the plane. Tensile strength, impact strength, and resistance to stress-cracking agents are all strongly enhanced, compared with those of unoriented polystyrene.[5]

"Crossed" molecular orientation was employed by Cleereman, who rotated the core of the mold during injection molding of thin-walled polystyrene tumblers.[6] The direction of molecular orientation varied through the wall, in a fashion similar to that of cross-laminated plywood. This led to a 3-fold increase in hoop strength, and a 10,000-fold increase in time to fail in a stress-crazing test in which tumblers were filled with corn oil or motor oil and pressurized.

Some thermoplastic polymers can be fabricated at relatively low temperatures, by methods similar to those used in metal fabrication (crystalline polymers below their melting points and amorphous polymers below their glass temperatures). These low-temperature operations, such as cold rolling[7], drawing[8], and forging[9], invariably introduce large orientations which strongly influence the mechanical performance of fabricated parts. Whereas unidirectional stretching of a sheet imparts uniaxial orientation, with marked loss of properties in the transverse direction, unidirectional *rolling* (because of the restraint against lateral reduction) often results in improved machine-direction properties with essential retention of transverse properties in the plane of the sheet. In low-temperature forming, the polymer must yield in a ductile manner, rather than fracture. This eliminates many thermoplastic materials; but, sometimes, mechanical pretreatment can convert an unusable thermoplastic to a usable condition. For example, polystyrene is too brittle to cold roll, but *oriented* polystyrene can be cold rolled and further oriented.

Another example of the role of fabrication in obtaining desired mechanical performance is the coextrusion of multilayer thermoplastic laminates. It is well known that multicomponent laminates often exhibit mechanical properties superior to those of the individual materials which constitute the separate layers. Some sword makers of early times hammered down alternate layers of hard and soft steel, obtaining blades which would take a fine cutting edge and yet were strong and tough.[10] Today, multilayer plastic-film laminates are manufactured in wide variety; in each case, the composite, multilayer laminate exhibits some property or combination of properties which cannot be matched by any one of the constituent materials. In particular, we wish to mention here an effect, which we have labeled "mutual interlayer reinforcement", which markedly alters the fracture behavior of some components of the composite.

The effect is this: thin layers of a high-modulus, low-elongation material are alternately sandwiched between thin (adhering) layers of high-elongation material. The film is tested to failure in tension, and it is observed that the high-modulus material undergoes a large ductile deformation before failing – in striking contrast to the free-film tensile behavior of this same material.

The high-elongation layers operate to prevent the propagation of transverse cracks across the brittle layers. With crack propagation so blocked, the stress in the brittle layers can reach the ductile yield point, and all layers can stretch out, together, to large elongations. Such a composite film has both a high modulus and a high elongation. It has a much higher work-to-fracture than would be exhibited by any of the individual layers.

Mutual interlayer reinforcement effects have been described in detail for Mylar-aluminum-Mylar composites, polyethylene-polystyrene-polyethylene composites, and 125-layer polystyrene-polypropylene films.[11]

Finally, this whole subject can be approached from various points of view, e.g., the physical-chemical mechanisms of mechanical response versus mechanical behavior from the viewpoint of materials selection and design. Horsley[12] has discussed this viewpoint in detail. Mechanical tests carried out on organic polymers can be crudely classified into two categories – those yielding numerical values which can be directly employed in design calculations, and those yielding numerical values which can be utilized only in a qualitative manner. Examples of the former type are elastic moduli (including time dependence) and time-dependent strength values. In contrast, the numerical value of a notched impact test cannot be used directly in design calculations. Horsley[12] comments: "In this context by good impact properties we really mean materials that exhibit *ductile* behavior . . . . From this viewpoint, the actual *energy* required to produce a ductile failure is of secondary importance, the main *performance* criterion being the position of the brittle-ductile transition . . . . . far too often, emphasis is placed in exactly the reverse order . . . . . avoidance of brittle fracture is the main service requirement."

This contrast between the search for mechanistic understanding and the utilization of property data in design and material selection should be kept in mind in any attempt to survey the current state of the field and provide guidelines for future developments. Both viewpoints are necessary, and to a large extent, it is the coordination between them which will determine the rate of significant progress.

2 HISTORICAL, TO 1940

Since the dawn of history, man has utilized natural polymers as materials for clothing, shelter, tools, and weapons. By the beginning of this century, the practical exploitation of these materials – wood, fibers, rubber, adhesives – had developed into myriad sophisticated arts which, to a large degree, form the basis for our current technology. The scientific study of these materials was also active, and in a phenomenological sense, most astute. The key to a *molecular* understanding, however, was the firm recognition of the high-molecular-weight chain of covalently bonded atoms, which was yet two decades in the future. This concept was not only the basis for all our modern molecular polymer science; it was also the key to rational polymer synthesis, and thus the foundation of the synthetic polymer industries.

By 1940, the exploitation of natural polymers was accompanied by a small, but rapidly growing, family of commercially produced synthetic polymers. Among others, polystyrene, phenolics, polyvinyl acetate, acrylates, methacrylates, polyvinyl chloride, synthetic rubbers, artificial fibers, safety glass, and laminated plastics were commercial realities.

In terms of polymer science, by 1940, many of the central problems with which we concern ourselves today were well recognized and subjected to serious analysis. A partial list of such problems would include: the molecular mechanism of rubber elasticity; the importance and nature of the glass transition and the crystalline-amorphous transition; the role of plasticizers; the influence of molecular orientation on mechanical properties; the time dependence of viscoelastic response; the Boltzmann Superposition Principle and to some extent the idea of time-temperature equivalence; the non-Newtonian flow of polymeric melts; and the relationship between molecular-weight distribution and tensile and impact strength. The study of dilute polymer solutions to determine polymer structures and the mechanisms and kinetics of polymerization reactions were also the subjects of scientific investigation.

3 HISTORICAL, 1940-1972

To a large extent, the advances of the last 30 years have been developments and elaborations of the basic concepts already established by 1940. However, this period has also seen some decisively new concepts, which have already had important effects on polymer science and polymer technology. Outstanding among these new concepts are the following three:

1. Recognition of the nature, the significance, and the control of *stereoregularity*
2. Discovery of *folded chain* laminar crystal morphology, and related derivative morphological features
3. Revelation of novel, subtle, and significant aspects of *fracture* mechanisms, and crazing.

(The most spectacular new development of this period – the elucidation of protein and nucleic acid structures, and the complementary roles of nucleic acid and proteins in biosynthesis – is only remotely connected with our topic of deformation and fracture, although, in the future, improved insights into motility, meiosis, and biological motions in general may bring these disparate fields into closer harmony.)

Developments in composite materials during this period deserve mention. Although the use of composites goes back to very early times, and current technology is based largely on earlier art, the scientific analysis of deformation and fracture of composites is a major achievement of recent origin.

Overall, the period from 1940 to 1972 has been one of significant advances on many fronts in the field of deformation and fracture of polymers. To some extent, the advances have suffered from fragmentation – with incomplete coordination between the efforts of those interested in deformation *mechanisms* and those concerned with mechanical response as it relates to material *selection* and *design*. Full integration of these various viewpoints is an important continuing task.

REFERENCES

1. Smith, T. L., *Pure Appl. Chem.*, **23**, 235 (1970).
2. Ferry, J. D., *Viscoelastic Properties of Polymers,* 2nd Edition, John Wiley and Sons (1970).
3. Williams, J. L., Cleereman, K. G., Karam, H. J., and Rinn, H. W., *J. Poly. Sci.*, 8, 345 (1952).
4. Broutman, L. J., and McGarry, F. J., *J. Appl. Polymer Sci.*, **9**, 589, 609 (1965).
5. Thomas, L. S., and Cleereman, K. J., *SPE Journal,* **28** (4), 2; (6), 9 (1972).
6. Cleereman, K. J., *SPE Journal,* **23**, 43 (1967); **25**, 56 (1969).
7. Broutman, L. J., and Patil, R. S., *Poly. Eng. and Sci.*, **11**, 165 (1971).
8. Li, H. L., Koch, P. J., Prevorsek, D. C., and Oswald, H. J., *J. Macromol. Sci.-Phys.*, **B4** (3), 687 (1970).

9. Wissbrun, K. F., *Poly. Eng. and Sci.*, **11**, 28 (1971).
10. Slayter, G., *Sci. Am.*, **206** (1), 124 (1962).
11. Schrenk, W. J., and Alfrey, T., Jr., *Poly. Eng. and Sci.*, 9, 393 (1969).
12. Horsley, R. A., *Appl. Poly. Symposia*, **17**, 117 (1971).

DISCUSSION on Paper by T. Alfrey

KNAPPE: In discussing the effect of processing conditions on the properties, one should also regard the effects of internal stresses. Like orientation, it is nearly impossible to avoid internal stresses since, during processing, parts of the polymer cool down or cure (cross-link) in an inhomogeneous temperature distribution. In crystallization, for instance, the growing of spherulites will cause internal stress of a higher order at the edges of the spherulite that may cause crazing and embrittlement. Most of the trouble observed in processing arises from internal stresses in combination with uncontrolled orientation.

ALFREY: I agree that residual stresses as well as frozen-in orientation profoundly affect the mechanical performance of polymeric articles. Large residual stresses can also exist in thermosetting resins, as the result of differential curing rates.

MAYER: Considering the possibility of a glassy polyethylene, do you know whether anyone has tried to splat cool (ref. the technique widely used now in metallurgy to obtain very large cooling rates) the material to suppress crystallization? Is a critical crystallization rate estimable?

ALFREY: I do not know whether the metallurgical splat-cooling technique has been attempted with molten polyethylene, but some extremely rapid cooling experiments have been made. In all cases, crystallization resulted.

HASSELL: What is splat cooling?

JAFFEE: Splat cooling is accomplished by impacting molten droplets against a cold metal surface thus producing a thin disk solidified at thousands of degrees per second with very fine grain size or even an amorphous structure. It is of interest in metals as a means of strengthening by grain refinement, e.g.,

$$\sigma_{ys} = \sigma_o + k_y d^{-1/2} \quad ,$$

where d is the average grain diameter and σ_o and k_y may be taken as empirical parameters, although they have significance in dislocation theory.

Splat cooling is of doubtful value in those polymers where it is easy to produce amorphous structures without the high cooling rates obtained by splat cooling.

HASSELL: Many workers have tried to supercool polyethylene from the melt by a rapid quench in liquid nitrogen. In all cases crystallization has occurred. However, the crystallization is not the maximum obtainable and further crystal growth and change will occur as the polyethylene is heated to 60 C. The interpretation of the dielectric and dynamic mechanical-property change due to this further crystallization is still a controversy. Some believe the changes in the α, β and γ transition peaks are due to the change in amorphous state, while others believe the changes to be due to change in crystal structure and its influence on the amorphous material.

MARK: Dr. Alfrey indicated that it has not yet been possible to prepare a sample of amorphous polyethylene. This can be supported by the fact that Dr. T. T. Wang of Bell Laboratories quenched a film of polyethylene (density 0.96, MI 0.2) at a rate faster than 10,000 C per second under high shear. He obtained a completely transparent, highly crystalline, and highly oriented film which did not become turbid by annealing at 110 C for 21 days. Its permanent transparency seems to be due to the presence of very thin, highly oriented crystallites which also cause an unusually high modulus (twice that of a normally oriented film) of this sample. Professor R. S. Porter has obtained filaments of polyethylene which are completely transparent, apparently had a similar structure, and showed a tensile strength six times higher than that of normally extruded and oriented samples.

HASSELL: In Professor Porters case, however, the filaments were made at a temperature close to the melting point and under high pressure. The similarity in properties again illustrates the rapid crystallization of polyethylene.

SHEN: I would think that if the molten polyethylene can be quenched in the shape of very fine droplets or extremely thin films, amorphous polyethylene could be made. However, since the glass transition temperature is very low, the amorphous state cannot be expected to persist at room temperature. It would, of course, be interesting to study the properties of glassy polyethylene if possible. We have made amorphous polyethylene by polymerization in ionized plasma, but the material is highly cross-linked and does not resemble the conventional polyethylene.

RADCLIFFE: The emphasis given here to the need for delineation of the mechanical processing history of polymers in trying to understand and predict their behavior is indeed important. However, this topic is only one aspect (though a major one) of the general question of characterization, i.e., the full definition of those chemical and structural aspects of the polymer that contribute (or may do so) to the particular property or behavior of interest. Other factors are, for example, chain length, degree of crystallinity or vitrification, purity, impurity particles, cavities, and thermal processing history. Unfortunately, it is not yet sufficiently common in the polymer literature to find adequate recognition of the importance of knowledge of these factors. The early history of inorganic solids – metals, ceramics, and electronic solids – was likewise confounded by similarly limited recognition. Hopefully, polymer research will increasingly pay more attention to defining the significance of the various characterization parameters important for specific properties and behavior.

MARK: Most commercial samples contain *accidental impurities* of up to more than 1 percent in the form of stabilizers, antistatic agents, lubricants, and remnants of catalysts. These impurities usually vary from lot to lot and are not specifically mentioned to the customers. Specially made samples can be very pure (less than a few parts per million) in terms of *external materials* but are usually ill defined in terms of *molecular-weight distribution*.

There exist, however, a number of fractions (or otherwise sharply characterized) of polymers such as polystyrene, PVC, polyethylene, etc., which are reasonably uniform in respect to molecular weight ($\overline{M}_w/\overline{M}_n$ as low as 1.03).

But even individual samples (fibers, films, or test bars) of these materials can (and do) still differ from each other in respect to their *morphology*. It is well known that the individual chain molecules may belong to an area in which they exist in the form of substantially random coils (amorphous or disordered area), to an area in which they "crystallize on themselves" (folded-chain crystallites), or, finally, to an area in which they "crystallize on each other" (extended-chain crystallites). It is almost impossible, at present, to establish quantitatively in given samples the weight percentages of the material which belongs to each of these three morphological (supermolecular) species. But, apparently, the morphology of a test piece has noticeable influence on its mechanical and thermal properties.

ANDREWS: The *importance* of precise characterization of experimental materials is illustrated by the fact that only 10 percent of *trans* units in otherwise *cis*-polyisoprene reduces the rate of lamellar crystal growth by a factor of 10^3. The *difficulties* of characterization are illustrated by the impossibility, as yet, of measuring molecular perfection (e.g., degree of isotacticity) to better than a few percent, when in the above example, 3 percent of noncrystallizable units can change crystallization rates by a factor of ten.

It would not be fair, however, to say that these problems are not recognized. Much current research is directed towards the elucidation of the effects of variables of composition upon the properties and reactions of polymers.

LANDEL: Can you comment on molecular features which control the effects of amplitude and axiality of orientation? For example, Professor Ferry has shown the central role of a molecular friction factor, which depends on monomer structure, in linear viscoelasticity. Is there an analogous parameter here?

ALFREY: I do not know of any study comparing different chemical species of polymer which answers the question of what molecular features determine the degree of response to molecular orientation. In the case of polystyrene, we know that the effects on the mechanical properties are strongly dependent upon which "viscoelastic modes" are employed in the orientation. The low-τ modes are much more effective than the high-τ modes.

DEFORMATIONS AND MOLECULAR MOTIONS IN POLYMERS ABOVE THE GLASS TRANSITION TEMPERATURE

John D. Ferry

Department of Chemistry and Rheology Research Center
University of Wisconsin
Madison, Wisconsin

ABSTRACT

Time-dependent deformations of polymers above the glass transition temperature are related to spectra of relaxation times which reflect certain modes of configurational motion of the macromolecules. The spacings and magnitudes of these characteristic times, even as determined from experiments in small deformations, are important in understanding large deformations and processes of rupture, since the rates of molecular motions will set the time scale for these as well. In very dilute solution, all relaxation times (except perhaps the shortest) are proportional to solvent viscosity; their spacings have been derived theoretically and checked experimentally for a variety of linear and branched molecules. In concentrated and undiluted systems, the presence of entanglement coupling sharply differentiates the time or frequency scale into characteristic zones of viscoelastic behavior. In the transition zone, the relaxation times reflecting local motions are proportional to a monomeric friction coefficient. The dependence of the friction coefficient on temperature, pressure, diluent concentration, and other variables can be described and formulated with an auxiliary parameter, the fractional free volume. The spacing of the transition-zone relaxation times in concentrated systems is less well understood, though the Rouse spectrum is a first approximation. In the terminal zone, relaxation times of uncross-linked polymers are enormously prolonged by entanglement coupling and appear to be quite closely spaced. In lightly cross-linked systems, effects of trapped and untrapped entanglement loci may be distinguished. The dependence of contributions to modulus or compliance on concentration, molecular weight, and other variables must be considered separately for each zone of viscoelastic behavior.

1 INTRODUCTION

In polymers above the glass transition temperature, thermal energy causes the macromolecules to perform configurational changes, at rates that depend on interactions with their neighbors – interactions which in some respects can be described by an effective frictional resistance to translational motion. The rates of molecular motion influence the manner of response to time-dependent mechanical stresses and the manner in which mechanical work is stored or dissipated. Thus they play an important role in time-dependent mechanical deformations, whether small or large, and in processes of rupture.

Measurements of linear viscoelasticity provide information about rates of molecular motion through the identification of spectra of relaxation times which can, to some degree, be related to characteristic motional modes. We first review here, briefly, the current state of knowledge for dilute polymer solutions, which are instructive because molecular theory is comparatively well developed, although the results are not directly applicable to most practical situations. We then consider the transition, plateau, and terminal zones in concentrated or undiluted uncross-linked polymers of high molecular weight, and cross-linked rubbery networks, discussing the magnitudes and spacings of relaxation times and their dependence on temperature, concentration, molecular weight, and other variables.

2 DILUTE POLYMER SOLUTIONS

In describing the viscoelastic properties of very dilute polymer solutions, the bead-spring molecular model has been quite successful. The molecule is represented as arbitrarily divided into submolecules; the frictional resistance to translational motion is concentrated at the junctions between them (beads); the submolecules are long enough to have a Gaussian distribution of configurations, and distortion of this distribution increases the free energy (entropy springs). The magnitude of the hydrodynamic interaction between the beads is gaged by a parameter $h^* = \zeta/(12\pi^3)^{1/2}\eta_s b$, where ζ is the friction coefficient of a bead (force/relative velocity), η_s is the solvent viscosity, and b is the root-mean-square distance between neightboring beads; it can be regarded as a measure of the ratio of bead "size" to interbead distance, and ranges typically between 0.1 and 0.3. In the theory of Zimm[1] for linear polymer molecules, a sequence of relaxation times τ_p is determined by a set of eigenvalues λ_p of a matrix which describes the forces acting on the beads; the index p runs from 1 to N, the number of submolecules. These eigenvalues, which correspond to characteristic modes of motion of the entire molecule, have been evaluated exactly by Lodge and Wu[2] for various values of h*. For branched molecules, the theory of Zimm and Kilb[3] describes the modes of motion by an appropriate matrix which has been evaluated to determine exact eigenvalues for star-shaped geometry by Osaki and Schrag[4] and for comb geometries with certain specific dimensions by Osaki.[5] In all cases, the results depend on N, but the relaxation time ratios τ_p/τ_1 approach independence of N for N large and $p << N$.

In Fig. 1, relaxation time spectra calculated in this manner are displayed for a linear polymer and for three types of branched molecules with eight branches of equal length. The heights are proportional to n k T, where n is the number of polymer molecules per cc. For star molecules with f arms, the odd-numbered relaxation times are (f - 1) degenerate; for comb molecules, some of the times are so closely spaced as to be practically degenerate and have been combined in the display for convenience.

These relaxation times govern the response of a very dilute solution to small time-dependent stresses. Because the times are usually too short to be conveniently observed by transients, the usual experiment is a sinusoidally oscillating stress, from which the storage and loss shear moduli G' and G'., respectively, are determined. For comparison with the theory, which ignores intermolecular interactions, it is necessary to extrapolate data to infinite dilution to obtain the intrinsic moduli

$$[G'] = \lim_{c \to 0} G'/c \tag{1}$$

$$[G''] = \lim_{c \to 0} (G'' - \omega\eta_s)/c \ , \tag{2}$$

where ω is radian frequency and c is concentration in g/cc. It is con-

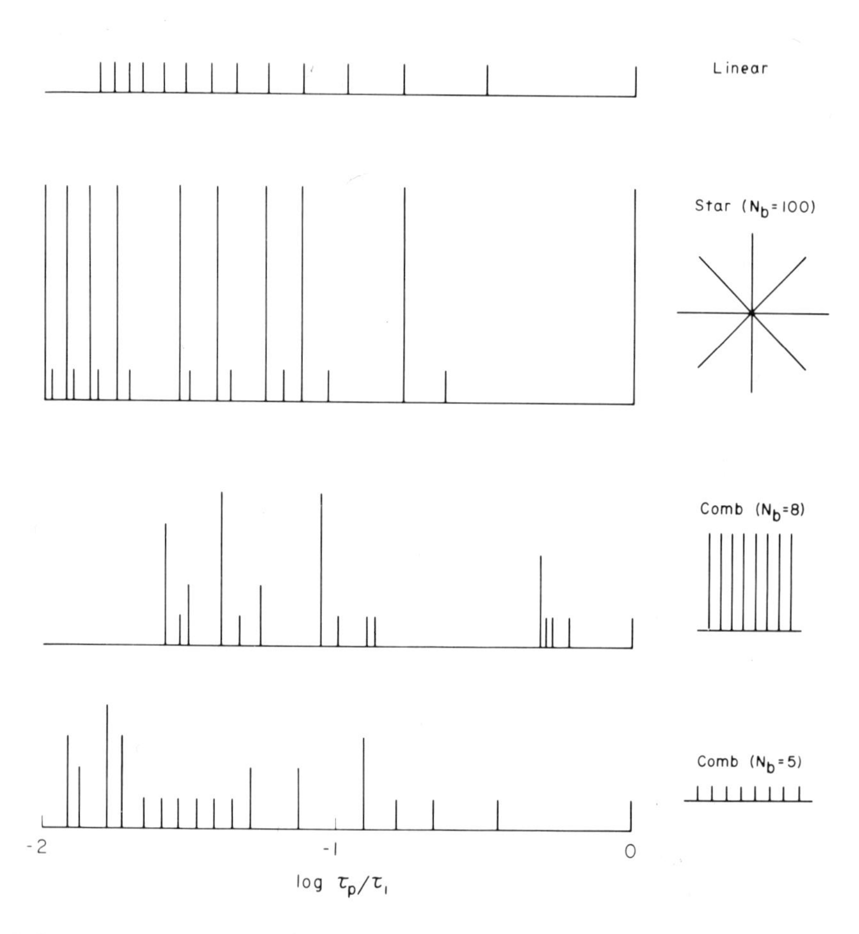

Fig. 1. Relaxation time spectra for polymers in very dilute solution. Linear, Zimm theory evaluated by Lodge and Wu; branched, Zimm-Kilb theory evaluated by Osaki and Schrag. For the comb polymers, some very closely spaced mechanisms have been arbitrarily combined. N_b is the number of submolecules per branch. Minimum height represents a modulus contribution of cRT/M.

venient to reduce these to the dimensionless moduli $[G']_R = [G']M/RT$ and $[G'']_R = [G'']M/RT$, where M is molecular weight. The frequency dependences of the dimensionless moduli are expressed as follows:

$$[G']_R = \sum_{p=1}^{N} \frac{(\omega\tau_1)^2(\tau_p/\tau_1)^2}{1+(\omega\tau_1)^2(\tau_p/\tau_1)^2} \tag{3}$$

$$[G'']_R = \sum_{p=1}^{N} \frac{(\omega\tau_1)(\tau_p/\tau_1)}{1 + (\omega\tau_1)^2(\tau_p/\tau_1)^2} \tag{4}$$

$$\tau_1 = [\eta]\eta_s M/RTS_1 \tag{5}$$

$$S_1 = \sum_{p=1}^{N} (\tau_p/\tau_1) \quad , \tag{6}$$

taking into account degeneracy where appropriate. In Fig. 2, the frequency dependences of $[G']_R$ and $[G'']_R$ are plotted logarithmetically for the relaxation spectra of Fig. 1. The abscissa scale is determined by the product $[\eta]\eta_s M/RT$, where $[\eta]$ is the intrinsic viscosity in cc/g.; these quantities are all measurable so there are no adjustable parameters except h*, provided N is sufficiently large and the frequency is not too high. The properties of the fictional submolecule do not appear. The curves are drawn for h* = 0.25, which is appropriate for a Θ-solvent, at least for the linear case.[6,7]

Experimental data extrapolated to infinite dilution confirms these calculations quantitatively for linear molecules[6-10] and for 4-arm and 9-arm stars[10,11] of high molecular weight. For comb geometries, qualitative agreement has been found, but experimental data have not yet been fully analyzed.[12] Because of computer limitations, it has not yet been possible to make calculations for combs with adequately large values of N.

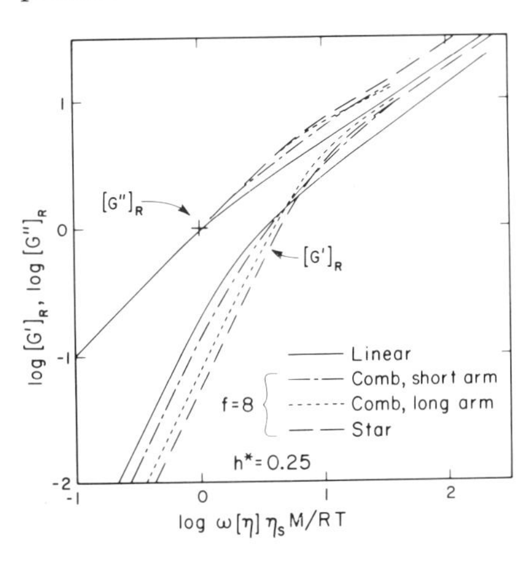

Fig. 2. Frequency dependence of intrinsic storage and loss shear moduli calculated from the spectra of Fig. 1.

It is evident that the bead-spring model ignores the chemical structure and specificity of the polymer, and the observed linear viscoelastic behavior does appear to depend only on the size and connectivity of the molecule for frequencies below $10\tau_1^{-1}$, where τ_1 is the longest relaxation time. The resistance to motion is provided solely by the environment, all relevant relaxation times being directly proportional to solvent viscosity; the polymer backbone is effectively completely limp. At higher frequencies and/or lower molecular weights, however, the value of N is no longer without influence; the bead-spring model is inadequate, and an additional mechanism of energy dissipation appears which depends on specific chemical structure.[13,14] The frequency dependence of G′ and G″ over an extended range can be described quite well by a theory of Peterlin[15] (Fig. 3), but the physical aspects of behavior at high frequencies are still uncertain.

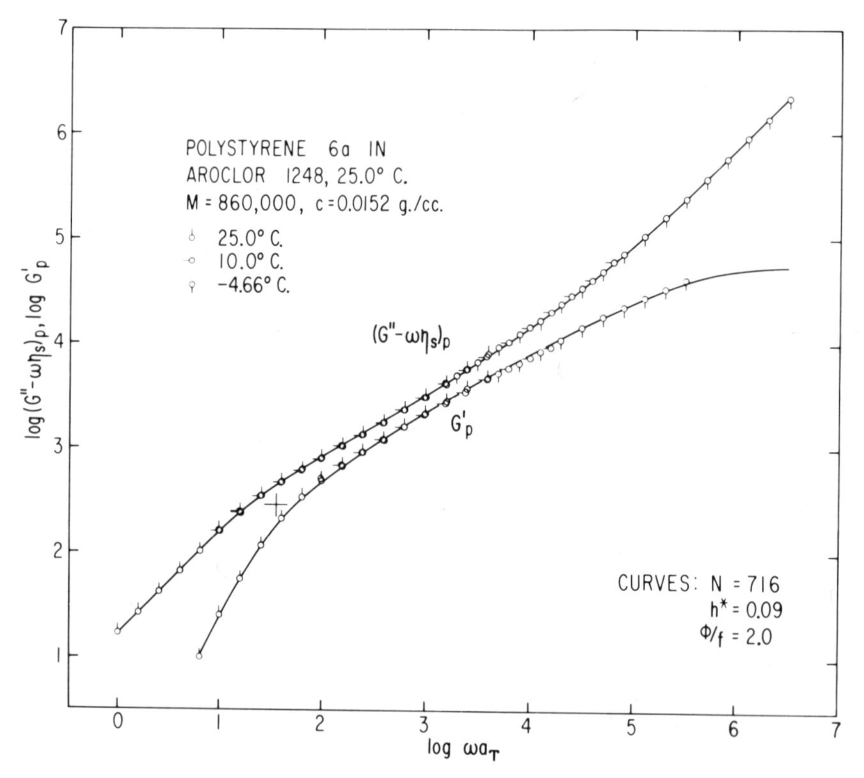

Fig. 3. Polymer contributions to storage and loss moduli in a dilute polystyrene solution in Aroclor 1248 (M = 860,000, c = 0.0152 g/cc) reduced to 25 C from measurements at different temperatures over an extended frequency range. Curves drawn from the theory of Peterlin with parameters indicated.

3 CONCENTRATED AND UNDILUTED POLYMER SYSTEMS

Although the motions of isolated polymer molecules that contribute to mechanical properties are fairly well understood, practical applications of this knowledge are rather limited. In most situations of practical interest, the molecules entwine and interpenetrate each other's domains. Their motions are then far slower and the relaxation spectra may have a quite different character.

If the molecular weight is below something like 10,000, the relative spacings of the relaxation times may not be too different from those in dilute solution, even for undiluted polymer.[16,17] However, such materials above the glass transition are also of somewhat limited practical interest. Of far greater importance are materials with high molecular weight whose rheological properties exhibit the complications associated with entanglement coupling.[18] For these, the linear viscoelastic behavior is sharply differentiated into separate zones on the time or frequency scale, as illustrated in Fig. 4 for the storage and loss shear moduli of a poly(n-octyl methacrylate)[19,20] of molecular weight 3.6 x 10^6. We consider, in turn, the transition, plateau, and terminal zones. The glassy zone will not be discussed in this review.

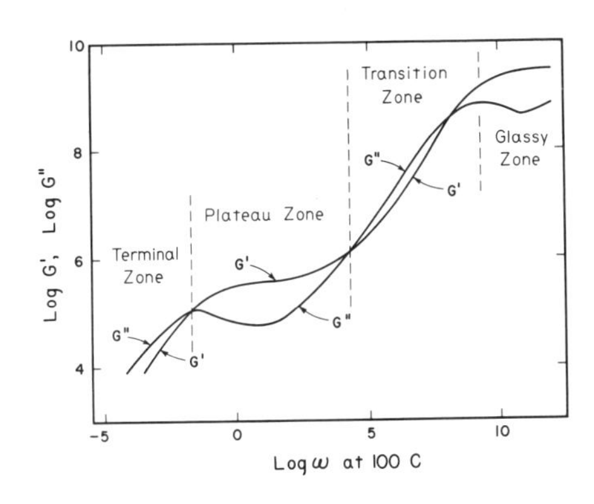

Fig. 4. Zones of viscoelastic behavior illustrated by the storage and loss shear moduli of poly-n-octyl methacrylate (M = 3.6 x 10^6) at 100 C.

3.1 Transition Zone

In the transition zone, linear viscoelastic properties reflect motions of molecular segments which run between enganglement loci and are almost indistinguishable from those of lightly cross-linked networks. They are often compared with the predictions of the Rouse theory, or more accurately, the Mooney modification thereof[21] which includes one infinite relaxation time to correspond to the very slowly slipping entanglement loci:

$$G' = (\rho RT/M_e) \left[1 + \sum_{p=1}^{N} \omega^2 \tau_p{}^2/(1 + \omega^2 \tau_p{}^2) \right] \tag{7}$$

$$G'' = (\rho RT/M_e) \sum_{p=1}^{N} \omega \tau_p/(1 + \omega^2 \tau_p{}^2) \tag{8}$$

$$\tau_p = a^2 M_e^{\ 2} \zeta_o / 6\pi^2 \rho^2 kTM_o^{\ 2} \ , \tag{9}$$

where ρ is the density of the (undiluted) polymer, M_e the average molecular weight between entanglement loci, M_o the molecular weight of a monomer unit, a^2 the mean square end-to-end separation per monomer unit, and ζ_o the monomeric friction coefficient. The relaxation spectrum of the Rouse theory differs somewhat from that of the Zimm theory with $h^* = 0.25$ shown at the top of Fig. 1; the spacings are greater, corresponding to a negligibly small value of h^*. In Fig. 5, the transition zone is compared with the frequency dependence calculated from Eqs. 7 and 8 with $M_e = 87{,}000$ as estimated from integration of the loss compliance.[22] The abscissa scale is adjusted arbitrarily and its position determines ζ_o in Eq. 9. The fit is satisfactory at the low frequency end of the transition zone; at higher frequencies, deviations appear for all polymers and are specific to chemical structure in a manner not well understood.

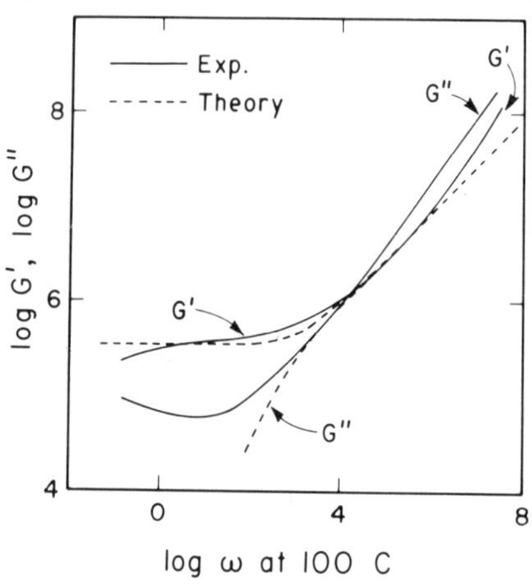

Fig. 5. Transition zone from Fig. 4 compared with Rouse-Mooney theory for a cross-linked network.

The most important feature of Fig. 5 is that it permits determination of ζ_o, the force per unit velocity per monomer unit required to push a chain through its surroundings. Although the theory is somewhat dubious, relative values of ζ_o serve to specify where on the time or frequency scale the transition zone will appear for a particular temperature, diluent concentration, etc., and since all the relevant relaxation times are proportional to ζ_o, it is important in setting the time scale for various mechanical properties.

Comparisons of ζ_o for two dozen polymers each at 100 C above its glass transition temperature[23] show a considerable range of magnitudes not readily correlatable with chemical structure (except that chains with disubstituted backbone atoms tend to have higher values). However, the dependence of ζ_o on temperature, pressure, diluent concentration, and molecular weight can be rather simply formulated in terms of the fractional free volume f by the generalized Doolittle equation[24]:

$$\log \zeta_o = \log \zeta_{oo} + \frac{B}{2.303}\left(\frac{1}{f} - \frac{f}{f_o}\right) , \tag{10}$$

where ζ_o and ζ_{oo} correspond to two states with fractional free volumes f and f_o, respectively, the latter being a reference state, and B is a numerical coefficient usually taken as unity. The dependence of f on different variables can be expressed as follows:

$$\text{Temperature } (T)^{25}: f = f_o + \alpha_f (T - T_o) \tag{11}$$

$$\text{Pressure } (P)^{26}: f = \pi f_o/(P - P_o + \pi) \tag{12}$$

$$\text{Volume Fraction of Diluent } (v_1)^{27}: f = f_o + \beta' (v_1 - v_1^o) \tag{13}$$

$$\text{Molecular Weight } (M)^{28}: f = f_o + A/\overline{M}_n , \tag{14}$$

where the subscript (or, for v_1, superscript) o refers to the reference state, usually respectively the glass transition temperature, atmospheric pressure, undiluted polymer ($v_1 = 0$) and infinite molecular weight; α_f, π, β', and A are parameters whose physical significance is generally obvious, and judicious guesses as to their magnitude can often be made if definite information is lacking. (Since A is of the order of 100 g/mole, molecular-weight dependence is trivial except for quite low molecular weights.) When mechanical properties – even fairly complicated ones like adhesion[29] or abrasion[30] – are governed by local molecular mobility, the dependence of the appropriate time, frequency, or velocity scales on temperature and other variables can often be successfully formulated and predicted by Eqs. 10 through 14.

In polymers strained with large deformations, the friction coefficient may be affected by molecular orientation, but there is as yet very little information available concerning this question.

3.2 Plateau Zone

In the plateau zone, molecular segments between entanglement loci rearrange rapidly within the experimental time scale, while entanglement slippage is slight. There is very little stress relaxation or energy dissipation.

3.3 Terminal Zone

In the terminal zone, linear viscoelastic properties reflect configurational motions of entire molecules which necessitate slippage through many entanglements. With increasing molecular weight, the relaxation times increase rapidly, shifting the terminal zone to longer times or lower frequencies as illustrated in Fig. 6 where the storage shear modulus of polystyrenes with narrow molecular-weight distribution is plotted logarithmically against frequency.[31] In the low-frequency limit, G′ and G″ are proportional to ω^2 and ω, respectively. The transition from this behavior to the plateau zone where G′ is virtually independent of frequency is abrupt, without an intermediate "power-law" region such as a Rouse relaxation spectrum produces; Graessley[29] has emphasized that this means a densely bunched spectrum, and his analysis of the drag interactions from extensive entanglement coupling leads to the spectrum portrayed in Fig. 7. The actual motions through the entanglement loci are very difficult to describe[30]; Graessley's calculation is based on the slippage of the shortest of four chains radiating from an entanglement locus. It reproduces quite well the frequency dependence of G′ in the terminal zone as portrayed in Fig. 6. Evidently, the long-range modes of motion in entangled systems *cannot* be represented as similar to those in dilute solution with abnormally large friction coefficients, as was formerly suggested.[32]

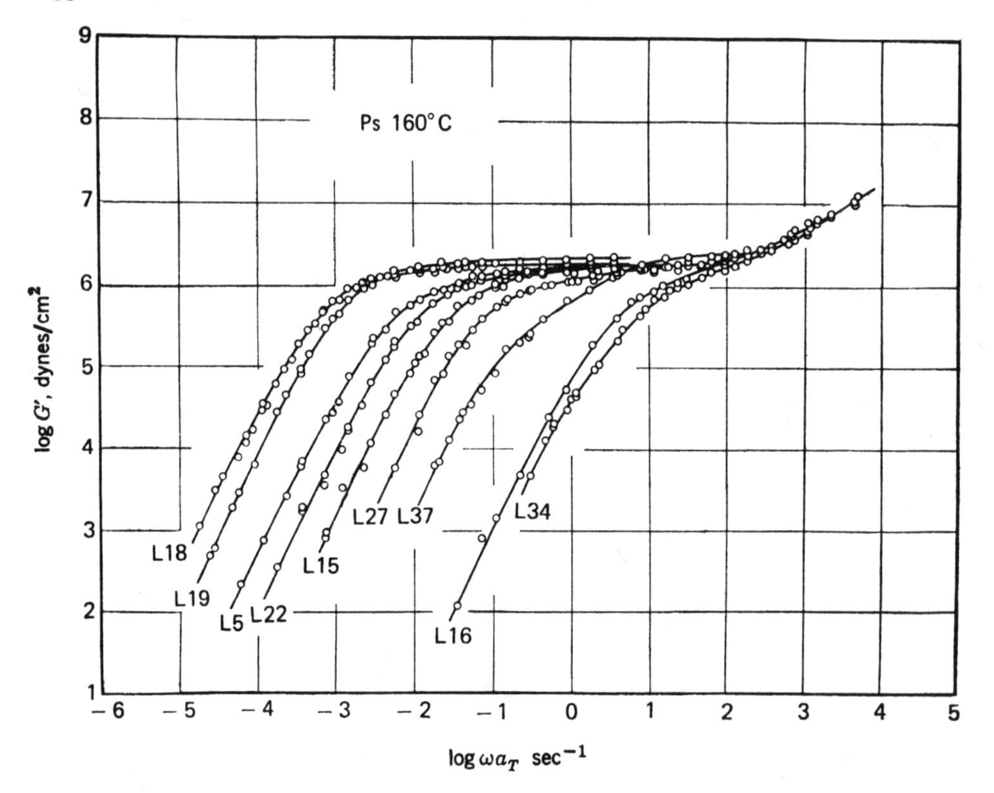

Fig. 6. Storage modulus of narrow-distribution polystyrenes, plotted logarithmically against frequency reduced to 160 C.[31] Viscosity-average molecular weights, from left to right, x 10^{-4}: 58, 51, 35, 27.5, 21.5, 16.7, 11.3, 5.9, 4.7.

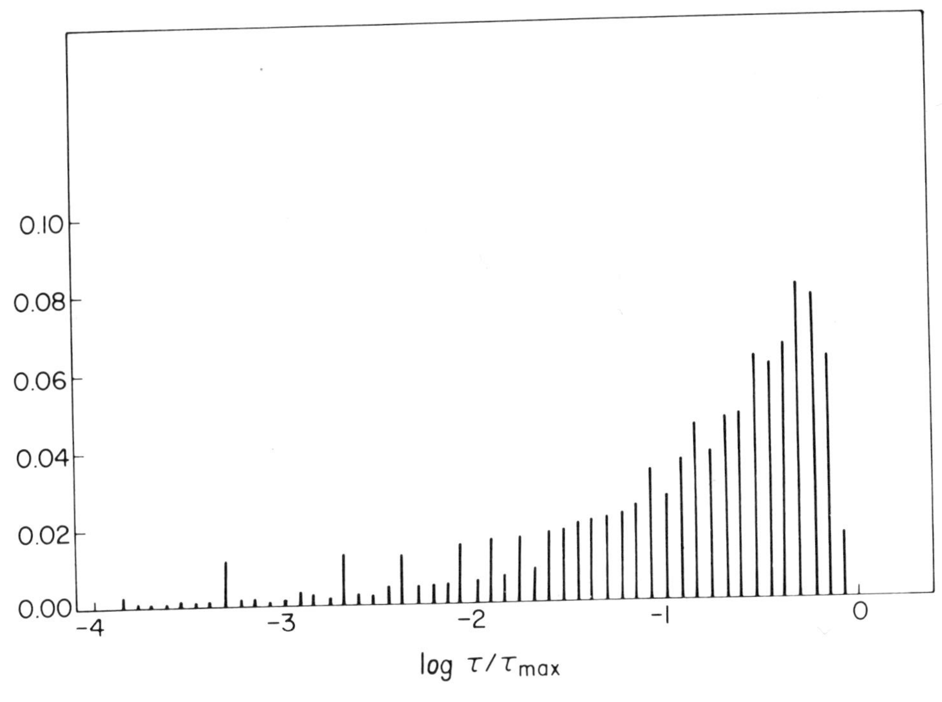

Fig. 7. Relaxation time spectrum accumulated in logarithmic intervals for molecules with 400 entanglement loci, calculated by Graessley.[29] Amplitudes are normalized so their sum is unity.

At high rates of deformation, the relaxation times in the terminal zone are shortened, as demonstrated by many experiments in which steady and oscillatory shear are combined.[33,34] Qualitatively, this probably corresponds to a reduction in the density of entanglement loci; it is probably important in describing processes of rupture.

In certain polymers, such as polymethyl methacrylate[35], polyethylene oxide[36], and polyacrylonitrile, relaxation times in the terminal zone appear to be abnormally long, as though the entanglements were especially stable, perhaps enhanced by specific attractive forces in appropriate intermolecular configurations; these couplings might be called hyperentanglements.

Since even the complicated motions slipping through entanglements involve the short-range frictional resistance of the environment, relaxation times in the terminal zone are proportional to ζ_o, and their dependences on temperature and other variables include as factors the effects on ζ_o summarized in Eqs. 10 through 14. In addition, there are factors which express the effects of these variables on the entanglement coupling. The ratio τ_1/ζ_o (where τ_1 is the longest relaxation time) appears to be almost independent of temperature except for the case of hyperentanglements mentioned above. The molecular-weight dependence of this ratio is expressed by the factor M^{ϵ} where ϵ is found experimentally to be between 3.3 and 3.7 in many recent experiments on polymers with narrow molecular-weight distribution.[37-42] The dependence of τ_1/ζ_o on diluent concentration probably cannot be expressed so concisely.

3.4 Dependence of Viscoelastic Constants in the Terminal and Plateau Zones on Molecular Weight and Concentration

In the plateau zone, the most important parameter describing linear viscoelastic behavior is the entanglement compliance J^o_{eN} or its reciprocal the modulus G^o_{eN}, which represents the pseudoequilibrium response of the entanglement network and can probably be best defined experimentally by a partial integration over the retardation spectrum or of the loss compliance. In numerous recent studies[37-46] it has been found to be essentially independent of M and proportional to c^{-2} as would be expected for a constant probability that two neighboring monomer units on different molecules will participate in an entanglement locus.

In the terminal zone, the most important viscoelastic parameters are the longest relaxation time τ_1, the steady-flow viscosity (at vanishing shear rate) η, and the steady-state compliance J^o_e. The dependence of τ_1/ζ_o, η/ζ_o, and J^o_e, as well as of J^o_{eN}, on M and c can be described by proportionality to powers of M and c expressed by the exponents summarized in Table I. The values apply to highly entangled systems and are therefore

Table I. Exponents Describing Dependence of Viscoelastic Constants on Molecular Weight and Polymer Concentration in Highly Entangled Systems

	M	c
J^o_{eN}	0[a]	-2[a]
τ_1/ζ_o	3.3 to 3.7[b]	
η/ζ_o	3.3 to 3.7[b]	3.3 to 3.7[b]
J^o_e	0[a]	-2 to -3[c]

(a) For $M > M_b$ when $v_2 = 1$.
(b) For $v_2 M > M_C$.
(c) For v_2 quite large (e.g., > 0.5); otherwise dependence may be quite complicated.

subject to certain restrictions on M and v_2 (volume fraction of polymer, viz., c/ρ_2 where ρ_2 is density of pure polymer) as stated in terms of M_C, the critical molecular weight for the effect of entanglement coupling, and M_b, another critical molecular weight which appears to be higher than M_C by a factor depending on the chemical nature of the polymer. These exponents are based on recent measurements mostly on polystyrene, poly(α-methyl styrene), polydimethyl siloxane, and poly-cis-isoprene. For systems with hyperentanglements, they may be different; in particular, the dependence of viscosity on concentration is greatly enhanced.[47]

3.5 Effects of Branching

Whereas in dilute solution the effects of branching on the spectrum of relaxation times can be described by the Zimm-Kilb theory as outlined above, in concentrated or undiluted systems the role of branch points may be entirely different because of the combined effects of branch points and entanglements. The steady-flow viscosity of a branched polymer is often, but not always, observed to be much smaller than that of a linear polymer of the same molecular weight[48-51], and the steady-state compliance is found to be markedly larger than that of the corresponding linear polymer[48,49] (in contrast to the case for dilute solutions, where it is invariably smaller). The entanglement compliance J_{eN}° is slightly higher than that of a linear polymer. It appears that even qualitative aspects of behavior will depend on the specific topology of the branched molecule.

4 CROSS-LINKED NETWORKS

In a lightly cross-linked network, the spectrum of relaxation times may be approximated by the Mooney modification[21] of the Rouse theory which leads to the frequency dependence of viscoelastic properties given in Eqs. 7 and 8, or by embellishments thereof.[52,53] If, however, the cross links have been introduced in a highly entangled system as is usually the case, the entanglements cannot be ignored. It is convenient to distinguish between trapped and untrapped entanglements; the criterion of entrapment is that all four strands radiating from an entanglement locus must be connected to the three-dimensional network.[54]

Statistical calculations show[54] that in a lightly cross-linked network a significant fraction of the entanglements may be untrapped, although the one or more loose strands radiating from them may be branched and entangled in the network so that configurational changes are very slow. Such networks do exhibit relaxation mechanisms with extremely long relaxation times in addition to those which can be approximated by the Rouse-Mooney spectrum, as illustrated in Fig. 8 for styrene-butadiene rubber cross linked to different extents.[55] The slow relaxation mechanisms have been attributed to dangling, branched, entangled structures[56], although alternative explanations in terms of cross-linking inhomogeneity have also been advanced[57].

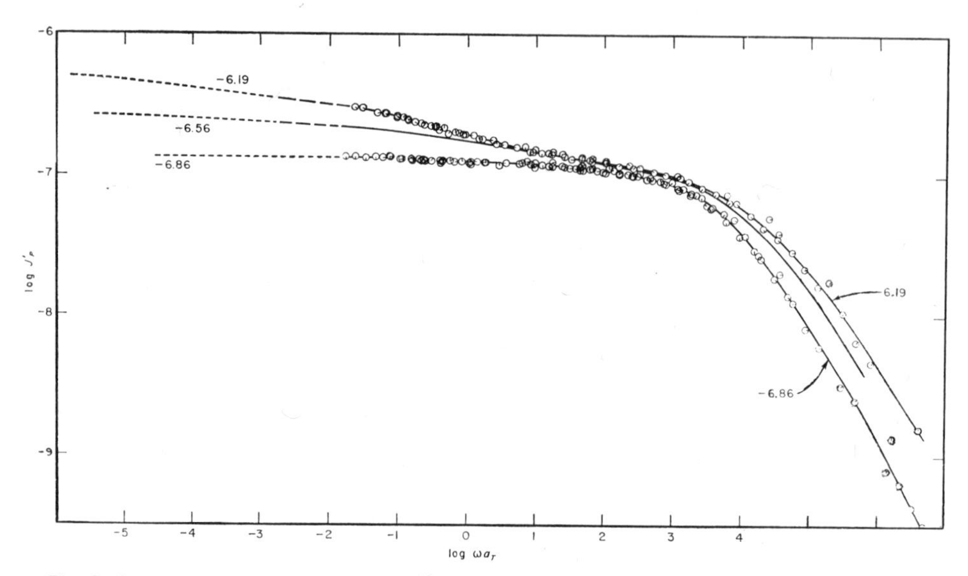

Fig. 8. Storage compliance, reduced to 15 C, plotted logarithmically for styrene-butadiene copolymer cross linked to three different extents as gauged by the values of log J_e shown at left.[55] *Frequency dependence at the right of 2 on the abscissa scale corresponds to transition zone; at the left, to additional slow relaxation mechanisms.*

The trapped entanglements must also influence viscoelastic properties but their role is uncertain. Presumably they contribute to the storage of free energy in deformation, but perhaps less than permanent cross links do and to an extent that may depend on both time scale and magnitude of strain. Some correlations between trapped entanglements and deviations from neo-Hookean stress-strain relations are evident[58], but behavior may be qualitatively different depending on whether the entanglements are more numerous than the cross links or vice versa. Recently, experiments on the cross-linking of entanglement networks in states of strain[59] have provided direct evidence of trapped entanglements and may clarify some of the above uncertainties.

In large deformations, the distribution of both trapped and untrapped entanglements will surely be different from that at rest, so relaxation spectra may be greatly distorted at high strains (as they are in noncross-linked systems at high strain rates). For example, studies of uniaxial extension of styrene-butadiene rubber over a wide range of constant strain rates[60] reveal certain rather slow time-dependent processes which appear only at large strains.

REFERENCES

1. Zimm, B. H., *J. Chem. Phys.*, **24**, 269 (1956).
2. Lodge, A. S., and Wu, Y. J., to be published.

3. Zimm, B. H., and Kilb, R. W., *J. Polymer Sci.*, **37**, 19 (1959).
4. Osaki, K., and Schrag, J. L., *Polym. Phys. Ed.*, **11**, 549 (1973).
5. Osaki, K., Mitsuda, Y., Schrag, J. L., and Ferry, J. D., *Trans. Soc. Rheol.* (to be submitted).
6. Osaki, K., Schrag, J. L., and Ferry, J. D., *Macromol.*, **5**, 144 (1972).
7. Osaki, K., *Macromol.*, **5**, 141 (1972).
8. Sakanishi, A., and Tanaka, H., *Zairyo*, **16**, 528 (1967).
9. Johnson, R. M., Schrag, J. L., and Ferry, J. D., *Polymer J. (Japan)*, **1**, 742 (1970).
10. Osaki, K., Mitsuda, Y., Johnson, R. M., Schrag, J. L., and Ferry, J. D., *Macromol.*, **5**, 17 (1972).
11. Mitsuda, Y., Osaki, K., Schrag, J. L., and Ferry, J. D., *Polymer J.*, **4**, 24 (1973).
12. Mitsuda, Y., Schrag, J. L., and Ferry, J. D., *Polymer J.* (submitted).
13. Massa, D. J., Schrag, J. L., and Ferry, J. D., *Macromol.*, **4**, 210 (1971).
14. Osaki, K., and Schrag, J. L., *Polymer J. (Japan)*, **2**, 541 (1971).
15. Peterlin, A., *J. Polymer Sci.*, Part A-2, **5**, 179 (1967); Part B, **10**, 101 (1972).
16. Bueche, F., *J. Chem. Phys.*, **20** 1959 (1952).
17. Barlow, A. J., Harrison, G., and Lamb, J., *Proc. Roy. Soc. (London)*, **A282**, 228 (1964).
18. Ferry, J. D., *Proc. 5th Intern. Cong. Rheol.*, **1**, 3 (1969).
19. Dannhauser, W., Child, W. C., Jr., and Ferry, J. D., *J. Colloid Sci.*, **313**, 103 (1958).
20. Berge, J. W., Saunders, P. R., and Ferry, J. D., *J. Colloid Sci.*, **14**, 135 (1959).
21. Mooney, M., *J. Polymer Sci.*, **34**, 599 (1959).
22. Sanders, J. F., and Ferry, J. D., *Macromol.*, **2**, 440 (1969).
23. Ferry, J. D., *Viscoelastic Properties of Polymers*, 2nd Ed., p 368, Wiley, New York, 1970.
24. *Ibid.*, p 320.
25. Williams, M. L., Landel, R. F., and Ferry, J. D., *J. Amer. Chem. Soc.*, **77**, 3701 (1955).
26. O'Reilly, J. M., *J. Polymer Sci.*, **57**, 429 (1962).
27. Fujita, H., and Kishimoto, A., *J. Chem. Phys.*, **34**, 393 (1961).
28. Fox, T. G., and Flory, P. J., *J. Appl. Phys.*, **21**, 581 (1950).
29. Graessley, W. W., *J. Chem. Phys.*, **54**, 5143 (1971).
30. Ziabicki, A., *Pure Appl. Chem.*, **26**, 481 (1971).
31. Onogi, S., Masuda, T., and Kitagawa, K., *Macromol.*, **3**, 109 (1970).
32. Ferry, J. D., Landel, R. F., and Williams, M. L., *J. Appl. Phys.*, **26**, 359 (1955).
33. Osaki, K., Tamura, M., Kurata, M., and Kotaka, T., *J. Phys. Chem.*, **69**, 4183 (1965).
34. Booij, H. C., Thesis, Leiden, 1970.
35. Masuda, T., Kitagawa, K., and Onogi, S., *Polymer J. (Japan)*, **1**, 418 (1970).
36. Yin, T. P., Lovell, S. E., and Ferry, J. D., *J. Phys. Chem.*, **65**, 534 (1961).
37. Tobolsky, A. V., Schaffhauser, R., and Böhme, R., *Polymer Letters*, **2**, 103 (1964).
38. Fujimoto, T., Ozaki, N., and Nagasawa, M., *J. Polymer Sci.*, Part A-2, **6**, 129 (1968).
39. Prest, W. M., Jr., *J. Polymer Sci.*, A-2, 8, 1897 (1970).
40. Odani, H., Nemoto, N., Kitamura, S., and Kurata, M., *Polymer J. (Japan)*, **1**, 356 (1970).
41. Nemoto, N.: *Polymer J. (Japan)*, **1**, 485 (1970); Nemoto, N., Moriwaki, M., Odani, H., and Kurata, M., *Macromol.*, **4**, 215 (1971).
42. Zosel, A., *Rheol. Acta*, **10**, 215 (1971); *Kolloid-Z. u. Z. Polymere*, **246**, 657 (1971).
43. Graessley, W. W., and Segal, L., *Macromol.*, **2**, 49 (1969).
44. Einaga, Y., Osaki, K., Kurata, M., and Tamura, M., *Macromol.*, **4**, 87 (1971).
45. Osaki, K., Kurata, M., and Tamura, M., *Polymer J. (Japan)*, **1**, 334 (1970).
46. Nemoto, N., Ogawa, T., Odani, H., and Kurata, M., *Macromol.*, **5**, 641 (1972).
47. Newlin, T. E., Lovell, S. E., Saunders, P. R., and Ferry, J. D., *J. Colloid Sci.*, **17**, 10 (1962).
48. Fujimoto, T., Narukawa, H., and Nagasawa, M., *Macromol.*, **3**, 57 (1970).
49. Masuda, T., Ohta, Y., and Onogi, S., *Macromol.*, **4**, 763 (1971).
50. Kraus, G., and Gruver, J. T., *J. Polymer Sci.*, Part A, **3**, 105 (1965).
51. Meyer, H. H., and Ring, W., *Kauts. Gummi*, **24**, 526 (1971).
52. Ref. 23, pp 263-265.
53. Chompff, A. J., and Duiser, J. A., *J. Chem. Phys.*, **45**, 1505 (1966).
54. Langley, N. R., *Macromol.*, **1**, 348 (1968).
55. Mancke, R. G., and Ferry, J. D., *Trans. Soc. Rheol.*, **12**, 335 (1968).
56. Langley, N. R., and Ferry, J. D., *Macromol.*, **1**, 353 (1968).

57. Labana, S. S., Newman, S., and Chompff, A. J., *Polymer Networks,* A. J. Chompff and S. Newman (Eds.), Plenum Press, New York (1971), p. 453.
58. Janácek, J., and Ferry, J. D., *J. Polymer Sci.,* A-2, **10**, 345 (1972).
59. Kramer, O., Ty, V., and Ferry, J. D., *Proc. Nat. Acad. Sci. U.S.*, **69**, 2216 (1972).
60. Smith, T. L., and Dickie, R. A., *J. Polymer Sci.,* A-2, 7, 635 (1969).

DISCUSSION on Paper by J. D. Ferry

TSCHOEGL: I would first like to make a comment and then ask a question. We have recently carried out a study in which we used block copolymer blends as model substances to elucidate the role of terminal chains in elastomeric systems. There are no terminal chains in a triblock copolymer rubber because all the rubbery (center) chains begin and end in glassy domains. By blending the triblock with known amounts of an appropriate diblock one can incorporate a controlled amount of terminal chains of known length and length distribution. Our measurements on the triblock as a function of frequency reveal a broad peak in $J''(\omega)$ which we ascribe to the trapped entanglements which exist in this material. In the triblock-diblock blends, this maximum is very much larger and shifts to lower frequencies. These measurements are in qualitative agreement with your observations.

Because of the mathematical complexity of calculating the behavior of polymer chains in dilute solution, one substitutes the well-known bead-and-spring model for the chain in the theories of Rouse and Zimm. In this model, each bead-and-spring represents several of the monomeric units of the chain. The sequence of these monomeric units is commonly referred to as the Rouse segment. The appealing feature of the original formulation of the theories was the fact that the Rouse segment appeared only as a working concept. No assumption needed to be made concerning its size because this dropped out of the final results. Since the discovery by Osaki of a computational error in the theory of Zimm, the size of the segment appears explicitly when the hydrodynamic interaction is neither dominant, nor vanishing, but takes on intermediate values. It is even possible in principle to determine the size of the segment. My question then is: does the Rouse segment have physical significance?

FERRY: In response to your comment, it is clear that your block copolymer systems provide a very promising means of investigating the roles of trapped and untrapped entanglements. I can answer your question as follows:

It is true that, in principle, the relaxation times are now functions of the number of submolecules, N, and hence of the number of monomer units per submolecule, P/N (P = degree of polymerization). However, if N is taken large (e.g., 200) and one limits consideration to behavior at not too high frequencies (e.g., $\omega < 10\tau_1^{-1}$), their dependence on N is very slight. At higher frequencies, application of the theory of Peterlin to experimental data necessitates assigning a value of P/N, and this parameter can be estimated in this way – it is 11 for polystyrene, and 16 for poly-α-methyl styrene. The physical significance of P/N is uncertain at present but may be clarified in the future.

ZAPAS: On your description of experimental results of small sinusoidal deformations superposed on constant shearing flow, you have suggested that the relaxation times tend to become smaller and sometimes zero. Recent experiments have shown that the superposed modulus $G'_{11}(\dot{\gamma},\omega)$ at high shear rates ($\dot{\gamma}$) and low frequencies (ω) becomes negative. This would suggest, according to your explanation, negative relaxation times. If that is the case, do you have any explanation regarding the meaning of those negative relaxation times?

FERRY: R. B. Bird and collaborators have introduced a constitutive equation which predicts negative values of G', but I cannot give any physical interpretation of this feature.

ONOGI: As you mentioned, branched polymers in the undiluted state show very complicated behavior in the terminal zone. Star-shaped polymers studied by us and comb-shaped polymers studied by Nagasawa and his co-workers show about 10 times larger J_e°, but we have difficulties in explaining this difference well. Could you give your opinion about this?

FERRY: The statements which I made about J_e° of branched polymers were in fact based on your work and Professor Nagasawa's. I do not know how to explain this striking behavior. Evidently the entanglements and the branch points interact in a complicated way.

KNAPPE: A problem also of practical interest is the influence of plasticizer concentration on viscosity and elasticity of PVC compounds in the temperature range of processing. Experimental results are scarce and sometimes in contradiction. Complications might arise from morphology and from specific interaction between the polymer and the plasticizer molecule. What can theory predict in this case?

FERRY: I believe that plasticized polyvinyl chloride is a somewhat pathological system; it appears to have the hyperentanglements which I mentioned and perhaps small crystalline regions as well. The simple principles which I have outlined cannot easily predict the behavior of such a material.

TSCHOEGL: The lack of superposition on the low-frequency side of the main transition peak in the loss compliance, which you observe in the polymers you have called "superentangled", also appears in our triblock-diblock copolymer blends, provided that the length of the terminal chains exceeds a certain minimum required for the formation of entanglements between these chains. We feel, in accord with Dr. Alfrey's comment, that this lack of superposition may be due to the presence of microcrystalline regions in the one system, and to glassy domains in the other. However, the reasons why the presence of such regions or domains should reveal entanglement features which are not observed in other amorphous polymers are not clear to us.

FERRY: Nor to me.

SHEN: You mentioned that the number of monomeric units per sub-molecule, N, is 11 for polystyrene and 16 for poly(α-methyl styrene). We have modified the Rouse theory for block copolymers and performed experiments on the block copolymer of styrene and α-methyl styrene. It is possible to calculate N with this treatment from maximum relaxation times. We found that N is of the order of 10 for poly(α-methyl styrene). The agreement with your results is interesting.

FERRY: This agreement is gratifying, though its significance is subject to careful scrutiny of the respective calculations.

SHEN: Professor Tschoegl just mentioned that he noted extra relaxations in polyblends of diblock and triblock copolymers. We have recently made measurements of the viscoelastic properties of polyblends of a triblock copolymer (SBS) with a homopolymer (polybutadiene). It was found that an additional loss peak can be seen, which is located between the loss peaks of the pure domain. The exact origin of this new peak is still obscure at present.

FERRY: Your observation of an additional loss peak is very interesting and certainly deserves further study.

HALPIN: Please comment about:

(1) The potential development of a reduced variable treatment for the glassy state, recognizing that glassy response is qualitatively different than rubberlike response.

(2) Your presentation of reduced variables explicitly states additivity of free volume for diverse sources. What is the state of experimental verification for this assumption?

FERRY: (1) I will leave discussion of the glassy state to others at this Colloquium.

(2) One way of testing additivity is to use temperature-dependence measurements to deduce the value of f_0 in a particular reference state and then to use concentration-dependence measurements (or, alternatively, pressure dependence) to obtain an independent value of f_0. In practice, they agree rather well, though not exactly. Such agreement provides important evidence of the utility of the free-volume treatment.

SCHWARZL: 1. Concerning the question of whether a time-temperature superposition principle could be formulated which works for polymers in the entire temperature region, I would like to make the following remarks: It is well known that relaxation processes occurring in the glassy state follow completely different shifting laws than the glass-rubber transition does. If, therefore, a generalized time-temperature superposition principle could be formulated by combining our knowledge of the shifting laws of the various relaxation processes occurring in polymers, this principle would become very complicated and it would depend on the chemical structure of the polymer in question. I therefore believe that such a principle would not be of much use for engineering applications.

2. The intimate connection of the free volume and the occurrence of the glass-rubber transition in shear was demonstrated at our institute by L.C.E. Struik. By prescribing an appropriate temperature history, the specific volume of a polymer may either decrease steadily or increase steadily or even go through a maximum as a function of time at constant temperature afterwards. It was shown that the position of the onset of the glass-rubber transition in shear shifted in a corresponding manner along the time axis for specimens with the same temperature pretreatment.

ON THE INTERCONVERSION OF LINEAR VISCOELASTIC FUNCTIONS WITH SPECIAL APPLICATION TO SHEAR BEHAVIOR IN THE GLASS-RUBBER TRANSITION REGION

F. R. Schwarzl

Centraal Laboratorium TNO
Delft, The Netherlands

ABSTRACT

The problem of interconverting measurable functions for linear viscoelastic materials is reviewed. A system of numerical formulae for such interconversions is presented, together with bounds for the errors. The information needed for this purpose is discussed.

The formulae are applied to data on the shear behavior of several amorphous high polymers in the glass-rubber transition region over wide ranges of time and temperature; the validity of the time-temperature superposition principle is discussed. The behavior of the viscoelastic material functions in the transition region is compared with that predicted by the diffusion theory of macromolecular networks.

1 INTRODUCTION

It is well-known[1,2] that the linear viscoelastic behavior of materials that obey the superposition principle may be characterized by various material functions, as for instance:

(a) *Creep compliance,* J(t), defined as the strain, as a function of time t, produced by a unit step in stress at time zero.

(b) *Relaxation modulus,* G(t), defined as the stress as a function of time, effected by a unit step in strain at time zero.

(c) *Storage compliance,* $J'(\omega)$, and *loss compliance,* $J''(\omega)$, defined as functions of angular frequency ω; these are the amplitudes of the in-phase components and the out-of-phase component of strain under conditions of steady-state response to a harmonic stress of angular frequency ω and unit amplitude.

(d) *Storage modulus,* $G'(\omega)$, and *loss modulus,* $G''(\omega)$, defined as the amplitudes of the in-phase component and the out-of-phase component of stress under conditions of steady-state response to a harmonic strain of angular frequency ω and unit amplitude.

(e) *Retardation spectrum,* $f(\tau)$, as a function of retardation time, τ; it is defined by the equation:

$$J(t) = J_o + \int_0^\infty f(\tau)\,[1-\exp(-t/\tau)]\,d\tau + t/\eta \quad , \qquad (1)$$

where J_o is the limit of the creep compliance for $t \to 0$, and $1/\eta$ is the limit of the rate of the creep compliance for $t \to \infty$; $f(\tau)$ is assumed to be a nonnegative function of the retardation time. This assumption is supported by a wealth of experimental evidence; its correctness follows from the experimental fact that the rate of creep is a completely monotonic function of time:

$$\dot{J}(t) \geqslant 0;\ \ddot{J}(t) \leqslant 0;\ \dddot{J}(t) \geqslant 0,\ \text{etc., for all } t > 0 \quad .$$

(f) *Relaxation spectrum,* $g(\tau)$, as a function of relaxation time τ; it is defined by the equation:

$$G(t) = G_\infty + \int_0^\infty g(\tau) \exp(-t/\tau) d\tau \quad , \qquad (2)$$

where G_∞ is the limit of the relaxation modulus for $t \to \infty$. If the retardation spectrum is a nonnegative function of the retardation time, the relaxation spectrum will be a nonnegative function of the relaxation time, and vice versa.[3]

Now, one of the problems frequently encountered in the investigation of relaxation behavior of polymers is to convert these characteristic material functions one into another. These problems have formally been solved by the theory of linear viscoelastic behavior, according to which the functions can be interconverted by means of linear integral transforms. As has been shown elsewhere[2], however, the actual application of these integral transforms to experimental data gives rise to basic difficulties and, in addition, to tedious calculations.

This prompted us to study the questions outlined below under the assumptions that one of the measurable characteristic material functions, defined under headings (a) through (d) above*, has been measured; that a finite number of measuring points is available at discrete times (angular frequencies) extending over a finite range on the time axis (frequency axis); neither the behavior at very short times (high frequencies) nor that at very long times (low frequencies) will be known; each measurement has a finite experimental error.

1. Is it, under these assumptions, still possible to convert the function measured into other measurable material functions?
2. How much experimental information is actually needed for this conversion?
3. What is the simplest numerical procedure for this purpose?
4. What are the truncation errors, i.e. the errors due to the fact that information is available only over a finite range of the time (frequency) scale?
5. What are the approximation errors, i.e. the errors due to the use of simple approximation formulae instead of integral transforms, even within the limited range in which the behavior has been measured?
6. How is the experimental error transmitted to the new function during the conversion process?

Answers to these questions have been obtained for all but two of the conversion problems mentioned above. They will be shortly reviewed.

2 DERIVATION OF APPROXIMATION FORMULAE AND ERROR BOUNDS

The method for deriving approximation formulae was first introduced by Ninomiya and Ferry.[4] It was supplemented with an estimation of the magnitude of the errors.[2] The method is illustrated here by a discussion of

*Note that we exclude the spectra!

the following simple approximation formula for calculating the creep compliance from values of the storage and loss compliances:

$$J(t) \sim A(t) = J'\,(1/t) + d \cdot J''\,(1/2t) + e \cdot J''\,(1/t) \qquad (3)$$

with d = 0.566 and e = −0.203.

It was derived from the representation of the viscoelastic material functions as integral transforms of the retardation spectrum[1], given by Eq. 1, and by

$$J'\,(\omega) = J_o + \int_0^\infty f(\tau)\,\frac{1}{1+\omega^2\tau^2}\,d\tau \qquad (4)$$

$$J''\,(\omega) = \int_0^\infty f(\tau)\,\frac{\omega\tau}{1+\omega^2\tau^{\cdot 2}}\,d\tau + 1/\omega\eta\ . \qquad (5)$$

The integrands of these expressions consist of the retardation spectrum times a function of $x = t/\tau$ or $x = 1/\omega\tau$, which is called the intensity function of the corresponding expression. The intensity functions of $J(t)$, $J'(\omega)$, $J''(\omega)$, and $A(t)$ are, respectively,

$$\chi(x) = 1 - e^{-x} \qquad (6)$$

$$\chi'(x) = x^2/\,(1 + x^2) \qquad (7)$$

$$\chi''(x) = x/(1 + x^2) \qquad (8)$$

$$\phi(x) = \chi'(x) + d \cdot \chi''(2x) + e \cdot \chi''(x)\ . \qquad (9)$$

The constant coefficients d and e in Eq. 3 have been determined in such a way that the intensity function $\chi(x)$ is approximated by $\phi(x)$ as closely as possible. Its error is defined by:

$$E(t) = A(t) - J(t) = \left[2d + e - 1\right]\frac{t}{\eta} + \int_0^\infty f(\tau)\,\Delta(x)d\tau\ . \qquad (10)$$

The corresponding intensity function is found to be

$$\Delta(x) = \phi(x) - \chi(x) = \chi'(x) - \chi(x) + d \cdot \chi''(2x) + e \cdot \chi''(x)\ . \qquad (11)$$

Next, we compare the intensity function of the error with that of either $J(t)$ or $J''(\omega)$. The behavior of $\{\Delta(x)/\chi(x)\}$ and of $\{\Delta(x)/\chi''(x)\}$ as functions of x will depend on the fixed values chosen for the constants d and e. The optimum values given in Eq. 3 were found by trial and error. The functions $\{\Delta(x)/\chi(x)\}$ and $\{\Delta(x)/\chi''(x)\}$ are plotted for these values in Fig. 1. The quotient $\Delta(x)/\chi''(x)$ tends to -0.07 for $x \to 0$; it has a maximum of 0.08 at x = 0.2 and a minimum of -0.08 for x = 4; it tends to 0.08 for $x \to \infty$. Simultaneously, the quotient $\Delta(x)/\chi(x)$ tends to -0.07 for $x \to 0$, and shows a maximum equal to 0.076 at x = 0.2 and some smaller extrema at higher values of x; it tends to zero for $x \to \infty$.

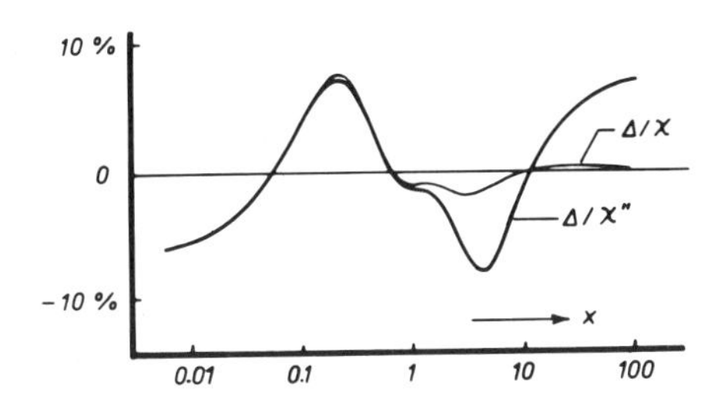

Fig. 1. $\Delta(x)/\chi(x)$ and $\Delta(x)/\chi''(x)$ as functions of x for Eq. 3.

We thus arrive at the following inequalities, valid for all positive values of x

$$-8\% \leqslant \{\Delta(x)/\chi''(x)\} \leqslant +8\% \tag{12}$$

$$-7\% \leqslant \{\Delta(x)/\chi(x)\} \leqslant +7.6\% \quad . \tag{13}$$

These yield the following bounds for the error of the approximation

$$-8\%\ J''(1/t) \leqslant E(t) \leqslant +8\%\ J''(1/t) \tag{14}$$

$$-7\%\ J(t) \leqslant E(t) \leqslant +7.6\%\ J(t) \quad . \tag{15}$$

By dividing these inequalities by J(t) and using the well-known inequality

$$J''(1/t) = J'(1/t) \cdot (\tan\delta) \leqslant J(t) \cdot (\tan\delta) \quad , \tag{16}$$

we obtain the following bounds for the relative error

$$-8\% \tan\delta \leqslant \frac{E(t)}{J(t)} \leqslant 8\% \tan\delta \tag{17}$$

$$-7\% \leqslant \frac{E(t)}{J(t)} \leqslant 7.6\% \quad . \tag{18}$$

These bounds obviously depend on the value of the damping at the point where the approximation is applied. For small damping values, they increase proportionally with the value of the damping, and for large damping values they approach an upper limit independent of the damping.

It is possible to derive slightly more restrictive bounds for Eq. 3. These are plotted in Figs. 2 and 3.*

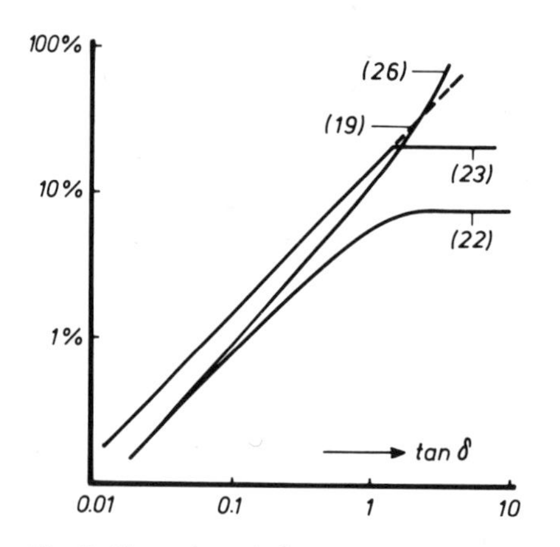

Fig. 2. Upper bounds for the relative error of Eqs. 19, 22, 23, and 26 as functions of the value of the damping.

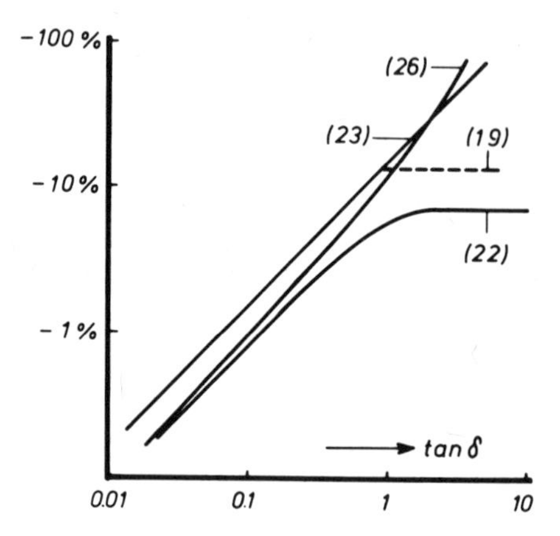

Fig. 3. Lower bounds for the relative error of Eqs. 19, 22, 23, and 26 as functions of the value of the damping.

3 CLASSIFICATION OF CONVERSION PROBLEMS

Using the above method we have investigated most of the conversion problems occurring in the theory of linear viscoelasticity. For each problem, it was assumed that the available measuring points of the respective function were evenly spaced on a logarithmic time or frequency scale, their distance corresponding to a factor of two for each ratio of successive times or frequencies. A set of numerical equations of increasing complexity and accuracy was derived, together with bounds for the relative error of each equation. The experimental information needed for a certain conversion problem depends on the value of the damping, tan δ, at the angular frequency where the conversion is to be performed. According to the information needed, we may classify the various problems as indicated in Table I.

We will restrict ourselves to a simple system of numerical equations (see Table II) which may often be applied successfully.
All these equations are extremely simple. Most of them do not require more than the knowledge of two measuring points at successive frequencies (times) for calculating one point of the desired function. If the system is applied to a series of n logarithmically equidistant measuring points so spaced that their distance corresponds to a factor of two in frequency or time, one arrives at a series of (n-1) logarithmically equidis-

*In these figures, Eq. 3 has been changed to Eq. 22. See also Table 2.

Table I. Classification of Conversion Problems

	Conversion	Remarks	Extensively Treated in Ref.
I	$J(t) \rightarrow J'(\omega)$	Easy for all values of tan δ	5
II	$J(t) \rightarrow J''(\omega)$	Difficult, and particularly tedious for low tan δ values	5
III	$J''(\omega), J''(\omega) \rightarrow J(t)$	Easy for all values of tan δ	6
IV	$G(t) \rightarrow G'(\omega)$	Easy for all values of tan δ	7
V	$G(t) \rightarrow G''(\omega)$	Very difficult for all tan δ values	7
VI	$G'(\omega), G''(\omega) \rightarrow G(t)$	Easy for small values of tan δ; difficult for large tan δ	8
VII	$J'(\omega) \leftrightarrow J''(\omega)$	Not yet investigated	
VIII	$G'(\omega) \leftrightarrow G''(\omega)$		

Table II. A Set of Simple Numberical Equations for the Conversion of Viscoelastic Functions

Numerical Equations ($\omega = 1/t$)	Bounds for Relative Error, %	Eq.
$J'(\omega) \sim J(t) - 0.855\ [J(2t)-J(t)]$	± 15 tan δ	19
$J''(\omega) \sim 2.12\ [J(t)-J(t/2)]$	Very unreliable	20
$J''(\omega) \sim 0.125\ J''(8\omega) + 1.93\ [J(t)-J(t/2)]$	± 25	21
$J(t) \sim J'(\omega) + 0.566\ J''(\omega/2) - 0.203\ J''(\omega)$	± 8 tan δ; ± 7.6	22
$G'(\omega) \sim G(t) + 0.855\ [G(t)-G(2t)]$	± 15 tan δ	23
$G''(\omega) \sim 2.12\ [G(t/2)-G(t)]$	Very unreliable	24
$G''(\omega) \sim 0.125\ G''(8\omega) + 1.93\ [G(t/2)-G(t)]$	± 25	25
$G(t) \sim G'(\omega) - 0.566\ G''(\omega/2) + 0.203\ G''(\omega)$	± 8 tan δ/ [1 - tan δ]	26

tant points for the resulting function. This series may then serve as a starting point for the next conversion. By applying the system of Table II to a series of equidistant points of a single viscoelastic function, all other viscoelastic functions can be derived without any further interpolation. With each step of the conversion, one measuring point is lost, and the experimental window becomes narrower by a factor of two.

In view of their simplicity, the equations cannot be expected to be very accurate. Estimates of their error bounds are indicated in the second column of Table II. More restrictive error bounds for Eqs. 19, 22, 23 and 26 are plotted versus the damping in Figs. 2 and 3. All four approximations are seen to be sufficiently accurate for low values of the damping ($\tan \delta < 0.2$); they can be safely applied in the entire glassy region of amorphous polymers, and almost everywhere for semicrystalline polymers. They might fail in the glass-rubber transition region of amorphous polymers, where damping values may reach the order of magnitude of unity, and in the onset of the flow region of noncrosslinked polymers.

More accurate formulae can be derived for Conversions I, III, IV, and VI (see Table I). When eight or nine equidistant measuring points are used instead of two, formulae may be derived for Conversions I, III, and IV, which are accurate to within 1 percent for all damping values. A more refined formula, involving eight equidistant terms for Conversion VI, will be accurate to within 1.5 percent for damping values up to unity. It will fail, however, for higher damping values. We therefore classified this conversion problem as easy for small damping values but difficult for high damping values.

More serious difficulties are met in Conversions II and V. Equations 20 and 24 were designated as unreliable for the following reason. Equation 20, e.g., involves only the difference $J(t)$-$J(t/2)$, which is proportional to the logarithmic creep rate in the vicinity of the point $t = 1/\omega$. An analysis of the problem showed[5], however, that one reliable formula for $J''(\omega)$ should be:

$$
\begin{aligned}
J''(\omega) \sim A''(t) = &-0.479\,[J(4t) - J(2t)] + 1.674\,[J(2t) - J(t)] \\
&+0.198\,[J(t) - J(t/2)] + 0.620\,[J(t/2) - J(t/4)] + 0.012\,[J(t/4) - J(t/8)] \\
&+ 0.172\,[J(t/8) - J(t/16)] + 0.0433\,[J(t/32) - J(t/64)] + \\
&+0.0108\,[J(t/128) - J(t/256)] + \ldots
\end{aligned} \tag{27}
$$

This formula represents an infinite series. The terms following the one with the coefficient 0.0108 will be shifted by a factor of four on the time scale in the direction of shorter times, each relative to its predecessor, and will have a coefficient which is exactly one-fourth of the coefficient of its predecessor. Equation 27 is very accurate. Its relative error is bounded by -2.7 percent and +2.7 percent for all damping values.

The main difference between Eqs. 20 and 27 is the presence of the short time tail in the latter formula. This shows that the value of the loss compliance at the angular frequency $\omega = 1/t$ is strongly influenced by the logarithmic creep rate around the time t, as well as by the logarithmic creep rate in the time region to the left of point t. It will therefore depend on the situation whether or not Eqs. 20 and 27 will yield about the same result or not. If the main contribution to $J''(\omega)$ comes from the logarithmic creep rate around t (i.e., from the principal terms 1.6 [J(2t) - J(t)], 0.6 [J(t/2) - J(t/4)] in Eq. 27), both formulae will yield about the same and the correct result. In this case, the simple formula (Eq. 20) may be safely applied. If, however, the main contribution to $J''(\omega)$ comes from the short time tail of Eq. 27, only Eq. 27 will yield the correct result, while application of the truncated formula Eq. 20 would yield values for $J''(\omega)$ that are far too low. This will be the case when the logarithmic creep rate in the short time tail is much higher than in the vicinity of point t. This situation will be met when the calculation is performed just at the end of a transition region in J(t). In this case, the calculation of $J''(\omega)$ from J(t) or of $G''(\omega)$ from G(t) will be a very difficult problem.

This "short time truncation problem" may be circumvented when information is available on the magnitude of the short time tail from other sources. In practice, data are often available from dynamic measurements, as well as from creep measurements. The calculation of the loss compliance from creep data can then be considerably simplified. Instead of the unreliable formula (Eq. 20) we can use Eq. 21 which is much more reliable. Its error bounds are ±25 percent for all damping values. This is due to the presence of the term 0.125 $J''(8\omega)$ in Eq. 21, which gives an estimate of the contribution of the short time tail to $J''(\omega)$. In Eq. 21, the term 0.125 $J''(8\omega)$ is known from dynamic measurements at somewhat higher frequencies, while the term 1.93 [J(t) - J(t/2)] is calculated from the results of creep measurements. Again, more accurate formulae of this type are available which involve more measuring points of the creep curve.[7]

4 APPLICATION TO MEASUREMENTS IN THE GLASS-RUBBER TRANSITION REGION

A combined torsional pendulum and torsion creep apparatus was used to measure the shear behavior of amorphous polymers in the glass-rubber transition region over an experimental window of 5 decades on the time or frequency scale. The apparatus has been described in more detail elsewhere.[9] It is sufficient here to mention that torsional vibration and torsional creep measurements can be performed on the same specimen immediately after each other without disturbing the temperature equilibrium in the thermostat. The temperature history of the specimen could be accurately prescribed by means of a thermostatic device using dry nitrogen. The shear behavior of polymers in the low-temperature tail of the glass-rubber transition region is known[10] to depend not only on the temperature of the specimen and the loading time (or vibration frequency), but also on the details of the previous temperature history of the sample. We have, therefore, carefully standardized the complete temperature history before and during the measurement in the following way: the specimen was first heated to about 15 C above its glass-transition temperature, T_g, and next annealed by cooling between glass plates at a rate of 1°C/hr to the lowest measuring temperature in the glassy state. It was conditioned at this temperature for 30 minutes. Vibration measurements were then performed at three frequencies, followed by a creep measurement during 2,048 seconds. Next, the specimen was unloaded, heated to the next higher measuring temperature, and allowed to recover at this temperature for another half hour. The measuring cycle was then repeated at this temperature. With this procedure it took some days to measure the complete shear behavior of one specimen. During this period, the specimen remained clamped in the apparatus, and the thermostat was kept overnight at the temperature of the previous recovery period. The following materials were investigated*:

(a) A cross-linked polyurethane rubber (PU) prepared at this laboratory. It has a rubbery network structure with trifunctional bulky cross links. The polypropylene ether chains that interlink the cross links have a mean molecular weight of 2000; its glass transition temperature, T_g, has been measured dilatometrically as -52 C.

(b) A polymethyl methacrylate (PMMA) polymerized at this laboratory. Its glass transition temperature was measured as 105 C.

(c) A rigid polyvinyl chloride (PVC) of the suspension type (Solvic 229) processed by means of extrusion at 190 C, whose glass transition temperature was unknown. In deviation from both other materials, PVC is assumed to be not completely amorphous; it shows a certain degree of crystallinity.

Figures 4, 5, and 6 show double logarithmic plots for a number of temperatures of the creep compliance of these materials versus the creep time. Open circles denote measuring points obtained directly by the digital creep method; crosses denote the values of the creep compliance as calculated from the free-vibration technique at the three frequencies of the vibration experiment. For this purpose, the equations of Table II were applied. The agreement between the results of both methods is very good. This, in effect, proves the applicability of the conversion methods used to arrive at these results.

The plots cover three orders of magnitude on the compliance scale and five orders of magnitude on the time scale. In spite of this, it is

---

*The author is indebted to Dr. Heijboer, Mr. Bree, and Dr. Den Otter for providing samples of PMMA, PU, and PVC.

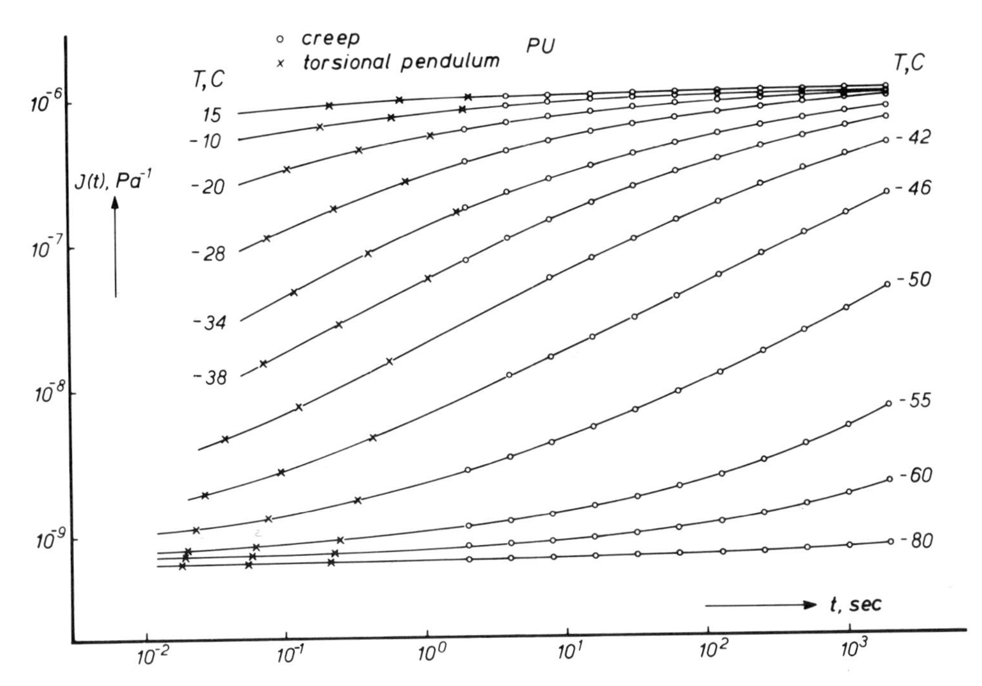

Fig. 4. Creep compliance in shear, versus time, for polyurethane rubber at various temperatures.

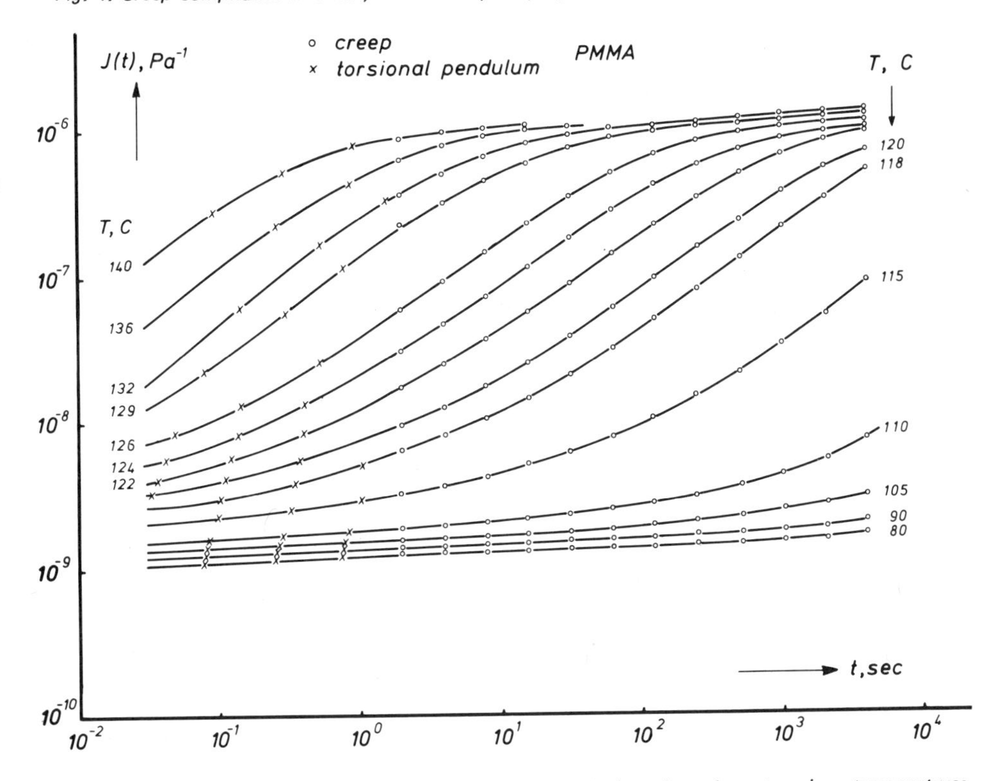

Fig. 5. Creep compliance in shear, versus time, for polymethyl methacrylate at various temperatures.

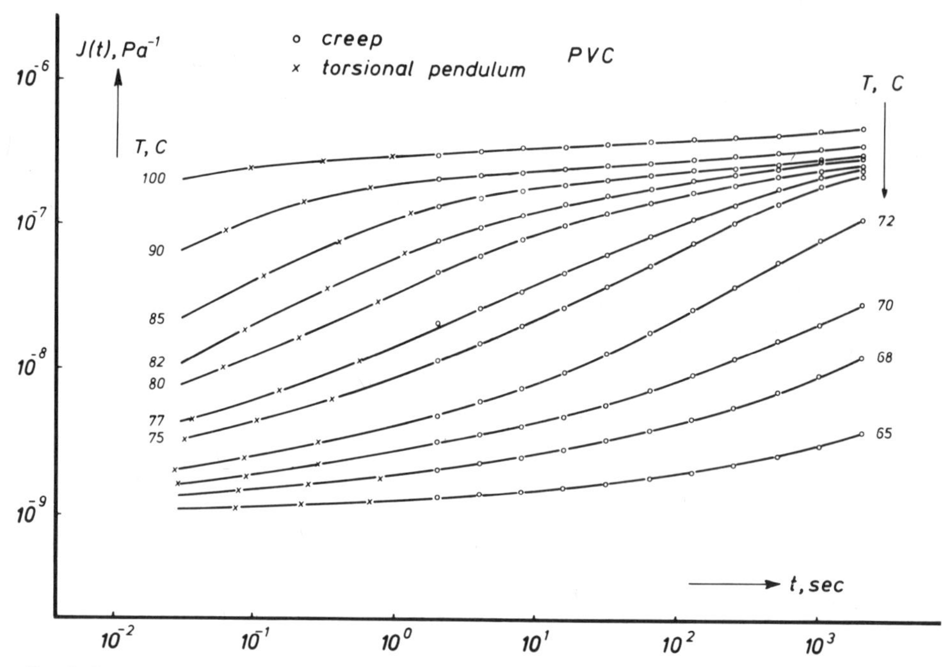

Fig. 6. Creep compliance in shear, versus time, for rigid polyvinyl chloride at various temperatures.

impossible to cover the complete glass-rubber transition at a single temperature. In all cases the strong influence of time and temperature on the compliance in the transition region is evident. However, significant differences are observed in the behavior of these three materials in the glass-rubber transition. Some of these differences are illustrated in Table III, where the following characteristics are listed: the height (H) of the step of the dispersion, in decades; the maximum value (n) of the double logarithmic slope of the compliance-time curves; the width (w) of the dispersion of J(t) in temperature scale at a characteristic time of 16 seconds; the characteristic temperature (T_s) where the compliance at 16 seconds reaches its logarithmic half-value of the dispersion step; and the glass-transition temperature.

Table III. Some Characteristics of the Glass-Rubber Transition

Material	H, decades	n	w, C	T_s, C	T_g, C
PU	3.1	0.48	60	-45	-52
PMMA	2.7	0.65	30	121	105
PVC	2.4	0.54	30	73	

The courses of the creep compliances (at 16 seconds) versus temperature for the three materials are compared in Fig. 7. Note that the unit of the temperature scale is the same for the three materials. A constant shift of the temperature scale has been applied only in the case of PU to facilitate comparison. It is very remarkable that the slope of the logarithm of compliance versus temperature is the same in the transition zone of all three materials, viz., 1.5 decades/10 C. In this figure also, the temperatures, T_s at the midpoints of the transition are indicated by arrows.

The curve for PVC consists of two parts of different slopes. Only the part with the higher slope is considered to be the proper glass-rubber transition. The other part in the temperature region between 90 C and 140 C is considered to be connected with a gradual melting or recrystallization process of the crystallite regions which are believed to be present in this PVC.

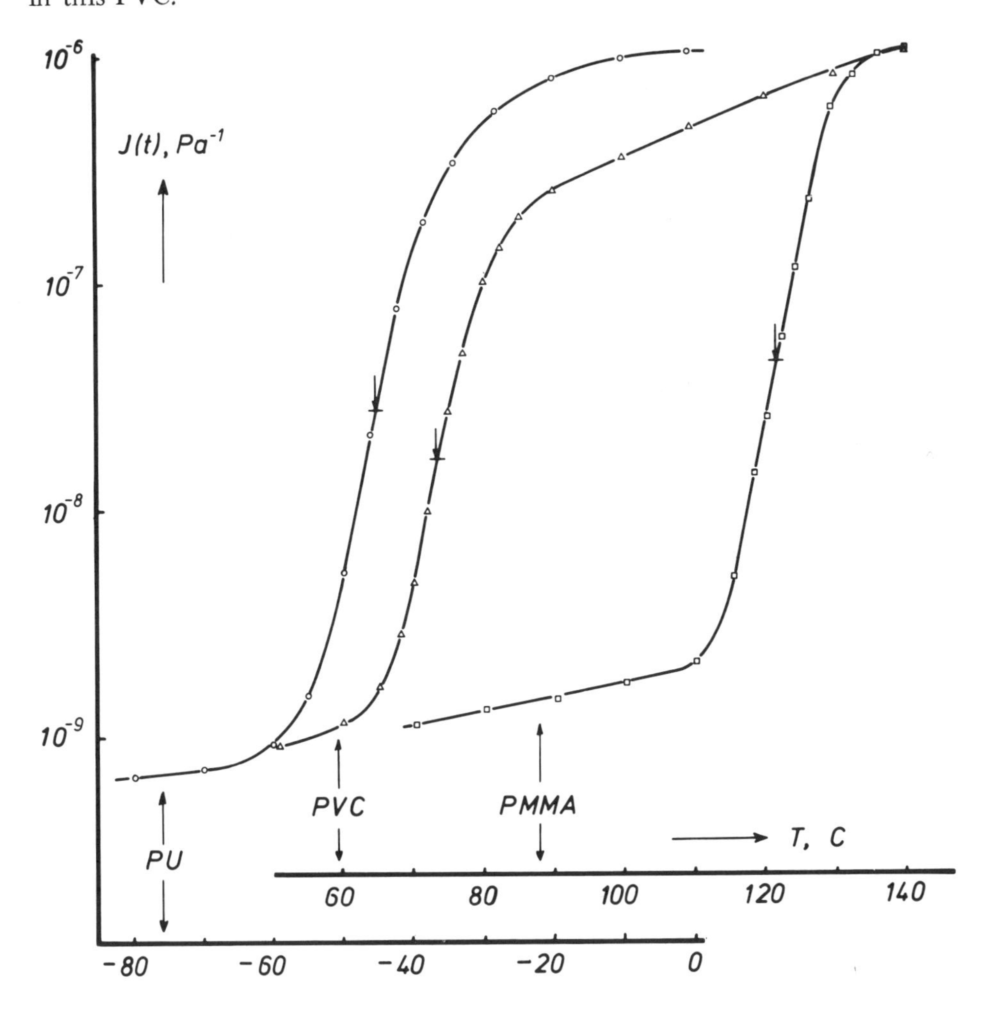

Fig. 7. Creep compliance in shear after 16 seconds versus temperature, for PU, PVC, and PMMA in the transition regions. The temperatures, T_s, at the midpoints of the transition are indicated by arrows.

5 APPLICABILITY OF THE TIME-TEMPERATURE SHIFT METHOD

The data were transformed by the conversion procedures into dynamic storage and loss moduli as functions of temperature and frequency, υ. Storage and loss shear moduli were divided by the absolute temperature, and plotted against frequency with temperature as a parameter. The resulting curves for PMMA are shown in Figs. 8 and 9. Curves for the storage moduli versus frequency for PVC are shown in Fig. 10.

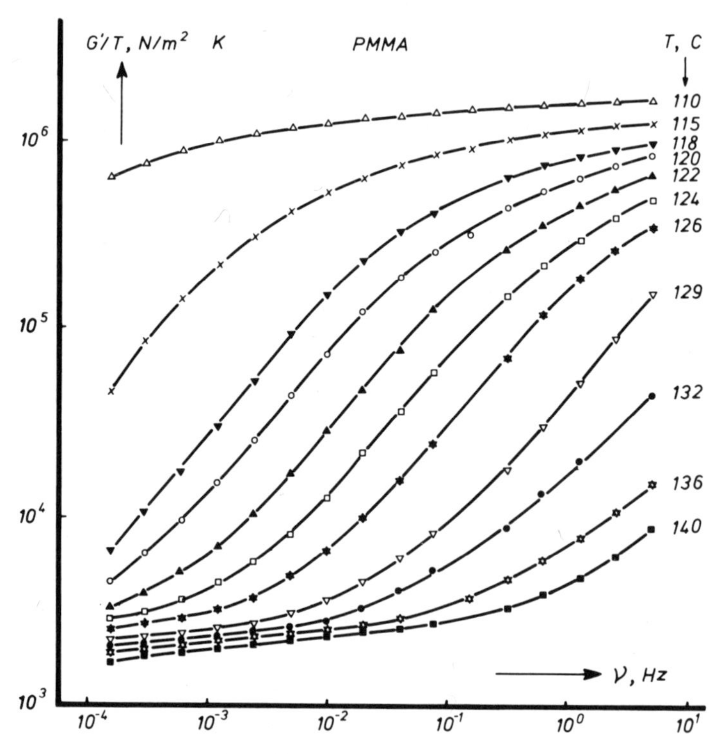

Fig. 8. Storage modulus in shear divided by absolute temperature, versus frequency, for polymethyl methacrylate at various temperatures.

It is generally assumed that data concerning the shear behavior of amorphous high polymers in the glass-rubber transition region obey the WLF time-temperature superposition principle.[11] This principle implies that storage and loss shear moduli, after division by density and absolute temperature, are unique functions of the reduced frequency, z, in such a way that:

$$\frac{G'(\upsilon,T)}{\rho T} = F'(z); \; \frac{G''(\nu,T)}{\rho T} = F''(z) \quad . \tag{28}$$

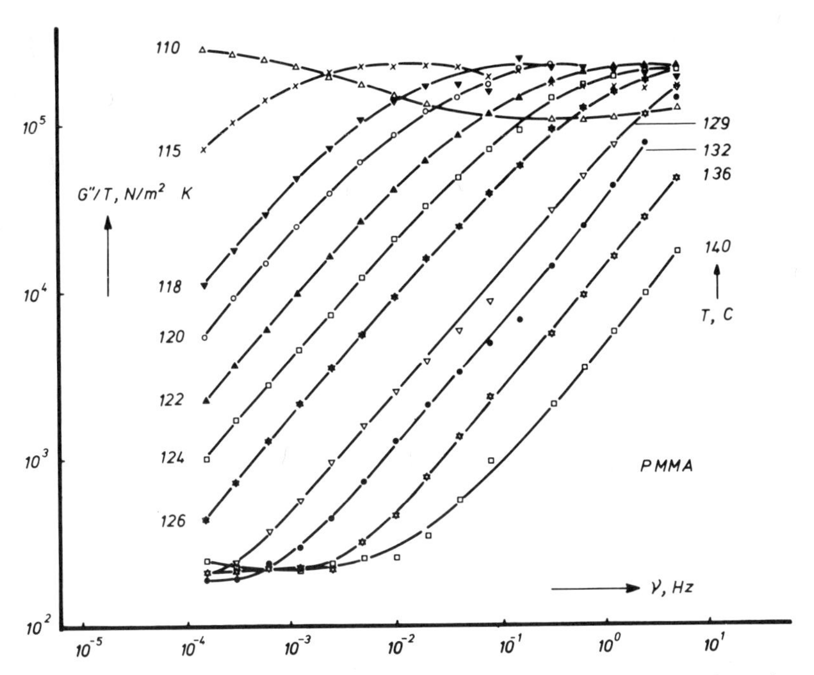

Fig. 9. Loss shear modulus divided by absolute temperature, versus frequency, for polymethyl methacrylate at various temperatures.

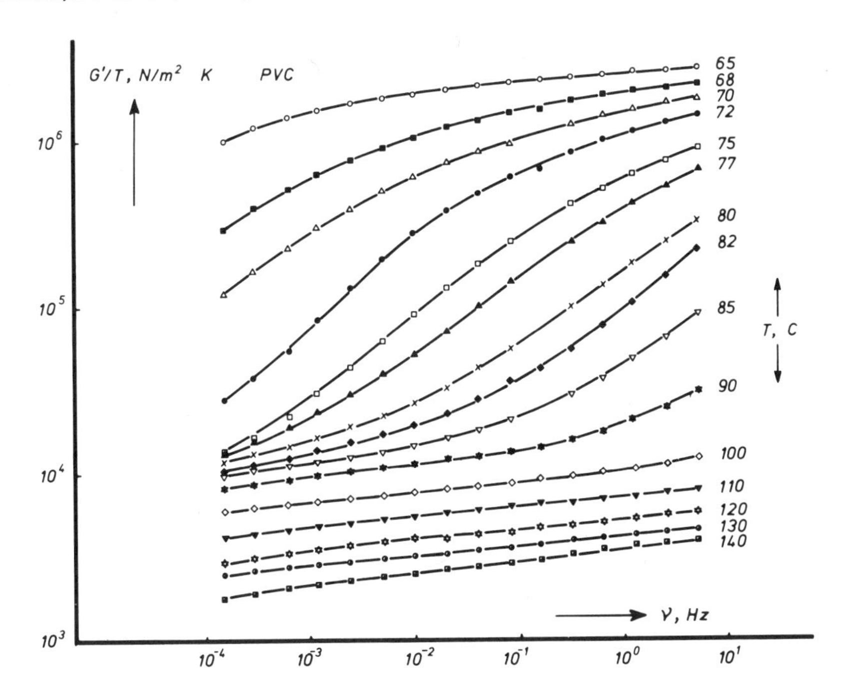

Fig. 10. Storage modulus in shear divided by absolute temperature, versus frequency, for PVC at various temperatures in the transition region and above.

Moreover, the reduced frequency should be given by:

$$\log z = \log \upsilon + \log a(T,T_o) \tag{29}$$

with

$$\log a(T,T_o) = -\frac{c_1(T-T_o)}{c_2 + T - T_o}\ , \tag{30}$$

where c_1 and c_2 are two constants that depend on the nature of the polymer and on the value, T_o, chosen for the reference temperature.

In view of these equations, it should be possible to reduce the data of Figs. 8 and 9 by a horizontal shift along the logarithmic frequency axis to unique master curves. A prior division of the curves of Figs. 8 and 9 by the appropriate densities proved to be without significant effect.

A reference temperature of T_o = 126 C was chosen, and curves for other temperatures were shifted on the logarithmic frequency scale until single master curves for G'/T and G''/T were obtained. This shifting procedure did not yield perfect superposition of the data. Significant deviations remained for G'/T curves at temperatures of 132 C and 136°C, and for G''/T curves at temperatures of 110 C and 115 C (see Fig. 12).

Equal shift values were applied to curves for G'/T and G''/T. The magnitude of the shift was determined from the curves for G'/T for temperatures between 110 C and 126 C. Between 129°C and 140°C, the magnitude of the shift could be determined more easily from the G''/T versus frequency curves.

A similar procedure could be applied to the data for polyurethane rubber. The values of log a, the magnitude of the shift, are listed for PU and PMMA in Table IV.

The shift function is plotted versus the temperature in Fig. 11. To facilitate comparison between the shift functions of these materials, the shift has been recalculated with the characteristic temperature, T_s, as the new reference temperature (cf. Table III). Next, the shift $\log a(T,T_s)$ was plotted against $T-T_s$. It is seen from Fig. 11 that no significant differences between the shift functions of the materials are obtained, if the reference temperature is defined appropriately. This is remarkable since the T_g values of these materials differ by more than 150 C. It is possible to describe the shift function by the "universal" time-temperature shift relation

$$\log a(T,T_s) = -\frac{d_1(T-T_s)}{d_2 + T - T_s}\ , \tag{31}$$

with d_1 = 15.52 C and d_2 = 58.0 C, which was proposed by Williams, Landel, and Ferry.[11] In this relation, the only material parameter is the characteristic temperature T_s.

The universal time-temperature shift relation was proposed originally in the form

$$\log a\,(T,T_u) = -\frac{f_1\,(T-T_u)}{f_2 + T - T_u} \tag{32}$$

Table IV. The Time-Temperature Shift Functions as Observed for Polyurethane Rubber and Polymethyl Methacrylate

Material			
PU, T_o = -42 C		PMMA, T_o = 126 C	
T, C	log a	T, C	log a
-70	8.41	110	5.40
-60	6.07	115	3.14
-55	4.42	118	1.94
-50	2.52	120	1.51
-46	1.10	122	0.97
-42	0.00	124	0.49
-38	-0.95	126	0.00
-34	-1.75	129	-0.77
-28	-2.81	132	-1.23
-20	-3.93	136	-1.75
-10	-5.12	140	-2.34
0	-6.13		
15	-7.41		
30	-8.36		

where f_1 = 8.86 and f_2 = 101.6.

It is easily seen that Eqs. 31 and 32 are completely equivalent, their only difference being the choice of the reference temperatures, which are connected by

$$T_u - T_s = 43.6 \text{ C} \;. \tag{33}$$

Master curves of G′/T and G″/T are shown for PMMA in Fig. 12, and for PU in Fig. 13. It was impossible to reproduce all experimental points on the master curves. To prevent overcrowding, about two thirds of them had to be omitted. The fit of the experimental points to the master curves is better for the cross-linked rubber (PU) than for the noncross-linked PMMA. The transition for PMMA is shorter in the log frequency scale than that for PU.

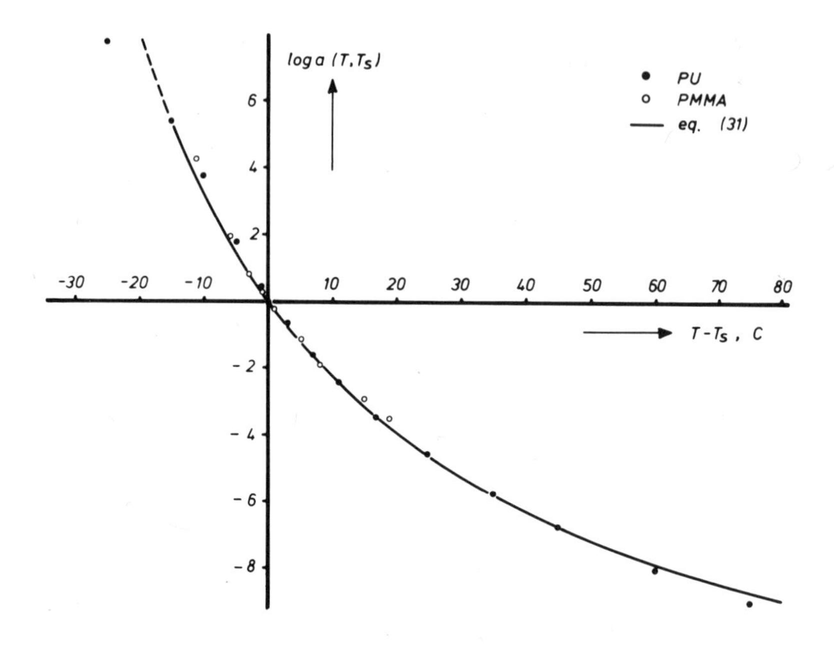

Fig. 11. Time-temperature shifts relative to the reference temperature, T_S, as functions of $T - T_S$ for PU and PMMA. Also shown is the theoretical line of the WLF equation according to Eq. 31.

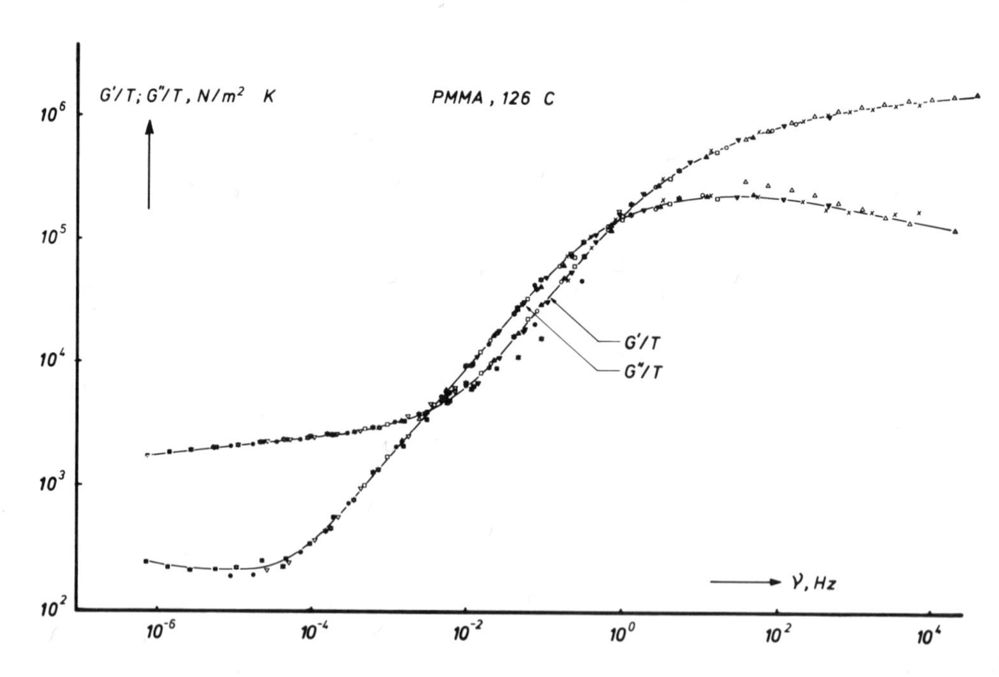

Fig. 12. Master curves of G'/T and G''/T for polymethyl methacrylate, reduced to a temperature of T_o = 126 C. Key to the symbols of the measuring points is the same as that in Figs. 8 and 9.

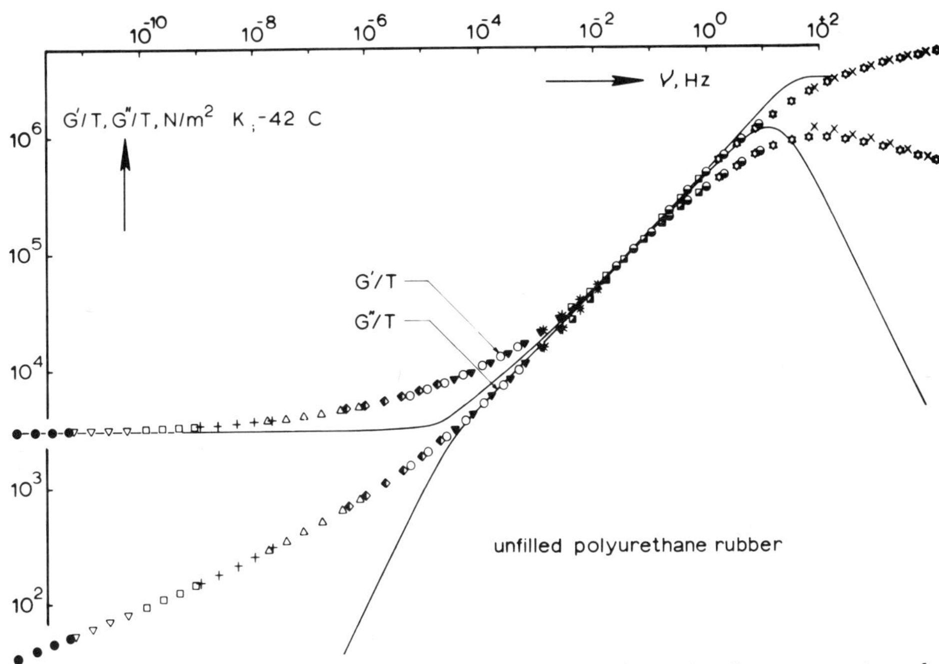

Fig. 13. Master curves of G′/T and G″/T for polyurethane rubber, reduced to a temperature of T_o = -42 C.

The master curves of G′/T and G″/T for PU are compared in Fig. 13 with the predictions of the modified theory of Rouse[12] for a cross-linked rubbery network. For this purpose, the value of the modulus in the rubbery region has been adjusted to the level found experimentally, and the number of Gaussian subchains has been arbitrarily chosen as N = 1000 (for details see Ref. 9).

Reasonable qualitative agreement is observed between theoretical and measured values of the storage shear modulus. The value of the double logarithmic slope in the transition region, namely 0.5, is particularly characteristic of the theory; it follows directly from the assumption of free draining, which, in this case, appears to hold. That this feature is not a general one is seen from the value of the double logarithmic slope of the transition curve of PMMA, which is 0.65, and therefore contradicts the assumption of the free draining.

It should finally be mentioned that our attempts to apply the time-temperature superposition procedure to the data for PVC failed. This may be seen from the curves in Fig. 10. Shear-modulus curves between 65 C and 75 C might be brought to approximate superposition by a horizontal shifting procedure. This does not hold, however, for curves in the temperature region above 75 C. Here a horizontal and a vertical shift would be necessary to effect superposition of the curves. We believe that in this temperature region, crystallization proceeds during the measuring cycle, changing the effective stiffness of the network in its rubbery state by an increase in crystallinity. Therefore, the shifting procedure is not applicable in its original form. It seems to be too early to try to correct for this effect by a simultaneous horizontal and vertical shift.

CONCLUSIONS

1. The interconversion formulae given in Table II have been shown to be valuable tools for determining the shear properties of polymers in the glass-rubber transition region over a wide experimental window.
2. The time-temperature superposition procedure was found to be applicable to the amorphous polymers PU and PMMA; it failed for the slightly crystalline PVC.
3. The position of the midpoint of the glass-rubber transition on the time-temperature plane obeyed the universal WLF equation.
4. The shapes of the master curves as functions of frequency depend on the topology of the molecular network of the polymer.

REFERENCES

1. Ferry, J. D., *Viscoelastic Properties of Polymers,* Wiley, New York (1961).
2. Schwarzl, F. R., and Struik, L.C.E., *Adv. Mol. Rel. Proc.,* **1** 201-255 (1968).
3. Gross, B., and Pelzer, H., *J. Appl. Phys.,* **22,** 1053 (1951).
4. Ninomiya, K., and Ferry, J. D., *J. Colloid Sci.,* **14,** 36 (1959).
5. Schwarzl, F. R., *Rheol. Acta,* **8,** 6-17 (1969).
6. Schwarzl, F. R., *Rheol. Acta,* **9,** 382-395 (1970).
7. Schwarzl, F. R., *Rheol. Acta,* **10,** 166-173 (1971).
8. Schwarzl, F. R. (in preparation).
9. Schwarzl, F. R., van der Wal, C. W., and Bree, H. W., *Chim. Ind. (Milan),* **54,** 51-58 (1972).
10. Struik, L.C.E., Centraal Laboratorium TNO, private communication.
11. Williams, M. L., Landel, R. F., and Ferry, J. D., *J. Am. Chem. Soc.,* **77** 3701 (1955).
12. Rouse, P. E., *J. Chem. Phys.,* **21,** 1272 (1953).

DISCUSSION on Paper by F. R. Schwarzl

TSCHOEGL: Your approach is based on the matching of intensity functions. With only a few coefficients, the error will be smaller in regions of the time scale where the match is more perfect and larger where it is less perfect. Would it not be possible to achieve smaller overall errors by changing the coefficients in different regions of the time scale to improve the match as one goes along? Since the calculations would normally be done on a computer anyway, the increase in computational complexity would be largely irrelevant.

SCHWARZL: In a few cases it was possible to give more than one simple formula for a conversion problem. Examples of pairs of formulae have been derived (see Ref. 6 above) for the calculation of the creep compliance from dynamic measurements. These formulae use different information, viz., from the course of the storage compliance and from the course of the loss compliance. One is more accurate in the case of low damping, the other in the case of high damping. For the calculation of dynamic properties from creep or stress relaxation, this procedure was not possible.

However, there should be another way of deriving more restrictive error bounds for the formulae which I have presented. If these formulae are applied to the conversion of viscoelastic properties of polymers, they will be much more accurate than one would expect from the given error bounds.

For the derivation of the error bounds, use was made only of the assumption that the spectra are nonnegative functions without any further knowledge of the viscoelastic behavior. The error will attain the given bounds only in the case that the spectrum consists of one sharp line which is situated at the most unfavorable place. With polymers, we deal mostly with broadly distributed spectra. Therefore, the positive and negative contributions to the integral representation of the error (cf. Eq. 10) will partially cancel. This will lead to an error with an absolute value which is perhaps a factor 10 smaller than the bounds.

This, however, should be proven in a certain case. It should be possible to give an estimate of the error from the measured values of the function which is to be converted. These estimates for the error have not yet been derived.

SHEN: I should like to comment that most of the techniques in the interconversion of viscoelastic functions involve integral equations. In fact, the original equations were derived as discrete sums. We have recently been able to numerically interconvert $G(t)$, G', G'', $J(t)$, J', J'', etc., in discrete form through linear programming. This technique automatically produces upper and lower bounds of the computed values without recourse to actual relaxation or retardation functions. Our results are in agreement with your conclusions in that loss quantities are more difficult to calculate than the storage ones.

SCHWARZL: I do not know enough of the method of linear programming in order to see whether this is equivalent to the numerical method which we have proposed. The error bounds obtained in our method automatically take advantage of the fact that the creep rates (or relaxation rates) are completely monotonous functions of the time. Owing to this fact, the truncation errors become much smaller. It would be interesting to know whether the method of linear programming also takes account of the total monotony of these functions, and how your upper and lower values compare with the error bounds I have given.

WILLIAMS: Would you care to comment on the practical advantages of your proposed interconversion technique compared with the Schapery interconversion using Prony Series? It would appear the two methods might, as a practical matter, complement each other because it is rather straightforward to deduce the complex moduli from the relaxation modulus using the Schapery approximation, whereas I gather your proposal might be less direct, and vice versa.

SCHWARZL: The collocation method of Schapery is a technique similar to that proposed by us. A comparison of the accuracies of both methods would be possible if error bounds were known for the collocation method. I do not know whether they have been given.

It seems to me that the method proposed by us is more straightforward and less time consuming than the collocation method, as it does not involve any curve fitting at all. You get directly the points of the converted curve as linear combinations of the points of the measured curve.

A further advantage of our method is that the way in which the experimental error is propagated to the converted quantity can be recognized easily. This might be more difficult with either the collocation method or with the method of linear programming.

TSCHOEGL: I do not think that this is a question of time saving. Schapery's collocation method derives an analytical expression from the data. This expression is then inverted by the methods of transformation calculus. The trouble is that there is no guarantee that the analytical expression which appears to represent, say, a relaxation-modulus curve adequately will, upon inversion, lead to an analytical expresssion which will also adequately describe the creep curve. The difficulties actually increase as the number of collocation points is increased. Some of the retardation times obtained from an inversion of the analytical expression obtained by collocation to a relaxation-modulus curve generally become negative, which is physically not correct. Moavenzadeh has shown that the relaxation or retardation times can be made positive if both relaxation and creep curves are used to derive the analytical expression. This, however, largely begs the question.

KOVACS: By investigating the temperature and time dependence of volume of PVC upon the thermal history of the sample, we found [Kovács, A. J., *Compt. Rend. Acad. Sci. (Paris)* (1956)], as you did, that this material is generally not in equilibrium above its glass transition temperature. Furthermore, the rate of isothermal volume contraction of the freshly quenched samples passes through a maximum between 80 and 90 C, which is just the glass transition range of PVC. This seems also to agree with your creep-compliance data.

ALFREY: There is still another feature of linear viscoelasticity which deserves some comment – namely, the applicability to multicomponent and *nonhomogeneous* time-dependent stresses, and even viscoelastic wave propagation. The transformations discussed by Dr. Schwarzl permit the calculation of shear (or tensile) creep response from shear (or tensile) stress-relaxation curves, etc. The superposition principle also permits the calculation of response to arbitrary multicomponent stresses if two material functions are known (one for volumetric behavior and one for shear behavior). In many cases, the viscoelastic polymer acts essentially as an *incompressible* material, and a *single* material function (specifying the shear viscoelasticity) is sufficient to allow prediction of multicomponent and nonhomogeneous responses.

In a formal sense, this extension is most concisely expressed in terms of operators (P and Q). For example, the differential equation governing the deflection of a loaded elastic beam,

$$EI \frac{d^4u}{dx^4} = w(x),$$

can be converted to a partial differential equation for a loaded *visco-elastic* beam by replacing the elastic modulus (E) by an appropriate ratio of operators:

$$QI \frac{\partial^4 u}{\partial x^4} = Pw(x,t),\ P = \sum_{n=o}^{N} p_n \frac{\partial^n}{\partial t^n},\ Q = \sum_{m=o}^{M} q_m \frac{\partial^m}{\partial t^m}.$$

where $P_\sigma = Q_\epsilon$.

Further, the solutions to such a governing partial differential equation are frequency factorable into a purely spatial function (identical in form with that for the corresponding *elastic* problem) and a temporal function. Actual numerical calculations are usually best done using one of the viscoelastic functions discussed by Dr. Schwarzl, rather than an operator formulation.

MATERIALS WITH MEMORY

R. S. Rivlin

Center for the Application of Mathematics
Lehigh University, Bethlehem, Pa.

1 INTRODUCTION

The characteristic property of viscoelastic solids which distinguishes them from perfectly elastic solids is the fact that, if they are subjected to a deformation which varies with time, the stress measured at time t, say, depends not only on the instantaneous value of the deformation gradients, but also on the whole previous history of the deformation gradients. In a series of papers, Green and Rivlin[1,2] and Green, Rivlin, and Spencer[3] have developed constitutive equations for such materials, in which the stress is expressed in terms of the deformation in the form of series of multiple integrals. It is the object of this paper to recapitulate this development, with particular emphasis on the physical assumptions regarding the material which are implied by the mathematical assumptions made in the theory. Such a development is perhaps timely in view of the extensive attempts in recent years to represent the behavior of actual materials in the form given by the theory.

In Section 2, we start with the assumption that the Cauchy stress is a functional of the deformation gradients for times up to and including t and express the restrictions imposed on the form of this dependence by the consideration that, if the body and the force system applied to it, including inertial forces, are simultaneously subjected to a time-dependent rigid rotation, the stress field at time t is correspondingly rotated by the amount of this rotation at time t. The functional dependence on the history of the deformation gradients up to and including time t is thus replaced, with any desired approximation, by functional dependence on the reduced Cauchy strain (i.e., the Cauchy strain less the unit tensor). The nature of this dependence can be further restricted if the material has some symmetry, but the explicit determination of the form of this restriction will not be discussed here. Rather, we shall be concerned with the assumptions regarding the functional dependence of the stress on the reduced Cauchy strain which enable us to obtain explicit representations.

Since much of the apparent complexity of the theory, as originally presented, stems from the tensor character of the equations involved, we first discuss the simple case in which only simple extension of a rod of the material under tensile force is considered. The stress is then considered to be a functional of the history of the fractional extension of the rod up to and including time t. The assumption is made that, for two extension histories for which the extensions at each instant of time are infinitesimally different, the corresponding stresses at time t are infinitesimally different. This assumption that the dependence of the stress on the extension history is continuous, enables us to obtain, with any desired approximation, representations for the stress at time t as the sum of a series of multiple integrals of the extension history. It is shown that the kernels in these multiple integrals must necessarily be such as to yield the property that the memory of the stress for extension history is a fading memory – the stress forgets the extension history in the distant enough past. This results from the continuity assumption and from the assumption that, whatever the extension history, the stress does not become infinite at infinite time.

It is pointed out that the basic assumption that the stress at time t is a functional of the extension history up to and including time t is strictly not broad enough to accommodate the behavior of materials which have internal friction, if deformations involving instantaneous changes of the velocity gradient are allowed. In such cases, the stress in the rod at time t must be assumed to depend not only on the history of the extension up to and including t, but also on the instantaneous value of the rate of extension at time t.

Finally in Section 4, the various types of behavior considered in Section 3 in the case of time-dependent simple extension of a rod are generalized to the case of arbitrary deformations of the material.

2 BASIC THEORY

In this section, the principles underlying the continuum mechanical theory of nonlinear viscoelastic solids are briefly summarized. We start by delimiting in mathematical terms the class of materials to which the theory is to apply. This class is not the most general with which we are concerned in this paper. However, it serves to illustrate the fundamental principles involved.

The deformation is described by specifying the vector position $\mathbf{x}(\tau)$ of a generic particle of the body at time τ, with respect to a fixed origin, as a function of its vector position $\mathbf{X}$ with respect to the same origin at some reference time t_r, say, which is conveniently taken to be a time at which the body is undeformed. Thus,

$$\mathbf{x}(\tau) = \mathbf{x}(\mathbf{X},\tau) \quad . \tag{2.1}$$

The displacement vector $\mathbf{u}(\tau)$, defined by

$$\mathbf{u}(\tau) = \mathbf{x}(\tau) - \mathbf{X} \quad , \tag{2.2}$$

may also be regarded as a function of $\mathbf{X}$ and τ, thus:

$$\mathbf{u}(\tau) = \mathbf{u}(\mathbf{X},\tau) \quad . \tag{2.3}$$

In terms of the components $x_p(\tau)$, X_A, and $u_p(\tau)$ of $\mathbf{x}(\tau)$, $\mathbf{X}$, and $\mathbf{u}(\tau)$, respectively, in a rectangular cartesian coordinate system x, Eq. 2.1 may be written

$$x_p(\tau) = x_p(X_A,\tau) \tag{2.4}$$

and Eq. 2.3 may be written

$$u_p(\tau) = u_p(X_A,\tau) \quad . \tag{2.5}$$

It is evident that if the nine deformation gradients $\partial x_p(\tau)/\partial X_A$ are specified at a point, then the amount by which any linear element at that point is stretched is determined. For brevity, we use the so-called comma notation, in which the subscript $,A$ denotes the operator $\partial/\partial X_A$. The nine deformation gradients $\partial x_p(\tau)/\partial X_A$ are then written $x_{p,A}(\tau)$.

We shall be concerned with materials in which the stress $\sigma_{ij}(t)$ at a generic particle measured at time t, which, for brevity, will be denoted σ_{ij}, may depend on the history of the deformation gradients $x_{p,A}(\tau)$ at that particle for all times from $\tau = -\infty$ to t. Thus, unlike the situation which exists in an elastic material, in which the stress at time t is determined uniquely by the deformation gradients at time t, in a viscoelastic material the stress at time t "remembers" the deformation gradients at all previous times and the material is said to possess "memory". We can express this physical idea in a mathematical statement – the *constitutive assumption* for the material – that the stress σ_{ij} at time t is a tensor-valued *functional* of the deformation gradient history $x_{p,A}(\tau)$ for $\tau = -\infty$ to t, thus:

$$\sigma_{ij} = \mathop{F_{ij}}_{\tau=-\infty}^{t} [x_{p,A}(\tau)] \quad . \tag{2.6}$$

This means simply that if the dependence of $x_{p,A}(\tau)$ on τ is known for a generic particle, then the stress at that particle at time t is determined.

We recall that, in classical elasticity theory, if we make the constitutive assumption that the stress at time t depends on the values of the nine displacement gradients $\partial u_i/\partial x_j$ at time t, it is shown that it must, in fact, do so through the six infinitesimal strain components e_{ij} defined by

$$e_{ij} = \frac{1}{2}\left(\frac{\partial u_i}{\partial x_j} + \frac{\partial u_j}{\partial x_i}\right) \quad . \tag{2.7}$$

The question arises – are there any corresponding restrictions which, in the case of the constitutive assumption (Eq. 2.6) for a nonlinear viscoelastic material, can be placed on the manner in which σ_{ij} depends on the deformation-gradient history? That such restrictions do, in fact, exist follows from a consideration somewhat similar to that used in classical elasticity theory. We consider that the body, together with the force system associated with it, including intertial forces, is subjected to a time-dependent rigid rotation. The stress field at time t, referred to the system x, differs from that which obtains in the absence of the superposed rotation only to the extent that it undergoes a rotation the amount of which is that of the superposed rotation at time t. From this consideration, it follows, by a purely mathematical argument, that the dependence of σ_{ij} on the deformation-gradient history must be of the following form:

$$\sigma_{ij} = x_{i,P}\, x_{j,Q} \mathop{F_{PQ}}_{\tau=-\infty}^{t} [E_{AB}(\tau)] \quad , \tag{2.8}$$

where $E_{AB}(\tau)$ is defined by

$$E_{AB}(\tau) = x_{k,A}(\tau) x_{k,B}(\tau) - \delta_{AB} \quad , \tag{2.9}$$

and the abbreviation $x_{i,P}$ is used for the deformation gradients $x_{i,P}(t)$ at time t. The nine quantities $E_{AB}(\tau)$ are called the components of the reduced Cauchy strain tensor at time τ and we note that $E_{AB}(\tau) = E_{BA}(\tau)$. Accordingly, only six of the quantities $E_{AB}(\tau)$ are independent. Thus, in passing from Eq. 2.6 to Eq. 2.8, we have replaced arbitrary dependence of σ_{ij} on the nine functions $x_{p,A}(\tau)$ by arbitrary dependence on the six independent components of the reduced Cauchy strain.

So far, we have made no assumption regarding the symmetry of the material. We recall that in classical elasticity, if the material possesses some symmetry, restrictions can be placed on the manner in which the stress depends on the infinitesimal strain components. In the case when the material is isotropic, these restrictions become particularly strong.

In principle, we can, by similar considerations, place restrictions on the form of the tensor-valued functional F_{PQ} in Eq. 2.8 if the nonlinear viscoelastic material has some symmetry. It emerges that the functional F_{PQ} must satisfy the condition

$$\mathop{F_{PQ}}_{\tau=-\infty}^{t} [\bar{E}_{AB}(\tau)] = a_{PM}\, a_{QN} \mathop{F_{MN}}_{\tau=-\infty}^{t} [E_{AB}(\tau)] \ , \tag{2.10}$$

where

$$\bar{E}_{AB}(\tau) = a_{AM}\, a_{BN}\, E_{MN}(\tau) \ , \tag{2.11}$$

for all a_{AM} belonging to the group of transformations describing the symmetry of the material. In the case when the material is isotropic, this group is the rotation group, so that, for an isotropic viscoelastic material, the tensor-valued functional F_{AB} must satisfy the condition of Eq. 2.10 for all rotations a_{AM}. This implicit restriction on the form of F_{AB} can be made explicit for any specified material symmetry by a method developed by Green and Rivlin[1] and Wineman and Pipkin[4]. However, we shall not pursue this here.

If the material considered is incompressible, specification of the deformation gradient history does not determine the stress completely, since the superposition on any force system of a hydrostatic pressure does not alter the deformation. Accordingly, the constitutive assumption of Eq. 2.6 must be replaced by

$$\sigma_{ij} = \mathop{F_{ij}}_{\tau=-\infty}^{t} [x_{p,A}(\tau)] - p\delta_{ij} \ , \tag{2.12}$$

where p is undetermined if the deformation-gradient history is specified. From this constitutive assumption, the constitutive equation

$$\sigma_{ij} = x_{i,P} x_{j,Q} \mathop{F_{PQ}}_{\tau=-\infty}^{t} [E_{AB}(\tau)] - p\delta_{ij} \tag{2.13}$$

follows, replacing Eq. 2.8. At the same time, the fact that in an incompressible material the volume of a material element must remain unchanged during deformation imposes on the reduced Cauchy strain the restriction

$$\det |E_{AB}(\tau) + \delta_{AB}| = 1 \quad . \tag{2.14}$$

There is one further restriction which can be placed on the constitutive Eq. 2.8 or 2.13 from considerations of a very general character. This involves the concept of *hereditary* material as one which remains unchanged in properties so long as it rests undeformed. For such a material, a shift in time, by an amount t_0, say, of the deformation history will shift the dependence of the stress versus time curve by an equal amount t_0 parallel to the time axis.

In the remainder of this paper, we shall restrict ourselves to hereditary materials.

3 NONLINEAR CONSTITUTIVE EQUATIONS

In Section 2 we have obtained certain restrictions which must be imposed on the manner in which the Cauchy stress depends on the deformation-gradient history. These are expressed by Eq. 2.8 in which, if the material possesses some symmetry, the tensor-valued functional F_{PQ} must satisfy Eq. 2.10.

In order to obtain more explicit forms for the constitutive equation, which reflect physically nonpathological behavior on the part of the material described, various representations of the functional F_{PQ} in terms of series of multiple integrals have been developed. Green and Rivlin[1,2] and Green, Rivlin, and Spencer[3] considered materials in which infinitesimal changes in the deformation history result in infinitesimal changes in the stress at time t. They showed that a sequence of approximations to F_{PQ} in the form of series of multiple integrals can be constructed which tend, in the limit, to F_{PQ}. In order to highlight the physical content of this development, free from the mathematical complications which arise from the tensorial character of the constitutive equations, we will discuss it in the context of time-dependent simple extension of a rod.

Accordingly, let σ be the stress at time t which results from a time-dependent fractional extension $e(\tau)$. Then, since σ is determined if $e(\tau)$ is specified for $-\infty < \tau \leqslant t$, we may say that σ is a functional of $e(\tau)$, thus:

$$\sigma = \underset{\tau=-\infty}{\overset{t}{F}} [e(\tau)] \quad . \tag{3.1}$$

It is convenient for the purposes of our discussion to replace the time τ by a timelike variable s, defined by

$$s = \frac{1}{(t-\tau+1)^{\rho}} \quad , \tag{3.2}$$

where ρ is positive. Then, for a specified t, the extension may be regarded as a function of s rather than of τ. Noting that when $\tau = -\infty$, $s = 0$ and when $\tau = t$, $s = 1$, we may rewrite Eq. 3.1, for a specified t, in the form

$$\sigma = \underset{s=0}{\overset{1}{F}} [e(s); t] \quad , \tag{3.3}$$

σ being a functional of e(s) and an ordinary function of t. For hereditary materials, we may omit the dependence of σ on t, and we then have

$$\sigma = \underset{s=0}{\overset{1}{F}} [e(s)] \quad . \tag{3.4}$$

Now, suppose that e(s) can be represented by a Fourier series (see Section 5) thus:

$$e(s) = \sum_{n=0}^{\infty} A_n \cos n\pi s \quad , \tag{3.5}$$

where

$$A_n = 2 \int_0^1 e(s) \cos n\pi s \, ds \quad ,$$

$$A_0 = \int_0^1 e(s) ds \quad . \tag{3.6}$$

The condition under which Eq. 3.5 for e(s) is valid is that e(s) be of bounded variation in the range s = 0, 1. If this condition is satisfied, then the sum of the series (Eq. 3.5) gives the value of e(s) at all points of the interval s = 0, 1 except possibly at points of discontinuity. Accordingly, if we assume that e(s) is a continuous function of s, then, if the coefficients $A_0, A_1, \ldots$ in the Fourier series (Eq. 3.5) are specified, the value of e(s) is determined at all points of the interval s = 0, 1, and σ may be regarded as a function of $A_0, A_1, \ldots,$ rather than as a functional of e(s).

We now suppose that in Eq. 3.4, σ is a *continuous* functional of e(s); i.e., an infinitesimal change* in e(s) results in an infinitesimal change in σ. Then, σ may be regarded as a continuous function of $A_0, A_1, \ldots$. To some degree of approximation, we may regard σ as a continuous function of the first N + 1 Fourier coefficients $A_0, A_1, \ldots, A_N$. This approximation may be made as close as we please by taking N sufficiently large. We denote by σ_N such an approximation to σ. Now, from Weierstrass's theorem, it follows that we can approximate σ_N as closely as we please by a polynomial in $A_0, A_1, \ldots, A_N$. Accordingly, we can approximate σ with any desired accuracy by a polynomial in $A_0, A_1, \ldots, A_N$. We note from Eq. 3.6 that the product of r, say, A's is an r-tuple integral and that the kernel in this integral is a continuous function of its arguments. For example,

$$A_1 A_2 = 4 \int_0^1\int_0^1 \cos \pi s_1 \cos 2\pi s_2 \; e(s_1)e(s_2)ds_1 ds_2 \quad . \tag{3.7}$$

Accordingly, σ may be approximated to any desired accuracy by the sum of a number of multiple integrals, thus:

$$\sigma = \sum_\mu \int_0^1 \ldots \int_0^1 f_\mu(s_1, s_2, \ldots, s_\mu)e(s_1)e(s_2)\ldots e(s_\mu)\, ds_1 ds_2 \ldots ds_\mu = \bar{\sigma}(\text{say}) , \tag{3.8}$$

where $\mu = 0, 1, 2, \ldots$, and the kernels in the multiple integrals are continuous functions of $s_1\ s_2, \ldots, s_\mu$. The term in Eq. 3.8 which corresponds to $\mu = 0$ is a constant. This may be taken to be zero if we assume that the stress is zero for zero strain history [i.e., $\sigma = 0$ when e(s) = 0]. The result (Eq. 3.8) is that obtained from the theory of Green and Rivlin[1] by specializing it to the case of simple extension of a rod.

*More precisely, if $e_1(s)$ and $e_2(s)$ are two extension histories and σ_1 and σ_2 are the corresponding values of stress at time t, then for $\epsilon > 0$ there exists a value of $\delta > 0$, such that

$$|\sigma_1 - \sigma_2| < \epsilon$$

provided that

$$\max |e_1(s) - e_2(s)| < \delta \ .$$

Using Eq. 3.2, we can rewrite the expressions of Eq. 3.6 for A_n as

$$A_n = 2 \int_{-\infty}^{t} \frac{\rho}{(t-\tau+1)^{\rho+1}} e(\tau) \cos \frac{n\pi}{(t-\tau+1)^{\rho}} d\tau ,$$
$$A_0 = \int_{-\infty}^{t} \frac{\rho}{(t-\tau+1)^{\rho+1}} e(\tau) d\tau . \tag{3.9}$$

Correspondingly, the approximate expression for σ in Eq. 3.8 may be rewritten as

$$\sigma \approx \bar{\sigma} = \sum_{\mu} \int_{-\infty}^{t} \bar{f}_{\mu}(\tau_1, \tau_2, \ldots, \tau_{\mu}) e(\tau_1)e(\tau_2) \ldots e(\tau_{\mu})d\tau_1 d\tau_2 \ldots d\tau_{\mu} , \tag{3.10}$$

where

$$\bar{f}_{\mu}(\tau_1,\tau_2, \ldots, \tau_{\mu}) = \frac{\rho^{\mu} f_{\mu} (s_1, s_2, \ldots, s_{\mu})}{[(t-\tau_1 + 1)(t-\tau_2 + 1) \ldots (t-\tau_{\mu} + 1)]^{\rho+1}} . \tag{3.11}$$

Since $f_{\mu}(s_1, s_2, \ldots, s_{\mu})$ is a bounded function of $s_1, s_2 \ldots, s_{\mu}$, and hence of $\tau_1, \tau_2, \ldots, \tau_{\mu}$, it follows that $\bar{f}_{\mu} (\tau_1, \tau_2, \ldots, \tau_{\mu})$ becomes vanishingly small as any of the variables $\tau_1, \tau_2, \ldots, \tau_{\mu}$ tends to $-\infty$. This implies that the material has fading memory in the usual sense that, as $t - \tau \to \infty$, the effect on σ of the extension, in a finite time interval about time τ, becomes vanishingly small.

From a purely mathematical point of view, if e(s) possesses one or more discontinuities in the interval s = 0, 1, then in order to determine e(s) at *all* points of this interval, we must specify not only the values of the Fourier coefficients given by Eq. 3.6, but also the values of e(s) at the various values of s at which the discontinuities occur. Let us suppose that discontinuities occur at $s = S_{\lambda} (\lambda=1, 2, \ldots, \nu)$ and let $e_{\lambda} = e(S_{\lambda})$. Then, if σ is a continuous functional of e(s), we may regard it as a continuous function of $A_0, A_1, \ldots$ and of $e_{\lambda}(\lambda=1, 2, \ldots, \nu)$, the nature of this function depending on the values S_{λ} of s at which the discontinuities occur. Approximate expressions $\bar{\sigma}$ for σ may then be obtained in the form of Eq. 3.8, where f_{μ} is a function of e_{λ} and S_{λ}, as well as of $s_1, s_2, \ldots, s_{\mu}$. The term corresponding to $\mu = 0$ is now a function of e_{λ} and S_{λ}.

In considering the applicability of this conclusion to real materials, it is well to bear in mind that discontinuous changes in the extension cannot in fact be produced. The importance of considering a discontinuous change in e(s) at time $s = S_1$, say, resides only in the fact that it provides an idealized model for a very rapid, but continuous, change in the value of e(s) in the interval $S_1 - \epsilon$, $S_1 + \epsilon$, where ϵ is small. Consequently, we may, in general, omit consideration of discontinuities in e(s). An exception arises when this discontinuity occurs at the instant at which the stress is measured, i.e., at s = 1. In order to accommodate perfectly elastic materials and materials exhibiting instantaneous elasticity within the framework of the theory, we must then include, in the expression for σ, explicit dependence on the instantaneous value e of e(s) at s = 1. Approximate expressions $\overline{\sigma}$ for σ may then be obtained in the form of Eq. 3.8, where f_μ is a function of e, as well as of $s_1, s_2, \ldots, s_\mu$. The term corresponding to $\mu = 0$ is now a function of e. This result is essentially that obtained from the theory of Green, Rivlin, and Spencer[3], by specializing it to the case of simple extension of a rod.

It is perhaps worthwhile to underline the fact that, whether or not the f's depend on e, the approximation to σ represented by Eq. 3.8 is of the following type. If ϵ is any specified positive quantity, we can construct an approximation $\overline{\sigma}$, say, to σ of the form of Eq. 3.8, such that

$$|\sigma - \overline{\sigma}| < \epsilon \tag{3.12}$$

for all extension histories e(s) which have bounded variation.

The kernels occurring in the expression for $\overline{\sigma}$ cannot be uniquely determined. We can, however, construct a sequence of approximations of the form of Eq. 3.8, corresponding to smaller and smaller ϵ, which tends in the limit to σ. Corresponding kernels in this sequence do not necessarily tend to a limit, so we *cannot* say that σ may, *without error*, be represented by an expression of a form similar to that in Eq. 3.8. The *absolute* error involved in approximating σ by $\overline{\sigma}$, may, however, be made as small as we please. In the neighborhood of $\sigma = 0$, the *percentage* error may then be large.

It is, however, perfectly safe to use a representation for σ of the form of Eq. 3.8, provided that we are concerned only with changes of σ of magnitude much greater than ϵ. Since ϵ may be made as small as we please, it might at first sight appear that this does not impose a meaningful restriction on the use of the representation in Eq. 3.8. Let us therefore consider an example in which this might, in fact, impose a meaningful restriction. Suppose we consider extension histories αe(s), with e zero, where α may be made as small as we please. For small enough α, and taking $f_0 = 0$, we may approximate $\overline{\sigma}$ by the first term in the series in Eq. 3.8 thus:

$$\overline{\sigma} = \alpha \int_0^1 f_1(s_1)e(s_1)ds_1 \quad . \tag{3.13}$$

This approximation becomes increasingly good as α decreases. However, regarded as an approximation to σ, the expression on the right-hand side of Eq. 3.13 has the limitation that σ only approximates $\bar{\sigma}$ with absolute error less than ϵ. It may be that, in order for Eq. 3.13 to provide a good approximation to σ, the value of α must be so decreased that the stress is comparable with ϵ. The fact that ϵ may be made as small as we please does not save us from this limitation, since choice of a lower value of ϵ requires a new approximate representation of the form in Eq. 3.8, and with this we may be driven to lower values of α, and hence of the stress.

If, however, σ may be expressed exactly in the form of Eq. 3.8, then, for sufficiently small values of α, we may approximate σ by the expression in Eq. 3.13 and, for somewhat larger values, by

$$\sigma = \alpha \int_0^1 f_1(s_1)e(s_1)ds_1 + \alpha^2 \int_0^1\int_0^1 f_2(s_1, s_2)\, e(s_1)e(s_2)ds_1\, ds_2 \quad , \qquad (3.14)$$

and so on. In this way, we obtain a hierarchy of expressions for σ that are valid over larger and larger ranges of the extension prior to time t. This is, of course, also true if σ can be approximated by an expression of the form of Eq. 3.8, with an error which is of a degree greater than μ, say, in the extension. Of course, whether or not one of the approximations has significant value in the study of a particular material, to which it applies in principle, will depend on whether the range of extensions for which it is valid is one over which experiments can be carried out with reasonable accuracy.

Coleman and Noll[5] have obtained n successive approximations essentially of the forms of Eqs. 3.13 and 3.14 by assuming that the functional F in Eq. 3.3 is Fréchet differentiable n times about the zero history [i.e., the history $e(s) = 0$, $0 \leqslant s \leqslant 1$]. Their argument is basically circular in that the definition of nth order Fréchet differentiability is that the functional F shall be expressible as the sum of homogeneous functionals of degrees 1, 2, . . . ,n in e(s), with a residue of degree greater than n in e(s).

The discussion has, so far, been based on the assumption that the stress at time t is a continuous functional of the extension history $e(\tau)$ for $-\infty < \tau \leqslant t$. It is evident that this is not necessarily an appropriate assumption for many of the materials with which we are concerned. For example, let us consider two deformations in both of which the extension is maintained at zero up to and including time t. In one of these, we maintain the extension zero after time t, while in the other, the rate of extension is changed discontinuously at time t from zero to a finite value. In many materials (e.g., in Newtonian fluids), the stress at time t will be quite different in the two cases.

In order to accommodate such materials in the framework of our mathematics, we may take our constitutive assumption in the following form: the stress is a functional of $e(\tau)$ up to and including time t and an ordinary function of $\dot{e}$ (the rate of extension at time t), thus:

$$\sigma = \mathop{F}_{\tau=-\infty}^{t} [e(\tau); \dot{e}] \quad . \tag{3.15}$$

Generalizing this concept to accommodate materials for which discontinuous changes at time t in the higher time derivatives $\ddot{e}, \dddot{e}, \ldots$ of the extension affect the stress at time t, we may make our initial constitutive assumption in the form: the stress at time t is a functional of $e(\tau)$ in the interval $-\infty < \tau \leqslant t$ and an ordinary function of $\dot{e}, \ddot{e}, \ldots$, thus:

$$\sigma = \mathop{F}_{\tau=-\infty}^{t} [e(\tau); \dot{e}, \ddot{e}, \ldots] \quad . \tag{3.16}$$

This is essentially the constitutive assumption used by Green and Rivlin[2], specialized to the case of simple extension of a rod.

Alternatively, we could accommodate such constitutive assumptions in the single assumption that the stress σ at time t is given by

$$\sigma = \mathop{L^t}_{\epsilon\to 0} \mathop{F}_{\tau=-\infty}^{t+\epsilon} [e(\tau)] \quad , \tag{3.17}$$

for small positive ϵ, i.e., stress at time t is the limit as $\epsilon \to 0$ of a functional of $e(\tau)$ over the range $-\infty < \tau \leqslant t + \epsilon$. To see this we note that

$$\dot{e} = -\mathop{L^t}_{\epsilon\to 0} \int_{-\infty}^{t+\epsilon} \delta'(\tau - t) e(\tau) d\tau \quad ,$$

$$\ddot{e} = \mathop{L^t}_{\epsilon\to 0} \int_{-\infty}^{t+\epsilon} \delta''(\tau - t) e(\tau) d\tau, \text{ etc.,} \tag{3.18}$$

where $\delta(\ )$ denotes the Dirac delta function and $\delta'(\ )$ denotes its derivative. It should be noted that even if, in Eqs. 3.15 and 3.16, σ is a continuous functional of $e(\tau)$ and a continuous function of the instantaneous values of the time derivatives of e at time t, the functional dependence on $e(\tau)$ expressed by Eq. 3.17 is not, in general, continuous.

Of course, if we limit ourselves to deformations in which the time derivatives of $e(\tau)$ are continuous at t, then the constitutive assumptions of Eq. 3.15 and Eq. 3.16 can be replaced by Eq. 3.1.

Approximations to σ in the form of series of multiple integrals may be made in the cases when σ is given by Eq. 3.15 or 3.16. The approximation takes the form of Eq. 3.10, with the kernels dependent on $\dot{e}$ in the case of Eq. 3.15 and on $\dot{e}$, $\ddot{e}$, . . . in the case of Eq. 3.16.

It should be appreciated, in interpreting the above remarks, that if the "slope" of $e(\tau)$ versus τ changes discontinously at time t, then, strictly, its time derivative at time t does not exist. By $\dot{e}$ we mean the right-hand derivative of $e(\tau)$ at time t. This is in accord with the usual convention in mechanics when we apply the Navier-Stokes equation, say, to situations in which the rate of deformation changes discontinuously at time t.

4 GENERALIZATION TO ARBITRARY DEFORMATIONS

In the previous section, we have attempted to show how mathematical expression may be given to a variety of possible types of mechanical behavior which may be exhibited by viscoelastic materials. We have done this in the relatively simple context of simple extension.

For each type of behavior we may obtain analogous equations which are valid for deformations of a general character. Thus, if we take our initial constitutive assumption in the form of Eq. 2.6, we have seen that the Cauchy stress must necessarily be expressible in the form of Eq. 2.8. Replacing the time τ by the timelike variables s, defined by Eq. 3.2, we see that the stress σ_{ij}, referred to a rectangular Cartesian system x, must be expressible in the form

$$\sigma_{ij} = x_{i,P}x_{j,Q} \underset{s=0}{\mathcal{F}_{PQ}} [E_{PQ}(s)] \quad , \tag{4.1}$$

where $E_{PQ}(s)$ is the reduced Cauchy stress at time s. This can be rewritten more succinctly in matrix notation, thus:

$$\sigma = \mathbf{F} \underset{s=0}{\overset{1}{\mathcal{F}}} [\mathbf{E}(s)]\mathbf{F}^T \quad , \tag{4.2}$$

where

$$\sigma = \| \sigma_{ij} \| , \ \mathbf{F} = \| F_{iA} \| = \| x_{i,A} \|, \ \mathbf{E}(s) = \| E_{AB}(s) \| \tag{4.3}$$

and $\mathbf{F}^T$ denotes the transpose of $\mathbf{F}$.

We now suppose that $\mathbf{E}(s)$ can be represented by a Fourier series thus:

$$\mathbf{E}(s) = \sum_{n=0}^{\infty} \mathbf{A}_n \cos n\pi s \quad , \tag{4.4}$$

where $\mathbf{A}_n$ is given by

$$\mathbf{A}_n = 2 \int_0^1 \mathbf{E}(s) \cos n\pi s \, ds \ (n \geqslant 1)$$
$$\mathbf{A}_0 = \int_0^1 \mathbf{E}(s) ds \quad ; \tag{4.5}$$

i.e., each of the components of the tensor $\mathbf{E}(s)$ can be expressed as a Fourier series. This is possible under the same conditions as apply in our discussion of the case of simple extension. Each of the components of $\mathbf{E}(s)$ must have bounded variation in the interval $[0, 1]$. Paralleling the discussion in Section 3, we assume that each of the components of $\mathbf{E}(s)$ is continuous everywhere, except possibly at $s = 1$. Then, in order to represent $\mathbf{E}(s)$ over the complete range $s = [0,1]$, we must specify not only the coefficients $\mathbf{A}_n$ in Eq. 4.4, but also the value $\mathbf{E}$ of $\mathbf{E}(s)$ at $s = 1$.

The tensor F in Eq. 4.2 is, of course, a tensor-valued functional, and we suppose that it is such that infinitesimal changes in the components of $\mathbf{E}(s)$ lead only to infinitesimal changes in the components of F. The tensor-valued functional F is then said to be a continuous functional of the tensor $\mathbf{E}(s)$. Analogously with the case of simple extension, we may approximate each of the components of F to any desired accuracy by a polynomial in the components of the tensors $\mathbf{E}$ and $\mathbf{A}_n$. Thus, each of the components F_{PQ} of F may be expressed, with any desired accuracy, in the form

$$\begin{aligned} F_{PQ} &\approx \sum_{\mu} \alpha_{PQA_1B_1 \ldots A_\mu B_\mu} A^{(n_1)}_{A_1B_1} A^{(n_2)}_{A_2B_2} \ldots A^{(n_\mu)}_{A_\mu B_\mu} \\ &= \bar{F}_{PQ} \text{ (say)} \ , \end{aligned} \tag{4.6}$$

where, for specified n, $A^{(n)}_{AB}$ are the components of $\mathbf{A}_n$; also $n_1, n_2, \ldots$ take integral values and the α's are functions (or, if we like, polynomial functions) of the components of $\mathbf{E}$.

Using Eq. 4.5, we may rewrite Eq. 4.6 in the form

$$F_{PQ} \approx \bar{F}_{PQ} = \sum_{\mu} \int_0^1 \ldots \int_0^1 f_{PQA_1B_1 \ldots A_\mu B_\mu}(s_1, \ldots s_\mu)$$
$$E_{A_1B_1}(s_1) \ldots E_{A_\mu B_\mu}(s_\mu) ds_1 \ldots ds_\mu \ , \tag{4.7}$$

where the kernels $f_{PQA_1B_1 \ldots A_\mu B_\mu}$ are continuous functions of $s_1, \ldots, s_\mu$ and functions (or polynomial functions) of the components E_{AB} of $\mathbf{E}$.

Using Eq. 3.2, we can, of course, rewrite Eq. 4.7 as

$$F_{PQ} \approx \bar{F}_{PQ} = \sum_\mu \int_{-\infty}^{t} \cdots \int_{-\infty}^{t} \bar{f}_{PQA_1B_1 \ldots A_\mu B_\mu}(\tau_1, \ldots, \tau_\mu)$$

$$E_{A_1B_1}(\tau_1) \ldots E_{A_\mu B_\mu}(\tau_\mu) d\tau_1 \ldots d\tau_\mu \;, \qquad (4.8)$$

where

$$\bar{f}_{PQA_1B_1 \ldots A_\mu B_\mu} = \frac{\rho^\mu f_{PQA_1B_1 \ldots A_\mu B_\mu}}{[(t-\tau_1+1) \ldots (t-\tau_\mu+1)]^{\rho+1}} \;. \qquad (4.9)$$

If the material considered has some symmetry, the tensor-valued functional F must satisfy the condition of Eq. 2.10. In approximating F by $\bar{F}$, to any desired accuracy, this can be done by an expression of the form of Eq. 4.7 which satisfies the restriction in Eq. 2.10 imposed by material symmetry.

We can develop constitutive equations for arbitary deformations corresponding to the simple extensional case presented in Eq. 3.16. We start with the constitutive assumption that the Cauchy stress components at time t are functions of the histories of the deformation gradients $x_{p,A}(\tau)$ for $-\infty < \tau \leqslant t$ and functions of the velocity gradients $\dot{x}_{p,A}$, acceleration gradients $\ddot{x}_{p,A}$, and so on, at time t. Thus,

$$\sigma_{ij} = F_{ij}\,[x_{p,A}(\tau);\, \dot{x}_{p,A},\, \ddot{x}_{p,A}, \ldots] \;. \qquad (4.10)$$

Paralleling the passage from Eq. 2.6 to Eq. 2.8, we reach the conclusion, in this case, that the stress must be expressible in the form

$$\sigma_{ij} = x_{i,P} x_{j,Q} F_{PQ}\left[E_{AB}(\tau); \frac{dE_{AB}}{dt}, \frac{d^2E_{AB}}{dt^2}, \ldots\right] \;. \qquad (4.11)$$

Using the notation of Eq. 4.3, we can, of course, rewrite Eq. 4.10 as

$$\sigma = \mathbf{F}F\left[\mathbf{E}(s); \frac{d\mathbf{E}}{dt}, \frac{d^2\mathbf{E}}{dt^2}, \ldots\right]\mathbf{F}^T \;. \qquad (4.12)$$

If the dependence of F_{PQ} on $E_{AB}(\tau)$ is continuous, then we can approximate F_{PQ} with any desired accuracy by a series of multiple integrals in the form of Eq. 4.8, the kernels now being functions of the components of the reduced Cauchy strain and its time derivatives at time t. If the dependence of F_{PQ} on $\mathbf{E}$, $d\mathbf{E}/dt$, $d^2\mathbf{E}/dt^2 \ldots$ is continuous, then the dependence of the kernels on these may be taken as polynomial, with any desired accuracy.

Just as in the previous cases discussed, if the material considered has some symmetry, the tensor-valued functional in Eq. 4.10 must satisfy the restrictions implied by Eq. 2.10.

5 APPENDIX

In order to arrive at the Fourier representation of Eq. 3.5 for e(s), we proceed in the following manner. We first form a function $\bar{e}(s)$ given by

$$\begin{aligned} \bar{e}(s) &= e(s) \qquad s = [0,1] \;, \\ \bar{e}(s) &= e(-s) \qquad s = [-1,0] \;. \end{aligned} \tag{5.1}$$

$\bar{e}(s)$ is thus an even function of s, defined in the range [-1,1] by Eq. 5.1.

We now define $\bar{e}(s)$, not just for s = [-1,1], but for all values of s, as a periodic function with periodicity 2, which is given by Eq. 5.1 in the range [-1,1]. We express this as a cosine Fourier series thus:

$$\bar{e}(s) = \sum_{n=0}^{\infty} A_n \cos n\pi s \;, \tag{5.2}$$

where, using the first of Eqs. 5.1,

$$\begin{aligned} A_n &= \int_{-1}^{1} \bar{e}(s) \cos n\pi s \, ds = 2 \int_0^1 e(s) \cos n\pi s \, ds \quad (n \geqslant 1) \;, \\ A_0 &= \frac{1}{2} \int_{-1}^{1} \bar{e}(s) \, ds = \int_0^1 e(s) \, ds \;. \end{aligned} \tag{5.3}$$

In view of the first expression in Eqs. 5.1, we obtain Eq. 3.5.

ACKNOWLEDGMENT

This paper was written with the support of the Office of Naval Research under Contract No. N00014-67-0370-0001 with Lehigh University. My thanks are due to Dr. D.G.B. Edelen for interesting discussions in connection with this work.

REFERENCES

1. Green, A. E., and Rivlin, R. S., *Arch. Rat'l Mech. Anal.*, **1**, 1 (1957).
2. Green, A. E., and Rivlin, R. S., *Arch. Rat'l Mech. Anal.*, **4**, 387 (1960).
3. Green, A. E., Rivlin, R. S., and Spencer, A.J.M., *Arch. Rat'l Mech. Anal.*, **3**, 82 (1959).
4. Wineman, A. S., and Pipkin, A. C., *Arch. Rat'l Mech. Anal.*, **17**, 184 (1964).
5. Coleman, B. D., and Noll, W., *Arch. Rat'l Mech. Anal.*, **6**, 355 (1960).

DISCUSSION on Paper by R. S. Rivlin

ZAPAS: According to your presentation, am I right to think that the Coleman-Noll simple fluid theory does not describe Newtonian fluids?

RIVLIN: I take it that the simple fluid theory to which you refer is that based on a constitutive assumption of the form

$$\sigma_{ij} = \underset{\tau=-\infty}{\overset{t}{F_{ij}}} \left[\partial x_p(\tau)/\partial x_q\right] \quad ,$$

where $x_q = x_q(t)$. Strictly, the Newtonian fluid does not fall within the scope of this constitutive assumption, if we allow discontinuous changes of velocity gradient at time t. However, if we do not allow such changes, the Newtonian fluid does fall within the scope of this constitutive assumption.

LEE: In today's presentation you introduced the influence of strain, strain rate, strain acceleration, etc., at time t on the functional law which introduces the influence of past history. As I recall, in your 1957 paper with Green, you showed that the earlier work with Ericksen fell within the scope of the functional integral theory. Perhaps it would be helpful to explain the situation in the light of the development you presented today.

RIVLIN: In the context of time-dependent simple extension, the relation between the functional constitutive assumption and the Rivlin-Ericksen theory is as follows. Starting with the relation

$$\sigma = \underset{\tau=-\infty}{\overset{t}{F}} \left[e(\tau)\right] \quad ,$$

we consider only extension histories such that $e(\tau)$ can be expressed as a Taylor series about time t, thus

$$e(\tau) = e(t) + (\tau - t)\, e(t) + \frac{1}{2!} (\tau - t)^2\, \ddot{e}(t) + \ldots$$
$$+ \frac{1}{\mu!} (\tau - t)^{\mu} \overset{(\mu)}{e}(t) + R_{\mu} \quad ,$$

where R_{μ} is the remainder. For this limited class of deformations, if R_{μ} is small enough, we can regard σ as a function of $e(t), e(t), \ldots, \overset{(\mu)}{e}(t)$, rather than as a functional of $e(\tau)$, in accordance with the Rivlin-Ericksen theory. Whether R_{μ} is, in fact, small enough will depend on both $e(\tau)$ and the functional F.

HALPIN: Please comment on the nature of the symmetry restrictions you have placed upon the constitutive response to represent material symmetry.

RIVLIN: The symmetry restrictions precisely parallel, for the nonlinear time-dependent situation, the restrictions which are placed on the linear constitutive equations of classical elasticity theory when we wish to specialize them to an isotropic material or to a material possessing one of the crystal symmetries.

HALPIN: The necessary test and utilization of constitutive theory is the characterization of a response under one test history and the prediction of a different history. Do the current nonlinear constitutive equations provide the same facility as is available with the Boltzmann superposition equation of linear analysis?

RIVLIN: The theory I have presented substantially parallels, in the general nonlinear case, the usual classical Boltzmann-type theory. However, it is much more complicated in that, to characterize a material, a large number of kernels is necessary and these may depend on more than one timelike variable. They will therefore, in general, be much more difficult to determine experimentally and, depending on the material and the class of deformations considered, may be impossible to determine in practice.

LEE: Is it possible to make some cogent and brief comments on how the theory you presented today would be modified by including the effect of varying temperature?

RIVLIN: Presumably, if we want to take account of varying temperature, the kernels in the various multiple integrals would be dependent on temperature. However, I do not feel that we are far enough along in the isothermal case to justify our embarking on this further complication.

ZAPAS: I was under the impression that in nonlinear viscoelastic materials, the stress history does not determine uniquely the strain at time t; e.g., the problem of squeezing a sphere.

RIVLIN: If the stress history is known at each point of the sphere, I think it is unlikely that the deformation will not be uniquely determined. However, if you merely specify a time-dependent hydrostatic pressure on the surface of the sphere, presumably there is some possibility of an instability which allows the sphere to take up some configuration other than the spherically symmetric one.

WILLIAMS: Does the author consider that it is possible to say whether stress or deformation may be regarded as imposed or are they simply interchangeable?

RIVLIN: If, in our functional relation between stress and strain, we wish to take the strain as the dependent variable and the stress as the independent variable, the problem arises as to which stress we shall take. That the Cauchy stress is not the appropriate one can be seen if we take the constitutive equation in the form

$$\sigma_{ij} = \frac{\partial x_i}{\partial X_A} \frac{\partial x_j}{\partial X_B} \mathop{F_{AB}}_{\tau=-\infty}^{t} [E_{PQ}(\tau)] \quad . \tag{1}$$

This may be rewritten as

$$P_{AB} = \mathop{F_{AB}}_{\tau=-\infty}^{t} [E_{PQ}(\tau)] \quad , \tag{2}$$

where P_{AB} is the Piola stress at time t defined by

$$P_{AB} = \frac{\partial X_A}{\partial x_i} \frac{\partial X_B}{\partial x_j} \sigma_{ij} \quad . \tag{3}$$

This would lead us to take the inverse relation in the form

$$E_{PQ} = \mathop{F_{PQ}}_{\tau=-\infty}^{t} [P_{AB}(\tau)] \quad , \tag{4}$$

where

$$P_{AB}(\tau) = \frac{\partial X_A}{\partial x_i(\tau)} \frac{\partial X_B}{\partial x_j(\tau)} \sigma_{ij}(\tau) \quad . \tag{5}$$

A formulation of this kind would be quite a reasonable one, but would lead to the difficulty that when boundary value problems are considered, in place of the usual equations of motion, we would have to use the relatively complicated equations of compatibility for the reduced Cauchy strain. In addition to their complexity, it should be noted that there are six equations of compatibility, while there are only three equations of motion. This reflects the fact that in formulation (1) there are really only three independent variable functions – the histories of the displacement components – while in formulation (4) there are six – the histories of the Piola stress components.

METHODS OF COMBUSTION THEORY IN THE MECHANICS OF DEFORMATION, FLOW, AND FRACTURE OF POLYMERS

G. I. Barenblatt

Institute of Mechanics
Moscow University
Moscow B-234, USSR

ABSTRACT

The mathematical combustion theory – aerothermochemistry – has developed a variety of concepts and approaches asymptotic in their nature which have enabled one to construct, for some important problems, a qualitative theory of the phenomenon and have formed the basis for directed experiment. It has been found recently that there exists an intimate relationship between the combustion theory and the mechanics of polymers. Because of this relationship it is possible to use directly and successfully the concepts and approaches of the combustion theory in the mechanics of polymers. In the present work, the relationship between the mechanics of polymers and the combustion theory is discussed from the general point of view; a survey of the most instructive work is given, in which the concepts and methods of the combustion theory are applied to the problems of deformation, flow, and fracture of polymers.

1 INTRODUCTION: RELATIONSHIP BETWEEN MATHEMATICAL COMBUSTION THEORY AND MECHANICS OF POLYMERS

Shortly after its origin, the hydrodynamics of reacting systems has developed as a branch of fluid mechanics which is of deep fundamental and applied significance. One of the first mechanicians to appreciate the cardinal importance of the new field was von Karman[1] who named this branch of hydrodynamics aerothermochemistry. The mathematical description of aerothermochemical processes is provided by the equations of gas dynamics which are complicated by the transformation of substance, heat generation, diffusion, and heat transfer. The salient feature is a strong temperature dependence of the reaction rate due to the large magnitude of the reaction activation energy U which, as a rule, is of the order of tens of kilocalories per mole. Because of the strong temperature dependence of the reaction rate, the kinetic processes are concentrated in very narrow regions, viz., at combustion fronts. In the remaining region, far larger in volume, the reaction rate is negligibly small and can be taken as zero (since the mixture is either cold or the transformation has been completed), transfer processes are absent, and we have an ordinary flow of an untransformed mixture or, conversely, of a fully transformed mixture. This makes it possible to describe combustion processes by asymptotic methods. To a first approximation, the "external solution"* neglects the front structure and leads to the gas dynamic combustion theory. To this approximation, the front is considered to be a discontinuity surface whose normal velocity V remains undetermined, i.e., in the pure gas dynamic combustion theory, the normal velocity of flame must be assigned independently, say, from experimental data.[3] This velocity can be determined theoretically by considering the front structure, i.e., the "internal solution". To a first approximation, the internal solution gives the thermal flame theory[4,5] which, on the assumption of one leading mixture component vanishing during the reaction, leads to the consideration, within the front, of the system of equations

$$\rho c \partial_t T = \partial_x (\lambda \partial_x T) + Q\Phi(n,T)$$

$$\rho \partial_t n = \partial_x (\rho D \partial_x n) - \Phi(n,T) \ . \qquad (1.1)$$

Here T is the temperature; n is the concentration of the leading mixture component; ρ, c, λ, and D are the density and the coefficients of thermal capacity, heat conduction, and diffusion, respectively; $\Phi(n,T)$ is the reaction rate; Q the reaction thermal effect; x is the coordinate measured

*Here use is made of concepts and approaches of the modern theory of singular perturbations.[2]

along the normal to the front, and t is the time. The characteristic linear scale is of the order of the front thickness, and is small compared with the characteristic dimensions of flow, as is the characteristic time. Of particular interest, therefore, is the asymptotic solution

$$T = T(\xi), \quad n = n(\xi), \quad \xi = x + Vt \quad , \tag{1.2}$$

which sets in at large times and corresponds to a uniformly propagating flame. Substitution of Eq. 1.2 in Eq. 1.1 leads to the boundary-value problem for a system of ordinary equations with boundary conditions at points $\xi = \mp\infty$ which correspond, on the scale of the front internal structure, to its boundaries. The problem is overdetermined, i.e., the number of boundary conditions is one unit more than necessary for its solvability at an arbitrary V. The flame velocity is therefore found as the eigenvalue from the condition of existence of a solution of the form of Eq. 1.2, in the main. The existence of this solution (Eq. 1.2) requires that the reaction rate in the cold mixture ahead of the front be zero. It would appear that this is inconsistent with the Arrhenius law, according to which the reaction rate is proportional to exp $(-U/RT)$ (R is the gas constant); this factor does not vanish at any finite T. In actual fact, this inconsistency is illusory since, at large activation energies, this factor in the cold mixture is many orders less than within the front, and to the approximation adopted, the reaction rate in the cold mixture ahead of the front must be taken equal to zero.

In addition to the above problems of flows with flames already generated, of great importance is the problem of ignition which, under the simplest assumptions, is stated as follows.[6,7] Let a vessel be filled with a reacting gas mixture. The reaction is described by system of the type Eq. 1.1; the boundary conditions at the vessel walls are taken as $T \equiv T_0 =$ const; the normal derivative of the concentration is zero. In principle, two types of reaction processes are possible. In the first type, the heat released during the reaction has no time to leave the vessel through its walls, the mixture is heated, and the reaction rate increases until the leading component of the mixture burns out perceptibly; this case is referred to as inflammation of mixture. In the second type of reaction, equilibrium sets in rapidly between the amount of heat released during the reaction and the amount of heat removed through the walls of the vessel. The problem is stated as follows: under what conditions may the second type of reaction be realized? If these conditions are not fulfilled, inflammation occurs.

In the second type of reaction, we may, over a certain interval of time, neglect burning-out and take $n \equiv n_0 =$ const, where n_0 is the initial concentration of the leading mixture component. In these conditions, the temperature distribution must settle down and hence tend, at large values of time, to the solution of the equation

$$\operatorname{div}(\lambda \operatorname{grad} T) + Q\Phi(n_0, T) = 0 \tag{1.3}$$

which satisfies the condition $T = T_0$ at the boundary. Thus, for the reaction to follow the second type, it is necessary that this solution exist; its nonexistence is a sufficient condition for inflammation. It is found that this solution exists only when the size of the vessel is sufficiently small; if a certain critical size of the vessel is exceeded, thermal inflammation of the mixture occurs.

The concepts of thermal inflammation, uniform flame propagation, and zero reaction rate ("cutting off" of the reaction rate) outside the front are fundamental for aerothermochemistry.

Polymeric materials (both amorphous and crystalline in solid state), as a rule, are bodies possessing pronounced microstructure of widely different scale, ranging from molecular to essentially supermolecular.[8,9] During deformation, flow, and fracture of polymeric materials, the elements of microstructure undergo transformations, also at different levels; these transformations are described by kinetic relations similar to the kinetic equations of ordinary chemical reactions at molecular level, with the essential difference, however, that the activation energy is strongly dependent on the mechanical stress. Besides, during the above transformations, heat is generally released, which in turn affects kinetics since the strong temperature dependence of the reaction rate is preserved.

For these reasons, in the analysis of deformation, flow, and fracture of polymers, the situation is basically the same as in the case of flow of chemically active gases, the study of which is the subject matter of aerothermochemistry. As in aerothermochemistry, the complete description of the processes of deformation, flow, and fracture of polymers is given by a set of equations of continuum mechanics and kinetic equations. It is found that the above-mentioned fundamental concepts of aerothermochemistry work directly in the mechanics of polymers. As always, the expediency of considering the general theory is ascertained after the analysis of some instructive problems, to which we now proceed.

2 NECK PROPAGATION

This phenomenon, discovered by Carothers and Hill[10], has been the subject of numerous, principally experimental, investigations. It resides in the fact that when a cylindrical polymeric specimen is stretched with a constant velocity, a sudden reduction of area (neck) occurs which does not progress to cause a complete rupture of the specimen but, assuming a certain shape, begins to propagate with a constant velocity along the specimen (Fig. 1). The force acting on the specimen remains constant the whole time the neck is propagating along the specimen (and is somewhat smaller than the force which acted on the specimen at the time immediately preceding the onset of neck propagation). At the neck, an orientational transformation of microstructural elements takes place which results in an appreciable strengthening of the material; the rate of this transformation depends strongly on the mechanical stress and temperature. Similarly to the rate of chemical reactions, the expression proposed for the rate of orientational transformation is proportional to $\exp[-U-\gamma\sigma)/RT]$, where U is the activation energy at zero stress, γ is a structure-sensitive constant of the material, and σ is the stress.

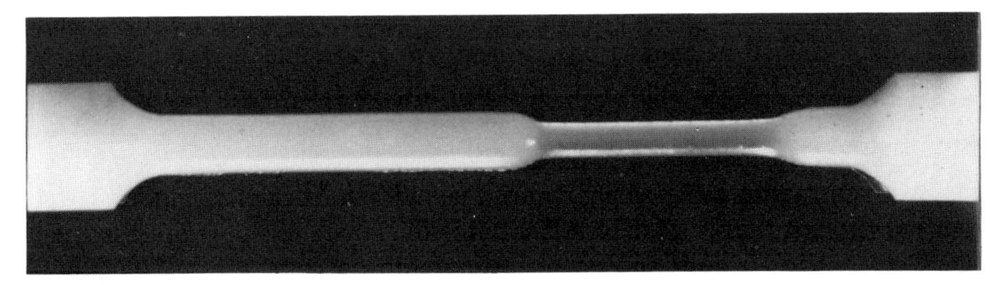

Fig. 1. Propagation of the neck along a specimen.

An analogy exists between the neck propagation and the uniform flame propagation considered in Section 1.[11,12] In the development of the theory of the phenomenon, the quantities involved are averaged over the section of the specimen and the following considerations are used. The microstructural elements are divided (to a certain extent arbitrarily) into transformed and untransformed ones; as experiments show, during transformation the density changes only slightly for many materials over a wide range of rates of extension. We neglect the change in density and assume that, during orientational transformation, the cross-sectional area of a microstructural element is reduced by a factor of α and the element elongates by an amount corresponding to a zero change in volume; the number of elements at the section remains constant. Introducing a naturally defined concentration of transformed elements n(x,t), we obtain the relations (x is the coordinate measured along the axis of the specimen, and t is the time)

$$\partial_t S + \partial_x Su = 0, \quad S/S_0 = 1 - (1-\alpha)n \quad . \qquad (2.1a,b)$$

Here S is the current and S_0 the original cross-sectional area of the specimen, and u is the velocity of the substance. To Eqs. 2.1a,b is added the energy balance equation

$$\partial_t(S\rho cT) + \partial_x(\rho cTSu) = \partial_x \ \lambda S\partial_x T + QqS - h\sqrt{S}\,(T-T_o) \quad . \qquad (2.2)$$

Here ρcT is the internal energy per unit volume, λ is the coefficient of heat conduction, h is the coefficient of heat transfer from the specimen surface, q is the rate of substance transformation, and Q is the thermal effect of transformation. Further, the equation of kinetics of transformation taking account of the nonlocal nature of transformation is written as:[13]

$$\partial_t Sn + \partial_x Snu = S\int_{-\infty}^{\infty} q[n(\xi,t),\ \sigma(\xi,t),\ T(\xi,t)]\ \rho(x-\xi)d\xi \quad , \qquad (2.3)$$

where $q(n,\sigma,T)$ is the rate of lengthwise uniform substance transformation, $\rho(x-\xi)$ is a correlation function which we assume to be symmetric and very rapidly decreasing, so that the following relations hold

$$\int_{-\infty}^{\infty} \rho(x-\xi)d\xi = 1, \quad \int_{-\infty}^{\infty} (x-\xi)\rho(x-\xi)d\xi = 0,$$

$$L^2 = \frac{1}{2}\int_{-\infty}^{\infty} (x-\xi)^2\,\rho(x-\xi)d\xi \quad , \qquad (2.4)$$

where L, the correlation scale, may have the values of the order of one or several characteristic sizes of the supermolecular structure. The correlation function $\rho(x-\xi)$ decreases rapidly and we may therefore restrict ourselves, in integral of Eq. 2.3, to the first terms in the expansion of the function q near the point $\xi = x$. Taking into account the strong stress dependence of the rate of transformation $q(n,\sigma,T)$ and neglecting the nonlocality due to concentration and temperature, it is possible to represent the right-hand side of Eq. 2.3 as

$$S\partial_x \left\{(L^2\partial_\sigma q)\partial_x\sigma\right\} \quad , \qquad (2.5)$$

so that the complete system of equations describing the process finally becomes

$$\sigma S = \sigma_o S_o = P, \; S/S_o = 1\text{-}(1\text{-}\alpha)n, \; \partial_t S + \partial_x Su = 0$$

$$S\partial_t n + Su\partial_x n = q(n,\sigma)S + S\partial_x \left\{ (L^2 \partial_\sigma q)\partial_x \sigma \right\} \tag{2.6}$$

$$S\partial_t \rho c T + Su\partial_x \rho c T = \partial_x \lambda S \partial_x T + QqS - h\sqrt{S}\,(T-T_o) \quad .$$

The uniformly propagating neck corresponds to the solution of the system (Eq. 2.6) of the "uniformly propagating wave" type for which all the unknowns depend only on the combination $\xi = x + Vt$. The number of boundary conditions at the boundaries of the neck front $\xi = \mp\infty$ is one unit more than necessary for the solution to exist for an arbitrary combination of the velocity of neck propagation V and the stress in the undeformed portion of the specimen σ_o, so that the solution exists only when a particular relation between V and σ_o is fulfilled. Under rather general assumptions concerning the rate of transformation $q(n,\sigma T)$, it is proved that the critical stress σ_o corresponding to each value of V exists and is unique.

For the important case of isothermal neck propagation, on the natural assumption that

$$q = A(1-n)\exp\left(-\frac{U-\gamma\sigma}{kT}\right) \tag{2.7}$$

where A is a constant, and k is the Boltzmann constant, the relation between σ_o and V is obtained in finite form[13]

$$\ln\frac{V}{V_*} = -\frac{U}{kT} + \frac{\gamma\sigma_o}{\alpha kT} \quad . \tag{2.8}$$

Here $V_* = AL\alpha/\sqrt{2}\ln(1/a)$ and it is taken that $\gamma\sigma_o/\alpha kT >> 1$. The linear relation between the stress and the logarithm of the velocity of neck propagation is well supported by experiments with capron fibers (Fig. 2).

In the contrary case of purely thermal neck propagation, of particular interest are the instability of uniform neck propagation and the occurrence of self-excited vibrations discovered by Hookway[14] and independently by Andrianova, Karguin, and Kechekyan[15]. Namely, in the same experimental conditions as described above but at higher rates of extension, the neck sometimes begins to propagate (Fig. 3) not with a constant velocity but by jumps which soon become regular, and the stress (in the undeformed portion of the section) varies periodically with time (Fig. 4). The available theory of this process[16] is based on the consideration of the interaction of elasticity of the specimen and the thermal processes at the propagating neck. The process at the neck itself is assumed to be quasi-steady, i.e., the

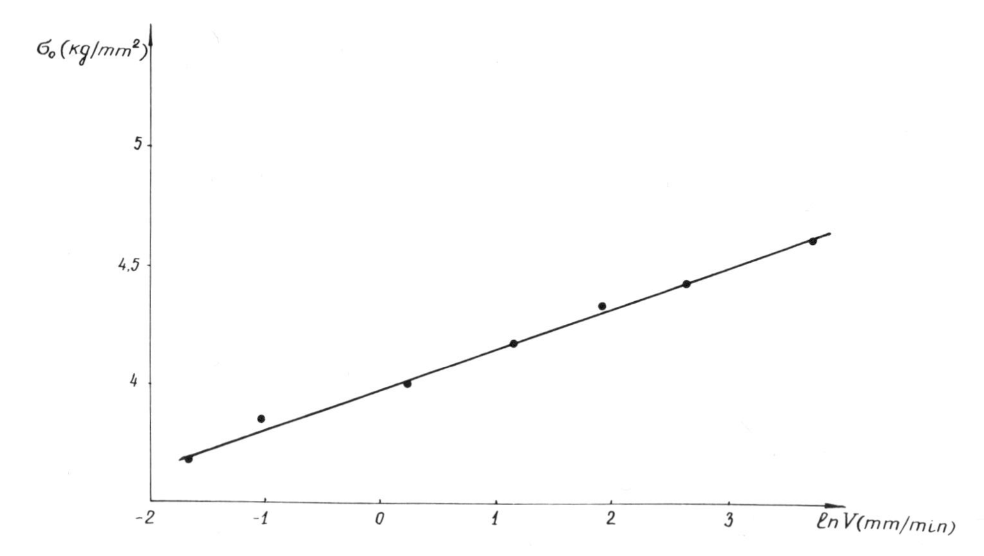

Fig. 2. Relation between the stress and propagation velocity of the neck.

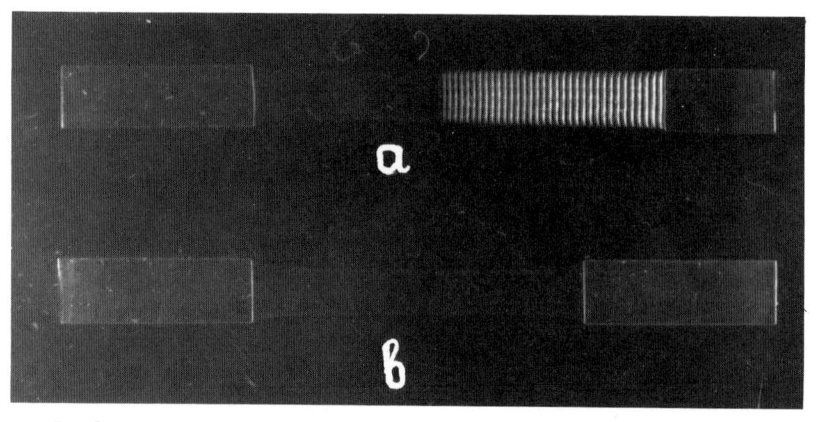

Fig. 3. a. Unstable propagation of neck region. b. Stable neck propagation.

instantaneous stress σ_o in the undeformed portion of the specimen, the instantaneous velocity of neck propagation V, and the instantaneous temperature in the neck zone T are assumed to be related by the same equation as for a uniformly propagating neck; this is also true for the total heat generation in the neck zone q(V,T). The resulting system of basic relations is of the form

$$\frac{dV}{dt} = (\partial_V \sigma_o)^{-1} \left\{ \frac{1}{\lambda} (V_o - V) - \frac{1}{\rho c} (\partial_T \sigma_o)[q(V,T) - h(T - T_o)] \right\}$$

$$\rho c \frac{dT}{dt} = q(V,T) - h(T - T_o) \quad , \qquad (2.9)$$

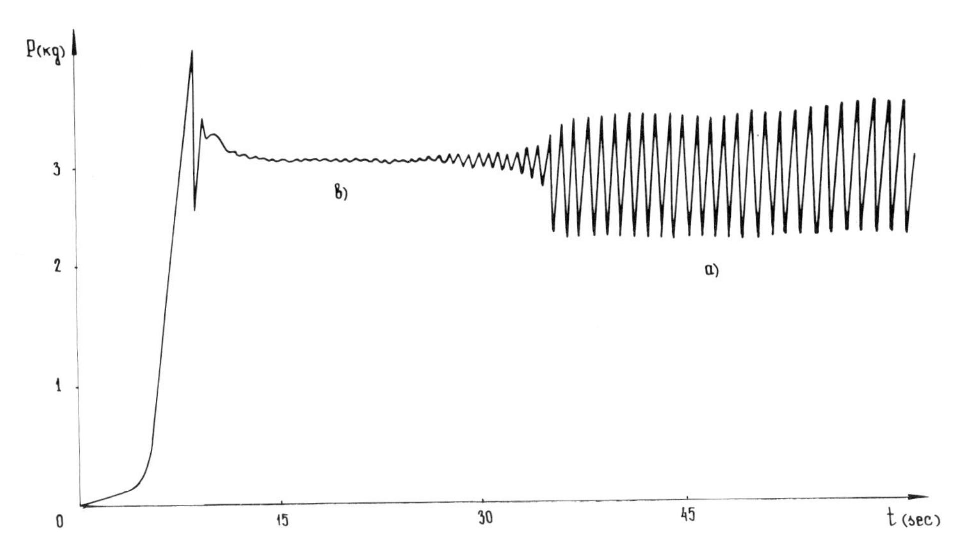

Fig. 4. Plot of stress in unstable neck propagation with time for sample in Fig. 3a.

where λ is the elastic compliance of the specimen, and V_0 is the speed of the grips. The analysis of this system (Eq. 2.9) shows that, under certain conditions, the uniform neck propagation ceases to be stable and stationary self-excited vibrations occur. The self-excited vibrations disappear when the heat removal becomes rather appreciable and/or the elastic compliance of the specimen as a whole is reduced (it increases during neck propagation because of the elongation of the specimen); the uniform neck propagation then maintains stability (Fig. 3b shows a neck developing under the same conditions as the neck of Fig. 3a but when the specimen is severely air cooled).

It is curious that in the case of the self-excited neck propagation, the mechanics of polymers forestalled the combustion theory; the work on the mathematically equivalent phenomenon of self-excited flame propagation in gasless compositions was done independently somewhat later.[17]

A detailed analysis of the microstructure in the neck zone shows that in sufficiently wide specimens, the straight-line front loses stability and the pattern becomes complicated (Fig. 5). Namely, the transformation at the neck is extremely nonuniform and is accompanied by the formation of numerous undertransformed pillars which gradually disappear as the neck progresses. The total surface of the transformation front increases sharply; this phenomenon closely resembles turbulent combustion with its characteristic islets of combustible mixture which burn away within a wide transition zone.

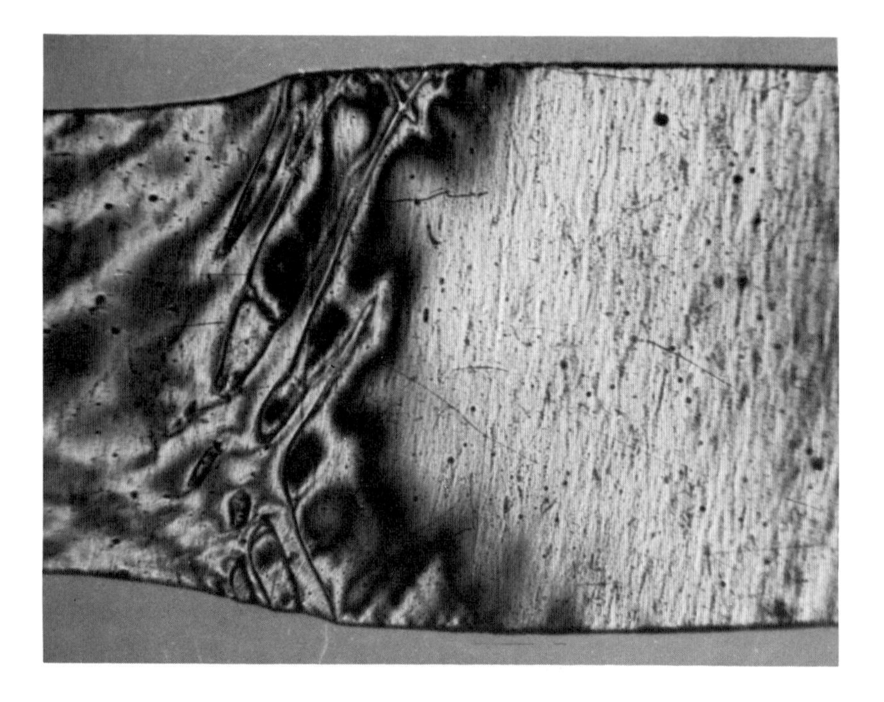

Fig. 5. Microstructure in the interface of neck region.

3 CRACK PROPAGATION IN POLYMERS

A systematic microscopic examination of crack propagation in polymeric materials of the polymethylmethacrylate type was carried out by Kambour[18], Bessonov[19], van den Booghart[20] and many other authors. On the basis of the results of these studies, the crack propagation in such materials may be represented as follows.[21] Under the action of applied loads, the cracks do not remain unchanged but slowly spread out; at the crack tips the elements of supermolecular structure undergo orientational transformation and considerably elongate, forming a system of bonds (Fig. 6) which deform and rupture owing to high stress concentration. The bonds acting on the crack surface give rise to a system of forces – cohesive forces – which prevent crack propagation. The orientational transformation and fracture of microstructural elements of polymers take place throughout the volume of the specimen, but the rates of this process near the crack edge, where stress concentration occurs, and in the bulk of the material differ by many orders. In problems of crack propagation in polymers, transformation and fracture of microstructural elements are therefore neglected everywhere except for a small crack-tip region, just where the fracture occurs. The size of the crack-tip region, denoted by d, is considered to be small compared with the crack length 2ℓ (the

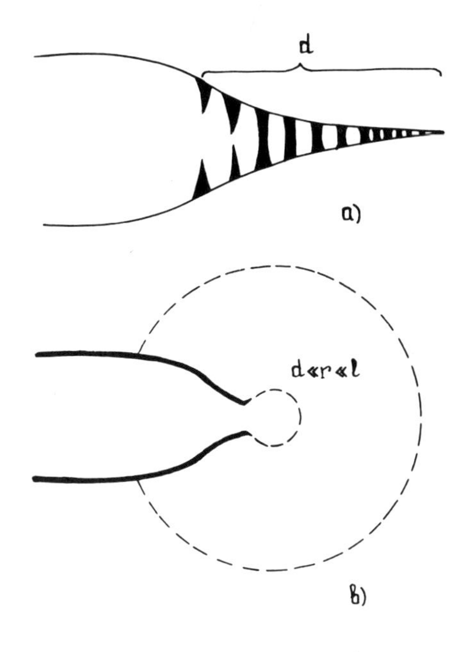

Fig. 6. Propagation of crack tip.

hypothesis of smallness of the crack-tip region). The asymptotic approach applied in the crack theory is close in idea to the approach of the combustion theory described in Section 1. Namely, the "external solution" is constructed, i.e., the solution of the problem of elasticity (or visco-elasticity if viscous effects are taken into account throughout the volume) for a cracked body subjected to applied loads. This solution holds everywhere except for a neighborhood of the crack-tip region having the size of the same order as the crack-tip region itself (the simple model of medium which is valid outside this neighborhood does not apply in it). Consider an "intermediate" asymptotic of the elastic field at distances r from the crack tip large compared to d but small compared to ℓ, i.e., $d \ll r \ll \ell$; because of smallness of the crack-tip region, the neighborhood where this asymptotic acts does exist. As can be shown[22] (for simplicity, we restrict ourselves to the case of plane strain, i.e., of cracks in thick plates), the intermediate asymptotic representation of the stress field is of the form

$$
\begin{aligned}
\sigma_{xx} &= \frac{N}{\sqrt{r}} \cos\frac{\theta}{2}\left\{1 - \sin\frac{\theta}{2}\sin\frac{3\theta}{2}\right\} - \frac{T}{\sqrt{r}}\sin\frac{\theta}{2}\left\{2 + \cos\frac{\theta}{2}\cos\frac{3\theta}{2}\right\} \\
\sigma_{yy} &= \frac{N}{\sqrt{r}} \cos\frac{\theta}{2}\left\{1 + \sin\frac{\theta}{2}\sin\frac{3\theta}{2}\right\} + \frac{T}{\sqrt{r}}\sin\frac{\theta}{2}\cos\frac{\theta}{2}\cos\frac{3\theta}{2} \qquad (3.1) \\
\sigma_{xy} &= \frac{N}{\sqrt{r}} \cos\frac{\theta}{2}\sin\frac{\theta}{2}\cos\frac{3\theta}{2} + \frac{T}{\sqrt{r}}\cos\frac{\theta}{2}\left\{1 - \sin\frac{\theta}{2}\sin\frac{3\theta}{2}\right\} .
\end{aligned}
$$

Here r and θ are the polar coordinates with center at the crack tip, the angle being reckoned from the tangent to the crack at its tip; the stress intensity factors N and T depend on the shape of the body and cracks in it and depend linearly on the magnitude of applied loads. The second basic hypothesis in the crack theory is the hypothesis of autonomity according to which the state of stress near the tips of all cracks (cracks proper and not notches) in a given material is the same. The hypothesis of autonomity extends not only to the crack-tip region itself but also to the region surrounding it, where the intermediate asymptotic (Eq. 3.1) acts. Hence, there are two important corollaries.

(1) The stress-intensity factor N represents a function of the single external parameter, the velocity of crack propagation u, which is universal for a given material under given ambient conditions:

$$N = \frac{1}{\pi} K(u) \quad . \tag{3.2}$$

(2) The stress-intensity factor T is zero, i.e., the stress field is locally symmetric near the edge of a crack (curvilinear in general). This condition determines the direction of propagation of curvilinear cracks.*

Thus, if the universal function K(u) is known for a given material, say, from experiment, the problem of development of cracks in a body under the action of a given load is easily solved in principle. For example, in the case of a single straight crack developing under the action of a system of loads proportional to one parameter P, $u = d\ell/dt$ (ℓ is the half-length of the crack), $N = N(P,\ell)$, and relation in Eq. 3.2 leads to the ordinary differential equation

$$N(P,\ell) = \frac{1}{\pi} K\left(\frac{d\ell}{dt}\right) \tag{3.3}$$

which determines the crack length as a function of time, and in the final analysis the lifetime, i.e., the time to fracture of a body as the time of existence of the solution of this equation. Basically, such an approach is analogous to the gas dynamic combustion theory (see Section 1) in which the internal front structure is not considered, the flame front is taken to be a surface, and the velocity of flame propagation is assumed to be a given function of pressure, or simply a physicochemical constant of mixture.

As in the combustion theory, of particular interest here is the consideration of the "internal solution". The characteristic linear scale is

---

*Deep investigation of the propagation of curvilinear cracks is performed recently by R. V. Goldstein and R. L. Sazganik.

now the scale of the crack-tip region d, so that the internal solution corresponds to a semi-infinite (or infinite) symmetrically loaded crack. We assign material properties in the crack-tip region and a certain model of kinetics of deformation and fracture of microstructural elements, i.e., bonds. The function K(u), therewith, is determined. Comparison with the data of macroexperiments may then confirm or refute the proposed model; this enables one to find the necessary constants of the model and thus to predict microprocesses taking place in the fracture zone from the data of macroscopic experiments. This approach was successively developed in a series of papers.[23,24] The simplest model assumes a crack to be semi-infinite (i.e., the initial bond size negligibly small), the material perfectly elastic everywhere outside the crack, the bonds themselves perfectly elastic up to fracture, and fracture proceeding by a thermo-fluctuation mechanism

$$\frac{dn}{dt} = -kn, \; k = \frac{1}{\tau_o} \exp\left(-\frac{U - \gamma\sigma}{kT}\right) \quad . \tag{3.4}$$

Here n is the density of bonds per unit surface area, t_o is the characteristic time whose order of magnitude coincides with the period of thermal fluctuations, U is the energy of rupture of an unloaded bond, and γ is a material constant which determines the transfer of stress σ to an individual bond and depends on the material structure. Thus, the same Arrhenius relation is taken for the constant of the rate of "bond-rupture reaction", but the activation energy is assumed to be dependent on the bond tension, i.e., the stress.

On the basis of the concept of the main crack developing in a specimen, this simplest model enables one to obtain the well-known formula of S. N. Zhurkov[25],

$$\tau = \tau_o \exp\left[\frac{U-\gamma\sigma}{kT}\right] \quad , \tag{3.5}$$

for the life τ of the specimen under stress σ. The paradox in the usual interpretation of this formula resides in the fact that, as would appear, fracture should proceed uniformly throughout the specimen, and after the lapse of lifetime, the whole specimen should instantaneously turn into powder everywhere. But this does not happen; because of the strong stress dependence of the reaction rate and the instability of the process of uniform fracture, the specimen fractures with the formation of the main crack, just as gas combustion occurs by flame propagation through a vessel and not by a simultaneous flash at all points.

4 HYDRODYNAMIC THERMAL EXPLOSION IN THE FLOW OF POLYMER MELTS

As pointed out in Section 1, thermal explosion, as one of the fundamental concepts of aerothermochemistry, has many various applications in the mechanics of polymers. Bostandzhiyan, Merzhanov, and Khudyaev[26] discovered and investigated the remarkable phenomenon of hydrodynamic thermal explosion (in this connection mention should also be made of the investigations of Gruntfest[17]), important for the analysis of flows of polymer melts.

The foregoing may be illustrated as follows. We consider a flow in a circular infinite tube of radius r_o. The thermal resistance of the tube walls is low, so that a constant temperature equal to the ambient temperature T_o is maintained at the walls. The equations of motion and energy are of the form (v is the velocity in the direction of the tube axis x, T is the temperature, J is the mechanical equivalent of heat, and λ is the coefficient of heat conduction)

$$\frac{1}{r}\,\partial_r(r\mu\partial_r v) - \partial_x p = 0,\ \partial^2_{rr}T + \frac{1}{r}\partial_r T + \frac{\mu}{\lambda J}(\partial_r v)^2 = 0\ ; \qquad (4.1)$$

for the viscosity μ, the Arrhenius relation holds to a sufficient approximation

$$\mu = \mu_o \exp(U/RT)\ , \qquad (4.2)$$

where the order of magnitude of the activation energy U is tens of kilocalories per mole, and the heating, i.e., the temperature drop in the flow is small, so that the Frank-Kamenetsky[6] approximation is fully valid for the temperature dependence of viscosity

$$\mu = \mu_o \exp(U/RT_o) \times \exp[-U(T-T_o)/RT_o^{\,2}]\ . \qquad (4.2a)$$

Introducing the dimensionless quantities

$$\xi = \left(\frac{r}{r_o}\right)^2,\ \theta = \frac{(T-T_o)U}{RT_o^2},\ \kappa = \frac{(\partial_x p)^2 r_o^{\,4}}{16\mu_o\,\lambda J}\,\frac{U}{RT_o^{\,2}}\exp\left(-\frac{U}{RT_o}\right), \qquad (4.3)$$

we obtain the boundary-value problem for the dimensionless excessive temperature θ:

$$\frac{d^2\theta}{d\xi^2} + \frac{1}{\xi}\frac{d\theta}{d\xi} + \kappa e^{\theta} = 0,\ \theta(1) = 0,\ \left(\frac{d\theta}{d\xi}\right)_{\xi=0} = 0\ , \qquad (4.4)$$

i.e., the situation of thermal explosion in a cylindrical vessel, familiar after the classical work of D. A. Frank-Kamenetsky.[6] The solution of this problem exists only when $\kappa < \kappa_{cr} = 2$; the condition $\kappa = 2$ determines the critical values of the parameters beyond which the steady flow of polymer through the tube becomes impossible. In contrast to a true thermal explosion, here no burning-out actually occurs, so that the analysis of the nonstationary postcritical flow is of special interest and it should be made in a less idealized statement. In particular, one must take into account the finiteness of the tube. The resulting self-excited vibrations may be used to explain peculiar vibratory regimes of extrusion of polymer melts from extruders.

5 VIBRATIONAL HEATING OF POLYMERS AND THERMAL VIBROCREEP

In the mechanics of polymers there is one more interesting analog of thermal explosion discovered independently by Shapery[28] and by Korobov and Ratner[29]. We consider here a somewhat complicated characteristic situation.[30] A cylindrical specimen of a polymeric material is subjected to combined action of a static basic axial load σ and a vibratory axial load $p = \sigma_o \sin\omega t$. Polymers display pronounced viscoelasticity which causes heat release under the action of the vibratory additional load. The intensity of heat generation in a unit volume per unit time is $W = \frac{1}{2}\sigma_o^2 \omega J''$, where J'', the so-called loss compliance, is a material characteristic which depends on temperature and vibration frequency. Over a certain range of parameters the quantity J'' may be represented as

$$J'' = \frac{K}{\omega^n} \exp\,[\beta T - T_o)] \quad ; \tag{5.1}$$

[for instance, for plasticized poly(n-butyl methacrylate) over a certain range of temperatures, $K = 5 \cdot 10^{-6}$ cgs, $n = 0.7$, $\beta = 0.26$ 1/°C], so that the energy equation for a long cylindrical specimen is of the form

$$\rho c \partial_t T = \lambda \frac{1}{r} \partial_r r \partial_r T + \frac{K}{2} \sigma_o^2 \omega^{1-n} \exp\,[\beta T - T_o)] \quad , \tag{5.2}$$

with the boundary and initial conditions

$$\partial_r T + \frac{h}{\lambda}(T - T_o) = 0 \ (r = r_o), \ \partial_r T = 0 \ (r = 0), \ T = T_o \ (t = 0) \ . \tag{5.3}$$

It is assumed that Newtonian heat exchange takes place at the boundary of the specimen. The temperature of the specimen grows with time (the strain rate under the action of the basic load also increases) and approaches the distribution defined by the stationary solution of the boundary-value problem:

$$\frac{d^2\theta}{d\xi^2} + \frac{1}{\xi}\frac{d\theta}{d\xi} + \kappa e^{\theta} = 0; \quad \left(\frac{d\theta}{d\xi}\right)_{\xi=0} = 0; \quad \frac{d\theta}{d\xi} + \eta\theta = 0 \ (\xi = 1)$$

$$\theta = \beta(T - T_o), \quad \xi = r/r_o, \quad \eta = h/\lambda r_o \tag{5.4}$$

$$\kappa = (\beta K/2\lambda)\sigma_o{}^2\omega^{1-n}r_o{}^2$$

if this stationary solution exists. As is seen, the situation arising here is again similar to thermal explosion. According to Barzykin and Merzhanov[31] who considered this problem for usual thermal explosion, the solution of the stationary problem exists when

$$x < x_{cr} = \frac{8s}{(1+s)^2}\exp\left[-\frac{4s}{\eta(1+s)}\right], \quad s = \frac{2}{\eta}\left(\sqrt{1+\eta^2/4^{-1}}\right), \tag{5.5}$$

so that the condition $x = x_{cr}$ determines the critical values of the parameters (for instance, of the radius r_o for a given amplitude and frequency) beyond which the temperature in the specimen begins to rise, there occurs softening, thermal destruction of the material, etc. The vibrational heating may have very serious consequences even for small amplitudes if the specimen is acted on by a large basic static load. If the basic static load, the stress σ, is considerably (say, one order) larger than the amplitude σ_o of the vibratory load, the direct, "force", effect of the vibratory additional load (for instance, the effect on the creep rate of a constant additional load equal in magnitude to the amplitude of the vibratory load) is small. This is seen, in particular, from the well-known Sherby-Dorn relation[32] for the creep deformation of polymers under a constant load as a function of time:

$$\epsilon_c = \psi\left[t \exp\left[-\frac{U-\gamma\sigma}{RT}\right]\right] \tag{5.6}$$

The same formula (Eq. 5.6) shows, however, that because of the large magnitude of U, the superposition of small vibrations which cause even a slight rise in temperature may appreciably and sometimes sharply increase the creep rate. This increasing of creep of polymers under the action of small vibrations which cause heating of the material is termed thermal vibrocreep. Naturally, it is most pronounced in regions of high stress concentration, for instance, near cracks. Namely, vibrations may be small in amplitude and the heating caused by them may be insignificant in the specimen as a whole. However, as first pointed out by Bartenev et al.[33], because of stress concentration and low heat conduction of polymers, there may be local heating near the crack edges which accelerates kinetic processes at the crack tip, accelerates crack propagation, and thereby decreases the lifetime of the specimen. The quantitative theory of this phenomenon was proposed in Ref. 34; the presence of local heating near the crack tips in a polymeric material (polyvinyl chloride) when vibrations were superimposed was directly observed by means of thermovision technique by Attermo and Oestberg.[35]

6 CONCLUSION

The foregoing examples show that the relationship between the mechanics of polymers and aerothermochemistry is based on the deep physical generality of the processes under study. The mechanics of polymers leads to a system of simultaneous equations of the transformation kinetics and the continuum mechanics; the kinetic laws of transformation of microstructural elements of polymers and chemical reactions in combustion are close; hence, in both cases, one obtains similar asymptotic approaches and concepts. Both in the combustion theory and in the mechanics of polymers, the kinetics of transformation is important only in regions of small extension, which greatly simplifies the intermediate asymptotic consideration.

REFERENCES

1. Karman, Th. von, and Penner, S. S., *Selected Combustion Problems,* AGARD, Buttersworths, London (1954).
2. Van Dyke, M., *Perturbation Methods in Fluid Dynamics,* Academic Press, New York-London (1964).
3. Landau, L. D., and Lifshitz, E. M., *Fluid Mechanics,* Pergamon Press, London, Addeson-Wesley Publishing Co., Reading, Mass. (1959), 536 pp.
4. Zeldovich, Ya. B., and Frank-Kamenetsky, D. A., *Compt. Rend. Acad. Sci. USSR, (Dokl. Akad. Nauk SSSR),* **19** (9), 693 (1938) (in English).
5. Zeldovich, Ya. B., *Zh. Fiz. Khim.,* **22** (1), 27 (1948); English translation available as 63-20808 from National Translation Center, The John Crerar Library, 35 West 33rd St., Chicago, Ill. 60616.
6. Frank-Kamenetsky, D. A., *Diffusion and Heat Exchange in Chemical Kinetics,* Princeton University Press, Princeton, N. J. (1955), 370 pp.
7. Gelfand, I. M., *Usp. Matem. Nauk,* **14** (2), 86 (1959); English translation available as AMST S2 V.29 p 295-381 from American Mathematical Society, 190 Hope Street, P. O. Box 6248, Providence, R. I. 02904.

8. Kargin, V. A., Kitaigorodsky, A. I., and Slonimsky, G. L., *Colloid. J. (Kolloidny Zh.)*, **19** (2), 141 (1957) (in English).
9. Keller, A., *Phil Mag.*, **2:** 1171 (1957).
10. Carothers, W. H., and Hill, J. W., *J. Amer. Chem. Soc.*, **54** (4) (1932).
11. Barenblatt, G. I., *J. Appl. Math. & Mech. (Prikl. Matem. i Mekhan.)*, **28** (6), 1264 (1964) (in English).
12. Barenblatt, G. I., Entov, V. M., and Segalov, A. E., *Proc. IUTAM Symposium on Thermo-inelasticity*, East Kilbride, June, 1968, Bruno A. Boley (Ed.), Springer-Verlag, Wien-New York (1970).
13. Barenblatt, G. I., *Izv. Akad. Nauk SSSR, Mekhan. Tverd. Tela*, No. 6 (1972); to be published in English in *Mech. Solids*, No. 6 (1972).
14. Hookway, D. C., *Proc. Textile Inst.*, **49** (7) (1958).
15. Kechekyan, A. S., Andrianova, G. P., Kargin, V. A., *Vysokomolekul. Soedin.*, **12** (11), 2424 (1970); *Polymer Sci. USSR*, **12** (11), 2743 (1970) (in English).
16. Barenblatt, G. I., *Izv. Akad. Nauk SSSR, Mekhan. Tverd. Tela.*, No. 5, 121 (1970); *Mech. Solids*, **5** (5), 110 (1970) (in English).
17. Shkadinsky, K. G., Khaikin, B. I., and Merzhanov, A. G., *Fizika gorenia i vzryva*, No. 1 (1971); to be published in English in *Combustion, Explosion, and Shock Waves*, Consultants Bureau.
18. Kambour, R. P., *J. Polymer Sci.*, Part A-2, **4**, 349 (1966).
19. Bessonov, M. I., *Usp. Fiz. Nauk.*, **83** (1), 107 (1964); Soviet Phys. Usp., **7** (3), 401 (1964) (in English).
20. Van den Booghart, A., *Conference Proceedings, Physical Basis of Yield and Fracture*, Oxford University Press (1966).
21. Barenblatt, G. I., Entov, V. M., and Salganik, R. L., in *Inelastic Behavior of Solids*, M. F. Kanninen, W. F. Adler, A. R. Rosenfield, and R. I. Jaffee (Eds.), McGraw-Hill Book Co., New York (1970).
22. Irwin, G. R., *Handbuch der Physik*, Springer-Verlag, Berlin (1958), Vol 6, p 551.
23. Barenblatt, G. I., Entov, V. M., and Salganik, R. L., *Inz. Zh. Mekhan. Tverd. Tela*, No. 5, 82 (1966), No. 6, 76 (1966), No. 1, 122 (1967); *Mech. Solids*, **1** (5), 53 (1966), **1** (6), 49 (1966), **2** (1), 79 (1967) in English).
24. Salganik, R. L., *Problemy Prochnosti*, No. 2 (1971); *Strength of Materials*, **3** (2), 200 (1971) (in English).
25. Zhurkov, S. N., *Internat. J. Fracture Mech.*, **1** (4) (1965).
26. Bostandzhiyan, S. A., Merzhanov, A. G., and Khudyaev, S. I., *Dokl. Akad. Nauk SSSR*, **163** (1), 133 (1965), Proc. Acad. Sci. USSR, **163** (1-6), 504 (1965) (in English); *Zh. Prikl. Mekh. i Tekhn. Fiz.*, No. 5, 45 (1965), No. 5, 38 (1968), *J. Appl. Mech. and Tech. Phys.*, No. 5, 30 (1965), No. 5, 552 (1968) (in English).
27. Gruntfest, I. J., and Young, J. P., *J. Appl. Phys.*, **35** (1) (1964).
28. Shapery, R. A., *AIAA Journal*, No. 5 (1964); No. 2 (1965); *J. Appl. Mech.*, **32** (3) (1965).
29. Ratner, S. B., and Korobov, V. I., *Dokl. Akad. Nauk SSSR*, **161** (4), 824 (1965); *Soviet Phys.-Doklady*, **10** (4), 361 (1965).
30. Barenblatt, G. I., Kozyrev, Yu. I., Malinin, N. I., Pavlov, D. Ya., and Shesterikov, S. A., *Zh. Prikl. Mekh. i Tekhn. Fiz.*, No. 5; 68 (1965), *J. Appl. Mech. and Tech. Phys.*, No. 5, 44 (1965) (in English); *Dokl. Akad. Nauk SSR*, **166** (4), 813 (1966), *Soviet Phys.-Doklady*, **11** (2), 117 (1966) (in English).
31. Barzykin, V. V., and Merzhanov, A. G., *Dokl. Akad. Nauk SSSR*, 120 (6), 1271 (1958); *Proc. Acad. Sci. USSR, Phys. Chem. Section*, **120** (1-6), 441 (1958) (in English).
32. Sherby, O. D., and Dorn, J. E., *J. Mech. Phys. Solids*, **6** (2) (1958).
33. Bartenev, G. M., et al., *Izv. Akad. Nauk SSSR, Otd. Tekhn. Nauk, Mekhan. i Mashinostr.*, No. 5, 176 (1960); Translation available as 65-13343 from National Translations Center, The John Crerar Library, 35 West 33rd St., Chicago, Ill. 60616. (in English).
34. Barenblatt, G. I., Entov, V. M., and Salganik, R. L., *Proc. IUTAM Symposium on Thermo-inelasticity*, East Kilbride, June 1968, Bruno A. Boley (Ed.), Springer-Verlag, Wien-New York (1970).
35. Attermo, R., and Oestberg, G., *Internat. J. Fracture Mech.*, 7 (1) (1971).

DISCUSSION on Paper by G. I. Barenblatt

LANDEL: Professor Barenblatt, you point out the existence of a similarity in the mathematics of these mechanical processes and flame propagation. What are the advantages of the flame-propagation approach?

BARENBLATT: Flame propagation belongs to a definite class of phenomena with a rather peculiar and well-elaborated mathematical structure. The similarity between neck propagation and flame propagation makes it advantageous to use this approach. I would like to stress that we have here a similarity but not a complete mathematical equivalence; i.e., while we could use the general concept of this approach in studying neck propagation, the existing formulae of the theory of flame propagation could not be used directly.

KOBAYASHI: Was this model used to describe sustained stress-crack growth in polymers? I surmise from your paper that this may be possible, but no explicit statement was made on this phenomenon.

BARENBLATT: This model has been used to describe crack growth in polymers. You could find the detailed explanation of our approach in References 21, 23, and 24.

MULLER: When we were occupied some years ago with the phenomenon of necking during cold stretching, we could show by thermosensitive fluorescent substances that the necking is, in the steady state, a hot zone migrating towards the still cold and still unstretched part of sample. This seems to be analogous to propagation of a flame. Indeed, employment of flame theory (theory of irreversible processes) has led to his reasonable unpublished results.

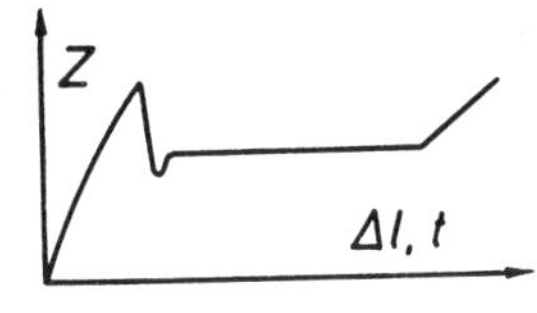

The force-elongation curves of materials which show necking possess a negative slope behind the maximum. This corresponds in the electrical case to a negative resistance. A negative resistance in a system being able to oscillate leads to initiation of oscillations – in our case to alternate strips of white and transparent character. We have seen this in the case of thin foils of Terylene. The vibrating system works only if there exists a spring, either a steel

spring or a sufficiently long piece of already stretched Terylene sample in series with the piece of Terylene to be stretched. Therefore, we were sure that the picture of exciting vibrations by a negative characteristic is alright.

BARENBLATT: Thank you for your interesting comment which, in fact, corresponds closely to our point of view. I want only to stress that the similarity between neck propagation and flame propagation exists even in the isothermal case and not only when thermal conductivity is the leading mechanism of neck propagation. On the contrary, the periodic neck propagation (in Terylene, for instance) is essentially nonisothermal; in analyzing this process, we have to take into account the heat-balance equation, together with the equation for elastic deformations. Therefore, the negative slope of some parts of the force-elongation curve and the large length of the specimen part already stretched is not enough for self-oscillations; the ceasing of oscillations due to blowing of the specimens shows it definitely.

MARK: About 15 years ago, Thompson and Tuckett studied the drawing of Terylene in considerable detail and arrived at the following interpretation: Neck drawing represents a phase transition from the disordered and unoriented state in the undrawn part to the ordered (crystalline) and oriented state in the drawn part. If an undrawn fiber or film is stretched, there exists a critical stress at which the sample separates in two phases: the undrawn (amorphous) and the drawn (crystalline) phases, with the neck as boundary. At a given temperature, each material has a "natural draw ratio", i.e., the reduction of the cross section at the neck is a characteristic quantity which does not depend on the conditions of the test, but only on the nature of the sample – molecular weight and chemical composition (polyethylene, polypropylene, polyamide, polyester) – and on temperature. If one stretches a material of this type, the stress increases up to a certain point of the stress-strain curve and then neck drawing occurs at constant stress. The characteristic stress-strain curve of drawable materials is shown in Fig. D-3; at point A drawing sets in and the sample elongates at constant stress until all is drawn at point B; from then on, the drawn material is being stretched. The dotted part of the curve is attainable only under nonequilibrium conditions (rapid stretching). Thompson and Tuckett point out that there exists a complete

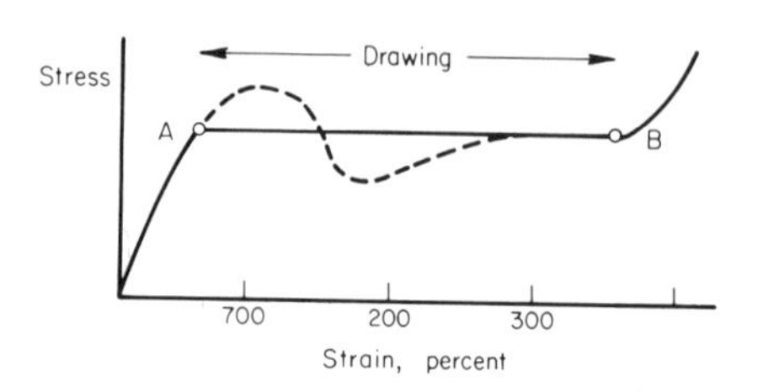

analogy of neck drawing (as a phase transition) with the isothermal condensation of a gas which is also a phase transition. In fact, the stress-strain curve of neck drawing is completely analogous to the van der Waals compression curve (see Fig. D-4) if one replaces the stress by the pressure and the strain by the volume. They derived for the neck-drawing process a corresponding equation which permits a quantitative description of the phenomenon.

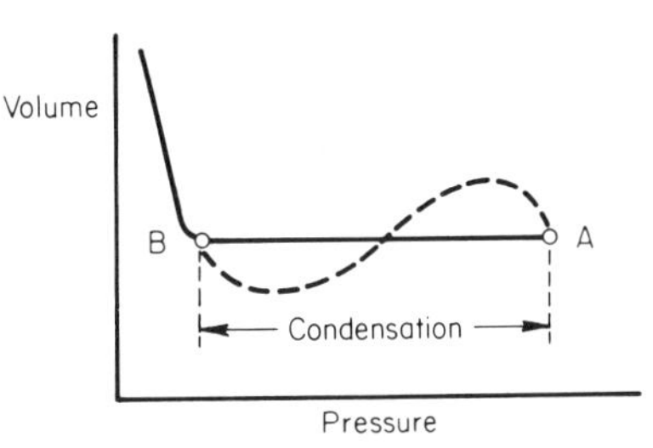

Part Two

PHENOMENOLOGY

ON THE NONLINEAR BIAXIAL STRESS-STRAIN BEHAVIOR OF RUBBERLIKE POLYMERS

G. W. Becker and O. Krüger

Bundesanstalt für Materialprüfung
Berlin, Germany

ABSTRACT

The nonlinear biaxial deformation behavior of rubberlike polymers has been determined by means of a special device. Using specimens in the form of square, thin sheets, stress-relaxation measurements have been carried out at room temperature for three materials: an unfilled natural rubber, SBR, and a plasticized polyvinyl chloride. As can be shown, there are simple transformation equations, depending only upon characteristic quantities of the material, that give one single master curve for each material. These transform the various stress-strain curves measured for different ratios of the two stresses.

1 INTRODUCTION

The nonlinear mechanical behavior of materials has been investigated for many years. It is not only a problem of fundamental interest but is also important for practical purposes: in many cases, materials and constructions used in practical applications are deformed up to finite strains, reversible or irreversible. Most striking is the wide nonlinear stress-strain range of rubbers or rubberlike materials. Therefore, those substances have been investigated, in respect to finite deformations, more often than others, like metals or glasses. But, as far as a phenomenological description is concerned, the nonlinear stress-strain behavior of rubberlike materials may also be used as a model for the general theoretical concept.

As has been discussed earlier[1], many different equations for the description of the nonlinear, uniaxial, stress-strain behavior of rubbers have been proposed. But, all of them have been proved insufficient if the whole range of deformations is considered – even in cases in which two or three different constants could be adjusted. As is well-known, a general description of the nonlinear behavior of materials is possible only from experiments in which the three principal stresses, and therefore the corresponding principal strains, are different. This follows from the concept of the existence of a strain-energy function[2]; in general, this may be concluded for materials with time effects from the nonlinear field theories.[3] But, multiaxial experiments are of importance not solely for these theoretical reasons, but also because the necessity for dealing with practical applications, in which materials are very often under multiaxial stresses, demands such experiments.

The easiest test of this type is the biaxial sheet experiment which has been used by various investigators.[4-8] In this case, a specimen in the form of a square thin sheet is stretched in two perpendicular directions parallel to the surface, whereupon the stress perpendicular to the surface is zero. A new version of such a device has been developed and is described in Section 3. With this apparatus, three rubberlike polymers – an unfilled natural rubber, SBR, and a plasticized polyvinyl chloride – have been investigated in a wide range of strains. The results are used to form some general considerations which may be useful for practical applications. Additionally, they may give new insight into the phenomenological treatment of the nonlinear behavior of rubbery polymers.

2 THEORETICAL CONSIDERATIONS AND PREVIOUS RESULTS

A treatise on the most general nonlinear field theory has been given by Noll and Truesdell.[3] It leads for isotropic elastic materials to equations of the type

$$\left.\begin{aligned} t_i &= \kappa_o + \kappa f_i + \kappa_{-1} f_1^{-1} \quad , \\ t_i &= \mu_o + \mu_1 f_i + \mu_2 f_1^{2} \quad , \\ &\text{with } i = 1, 2, 3, \end{aligned}\right\} \quad (2.1a,b)$$

where, in the most general concept, the f_i are the diagonal components of an appropriate deformation tensor. The κ_j or μ_j respectively are scalar functions of the invariants of this tensor, or are also simple symmetric functions of the principal strains $\lambda_i = 1 + \epsilon_i$, where the ϵ_i are the corresponding Cauchy strains. The t_i are the principal stresses, based on the actual cross-section area. If the f_i have been chosen, then the κ_j or the μ_j can be determined, e.g., from biaxial stress-strain experiments. The main difficulty is in choosing the deformation tensor so that the functions κ_j or μ_j do not become too complicated.

A possible deformation tensor is the well-known Cauchy-Green tensor **B** with the diagonal components λ_i^2. Their introduction into Eq. 2.1a,b leads to equations which are identical with those of the theory given by Rivlin and co-workers.[2] This theory is based on the principle of the strain-energy function W, which should depend on the three principal invariants of the tensor **B**:

$$W = W(I_B, II_B, III_B) \quad ,$$

with the invariants

$$\left.\begin{aligned} I_B &= \lambda_1^2 + \lambda_2^2 + \lambda_3^2 \\ II_B &= \lambda_1^2\lambda_2^2 + \lambda_2^2\lambda_3^2 + \lambda_3^2\lambda_1^2 \\ III_B &= \lambda_1^2\lambda_2^2\lambda_3^2 . \end{aligned}\right\} \quad (2.3a,b,c)$$

Differentiating W with respect to the strains λ_i, with

$$\sigma_i = \partial W / \partial \lambda_i ,$$

leads to Eq. 2.1a, if use is made of the identity

$$\sigma_i = \frac{t_i \, III_B^{1/2}}{\lambda_i} \qquad (2.4)$$

and of the equations[9]:

$$\left.\begin{aligned} 2\frac{\partial W}{\partial I_B} &= \kappa_1 \, III_B^{1/2} \\ 2\frac{\partial W}{\partial II_B} &= -\frac{\kappa_{-1}}{III_B^{1/2}} \\ 2\frac{\partial W}{\partial III_B} &= \frac{1}{III_B^{1/2}}\left(\kappa_o + \kappa_{-1}\frac{II_B}{III_B}\right). \end{aligned}\right\} \quad (2.5a,b,c)$$

The σ_i are the principal stresses, based on the original cross-section area. The determination of the functions κ_i therefore is replaced by the determination of the gradients of W with respect to the invariants of **B**.

There have been various investigations in order to determine experimentally the κ_j or the gradients of W. In most cases, rubbers or rubberlike materials have been used because of the very small time dependence of their mechanical properties and because of their large strain range. Furthermore, for most deformation states, the assumption of incompressibility ($III_B = 1$) is approximated very well. This allows a remarkable simplification of the theoretical equations. In addition, if biaxial sheet experiments are considered, setting $t_3 = 0$ gives a further simplification.

The first biaxial sheet tests of this kind were carried out by Rivlin and Saunders.[4] Their results have been described by the simple approximations[10]

$$\frac{\partial W}{\partial I_B} = \text{const., and } \frac{\partial W}{\partial II_B} \sim \frac{1}{II_B} \quad . \qquad (2.6a,b)$$

Sometimes the approximation $(\partial W/\partial II_B) \approx$ const. is used, which leads to the Mooney-Rivlin equations[11]:

$$\left.\begin{aligned} \sigma_1\lambda_1 &= 2(\lambda_1^2 - \lambda_3^2)(C_1 + C_2\lambda_2^2) \\ \sigma_2\lambda_2 &= 2(\lambda_2^2 - \lambda_3^2)(C_1 + C_2\lambda_1^2) \end{aligned}\right\} \quad (2.7a,b)$$

for biaxial experiments, and

$$\sigma\lambda = 2(\lambda^2 - \lambda^{-1})(C_1 + C_2\,\lambda^{-1}) \qquad (2.8)$$

for uniaxial experiments.

Since it was found that $(\partial W/\partial II_B) \ll (\partial W/\partial I_B)$, often $\partial W/\partial II_B$ is considered small enough to be neglected. This approximation leads to the Eq. of the well-known kinetic theory of rubber elasticity[12]:

$$\left.\begin{aligned} \sigma_1\lambda_1 &= G\,(\lambda_1{}^2 - \lambda_3{}^2) \\ \sigma_2\lambda_2 &= G\,(\lambda_2{}^2 - \lambda_3{}^2) \end{aligned}\right\} \quad (2.9a,b)$$

for biaxial experiments, and

$$\sigma\lambda = G(\lambda^2 - \lambda^{-1}) \quad (2.10)$$

for uniaxial experiments, where G is the shear modulus.

In a later investigation, one of the authors could show[9] that the approximations in Eqs. 2.6a,b simplify the measured curves too much. As can be seen from Fig. 1, the curves of $\partial W/\partial I_B$ as functions of $I_B - 3$ or $II_B - 3$, respectively, are much more complicated, especially insofar as the range of smaller deformations is concerned. This follows very clearly from the semilogarithmical diagrams in Fig. 1b. Similar deviations from the approximation in Eq. 2.6b have been observed for $\partial W/\partial II_B$.[9]

These results are in agreement with more recent investigations[7,13]: they all show that there are no single proportionality constants in the two approximations (Eqs. 2.6a, b), if the whole range of strains is considered. Especially for very small deformations for which the classical theory of elasticity should be valid, these "constants" become functions of the strains or, in other words, the constants at small deformations are not the same as those at finite deformations.

There has been a number of attempts to obtain more general expressions. Very recently, Tschoegl[14] has shown that the expansion of the function W in terms of the invariants I_B and II_B leads to equations which are in agreement with those of other authors who have proposed at least three constants; e.g., instead of Eq. 2.8, one obtains:

$$\sigma\lambda = 2(\lambda^2 - \lambda^{-1})\,[C_1 - C_2\,\lambda^{-1} - 2C_3\,(\lambda - 1)^2(1 - 2\lambda^{-1})]\,. \quad (2.11)$$

A further possibility is the use of a more appropriate deformation tensor which leads to the simple functions κ_i in Eq. 2.1a,b at smaller strains. As has been remarked earlier[9], there is an infinite number of possibilities: every tensor with diagonal components given by $f_i \approx \epsilon_i$ at small deformations. This is true for all components of the type

$$f_i = (\lambda_i{}^m - \lambda_i{}^n)/(m - n), \text{with } m \neq n \quad (2.12)$$

and also for transcendental functions as

$$f_i = \ln\lambda_i \;, \quad (2.13)$$

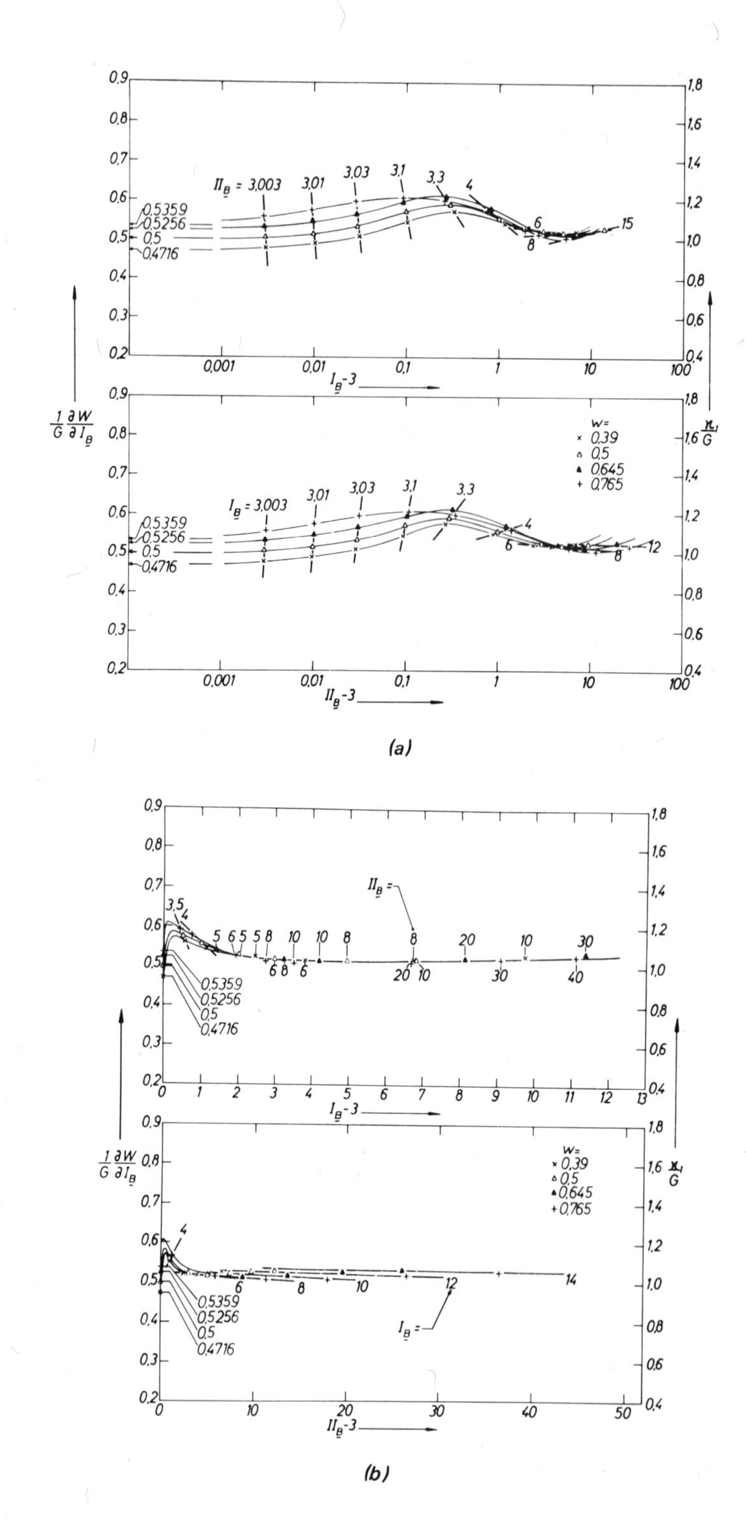

Fig. 1. (1/G) ($\partial W/\partial I_B$) as a function of I_B-3 and II_B-3. Abscissa (a) linear, (b) logarithmic.

which has been introduced as a strain measure by Hencky.[15] The Hencky strain measure has been used by Valanis and Landel.[16] There are various other possibilities, however, for obtaining stress-strain equations for finite deformations using Eq. 2.1a,b with a deformation tensor of the type given in Eq. 2.13. This may be treated elsewhere; but, it can be stated that presently known theoretical expressions fail to describe the measurements very well.

3 NEW BIAXIAL EXPERIMENTS

3.1 Measuring Device

The apparatus, which has been used for the investigation of the nonlinear, biaxial stress-strain behavior of rubberlike polymers, is shown in Fig. 2, for a general biaxial deformation test. It is built into a universal tensile testing machine. The specimen in the form of a square, thin sheet is held in its position parallel to the guide beams of two identical Y-shaped assemblies by hooks. The hooks slide on one end into reinforced holes of the specimen without stretching it before the test. The assemblies are clamped into jaws, from which the lower one is rigidly mounted to the movable crosshead of the machine, while the upper one is used for the load transfers. This jaw is connected with a vertical measuring load cell by

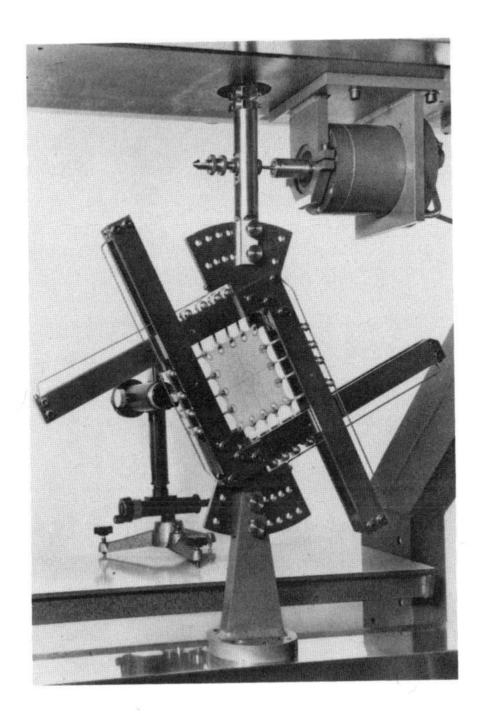

Fig. 2. Device for biaxial sheet experiments.

a flexible coupling blade and with the horizontal measuring load cell by a coupling steel wire.

Depending on the angle position of the two assemblies, the movement of the lower crosshead leads to a biaxial stretch of the specimen with a given ratio of the two strains or stresses, respectively. The ratio $w = \sigma_2/\sigma_1$ of these two stresses can be varied between 0 (uniaxial tension) and 1 (homogeneous biaxial tension) or if the subscript indices for the two directions are interchanged, up to values >1. A double circle target with cross lines is printed on the specimen and can be used for measuring the deformations with a cathetometer. Details of the entire device are described elsewhere.[17]

3.2 Specimens

Three different materials have been investigated: an unfilled natural rubber, SBR, and a plasticized polyvinyl chloride (PVC). The specimens of natural rubber and SBR were vulcanized in a special mold, which prepares the specimens already with reinforcing rings around the holes (Fig. 3). The PVC specimens were cut from fabricated sheets, using little ring pieces glued to the holes for their reinforcement; in this case, the sheets have been tempered to avoid internal stresses. The PVC had a plasticiser content of 32 volume percent dioctylphthalate. The target for the measurement of the deformations was printed on all specimens with the device shown in Fig. 4.

Fig. 3. Mold for the vulcanization of the rubbery specimens.

Fig. 4. Printing device for the specimens.

3.3 Results

With the biaxial sheet testing device, stress relaxation measurements have been carried out at room temperature. The time dependence of the stresses was very small for the two rubbers. In contrast, in the plasticized PVC, the stresses decreased about 30 percent within one decade of time. But, for this material it can be shown that the dependencies of the stresses on time and deformation are factorizable. Thus, general stress-strain curves at a fixed time, $t = 10$ minutes, are considered only.

For each angle position of the two Y-shaped assemblies, i.e., for a special stress ratio $w = \sigma_2/\sigma_1$, a complete series of experiments with deformations ϵ_1, and ϵ_2 from 0.5 to 300 percent, has been made. An example of a stress-strain curve $\sigma_1(\epsilon_1)$ is shown in Fig. 5 for natural rubber (NR) with $w = 0.58$.

In order to describe the measured biaxial stress-strain curves, one single master curve for each material was obtained by transforming the various curves measured for different ratios w of the two stresses. As can be shown, such transformations are possible. If the curves are plotted into log-log-diagrams, there are linear shifts parallel to the axis. As an example, the master curve for NR is shown in Fig. 6 for the reference ratio $w = 1$. Similar curves could be obtained for SBR and plasticized PVC.[17]

A complete survey of a whole set of average curves for different ratios w can be obtained from the general master curve. These curves have been plotted in Fig. 7 for positive and negative deformations ϵ_1. The ratios $w^* = 1/w$ are used in cases where the subscript indices for the two directions of the stresses have been interchanged.

The stress-strain curves of the plasticized PVC for $w = 1$ are shown in Fig. 8 for different times t ranging from 0.1 to 50 minutes. As it may be

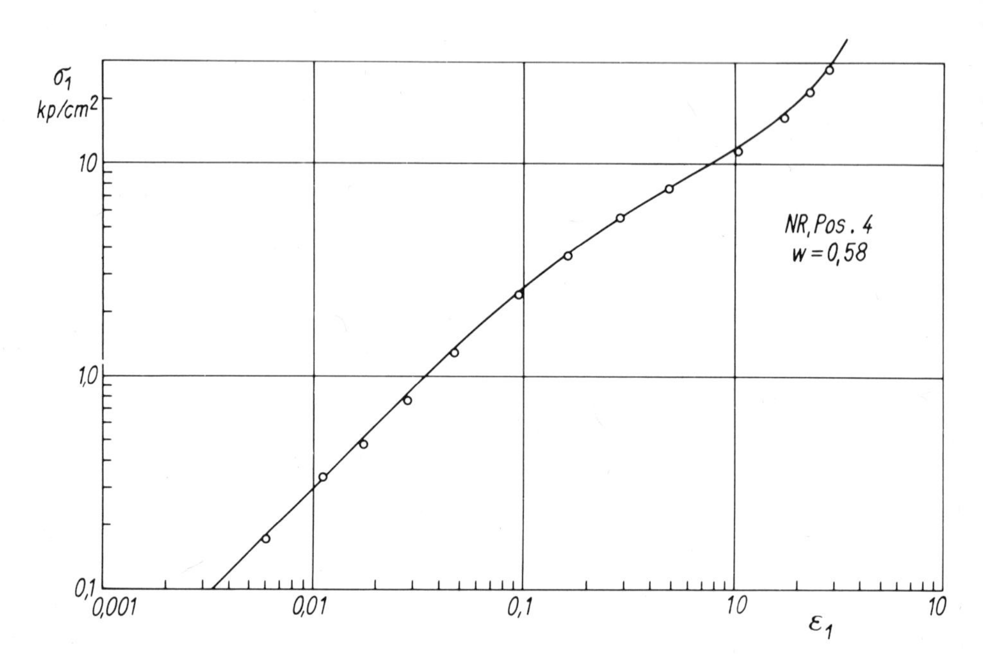

Fig. 5. Stress-strain data $\sigma_1(\epsilon_1)$ for natural rubber at a stress ratio w = 0.58.

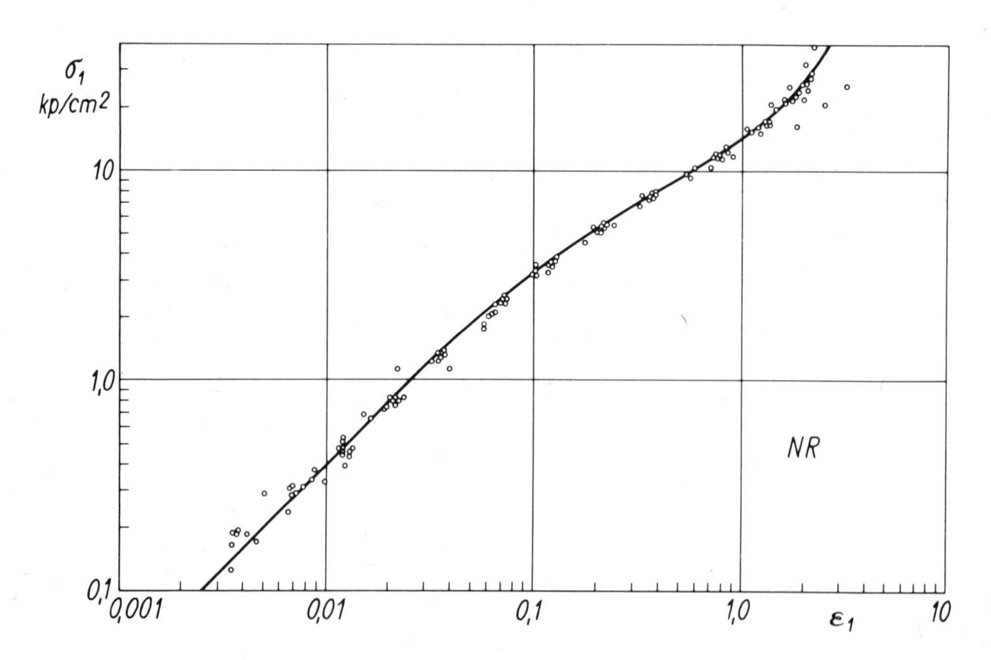

Fig. 6. Stress-strain master curve for natural rubber at a reference stress ratio w = 1.

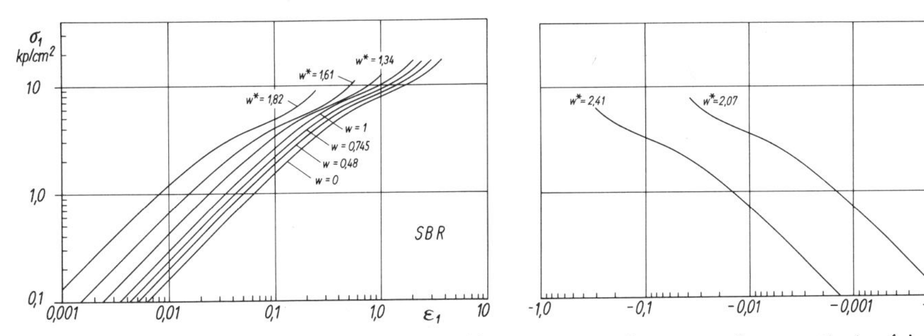

Fig. 7. Average curves $\sigma_1(\epsilon_1)$ for SBR at different stress ratios w or w, respectively; (a) positive and (b) negative deformations ϵ_1.*

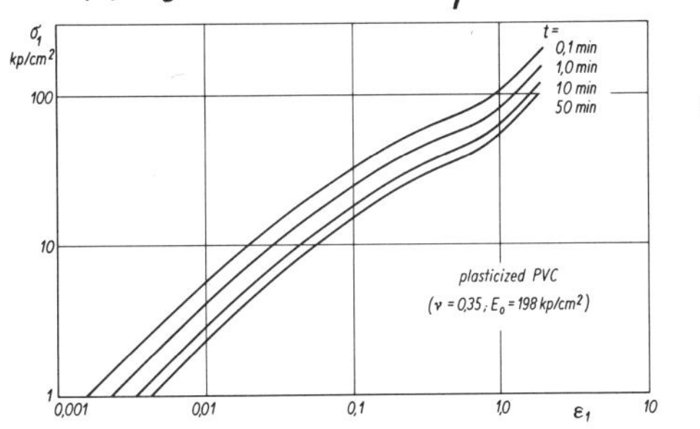

Fig. 8. Stress-strain curves for plasticized PVC at a stress ratio w = 1; parameter: measuring time t.

Fig. 9. Master curves $\sigma_1(\epsilon_1)$ for plasticized PVC, natural rubber (NR), and SBR, at a stress ratio w = 0.

seen from the parallelity of the curves, the dependencies on time and deformation can be separated, as has been mentioned above.

Finally, the master curves of the three investigated materials have been plotted together in Fig. 9 for the reference stress ratio w = 0 (uniaxial case). The curvature is very similar in all three cases. As is well-known for such materials, there are linear relations between the stresses and the strains up to values $\epsilon \approx 0.1$. For larger strains, on the other hand, the curves undergo the characteristic decrease of the tangent, followed by an increase up to a sudden break.

4 PHENOMENOLOGICAL DISCUSSION OF THE RESULTS

The shifts of the stress-strain curves (plotted into log-log-diagrams) parallel to the axis are mathematically equivalent to the multiplication of the stress and strain by fixed factors for each curve. The interpretation of the shift factors is possible through the use of a generalization of the simple equations of the classical theory of elasticity by an unknown function.

Let σ_1 and σ_2 be the two stresses, ϵ_1 and ϵ_2 the two deformations, E the Young's modulus of the special material, υ Poisson's ratio, and w =

σ_2/σ_1 the stress ratio of a special stress-strain curve. Then, the nonlinear biaxial stress-strain relations may be written as

$$\left.\begin{aligned} \frac{\sigma_1}{E} h\left(\frac{\sigma_1}{E}\right) &= \frac{\epsilon_1}{1-\upsilon} \\ \frac{\sigma_2}{wE} h\left(\frac{\sigma_2}{wE}\right) &= \frac{\epsilon_2}{w-\upsilon} \quad , \end{aligned}\right\} \tag{4.1a,b}$$

where, in general, $E = E(t)$ is time dependent, $h\,(\sigma_1/E)$ is an unknown function which generalizes the stress-strain curve subject to the condition

$$h\left(\frac{\sigma}{E}\right) = 1 \quad . \tag{4.2}$$

In this notation, for example, the uniaxial stress-strain relation is obtained from Eqs. 4.1a and 4.1b with $\sigma_2 = 0$ or $w = 0$ as

$$\frac{\sigma_1}{E} h\left(\frac{\sigma_1}{E}\right) = \epsilon_1 \quad , \tag{4.3}$$

and the homogeneous biaxial stress-strain relation with $\sigma_2 = \sigma_1$ or $w = 1$ as

$$\frac{\sigma_{1,2}}{E} h\left(\frac{\sigma_{1,2}}{E}\right) = \frac{\epsilon_{1,2}}{1-\upsilon} \quad . \tag{4.4}$$

From the experiments it could be verified that any other stress-strain relation $\sigma_1(\epsilon_1)$ for a general w is obtained from Eq. 4.3 by the multiplication of ϵ_1 with the factor

$$e = \frac{1}{1-\upsilon w} \quad . \tag{4.5a}$$

Since σ_1 is unchanged, this corresponds to the use of the factor

$$s = 1 \quad . \tag{4.5b}$$

On the other hand, stress-strain relations $\sigma_2\,(\epsilon_2)$ for a general w could be obtained from Eq. 4.3 by multiplication with the factors

$$\left.\begin{aligned} e &= \frac{1}{w-\upsilon} \quad , \\ s &= \frac{1}{w} \quad . \end{aligned}\right\} \tag{4.6a,b}$$

Equations 4.5a and 4.5b and 4.6a and 4.6b could be derived easily by comparing Eqs. 4.1a and 4.1b with Eq. 4.3.

An experimental verification of the shift factors of SBR is given in Figs. 10 through 13. Figure 10 shows the dependence of e on $1/(1-\nu w)$. Since here the reference position of the master curve has been taken for w = 1, the slope of the straight line is $1-\nu$, as follows if Eqs. 4.1a and 4.1b are compared with Eq. 4.4. Figure 11 shows the dependence of e on $1/[(1/w^*)-\nu]$, with $w = 1/w^*$. Here also the slope is $1-\nu$. Figure 12 shows the dependence of s on w or w*, which follows from Eq. 4.5b or 4.6b, respectively. Finally, Fig. 13 demonstrates the dependence of a general shift factor V = s/e on w or w*, respectively: a straight line with the slope $-\nu/(1-\nu)$, as follows if the reference position of the master curve has been taken for w = 1. The axial section for w = 0 equals, in this case, $1/(1-\nu)$.

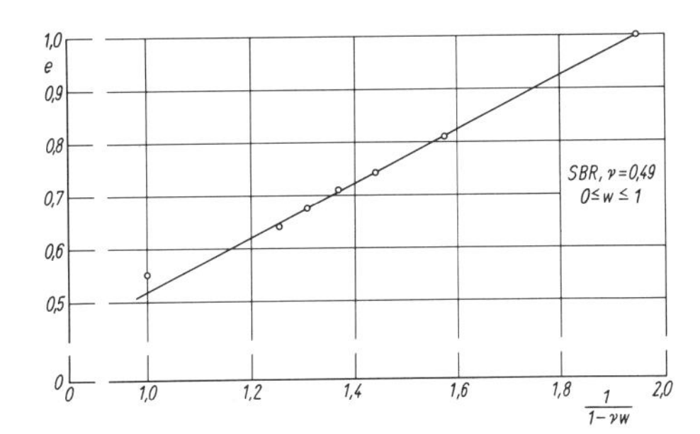

Fig. 10. Shift factor e as function of 1/(1-υw) for SBR - with respect to the reference position w = 1.

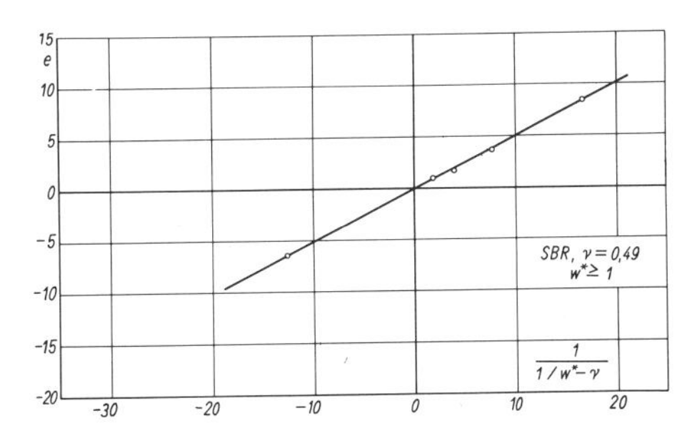

Fig. 11. Shift factor e as function of 1/[(1/w)-υ] for SBR - with respect to the reference position w = 1.*

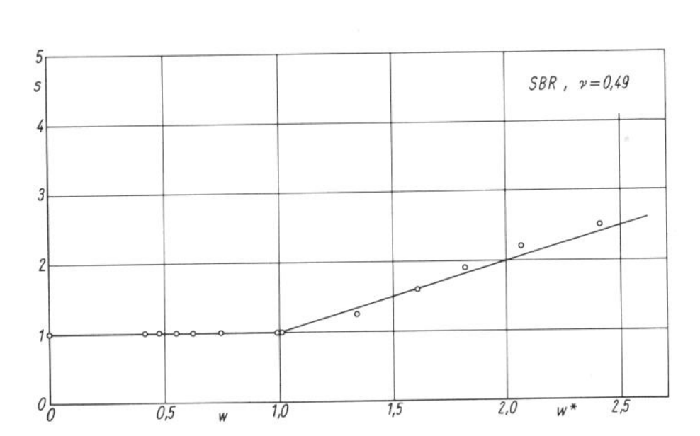

Fig. 12. Shift factor s as a function of w or w = 1/w, respectively, for SBR with respect to the reference position w = 1.*

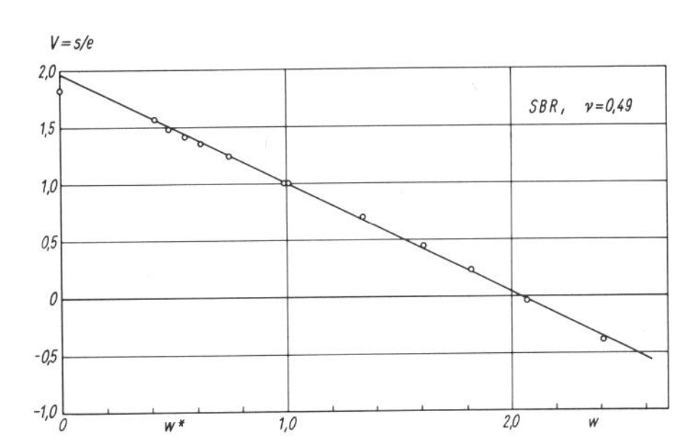

Fig. 13. Shift factor V = s/e as a function of w or w = 1/w, respectively, for SBR with respect to the reference position w = 1.*

Using these considerations, the experimentally determined shift factors permit the determination of the Poisson's ratio ν and by the simple Eqs. 4.1a and 4.1b, the Young's modulus E. The values given in Figs. 8 through 13 have been determined in this way.

Of course, instead of using Eqs. 4.1a and 4.1b, the possibility exists of describing the nonlinear biaxial stress-strain behavior by the relations

$$\left.\begin{aligned} \frac{\sigma_1}{E} &= \frac{\epsilon_1}{1-\upsilon w}\, g\left(\frac{\epsilon_1}{1-\upsilon w}\right) \\ \frac{\sigma_2}{wE} &= \frac{\epsilon_2}{w-\upsilon}\, g\left(\frac{\epsilon_2}{w-\upsilon}\right) \quad . \end{aligned}\right\} \qquad (4.7a,b)$$

But, in this case, the unknown function g depends upon Poisson's ratio υ and the stress ratio w in a more complicated form. On the other hand, there are many ways to derive the function h (σ_i/E) empirically from the measured curves. This also may be treated elsewhere.

5 SUMMARY

From the experimental data on biaxial sheet experiments for a natural rubber, SBR, and a plasticized PVC, it was shown that all curves for biaxial experiments under different stress ratios $w = \sigma_2/\sigma_1$ may be accurately calculated from a simple uniaxial stress-strain curve up to deformations of 300 percent. Furthermore, from the stress-relaxation curves of the plasticized PVC and the stress-strain curves at different times, it could be concluded that the dependencies of the stress on deformation and time can be separated in two different factorized functions. These results are of importance for the practical application of such materials in engineering because they show that uniaxial experiments only are necessary to predict the general biaxial stress-strain behavior.

ACKNOWLEDGMENT

The authors wish to thank the Deutsche Forschungsgemeinschaft (DFG) for their support of this investigation.

REFERENCES

1. Becker, G. W., and Rademacher, H.-J., *J. Polymer Sci.,* **58**, 621 (1962).
2. Green, A. E., and Adkins, J. E., *Large Elastic Deformations and Non-Linear Continuum Mechanics,* Oxford University Press, London, 1960.
3. Noll, W., and Truesdell, C., "The Non-Linear Field Theories of Mechanics", in *Encyclopedia of Physics,* S. Flügge (Ed.), Springer-Verlag, Berlin, 1965, Vol III/3.
4. Rivlin, R. S., and Saunders, D. W., *Phil. Trans. Roy. Soc. London,* **A243**, 251 (1950/51).
5. Blatz, P. L., and Ko, W. L., *Trans. Soc. Rheology,* **6**, 223 (1962).
6. Becker, G. W., *Testing Device for Biaxial Deformation of Elastomeric Materials,* Invention Report No. 30-616, Jet Propulsion Laboratory, Pasadena, California, 1964.
7. Obata, Y., Kawabata, S., and Kawai, H., *J. Polymer Sci.,* Part A-2, **8**, 903 (1970).

8. San Miguel, A., *Experimental Mechanics,* **12**, 155 (1972).
9. Becker, G. W., *J. Polymer Sci.,* Part C, **16**, 2893 (1967).
10. Gent, A. N., and Thomas, A. G., *J. Polymer Sci.,* **28**, 625 (1958).
11. Rivlin, R. S., *Phil. Trans. Roy. Soc. London,* **A240**, 459 (1948), **A241**, 379 (1949).
12. Treloar, L.R.G., *The Physics of Rubber Elasticity,* 2nd Ed., Clarendon Press, Oxford, 1958.
13. Bartenew, G.M., Nikiforow, W. P., and Awrustschenko, B.Ch., *Paste u. Kautschuk,* **17**, 37 (1970).
14. Tschoegl, N. W., *Rubber Chem. Technology,* **45**, 60 (1972).
15. Hencky, H., *Z. Techn. Phys.,* **9**, 215 (1928).
16. Valanis, K. C., and Landel, R. F., *J. Appl. Phys.,* **38,** 2997 (1967).
17. Becker, G. W., and Krüger, O., *Die Deformation hochpolymerer Werkstoffe im dreiachsigen Spannungszustand,* Research Report Be 391/1, Deutsche Forschungsgemeinschaft, 1972.

DISCUSSION on Paper by G. W. Becker and O. Krüger

RIVLIN: I would like to point out that the equation $t_i = \mu_o + \mu_i f_i + \mu_2 f_i^2$, which Dr. Becker has attributed to Truesdell and Noll, is certainly not due to them. It should probably be attributed to Reiner (1945). In any case, I and others have pointed out that unless the μ's are derivable from a strain-energy function – in which case it becomes the constitutive equation of finite elasticity theory for pure homogeneous deformations – it would represent a material from which energy could be drawn indefinitely, in violation of the law of thermodynamics. It is, however, meaningful in the context of stress-relaxing isotropic viscoelastic materials, if we consider the material to be subjected to a pure homogeneous deformation at some instant of time, and to then be held at constant deformation. The stress at time t after the deformation is produced is then given by the formula, but the μ's must, in this case, depend also on t (Rivlin, 1956).

Dr. Becker has mentioned the apparent scatter in the experimental values obtained for $\partial w/\partial I_1$ and $\partial w/\partial I_2$ in the experiments of Rivlin and Saunders. This is in large part due to the manner in which they are plotted and to the computation. If the values of the forces are calculated from the smooth lines drawn in the slide showing the results of Rivlin and Saunders, they do in fact agree very well with the measured values.

I do not understand why Dr. Becker considers that there is conflict between the results of Rivlin and Saunders and classical elasticity theory and would welcome his farther comments on this point. Also, I am surprised at the very low value he obtains for Poisson's ratio in the case of plasticized PVC.

BECKER: I thank Dr. Rivlin for his clarifying comments in respect to Eq. 4.3 of my paper, with which I fully agree.

Answering the last two questions of Dr. Rivlin, I would like to point out that the special approximations – Eqs. 2.6a and 2.6b of my paper – drawn from the results of Rivlin and Saunders oversimplify these results. So, especially at small deformations, the proportionality "constants" of these approximations, taken at finite strains, become functions of the deformations, or vice versa. The values of Poisson's ratio for the three investigated substances have not been measured directly, but only derived from the empirical shift factors using Eqs. 4.5a and 4.6a of my paper. This method is, of course, not very precise. So, the value of Poisson's ratio for plasticized PVC might be too low.

ON THE PHENOMENOLOGY OF RUBBERLIKE BEHAVIOR*

Robert F. Landel and Robert F. Fedors

Jet Propulsion Laboratory
California Institute of Technology
Pasadena, California

ABSTRACT

A simple functional statement of the stress-strain response can be used as the basis for a phenomenological investigation (or presentation) of the response, including rupture. Experimental observations on time-temperature or time-chain concentration superposition, time-strain factorizability, and the proportionality of stress with chain concentration can all be invoked to produce a simple, rather all-encompassing representation of the response. The results turn out to be very similar for all elastomers.

A semiquantitative molecular theory is reviewed which, though incomplete, accounts for much of the response in the rubbery region. Nine

*This paper presents the results of one phase of research carried out at the Jet Propulsion Laboratory, California Institute of Technology, under Contract No. NAS7-100, sponsored by the National Aeronautics and Space Administration.

molecular parameters are identified. Unsolved problems include the origin and molecular description of the slow relaxation processes associated with entanglements and a molecular theory of finite strain elastic and viscoelastic response.

1 INTRODUCTION

As previously noted[1] the tensile stress-strain response of a simple filled elastomer can be functionally stated as

$$\sigma = E(P_1, t, T, \phi, \ldots)\epsilon\ f(\epsilon, t, P_2, \ldots) \quad , \tag{1}$$

where σ is the stress, ϵ is the strain, P_1 and P_2 are one or more material parameters, ϕ is the volume fraction of filler, t is the time, and T is the temperature. The character of the function f is such that $f \to 1$ as $\epsilon \to 0$. If, as is usually true for rubbers, the character of the σ,ϵ response does not change near rupture, then the rupture points will be statistically distributed along the σ,ϵ curve. Hence, the interrelationships at rupture are given by the same or a very closely related functional statement.

These functional statements then permit the evaluation of the various relationships between the functions and the material or experimental parameters.[2-4] For example, in constant-strain-rate experiments (ϵ = Rt) and for ϕ = 0, P_1 = constant, at rupture $\epsilon_b = Rt_b$ and, noting that $\epsilon_b = g(t_b, P_2)$, the shape and location of the failure envelope in the σ_b, ϵ_b plane is given by[3]

$$\frac{d\sigma_b}{d\epsilon_b} = f\frac{\partial E/\partial t_b}{\partial g/\partial t_b} + \left[E\frac{\partial f}{\partial P_2} - f\frac{\partial g}{\partial P_2}\frac{(\partial E/\partial t_b)}{(\partial g/\partial t_b)}\right]\frac{dP_2}{d\epsilon_b} + E\frac{\partial f}{\partial \epsilon_b} \tag{2}$$

and the examination can be carried out with functions, such as E(t) or f(ϵ), which are empirical, theoretical, or both. For normal rubbers, this examination quickly reveals a number of important generalizations. Thus, for normally vulcanized rubbers, time and strain are factorizable over wide ranges of t and ϵ, and so

$$\sigma = E(P_1, t)\epsilon f(P_2, \epsilon) \quad . \tag{3}$$

Here, E(t) is the familiar modulus of linear viscoelastic theory. It can be seen that whatever affects the modulus, such as a change in P_1 or its time dependence, carries over directly to modify the stress at finite strains and at rupture. Consequently, for example, time-temperature superposition should apply to the rupture response and σ_b, like E(t), should be proportional to the concentration of effective network chains, υ_e, t, and T, as is observed experimentally. Further, the occurrence of multiple relaxation peaks (or "steps" in the E vs. t response) should give rise to peaks in ϵ_b vs. t_b plots or to steps in σ_b vs. t_b response. Such peaks have been found by Janacek and co-workers in crosslinked poly(hydroxy ethyl methacrylate).[5]

Similarly, if the function f becomes time dependent, then the character of the failure envelope should change. This has been demonstrated by Halpin[6] and put on a somewhat more formal basis by Smith[7]. (Actually, Smith formalized the ϵ_b-t_b relationship for the case of a Rouse model and used this to predict the occurrence of yielding rather than to obtain information about the failure envelope.)

Considering the effect of network-chain concentration, for example, the failure envelopes for different polymers might appear as in Fig. 1. As the temperature is lowered or the strain rate is increased, the points trace out a curve open toward the σ_b axis as ϵ_b first increases then decreases, while σ_b steadily increases. Since all of the rubbers have different crosslink densities or chain concentrations, the results should be normalized or reduced to a standard value of υ_e. According to Eq. 3, this can be done simply by taking σ_b/υ_e as the proper reduced variable. Figure 2 shows the results of such a reduction (including the temperature factor normally used with the modulus, too). It can be seen that the high-temperature or

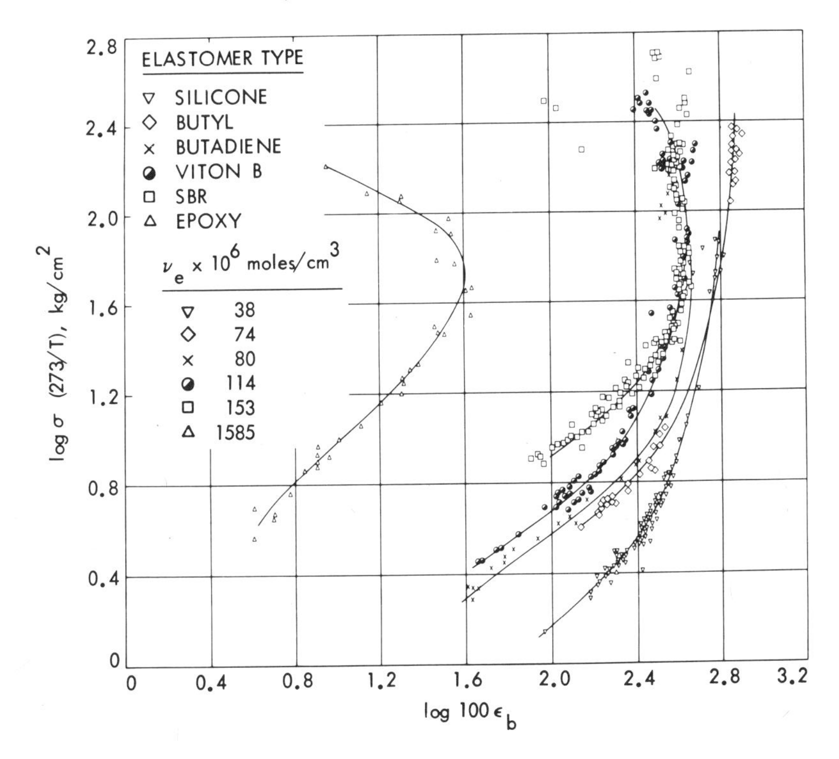

Fig. 1. Failure envelopes for five rubbers and a plastic.

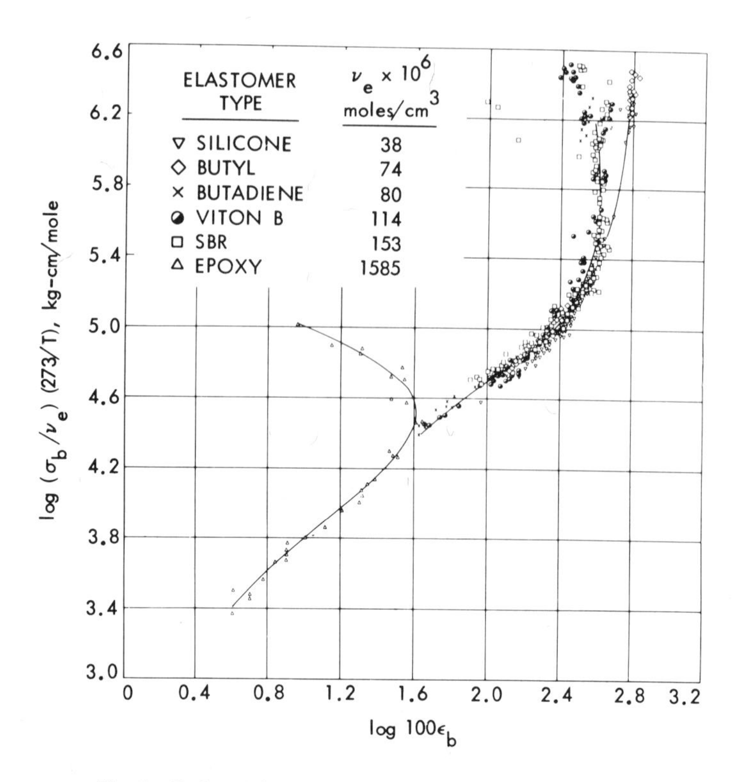

Fig. 2. Reduced failure envelope plot of the data of Fig. 1.

long-time portions of the envelopes superpose to a master curve. This curve is the *rupture stress per chain* and can be seen to be *independent of the chemical nature of the rubber*. Hence, in this region, we have another example of a response which, like the dilute-solution dynamic behavior described by Prof. Ferry, is unaffected by the nature of the molecular architecture. Therefore, the first question of this conference, "How does molecular structure affect the mechanical properties?" can be simply answered – it does not affect a major portion of the failure envelope. The break response as characterized by the failure envelope is largely controlled by the mere fact that a network is present. Hence, it is clear that it is not detailed molecular structure which must be modified to change the rupture response, but the nature, i.e. topology, of the network.

The reduced-failure-envelope concept applies to literature data on filled, swollen, and foamed polymers too[8], though the reduced envelopes differ from that of the pure gum material. Such data are difficult to treat quantitatively because reliable values of υ_e are not available.

One of the unexplored areas of this phenomenological description of the stress-strain response is that the parameters P_1 and/or P_2 could be (intentionally or unintentionally) strain dependent. Thus, a strain-dependent parameter P_2 has been invoked to describe the stress-strain response of glass-bead-filled SBR. Here the rubber separates from the beads as the specimen is strained, leaving a void. Using a simple analysis of the size of the void relative to the macroscopic specimen strain, Fedors[9] has shown that the actual volumetric loading ϕ can be replaced with an effective value which is a function of: the strain, the maximum possible value of ϕ (a parameter unique to each filler-rubber system), and an adhesion-related parameter. Both of these subsidiary parameters can be evaluated independently. The theory gave good agreement for *both* loading and unloading curves and for strains up to rupture. Thus, in answer to the second Conference question on the effect of an incoherent or incompatible phase, in one case at least (nonreinforcing filler) there is a straightforward and fairly complete answer, while for other systems (foams, reinforcing fillers) the reduced-variable scheme is applicable system by system. It is felt that an analogous approach would be very useful and instructive in describing the response of ionically linked and block copolymer-based rubbers, where the system has labile crosslinks. It is felt that the strain-dependent υ_e is a major source of the "increased" strength.

In contrast to this presentation, essentially phenomenological though guided by molecular concepts, there does exist a semiquantitative molecular theory for at least uniaxial deformation and rupture of elastomers. It requires only a combination of extant theories of E(t) and f(ϵ). Considering these in turn, E(t) originates from contributions from chain configurational changes, from entanglement slippage, and from the permanent network. The chain contribution can be calculated from the Marvin extension of the Rouse theory[10] by transforming[11] his $G'(\omega)$ and $G''(\omega)$ results, as tabulated by Ferry[12a], to E(t) by the Yagii-Maekawa approximation[13].*,** The molecular parameters required are the density ρ, the average monomeric friction factor ζ_0, an effective bond length along the chain backbone a, the molecular weight between entanglements M_{en}, the molecular weight of the sample M, and, to account for the temperature dependence, the glass transition temperature T_g, and the thermal expansion coefficient α. These are the same parameters mentioned in Prof. Ferry's lecture. As an example of such a calculation, Fig. 3 shows log E(t) vs. log t for five types of uncrosslinked elastomers. An initial molecular weight of 200,000 is assumed in each case; the values of the other parameters may be found in Prof. Ferry's book.[12b] Silicone, natural rubber (N), and butyl rubber illustrate the effect of increasing ζ_0 on the location of the transition zone. The increasing height of the entanglement

*Similar to the transformation indicated by Dr. Schwarzl in Session I of this Symposium.

**Prof. Vinogradov (private communication) feels that a recent treatment of entanglement effects, in his laboratory, is superior to that of Marvin, but we have not yet seen the paper.

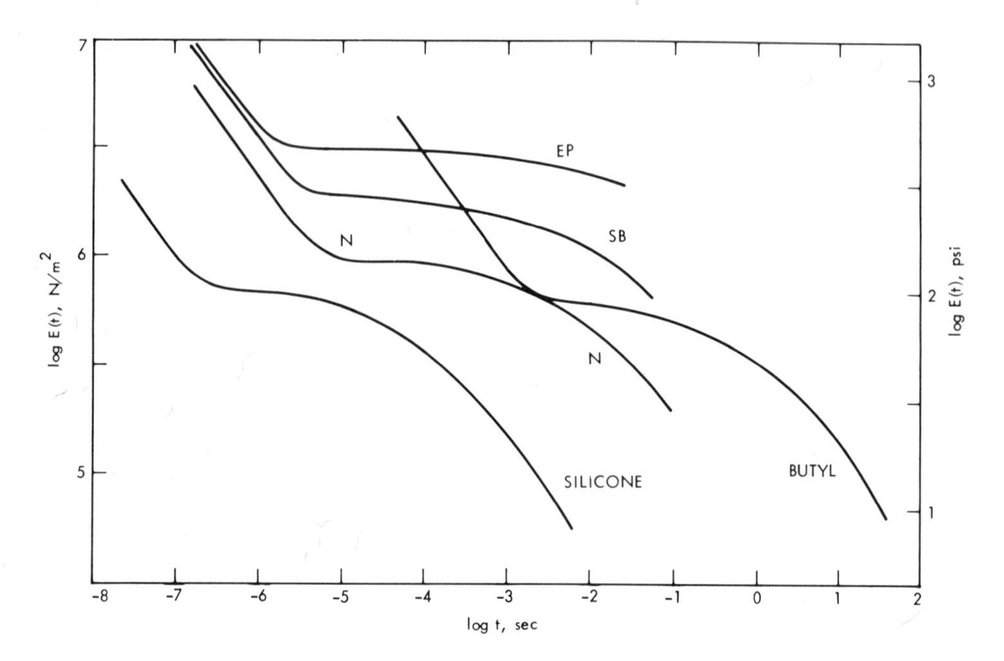

Fig. 3. Calculated stress-relaxation behavior at 298 K for five uncrosslinked elastomers of M = 200,000: ethylene-propylene (56/44) EP, styrene-butadiene (23.5/76.5) SB, natural rubber N, butyl, and dimethyl siloxane.

plateau in the series of hydrocarbons, natural rubber (N), styrene butadiene rubber (SB), and ethylene propylene rubber (EP) reflects their decreasing M_{en}. It can be seen that the two synthetic rubbers SB and EP have evidently had their copolymer composition adjusted (either intentionally or unintentionally) to match the ζ_0 value of natural rubber. Unfortunately, however, M_{en} values have not been matched, so that further adjustments in molecular architecture will be required in order to match the response of N.

On crosslinking the polymer, there is an additional contribution to the modulus, that of the permanent network. This can be separated from the chain contribution either directly[14] or on the basis of the Rouse-Mooney theory[15] discussed by Prof. Ferry.

The network contribution is $f'g\upsilon_e RT$, where f′ is the fraction of the network which is effective and includes a correction term for dangling chain ends, and g is the gel fraction. Lacking ideal systems where υ_e can be determined a priori from the chemistry of the reactants alone, υ_e must be determined for each sample in a separate experiment.

There is an additional relaxation process, however, ascribed to entanglement slippage. The treatment of chain slippage is deemed to be in an unsatisfactory state, so the empirical Plazek function[16] $\psi(a_x)$ is employed to assess this contribution. In effect, this gives the rate of decay of E(t) from the entanglement plateau to the equilibrium plateau. Plazek found the same functional form for this process for both N and SB measured in shear; Thirion[17] finds substantially the same form in tensile stress relaxation (also N and SB), as did Tsuge, Arenz, and Landel[18] in biaxial stress relaxation (SB). We believe this function to be independent of the polymer over wide ranges in υ_e, but suspect this will no longer be true at very high or very low chain densities.

If we take the modulus to be the simple sum of these three contributions,

$$E(t) = E(t)_{Ch} + E_{network} + E(t)_{en} \quad , \tag{4}$$

then E(t) can be calculated for most of the useful operating range of time or temperature with no adjustable parameters. The three dominant molecular parameters are T_g, ζ_o, and M_{en}. Figure 4 illustrates the effect of crosslinking silicone and SB to different network chain concentrations. It can be seen that when the entanglement plateau is appreciably larger than the equilibrium value, as in SB, the approach to equilibrium is very, very slow. Figure 5 shows a comparison of measured and calculated results for four of the rubbers mentioned previouly, again assuming the original

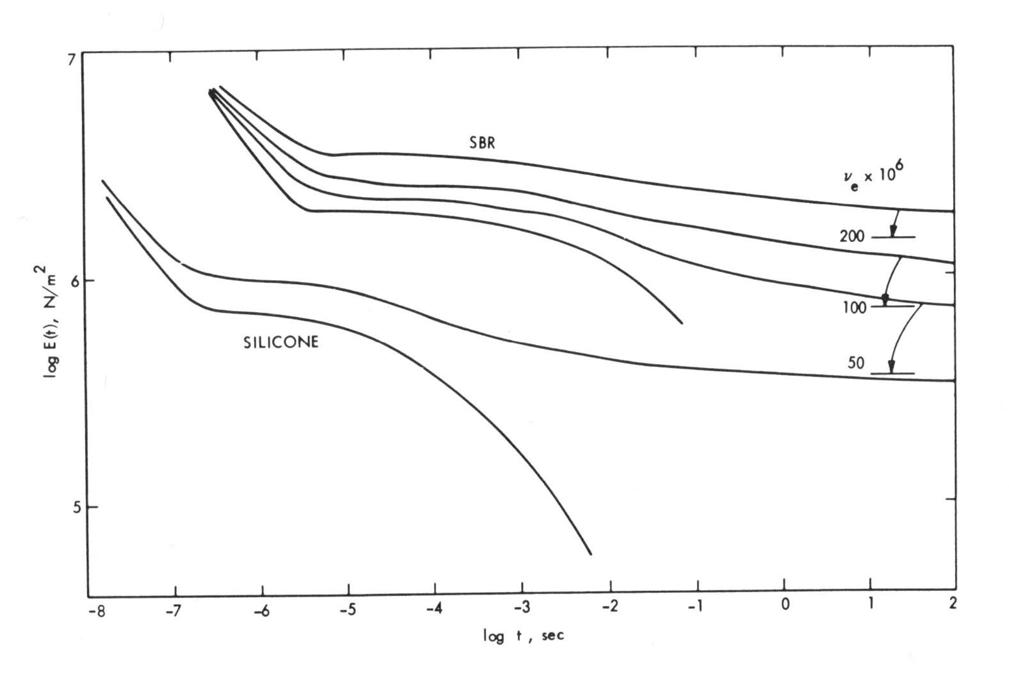

Fig. 4. Calculated stress-relaxation curves for silicone, uncrosslinked and crosslinked to υ_e = 50 x 10^6 moles/cm^3, and SBR crosslinked to υ_e = 50, 100, and 200 x 10^6 moles/cm^3. The horizontal bars show the location of the equilibrium modulus for M = 200,000, T = 298 K for both rubbers.

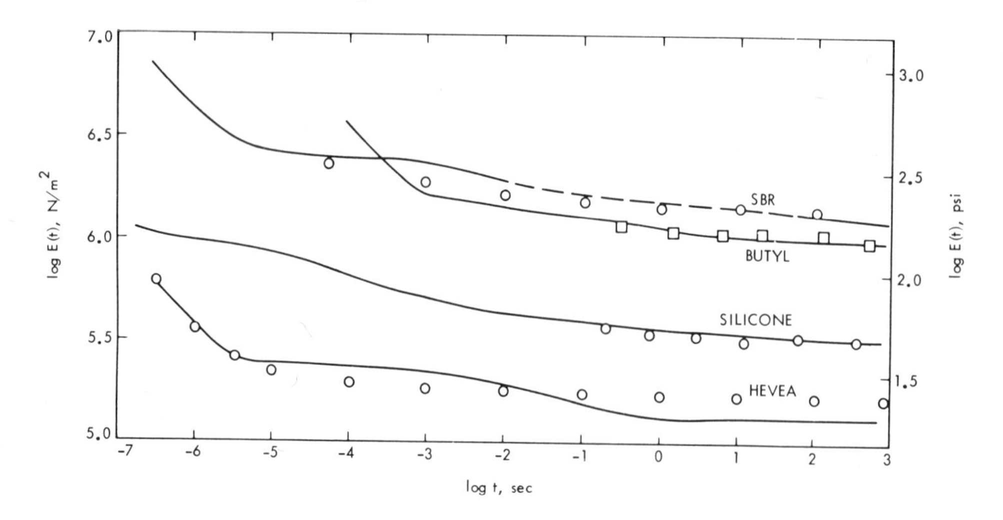

Fig. 5. Calculated vs. experimental modulus curves for four elastomers. M assumed 200,000 in each case; T = 298 K. The values of υ_e are: SBR 100, butyl 113, silicone 38, and HEVEA 168 moles/cm^3.

molecular weight to be 200,000 in each case. For silicone and butyl rubber, the experimental data represent the isochronal $2(C_1 + C_2)$ response as determined from constant-strain-rate tests by Smith[19]; for natural rubber, the creep data of Plazek[16] (his specimen J); and for SBR, our own tensile data[20]. In all cases the agreement is within 25 percent, and for the SB, where υ_e and M are more precisely known, the calculated curve agrees with experiment to better than 12 percent (except in the first half-decade of time). Thus, the modulus is shown to be calculable from molecular parameters which can be evaluated a priori (or are already known) over at least seven decades of reduced time. Put in another way, this corresponds to an essentially correct prediction over temperature ranges of -40 to +80 C for SB, -45 to +200 C for silicone, -20 to +150 C for butyl, and -50 to 90 C for N. It can be seen that these temperatures cover substantially all of the normal operating range of these materials.

The major problem is thus identified as the need for a molecular theory of entanglements and their relaxation processes. Such a theory should undoubtedly obviate the need for the admittedly crude but surprisingly effective additivity assumption of Eq. 4.

Considering the region of nonlinear response, when $f(\epsilon) \neq 1$ in Eq. 1, again there is no adequate molecular theory for finite-strain elastic response, much less the viscoelastic response. However, if factorizability holds, then the molecular theory should apply to isochronal data. Unfortunately, as well known, the best molecular theories for the σ,ϵ response are somewhat inadequate. Recent development in continuum mechanical analysis and corresponding finite-deformation multiaxial experiments indicate, however, that a simple cubic-lattice network model should suffice and, moreover, that up to quite large strains, the only material parameter required to describe the response is E(t) [strictly speaking, G(t)].

The large-strain response of a rubber has been treated in the past in terms of a stored-energy function W and its dependence on the strain invariants.[21-25] Experimentally, the strain dependence (or, specifically, the invariant dependence) of the derivative of the stored-energy function W with respect to the invariants must be determined. This is not easy, and no generally valid representation of W has yet been found, as the response at smaller strains is very complex.[23,24] This point is discussed more fully by Dr. Becker in Session II. Recently a different approach has been suggested for determining W, in which the independent variables remain the stretch ratios, rather than the strain invariants, i.e.,

$$W = W(\lambda_1, \lambda_2, \lambda_3) \tag{5}$$

instead of

$$W = W(I_1, I_2, I_3) \quad , \tag{6}$$

where each invariant

$$I = I(\lambda_1, \lambda_2, \lambda_3) \quad . \tag{7}$$

Under the hypothesis that W is a separable function of λ_i, such that

$$W(\lambda_1, \lambda_2, \lambda_3) = w(\lambda_1) + w(\lambda_2) + w(\lambda_3) \quad , \tag{8}$$

and taking rubber to be incompressible, it turns out[26] that the gradient of the stored-energy function ($\partial w / \partial \lambda$) can be established in graphical form with very little experimental difficulty.[18,25,27] The results, however, have not lent themselves to an analytical representation, though they can be approximated as[26,27]

$$w' = 2G \ln \lambda \tag{9}$$

or

$$w' = 2G\epsilon / \lambda \tag{10}$$

for λ's lying roughly in the range 0.8 to 2.0.

In more recent studies[18] it has been shown that the rate of stress relaxation is not only independent of the strain but also of its degree of biaxiality (for SB at room temperature and times up to 10^4 sec). Hence, factorizability has been confirmed for biaxial experiments. The stress-relaxation data could be extrapolated to equilibrium, and it was shown that $w'(\lambda)$ at equilibrium is directly proportional to υ_e. These facts form the basis for the statement made above that only one material parameter is required to describe even finite-strain behavior (in cases where the rubber can be considered incompressible).

These few paragraphs therefore indicate that, in answer to Conference Questions 3 and 4 on phenomenological approaches to developing constitutive equations and their relevance to fracture theory, the usual emphasis on the use of invariants is too restrictive experimentally. The new approach suggested has not been assessed as a means of characterizing failure, so this remains very much an open question.

Returning to the molecular picture of large strain response, it should be recognized that at very high extensions, finite extensibility of the average chains will begin to play a role in deformation and rupture. This is automatically accounted for in the continuum treatment just described.* In molecular theory, however, only the work of Treloar[29], with the inverse Langevin function treats this effect. Though the resulting stress-strain equation is inadequate at low and moderate uniaxial tensile strains, it does describe rather well *both* the region of the sharp upturn in tension *and* compression results. In this case, the molecular parameters P_1 and P_2 are υ_e (again) and the number of equivalent statistical segments per chain, n, respectively, e.g.,

$$\sigma = \frac{1}{3} \upsilon_e \, R \, T \, n^{1/2} \left[\mathcal{L}^{-1} \left(\frac{\lambda}{n^{1/2}} \right) - \frac{1}{\lambda^{3/2}} \, \mathcal{L}^{-1} \frac{1}{(\lambda^{1/2} \, n^{1/2})} \right] , \qquad (11)$$

where $\mathcal{L}^{-1}$ is the inverse Langevin function, defined by $\mathcal{L}^{-1}(x) = y$ and $x = \coth y - \frac{1}{y}$. The value of n is not unique to a given rubber, as is ζ_0 or T_2, but depends on f′, g, and υ_e. For a perfect network it can be determined in any of several related ways, e.g.,[30] from the average projected bond length of the backbone ℓ, the average molecular weight per backbone bond M_ℓ, and the mean square end-to-end separation of a chain per unit molecular weight $\langle r_0{}^2 \rangle / M$ (determined from measurements on dilute solutions of the uncrosslinked polymer):

$$n = \left(\frac{\ell}{M_\ell} \right)^2 \left(\frac{\langle r_0{}^2 \rangle}{M} \right)^{-1} \frac{\rho}{\upsilon_e} . \qquad (12)$$

It can also be estimated from birefringence measurements on the elastomers.[31] Figure 6 illustrates how changing υ_e or n affect the stress-

*Though Alfrey believes the underlying assumption of separability will break down at very high chain extensions.

strain response. The initial slope is determined by υ_e, but the rapidity of the upturn is controlled by n – the smaller the value, the more rapid the rise beyond the inflection.

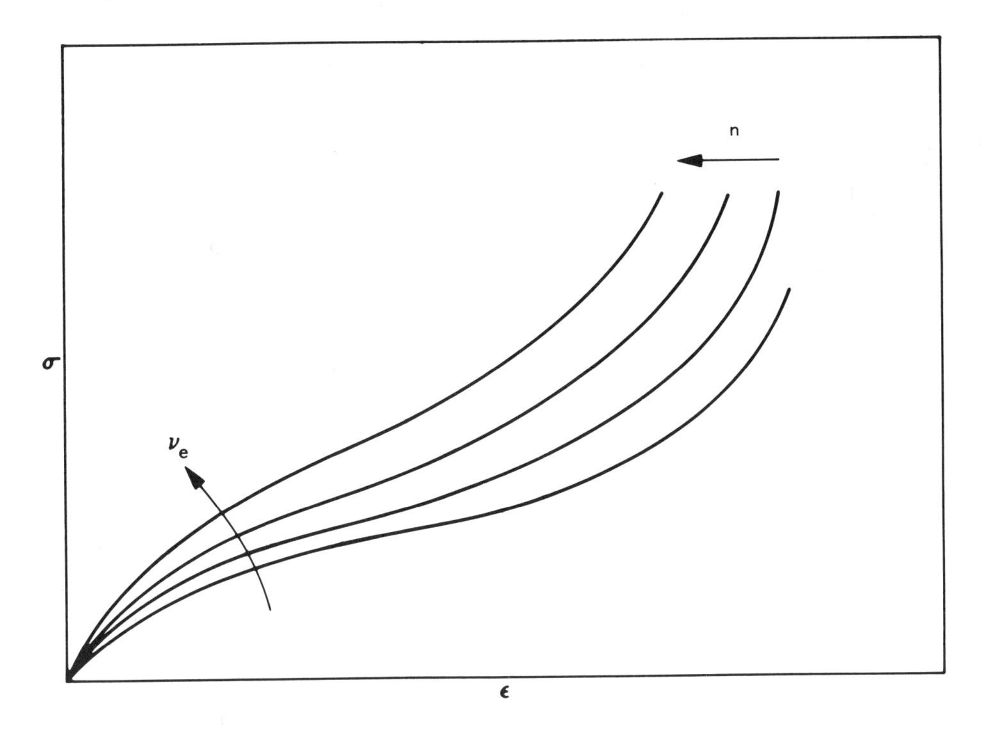

Fig. 6. Effect of increasing υ_e or n on the shape of the stress-strain curve. At fixed υ_e, n can be increased by lowering the chain flexibility or increasing the number of dangling chain ends.

Alternatively, stress-strain response based on the inverse Langevin function can be taken as a failure criterion in the sense of giving the failure envelope, σ_b vs. ϵ_b, for it contains an upper limit to σ.[29,32-34] Other criteria might be ϵ_b vs. t_b[35] or σ_b vs. t_b[36]. Equation 1, however, is more successful in describing rupture than in describing the response prior to rupture. Figure 7 shows that the description of the envelope based on Eq. 11 is very good indeed and that the υ_e values required for the fit shown agree well with those measured directly from swelling experiments. The values of n obtained from the fit are in fair agreement with those calculated via Eq. 12, but this calculation has to assume a perfect network. Where the necessary sol-gel data are available and a reasonable estimate of the chain end correction factor can be made, the envelope can be calculated very closely. Figure 8 compares the calculated responses for SB with υ_e values of 50 and 100μ moles/cm^3 (full curves) with the experimental data for a sample with υ_e = 100 μmoles/cm^3 (open circles).

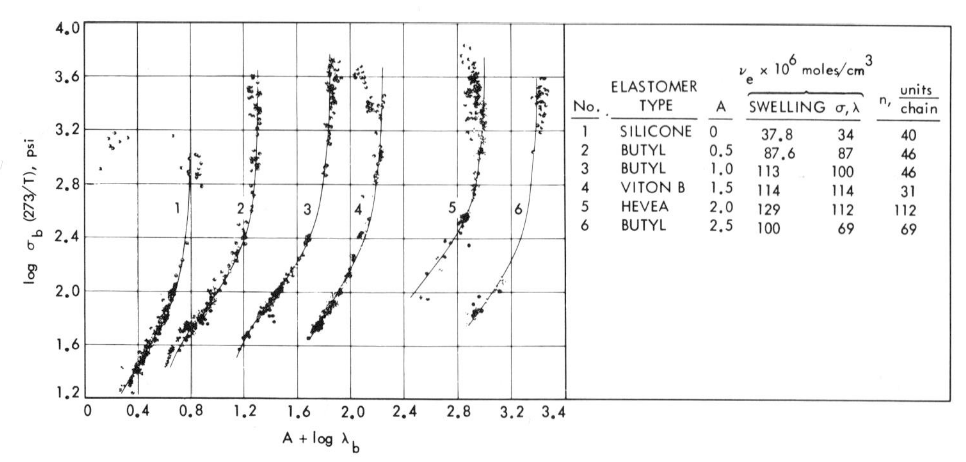

Fig. 7. Fit of the inverse Langevin function σ-ε relationship to failure envelopes of five rubber samples, comparison of υ_e values obtained from swelling and from the fit, and the values of n for each sample.

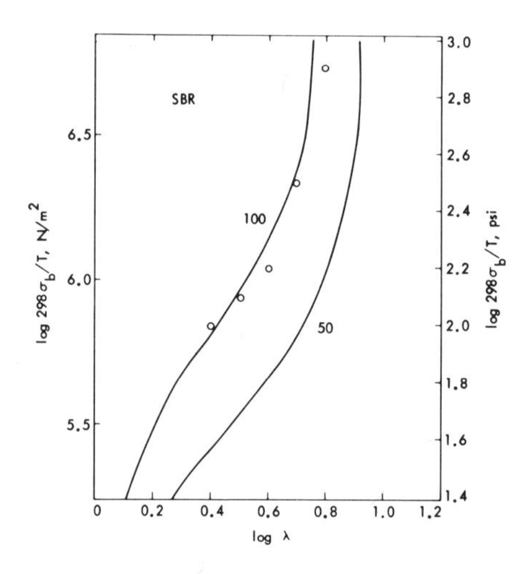

Fig. 8. Lower portion of SBR failure envelopes calculated for two values of υ_e x moles/cm^3 as shown. Open circles — experimental at υ_e = 100 x 10^{-6} moles/cm^3; full curves — at υ_e = 50 and 100 x 10^{-6} moles/cm^3.

As we have stated earlier, that which affects the modulus should affect the rupture response in the same way. Since Plazek has shown that a change in the crosslink density shifts the compliance along the time scale by a shift factor designated a_x, the same should be true for σ_b and ϵ_b. For ϵ_b, this shift should be valid only in the high-temperature region. Figure 9 shows the superposition obtained with several sets of data on SBR[3,37] using $a_x = (C/\upsilon_e)^{7.7}$, where C is a constant. The data of Smith[19], being the most extensive, are taken as the reference and shown as the solid line. The dashed lines indicate the spread in his data. These were obtained at one υ_e but varying test rate and temperature. The data of Baranwal[35] and Healy[39] were similarly obtained, but that of Taylor and Darin[40], Epstein and Smith[41], and Fedors and Landel[20] were ob-

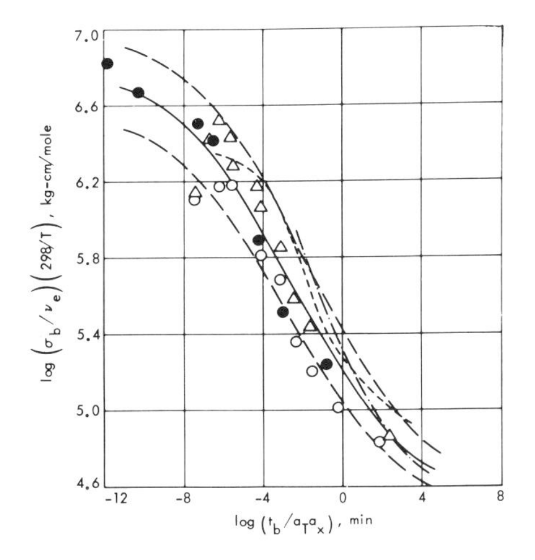

Fig. 9. Breaking stress per effective chain for SBR as a function of reduced time, i.e., a time scale corrected for changes in both temperature and effective chain concentration. Solid and long dashed lines indicate average and spread in Smith's data[19]; dot-dash, data of Baranwal[38]; short dash: Healy[39]; △: Taylor and Darin[40]; ○: Epstein and Smith[41]; and ●: Fedors and Landel[20].

tained in tests at a single rate and temperature, but using specimens of varying υ_e. Figure 10 shows the corresponding results for ϵ_b, where it is passing through its maximum at short reduced times, this new shift process superposes all data to within the normal experimental scatter. The same technique has been successfully applied to Smith's Viton data[42] and our own fluorosilicone data[43]. The reduced curves are of course displaced along the time scale, reflecting the different values of C, but all have the same shape. Thus the character of the time dependence of rupture, especially σ_b is again substantially independent of the nature of the rubber chain – hydrocarbon, fluorocarbon, or fluorosilicone.

This new technique offers an additional way of exploring long-time behavior, without having to go to extremely high temperatures, where degradation may set in. As an example of a possible use of this reduced time scale, Fig. 11 shows the time-to-rupture dependence of SBR specimens subjected to both static and dynamic fatigue as reported by Lake and Lindley.[45] The full curve is the response predicted from uniaxial constant-strain-rate data. These data represent the longest fatigue-lifetime studies in an unfilled elastomer that we know of, the maximum lifetime being 6 months. The agreement is excellent.* If only temperature-reduced times were employed, tests at 145 C would have been required for constant-strain-rate tests at unit strain rate.

*For the dynamic fatigue data, it was assumed simply that t_b is equal to the total elapsed time. Since this paper was written, fatigue studies on filled methyl silicone rubber have shown that for this material, data correlate with the number of cycles to failure, rather than the time to failure. The dynamically loaded (to constant amplitude) specimens fail sooner and at a time which depends on the frequency (0.1, 1, 10 cps). This underscores Dr. Thomas' remarks. The predicted lifetime represents an upper limit, probably valid for static loadings only in such cases.

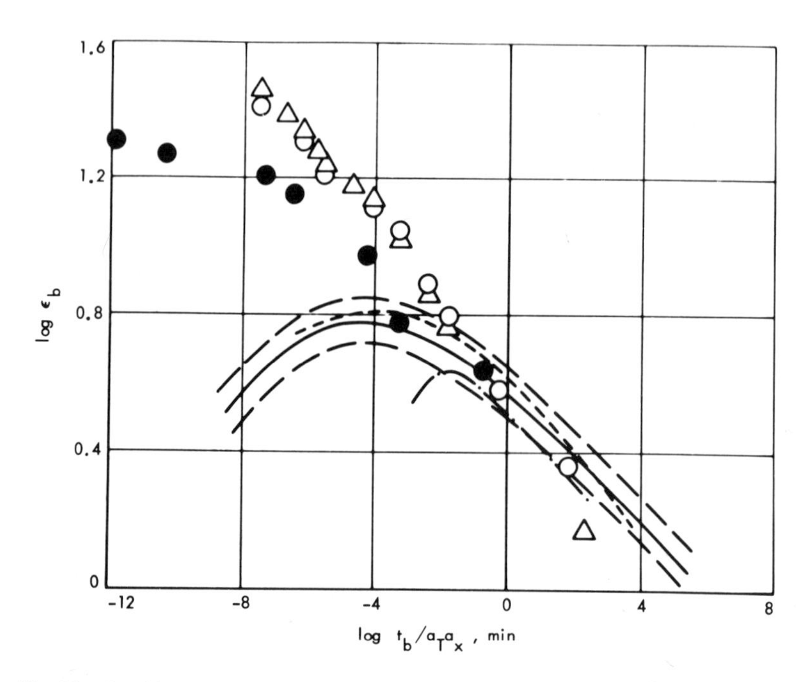

Fig. 10. Breaking strain versus time- and chain-concentration-reduced time for SBR. Key — same as Fig. 9.

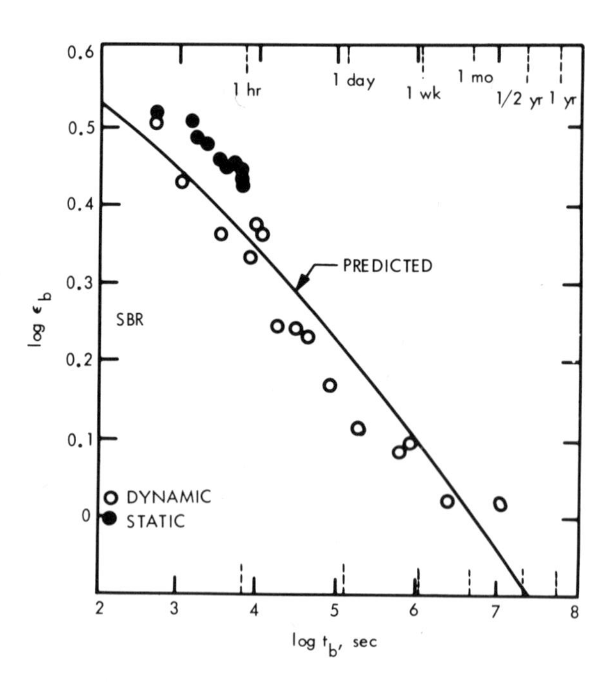

Fig. 11. Dependence of breaking strain on rupture time for SBR in fatigue. The solid curve is the response predicted from constant-strain-rate-rupture data; points, data of Lake and Lindley[45].

In conclusion, it should be noted that the overall treatment described here offers a straightforward framework for interpreting the effects of degradation on mechanical response. Rather than measuring the change in properties, such as E or σ_b, it would appear that a wider range in mechanical-behavior response could be assessed from a given effort if the rate of change of the fundamental parameters such as υ_e, M_{en}, T_g, and ζ_o were determined instead of direct measurement of properties.

REFERENCES

1. Landel, R. F., "Some Effects of Fillers on the Deformation and Rupture of an Elastomer", in *Mechanics and Chemistry of Solid Propellants,* A. C. Eringen, H. Liebowitz, S. L. Koh, and J. M. Crowley (Eds.); Pergamon Press, London (1967), 575 pp.
2. Fedors, R. F., and Landel, R. F., "Fracture of Amorphous Polymers", in *Proc. First Int. Conf. on Fracture,* T. Yokobori, T. Kawasaki, and J. L. Swedlow (Eds.), The Japanese Society for Strength and Fracture of Materials, Tokyo, Japan (1966), Vol 2, p. 1247; *Rubber Chem. and Techn.,* **40**, 1049 (1967).
3. Landel, R. F., and Fedors, R. F., "A Molecular Theory of Elastomer Deformation and Rupture", in *Mechanical Behavior of Materials,* The Society of Materials Science, Japan (1972), Vol III, p. 496.
4. Fedors, R., "Uniaxial Rupture of Elastomers", in *Elastomers: With Emphasis on the Stereospecific,* W. M. Saltman (Ed.) M. Dekker, New York (1972), in press.
5. Janacek, J., private communication(September 1972).
6. Halpin, J. C., *J. Polymer Sci.,* Part C, **16**, 1037 (1965); and *J. Appl. Phys.,* **36,** 2975 (1965).
7. Smith, T. L., Paper presented at the Sixth International Congress on Rheology, Lyon, France, September, 1972.
8. Fedors, R. F., and Landel, R. F., Unpublished results.
9. Space Program Summaries 37-41, **IV,** 97 (August, 1966); 37-54, **III**, 97 (December, 1968); and 37-55, **III,** 193 (February, 1969), Jet Propulsion Laboratory, Pasadena, California.
10. Marvin, R. S., *Viscoelasticity – Phenomenological Aspects,* J. T. Bergen (Ed.) Academic Press, New York (1960), p. 27.
11. Landel, R. F., Unpublished results.
12. Ferry, J. D., *Viscoelastic Properties of Polymers,* John Wiley & Sons, New York: (a) 1st Edition (1961), p. 462; (b) 2nd Edition (1970).
13. Yagii, K., and Maekawa, E., *Nippon Gomu Kyokaishi,* **40,** 46 (1967).
14. Landel, R. F., *J. Colloid Sci.,* **12**, 308 (1957).
15. Mooney, M., *J. Polymer Sci.,* **34,** 599 (1959).
16. Plazek, D. J., *J. Polymer Sci.,* Part A-2, **4,** 745 (1966).
17. Chausset, R., and Thirion, P., *Physics of Non-Crystalline Solids,* J. Prins (ed.), N. Holland Pub., Amsterdam (1965), p. 345.
18. Tsuge, K., Arenz, R. J., and Landel, R. F., "Finite Deformation Behavior of Elastomers: VI. Dependence of W on Degree of Crosslinking for SBR", in *Mechanical Behavior of Materials,* The Society of Materials Science, Japan (1972), p. 433.
19. Smith, T. L., *J. Polymer Sci.,* Part C, **16,** 841 (1967).
20. Fedors, R. F., and Landel, R. F., "A Test of the Predictability of the Properties of Filled Systems", presented at the AIAA Third Propulsion Joint Specialist Conference, Washington, D. C., AIAA Preprint No. 67-491 (July, 1967).
21. Rivlin, R. S., "Large Elastic Deformations" (and reference cited therein), in *Rheology,* Vol. I, F. Eirich (Ed.), Academic Press, New York (1956).
22. Green, A. E., and Zerna, W., *Theoretical Elasticity,* Oxford University Press, London 2nd Ed. (1954), 2nd Edition.
23. San Miguel, A., and Landel, R. F., *Trans. Soc. Rheol.,* **10,** 369 (1966).

24. Becker, G. W., *J. Polymer Sci.,* Part C, **16**, 2893 (1967).
25. Obata, Y., Kawabata, S., and Kawai, H., *J. Polymer Sci.,* **8**, 903 (1970).
26. Valanis, K. C., and Landel, R. F., *J. Appl. Phys.,* **38**, 2997 (1967).
27. Smith, T. L., and Dickie, R. A., *J. Polymer Sci.,* Part A-2, **7**, 635 (1969).
28. Alfrey, T., Jr., "Equations of State for Elastomers", in *Proc. of Conf. on Polymer Structures and Mechanical Properties,* U. S. Army Natick Laboratories, Natick, Massachusetts (1967).
29. Treloar, L.R.G., *The Physics of Rubber Elasticity,* Oxford University Press, London (1958), 2nd Edition.
30. Bueche, F., Kinzig, B. J., and Voen, C. J., *Polymer Letters,* **3**, 399 (1965).
31. Furukawa, J., Nishioka, A., and Kotani, T., *Polymer Letters,* **8**, 25 (1970).
32. Halpin, J. C., *J. Appl. Phys.,* **36**, 2475 (1965).
33. Smith, T. L., and Frederick, J. E., *J. Appl. Phys.,* **36**, 2996 (1965).
34. Fedors, R. F., and Landel, R. F., Space Programs Summary 37-36, **IV**, 137 (December, 1965), Jet Propulsion Laboratory, Pasadena, California.
35. Bueche, F., and Halpin, J. C., *J. Appl. Phys.,* **35**, 36 (1964).
36. Ninomiya, K., *Nippon Gomu Kyokaishii,* **41**, 893 (1968); *J. Soc. Mater. Sci. (Japan),* **19**, 282 (1970).
37. Fedors, R. F., and Landel, R. F., Space Programs Summary 37-58, **III**, 180 (August, 1969), Jet Propulsion Laboratory, Pasadena, California.
38. Baranwal, K. D., Ph.D. Thesis, University of Akron (1967).
39. Healy, J. C., Ph.D. Thesis, University of Akron (1967).
40. Taylor, G. R., and Darin, S. R., *J. Polymer Sci.,* **17**, 511 (1955).
41. Epstein, L. M., and Smith, R. P., *Trans. Soc. Rheol.,* **2**, 219 (1958).
42. Smith, T. L., *Proc. of the Fourth Internat. Congress on Rheol.,* E. H. Lee, (Ed.), Interscience, New York (1965), Part II, p. 525.
43. Fedors, R. F., Unpublished results.
44. Lake, G. J., and Lindley, P. B., *J. Appl. Polymer Sci.,* **8**, 707 (1964).

DISCUSSION on Paper by R. F. Landel

TSCHOEGL: Your approach assumes that strain-time factorization is applicable up to rupture. Is this assumption generally justified?

LANDEL: Yes. And some limits of its applicability can be readily stated. One must be working with a material which is not reacting chemically or undergoing a phase change. Moreover, within the class of amorphous unfilled polymers, it will hold from the region beyond the maximum value of λ_b (measured on t or T) and at all strains up to rupture. It is important to mention that $(\lambda_b)_{max}$ occurs in the region of the time scale which is *beyond* the transition zone and *into* the plateau zone as described by Professor Ferry.

ANDREWS: Is the shift (a_x), due to crosslinking, the same effect as the raising of T_g by crosslinking, or is it additional to this?

LANDEL: It is additional and very much larger.

WILLIAMS: I should like to reiterate one of my long-standing objections to extending, without qualification, the use of the one-dimensional failure envelope to specimens subjected to repeated and, especially, fatigue loadings. The failure envelope has been shown experimentally to be independent of rate effects, but only for essentially monotonic loadings. When oscillatory loadings are imposed, sufficient internal heating can arise to effectively change the temperature of the test. Because most data have been recorded for small-size tensile specimens, the heat generated can frequently be dissipated to the extent that thermo-mechanical coupling effects are not observed. In turn then, such apparent correlation of fatigue-failure tests with the failure envelope can generate a false sense of security and mislead designers who may then misapply the failure-envelope concept to larger size specimens where internal heating generated during fatigue can be a controlling factor not reflected in the failure envelope.

THOMAS: I would like to make two points concerning Dr. Landel's paper.

First, I have certain reservations about the significance of failure envelopes as a method of representing rupture behavior. They appear to be in large measure simply the stress-strain curves of the materials, and the influence of rate and temperature of test is primarily to vary how far up the curve one can proceed before failure occurs. Thus a rupture criterion hardly enters into the representation. In fact, the most interesting aspect of failure given by the measurements is the influence of rate and temperature on breaking stress (or strain) and the failure envelope representation suggests this influence.

Second, the failure of SBR rubber under repeated stressing is not solely dependent on time as such. If the frequency of cycling is more than about 1 Hz, there is a significant component associated with the cycling process itself. For example, the crack growth under repeated stressing is, in general, more rapid than would be expected simply on the basis of the time integral of the steady-state crack-growth behavior. Natural rubber, a strain-crystallizing elastomer, is an extreme example in that purely-time-dependent crack growth is virtually absent, and SBR can be considered as intermediate between this behavior and that of a hypothetical purely-time-dependent material.

LANDEL: It is just this suppression of the time effects which makes the failure envelope an attractive failure *criterion* as opposed to description. In this connection it is important to recall the tensile property surface – the envelope is the projection of the failure points on the σ,ϵ plane. Either of the other two projections to the σ,t or ϵ,t plane could also be used but whenever we can ascertain a time-independent (or nearly so) portion of the response it seems better to do so.

For example, time-temperature superposition could just as well be given in terms of a temperature-dependent activation energy instead of an a_T factor. While this would certainly present no essential difficulties in superposition, the usual separation into a time-dependent function (one of the spectra) and a time-independent function (log a_T) leads to a simpler, more economical and certainly more revealing representation.

Considered in this way, the time dependence of σ_b or ϵ_b arises from the time dependence of E and the slight but profoundly effective differences in the shapes of the stress-strain curve and the failure envelope.

As for the second question, you may well be correct and one could get a poor estimate of the lifetime of a gum rubber by this approach. Still, it is remarkably successful for these data, which are the longest-duration results we could find published on unfilled SBR. Moreover, if the material does fail sooner than predicted, as you indicate it should, this is in reality a benefit, since one is generally interested in the longest lifetime possible; hence the proposed method would give an upper bound (in the absence of chemical attack, of course).

FRACTURE AND FIBER FORMATION OF POLYETHYLENE CRYSTALS

H. Gleiter, E. Hornbogen, and J. Petermann*

Institut für Werkstoffe
Ruhr-Universität Bochum
Bochum, West Germany

ABSTRACT

Deformation experiments with substrate free polyethylene single crystals are employed to study the plastic deformation and the formation of fibers from polyethylene single crystals. The observations by transmission electron microscopy and electron diffraction indicate that substrate-free polyethylene crystals fail by brittle fracture at all temperatures between 77 K and 388 K and all strain rates between 10^{-3} and 10^{-1} sec^{-1}. The formation of fibers and their molecular structure were studied by drawing fibers from substrate-free polyethylene single crystals. The results suggest that the formation of fibers occurs by a two-step process. The first step is

*Now with: Lehrstuhl für Werkstoffphysik – Bau 2, Universität Saarbrücken, Saarbrücken, West Germany.

the breaking off of single blocks of folded chains from the single crystals so that a "string of pearls" structure is obtained. If the temperature is sufficiently high, this process is followed by the thermally activated rearrangement of the molecules in the drawn fibers so that a "bamboo structure" results.

The effect of lattice defects present in the drawn polyethylene single crystals on the formation of fibers was studied by introducing crosslinks into the single crystals. The crosslinks were generated by irradiating the polyethylene crystals with 60 kV electrons prior to fiber drawing. It was observed that the introduction of crosslinks results in a network of interconnected fibers the "mesh-size" of which decreases with increasing crosslink density so that, finally (at a dosage of about 200 Mrad), drawing results in a thin continuous film (approximately 15 A thick).

1 INTRODUCTION

Although many details on the structure and molecular processes in crystalline polymers have been deduced from the results of scattering experiments (scattering of electrons, X-rays, and light) and from the application of resonance methods (NMR and ESR measurements), the understanding of the relationship between the structure-sensitive properties and the molecular structure is still fragmentary. The application of refined experimental methods and high-resolution transmission electron microscopy have made possible, in recent years, considerable advances in our understanding of the deformation processes in crystalline polymers in terms of the molecular motion.

Examples of polymer single crystals (preferentially polyethylene single crystals) which were deformed during handling have been reported by several authors.[1] Other work describes crystals deformed by scratching with a needle[2], by collapsing or folding from the suspension onto a flat substrate[2,3] and by shearing between glass plates[4]. Systematic studies of the plastic deformation of polymer single crystals have been carried out by either embedding the crystals in bulk material[4-6] or supporting them by a substrate[7-11] that was deformed with the adhering crystal. Although important information was derived from those studies, the experiments suffered from the fact that the thin polymer single crystals were forced by a relatively thick substrate to deform in the same manner as the substrate. Furthermore, as long as the crystal adheres to the substrate, the fracture of the polymer crystals is limited to the formation of microcracks, since the growth process of the microcracks is inhibited by the adhesion forces between the substrate and the crystals. From the experiments with crystals adhering to a substrate, it is therefore hard to deduce the actual behavior of the crystals and the molecular processes that occur during plastic deformation and fracture. Hence, it appears promising to study the mechanical behavior (plastic deformation, fracture, and fiber formation) of polymer crystals by using substrate-free crystals. It is the purpose of this paper to report some of the results obtained from those investigations.

2 EXPERIMENTAL PROCEDURES

Single crystals of linear polyethylene (Lupolen 6011 L, $M_w = 10^4$) were grown by standard methods[12] from a solution of 0.005 percent of the polymer in xylene. The crystallization temperature was 353 K. The substrate-free polyethylene crystals were produced by the following method. Electron microscope grids were coated with a thin carbon film. The carbon-coated grids were deformed about 5 percent in a tensile device so that cracks normal to the axis of the tensile stress were obtained in the carbon film.

The width of the cracks was roughly between 1 and 5μ. In order to deposit polyethylene single crystals on the fractured carbon film, the carbon-coated grids were floated on a water surface (Fig. 1). Then a few droplets of the suspension of the polyethylene crystals in xylene were deposited on the water surface so that a thin layer of xylene on top of the water was obtained. The xylene was then evaporated, which resulted in the deposition of the polyethylene single crystals on the carbon-coated grids. The experiments showed that the cracks in the carbon film were bridged by the single crystals as indicated schematically in Fig. 1. Hence, between the two edges (A and B) of the carbon film, a "free" area of a polyethylene single crystal was obtained. After the xylene was evaporated, the specimens were removed from the water surface and were dried in vacuum.

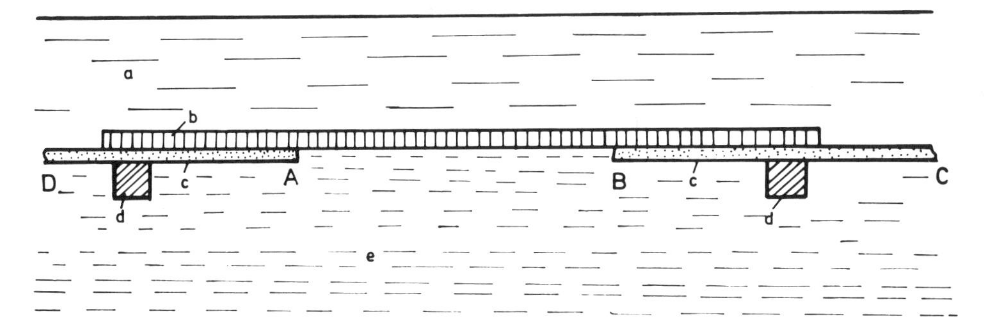

Fig. 1. Arrangement to produce substrate-free polyethylene crystals (schematically) [a: xylene, b: polyethylene single crystal, c: carbon film, d: electron microscope grid, e: water, A, B: crack in the carbon film bridged by the polyethylene crystal (b)].

In order to study the *deformation* processes and *fracture* in the substrate-free crystals, the specimens (Fig. 1) were deformed by straining them in a tensile device in the direction AD and BC. Since the crystals are flattened out when deposited on the carbon film, the strain axis is approximately normal to the [001] direction of the crystals. Hence, the results of these experiments are relevant only for the deformation behavior of the crystals under uniaxial strain in the (001) plane. The tensile experiments were carried out in air at various temperatures between 77 K and 388 K with deformation rates ($\dot{\epsilon}$) between 10^{-1} sec^{-1} and 10^{-3} sec^{-1}. Since the polyethylene crystals were deposited at random on the carbon film, there was no correlation between the tensile axis (AD and BC, Fig. 1) and the crystallographic orientation of the crystals. The deformed crystals were examined by transmission electron microscopy, employing bright- and dark-field methods. The molecular processes involved in the *formation of fibers* from the substrate-free crystals were studied by straining the polyethylene crystals (Fig. 1) in the direction AD and BC so that a crack was generated in the substrate-free area between A and B, resulting in the formation of fibers that were pulled across the crack.

In order to investigate athermal and thermally activated processes during fiber formation, fibers were drawn at 77 K, 293 K and 383 K. At 77 K, the athermal processes are dominant. At 293 K and 383 K, thermal activation becomes important. From the difference in the fiber morphologies and structures observed at the various temperatures, it was possible to deduce the role of the athermal and the thermally activated processes in the formation of fibers.

The relationship between the structure of the fibers and the presence of lattice defects in the polyethylene single crystals was investigated by drawing fibers at 77 K and 293 K from polyethylene single crystals that were irradiated in vacuum (10^{-5} torr.) at 293 K with 60 kV electrons prior to deformation. The electron irradiation introduces structural defects in thepolyethylene crystal lattice since it generates[13-16], preferentially, crosslinks between the molecular chains. The concentration of the crosslinks was controlled by means of the radiation dose. The radiation source used was the defocused beam of the electron microscope (Philips EM 300). The dosage was calculated from the beam current and the beam cross section measured. The specimens (both the irradiated and nonirradiated crystals) that were deformed at the lowest temperature (77 K) were directly transferred at 77 K from the tensile device into the cold stage of the electron microscope and studied by transmission electron microscopy at 130 K. The specimens deformed at the other temperature were investigated at 293 K. Both bright- and dark-field techniques were used. The dark-field micrographs were generated at 130 K by either [110] or [200] reflections. The micrographs of the specimens observed at room temperature were obtained by using [110] and [200] reflections together. All investigations were carried out with a Philips EM 300 electron microscope that was operated at 60 kV.

3 RESULTS

3.1 Deformation and Fracture

It was found that at all temperatures and all deformation rates, the polyethylene crystals fractured if they were strained in any direction parallel to the (001) plane. Figures 2 and 3 show bright- and dark-field electron micrographs of crystals fractured at high and low temperatures. At room temperature and below, the fracture path was frequently observed to consist of ledges (Fig. 3). At temperatures close to the melting point the fracture occurred normal to the direction of maximum strain (Fig. 2b). In order to find out whether or not there was any plastic deformation in the remaining unfractured substrate-free area of the polymer crystals (for example in the area a of Fig. 2b), the crystals were

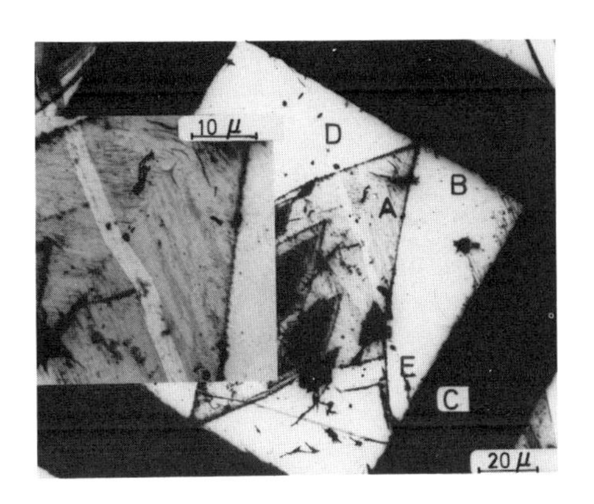

Fig. 2a. Bright-field electron micrograph showing a polyethylene crystal (A) on a carbon film (B) supported by a copper grid (C). The polyethylene crystal bridges of the crack (D-E) in the carbon film. The inserted part shows the crack (D-E) bridged by the polyethylene crystal (A) at the magnification of Fig. 2b.

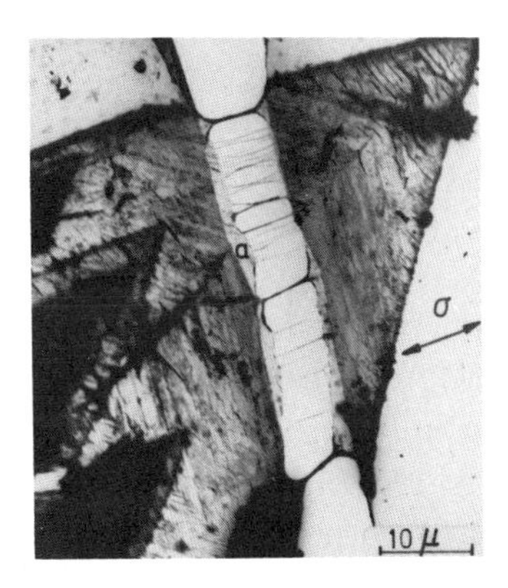

Fig. 2b. Tensile deformation at 388 K of the polyethylene crystal shown in Fig. 2a. The strain rate was $\epsilon = 10^{-3} s^{-1}$. The arrow indicates the direction of the strain. It may be seen that the crystal fractured in the substrate-free area. Fibers are pulled across the crack.

photographed before and after the tensile test was carried out (cf. Figs. 2a and 2b). Under the experimental conditions used (the temperatures and strain rates used), no plastic deformation of the crystals was observed. The plastic deformation of the crystals was checked by measuring the distance between two reference points (for example two small dust particles) on the crystal surface before and after the tensile experiments. By using dust particles with a separation of about 1μ or more, a plastic deformation of more than 1 percent can be detected. Since no change in the particle separation was observed, we are led to conclude that if there is any plastic deformation at all in the (001) plane of the crystals prior to fracture, it is less than 1 percent.

Frequently, in the dark-field micrographs (Fig. 3), a pattern of parallel lines and contrast effects due to [001] screw dislocations[17] were observed in the fractured crystals. The diffraction pattern of the deformed crystals indicated that the crystals had an orthorhombic structure. Diffraction spots corresponding to deformation twins[1] or a monoclinic crystal structure[1,19,20] were not observed.

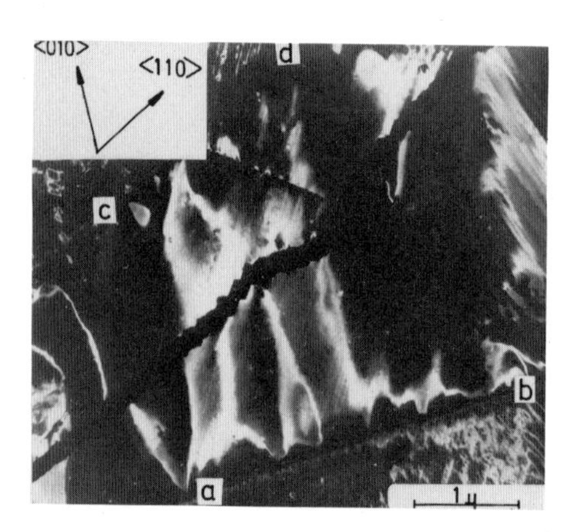

Fig. 3. Dark-field electron micrograph of a fractured polyethylene crystal deformed in tension at 293 K. The strain rate was 10^{-1} s^{-1}; a, b, and c, d are the edges of the carbon film; the area a, b, c, and d is the substrate-free area of the crystal. The dark region between the stepped edge of the bright areas represents the crack in polyethylene crystal.

3.2 Formation of Fibers

(a) Nonirradiated Crystals. Figure 4 shows the electron-diffraction pattern of fibers drawn at 77 K. The fiber direction is indicated by the arrow. The diffraction pattern is formed by a large number of diffraction spots that are arranged on rings around the origin, indicating that the fibers consist of crystalline regions. From the diameter of the rings it follows that the crystalline regions have an orthorhombic structure and the same lattice constant as the polyethylene single crystals. In some cases, diffraction spots were observed which may be attributed to the monoclinic crystal modification of polyethylene[1,19,20] present in the fibers. The fact

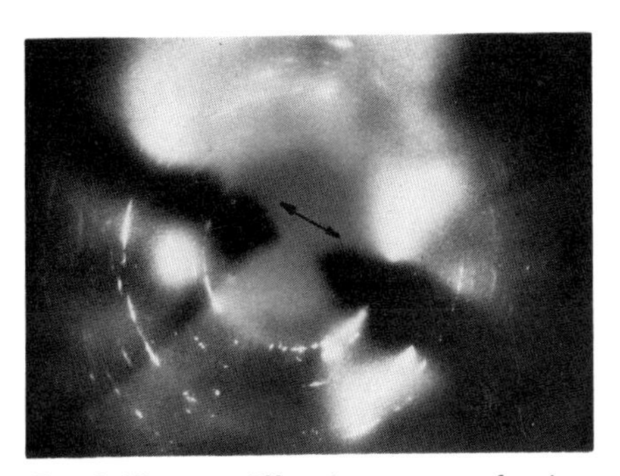

Fig. 4. Electron-diffraction pattern of polyethylene fibers drawn from free polyethylene single crystals at 77 K.

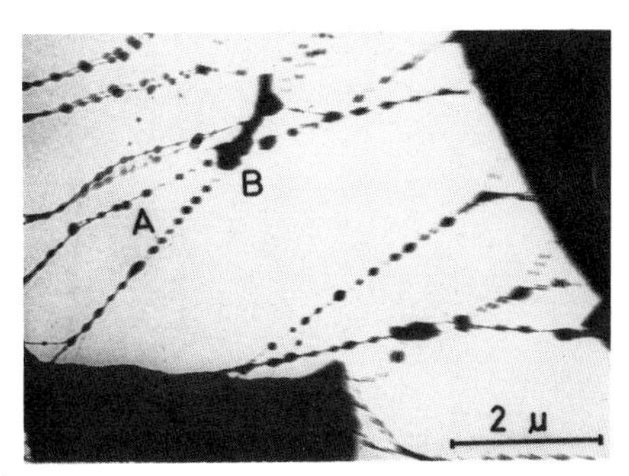

Fig. 5. Bright-field electron micrograph of polyethylene fibers drawn at 77 K from single crystals. The fibers have a "string of pearls" structure. Some fibers have moved during exposure and hence appear to be out of focus.

that the diffraction spots are arranged randomly on the rings suggests that the crystallographic axes of the crystalline regions are oriented at random with respect to the fiber direction. The shape and the arrangement of the crystalline regions in the fibers may be seen from Figs. 5 and 6. Fig. 5 shows a bright-field micrograph of fibers formed at 77 K. It may be seen (cf. for example, the fibers at A) that two parts of a fiber may be distinguished: a thin "thread" (diameter 100 to 500 A) that connects irregularly shaped blocks so that a "string of pearls" structure results. The size of the blocks varies between 10^2 and 10^3 A.

At the points at which several fibers converge, large blocks may be seen (cf. Fig. 5 at B). Figure 6 shows a dark-field micrograph of fibers drawn at 77 K. The blocks appear bright in dark field. Hence, it follows that the blocks are small single crystals. Selected-area electron diffraction from the crystalline blocks (Fig. 3) showed that the blocks have the well-known orthorhombic structure of polyethylene. The thin thread connecting the blocks appears rather dim in the dark field (Fig. 6), suggesting that there is weak coherent electron scattering by the molecules forming the thread. If the fibers were tilted with respect to the electron beam, some blocks became brighter whereas others became darker, indicating that the crystallographic directions of the blocks were not parallel. No such effect was noticed in the thread connecting the blocks. The effect of a temperature increase on the morphology of the fibers is shown in Figs. 7a and 7b. Figure 7a shows fibers (bright field) drawn at 77 K. After the electron micrograph (Fig. 7a) was taken, the specimen was heated to 293 K in the electron microscope and photographed again (Fig. 7b), at 293 K. It may be seen that heating resulted in the "dissolution" of the large crystalline blocks. The blocks were observed to "shrink" as the temperature increased and finally disappeared. The dissolution of the blocks seems not to be due to tensile stresses (e.g., thermal stresses) acting on the fibers, since the dissolution was also observed in fibers bulging out between their end points.

Fig. 6. Dark-field micrograph of polyethylene fibers drawn at 77 K from single crystals. The fibers consist of crystalline blocks (bright spots) connected by a thin thread (dim line between the bright spots).

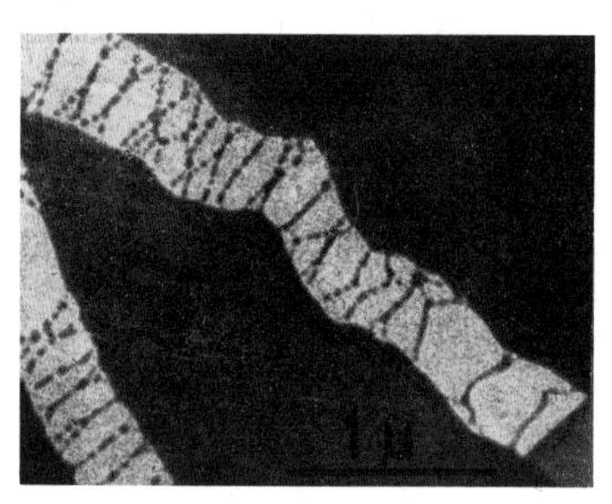

Fig. 7a. Fibers drawn at 77 K and photographed at 130 K. The "string of pearls" structure of the fibers is clearly visible.

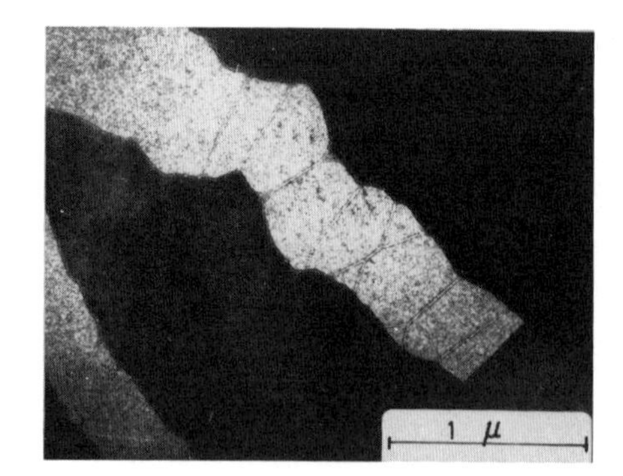

Fig. 7b. Structure of the fibers shown in Fig. 7a after heating the specimens from 130 K to 293 K. It may be seen that the crystalline blocks are removed so that fibers with a constant diameter are obtained. Some of the fibers fractured during heating.

The electron-diffraction pattern of the fibers shown in Fig. 7b was similar to the one shown in Fig. 4 with a superimposed fiber texture, suggesting that the fibers consist of crystalline regions some of which are crystallographically oriented at random (ring pattern) and other regions in which the molecules are aligned parallel to the fiber axis (fiber texture pattern).

Figure 8 shows a dark-field micrograph of fibers that were drawn at 77 K and subsequently annealed for 10 minutes at 293 K. The small bright areas are crystalline regions oriented so that the Bragg reflection condition is fulfilled.

Most of the crystalline regions are roughly spherical, with diameters between 50 and 200 A. Their crystallographic axes are, in general, not

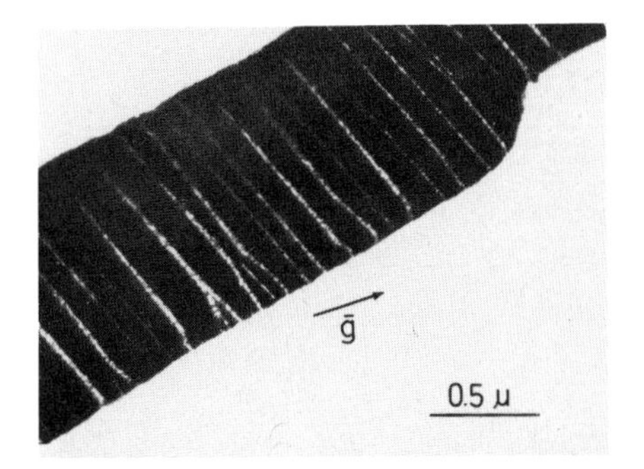

Fig. 8. Small crystalline regions in fibers that were drawn at 77 K and annealed for 20 minutes at 293 K.

parallel, since new regions became visible as the specimen was tilted whereas other regions disappeared. A few regions may be seen which have the shape of a line (disk) that is normal to the fiber axis and extends over the entire cross section of the fiber. The thickness of the lines (disks) is about 150 A.

A pronounced fiber texture was obtained (Fig. 9) after an annealing treatment at elevated temperatures. The number of statistically arranged diffraction spots decreased with increasing annealing times and temperatures, whereas the intensity of the fiber texture increased, suggesting that the number of molecules aligned parallel to the fiber axis increases during annealing. Figure 10 shows a dark-field micrograph of fibers drawn at 293 K with a subsequent annealing treatment (383 K) of 20 hours. It may be seen that at 383 K, the fibers have a "bamboo-structure" consisting of stacks of disk-shaped crystalline regions extending over the fiber cross section. In addition, a few irregularly shaped crystalline regions are visible. Fibers drawn at 383 K with a subsequent annealing treatment at the same temperature show a similar structure. It may be mentioned, however, that even after long annealing times (more than 100 hours at 383 K), the fiber

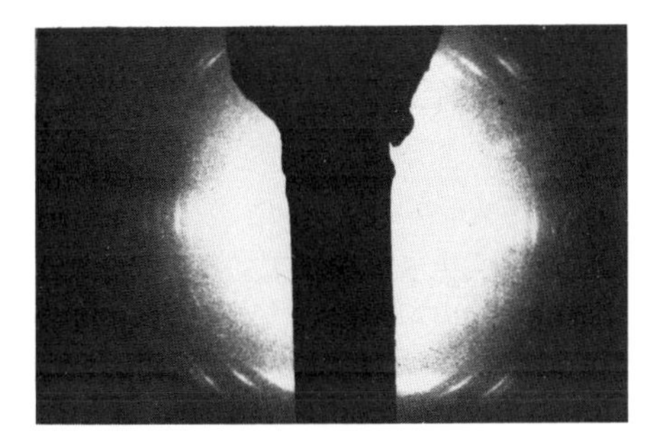

Fig. 9. Texture of fibers drawn at 383 K with a subsequent anneal at the same temperature for 15 minutes. The direction of the fibers is vertical.

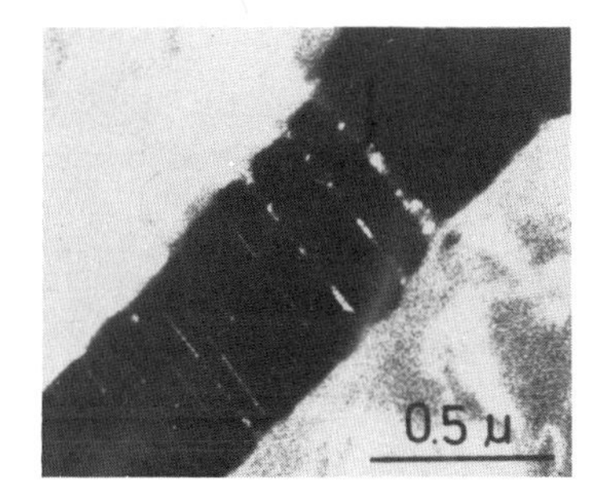

Fig. 10. "Bamboo-structure" of fibers that were annealed 20 hours at 383 K.

texture remained imperfect: a few diffraction spots outside of the fiber texture pattern were still visible and, furthermore, the texture pattern consisted of arcs, indicating an imperfect alignment of the molecules.

(b) Irradiated Crystals. If the crystals were irradiated with a dosage of 0.6 Mrad or less prior to low-temperature (77 K) deformation, the fibers consisted of statistically oriented blocks of folded chains just as in the case of nonirradiated crystals. The only change noticed was the decrease in the size of the blocks with increasing radiation dose.

If the dosage was above 0.6 Mrad, fiber formation ceased and the irradiated crystals underwent brittle fracture.

Fibers with a completely different morphology and molecular structure were obtained if the drawing was carried out at room temperature.

The following fiber morphologies as a function of the radiation dose were observed. Crystals irradiated with 11 Mrad or less formed branched interconnected fibers. Both the degree of branching and the fiber diameter increased with increasing dosage (Fig. 11). If the radiation dose was between 11 Mrad and 25 Mrad, wide bundles of fibers (Fig. 12 at A and B) were obtained. The width of these bundles increased with increasing radiation dose. Above 30 Mrad, practically 100 percent of the drawn material was incorporated into the wide bundles. The width of the bundles approached the width of the single crystal, so that drawing resulted in a single wide bundle, which may equally well be described as a thin continuous film that consists of many interconnected fibers. The formation of the thin continuous film is demonstrated in Figs. 13a and b. Both figures show the same irradiated crystal (35 Mrad) before (Fig. 13a) and after (Fig. 13b) deformation. At P, a continuous thin film has been formed during drawing. The plastic strain in the thin film was measured by comparing the distance between two reference points (for example dust particles) on the crystal before and after deformation. It was found that the plastic strain in the film was about 700 percent. Hence, if the density of the material does not drastically change as a result of the deformation, the thickness of the film is approximately 15 A.

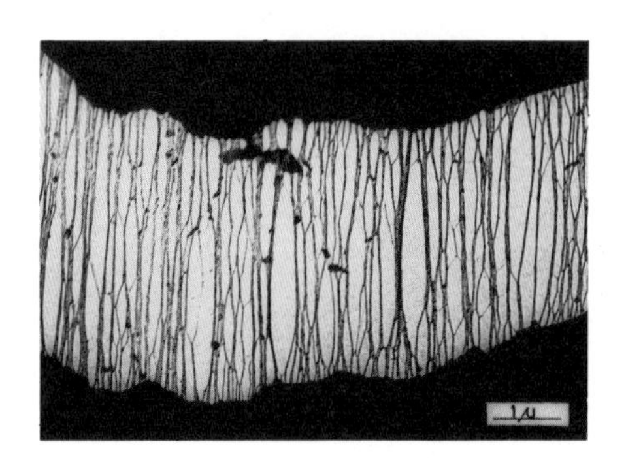

Fig. 11. Branched fibers drawn from a polyethylene single crystal irradiated with 60 kV electrons (dosage 10.5 Mrad).

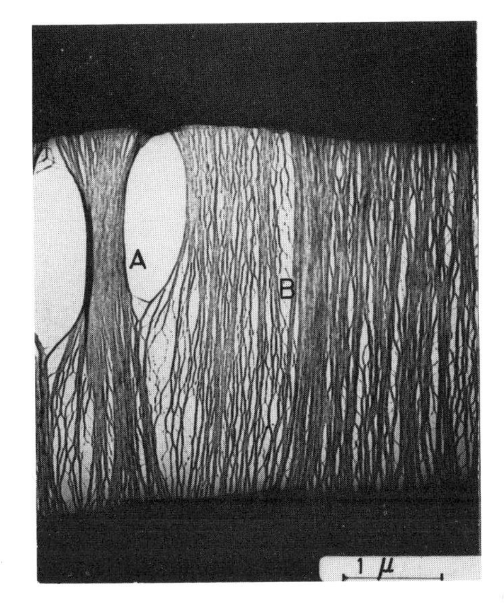

Fig. 12. Bright-field electron micrograph showing the formation of "wide bundles" (at A and B) and branched polyethylene fibers after irradiation with 15 Mrad.

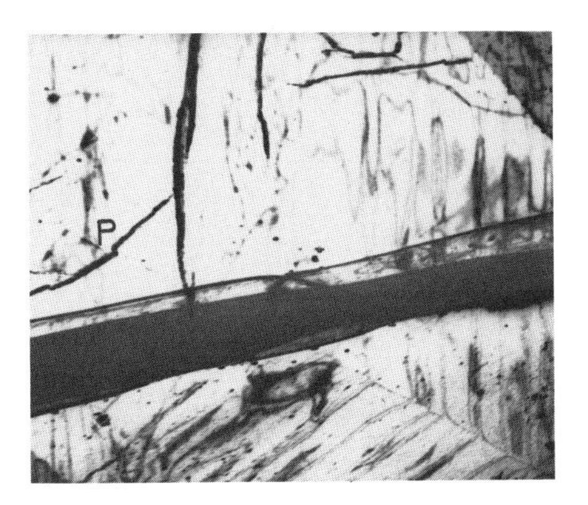

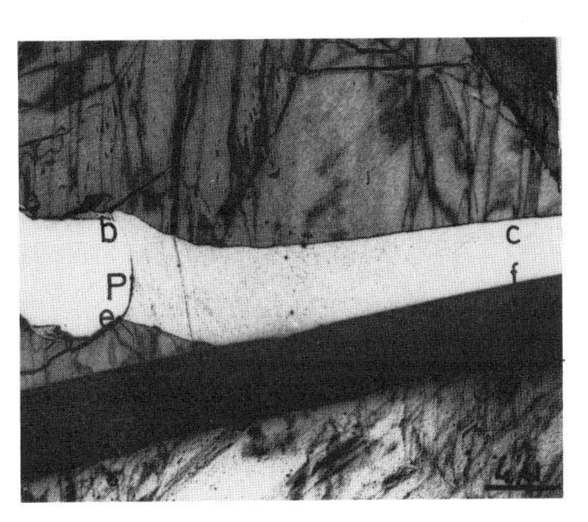

Fig. 13 a,b. Electron micrographs (bright field) demonstrating the formation of a continuous thin film from a polyethylene crystal irradiated with 3 Mrad prior to deformation. Figs. 13 (a) and (b) show the same crystal before and after deformation, respectively. The formation of the film commenced at P. The lines b, c and e, f are the boundaries between the thin film and the undeformed crystal area. The direction of drawing is normal to the direction ef.

If an already strained crystal, for example, the crystal shown in Fig. 13b, was elongated further, it was observed that the lines bc and ef sweep across the crystal. The regions swept by the lines are "transformed" into the thin film. The unswept crystal parts remained undeformed. This observation indicates that the formation of the thin film occurs by a discontinuous process in a narrow transition region (the regions of the lines bc and ef in Fig. 13b). If the radiation dose was above 200 Mrad, the formation of the thin film ceased and the material was uniformly strained during deformation.

Information about the molecular structure of the fibers (Fig. 11), of the wide bundles (Fig. 12), and of the thin film (Fig. 13b) was deduced from selected-area electron diffraction and from the results obtained by dark-field electron microscopy. The electron diffraction pattern (Fig. 14) of the fibers indicates that in the "as-drawn" state, the molecules are roughly aligned parallel to the fiber axis (fiber texture). The width and diffuseness of the diffraction spots suggests that the defect density in the crystalline regions may be high and the molecular alignment is imperfect. If the drawn specimens were annealed for 1 hour at 383 K, a more perfect fiber texture resulted (Fig. 15). The microcrystalline structure in the as-drawn and annealed fibers may be seen in Figs. 14 and 15. In the as-drawn state (Fig. 14) the fibers consisted of irregularly shaped crystalline regions the width of which is about 100 to 300 A. If the drawn material was annealed, the crystalline regions grew and finally extended over the entire fiber cross section (bamboo structure) (Fig. 15). The change in the diffraction pattern (narrow diffraction lines, Fig. 15) indicates that in addition to the lateral growth of the crystallites, the alignment of the molecules in the crystalline regions becomes more regular and perfect during annealing. The structure of the single fibers and the structure of the wide bundles and the thin film were found to be closely related. From Fig. 15 (regions B and C), it may be seen that the wide bundles consist of aligned small crystalline regions which grow during the annealing process. The size, the shape, and the molecular structure of the crystalline regions is basically the same in single fibers and bundles. The same applies to the thin film.

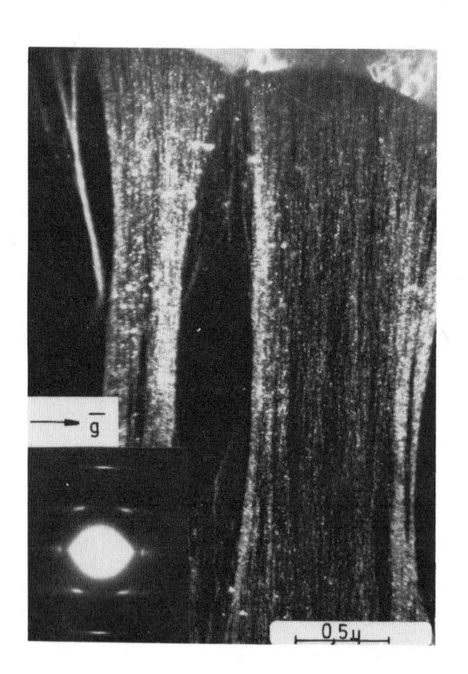

Fig. 14. Dark-field electron micrograph (g = <110> and <200>) of fibers drawn at room temperature from an irradiated-polyethylene crystal (15 Mrad). The fibers consist of small crystalline regions (bright spots) the orientation of which corresponds approximately to the fiber texture (cf. the inserted diffraction pattern of the fibers). The crystalline regions oriented with respect to the electron beam according to the Bragg reflection condition appear as bright spots.

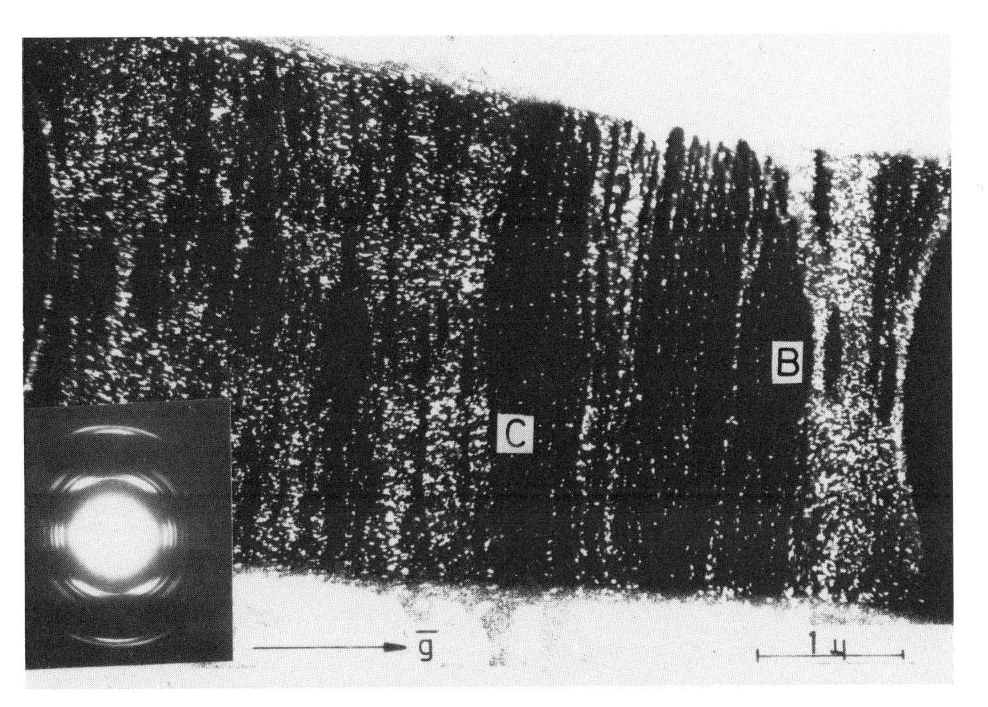

Fig. 15. Dark-field electron micrograph of fibers and wide bundles (at B and C) annealed 1 hour at 383 K after drawing. The irradiation and drawing procedure was the same as in Fig. 14. The fibers consist of crystalline regions (bright areas) extending over the entire cross section. The crystallographic orientation of the regions is such that a fiber texture results (cf. inserted electron diffraction pattern of the fibers). The structure of the wide bundles (at B and C) is similar to the fiber structure.

In order to check the temperature dependence of the fiber morphology at temperatures above 293 K, fibers were drawn from irradiated crystals at room temperature and at 383 K. No change in the fiber morphology was noticed: the same radiation dose led to the same fiber structure independent of the drawing temperature.

4 DISCUSSION

4.1 Deformation and Fracture

The experimental observations reported on the deformation of substrate-free crystals suggest that polyethylene crystals fail by brittle fracture at temperatures between 77 K and 388 K and strain rates between 10^{-1} sec^{-1} and 10^{-3} sec^{-1} if they are strained in tension in the (001) plane. If there was any plastic deformation at all, it was below 1 percent.

These results are at variance with the observations on polyethylene crystals deformed on substrates.[1-11] It appears that the discrepancy is due to the fact that in substrate-free crystals, fracture is possible, whereas in crystals deposited on a substrate, crack growth is limited by the substrate as long as the crystal adheres to it. Therefore, the brittleness of polymer single crystals may, in general, not be observed if the experiments are carried out with crystals adhering to a substrate.

The observed line pattern in the dark-field micrographs of the deformed crystals (Fig. 3) may be interpreted as slip traces produced by the movement of [001] dislocations, since the lines showed the contrast effects recently reported for slip traces in polyethylene crystals.[21] The movement of [001] dislocations in the strained polyethylene crystals may be understood in terms of crystal buckling. If the applied tensile stress were everywhere exactly normal to the [001] direction of the crystals, the force exerted on a [001] dislocation by the strain field[22] would be zero. However, if the c-direction of the crystals deviates locally from the perpendicular orientation, a shear stress parallel to [001] exists which may result in the movement of the [001] dislocations.

4.2 Formation of Fibers

(a) Nonirradiated Crystals. The observations reported suggest that both thermally activated and athermal processes are involved in the formation of fibers from polyethylene single crystals. The athermal processes may be deduced from the fiber structure observed at low temperatures, since thermal activation is negligible at 77 K. The structure of the fibers drawn at 77 K may be understood if it is assumed that the formation of the fibers occurs by the breaking off of single blocks of folded chains from single crystals[23,24] and incorporating those blocks into the fibers. The irregular shape and the different sizes of the blocks may be due to imperfections in the structure of the single crystals. It is well established[25] that polyethylene single crystals have a high density of structural imperfections, for example, small-angle boundaries between mosaic blocks. Some of the imperfections may cause a local stress enhancement in the single crystal if an external stress is applied during fiber drawing. The stress enhancement due to those imperfections may result in local fracture of the single crystal with a fracture path following approximately the line of maximum local stress so that a small crystal block is broken off. Fracture into single blocks of folded chains may partly unfold the molecules[26] connecting two adjacent blocks and may incorporate the partially unfolded sections in the thread between the blocks. This idea is in agreement with the observations that there was a strong electron diffraction in the crystal blocks but weak electron scattering from the thread without any indication of a crystalline structure. It also agrees with the observed different cross section of the thread and the blocks. If it is assumed that

the maximum tensile stress in the blocks during fiber drawing is approximately the stress necessary to fracture the van der Waals bonds between the folded chains, then the maximum stress in the fibers is roughly 10 to 100 times higher. A stress of that magnitude may be supported by the covalent bonds of partly extended chains in the thread, but it would be hard to understand if the thread had a large volume fraction of folded chains. Hence, the fibers drawn at two temperatures may have a structure as indicated schematically in Fig. 16.

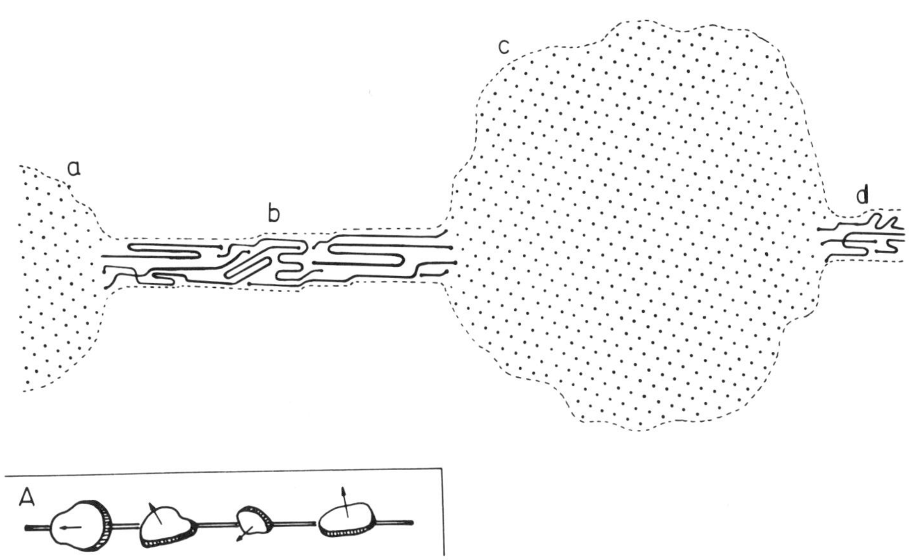

Fig. 16. Molecular structure of a fiber drawn at low temperatures (schematically) (a, c: blocks of folded chains; b, d: threads connecting blocks of folded chains). In both blocks (a and c) the molecular systems are drawn normal to the paper. In general, the crystalline blocks are randomly oriented with respect to the fiber axis as shown schematically by the inserted drawing A. The direction of the folded chains is indicated by an arrow.

The thermally activated processes observed during annealing at 293 K and 383 K (dissolution of the blocks, Figs. 7a and 7b; formation of small crystalline regions of random orientation in the fibers, Fig. 8; formation of the bamboo-structure, Fig. 10, and the fiber texture, Fig. 9) suggest a thermally activated rearrangement of the molecules in the fibers so that the free energy decreases. The rearrangement of the molecules appears to be a two-step process. The first step results in the dissolution of the blocks and the formation of a fiber with a rather uniform diameter consisting of randomly arranged crystallites (Fig. 8). A further reduction of the free energy is achieved by a second process: the molecules in the randomly oriented crystallites are aligned parallel to fiber axis (fiber texture). During the second process, disk-shaped crystalline regions are formed which extend over the entire cross section of the fiber. The formation of those crystallites results in the well-established periodic structure of annealed fibers.[7,18,23,26-31] Since both processes observed require a substantial rearrangement of the molecules, it appears likely that partly unfolded molecules formed during the drawing process refold during annealing.

If fibers are drawn at elevated temperatures, both the (athermal) fracture of single crystals into single blocks of folded chains and the thermally activated rearrangement of the molecules may occur simultaneously during the drawing process. Hence, the fibers obtained have a fiber texture which becomes more pronounced if the drawing (annealing) temperature is increased.

The observations reported on the mechanism of crystal fracture into single blocks of folded chains are in agreement with the ideas put forward by Peterlin.[23,24] The observations (Figs. 4, 6, and 8) have shown the blocks are oriented at random in the as-drawn fiber.This result is at variance with most models[23,24,26-29,32-34] of the as-drawn fiber structure which assume that the molecules of the crystalline regions are aligned parallel to the fiber axis. In fact, the result of the annealing experiments (Figs. 7a, 7b, and 10) suggest that the alignment is due to a thermally activated rearrangement process of the molecules (recrystallization) in the drawn fiber, resulting in the periodic fiber structure frequently reported. The alignment of the molecules in the recrystallized structure may be due either to the growth of aligned nuclei or to a growth-selection process. If the drawing process is carried out at high temperatures, the rearrangement of the molecules starts immediately after drawing, so that the statistical arrangement of the crystalline regions is hard to observe. The idea[26,35] that there is a considerable refolding of the molecules during the annealing of the fibers is in agreement with the observations reported. After annealing at room temperature, a fiber structure was obtained that consisted of many small crystalline regions (Fig. 8), suggesting that refolding of partly unfolded molecules takes place. This result agrees well with the observations by ESR and infrared-absorption results.[36-40]

(b) Irradiated Crystals. The formation of fibers from irradiated crystals at 77 K may be understood if it is assumed that at a radiation dose of 0.6 Mrad or less, the process of fiber formation remains basically unchanged and is modified only in the sense that the crosslinks introduced into the structure of the crystals increase the binding forces between the adjacent molecular chains. In nonirradiated polyethylene crystals, a small block is broken off when the local stress is sufficiently high to overcome the van der Waals forces between adjacent molecular chains. If a crystal has been irradiated prior to the drawing process, the stress necessary to break off a crystal block is increased and consists of two components: the stress required to break the van der Waals bonds

plus the stress necessary to break the covalent bonds due to the crosslinks. Hence, increasing crosslink density in the single crystals increases the force necessary to break off the crystal blocks. Finally, when the crosslink density is so high that this force becomes larger than the maximum strength of the thread (b and d in Fig. 16) connecting two crystal blocks of a fiber, the formation of fibers ceases, since it is easier to rupture the thread of the fiber than to break off crystal blocks.

Fibers drawn at room temperature from irradiated crystals show morphologies that are completely different from the ones observed at low temperatures. If the drawing process is carried out at room temperature, fiber formation (formation of bundles or thin films) was observed to occur at all crosslink densities. This result suggests that at room temperature, a thermally activated process is superimposed on the athermal fracture of the single crystals into single blocks of folded chains. By analogy to the processes observed in fibers drawn from nonirradiated crystals, it is suggested that the superimposed process is the thermally activated rearrangement of the molecules during drawing. In nonirradiated crystals, the rearrangement involves unfolding, refolding, and local chain movements. However, at high crosslink densities, the differing geometry of the molecular structure of the fibers and of the single crystals may require that some of the crosslinks formed during irradiation have to be broken to allow the thermally activated rearrangement that was observed during drawing. Since covalent bonds cannot be broken by thermal activation only, this implies that drawing results in local stress concentrations in the order of 0.1E*.[39,41]

The change in the fiber morphology as a function of the radiation dose may be interpreted in terms of the crosslinks between the molecules of the single crystals. If a nonirradiated material is drawn, the fibers form a network at the nodes of which are located crystal blocks (Figs. 5 and 6). The necessary requirement for this kind of structure to be stable is that the strength of the crystal block at the node is higher than the stress exerted by the fibers on the block. If the material is irradiated, crosslinks are formed between adjacent molecules. Those crosslinks increase the strength of the material. Hence, we obtain more and more single crystal blocks that are sufficiently strong to form a branching point between fibers. Thus, with increasing crosslink density, a network-type structure results (Fig. 11). If the density of the crosslinks is increased, the mesh size of the fiber network becomes smaller and, finally, the wide bundles and the continuous thin film are obtained (Figs. 13, 14, 15). If the radiation dose is above 200 Mrad, the formation of the continuous thin film ceases and drawing results in a uniform stretching of the crystal. This behavior

*E is the Young's modulus of the material.

may be interpreted in terms of the crosslink density of the material. Above 200 Mrad, the density of crosslinks is so high that the molecular structure corresponds to a highly crosslinked rubber and, therefore, a uniform straining of the material may be expected.

Just as in the case of fibers drawn from nonirradiated polyethylene crystals, the thermally activated rearrangement of the molecules during drawing results in the well-known fiber texture (Fig. 9). However, at room temperature, the fibers contain a high defect density and the alignment of the molecules parallel to the fiber axis is rather imperfect. Annealing at elevated temperatures allows recrystallization to occur (Fig. 15), so that the free enthalpy of the system is lowered. According to the observations, this is achieved by the following three processes: a more perfect alignment of the molecules parallel to the fiber axis, the lateral growth of the crystalline regions, and probably a decrease in the defect concentration of the crystallites of the fibers.

ACKNOWLEDGMENT

The financial support of the Deutsche Forschungsgemeinschaft is gratefully acknowledged. The polyethylene has been given by the Badische Anilin- und Sodafabriken (BASF).

REFERENCES

1. Geil, P. H., *Polymer Single Crystals,* Interscience Publishers, New York (1963).
2. Rennecker, D. H., and Geil, P. H., *J. Appl. Phys.,* **31**, 1916 (1960).
3. Geil, P. H., and Rennecker, D. H., *J. Polymer Sci.,* **51**, 569 (1961).
4. Fischer, E. W., unpublished, quoted in Ref. (1), pp 104 and 442.
5. Speerschneider, C. J., and Li, C. H., *J. Appl. Phys.,* **33**, 1871 (1962).
6. Geil, P. H., Anderson, F. R., Wunderlich, B., and Arakawa, T., *J. Polymer Sci.,* Part A, **2**, 3707 (1964).
7. Geil, P. H., *J. Polymer Sci.,* Part A, **2**, 3813, 3835, 3857 (1964).
8. Kiko, H., Peterlin, A., and Geil, P. H., *J. Polymer Sci.,* Part B, **3**, 157, 257, 263 (1965).
9. Sauer, J. A., *Ann. N. Y. Acad. Sci.,* **155**, 517 (1969).
10. Cerra, P., Morrow, D. R., and Sauer, J. A., *J. Macromol. Sci.-Phys.,* Part B, **3** (1), 33 (1969).
11. Gleiter, H., and Argon, A. S., *Phil. Mag.,* **24**, 71 (1971).
12. Holland, V. F., and Lindenmeyer, P. H., *J. Polymer Sci.,* **57**, 589 (1962).
13. Charlesby, A., *Atomic Radiation and Polymers,* Pergamon Press, London, 1960.
14. Salovey, R., and Bassett, D. C., *J. Appl. Phys.,* **35** 3216 (1964).
15. Chapiro, A., *Radiation Chemistry of Polymeric Systems,* Interscience Publishers, New York, 1962.
16. Orth, H., and Fischer, E. W., *Makromol. Chem.,* **88**, 188 (1965).
17. Petermann, J., and Gleiter, H., *Phil. Mag.,* **25**, 813 (1972).
18. Hay, I. L., and Keller, A., *Nature,* **204**, 862 (1964).
19. Pierce, R. H., Tordella, J. P., and Bryant, W. D., *J. Am. Chem. Soc.,*
20. Tanaka, K., Seto, T., and Hara, T., *J. Phys. Soc. Japan,* **17**, 873 (1962).

21. Petermann, J., and Gleiter, H., *J. Polymer Sci.*, Part A-2, **10**, 1731 (1972).
22. Peach, M., and Koehler, J. S., *Phys. Rev.*, **80**, 436 (1950).
23. Peterlin, A., *J. Mat. Sci.*, **6**, 490 (1971).
24. Sakaoku, K., and Peterlin, A., *Makromol. Chem.*, **108**, 234 (1967).
25. Hosemann, R., Cackovic, H., and Wilke, W., *Naturwissenschaften*, **54**, 278 (1967).
26. Dismore, P. F., and Statton, W. O., *J. Polymer Sci.*, **Part C**, 13, 133 (1966).
27. Fischer, E. W., Goddar, H., and Schmidt, G. F., *Makromol. Chem.*, **119**, 170 (1968).
28. Hearle, J.W.S., *J. Polymer Sci.*, Part C, **20**, 215 (1967).
29. Peterlin, A., *J. Polymer Sci.*, Part C, **9**, 61 (1965).
30. Peterlin, A., Kiko, H., and Geil, P. H., *Polymer Letters*, **3**, 151 (1965).
31. Peterlin, A., Ingram, P., and Kiko, H., *J. Makromol. Chem.*, **86**, 294 (1965).
32. Anderson, F. R., *J. Polymer Sci.*, Part C, **8**, 275 (1965).
33. Gubanov, A. I., and Chevychelov, A. D., *Sov. Phys. Solid State*, **4**, 4 (1962).
34. Pechold, A., *J. Polymer Sci.*, Part C, **32**, 123 (1971).
35. Statton, W. O., *J. Polymer Sci.*, Part C, **32**, 219 (1971).
36. Fischer, E. W., Goddar, H., and Schmidt, G. F., *Kolloid-Z. u. Z. Polymere*, **226**, 30 (1968).
37. Peterlin, A., and Olf, H. G., *J. Polymer Sci.*, Part A-2, **4**, 587 (1966).
38. Hyndman, D., and Origlio, G. F., *J. Polymer Sci.*, **39**, 556 (1959).
39. Gilman, J. J., *Chapter in Fracture*, B. L. Averbach (Ed.), M.I.T., Wiley, N.Y. (1959), p 193.
40. Statton, W. O., Koenig, J. L., and Hannon, M. J., *J. Appl. Phys.*, **41**, 4290 (1970).
41. Drowan, E., *Repts. Progr. Phys.*, **XII**, 185 (1948).

DISCUSSION on Paper by H. Gleiter

TAKAYANAGI: I want to know the thickness of the single crystals and whether they thicken before drawing at the highest temperature employed in your measurements. If thickened; the crystallite size is expected to be much larger than that of your observation.

GLEITER: There was no measurable thickening of the crystals before drawing. The thickness of the single crystals was about 100 A.

ANDREWS: You have deliberately avoided using a substrate in the work described and obtain results which differ from those when a substrate is present. However, would you agree that deformation of single crystals on a substrate is more likely to represent the case of a lamellar crystal deforming in the bulk polymer?

GLEITER: I agree with Professor Andrews that further experiments are necessary to check whether the results obtained from the solution-grown single crystals can be applied to the fiber formation in bulk polymers. I am not sure, however, that the results obtained by the deformation of single crystals on a substrate are more likely to represent the case of a lamellar crystal deforming in the bulk polymer, since various substrates influence the fiber structure in different ways.

ANDREWS: You suggest that the size of the folded-chain packets, broken off the crystal during fiber formation, is governed by some mosaic defect structure in the undeformed crystal. However, moiré patterns in undeformed crystals are usually highly perfect and do not reveal any such defect structure.

GLEITER: We have no experimental evidence that the size of the blocks is governed by mosaic blocks since we have not seen mosaic blocks. In suggesting that the mosaic structure may be important, I had in mind the mosaic structure proposed recently by Pechhold (to be published). It may well be, however, that the block size is governed by some other kind of defect located either in the fold surface or in the interior of the crystals. So far, we have not been able to identify the defects.

TAKAYANAGI: Professor Gleiter has shown the electron micrographs of broken fragments of lamellar crystals in the fiber.

I want to show our electron microscope observation of formation of crystal blocks in the lamellar crystals by using a special technique just before the start of fiber formation (Fig. D-1).

This figure shows that the block crystals with edge length ca. 300 A are generated uniformly in the field. The polyethylene film used was that which was had been elongated 15 percent at 55 C.

GLEITER: We have looked for mosaic blocks in solution-grown single crystals by means of moire fringes and high-magnification transmission electron micrographs using both dark- and bright-field techniques. We were not able to obtain any experimental evidence for the existence of mosaic blocks. These findings, however, may not necessarily be relevant to the findings of Professor Takayanagi, since his observations are made on thin films.

KOVACS: Did you observe differences in the morphology of the fibers drawn from multilayer crystals as compared with that you have shown in monolayer crystals?

GLEITER: Yes, we did observe differences in the morphology of the fibers drawn from multilayer crystals. The fibers formed a complex network and, in some cases, a kind of film formation was observed. However, at low temperatures (77 K), blocks of folded chains were seen to be broken off even if the multilayer crystal consisted of more than two layers.

HULL: There have been a number of reports of a martensitic-type transformation, from an orthorhombic to a monoclinic crystal structure, when polyethylene crystals are deformed using a Mylar substrate. Have you observed such a transformation in your experiments and, if not, can you account for the different behavior?

GLEITER: No, we have not observed such a transformation. From the observations about the stability of the monoclinic structure it follows that the monoclinic structure is stable only if the polymer crystal adheres to a substrate so that the transformation from the monoclinic structure to the orthorhombic one is prevented by the stresses between the substrate and the polymer crystals. It appears likely that the stresses in substrate-free crystals are not high enough to stabilize the monoclinic structure.

THE MICROSTRUCTURE AND PROPERTIES OF CRAZES

D. Hull

University of Liverpool
Liverpool, England

ABSTRACT

A brief résumé of the established characteristics of craze deformation and structure is given along with a summary of the significance of crazing in the bulk deformation and fracture of high polymers. This is followed by a more detailed consideration of recent work in two important areas of interest, viz (1) microstructure of crazes and (2) stress-strain response of crazes.

The microstructure of crazes has been studied using high-resolution transmission electron microscopy. The crazes were produced in bulk specimens, microtome thin sections, and solvent-cast thin films. The structure of crazes was similar in all specimens and consisted of highly drawn fibrils normal to the craze-matrix interface with linking, finer cross fibrils. In the thin films it was possible to obtain much larger craze strains and this resulted in a finer fibril structure. A model of the microstructure of a craze is presented.

The stress-strain characteristics of a craze have already been studied by Kambour and Kopp using a method which requires direct observation of the change in thickness of a thick craze during the application of a loading cycle. An alternative approach is described which is based on microstrain measurements of specimens containing a high density of very fine crazes. Quantitative microscopy is used to determine the craze density and average thickness. The results obtained on polystyrene are similar to Kambour and Kopp's observations on solvent induced crazes in polycarbonate. The results are discussed in the context of the observed craze microstructure.

1 CHARACTERISTICS OF CRAZING

In the last few years there has been some intense experimental work on the structure and properties of crazes and the relation between craze formation and fracture. A comprehensive review of the work carried out on glassy polymers has been published recently by Rabinowitz and Beardmore[1], and it is not intended here to cover all this ground again. However, it is necessary to emphasize some of the important characteristics of crazing since our understanding of the phenomenon is far from complete and there is still considerable confusion about crazing-like processes, particularly in "nonglassy" plastics.

The main characteristics of crazes in transparent, glassy, isotropic polymers, such as slowly cooled compression molded polystyrene or polymethyl methacrylate, are fairly well defined and in the main generally accepted. They are:

1. A craze is a highly localized region of plastic deformation in which the strains are of the order of unity (100 percent).
2. Crazes formed in a uniaxial tensile stress field have a similar shape to a crack, and the plane of the craze is at right angles to the stress axis. The planar dimensions of the craze are many orders greater than the thickness, which is typically less than 1 μm.
3. In a complex stress field, the plane of the craze is normal to the maximum principal tensile stress.[2-4]
4. Crazes form only in a tensile field and the criteria for visible crazing[3] in a biaxial stress field is:

$$\sigma_b = A + B/I_1 \quad , \qquad (1)$$

where σ_b is the difference between the principal stresses $(\sigma_1 - \sigma_2)$, I_1 is the first stress invariant $(\sigma_1 + \sigma_2)$, and A and B are parameters which depend on a number of testing and material variables such as temperature and molecular weight.

5. The craze volume has a lower density than the surrounding material, and the microstructure consists of a high density of interpenetrating micropores (see Reference 6 for a review of this work) surrounded by drawn polymer in a fibrillar form.[7-10] These structural features are responsible for properties of the craze such as (a) the lower refractive index and highly reflective craze-matrix interfaces, (b) the load-bearing capacity, (c) the porosity, and (d) the eventual breakdown of craze by cavitation processes.[11]

In addition to these well-defined characteristics, there is a range of other properties which have to be taken into account in any generalized model for the crazing process. Notably, these are the features which relate to mechanism and kinetics of craze nucleation and propagation which indicate that crazing is a thermally controlled stress-activated process involving local molecular motion.

Of the characteristics outlined above, those relating to the morphology and microstructure of the craze may be considered as defining a craze. In glassy polymers these characteristics appear to be changed in detail only by changes in the conditions under which crazes form (e.g., temperature, strain rate, and environment) and the molecular structure and conformation of the polymer (e.g., molecular weight, orientation, degree of crosslinking). When this approach is extended to other polymers, it is clear that crazing defined in this way is not restricted to glassy polymers. Thus, for example, crazes have been observed in semicrystalline spherulitic polymers such as polypropylene[10] and isotactic polybutene-1[12] and in oriented crystalline polymers such as poly(ethylene terephthalate)[13] and poly(4 methyl-pentene-1).[14] In addition, craze-type microstructures have been observed many times[15] during the deformation of spherulitic thin films and single crystals. The distinction between crazing and phenomena such as stress whitening on the one hand and pure fracture on the other is far less distinct in semicrystalline polymers than in glassy polymers. It is usually considered that stress whitening is associated with the formation of microvoids. Whether or not this is a crazing process, then, depends on the distribution of the voids. If they are in planar arrays bounded by a drawn fibrillar structure, then, according to the above analysis, the whitening is a crazing process. The problem reduces to one of scale. Thus, for example, the planar arrays of voids may cross the individual arms of a spherulite or may extend across many spherulites largely ignoring the spherulitic morphology. It seems likely that a number of possible types of craze morphology may exist, and some recent work by one of my collaborators[14] indicates that different types of crazing may develop under different regimes of molecular orientation. His work has been concerned with the fracture of injection molded T.P.X. [an I.C.I. copolymer based on poly(4-methyl-pentene-1)]. This material has about 40 to 50 percent crystallinity in the form of lamellar blocks. The orientation of the lamellae varies across the section owing to the molding conditions. Figure 1 shows a specimen which has been loaded in tension parallel to the injection-molding direction. A number of different modes of deformation are apparent. In Regions A and B, well-defined extensive crazes have formed in a manner entirely analogous to polystyrene. The planar dimensions of the crazes extended over 1 mm in the direction normal to the surface of the photograph, indicating that the craze region has embraced large volumes of lamellar blocks. In regions C and D, stress whitening has occurred. No structure can be seen at low magnifications, but at high magnifications, fine ragged crazes can be resolved (Fig. 2). In this case the morphology of the crazes appears to be related to the lamellar block structure.

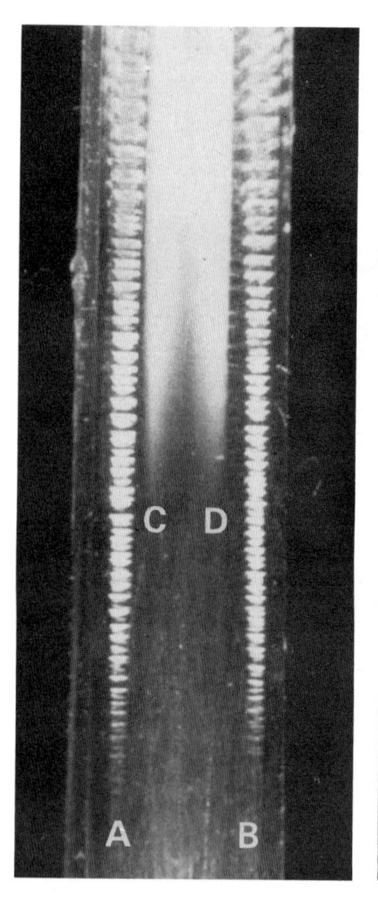

Fig. 1. Stress whitening and crazing in different parts of an injection-molded tensile specimen of TPX. The stress whtening has all the characteristics of fine crazes when observed at high magnifications.

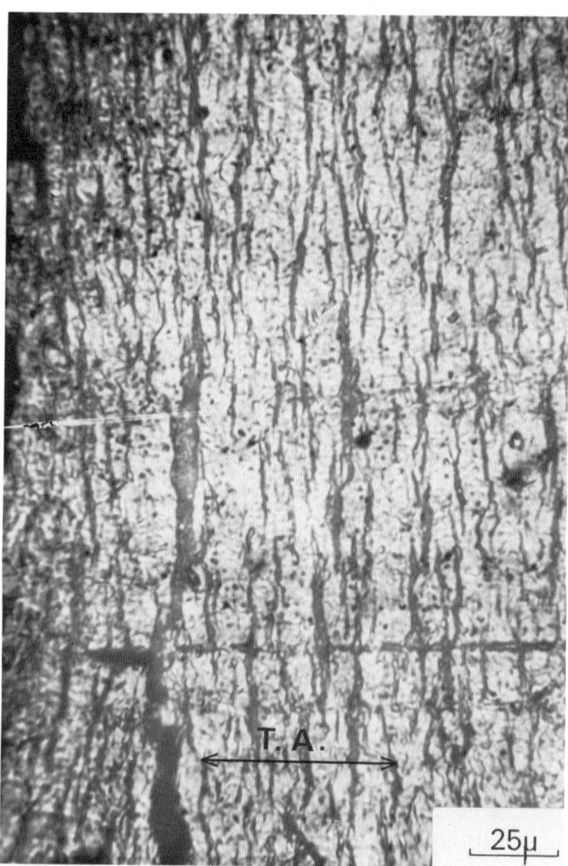

Fig. 2. Fine crazes observed in the stress-whitened regions C and D of the specimen shown in Fig. 1.

The characteristics of crazes are in many ways unique, and many important implications arise, some of which will be considered briefly. The first arises directly out of the distinction which can be made between crazing and other mechanisms of plastic deformation in polymers, notably shear yielding. The latter may occur (a) in narrow zones of intense shear (shear bands), (b) in large regions of homogeneous shear, and (c) as interlamellar or intralamellar sliding processes. Crazing involves an increase in volume and responds to the dilational components of the stress field, whereas shear processes respond primarily to the shear components. Thus, according to Sternstein et al.[3], the yield criteria for polymethyl methacrylate in the absence of crazing is

$$\tau = \tau_o - \mu \, \sigma_m \quad , \tag{2}$$

where τ is the observed shear yield stress, τ_o is the yield stress in pure shear and σ_m is the mean normal stress $I_1/3$. Following Sternstein and Ongchin[5], the criteria represented by Eq. 1 and 2 may be represented as shown in Fig. 3 for biaxial tensile tests. In this example, crazing will always precede fracture, but in other quadrants of the biaxial stress state, the two curves will intersect and there will be regions where crazing will not occur. It is well known, for example, that crazing does not occur in uniaxial compression. In addition, the relative positions of the two curves in Fig. 3 may change owing to the different dependence of crazing and shearing processes on, for example, temperature and molecular orientation. Thus, if increasing the temperature reduces the shear yield stress more rapidly than the crazing stress, there will be a temperature transition in the crazing behavior. Crazing and shearing are then identified as competitive deformation processes; this is particularly relevant to craze-induced

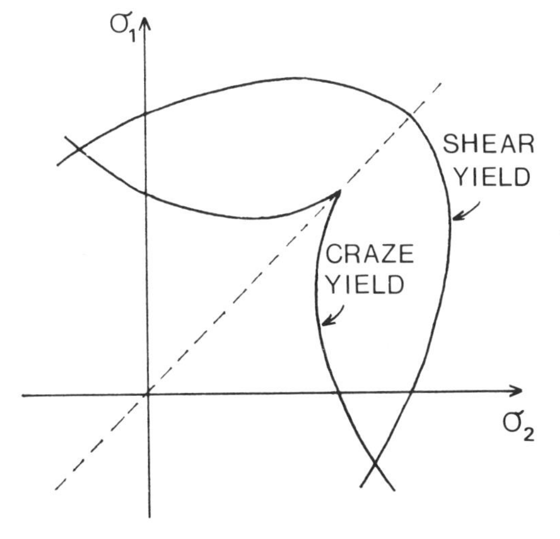

Fig. 3. Schematic representations of normal stress and shear stress yielding envelopes in biaxial tensile stress, after Sterstein and Ongchin.[5]

fracture, and brittle fracture may be suppressed by increasing the crazing stress or reducing the shear yielding stress.

The crazing response to dilational stresses has resulted in a number of models for craze nucleation and growth.[3,5,17-19] It is not possible to review these in this paper, but the main ideas which arise are, firstly, that the presence of a local high triaxial tensile stress results in a small volume of material in which the molecular mobility is increased and the material behaves as a rubber or plasticized thermoplastic, and, secondly, that this volume cavitates in a way which depends on the surface energy of the voids and the extent of orientation hardening associated with fibrillation.

The evidence for a close association between crazing and fracture is now incontrovertible, and a detailed mechanism of craze breakdown leading to the nucleation and growth of cracks has been developed on the basis of extensive microscopic observations of the fracture of polystyrene. The formation of a planar volume of microvoids is a necessary precursor to final craze breakdown which involves the failure of individual fibrils. This leads to the coalescence of the microvoids to form larger voids which, in turn, coalesce to form a planar cavity or crack within the craze. It follows that the fracture behavior is determined by the crazing characteristics of the material and that a modification of the fracture process arises primarily from a change in the crazing behavior. A rather interesting example of this is illustrated in Fig. 4. from work by another of my collaborators, Miss L. Camwell.[20] This shows the displacement to failure of a sheet specimen compressed end on between two anvils. Failure occurs by the formation of a craze parallel to the compression axis as a result of

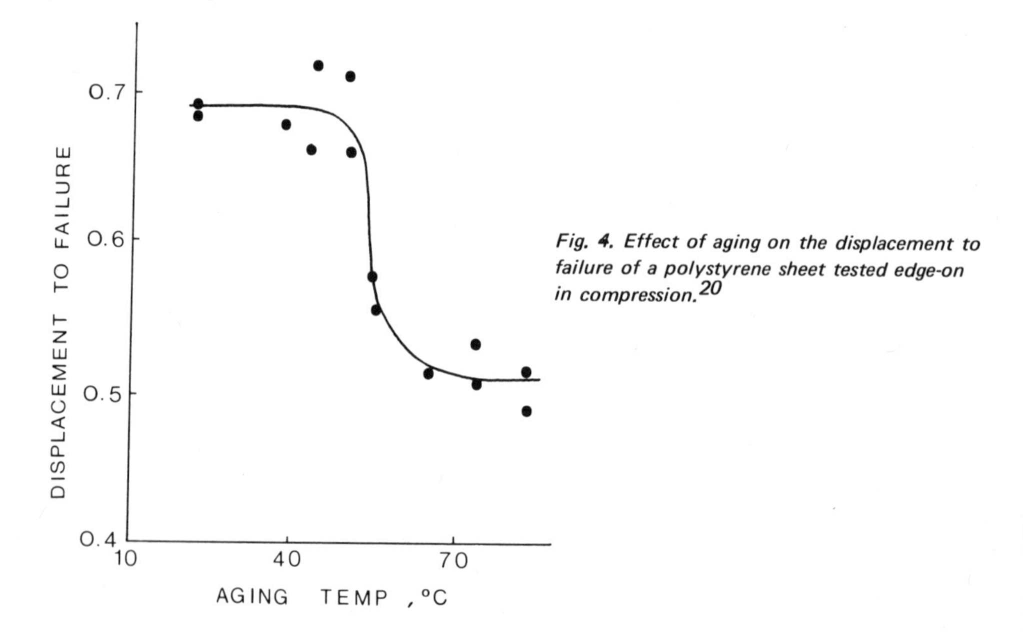

Fig. 4. Effect of aging on the displacement to failure of a polystyrene sheet tested edge-on in compression.[20]

secondary tensile stresses generated by shear zones. The interesting feature of these results is the change in displacement to failure arising from aging. All the specimens were initially water quenched from close to T_g and subsequently aged for 1 hour at the temperatures indicated. The sharp decrease in the displacement to failure after aging is due to the relative ease of shear-band and craze formation. The aging appears to modify the homogeneity of the shearing processes rather than the crazing process.

Finally, reference should be made to the fact that crazing is a high strain phenomenon, which means that it is possible that crazing could absorb a significant amount of energy during fracture and hence provide a toughening mechanism. However, because crazes are so thin, very high densities of crazes are required and, in addition, premature failure must be avoided. These two conditions are achieved in some rubber-modified polymers such as high-impact polystyrene.[21]

Although it is now possible, as has been demonstrated here, to outline a significant number of well-defined characteristics of crazes, there are still many aspects of crazing which are not understood. There is a real need for detailed experimental work on the microstructure and properties of crazes. The remainder of this paper is concerned with two aspects of this subject which have been selected from a wider range of problems we are studying.

2 MICROSTRUCTURE OF A CRAZE

The elegant work of Kambour and his colleagues[7,9] referred to earlier, and others, has established some important features of the microstructure. However, the techniques which have been used have either been indirect or, in the case of direct microscopic methods, subject to uncertainty because of the modifications to the microstructure which result from using the technique. To resolve the microvoided structure, electron-microscope techniques are essential. The main difficulty in preparing thin sections is in avoiding the collapse of the voids and relaxation of the craze. Kambour[22] achieved some success in polycarbonate by doping the craze with silver using aqueous silver nitrate, and later, Kambour and Holik[7,23] impregnated solvent-induced crazes in PPO poly(2,6-dimethyl-1, 4-phenylene oxide) by infusing liquid sulfur. The latter work clearly demonstrated the fibrillar nature of the microstructure and the technique was extended[9] to study other polymers using an iodine-sulfur eutectic.

Recently[24], we have been successful in producing thin sections of polystyrene containing crazes without the use of reinforcing impregnants or a special diamond knife for microtoming. In the initial work, two types of studies were made: (a) examination of thin sections cut from bulk material which had been deformed in tension at room temperature in air to produce crazing and (b) examination of thin sections of uncrazed material which were subsequently strained to induce crazing in the thin section.

For the first series of experiments, care was taken to ensure that the sections (600 to 1200 A thick) were cut normal to the plane of the craze and that no artifacts were produced by the cutting processes. It was not possible to avoid fine "chatter" marks normal to the cutting direction, but their effects were minimized by cutting in a direction 60 degrees to the craze. Prior to observation, the sections containing the craze were deformed on a microstraining device. For the second series of experiments, thin slices of uncrazed material were prepared and strained in a similar way to produce crazing.

Some typical observations of craze microstructure made using the above techniques are shown in Figs. 5 and 6. A number of features of the microstructure can be established from these observations, and may be summarized as follows:

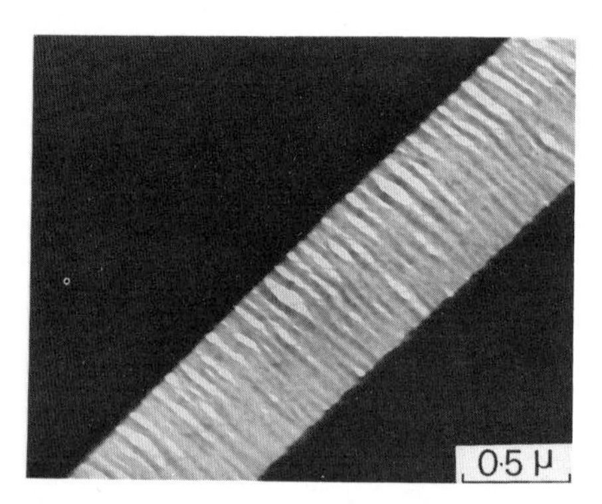

Fig. 5. Microstructure of a craze formed in a bulk specimen.

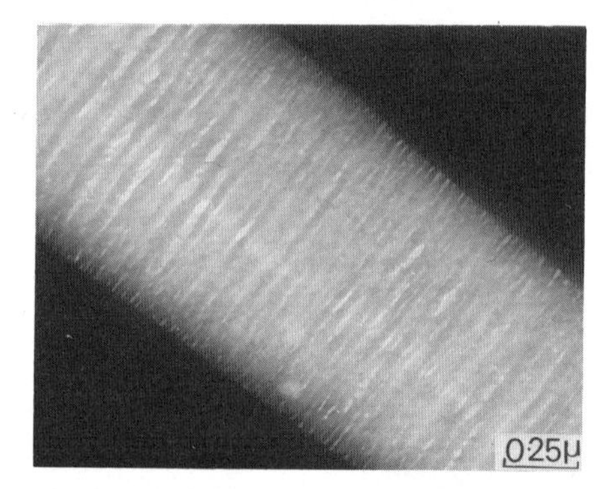

Fig. 6. Microstructure of a craze formed in a microtome thin section after straining.

1. The boundary between the craze and undeformed material is sharp and well defined.
2. The main feature of the structure in the early stages of craze growth is the development of an array of fibrils 250 A thick, which are joined together by finer fibrils less than 50 A thick. This produces an interconnecting three-dimensional array of fibrils and the structure has been described as an open-celled foam.[6] The size of the microvoids is comparable to that of the fibril thickness.
3. The fibrils form at right angles to the craze-matrix interface.
4. The fibrillar microstructure develops in the early stages of craze formation as indicated by Fig. 7. Thus, the fibrillar array extends to the tip of the craze and the thickness of the fibrils is similar

throughout the tapered craze section. This is an important observation in connection with the mechanism of craze growth since it implies that growth occurs by extension normal to the plane of the craze and the spreading of the fibril structure into adjacent material.

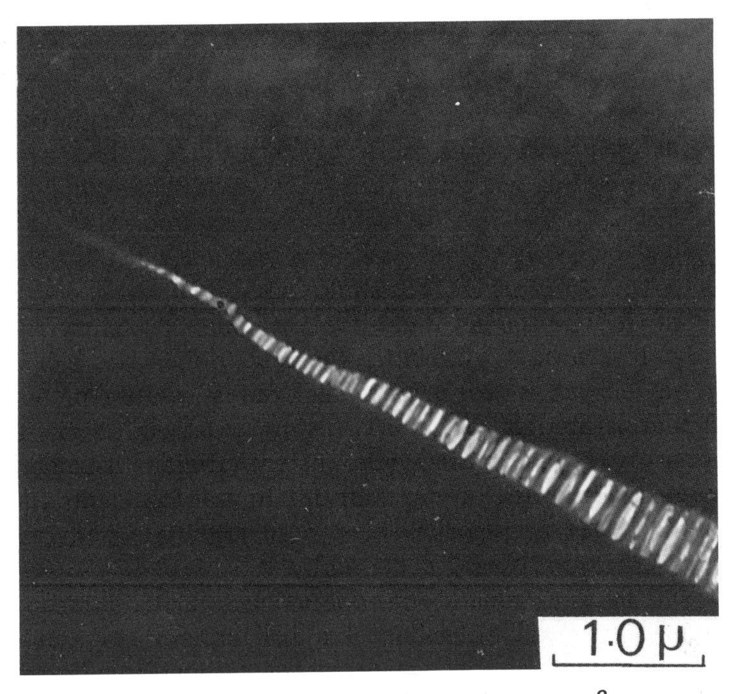

Fig. 7. Microstructure of the tip of a growing craze.[8]

A comparison of Figs. 5 and 6 shows that the microstructure formed in precrazed sections is very similar to that formed in stretched sections of uncrazed material. The distribution and dimensions of the fibrils are identical. This observation and the difficulties associated with accurate damage-free microtoming has led us to develop another technique for examining craze microstructure.[24] Thin films of polystyrene have been produced on glass slides by solvent evaporation of a 10 percent solution of a general-purpose grade of polystyrene ($\overline{M}_v = 2.03 \times 10^5$) in xylene. The films were then stretched in a number of ways. The most successful method involves mounting the film on a Mylar substrate which is then stretched. During stretching, the substrate buckles so that the polystyrene film separates and is stretched independently. The use of this technique offers a number of advantages to the previous method. Uniformly thick cast films can be produced which have a very smooth surface finish. This reduces premature failure, which often occurred in microtomed sections. In addition, the thickness of the film can be controlled and experiments are possible with much thicker samples. The use of the Mylar also allows the crazes to be stretched to much larger strains without failure.

Examples of the microstructures produced by the techniques outlined above are shown in Fig. 8 and a schematic representation of the various stages in the development of high-strain craze microstructure is shown in Fig. 9. The fibrillar structure characteristic of the microtomed sections develops in the early stages of craze deformation, and no distinction can be made between thin sections and films. In the thick films, large numbers of overlapping fibrils can be seen. There appears to be no effect of film thickness on the fibril structure. A new feature, which had not been recognized in our earlier work, is the development of a finer fibrillar structure as the craze thickens, leaving evidence of the coarse fibrillar

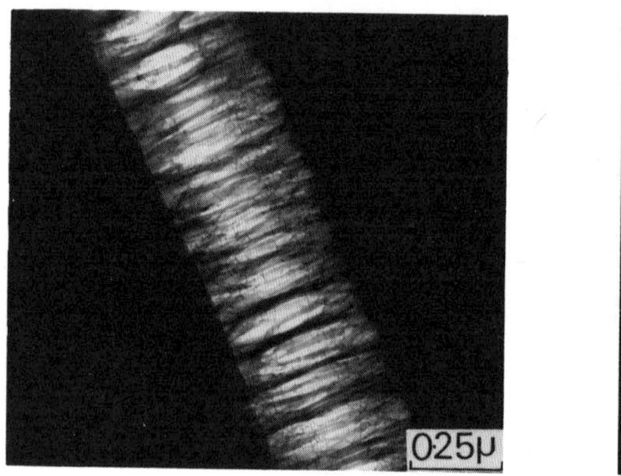

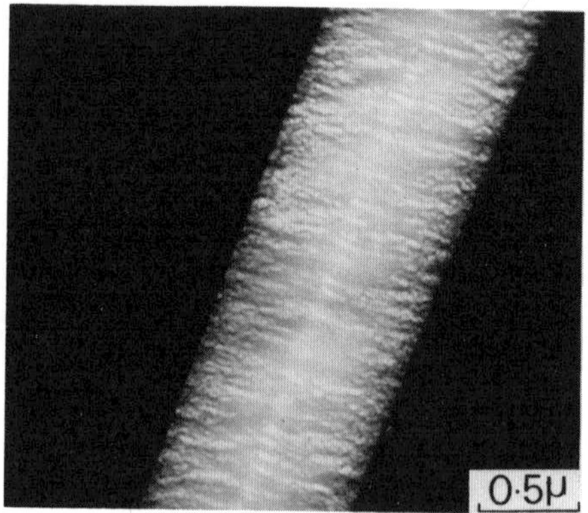

Fig. 8. Examples of microstructures of crazes formed in a solvent-evaporated thin film after straining.

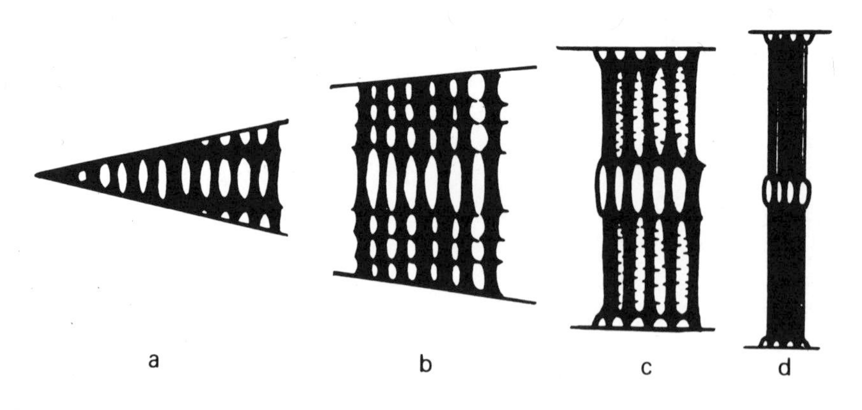

Fig. 9. Schematic illustrations of the variation in microstructure of a craze in polystyrene with increasing craze width. The dimensions of the craze tip are exaggerated and the width of the craze in c is normally much less than in d.

structure as voids at the craze matrix interface and along the center of the craze. This effect is more pronounced in some crazes than in others. The development of the larger voids occurs in the early stages of craze formation and remains at the interface as the craze thickens.

The microstructure shown in Fig. 8b and sketched in Fig. 9 is produced in crazes which have not failed prematurely and which have undergone large strains, causing drawing at the center of the craze. The coarse 250 A thick fibrils break down into finer elementary fibrils less than 100 A thick in a way analogous to the mechanism described by Peterlin[25] for fibrillation during the stretching of spherulitic thin films of polyethylene. The rows of larger microvoids remain at the craze matrix interfaces and in the center of the craze.

An important aspect of craze microstructure is the way in which it is related to the final failure of the craze and the fracture behavior of the material. Clearly, fracture must be described in terms of the failure of fibrils, and two types of experimental study indicate that this normally involves the necking and failure of the 250 A fibrils. The first type of observation has been made directly on thin sections and films. Two examples of the two main failure modes are illustrated in Figs. 10 and 11. The craze in Fig. 10 has failed by necking of the fibrils along the central region of the craze, leaving a layer of craze (fibrils) on each half of the specimen. In polystyrene this mode of failure is characteristic of relatively slow crack growth and the region associated with crack nucleation.[11] It is clearly recognized in the optical fractography by the colored fringes which are produced. Craze failure occurs gradually by the coalescence of microvoids due to failure of individual fibrils.

Fig. 10. Failure along the center of a craze.

The mode of failure shown in Fig. 11 is quite distinct from the center craze failure. Fracture occurs by progressive failure of fibrils at the craze-matrix interface. This process has also been identified in optical microscope studies[11]; one face of fracture shows a craze layer, whereas the other is devoid of craze. The fracture path tends to oscillate between

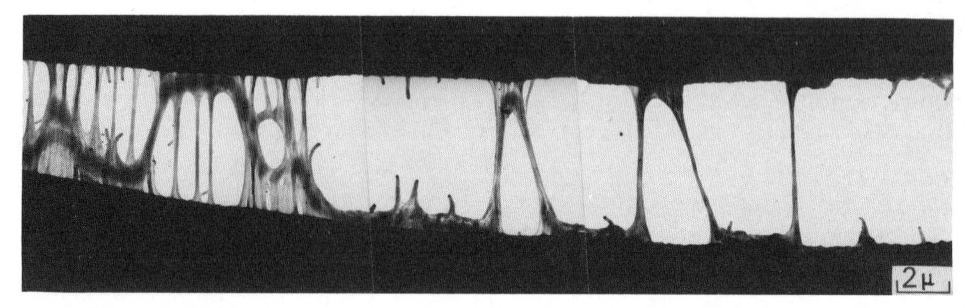

Fig. 11. Failure along the craze interface. The crack has propagated alternatively along the two interfaces.

one craze-matrix interface and the other, and this produces characteristic patterns of patch and mackerel.[11] This mode is associated with fast crack propagation. It will be noted in Figs. 10 and 11 that the fibrils which have failed are about 250 A thick and the craze has not developed the structure associated with large craze strains, Figs. 8a and 9d.

The second piece of evidence for failure of the 250 A-thick fibrils is obtained from fracture-morphology studies at high resolution. Bird, Rooney, and Mann[26] used replicas to study the fracture surfaces of injection molded polystyrene and reported that the central region of fracture was covered with fibrils 100 to 500 A thick. We have repeated and extended this work and shown that this fibril type of failure occurs over the whole fracture surface. The appearances of the fibrils in different parts of the fracture are entirely consistent with the failure modes shown in Figs. 10 and 11.

3 STRESS-STRAIN RESPONSE OF CRAZES

At the beginning of the paper, the large strain associated with crazing was noted as one of the main characteristics of the phenomenon. The microstructures described above are consistent with this, as are the measurements which have been made using optical techniques.[6] Little work has been done on the magnitude of the strains associated with crazing. However, Bucknall and Smith's work[21] on stress whitening in high-impact polystyrene provides strong evidence for large microscopic strains associated with very high craze densities, and Kambour and Kopp[27] have been able to measure directly the displacements across a solvent-induced thick craze formed in polycarbonate. No corresponding studies have been made on crazes formed in the absence of a solvent and it is unlikely that Kambour and Kopp's method can be used because the thickness of dry crazes precludes the accurate measurement of their thickness.

We have used an alternative approach[28] to determine the stress-strain behavior of crazes in tension. A grade of polystyrene (H R Carinex, Shell Chemicals Ltd., $\overline{M}_v = 2.03 \times 10^5$) was selected which crazes profusely when strained in air at room temperature. Parallel-sided specimens were prepared and all the surfaces were ground and hand polished with dry and wet gamma alumina. A typical stress-strain curve is shown in Fig. 12. Fine surface crazes, which do not penetrate into the bulk of the specimen, are observed at a stress slightly below that at which the first deviation of the stress-strain curve from linearity (a in Fig. 12) can be detected. The number of crazes increases as the stress increases, and at a much higher stress (b in Fig. 12), a dense zone of long, deeply penetrating crazes is nucleated at one side of the specimen (Fig. 13). The maximum in the stress-strain curve (c in Fig. 12) coincides with the extension of the craze zone entirely across the specimen. The zone then spreads along the specimen, and the load on the specimen drops. Extensive deformation is prevented by premature failure.

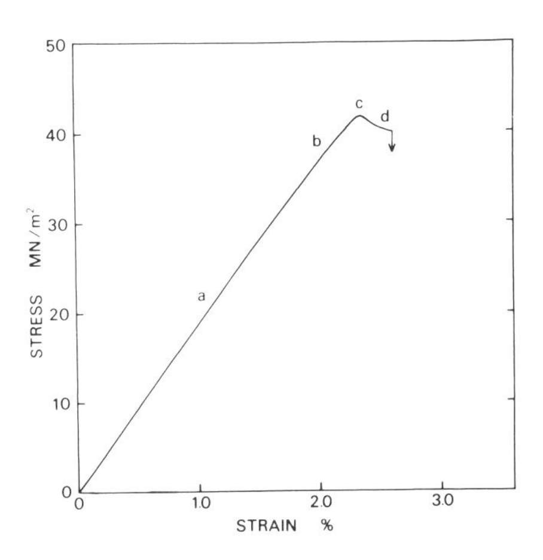

Fig. 12. Stress-strain curve of a specimen which has craze yielded.

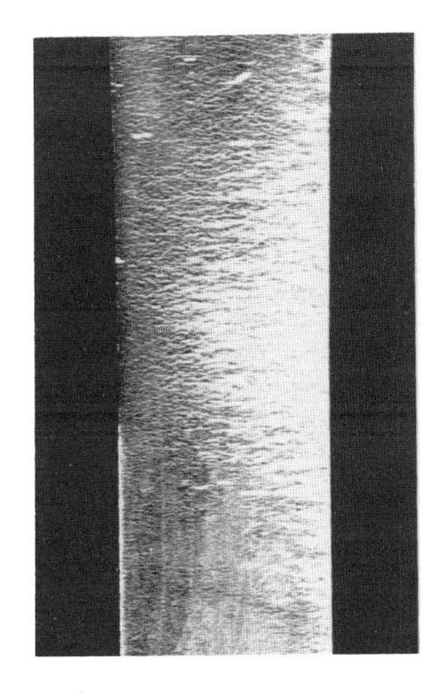

Fig. 13. Dense zone of deeply penetrating crazes formed at an early stage of craze yielding. Fine surface crazes which form at a much lower stress can be seen over the whole specimen. The deep crazes are long and straight. The ragged appearance is due to the angle of the illumination.

The form of yielding described above has the same characteristics as Lüders band formation and growth in mild steel. All the strain recorded on the stress-strain curve is the result of either elastic deformation or craze deformation. The shape of the curve depends on the crosshead speed of the tensile machine, the "elasticity" of the machine and specimen, and the rate of craze growth and multiplication. The onset of instability (c in Fig. 12) arises because the rate of increase in the length of the specimen due to crazing is greater than the rate of crosshead displacement, so that the load on the specimen is relaxed (in a hard-beam tensile machine). During the spreading of the zone of craze, the load continues to relax for the same reasons. The phenomenon is called "craze yielding".

The yielding process is of only secondary interest in the context of this paper, but is described in more detail elsewhere.[28] The important feature is the form of the specimen after craze yielding, which is illustrated schematically in Fig. 14. Crazes extend completely through the specimen cross section and are all parallel to each other, so that the specimen may be regarded as a multiple sandwich of crazed and uncrazed material. The stress-strain curve of the craze was determined by measuring the total strain of crazes in this composite specimen. The testing sequence is shown in Fig. 15. The specimens are first loaded to a stress which is one third of the craze-yielding stress and unloaded again to determine the elastic response of an uncrazed specimen. This cycle is followed by a high-stress cycle in which the specimens are craze yielded and unloaded before fracture. Finally, the stress is cycled to the same level as in the first cycle.

The elongation due to extension of the crazes is determined by subtracting the elastic strain (curve 1, Fig. 15) from the total strain (curve

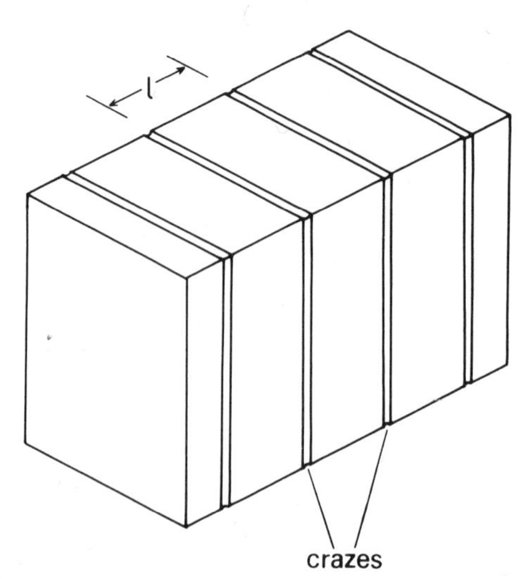

Fig. 14. Geometry of crazes after craze yielding.

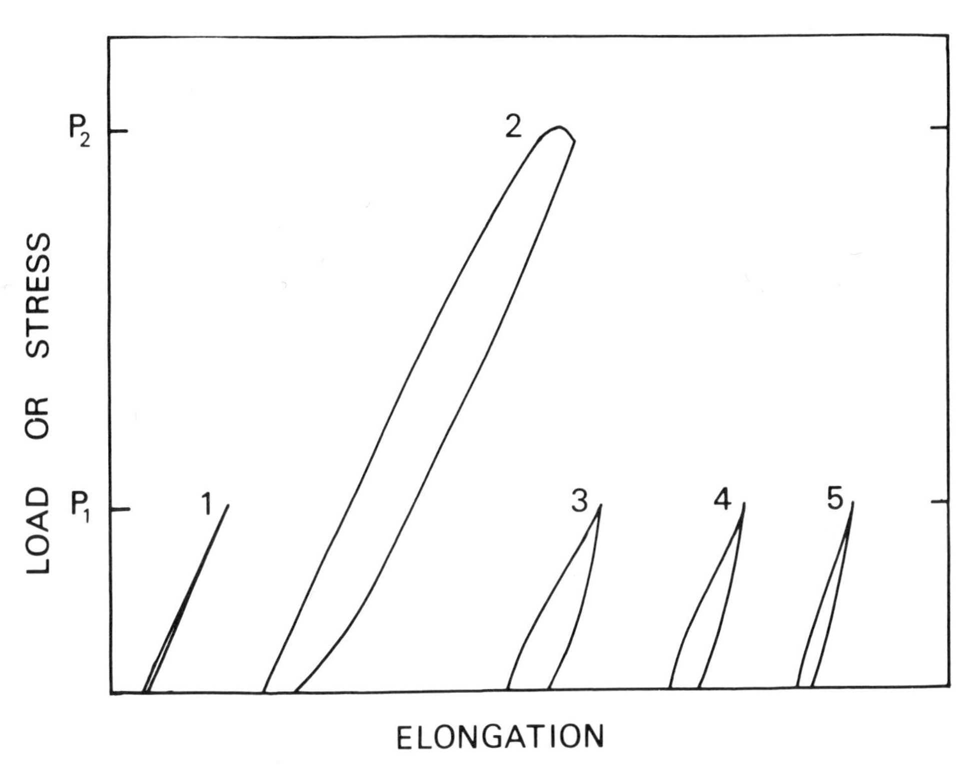

Fig. 15. Testing sequence used to determine the stress-strain of the craze. The diagram is not to scale.

3, Fig. 15). To achieve sufficient sensitivity, the displacements were measured using a microstrain technique with a sensitivity of 5 x 10^{-4} mm on a specimen gage length of 3 cm. To determine the strain it is necessary to measure the total thickness of craze. It is very difficult to obtain accurate values, mainly because the crazes are less than 1.0 μm thick and are usually in the range 0.1 μm to 0.5 μm. The following procedure was adopted. The total number of crazes in each specimen was measured by sectioning along the midsection parallel to the tensile axis and using a line-intercept method to determine the distribution and density of crazes along the whole gage length. The average thickness of the crazes was determined using carbon replicas and scanning electron microscopy. Neither of these two methods was entirely satisfactory. Typically, the mean value of craze thickness was 0.19 μm and the total craze thickness in a craze-yielded specimen was 0.08 mm.

The stress-strain curve of the craze determined from the total craze displacement and the total craze thickness is shown in Fig. 16. Curve a was produced immediately after the craze-yielding cycle and shows marked nonlinear behavior with a large hysteresis loop. Further stress cycles result

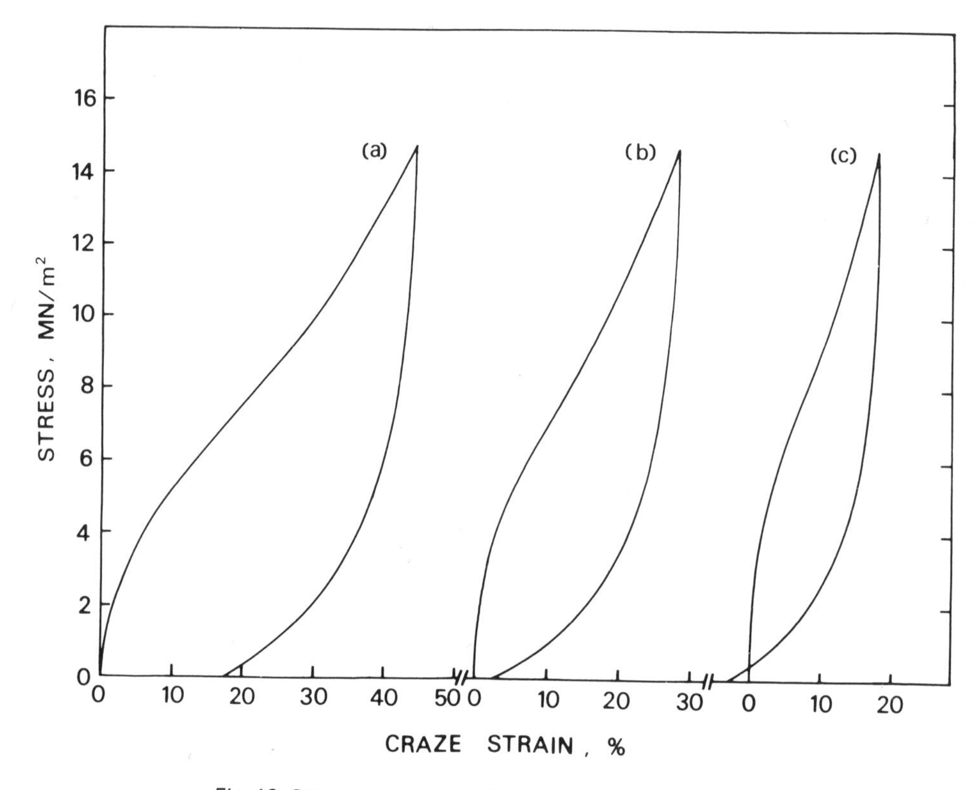

Fig. 16. Stress-strain curves of craze — see text for details.

in the curve becoming steeper and the hysteresis loop less pronounced. The size of the loop depended on the length of time the specimen was held at zero stress before the next cycle. Thus, curve b corresponds to a 20-second hold time and curve c to a zero hold time.

The agreement between the results shown in Fig. 16 and those reported by Kambour and Kopp[27] is remarkable. The general shape of the curve and the effect of cycling are reproduced. In Kambour and Kopp's work it was possible to show that the craze strain observed during the cyclic stress was due to stretching out of the material already in the craze, rather than to the formation of new crazed material at each side. This must also be true of our experiments. The stress required to produce the crazes was three times the stress used to determine the stress-strain characteristics of the craze. Indeed, after the initial craze straining, the work-hardening rate accelerates rapidly and approaches the elastic line of the uncrazed material.

4 RELATION BETWEEN CRAZE MICROSTRUCTURE AND PROPERTY

Kambour[29] has noted that the stress-strain characteristics of craze can be interpreted by considering the craze as a thin layer of variable-density, open-cell, polymer foam bonded between two rigid grips. The effect of stress on the preformed craze is, then, to open up new holes and produce a further reduction in density. The stress-strain response is likened to that of a polymer foam. The detailed observations of craze microstructure reported earlier indicate that significant changes in the size and distribution of the fibrillar structure occur at larger craze strains and that the final morphology cannot be compared directly with that of a foam.

Several difficulties have still to be resolved before an interpretation of the stress-strain response in terms of microstructure can be attempted. The most serious is that of achieving a direct correlation between the observed microstructure and the measured craze strains. The conditions under which crazes are induced in thin films are still rather arbitrary, and we do not know whether the structures which are observed correspond to those in a simply stressed craze. In addition, craze strain relaxation occurs very rapidly and the microstructure and the measured strain will clearly depend on the amount of relaxation which has occurred. Little is known about the relaxation processes, and we are currently modifying the microstrain equipment to measure the relaxation strains directly.

Correlation of the craze microstructure with fractographic observations is more straightforward, and we have achieved considerable success in this regard. However, here again, relaxation effects affect the interpretations which are made, as is evident from Fig. 11.

ACKNOWLEDGMENTS

I am indebted to my colleagues for their contributions to this work and for their permission to use unpublished results.

REFERENCES

1. Rabinowitz, S., and Beardmore, P., *Critical Reviews in Macromolecular Sciences,* CRC Press, Cleveland (1972), Vol. 1., No. 1, p 1.
2. Klemperer, W. B., *Theodore von Karman Anniversary Volume, Applied Mech.,* p. 328, Cal. Tech. (1941).
3. Sternstein, S. S., Ongchin, L., and Silverman, A., *Applied Polymer Symposia,* Interscience, New York (1968), No. 7, p 175.
4. Bevis, M., and Hull, D., *J. Mat. Sci.,* 5, 983 (1970).
5. Sternstein, S. S., and Ongchin, L., *ACS Polymer Preprints,* **10**, 1117 (1969).

6. Kambour, R. P., *Applied Polymer Symposia,* Interscience, New York (1968), No. 7, p 215.
7. Kambour, R. P., and Holik, A. S., *J. Polymer Sci.,* Part A-2, **7**, 1393 (1969).
8. Beahan, P., Bevis, M., and Hull, D., *Phil. Mag.,* **24**, 1267 (1971).
9. Kambour, R. P., and Russell, R. R., *Polymer,* **12**, 237 (1971).
10. Van den Boogaart, A., *Conference Proceedings, Physical Basis of Yield and Fracture,* Oxford University Press (1966), p 167.
11. Murray, J., and Hull, D., *Polymer,* **10**, 451 (1969).
12. Williams, D.R.G., *Applied Polymer Symposia,* John Wiley and Sons, Inc. (1971), No. 17, p. 25.
13. Harris, J. S., and Ward, I. M., *J. Mat. Sci.,* **5**, 533 (1970).
14. Owen, T. W., unpublished work.
15. Vadimsky, R. G., Keith, H. D., and Padden, F. J., *J. Polymer Sci.,* Part A-2, **7**, 1367 (1969).
16. Woodward, A. E., and Morrow, D. R., *J. Polymer Sci.,* Part A-2, **7**, 1651 (1969).
17. Gent, A. N., *J. Mat. Sci.,* **5**, 925 (1970).
18. Knight, A. C., *J. Polymer Sci.,* Part A, **3**, 1845 (1965).
19. Marshall, G. P., Culver, L. E., and Williams, J. G., *Proc. Roy. Soc.,* Part A, **319**, 165 (1970).
20. Camwell, L., unpublished work.
21. Bucknall, C. B., and Smith, R. R., *Polymer,* **6**, 437 (1965).
22. Kambour, R. P., *Polymer,* **5**, 143 (1964).
23. Kambour, R. P., and Holik, A. S., *ACS Polymer Preprints,* **10**, 1182 (1969).
24. Beahan, P., Bevis, M., and Hull, D., unpublished work.
25. Peterlin, A., *J. Mat. Sci.,* **6**, 490 (1971).
26. Bird, R., Rooney, J. G., and Mann, J., *Polymer,* **12**, 742 (1971).
27. Kambour, R. P., and Kopp, R. W., *J. Polymer Sci.,* Part A-2, **7**, 183 (1969).
28. Hoare, J., and Hull, D., *Phil. Mag.,* **26**, 443 (1972).
29. Kambour, R. P., *Polymer Eng. and Sci.,* **8**, 281 (1968).

DISCUSSION on Paper by D. Hull

KNAUSS: In measuring the stress-strain behavior of the craze material under repeated loading, it is extremely important to prevent the formation of additional fibrillar craze material from the craze wall. For, if even a small amount of such additional material is produced, the "stress-strain behavior" is affected through a change in the effective "gage length" of the craze. How did you ascertain whether or not any craze-wall material was transformed into fibrillar craze material during subsequent load cycles?

HULL: I agree entirely with your point. We do not have any experimental results which can be used to answer the question definitively. However, it will be noted that the stress-strain response of the craze was determined only at stresses up to 33 percent of the stress required to produce craze yielding. It seems unlikely that any craze growth will occur. Electron microscope observations on growing crazes support this conclusion.

WILLIAMS: Your talk implied that the "patch" and "mackerel" fracture modes in polystyrene, resulting from internal craze and interface crack growth, are due to inhomogeneities alone. I feel that such effects are often governed by the energy supplied during crack propagation, i.e., rougher surfaces for higher energies.

HULL: We have demonstrated that the "patch" or "mackerel" fracture modes are a characteristic feature of one of the failure processes of crazes in polystyrene. There is clear evidence that this is related to craze microstructure. However, the general fracture path is determined by the distribution of crazes around the propagating crack which is, in turn, dependent on the energy supplied during crack propagation. The rougher surfaces arise out of the changes in craze distribution and should not be confused with the failure mode of individual crazes.

ROSENFIELD: Is craze formation a necessary prerequisite for fracture of glassy polymers?

HULL: I take it that the question refers to brittle-type fracture. I know of no conclusive evidence for fracture without crazing. One or two workers have indicated that this has occurred, but I don't think the experimental observations were sufficiently rigorous to justify their conclusion. We have tested polystyrene in a wide variety of conditions and in all cases craze formation preceded fracture.

HALPIN: I think that what may be the most important point here is the question of how crazes form from bulk polymers; not so much how the developed crazes deform. There is a general principle developing out of this work which I feel needs emphasis. Bulk polymers generally exhibit low bulk or volume strength; consequently, the development of positive hydrostatic stress fields adjacent to points of stress concentration produces cavitation in local volume elements. In elastomers, this leads to the appearance of fibers, or ligaments at the notch root. In hard, brittle, plastics, catastrophic fracture may be coincident with cavitation. In plastics capable of cold drawing, the cavitated material is plastically drawn and stabilizes into fibers if the material is of the strain-hardening type. The mean dimensions of the fibrous craze structure is probably determined by the thermodynamic restrictions on void size. In each of the material classes, elastomeric, .brittle, and ductile solids, fracture is initiated by cavitation processes.

EFFECTS OF HYDROSTATIC PRESSURE ON THE DEFORMATION AND FRACTURE OF POLYMERS

S. V. Radcliffe

Case Western Reserve University
Cleveland, Ohio

ABSTRACT

Accurate observations under controlled conditions of the effects of increased environmental pressure on the mechanical behavior of polymers are relatively recent. From a critical analysis of these various pressure observations, phenomena that appear to be characteristic of such effects for the major polymer classes are identified, and the validity of hypotheses advanced for particular phenomena in specific polymers, together with their generality, is examined.

For the modulus (i.e., the preyield region of the stress-strain curve), the larger pressure dependence for semicrystalline compared with amorphous glassy polymers is associated with the pressure-induced increase in the temperature of subambient relaxation processes in the disordered component of the structure. In an analogous manner, increases in the glass transition temperature with pressure cause discontinuous increases in modulus for elastomers. For the yield stress, the measured pressure dependence – both from hydrostatic and from biaxial stress experiments

– conforms to a modified von Mises yield criterion. However, unlike the modulus, there is no clear differentiation in the behavior of crystalline and amorphous polymers. Although the "volume-change equivalence" hypothesis is found to be invalid, the temperature and pressure dependence of yield stress can be correlated in a manner analogous to the time-temperature superposition concept. Applications of rate theory to the pressure, strain rate, and temperature dependence of yielding appear promising, but have not yet elucidated the specifics of the molecular mechanisms involved in polymer yielding.

The brittle fracture and crazing of amorphous glassy polymers can be suppressed by pressure and shear yielding induced. For polystyrene, there is evidence that this brittle-ductile transition may result from changes in crack-propagation characteristics rather than from simply a suppression of crazing. In some normally ductile polymers, decreases in strain to fracture occur with increases in pressure; such effects appear to be associated with pressure-induced changes in the temperature of relaxation processes.

It is concluded that the development of experimental techniques for studying polymer behavior under pressure has reached the stage where substantial contribution can be made to the elucidation of the factors controlling the mechanical behavior of this class of materials.

1 INTRODUCTION

In the early 1950's, P. W. Bridgman[1] demonstrated for the first time for a polymeric material that, in a manner analogous to some normally brittle metals, ductile behavior in mechanical tests could be induced by increased hydrostatic pressure. A commercial plastic (a melamine formaldehyde resin – Melmac 404) that fractured catastrophically when stressed under uniaxial tension under ambient conditions was found to be capable of straining substantially and exhibiting yielding prior to shear failure when loaded similarly but subjected to a hydrostatic stress of some 24 kb (~353 ksi). However, in addition, unlike metals and other inorganic solids, the "initial stiffness" was reported to increase threefold at the higher pressure. Despite the crudity of these early experiments insofar as stress and strain measurements were concerned, the principal features of the inhibition of fracture and the increased modulus were sufficiently striking (and subsequently to be confirmed as characteristic of a number of polymers) that it is surprising to find a delay of some 10 years before the further extensive work that has developed in this area.[2-23]

Thus, it was not until 1964 that Ainbinder and co-workers[2] in the Soviet Union reported for tests up to only 2 kb, the additional important characteristic that the compressive yield maximum of a ductile polymer increased (by 150 percent at 2 kb) and that the modulus for several elastomers increased over the same pressure range from values typical of the rubbery state to those of the glassy state. In the same year, Paterson[3] in Australia interpreted similar modulus effects in elastomers in terms of a pressure-induced depression of the glass transition temperature to below ambient. Simultaneously, Holliday, Pugh, and co-workers[4] in Britain confirmed the effect of pressure in inducing ductile behavior by showing that the glassy polymer, polystyrene, would yield in tension under a hydrostatic pressure of 7 kb (~103 ksi) at stresses considerably greater than its normal tensile fracture stress. The next main stage of development came in 1968 with the measurement of true stress and true strain (through the use of apparatus with direct optical viewing of the specimen) during uniaxial mechanical tests at constant strain rate at up to 8 kb as applied by Sardar, Baer, and Radcliffe, to the semicrystalline polymer polyoxymethylene.[8] This latter development has permitted the detailed examination of the validity of possible macroscopic flow and fracture laws for polymers and the study of stress relaxation at constant pressure. (Note: A variety of types of apparatus for mechanical testing of polymers is now in use and it is particularly important in comparing data from different laboratories to be cognizant of the machine characteristics and the nature of the measurements made.) A significant further experimental advance is the ability to measure the sensitivity of the flow stress to changes in strain rate at constant pressure reported recently by Davis and Pampillo.[24]

At the present time, some 25 papers concerning the mechanical behavior of polymers under pressure have appeared in the literature – as summarized in Table I. These numerous investigations for a range of polymers from the different major classes have developed a variety of data, much of which are of sufficient accuracy and reproducibility that it is now feasible to discern phenomena that appear characteristic of pressure effects on the mechanical behavior of such materials and to seek a generalized approach to elucidating them. The present paper is directed toward exploring such questions, the validity of various hypotheses that have been advanced for particular phenomena, and the significance for the general understanding of polymer behavior. In the following sections, the nature of the pressure dependence of the sequential portions of the stress-strain curve – preyielding, yielding, large-scale plastic flow (work hardening), and fracture – is examined for the major polymer classes. One factor which still frequently makes exact comparisons difficult between results for the same polymer from different laboratories is the absence or inadequacy of characterization of the polymer and the limited recognition of the influence of preparation and processing history on properties. Where possible, such features have been taken into account here.

Table I. Summary of Experimental Research on the Effects of Pressure on the Mechanical Behavior of Polymers

Year	Authors	Reference	Polymers	Pressure Maximum, kb	Applied Stress
1953	Bridgman	1	Melamine-formaldhyde	24	Tension
1964	Ainbinder, et al.	2	Polymethyl methacrylate Polyvinyl chloride Ebonite	2	Tension Compression
1964	Paterson	3	Elastomers	10	Compression
1964	Holliday, et al.	4	Polystyrene	8	Tension
1965	Ainbinder, et al.	5	As for Ref. 2, plus: Polystyrene Nylon 6 Polytetrafluoroethylene Polyethylene Several filled plastics	2	Compression Tension
1967	Laka and Dzenis	7	Polymethyl methacrylate Polytetrafluoroethylene Polyethylene Several filled polymers	2	Tension
1968	Sardar, Radcliffe, and Baer	8	Polyoxymethylene	8	Tension (and stress relaxation)
1968	Pae, Mears, and Sauer	9	Polypropylene	7	Tension
1968	Pae and Mears	10	Polyethylene Polytetrafluoroethylene	5.6 4	Tension Tension
1969	Biglione, Baer, and Radcliffe	12	Polystyrene High-impact polystyrene Acrylonitrile-butadiene -styrene (ABS) Rubber-reinforced PMMA	6 6 6 2	Tension Tension Tension Tension
1969	Mears, Pae, and Sauer	13	Polyethylene Polypropylene	7	Tension and compression Tension
1969	Mears and Pae	14	Polycarbonate	5	Tension
1969	Vroom and Westover	15	Polystyrene Polymethyl methacrylate ABS Phenoxy ethylene copolymer Polytetrafluoroethylene	2	Tension (mercury used as pressure fluid

Table I. (Continued)

Year	Authors	Reference	Polymers	Pressure Maximum, kb	Applied Stress
1969	Weaver and Paterson	16	Filled elastomer	10	Compression
1969	Uy, McCann, and Hettwer	18	Polysulfone	10	Tension
1970	Sauer, Mears, and Pae	19	Polycarbonate Polytetrafluoroethylene	7 5	Tension Compression and tension
1970	Rabinowitz, Ward, and Perry	20	Polymethyl methacrylate Poly(ethylene terephthalate)	7.5	Torsion (thin-walled cylinders)
1971	Christiansen, Baer, and Radcliffe	21	Polycarbonate Polyethylene terephthalate Polychlorotrifluoroethylene Polytetrafluoroethylene	8	Tension
1971	Pugh, Chandler, Holliday, and Mann	21	Polystyrene Polymethyl methacrylate Polyethylene Nylon (cold drawn)	7.7	Tension
1972	Davis and Pampillo	23	Polyethylene	8.3	Tension (strain-rate dependence)
1972	Pae and Sauer	24	Polyvinyl chloride	7	Tension
1972	Sauer, Bhateja, and Pae	25	Polyimide Polysulfone	7	Tension and compression

2 PREYIELD BEHAVIOR

The characteristic nonlinearity of the extensive region, typically up to 10 percent strain of the stress-strain curve preceding macroscopic yielding in polymers makes it impossible to define a unique modulus from the slope, as for inorganic solids. Instead, it is common to define a nominal "Young's modulus" from the slope of or secant to the curve at low strain – typically in the region of 1 percent strain. Even such measurements are made more difficult at pressure owing to the additional experimental problems involved and the effects of hydrostatic compression on specimen dimensions. However, the general increase in the slope of the preyield region with increase in environmental pressure is sufficiently striking that several investigators have reported measurements of the magnitudes involved. Ainbinder and co-workers[2,5] showed various such increases for a range of polymers, although the experimental techniques and modest pressure range (2 kb) reduce the accuracy of their measurements and makes comparison difficult. Ainbinder[11,17] has proposed that such modulus effects (and likewise the increases in yield stress with pressure) can be accounted for in terms of equivalent temperature effects through analogous decreases in specific volume associated with either increase in pressure at constant temperature (compressibility) or decrease in temperature at constant pressure (thermal contraction). The incorrectness of this view will be discussed later in connection with yield phenomena.

The comparative Young's modulus measurements made over a wide pressure range by Baer and Radcliffe et al.[8,12,21], Pae and Sauer et al.[9,13,19,24,25], and Pugh and Holliday et al.[22] provide more reliable values for a number of amorphous (glassy) and semicrystalline polymers. Table II ranks the comparative modulus changes for a variety of polymers derived from the experimental data reported from these groups. While the measures of modulus change listed in Table II must be regarded as approximate only, the fact that the ranking distinctly separates the five semicrystalline polymers as having relative modulus values in the region of 2 or higher from the lower six, which all have values below 1.5, points to the significance of polymer structure in determining the pressure response.

Table II. Pressure Dependence of Young's Modulus

Polymer	Reference	Relative Modulus[(a)]
Polyethylene (PE)	13	4.5
Ditto	22	2.2[(b)]
Polychlorotrifluoroethylene (PCTFE)	21	3.4
Polytetrafluoroethylene (PTFE)	21	3.2
Ditto	19	2.6
Polyoxymethylene (POM)	21	2.0
Nylon 6.6	22	1.9[(b)]
Polyethylene terephthalate (PET)	21	1.4
Polycarbonate (PC)	21	1.2
Polymethyl methacrylate (PMMA)	22	1.25[(b)]
Polystyrene (PS)	12	1.1
Ditto	22	(close to 1.0)[(b)]
Polyimide (PI)	25	1.3
Polysulfone	25	1.2

(a) Ratio of modulus at hydrostatic pressure of 50,000 psi to that at atmospheric pressure (the increase in modulus is approximately linear for all the polymers within this range).

(b) Estimated from data for one pressure, assuming linearity.

It is well established that marked changes in elastic moduli and other mechanical properties can occur in polymers when a phase transition or glass transition occurs. Thus, in PTFE, a solid-solid phase transition takes place in the vicinity of 5.5 kb[27], and strong decreases in modulus and yield stress have been observed there[21]. In the case of elastomers, Paterson[3,16] has shown that a very striking increase in Young's modulus (1000-fold) for natural rubber in the region of 5 kb is associated with the pressure-induced raising of the glass transition temperature through ambient. Although corresponding to the latter results, the variation in the relative modulus values for the amorphous polymers in Table II might be expected to relate to their different glass transition temperatures, the table does not show such a correspondence consistently. Sauer et al.[19,25] have suggested that, to a first approximation, finite strain theory predicts that the pressure dependence of the modulus is given by:

$$m = 2\,(5\text{-}4\nu)\,(1\text{-}\nu) \quad ,$$

where m is the slope of the modulus-pressure curve and ν is Poisson's ratio. However, the degree of agreement with experimental values appears insufficient for this approach to be useful in distinguishing the different pressure effects in various polymers.

For the semicrystalline polymers (which typically exhibit lower moduli than amorphous glassy polymers), there is evidence in several cases that the application of pressure increases the modulus by raising the transition temperatures of relaxation processes through ambient. Sardar, Radcliffe, and Baer[8] first suggested this effect for POM on the basis of an established β transition in the vicinity of -75 C in the disordered regions of the polymers, and their evidence from stress-relaxation measurements of an appropriate pressure-induced shift in that transition. The extension of this hypothesis to other crystalline polymers[21] provides a logical basis for qualitative explanation of their pressure behavior. Pugh et al.[22] have pointed out that their modulus observation on polyethylene may be attributable to such pressure effects on established low-temperature transition in that polymer also. Although knowledge of the necessary transition phenomena and their pressure dependence is yet too limited to fully validate this hypothesis for semicrystalline polymers as a class, the available evidence points strongly to the pressure response of the disordered regions of such polymers being the controlling factor in changing the modulus. Whether the changes result from the pressure induction of an effective glass transition in these lower density regions of the polymer structure or from other transition phenomena remains uncertain. However, it appears clear that the introduction of any such modifications in the properties of the disordered regions must play a role in the observed increases in the stress required to induce subsequent plastic yielding in these polymers under pressure.

While there are no direct structural parallels in inorganic solids for the pressure effects on modulus discussed above, the pressure-dependence of their moduli, as might be expected, is influenced strongly by the onset of pressure-induced first-order transitions in phase. In addition, it is of interest to note that pressure-induced shifts in the Néel transition temperature in chromium (a second-order transition involving changes in electron-spin or magnetic ordering) induce correspondingly strong changes in Young's modulus.[28] While for inorganic solids in general the relative increases in modulus induced under pressure are much less than for even the amorphous polymers, it is important to take into account the much larger modulus values characteristic of the former relative to the hydrostatic stress levels applied. Thus, the modified Murnaghan theory as used in geophysical analysis expresses the pressure dependence of the modulus in the form:

$$E = E_o + a \cdot P \quad ,$$

and the order of magnitude of the observed pressure dependence for at least the amorphous polymers appears more in keeping with that for other solids when viewed in this way.

3 YIELD PHENOMENA

In recent years, theoretical and experimental interest in the mechanical behavior of polymers has extended from the previous focus on the linear viscoelastic region preceding macroscopic flow to the large strain phenomena of yielding, flow, and fracture. The complexity of the structural effects accompanying such large strains – in ductile polymers, plastic yielding itself is typically preceded by viscoelastic strains up to some 10 percent – has made the elucidation of the controlling mechanism especially difficult. The early evidence of strong effects of increased hydrostatic pressure on these phenomena[1-4], including that of the possibility of suppressing the brittle-fracture characteristic of many amorphous polymers and so permitting the examination of their plastic yielding, has been a major factor in stimulating the recent growth of research activity in this area. In addition to the theoretical implications, improved knowledge of such phenomena is of potential value in connection with the application of cold-forming processes to polymers.

In the case of yielding, the principal efforts have been directed to measurements of the pressure dependence of the yield stress with respect to the definition of yield criteria and the use of continuum plasticity theory analyses to account for the observed behavior. The latter approach necessarily ignores the fact that polymers do not typically approximate to the ideal "rigid-plastic" solid and that there is direct experimental evidence for some amorphous polymers that volume changes occur during straining in the preyield region and continue during the yield process itself.[29] In addition to facilitating the development of laws of macroscopic response to applied stress fields, the nature and analysis of high-pressure behavior has proved of value in elucidating the possible molecular processes involved. Both aspects will be examined here.

The historical development and interrelation of the various relationships proposed to describe the pressure dependence of yield stress in polymers have been reviewed in detail[8], and only the main points need attention here. Until relatively recently, the yield criteria for ductile, relatively incompressible inorganic solids have been used practically for polymers, namely the von Mises (or critical elastic shear strain energy density) criterion and the Tresca (or critical shear stress) criterion. For both, the flow stress depends only on the deviatoric and is independent of the hydrostatic component of the applied stress. Introduction of the effect of pressure or mean normal stress,

$$\sigma_m = (\sigma_1 + \sigma_2 + \sigma_3)/3$$

leads to the equation known as the modified Tresca or "Mohr-Coulomb" relationship (which was proposed initially for soils)

$$\tau_{max} = \tau_o - \mu' \sigma_m$$

or to the modified von Mises equation

$$\tau_{oct} = \tau_s - \mu \sigma_m \quad ,$$

where τ_o and τ_s are the maximum and pure shear yield stresses at atmospheric pressure and μ' and μ are materials constants defining the pressure sensitivity of the yield stress. These various parameters can be expected to be strain-rate and temperature dependent. The latter equation is the form suggested by Sternstein and Ongchin[30] for shear yielding on the basis of biaxial stress experiments in glassy polymers (where μ is considered as a "bulk" friction factor) in preference to the earlier application[29,32a] of the Mohr-Coulomb relationship to such data. This modified von Mises criterion was also developed independently by Bauwens[31] in connection with his application of the Eyring theory of non-Newtonian flow to polymers.

For the experimental testing of yield criteria, it is important to derive true stresses. Unfortunately, few of the reported pressure measurements are of this type, and approximate corrections must be applied. Table III compares the values of the parameter μ obtained directly or by correction from pressure experiments and includes values from biaxial stress experi-

Table III. Pressure Dependence of the Yield Stress in Polymers

Polymer	Reference	Pressure Factor
Polyethylene (PE)	13	0.046[(a)]
Ditto	20	0.075
"	22	0.08[(a)]
"	32b	0.05[(b)]
"	23	0.07
Polychlorotrifluoroethylene (PCTFE)	21	0.12
Polytetrafluoroethylene (PTFE)	21	0.048
Ditto	10	0.032[(a)]
Polyoxymethylene (POM)	21	0.10
Nylon 6:6	22	0.03[(a)]
Polyethylene terepthalate (PET)	21	0.054
Ditto	32b	0.09[(b)]
"	20	0.075
Polybisphenol A carbonate (PBAC)	21	0.072
Ditto	14	0.047[(a)]
"	48	0.075[(c)]
Polymethyl methacrylate (PMMA)	20	0.204 (torsion)
Ditto	22	0.17[(a)] (fracture)
"	33	0.15[(b)]
"	32b	0.158[(b)]
Polystyrene (PS)	21	0.084
Ditto	22	0.055[(a)]
"	32b	0.25[(b)]
Polyimide (PI)	25	0.06
Polysulfone	25	0.08
Polypropylene (PP)	13	0.092[(a)]
Polyvinyl chloride (PVC)	32b	0.11[(b)]
Epoxy I	32b	0.09[(b)]
Epoxy II	32b	0.19[(b)]

(a) Parameter computed from reported data.
(b) From biaxial measurements at atmospheric pressure.
(c) From uniaxial tension and compression measurements at atmospheric pressure.

ments at atmospheric pressure. The tabulated values of μ indicate an encouraging level of agreement in most cases from pressure measurements on a given polymer by different investigators. Indeed, some differences in response are to be expected from the different preparation and processing of the specimen material in each case. In addition, agreement is generally good between the values obtained from pressure and biaxial tests – with the exception of polystyrene, where the biaxial value appears anomalously high. Despite this agreement, the type of distinct separation in response between semicrystalline and amorphous polymers that was found for the modulus (Table II) is absent for the pressure constant μ. The first 11 polymers in Table III are the same as those listed in Table II and the lack of a comparable separation is immediately apparent. On the basis of an earlier comparison of a more limited number of pressure-derived μ values only[21], qualitative agreement was obtained between the ranking by magnitude of μ with the strength-limiting temperature (the melting point or glass transition temperature). The present analysis points to the probable greater importance of structural factors in determining the value of the pressure constant.

The generally strong dependence of the yield stress of polymers on pressure is in marked contrast to the general absence of detectable effects in metals, providing care is taken to avoid the conditions that give rise to pressurization phenomena.[34] The latter effects, which appear to be prominent only in the body-centered cubic transition metals, occur as a dramatic *decrease* in yield stress on testing under ambient conditions after subjecting a specimen to hydrostatic compression above a critical minimum pressure. The effect has been demonstrated to result from the generation of mobile dislocations due to differential compression of impurity particles and the metal matrix.[34] Pressurization experiments in various polymers studied in this laboratory have shown no influence on the subsequent mechanical behavior under ambient conditions.[8,12,21] Such experiments, with and without sheathing to prevent direct contact of the pressure fluid with the polymers, also confirmed that the contact did not influence their behavior.[21,22] However, there are some combinations of medium and polymer where the yield stress is affected – as for polyethylene, which exhibited different mechanical behavior in pentane from that in water.[23] Although not a major problem, it is clear that attention should be given to ensuring that a "neutral" pressure fluid is selected for a given polymer.

A consequence of the conformity of polymer yield behavior to the modified von Mises relationship is the prediction of a higher yield stress in tension than in compression at atmospheric pressure. Such a difference has long been recognized for a number of polymers (especially as it has some consequences for their engineering use), although the corollary of a pressure dependence of yield stress has not. Although differences in tensile and compressive yield stress are known not to be characteristic of metals, recently such effects – known as "strength-differential" or "S/D" effects – have been observed prominently in iron-carbon martensites.[35] On the basis of characteristics of polymer behavior, the pressure dependence of such alloys has been measured and shown to be substantial.[36] This result is consistent with plasticity theory and points to the likelihood that a volume or dilatational mechanism is associated with the yield process in this special class of alloys, as for polymers. It should be noted that the theory does not *require* the existence of a volume change, although some variants would give such a change.

The preceding discussion of the fit of the measured pressure dependence of polymers to the shear yield criterion supports the use of the latter for the prediction of the advent of yielding under any stress state in which this mode of yielding is dominant. At the same time, the conditions that a satisfactory yield mechanism must meet are specified. One approach to this question of mechanism is that of Ainbinder[11,17] who has argued that the controlling feature is specific volume (and the related free volume), so that a given volume change, whether induced by increase in pressure or decrease in temperature, will bring about the same increase in yield stress. However, direct measurements of the temperature and pressure dependence of yield stress and specific volume for polycarbonate[2] have established that there is no such direct relationship between volume and yield-stress changes, although it was shown that a simple "shift factor" involving only physical constants can be used to relate the two effects.

Principally on the basis of observations of strength differential and pressure effects in amorphous polymers, attempts have been initiated to develop appropriate analyses and mechanistic models for the kinetic processes involved in flow. Robertson's[37,38] initial application of the formalism of the Eyring model for non-Newtonian flow in which deformation is considered as a rate process has been modified by Duckett et al.[26] by an additional pressure-work term and applied to PMMA, but with only modest agreement between theory and experiment. A variant approach by Argon et al.[39] involving the translation of a "knotted" polymer chain between adjacent chains has proved only conceptually useful to date. Bauwens and co-workers[48] used a modified Eyring model to account for the observed change in activation energy in polycarbonate with temperature in terms of a shift to two different activated flow processes in this polymer at lower temperature (<-50 C). Davis and Pampillo[23] have also applied rate theory to determine the activation volume characteristic of the molecular processes involved in flow in the semicrystalline polymer polyethylene from measurements of the strain-rate sensitivity of yielding at pressure. It is of interest that the initial results point to the prominence of the disordered region in controlling flow in this polymer. Although promising, the rate-theory approach is subject to the same limitation as has often been found in its analogous application to metals – it defines some characteristics that a satisfactory model must meet, but it is difficult to compute such characteristics directly for any specific model. Furthermore, although it has been contended[23] that the ratio of yield stress to Young's modulus, σ/E, is essentially independent of temperature, strain rate, and pressure for polymers, it is clear from the different pressure dependence shown for the parameters in this ratio in Tables II and III that the ratio cannot be regarded as independent of pressure. Thus, the observation by Davis and Pampillo[23] that the ratio for polyethylene is "roughly constant" with pressure must be regarded as of questionable significance and, in any case, of no significance as a general characteristic of polymer flow mechanisms.

Very recently, Wu[40] has examined for polycarbonate and polyethylene in torsion experiments on tubular specimens, the effects of strain rate on plastic flow at pressures up to 1 kb at room temperature and between -196 C and 80 C at atmospheric pressure. Activation analysis was also applied to these measurements to determine shear activation volume, dilatation activation volume, and activation enthalpy. For both polymers, the results indicate that activation occurs in a region of a size comparable to that of the molecular units. However, for polycarbonate there appears to be no reasonable model to account for the observed changes in activation volume with temperature. In the case of polyethylene, there are indications that entropy effects are important in the rate mechanism of deformation.

The complexity of the polymer structures, whether amorphous or semicrystalline, makes it clear that the continued development of measurements (on well-characterized polymers) of the temperature, pressure, and strain-rate dependence of the principal parameters that describe plastic flow is essential for the elucidation of the microscopic flow mechanisms involved.

4 FRACTURE

The earliest observations on the influence of pressure on fracture in polymers were concerned with materials that normally failed in a brittle manner after little deformation when ubjected to tensile stress.[1,4] Since that time, the pressure dependence of fracture of the brittle amorphous polymers polystyrene[12,22], polymethyl methacrylate[22], and polysulfone[25] has been examined, but in detail only for polystyrene. A pressure-induced transition from brittle to ductile behavior in tension has been determined in the vicinity of 3 kb for both PS[12,22] and polysulfone[25]. However no transition was apparent for PMMA at the higher pressure examined, 7.7 kb, although the fracture stress and strain were both increased substantially. The shape of the stress-strain curve suggests that the polymer is close to the point of the onset of plastic flow. For the PS, the direct viewing of the specimen during straining at pressure demonstrated that increase in pressure within the brittle range resulted first in suppression of crazing* and the associated onset of a more brittle (smooth planar fracture surface) mode of fracture even though the stress at fracture increased.[12] Direct measurements of crack-propagation velocity for an artificially precracked specimen showed that the initial slow-crack-growth stage preceding high-speed growth at atmospheric pressure was suppressed with increase in pressure. Instead, the crack remained stationary until a critical applied stress level and then grew at high speed. With further increase in pressure into the transition region, the crack was found to grow only slowly before stopping completely, and failure occurred elsewhere in the specimen by a ductile tearing process after yielding and plastic flow. This cessation of crack growth appears analogous to the phenomenon observed in some metals of crack blunting and stress relaxation by plastic deformation ahead of the tip. Support for this mechanism in polymers is shown by Vincent's[41] observation of elongated plastic zones in glassy polymers at atmospheric pressure, which are situated at the tips of artificial cracks and which grow as the crack length increases. This demonstration of the suppression of crazing at pressure well prior to brittle-ductile transition invalidates the suggestion of Sternstein[42,43] that the transition is due simply to a suppression of cavitation (crazing) and consequently of "normal stress yielding" relative to shear yielding. Rather, it should be considered in terms of an increase in the applied stress required to initiate and propagate "Griffith"-type cracks (to which the crazes can be approximated[12]) to a level in excess of that required to initiate shear yielding. Currently, theoretical analyses are available for the pressure dependence of the analogous cleavage-fracture process that occurs in a number of metals[44] and have been found to predict the measured behavior closely for the normally brittle metal beryllium[45,46]. However, the models on which such analyses are based are specific to the structure of metals and cannot be applied directly to polymers. The development of suitable analyses to improve understanding of brittle-fracture processes for the latter appears to require a more extensive knowledge of crack-energy and -propagation characteristics and their pressure dependence for the various classes of polymers.

For normally ductile polymers, and for brittle polymers above their transition pressure, the pressure dependence of the *true* fracture stress is slightly higher than that for the yield stress, in marked contrast to metals where the only strong pressure effect is on the fracture stress. This change in fracture behavior under pressure cannot be interpreted at the present

*An interesting application of this ability to observe deformation at pressure is the demonstration that in glassy polymers toughened by the addition of dispersed rubber particles, the mechanism of toughening is the formation of microcrazes ("stress whitening") around the particles under applied stress. Pressure suppresses such stress-whitening and the polymer becomes brittle.[12]

underdeveloped stage of understanding of the mechanisms of "ductile" fracture in polymers, except insofar as indicating that dilatational processes are involved in fracture as well as yield mechanisms.

One distinctive feature of the fracture characteristics under pressure is that while the fracture stress always increases with pressure, the fracture strain increases or decreases from that observed under ambient conditions, depending on the specific polymer involved. An increase in fracture strain is the type of behavior that might be expected and it does appear to be the more general effect for the range of polymers that has been studied. A decrease in fracture strain with pressure will occur when the pressure-transmitting fluid acts as a stress-crazing agent for a given polymer – as noted, for example, by Sauer et al.[25] for polysulfone in heptane. Indeed, Vroom and Westover[15], on the basis of comparative tensile measurements at relatively low pressures using mercury as a "neutral" pressure fluid, have observed decreases in ductility and fracture stress and contended that increases in ductility by other investigators are associated with the fluids used acting as plasticizers. However, experiments on pressurization of several polymers, using protective specimen sheathing[21,22] have failed to show any such effects. It has been suggested[22] that the conflicting results of Vroom and Westover may originate from damage to their thin specimens during pressurizing of their high-density pressure medium.

Decreases in fracture strain with pressure under test conditions that preclude fluid effects have been examined in detail for PBAC, PCTFE, PTFE, and PET[21], and have been shown to occur in HDPE and Nylon 6.6.[22] This type of behavior is qualitatively in keeping with that observed with decreasing temperature. This correlation points to the possibility that the pressure effects may arise from the pressure-induced increase to ambient or lower temperature transitions. Boyer[47] has suggested that temperature-dependent changes in deformation behavior involving decreasing strain to fracture often appear to be associated with mechanical relaxations. Polycarbonate and PET do exhibit changes to the brittle behavior associated with such relaxations (at -110 C and -60 C, respectively). There is also evidence of similarly associated changes for PCTFE. In all these cases, the relaxation processes appear to involve changes in polymer-chain-segment mobility. The established general influence of pressure in shifting the temperature of transitions encourages the interpretation of the pressure-induced reductions in fracture strain in a like manner.

5 CONCLUSIONS

The principal characteristics of the effects of hydrostatic pressure on the successive regions of the stress-strain behavior – preyield, plastic yielding, and fracture – in the wide range of polymers from the major classes are shown to be the following:

1. Young's modulus increases monotonically with pressure for all polymers yet examined. The pressure dependence is several times larger for semicrystalline polymers than for amorphous glassy polymers. In the former class of polymers, the continuous increase with pressure of the temperature of subambient relaxations in the disordered component in the structure is associated with the observed changes in modulus.

2. Pressure-induced transitions, such as the solid-solid transition in PTFE and the glass transition in elastomers, give rise to discontinuity in the pressure dependence of the modulus.

3. The yield stress increases continuously with increase in pressure for ductile polymers, and also for normally brittle polymers above a critical pressure. In the absence of phase changes, the measured pressure dependence conforms to a modified von Mises yield criterion. Values of the constant in the pressure term of this linear relationship are preponderantly in the range from 0.05 to 0.1 and, unlike the modulus, there is no clear differentiation between the values for semicrystalline and amorphous glassy polymers.

4. The nonequality of tensile and compressive yield stresses characteristic of polymers arises from the normal stress dependence of the yield process.

5. The concept of a volume-change equivalence for temperature decrease or pressure increase as the basis of the associated change in yield stress has been shown to be in error. However, there does appear to be a direct correlation between the temperature and pressure dependence of yield stress, in a manner analogous to the time-temperature superposition concept.

6. A clear understanding of the molecular mechanisms involved in the yield process has not yet emerged from the various applications of the rate-theory approach to the analysis of the pressure-dependent behavior.

7. The brittle-fracture process associated with craze formation in amorphous glassy polymers is suppressed by pressure and, beyond a critical range, ductile behavior (shear yielding) is exhibited. In the case of polystyrene, this brittle-ductile transition is not only preceded by the suppression of crazing but is also associated with changes in the propagation characteristics of cracks.

8. The pressure-dependence of the fracture stress for ductile polymers and for brittle polymers above the transition is generally similar to that for the yield stress.

9. Depending on the polymer, increases or decreases in the strain to fracture occur with increase in pressure. For a number of polymers, such decreases appear to be associated with pressure-induced increases in the

temperature of relaxation processes.

It is concluded that the development of techniques for quantitative investigation of the pressure dependence of the mechanical behavior of polymers has now reached the stage of providing an important additional tool for assisting in the elucidation of the mechanisms involved in such behavior.

REFERENCES

1. Bridgman, P. W., *J. Appl. Phys.*, **24**, 560 (1953).
2. Ainbinder, S. B., Laka, M. G., and Maiors, I. Y., *Dokl. Akad. Nauk SSR,* **159,** 1244 (1964).
3. Paterson, M. S., *J. Appl. Phys.*, **35**, 176 (1964).
4. Holliday, L., Mann, J., Pogany, G. A., Pugh, H.L1.D., and Gun, D. A., *Nature,* **202**, 381 (1964).
5. Ainbinder, S. B., Laka, M. G., and Maiors, I. Y., *Mekh. Polim.,* **1** (1), 65 (1965); [*Poly. Mech.,* **1** (1), 50 (1965)].
6. Holliday, L., *J. Chem. Ind.*, p 970 (June 1967).
7. Laka, M. G., and Dzenis, A. A., *Mekh. Polim.,* **6**, 1043 (1967).
8. Sardar, D., Radcliffe, S. V., and Baer, E., *Polymer Eng. Sci.,* **8**, 290 (1968).
9. Pae, L. D., Mears, D. R., and Sauer, J. A., *J. Polymer Sci.,* Part B, **6**, 773 (1968).
10. Pae, K. D., and Mears, D. R., *J. Polymer Sci.,* Part B, **6**, 269 (1968).
11. Ainbinder, S. B., *Mekh. Polim.,* **6**, 986 (1968).
12. Biglione, G., Baer, E., and Radcliffe, S. V., *Fracture,* 1969, (Proc. 2nd Inter. Conf., Bristol, April 1969), P. L. Pratt, et al. (Eds.), Chapman and Hall, London (1969), p 520.
13. Mears, D. R., Pae, K. D., and Sauer, J. A., *J. Appl. Phys.,* **40**, 4229 (1969).
14. Mears, D. R., and Pae, K. D., *J. Polymer Sci.,* Part B, **7**, 349 (1969).
15. Vroom, W. I., and Westover, R. F., *Soc. Plastics Engrs. J.,* **25**, 58 (1969).
16. Weaver, C. W., and Paterson, M. S., *J. Polymer Sci.,* Part A-2, **7**, 587 (1969).
17. Ainbinder, S. B., *Mekh. Polim.,* **3**, 449 (1969).
18. Uy, J. C., McCann, D. R., and Hettwer, P. F., *Ocean Eng.,* **1**, 573 (1969).
19. Sauer, J. A., Mears, D. R., and Pae, K. D., *European Polymer J.,* **6**, 1015 (1970).
20. Rabinowitz, D., Ward, I. M., and Parry, J.S.C., *J. Mater. Sci.,* **5**, 29 (1970).
21. Christiansen, A. W., Baer, E., and Radcliffe, S. V., *Phil. Mag.,* **24**, 188 451 (1971).
22. Pugh, H.D.L1., Chandler, E. F., Holliday, L., and Mann, J., *Polymer Eng. Sci.,* **11,** 463 (1971).
23. Davis, L. A., and Pampillo, C. A., *J. Appl. Phys.,* **42**, 12, 4659 (1971).
24. Pae, K. D., and Sauer, J. A., *Mech. Eng.* (in press).
25. Sauer, J. A., Bhateja, S. K., and Pae, K. D., *Proc. Third Inter-American Conf. on Materials Technology,* Rio de Janeiro, Brazil, 1972 (in press).
26. Duckett, R. A., Rabinowitz, S., and Ward, I. M., *J. Mater. Sci.,* **5**, 909 (1970).
27. Weir, C. E., *J. Res. Nat. Bur. Std.,* **46**, 207 (1951).
28. Robertson, J. A., and Kipsitt, H. A., *J. Appl. Phys.,* **36**, 2843 (1965).
29. Whitney, W., and Andrews, R. D., *J. Polymer Sci.,* **16**, 2981 (1967).
30. Sternstein, S. S., and Ongchin, L., *Amer. Chem. Soc. (Polymer Preprints),* **10,** 117 (1969).
31. Bauwens, J. C., *J. Polymer Sci.,* Part A-2, **5**, 1145 (1967); Part A-2, **8**, 893 (1970).
32. Bowden, P. B., and Jukes, J. A., *J. Mater. Sci.,* **3**, 183 (1968); **7**, 52 (1972).
33. Sternstein, S. S., Ongchin, L., and Silverman, A., *Applied Polymer Symposia,* **7**, 175 (1968).
34. Radcliffe, S. V., "Pressure-Induced Effects on Defect Structure and Properties", Chapter 12 in *Mechanical Behavior of Materials Under Pressure,* H.L1.D. Pugh (Ed.), Elsevier Publishing Company, Ltd., New York (1970).
35. Rauch, G. C., and Leslie, W. C., *Metallurgical Trans.,* **3**, 373 (1972).
36. Daga, R., and Radcliffe, S. V., to be published.
37. Robertson, R. E., *J. Appl. Polymer Sci.,* **7**, 443 (1963); *J. Chem. Phys.,* **44,** 3590 (1966).
38. Robertson, R. E., *Applied Polymer Symposia,* **7**, 201 (1968).
39. Argon, A. S., Andrews, R. D., Godrick, J. A., and Whitney, W., *J. Appl. Phys.,* **39**, 1899 (1968).

40. Wu, W., "Plastic Deformation of Polymers", Doctoral Thesis, Massachusetts Institute of Technology, May, 1972.
41. Vincent, P. I., *Polymer,* **12**, 534 (1971).
42. Ongchin, L., and Sternstein, S. S., *Bull. Amer. Phys. Soc.*, **14**, CK8-CK9 (1969).
43. Sternstein, S. S., and Ho, T. C., *J. Appl. Phys.* (1972) (in press).
44. Francois, D., and Wilshaw, T. R., *J. Appl. Phys.*, **39**, 4170 (1968).
45. Aladag, E., Pugh, H.Ll.D., and Radcliffe, S. V., *Acta Met.*, **17**, 1467 (1969).
46. Bedere, D., Jamard, C., Jarlaud, A., and Francois, D., *Acta. Met.*, **19**, 973 (1971).
47. Boyer, R. F., *Polymer Eng. Sci.*, **8**, 101 (1968).
48. Bauwens-Crowet, C., Bauwens, J-C., and Hómes, G., *J. Mater. Sci.*, **7**, 176 (1972).

DISCUSSION on Paper by S. V. Radcliffe

TSCHOEGL: We have carried out constant extension-rate measurements under superposed hydrostatic pressure on a glass-bead-filled rubber. This is a "dewetting" system in which voids form around the glass beads upon deformation. The pressures involved were quite low, the effect of pressure being quite noticeable at pressures of the order of 50 to 150 bars. The broad features are very similar to those you have detailed. With an increase in pressure, the initial modulus increases, the elongation at break decreases, and the stress at break increases. Simultaneous measurements of the volume changes accompanying the deformation show that the superposed hydrostatic pressure prevents void formation just as it prevents the formation of crazes in your materials.

RADCLIFFE: The results that you cite are in keeping, as you suggest, with our results on polystyrene. In addition, our other work on this material toughened by rubber-particle additions, high impact polystyrene,* showed analogous behavior to that observed for the glass particles in rubber. Thus, increase in pressure initially embrittled the material by suppressing the craze formation (which would result in volume dilatation) that normally occurs in the polystyrene adjacent to the rubber particles.

VINCENT: You have suggested that the pressure-dependence of the modulus is less for noncrystalline than for crystalline polymers. The data could also be interpreted as showing that the pressure-dependence is less when the material is tested below rather than above the glass transition. These two possibilities may be distinguished by studying crystalline samples of polymers, such as poly(ethylene terephthalate) or poly(tetramethylene terephthalate), whose glass transitions are above room temperature.

---

*Biglione, G., Baer, E., and Radcliffe, S. V., *Fracture, 1969*, P. L. Pratt, et al. (Eds.), Chapman and Hall, London, 1969, p 520.

RADCLIFFE: I should have been more specific in using the phrase amorphous polymers, since I had intended to indicate that such polymers in the glassy state are less compressible as a class than the semicrystalline polymers. The experiment that you suggest should indeed be effective in checking this view.

HULL: In connection with your observation of brittle fracture under pressure in the absence of crazing, I wonder whether or not you have any evidence for the effect of pressure *firstly* on the nucleation of crazing and *secondly* on the propagation of crazes. It seems to me that the pressure dependence may be different for the two processes and provide an explanation for the apparent occurrence of crazeless fracture.

RADCLIFFE: We have taken the view* that fracture in polystyrene in the pressure-regime between the suppression of visible crazing and the brittle-ductile transition is due to the catastrophic growth of an inherent flaw, although the possibility of one developed during the increase in applied stress (i.e. a craze) cannot be excluded on the basis of present evidence. More detailed experimental analysis over this regime is needed to clarify this point.

SHEN: Do the polymers that have been exposed to high hydrostatic pressure still exhibit increased yield stress at atmospheric pressure (particularly for highly crystalline polymers or polymers with high T_g's that have long volume relaxation times)?

RADCLIFFE: We have conducted such pressurization experiments for a number of polymers and find that, providing the pressure remains hydrostatic and there is no reaction between the fluid and the polymers or phase change in the polymers, the relaxation of the polymer back to its initial dimensions after hydrostatic compression is complete and it exhibits no change in its mechanical behavior at atmospheric pressure.

*Biglione, et al., *op. cit.*

ATTEMPT TO EXPLAIN CRAZING IN AMORPHOUS THERMOPLASTICS AND ADHESION FRACTURES IN SEMICRYSTALLINE THERMOPLASTICS AND FILLED POLYMERS

G. Menges

Institut für Kunststoffverarbeitung
Aachen, West Germany

ABSTRACT

In stressed amorphous thermoplastics, cracklike phenomena are observed to develop perpendicularly to the highest normal strain. In semicrystalline thermoplastics, adhesive cracks occur mainly at spherulite boundaries which are positioned perpendicularly to the highest normal strain. Corresponding results were found with multiphase materials, e.g., filled elastomers, thermosets, and thermoplastics.

From the phenomenological similarity of all these observations, the conclusion is drawn that crazes in amorphous thermoplastics are adhesion fractures at particle boundaries. Such particle boundaries are, for instance, the boundaries of raw-material grains which are not destroyed during the plasticizing and processing of a material. Consequently, similar conditions exist as with multiphase materials; the properties of the particles, however, differ less. As "craze material", "soft" particles must be regarded as those which are stretched during the propagation of the craze.

The surprising existence of constant threshold values of normal strain for the formation of adhesion cracks in all the plastic materials mentioned is explained by the power-law dependence of secondary valence forces (adhesion forces) on distance. After a certain deformation, adhesion cracks develop at numerous places on account of the statistical distribution of critical phase boundaries. This leads to the formation of microcracks. These cracks grow until they reach particles which stop them. These considerations suggest that the strain limit for debonding at critical phase boundaries be estimated by means of an equation which is based on Griffith's theory and solved with respect to strain. Also, an attempt is made to explain the dependence of strain at fracture on strain rate.

1 PRESENT KNOWLEDGE

With unfilled, transparent, amorphous plastomers such as polymethyl methacrylate, polyvinyl chloride, polystyrene, and polycarbonate, it can be observed that at low mechanical-deformation rates, cracklike formations develop.

For uniaxial tensile stress, these crazes are oriented perpendicularly to the direction of deformation. As shown in Fig. 1, they adopt an almost regular pattern; their distance from each other as well as their length depends on the rate of deformation. Kambour[1] observed that these cracks are bridged by batches of material.

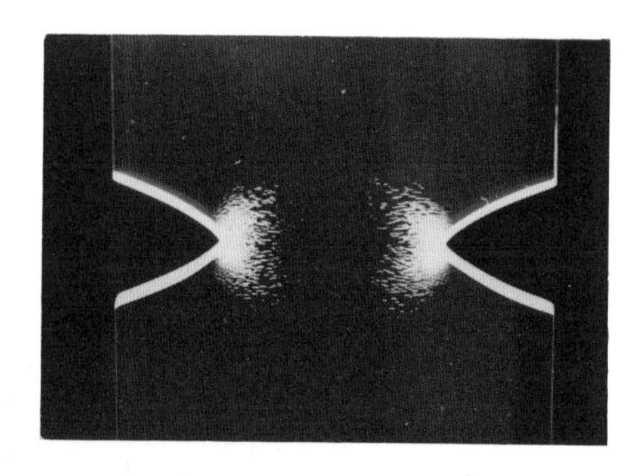

Fig. 1. Crazes in notched polymethyl methacrylate (PMMA) bars.

Morbitzer and Holm[2] who observed the creep behavior of various plastomers by means of scanning electron microscopy, discovered in recent investigations that

1. Crazes always develop before fracture
2. The propagation of cracks always follows along existing crazes.

Therefore, we seem to be justified in regarding the formation of crazes as a first irreversible deformation and a phase of failure. Consequently, we gave particular attention to this phenomenon and studied it extensively with greatly differing plastic materials.[3-13] In our opinion, it is more important for technical applications to know details of the formation of crazes than the details of the final fracture.[4]

We observed that in the unfilled amorphous plastomers, which were tested by us, crazes of the same type develop. Their number, rate of growth, and the strain under which they can be seen for the first time increase with increasing deformation rate, while their distance from each other decreases. Even after long times, crazes will not develop during creep-relaxation tests when strain is kept below a definite limit.[3-5]

Through further investigations we learned that in semicrystalline plastomers and in filled resins, there will be no crazing, but instead cracks will form.[6-8] These cracks have, however, the following in common with crazes: they originate only after certain threshold values of strain are exceeded and they are always oriented perpendicularly to the direction of highest strain. For all materials tested, unfilled and filled, amorphous and semicrystalline plastomers as well as filled cured resins, the following result was obtained: the strain limits below which no crazes or cracks are observed are independent of the state of stress, whether uniaxial or multiaxial, homogeneous, or inhomogeneous.[4] The decisive influence for the occurrence of these phenomena is exclusively the state of deformation of the material in the critical area.

Finally, we also found that at lower deformation rates and under the influence of different surrounding media, similar phenomena always occurred prior to the final failure. Failure could be of brittle character in one case and of a tenacious one combined with high deformation in another.[9] The strain limits were not influenced so long as no medium was absorbed by test specimens and no solution – detectable by loss of weight – had started (Fig. 2). With test specimens showing no absorption of liquid medium in the unstressed state it was observed that the crazes were filled by the surrounding agent. This was followed by the development of new crazes deeper within the material, which were also filled later by the agent.[9]

2 HYPOTHESIS FOR THE DEVELOPMENT OF CRACKS AND CRAZES

From the test results at hand, the conclusion is drawn that cracks in filled resins and plastomers as well as crazes in amorphous plastomers are, in principle, the same phenomenon. In filled plastic materials and semicrystalline plastomers, cracks are definitely adhesion fractures at phase boundaries the orientation of which is disadvantageous, i.e., perpendicular to the direction of strain. This is proved by numerous observations on

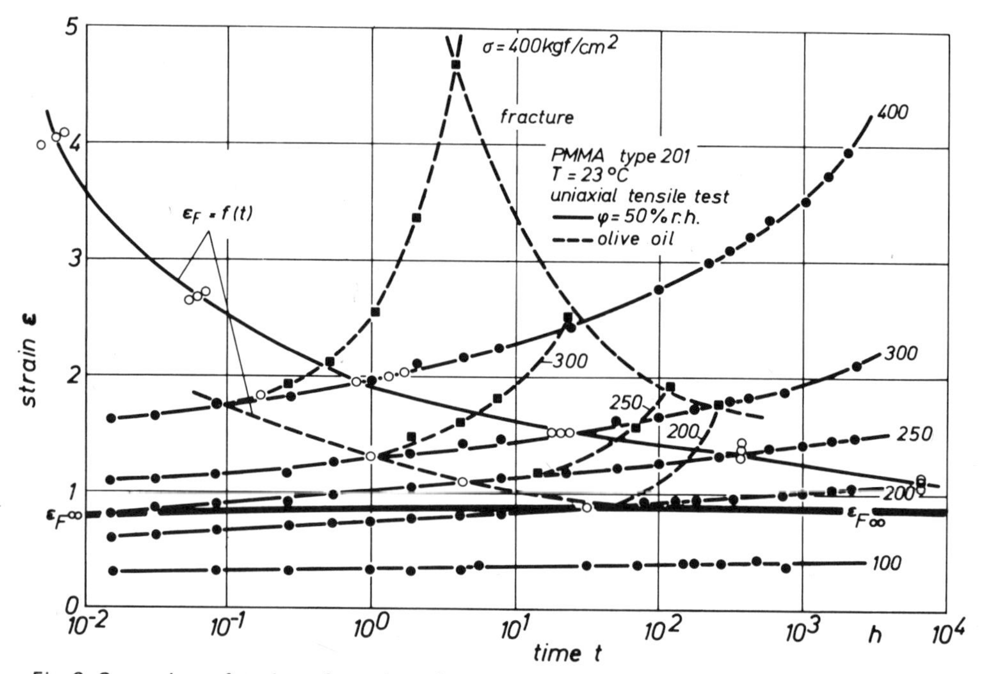

Fig. 2. Comparison of strain at formation of crazes and elongation at fracture during tensile creep tests with PMMA in air and olive oil[10]; ϵ_F = point of appearance of crazes; $\epsilon_{F\infty}$ = strain limit for production of voids.

various multiaxially loaded samples. Fig. 3 may serve as an example. Fig. 3 is a photomicrograph of a polypropylene test bar which was strained in creep until a first dimming showed up, and which was, in the stretched state, briefly dipped in a crack detecting agent. The adhesion cracks are stopped at phase boundaries, which are situated more or less perpendicularly to the direction of their propagation. This can be seen clearly in Fig. 3.

Fig. 3. Photomicrograph of adhesive fracture between spherulite boundaries in polypropylene test bar. Direction of fracture is perpendicular to direction of strain. The fracture was marked in the strained state by infusion of a crack-checking agent.[8]

It is well known that substructures or "domain structures" exist in amorphous plastomers.[14] Often, these substructures are very coarse and can be seen with the naked eye, for example, in the technological, so-called "solvent test". This test is used with rigid PVC to prove the quality of the extrusion process. Under the influence of the solvent, the material swells. The particles from which the material is welded together by the extrusion process relax to their original shape and the particle joints break. After the solvent is evaporated, an accumulation of rather disconnected particles is left.[11]

Hattori, Tanaka, and Matsuo[16] found that PVC powder grains (suspension and bulk polymerized PVC) of about 100-μm diameter consist of quasi-round particles of about 0.5 to 1.5-μm diameter (Fig. 4). In these microparticles, a further substructure was detected which these authors term lattice structure or fibrils. This structure has characteristic dimensions of about 300 A. At melt temperatures of more than 190 C, the boundaries of these microparticles vanished, while the substructure of 300 A remained.

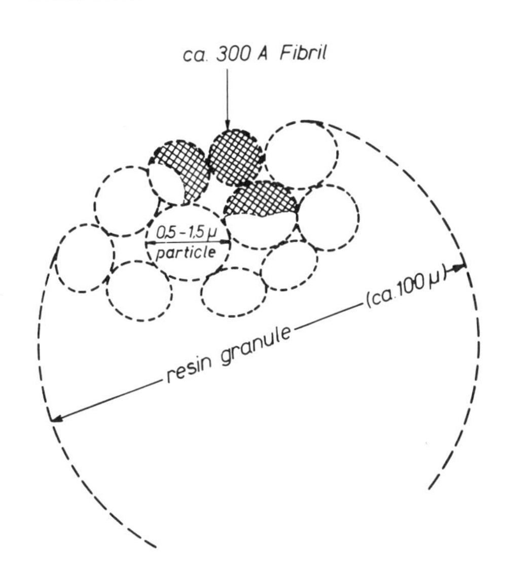

Fig. 4. Schematic graph of multiphase structure of E-PVC granules.[16]

These observations also prove the existence of microscopic or sub-microscopic phase boundaries, the strength of which is without doubt less than that of homogeneous material. From this it is concluded that crazes are initiated by the debonding of phase boundaries, i.e., by adhesion fractures. Compared with filled plastics or semicrystalline plastomers, the boundaries are much smaller owing to the size of the molding-compound particles or other "domain structures". Also, the differences in strengths of the particles are smaller, as expected.

Because of the low level of strength of the particles, an already existing adhesive crack can stretch a particle which tries to stop its growth. Such stretched particles form the "craze material" observed by Kambour and others. A schematic drawing may serve to explain this conception (Fig. 5).

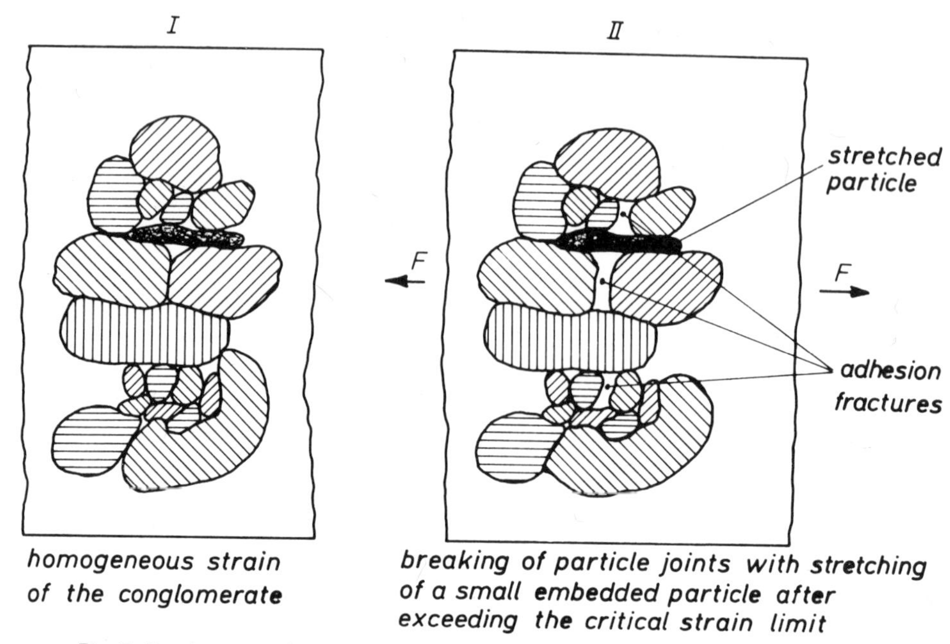

Fig. 5. Development of crazes in unfilled amorphous plastomers consisting of particles "welded together".

3 INFLUENCE OF PHASE BOUNDARIES ON THE ADHESION FRACTURE (CRAZES AND CRACKS)

Unfilled cross-linked polymers, elastomers, as well as thermosets, show almost no creeping prior to fracture. Deformation is elastic and reversible. After a certain load time, brittle failure occurs almost spontaneously. If, however, a dispersion of resin with 10 percent or more filler is stressed in the same way, deformation rate is low and almost constant in the beginning but later suddenly increase. This sudden creep is irreversible, and, if the resins are transparent, the test specimens can be observed to have changed color, i.e., they have become "milky" and nontransparent. For this the term "stress whitening" is often used. The matrix has separated from filler particles. Schwarzl[15] named this phenomenon "adhesion fracture", during his investigations of noncrystalline elastomers containing different percentages of filler material. We ourselves found identical behavior with fiber-reinforced thermosets.[13]

It is remarkable that the investigations reported by Schwarzl reveal an identical strain limit for the formation of adhesion fractures (Fig. 6) in elastomers containing different percentages of filler material. In the course of these investigations, salt crystals of greatly differing sizes were dispersed in the elastomer matrix at always the same percentage by volume. The tensile-strain curves in Fig. 6 show that the smaller the particle size of the dispersed phase, the higher is the modulus of elasticity. While adhesion fractures occur with all dispersions at about the same strain, the corresponding stress at adhesion fracture increases with the modulus of elasticity of the composite body.

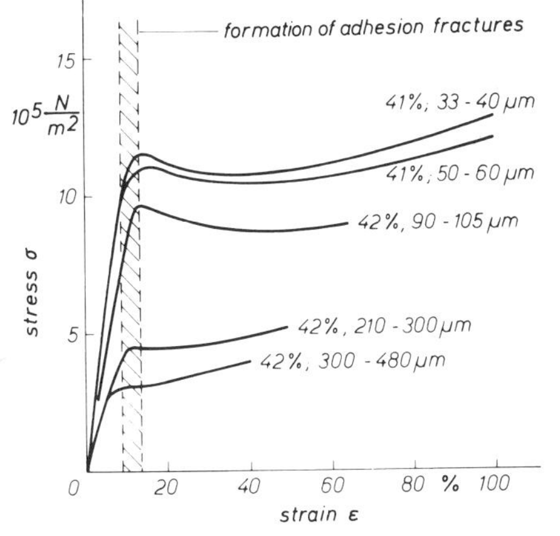

Fig. 6. Stress-strain diagrams for a strain rate of 300 percent/min at 20 C for dispersions of about 41 percent common-salt fractions of different particle size in polyurethane rubber.[15]

The higher modulus of elasticity can be explained by the number of particles of the dispersed phase. In all cases, the concentration of filler material was identical: 41.5 ± 0.5 percent by volume. The number of particles, however, increased with decreasing particle size. This is the reason for the stiffening of material, as evidenced by the increasing modulus of elasticity. Even the elongation at rupture increased with increasing number of particles by a factor of 2.5. This surprising increase of energy of fracture with increasing number of particles is certainly due to the size of the matrix bridges between the filler particles, which have become smaller by a factor of about 100 for the smaller grains. This means that the smaller the size of the resin bridges, the less influence the failure of single-matrix bridges, which occurs more and more once the adhesion fracture strain is exceeded, has on the final failure. The frequent, qualitative observation that parts with a finely crystalline structure absorb higher fracture energies than do those with coarsely crystalline structures corresponds with this.

The most important conclusion, which is drawn from Schwarzl's work[15], is that the critical strain for adhesion fractures is independent of the number of filler particles. This is confirmed by our own observations. With glass-fiber-mat-reinforced polyester test specimens, no correlation was obtained between glass content and debonding strain.

These results also explain the reinforcing effect of finely dispersed carbon or silica particles in elastomers and plastomers. These filler materials are not only the most finely dispersed ones in use, but they also have the most highly structured and, consequently, largest surface. It can be assumed that the reinforcing effect, i.e., the relation of strength between reinforced and unreinforced material, is influenced not only by the number of filler particles, but also by their surfaces, since the fracture propagation is impeded by transversely positioned phase boundaries. It is obvious that, apart from the mere adhesive strength between filler and matrix, the structure of the filler surface is important. A highly structured surface will have a much higher resistance against debonding of matrix and filler, because there are always disadvantageously oriented new phase boundaries which hinder fracture propagation.

The influence of impact stress and low temperatures on the behavior of the glass-fiber-mat-filled polyesters[13] is now considered. With a composite system, glass-fiber-mat-reinforced polyester resin, no alteration of the strain where adhesion fracture begins could be observed for the whole range of practical deformation rates now in existence, viz., from 10^{-2} to 10^{3} percent/min. This result was to be expected. The elongation at rupture, however, decreases with increasing deformation rates. The same tendency exists, of course, for decreasing test temperatures (Fig. 7). In both cases the matrix becomes more brittle. With a completely brittle matrix, i.e., at low temperatures and high deformation rates, cracks in the matrix bridges will occur simultaneously with the adhesion fractures. We noticed that these cracks are longer. This means that an adhesion fracture spreads spontaneously to both sides and thus penetrates phase boundaries without being stopped.

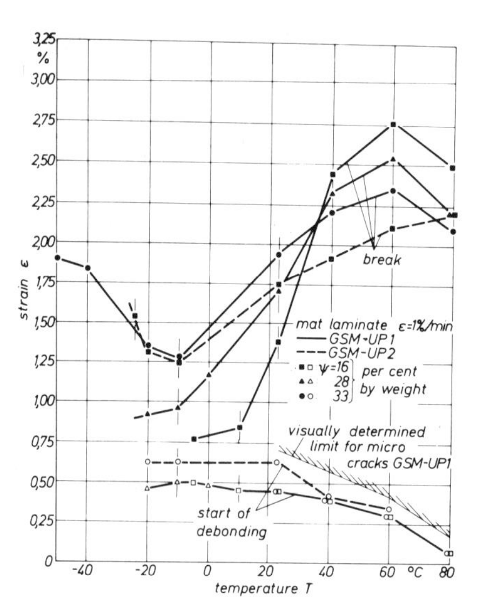

Fig. 7. Influence of temperature on the critical strain limit at which adhesion fractures are observed and on elongation at rupture of glass-fiber-mat-reinforced polyester resins.[13]

Contrary conditions exist for a hard matrix and a soft filler material, of which the most extreme example is a foam. An important example with a less extreme ratio of moduli is ABS. ABS systems have been examined for crazing and impact strength by Pohl, Gruber, and Taeger.[17]

They observed the best behavior when the elastomer particles had a diameter of 5 μm. It was previously known that numerous small cracks develop around the elastomer particles. Consequently, the elastomer particles serve to initiate crazes and thus to absorb energy. This observation confirms our conception. On the one hand, the soft particles act as predetermined breaking points, while on the other, as in the case of a good adhesion, they will carry a bigger part of load the more they are stretched. Furthermore, they will try to close the cracks again after the stress is released. Indeed, elastomer particles which are particularly well jointed to the matrix by grafting are the most effective ones.[2,17]

In our own investigations concerning static and fatigue strength and development of crazes in ABS during long-term tests, this mechanism was confirmed. The cracks or crazes are also oriented perpendicularly to the direction of highest normal strain.

From the behavior of resins reinforced with coarse fillers, we learn that the strain at which the first adhesive fracture begins is independent of temperature and strain rate when the state of the matrix is not changed. Therefore, we can conclude that in materials with finer phase boundaries, namely, unfilled amorphous and semicrystalline plastomers, we should find the same behavior. Since the particles are very fine, the first adhesion fracture at phase boundaries is not visible. There seems to exist only a dependence on deformation rate. We state that all materials with domain structures or phase boundaries from filler particles have a constant strain limit at which adhesion fracture on phase boundaries develop. That is shown schematically in Fig. 8.

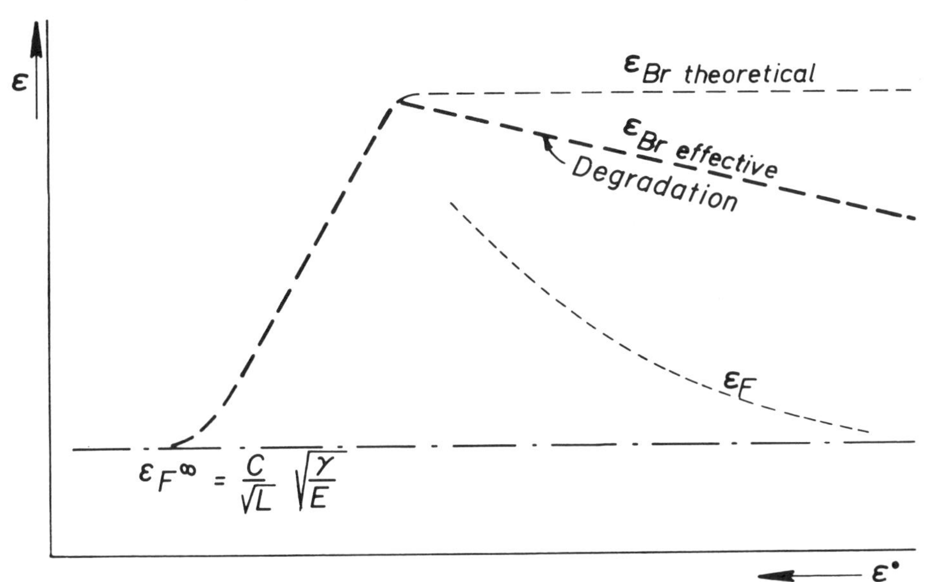

Fig. 8. Schematic plot of strains as a function of deformation rate for various plastic materials. $\epsilon_{F\infty} = \epsilon_{crit}$ = strain threshold for microcracking; ϵ_F = visibility limit of crazes.

4 EXPLANATION OF THE PHENOMENON OF CONSTANT STRAIN LIMITS FOR DEBONDING (DEVELOPMENT OF ADHESIVE FRACTURES)

As we have seen (Figs. 7 and 8), crazes and adhesive fractures occur in all polymers when certain strain limits are exceeded. On this basis, we established a new kind of strength analysis, as these strain limits proved to be independent of the state of stress (uni- or multiaxial, homogeneous, or inhomogeneous), of the kind of load (static, cyclic, dynamic, or intermittent), and of environmental conditions (temperature and surrounding medium) so long as the state of the material itself does not change. It is undeniable that bodies are held together by forces of attraction between their atoms and molecules. In a body consisting of particles welded together, various kinds of adhesive forces will be in effect which decrease at least with the second power of the distance of the boundaries. So long as the body is deformed homogeneously, only small alterations of distance will occur. But if the body is not completely homogeneously deformed, which happens to be the case with all real bodies, the first weak point will show a higher deformation and break open. The resultant strong reduction of adhesive force will certainly initiate a chain reaction: the developing debonding proceeds to the nearest phase boundary where it will be stopped if the adjacent particle possesses a higher strength in the direction of deformation. If this strength is sufficient, this particle will bear the additional load which has become effective because of the adhesion fracture, but it will be strained correspondingly, In consequence of this local deformation, additional weaker areas around the particle will break open and these breaks will occur chiefly in the plane of the first adhesion fracture. Thus, the first adhesion fracture will initiate a sequence of further ones in equally stressed cross sections.

An explanation can be deduced from a hypothesis presented by Szabo.[18] Szabo starts from the basis that real materials consist of particles which attract each other and are pressed together by the neighboring ones. A tensile force, applied to the body externally, will thus reduce the compression of the particles. When the tensile force is sufficient to neutralize the adhesive forces, the particles will separate from each other. Consequently, the deformation at separation should have a constant value. If the elastic properties of all the particles of a body are in every direction identical, separation will occur simultaneously in all particles at the same strain; the body breaks.

Bodies consisting – as most real materials – of particles having different properties in different directions, however, will show a partial debonding of the least coherent particles. This debonding will take place at that strain at which the surface of the particles in question experiences equilibrium of retractive and attractive forces.

Furthermore, from this theory it can be concluded that the surface of separation in isotropic materials will always be oriented perpendicularly to the direction of the tensile force; with multiaxial stresses, the debonding should occur perpendicularly to the highest normal strain. Under shear stress, for example the debonding should be oriented perpendicularly to the tensile component. This we could confirm.[5]

Finally, it remains to be made clear why crazes do not reach their final length at once, as other adhesion fractures do (Fig. 9). Crazes grow slowly to a certain length which obviously has the same average value for one material and for different deformation rates, i.e., different creep stresses.[12] As the particles and the joints between them are staggered in space, a first adhesive fracture developing at such a joint will tear other particles lying in its way out of the material and stretch or tear them apart. Thus, the growth of the adhesion fracture to an average length will take time.

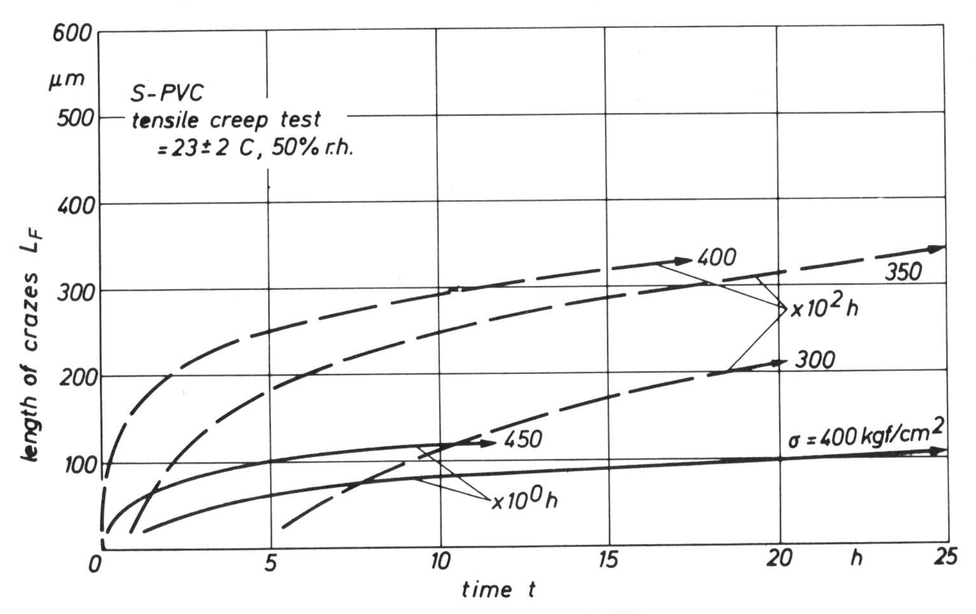

Fig. 9. Growth of crazes in S-PVC.

5 ESTIMATION OF THE STRAIN CAUSING THE FIRST DEBONDING OF PHASE BOUNDARIES

Griffith's theory may be modified for an estimation of the formation of first adhesion fractures at phase boundaries. According to Griffith, the following equation is valid

$$\sigma_{crit} = C \left(\frac{\gamma \cdot E}{L} \right)^{1/2} .$$

Because of the dependence of stress on deformation rate, it is advantageous to change over to strain using Hooke's law to substitute strain for stress:

$$\epsilon_{crit} = \sigma_{crit}/E = C\left(\frac{\gamma}{L \cdot E}\right)^{1/2} ,$$

where

C is a constant

$\epsilon_{crit} = \epsilon_{F}\infty$ is the strain at which first debonding phenomena of particle boundaries is observed at strain rates below the relaxation rates of the particles after a load release

γ is interfacial tension under the existing conditions

L is average length of particles which must be regarded as critical ones

E is modulus of "elasticity" governing the relaxation of particles after tension release (not the modulus of bulk material).

Without giving discrete values, it can be estimated, by means of this formula, that ϵ_{crit} varies only with γ and E, as the length L is constant for a given material and equivalent to the average diameter of critical particles. As the interfacial tension is rather independent of temperature, the debonding strain ϵ_{crit} will – for one surrounding medium – change significantly only when the modulus of elasticity changes significantly, e.g., owing to changes of state.

If the material is stressed in a different medium, the interfacial energy changes. Where the medium diffuses not only along particle boundaries, but also into the particles themselves – as for example in spherulites of semicrystalline polymers – the modulus of elasticity is changed as well. Thus, it can be realized that the given formula is an adequate approach to describe in a qualitative way the experimental observations.

6 INTERPRETATION OF THE MATERIAL'S BEHAVIOR BETWEEN THE FORMATION OF CRAZES AND RUPTURE

When all particles in a material are "frozen", i.e., under high deformation rates or at low temperatures, the first adhesive fracture spreads through the specimen causing it to rupture. The material's behavior is brittle. When some particles or domain structures are softened, the adhesive fracture is stopped and particles between cracks are stretched. We studied this stretching process[5] and found that it is quite different from the one prior to the formation of the adhesive fractures. The deformable particle, which is clamped at its ends (Fig. 5) and struck by the crack is now distorted by displacements which start at the upper boundary of the crack (Fig. 10). Under further deformation, displacements of adjacent cracks (crazes) unite and later form Lüders bands (flow ribbons) at an angle of 30 to 40 degrees to the highest normal strain direction.

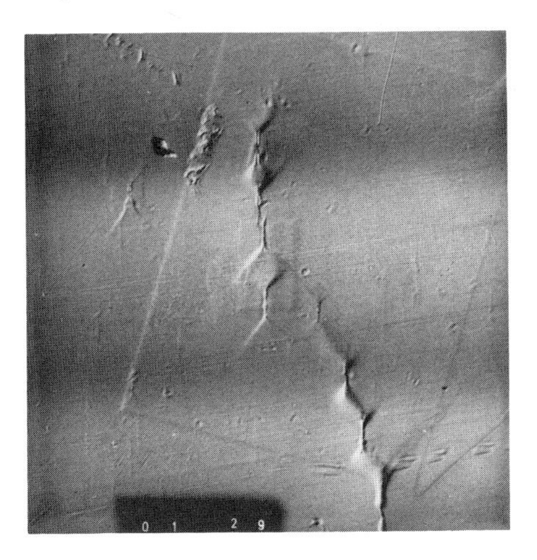

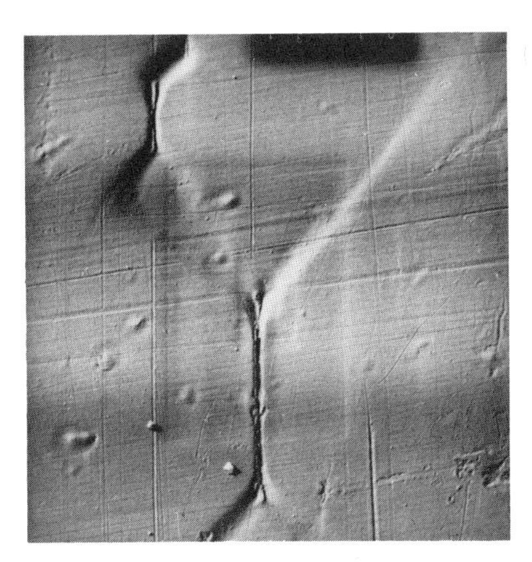

Fig. 10. Displacements starting from the tip of a craze formed under creep.[5]

The final state of deformation then is necking, which begins at two crossing flow ribbons.

Under very low deformation rates, the surrounding medium enters the crazes or cracks and diffuses from there into the adjacent material, and this may lower the interfacial tension. We observed a very high deformation of a small part of the material (Fig. 11). The attack on the stretched material accelerates and rupture occurs if loading of the sample is continued.

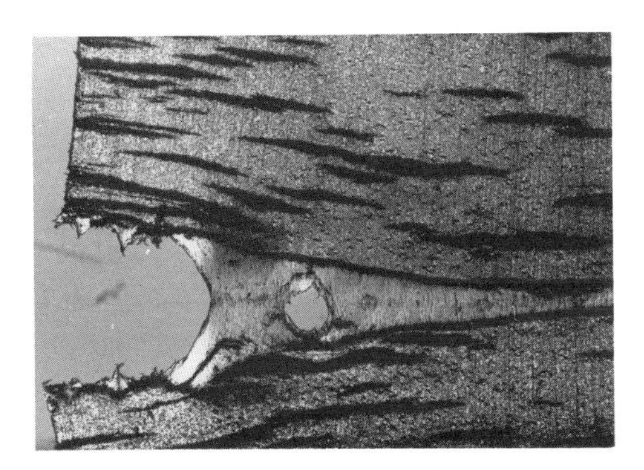

Fig. 11. Stretching of spherulites (polypropylene) in creep in water after the surrounding medium had entered the adhesion cracks and had decreased the interfacial tension.

REFERENCES

1. Kambour, R. P., *Polymer Eng. Sci.*, **8** (4), 281-289 (1968).
2. Morbitzer u. Holm, unpublished work of Farbenfabriken Bayer Leverkusen (1972).
3. Menges, G., and Schmidt, H., *Kunststoffe,* **57** (11), 885-890 (1967).
4. Menges, G., "Erleichtertes Verständnis des Werkstoffverhaltens bei verformungsbezogener Betrachtungsweise", *Fortschrittsbericht VDI-Z,* Series 5, No. 12.
5. Schmidt, H., "Untersuchungen der Fließzonenbildung und des mechanischen Langzeitverhaltens von thermoplastischen Kunststoffen bei ein- und zweiachsig wirkenden Zugspannungen", Thesis RWTH, Aachen (1971).
6. Menges, G., Dolfen, E., Papers of Haus der Technik H 159, Essen, 1967.
7. Dolfen, E., "Bemessungsgrundlagen für tragende Bauelemente aus glasfaserverstärkten Kunststoffen, insbesondere durch Glasseidenmatten bewehrte Polyesterharze", Thesis RWTH, Aachen (1969).
8. Menges, G., Alf, E., *Kunststoffe,* **62** (4), 259-267 (1972).
9. Menges, G., Riess, R., and Taprogge, R., Paper read at the Sixth Kunststofftechnisches Kolloquium of the IKV, Aachen, West Germany, 1972, Institut für Kunststoffverarbeitung; see also *Materialprüfung,* **14** (4), 141-146 (1972).
10. Menges, G., Schmidt, H., and Berg, H., *Kunststoffe,* **60** (11), 868-872 (1970).
11. Roberg, P., "Dimensionierungsgrundlagen für dynamisch beanspruchte Kunststoff-Rohrsysteme, dargestellt am Beispiel einer PVC-hart-Rohrleitung", Thesis RWTH, Aachen (1971).
12. Alf, E., Thesis RWTH, Aachen (1972).
13. Menges, G., and Roskothen, H.-J., *Kunststoff-Rundschau,* No. 9, 479-487 (1972).
14. Bonart, R., *Kolloid-Z. u. Z. Polymere,* **231** (1-2), 438-456 (1969).
15. Schwarzl, F. R., *Kunststoff-Rundschau,* **18** (1), 49-55 (1971).
16. Hattori, T., Tanaka, K., and Matsuo, M., *Polymer Eng. Sci.,* **12** (3), 119-203 (May 1972). (cf. for further literature).
17. Pohl, E., Gruber, T., and Taeger, F., *Plaste und Kautschuk,* **19** (3), 438 (1972).
18. Szabo, C., *Grundlagen einer neuen Festigkeitstheorie,* Vols 1 and 2, Bauverlag GmbH, Wiesbaden and Berlin, Germany (1971).

DISCUSSION on Paper by G. Menges

KAUSCH: The concept of introducing energetic considerations into analysis of craze formation has promised to be very fruitful. In your paper you stress the point that – in PVC – the boundaries of granules are the locus of cavitation, i.e., craze initiation. I wonder whether this condition is not too restrictive and whether craze initiation within granules is also possible as a consequence of molecular stress concentration?

MENGES: The boundaries of granules are the coarsest domain structure in unfilled PVC. It is well known that these boundaries are very weak, as it can be seen in solvent tests. Such tests are regular technological tests in PVC-pipe factories. And if it is possible to produce cracks by this simple method, this is a clear proof of the weakness of these boundaries. So we concluded that they also break under certain deformation rates. We know that in an annealed PVC of good quality, this critical strain is about 0.8 percent. I agree with you that this strain limit is shifted by frozen stresses and by orientation, as we have just found.

MAYER: Would you describe the nature of the domain structures to which you have referred? They seem to be somewhat between the size of spherulites and the size of the preextruded pellets used in forming the polymer but I have been unaware of such structures; have you evidence for them?

MENGES: In semicrystalline materials, these domain structures are spherulites and fibrils within spherulites. In amorphic unfilled materials, we also have different kinds of domain structures. The coarsest are particles of the raw material as the material comes out of suspension or emulsion reactors. Inside the particles are finer structures. These have quite recently been investigated by Hatori et al.* The finest structure they found in PVC was a lattice with distances of 300 A. We find such domain structures by solvent testing, in which a few drops of a solvent are placed on the material.

KNAPPE: As was pointed out in Menges' paper, the boundaries of spherulites are very critical in crazing. It should be kept in mind that at this site we have a concentration of defects and internal stress. This situation can lead to crazing during slow cooling or annealing not too far from melting point. Therefore, one should expect that crazing and stress cracking would be strongly dependent on thermal history. It seems doubtful that a critical strain value independent of thermal history and environment would exist, as in the case of homogeneous amorphous polymers.

MENGES: The strain dependence of crazing and cracking is effectively influenced by production parameters, especially orientation and thermal stresses. The unique dependency on strain is valid only for materials with identical history. As is well known, frozen stresses (prestrains) may be so high that crazing is produced instantly after a part is molded.

*Hattori et al., *Polymer Eng. Sci.*, **12** (3), 119 (1972).

AGENDA DISCUSSION: PHENOMENOLOGY

*N. W. Tschoegl**

Division of Chemistry and Chemical Engineering
California Institute of Technology
Pasadena, California

*J. Becht***

Battelle-Institut e. V.
Frankfurt/Main 90, Germany

The Agenda Discussion on Phenomenology can conveniently be divided into three sections. The first deals with the discussion based on the presentation by Rivlin (Session I) and those by Becker and by Landel (Session II), all of which consider the problem of the phenomenological description of the behavior of polymeric materials in large deformations at atmospheric pressure.

The second section deals with the discussion centered on the paper by Radcliffe (Session II) which describes the effect of pressure on the flow and fracture of polymers.

---

*Chairman.
**Secretary.

The third section is concerned with problems of crazing in polymers.

1 PHENOMENOLOGICAL DESCRIPTION OF POLYMER BEHAVIOR

The Chair opened the discussion by summarizing the current state of the phenomenological description of polymer behavior. In principle, the continuum mechanical theory of large, purely elastic deformations of isotropic incompressible materials as developed mainly by Rivlin, Green, and others allows one to describe large elastic deformations of rubbery materials in a satisfactory way if the strain-energy-density function for the material is known. Unfortunately, strain-energy-density functions which describe large deformations in all deformation fields (simple tension, equibiaxial tension, pure shear, etc.), with the same values of the material parameters, are valid only for moderate deformations. The well-known function

$$W = \frac{1}{2} G \left(\lambda_1{}^2 + \lambda_2{}^2 + \lambda_1{}^{-2} \lambda_2{}^{-2} - 3\right) \qquad (1)$$

in which the shear modulus, G, is the only material parameter, is an example. Equation (1), valid only for rather small deformations, can be derived from continuum mechanical as well as statistical mechanical considerations. Other strain-energy-density functions based on G as the sole parameter, but valid to much higher deformations, have been proposed by Valanis and Landel[1], and by Dickie and Smith[2]. The also well-known equation

$$W = c_1 \left(\lambda_1{}^2 + \lambda_2{}^2 + \lambda_1{}^{-2}\lambda_2{}^{-2} - 3\right) + c_2 \left(\lambda_1{}^{-2} + \lambda_2{}^{-2} + \lambda_1{}^2\lambda_2{}^2 - 3\right) \qquad (2)$$

does not generally represent behavior in different deformation fields with the same material parameters. Such a strain-energy-density function does not allow one to formulate true constitutive behavior. Thus, the quest continues for strain-energy-density functions which lead to true constitutive equations for at least some classes of materials up to the point of rupture.

In the absence of a constitutive equation it is necessary to resort to the experimental determination of $\partial W/\partial I_1$ and $\partial W/\partial I_2$, the derivatives of the strain-energy-density function with respect to the first and second principal invariants of the deformation tensor. Both derivatives are functions of both invariants. Hence, two three-dimensional mappings of properties are required. As shown recently by Kawabata[3], this is a formidable task that, at the present state of our knowledge, must be repeated on each individual material, since the effect of structural features such as molecular architecture, crosslink density, network imperfections, sol fraction, etc., on $\partial W/\partial I_1$ and $\partial W/\partial I_2$ are not understood.

Valanis and Landel[1] have proposed an approach which is based on the assumption that the free-energy changes associated with the deformations of an isotropic material in the three principal directions are additive. It follows from this assumption that the strain-energy-density function for an incompressible material can be written in the form

$$W = w(\lambda_1) + w(\lambda_2) + w(\lambda_3) \quad , \tag{3}$$

and it suffices to determine $\partial w(\lambda)/\partial\lambda$, i.e., only one two-dimensional mapping is needed. Various workers have since shown that the assumption of the additivity of the free energy of deformation is a good one for natural rubber[1], for styrene-butadiene rubber[2,3] and for acrylonitrile-butadiene rubber[4]. The material function, $\partial w(\lambda)/\partial\lambda$, can be obtained from measurements in pure shear[1] or from measurements in simple tension[5]. The approach of Valanis and Landel lightens the task considerably whenever its applicability can be demonstrated.

When the material cannot be regarded as incompressible, the strain-energy-density function becomes a function also of $\partial W/\partial I_3$, where I_3 is the third principal invariant of the deformation tensor. In that case, both the theory and its experimental application become rather complicated.

The theory of large, purely elastic deformations is at best a good approximation in the case of slow deformations of rubbers. Plastics are not only generally compressible but also show considerable viscoelasticity, i.e., their behavior is time dependent. Again, the continuum mechanical description of time-dependent behavior may be regarded as solved in principle. However, even the relatively simple theory of Coleman and Noll[6] requires twelve material functions and three material constants to describe the behavior. Such theories are useless from a practical point of view. Attempts are currently under way to ascertain whether there are classes of materials for which some of the material functions contribute negligibly. It has been found in some cases[7,8] that an adequate description can be obtained with a limited number (say about four) of material functions. So far, it is not clear whether the same functions will yield a useful description over wide ranges of time and temperature and in all modes of deformation.

Thus, although tremendous progress has been made in formulating the phenomenological (i.e., continuum mechanical) description of the behavior of polymeric materials theoretically, much remains to be done to arrive at solutions that are useful for the practicing polymer engineer. From a practical point of view, perhaps the most useful development in recent years is the realization that the effects of strain and time can often be factored up to quite large deformations. This approach has been pursued particularly by Thor Smith[9] and by Landel[10]. Up to the limit to which strain-time factorization is applicable, the stress in a constant-rate-of-strain equation in simple tension is given by

$$\sigma(t, \lambda) = F(t)\ \Gamma(\lambda) \quad , \tag{4}$$

where F(t) contains all the time dependence and $\Gamma(\lambda)$ all the strain dependence of the stress. The relation between the constant-strain-rate modulus, F(t), and the relaxation modulus, E(t), in infinitesimal deformation is given by

$$E(t) = F(t) + \frac{d\ F(t)}{d\ \ell n\ t} \quad . \tag{5}$$

The strain function, $\Gamma(\lambda)$, can be determined experimentally but should also be accessible through the Rivlin-Green theory.

Thus, for rubbery materials in which mechanical properties do not strongly depend on time, a combination of the Valanis-Landel approach with strain-time factorization yields a practical way of describing and predicting time-dependent material behavior in large deformation. When such simplification is not applicable, the situation appears quite hopeless at this stage.

The Chair invited the participants to voice their views with respect to the directions future research in this area should take. Rivlin suggested that research should concentrate on carefully selected materials which behave in a relatively simple fashion to obtain a thorough understanding of at least some materials. He referred to work he did in collaboration with Bergen and Messersmith of the Armstrong Cork Company over 15 years ago. Stress-relaxation experiments were made on unfilled PVC and on clay-filled PVC at various filler concentrations. The deformations were small enough so that classical infinitesimal-strain theory could be used. A fairly complete set of experiments on simultaneous extension and torsion of a circular cylinder could be interpreted in terms of a constitutive equation of the form

$$\underline{\sigma} = \alpha \cdot \theta \cdot \underline{\epsilon} \quad , \tag{6}$$

where $\underline{\sigma}$ and $\underline{\epsilon}$ are the stress and strain tensors, respectively, θ is a function of the strain invariants, and α is a function of time. For the unfilled PVC, θ was found to be constant, i.e., the material was linear. For the filled PVC, the dependence of θ on the strain invariants became increasingly stronger as the concentration of the filler increased, i.e., the material became increasingly nonlinear.

Zapas felt that it is necessary to distinguish elastic and viscoelastic solids as carefully as one distinguishes simple and nonsimple materials. For the simple solids we have the elegant theory of Green and Rivlin. Some special cases of this theory were presented in the session, where either a time-dependent strain energy or a strain-potential function was empirically represented by curve fitting experimental data. The separation of these functions into a product of a Young's modulus and a function of strain should be treated as no more than a good approximation for many materials and should be used only for engineering purposes. His data on plasticized PVC, contrary to the findings of Becker, could not be represented by this approximation. He thought that strain-time factorization would not be useful for materials such as polyisobutylene and vulcanizates of butyl rubber, nor for natural rubber either, although it might be more applicable to these materials than to PVC.

Similar cautions were restated by Schwarzl with concurrence by Landel. Schwarzl thought that a theory based on the assumption of factorizability of time and strain dependence should be applied only to a special class of materials (such as, e.g., the crosslinked elastomers discussed by Landel) in a special domain of time and temperature (e.g., in the rubberlike state). He did not expect this theory to remain valid in cases where the damping becomes higher than, e.g., 0.05; and he expected it to break down at the onset of the rubber-glass transition. He did not, however, challenge the practical usefulness of a theory which – at least for a certain class of materials – is able to describe stress-strain behavior adequately, if only in the rubbery region.

Tschoegl emphasized that there exist cases where strain-time factorization has been found to apply experimentally. Theoretical justification for these observations, on a continuum mechanical basis, were presented by Valanis at the International Rheology Congress in Lyon.

Halpin pointed out the contrast between the general approach presented by Rivlin and the contributions of Becker and of Landel. Rivlin defined a permissible constitutive equation with constraint upon the number of independent material coefficients. For a restricted class of materials, no additional material functions or constants may appear in the time-dependent case than are required to describe the equilibrium problem. The validity of this assumption is suggested by the apparent utility of factorization techniques. If the assumption could be formalized, it would provide an exclusion principle to limit the number of material functions required to describe a given material.

J. G. Williams reiterated that there are no fundamental theoretical problems in finite strain elasticity and only one in nonlinear viscoelasticity. Difficulties do arise in specific applications, such as defining a suitable strain-energy function for a particular material or interconverting a set of time data to another form, but these difficulties are not of principle but of method. The only remaining unsolved theoretical problem lies in converting time histories for nonlinear viscoelastic materials. Existing theories are particularly inept, and a new formulation, or at least a significant modification of existing formulations, is needed.

Becker remarked that no matter how well developed the theory is, the problem of describing the material behavior experimentally still remains. He felt that the simplest way was the best. Since the elastic theory is simpler than the viscoelastic one, modification of the elastic theory to a viscoelastic one through strain-time factorization is the best expedient from an experimental point of view.

Zapas queried Becker's use of a value of 0.47 for Poisson's ratio for rubber. He pointed out that the value used by other workers is closer to 0.4999. The difference is a significant one. Schwarzl suggested that exact values of Poisson's ratio for rubberlike materials can be obtained only by simultaneous measurements of the shear and the bulk modulus at the same temperature and frequency. When this is done, one obtains values between 0.4990 and 0.4995. Accurate values of Poisson's ratio cannot be obtained from measurements in two-dimensional states of stress (or strain).

Onogi recalled Landel's paper pointing out the significant effect of molecular weight between crosslinks on the fracture properties of amorphous elastomers. In his laboratory, a similar effect is being studied on uncrosslinked narrow-distribution polystyrenes having different molecular weights ($M_w = 1.31 \sim 8.43 \times 10^5$). Uniaxial stress-strain measurements under constant rate of extension were carried out on films of the polystyrenes at temperatures ranging from 106.5 C to 178 C and at extension rates ranging from 0.96×10^{-3} to 4.81×10^{-3} cm/sec. At temperatures lower than a critical value or at rates of extension higher than a critical value, the specimens could be elongated rather uniformly. Above the critical value, thinning could be observed. The stress and strain at which this thinning started, σ_f and ϵ_f, were functions of temperature and molecular weight. When σ_f and ϵ_f were plotted against the strain rate at different temperatures, the data could be superposed according to the time-temperature superposition principle, and the shift factors could be well represented by the WLF equation. From the molecular-weight dependence of ϵ_f, the molecular weight between entanglement loci, M_e, could probably be evaluated, assuming affine deformation of the network structure due to entanglements. Onogi hoped that this type of analysis would give more information on the fracture behavior of amorphous polymers not only above but also below the glass transition temperature.

Landel commented on Onogi's use of two measures of strain: the stretch ratio λ and the natural strain $\ell n\ \lambda$. The former was used to describe the "elastic" part of the response and the latter for the "viscous" part. Landel stated that in Onogi's case there was a clean-cut separation, since Onogi also changed from a constant-strain-rate experiment to one with an exponential strain history when the specimen thinned. In Landel's opinion, these were especially convenient experiments and strain measures for the two types of response. By contrast, the experimental response of a viscoelastic fluid is very difficult to analyze, if a constant Cauchy strain rate is employed. Similarly, in a highly crosslinked rubber, usually tested at a constant (macroscopic) Cauchy strain rate, there will be a large flow contribution as the chains disentangle and/or entanglements slip. Landel felt that it was an open question as to what measure of strain would be appropriate in this case.

2 THE EFFECT OF PRESSURE ON POLYMER BEHAVIOR

The first, and most substantial, part of the Agenda Discussion centered on the behavior of polymers in large deformation at atmospheric pressure. While the effect of temperature on polymer behavior is fairly well understood, and while time effects can be treated, at least theoretically, the effect of pressure on polymer behavior has been much less well explored. Radcliffe, in his presentation (Session II), briefly reviewed the work done in this area so far. His paper formed the basis of the discussion reported in this section.

In amplification of his paper, Radcliffe pointed out that experimental comparison of the respective volume changes for polycarbonate, associated with a given increase in tensile yield stress, as effected either by an increase in pressure at room temperature or by a decrease in temperature at atmospheric pressure shows that they are not the same – in contrast to the hypothesis proposed by Ainbinder. However, it can be demonstrated[11] that a simple shift factor can be developed, analogous to the WLF relation, to permit superposition of pressure and temperature behavior at constant strain rate. The resulting relationship involves only physical constants of the material and provides a useful correlation between pressure and temperature effects.

Ferry commented on the importance of considering free volume in pressure-temperature superposition. He suggested that Radcliffe's two curves of yield stress against $\Delta V/V$ might more nearly coincide if the plot were made against the change in fractional free volume rather than against the relative change in total volume. The compression of free volume under pressure is a relatively *small* proportion of the total compression, whereas the contraction of free volume with decreasing temperature is a relatively *large* proportion of the total contraction. The free-volume changes can be estimated by taking the temperature and pressure coefficients as the differences between the macroscopic coefficients above and below the glass transition temperature. Phenomena involving molecular mobility are related to free volume, not to total volume.

With reference to Radcliffe's statement that rubber-reinforced high-impact polystyrene (HIPS) becomes brittle when subjected to moderate hydrostatic pressure because of suppression of crazing, Alfrey pointed out that *biaxially oriented* HIPS responds to tensile loading by shear yielding without crazing. He therefore suggested that biaxially oriented HIPS might not exhibit a brittle regime under hydrostatic pressure.

Bauwens made some comments concerning the influence of temperature and strain rate on the effect of hydrostatic pressure on the yield stress. His comments were based on his recent experiments on bisphenol-A-polycarbonate.[12] Bauwens had previously proposed[13,14] the relation

$$\tau_o + \mu p = f(\dot{\epsilon}, T) \quad , \tag{7}$$

where τ_o is the octahedral shear stress, μ is a constant, p is the hydrostatic stress, and ϵ and T are the strain rate and the temperature, respectively. From this he deduced the relation

$$\frac{|\sigma_c|}{\sigma_t} = \frac{\sqrt{2} + \mu}{\sqrt{2} - \mu} \quad , \tag{8}$$

where σ_c and σ_t are the compressive and tensile stresses for tests performed at the same temperature and strain rate. It is thus possible to calculate the influence of pressure on yielding by the measure of σ_c and σ_t.

In Bauwens' laboratory, tensile and compression tests were performed on polycarbonate over a wide range of temperatures and strain rates. It was found that μ is constant between -50 and +120 C and has a value in agreement with data presented by Radcliffe.[15] At temperatures below -50 C, μ increases. Bauwens believes that this increase may be related to the β-transition. Below a temperature which depends on the strain rate, the yield phenomenon implies two rate processes, α and β; thus, the yield stress becomes

$$\sigma = \sigma_\alpha + \sigma_\beta \quad . \tag{9}$$

The parameters related to each process have different values. In particular,

μ_α and μ_β are quite different. According to Bauwen's results on polycarbonate, $\mu_\beta > \mu_\alpha$. For this reason, μ must increase when the β contribution to the yield stress becomes significant.

3 CRAZING IN POLYMERS

The final part of the Agenda Discussion on Phenomenology concerned itself with a rather different area, viz., the significance of crazing in the bulk deformation and fracture of glassy polymers.

Knappe pointed out that stress-cracking under the influence of solvents is strongly dependent on the molecular weight. Very high molecular weights, as in the case of cast poly(methyl methacrylate), can prevent stress cracking. Slight crosslinking, e.g., in irradiated polyethylene, can reduce stress cracking and raise the time to rupture under constant load by several orders of magnitude. Since crazing and cracking are related phenomena, he raised the question as to whether the phenomenology of crazing could lead to an explanation of stress cracking.

Hull thought that the general morphology and microstructure of crazes was similar in a wide variety of materials and under a variety of conditions. There is a growing amount of evidence for changes in the fine detail of the morphology and microstructure. These changes could well account for the changes observed in the fracture behavior. He had recently studied the effect of gamma radiation on the morphology and microstructure of crazes using the techniques described in his paper. After irradiation, the crazes were much shorter and more irregular but the microstructure remained the same. Molecular weight and, in particular, molecular-weight distribution can have a significant effect on crazing phenomena, but once again, at high molecular weights the morphology and microstructure were little affected. These factors influenced mainly the amount of crazing and the stress levels at which crazes nucleate and grow. In confirmation of this, at the Gordon Conference on Polymer Physics this year, Baer reported on some recent observations on craze microstructure using Hull's techniques. Baer's group found that in low-molecular-weight polystyrene, the density of fibrils in the craze is reduced markedly.

Andrews remarked that the distance between a craze tip at which the extending craze filaments strain harden strongly will be related to the stress concentration at the craze tip via the "equivalent crack" concept recently described.[16] The faster the strain hardening, the lower is the stress concentration and the more craze resistance is exhibited by the polymer. This may explain many effects, such as that of molecular weight on crazing. Rivlin pointed out that a rather similar question with regard to thread formation arises in the case of pressure-sensitive adhesives of the type used in Scotch tape.

REFERENCES

1. Valanis, K. C., and Landel, R. F., *J. Appl. Phys.*, **38**, 2997 (1967).
2. Dickie, R. A., and Smith, T. L., *Trans. Soc. Rheol.*, **15**, 91 (1970).
3. Kawabata, S., *J. Polymer Sci.*, Part A-2, **8**, 903 (1970).
4. Obata, Y., Kawabata, S., and Kawai, H., *Zairyo*, **19**, 330 (1970).
5. Peng, T. J., and Landel, R. F., *J. Appl. Phys.*, **43**, 3064 (1972).
6. Coleman, B. D., and Noll, W., *Rev. Mod. Phys.*, **33**, 239 (1961).
7. McGuirt, C. W., and Lianis, G., *Trans. Soc. Rheol.*, **14**, 117 (1970).
8. Valanis, K. C., and Landel, R. F., *Trans. Soc. Rheol.*, **11**, 243-256 (1967).
9. Smith, T. L., *J. Polymer Sci.*, Part A-2, **7**, 635 (1969); *Proceedings of the International Conference on the Mechanical Behavior of Polymers*, The Society of Materials Science, Japan (1972), Vol III, pp 431-442.
10. Landel, R. F., and Fedors, R. F., *Rubber Chem. Tech.*, **40**, 1049 (1967).
11. Christiansen, A., Baer, E., and Radcliffe, S. V., *Phil. Mag.*, **24**,(188), 451 (1971).
12. Bauwens-Crowet, C., Bauwens, J.-C., and Homes, G., *J. Mater. Sci.*, **7**, 176-183 (1972).
13. Bauwens, J.-C., *J. Polymer Sci.*, Part A-2, **5**, 1145-1156 (1967).
14. Bauwens, J.-C., *J. Polymer Sci.*, Part A-2, **8**, 893-901 (1970).
15. Christiansen, A., Radcliffe, S. V., and Baer, E., Conference on The Yield, Deformation and Fracture of Polymers, Cambridge, April, 1970.
16. Andrews, E., *Polymer*, **13**, 337 (1972).

Part Three

MOLECULAR DESCRIPTIONS OF DEFORMATION

RATE PROCESSES IN THE PLASTIC DEFORMATION OF POLYMERS

J. C. M. Li

Department of Mechanical and Aerospace Sciences,
University of Rochester,
Rochester, New York

C. A. Pampillo and L. A. Davis

Materials Research Center
Allied Chemical Corporation
Morristown, New Jersey

ABSTRACT

Some recent efforts in the application of rate theory to the plastic deformation of polymers are reported. Concepts such as the activation strain volume are defined. The volumetric effects during the deformation of polymers are discussed in detail. Evidences are provided to show that plastic deformation is not a near-equilibrium process under usual conditions. The shear-strain volume is found to decrease with shear stress, obeying a general correlation for all materials. The nature of simultaneous processes is described briefly and applied to the transition region so as to map out domains for individual processes. Work hardening in tension is formulated by assuming a certain shear stress needed for sliding between molecules and the gradual orientation of chain directions toward the tensile axis.

1 INTRODUCTION

Ever since Einstein[1] derived his famous relation between viscosity and self-diffusivity in liquids, viscous flow has been treated as a molecular process. This microscopic concept of macroscopic flow has been extended to creep of metals by showing that the activation energy[2] and activation volume[3] for creep are comparable to those for self-diffusion.

Plastic deformation of solids was considered as a thermally activated rate process as early as 1925 when Becker[4] applied the Boltzmann principle to the nucleation of a slip region. Extensive measurements and studies have been done on crystalline materials and considerable progress has been made regarding the atomic or molecular processes involved. However, except for a limited early attempt of Sherby and Dorn[5], very little effort has been expended on polymers.

An understanding of rate-controlling mechanisms is essential for the improvement of material performance by providing relations between microscopic and macroscopic parameters. Such understanding is also helpful in suggesting reliable and meaningful constitutive equations for the prediction of behaviors under a variety of external conditions.

It is attempted here to report some recent efforts in the analysis of rate and volumetric data, in the determination of apparent activation parameters and the separation of individual processes, and in the development of a theory of work hardening in tension.

2 RATE-THEORY FORMULATION

No specific mechanisms will be proposed at this time except that plastic strain (shear or normal) is supposed to increase by the motion of certain conformational entities (dislocations, for example) through their activated states. The process is thermally as well as stress activated. The rate of increase of strain is given by

$$\dot{\epsilon} = \dot{\epsilon}_c \exp(-\Delta F/kT) \quad , \tag{1}$$

where $\dot{\epsilon}_c$ is the fastest attainable strain rate, ΔF is the standard free energy of activation, k is the Boltzmann constant, and T is the absolute temperature. The quantity $\dot{\epsilon}_c$ involves the number of all the mobile entities, the vibrational frequency or the number of attempts per unit time in reaching the activated state, and the strain contributed by each entity after it reaches the activated state. The exponential term is the probability of success of the attempts and is related, according to the rate theory, to the difference in free energy of the activated and normal configurations of the entity in their respective standard states. In other words, the exponential term is the equilibrium constant between the activated and normal states of the entity or the ratio of respective concentrations.

Equation 1 neglects the reverse reaction or the backward motion of the entities. This is justified under normal laboratory testing conditions. However, if the stress is very small so that the rate of backward motion cannot be neglected from that of the forward motion, an exponential term representing the probability of a successful attempt in the reverse direction should be included. This is discussed later.

Since ΔF is the standard-free-energy difference between the activated and the normal states of the mobile entity, it obeys all the relations of classical thermodynamics. Considering, for example, three independent variables: the temperature, T, the hydrostatic pressure, P, and the shear stress (corresponding to the plastic shear strain), τ. The temperature dependence gives the enthalpy of activation:

$$\Delta H = \left.\frac{\partial(\Delta F/T)}{\partial(1/T)}\right|_{\tau,P} = kT^2\left(\frac{\partial \ln\dot{\epsilon}}{\partial T}\right)_{\tau,P} , \tag{2}$$

or the difference in enthalpy between the activated and the normal configurations of the entity in their respective standard states. The second relation involves the assumption that the preexponential term $\dot{\epsilon}_c$ is not a function of temperature. Such assumption may be justified, for example, when the test is performed on the same specimen in the same conformational state at two different temperatures.

Similarly, the pressure dependence of ΔF gives the activation volume:

$$\Delta V = \left(\frac{\partial \Delta F}{\partial P}\right)_{\tau,T} = -kT\left(\frac{\partial \ln\dot{\epsilon}}{\partial P}\right)_{\tau,T} , \tag{3}$$

and the shear-stress dependence gives the shear-strain-volume[6] of activation:

$$\Omega_\tau = -\left(\frac{\partial \Delta F}{\partial \tau}\right)_{T,P} = kT\left(\frac{\partial \ln\dot{\epsilon}}{\partial \tau}\right)_{T,P} . \tag{4}$$

Although the activation volume, ΔV, sometimes called the dilatational activation volume, can be interpreted simply as the volume difference between the activated and the normal states of the mobile entity, the physical meaning of the shear-strain volume of activation is slightly involved. Briefly, it is the local additional shear strain in a volume element produced during the activation process integrated throughout the solid.[6] Like strain, it is a tensorial quantity of second rank. In order to fully describe the activated state, six components of strain volumes are needed. However, the present status of experimental information is not sufficient to justify such involved formulism. Instead only two components, the hydrostatic component and the shear component, are considered here.

As is known, there are two modes of plastic deformation, the shear mode and the tensile mode (crazing). The foregoing is for the shear mode of deformation. For the tensile mode, a set of similar quantities can be defined. Three independent variables, such as temperature, pressure, and the tensile stress, σ, can be used. A tensile-strain -volume of activation is given by:

$$\Omega_\sigma = -\left(\frac{\partial \Delta F}{\partial \sigma}\right)_{T,P} = kT\left(\frac{\partial \ln \dot{\epsilon}}{\partial \sigma}\right)_{T,P} . \tag{5}$$

Since the free energy of activation is a function of many variables, it is customary to refer to a value ΔF_o at zero stress and zero pressure so that

$$\Delta F = \Delta F_o + \int_0^P \Delta V dP - \int_0^\tau \Omega_\tau d\tau . \tag{6}$$

However, ΔF_o still may be a function of temperature:

$$\frac{\Delta F_o}{T} = \frac{\Delta F_o^o}{T_o} + \int_{T_o}^{T} \Delta H \, d\left(\frac{1}{T}\right) , \tag{7}$$

where ΔH is the enthalpy of activation. Since ΔF is a thermodynamic state function, the path of integration is immaterial.

3 PRESSURE AND VOLUME EFFECTS

3.1 Macroscopic Volume Effects

When a polymer deforms plastically, it may have a volume change associated with the process. Qualitatively this could be the cause for the effect of hydrostatic pressure on the plastic behavior of polymers. However, the detailed nature of the effect is somewhat involved.

First of all, there are several volume effects which can be identified in the process. An instantaneous volume change can be measured during deformation. It is usually positive for tension and negative for compression. When the change of volume is plotted against stress as shown in Fig. 1, it has a linear part which has elastic origin. The slope of this part is $(1-2\nu)/E$ where E is Young's modulus obtainable from the linear part of the stress-strain relation and ν is Poisson's ratio obtainable from this plot. At higher elastic strain equal to σ/E, the elastic volume change may deviate from linearity as shown, but can be calculated from σ, E, and ν. Thus, the instantaneous volume change can be separated into two parts, the elastic volume change and the plastic volume change. The plastic volume change is found[7] positive for PMMA and negative for Lexan polycarbonate (PC), both in the case of compression.

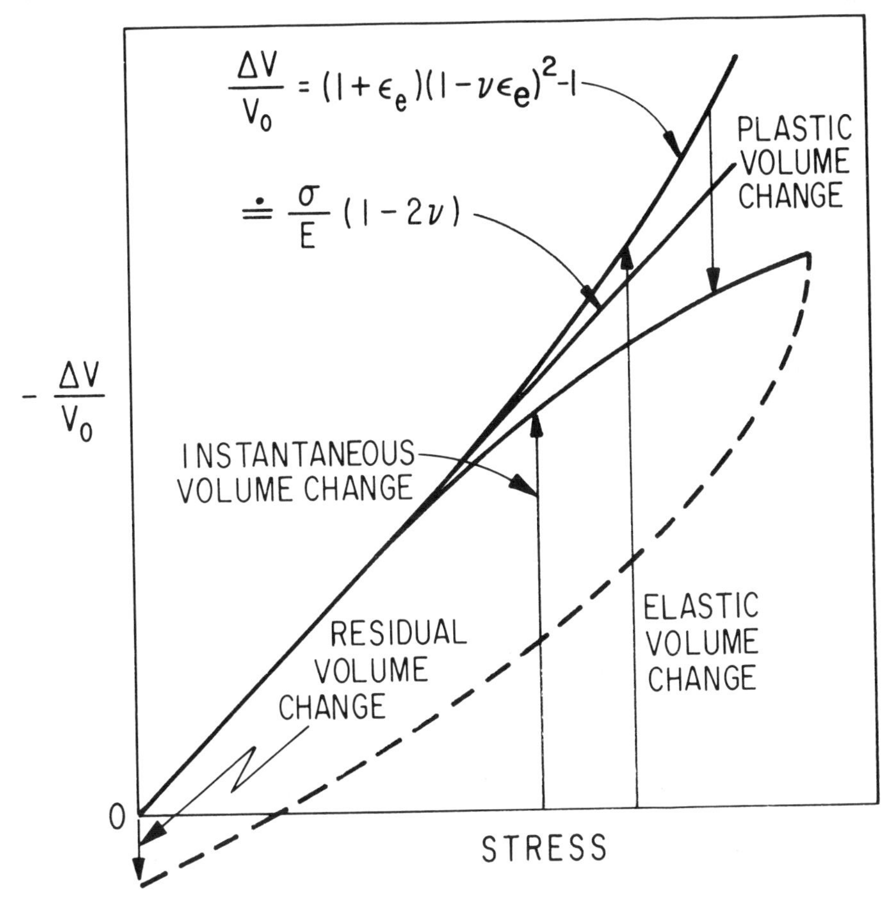

Fig. 1. Schematic representation of macroscopic volume effects in the compression of a polymer.

If the plastic volume change is actually due to plastic deformation, there should be a one-to-one relation between the plastic volume change and plastic strain. This is shown not to be the case in Fig. 2 for Lexan PC. Upon unloading the specimen, the plastic volume change becomes zero but the plastic strain remains finite. This observation suggests that either the volume change and plastic deformation are two independent processes or there is a mechanism which relaxes volume without affecting plastic strain. The term "plastic volume change" is then of questionable validity.

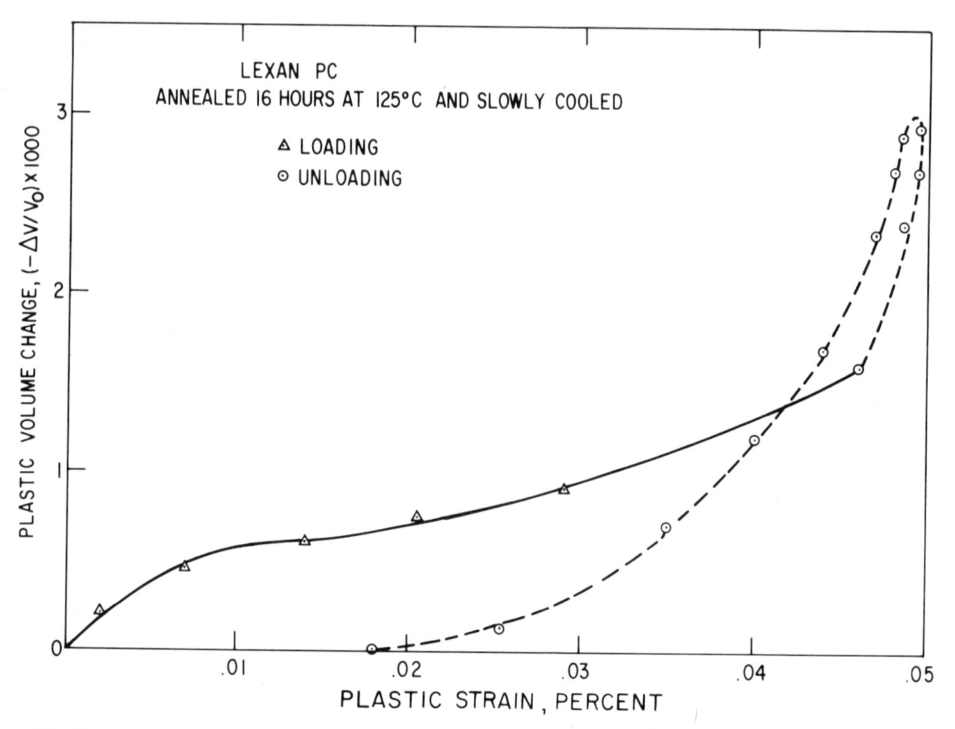

Fig. 2. Relation between the plastic volume change and plastic strain for Lexan polycarbonate.

Upon unloading the specimen, the volume-stress relation may behave as shown by the dotted line in Fig. 1. A residual volume change can be defined when the stress reaches zero. Ignoring all the transient volume changes, the residual volume change may be the quantity which should be associated with plastic deformation. However the residual volume change is found to be positive in PE[8] and negative in PMMA[8] and PC[7]. If the increase of flow stress with hydrostatic pressure[9] is going to be explained by the extra PV work, the residual volume change should be a positive quantity.

The fact that none of these volume changes, the instantaneous volume change, the elastic volume change, the plastic volume change, or the residual volume change, can explain the pressure effect of flow stress is demonstrated in Figs. 3 and 4 for Lexan PC. While the effect of pressure on the flow stress is positive, all these volume effects are negative.

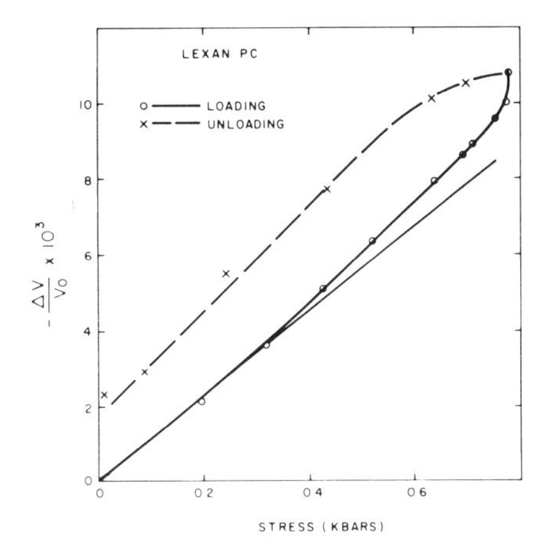

Fig. 3. Instantaneous volume change for Lexan polycarbonate (annealed at 165 C for 2 hours and quenched in air).

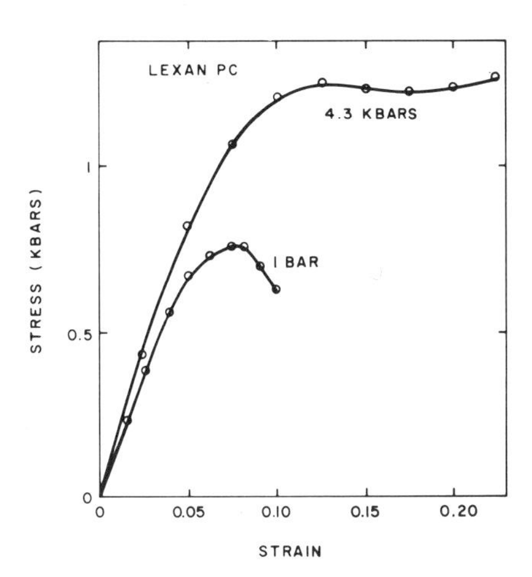

Fig. 4. Effect of hydrostatic pressure on the stress-strain relation for Lexan polycarbonate.

3.2 The Free Volume

Another way of looking at the volume effect is not from the viewpoint of "PV work" but from the viewpoint of microstructural changes. This is the "free-volume" concept used successfully in explaining the sharp viscosity change above the glass transition temperature of a glassy polymer. In view of the similarity between viscosity and plastic deformation, a suggestion[10,11] was made that tensile yielding occurs when the volume increase produced by the tensile stress supplies enough free volume for plastic deformation to take place. However, this concept fails to explain yielding in compression as indicated by Sardar, et al.[12] and in simple shear by Sternstein, et al.[13] Hence, the free-volume concept is not useful in understanding yielding, let alone the effect of pressure on yielding.

3.3 The Activation Volume

On the basis of the rate theory, the volume associated with the pressure effect is the activation volume given by Eq. 3. This activation volume, like activation enthalpy, is not macroscopically measurable except through the effect of pressure on the strain rate. Even if one of the macroscopic volume changes just discussed is associated directly with plastic deformation, it has very little relation with the activation volume as shown in Fig. 5. For example, the sign of the macroscopical volume change can be either positive or negative, independent of the sign of the activation volume.

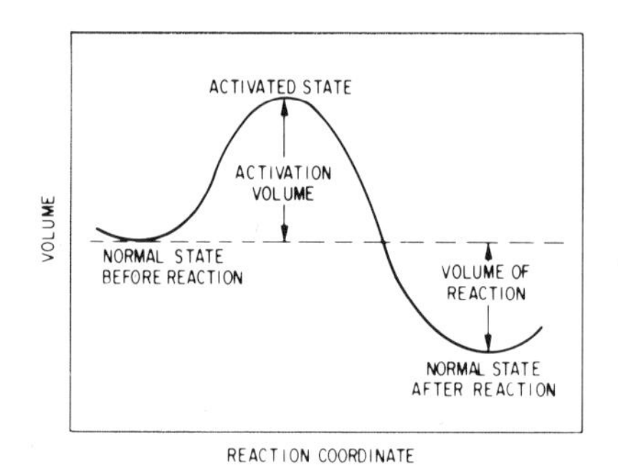

Fig. 5. Schematic representation of volumetric effects in a chemical reaction.

The macroscopic volume change becomes important only in the case of near-equilibrium processes in which the reverse reaction proceeds at a rate comparable to that of the forward reaction. In this case the free energies of activation for the forward and reverse reactions are, respectively, from Eq. 6:

$$\Delta F_f = \Delta F_o + P\Delta V_f - \tau\Omega_f \tag{8}$$

$$\Delta F_r = \Delta F_o + P\Delta V_r + \tau\Omega_r \ , \tag{9}$$

where ΔV and Ω are respective activation volumes of near-equilibrium values. From Eq. 1 the net strain rate is:

$$\dot{\epsilon} = \dot{\epsilon}_c \ [e^{-\Delta F_f/kT} - e^{-\Delta F_r/kT}]$$

$$= (\dot{\epsilon}_c/kT)\ e^{-\Delta F_o/kT}\ [P(\Delta V_r - \Delta V_f) + \tau(\Omega_f + \Omega_r)] \ . \tag{10}$$

When $P = 0$, let the stress required to maintain a certain strain rate be τ_o, it is seen that

$$\tau = \tau_o + \beta P \ , \tag{11}$$

where

$$\beta = (\Delta V_f - \Delta V_r)/(\Omega_f + \Omega_r) \ , \tag{12}$$

a quantity directly proportional to the macroscopic volume change.

However, a characteristic for near-equilibrium processes is a linear relation between stress and strain rate at zero pressure as shown in Eq. 10. This is seldom true under normal testing conditions at low temperatures. When the reverse reaction is negligible compared with the forward reaction, the stress-pressure relation can be obtained by differentiating Eq. 6:

$$d\Delta F = \Delta V dP - \Omega d\tau = 0 \tag{13}$$

for a certain strain rate. Hence, in the linear approximation of Eq. 11, the quantity β is given by

$$\beta = \Delta V/\Omega \quad , \tag{14}$$

namely, the ratio of two activation volumes.

The activation volume for AC1220 high-molecular-weight PE has been measured[14] at room temperature and found to be about 260 A^3 at 1 atmosphere. The activation volume decreases with increasing pressure and shows a transition at about 2.5 kbar. This is discussed later.

4 TEMPERATURE AND HEAT EFFECTS

The flow stress of polymers increases with decreasing temperature, as in the case of most other materials. Since, usually, plastic deformation produces heat, it may appear from the Le Chatelier principle that the flow stress should decrease with decreasing temperature. This proves once again that plastic deformation is not a near-equilibrium process. To illustrate, Eqs. 6 and 7 can be written near equilibrium for the forward and reverse reactions, respectively:

$$\frac{\Delta F_f}{T} = \frac{\Delta F_o^o}{T_o} + \Delta H_f \left(\frac{1}{T} - \frac{1}{T_o} \right) - \frac{\tau \Omega_f}{T} \tag{15}$$

$$\frac{\Delta F_r}{T} = \frac{\Delta F_o^o}{T_o} + \Delta H_r \left(\frac{1}{T} - \frac{1}{T_o} \right) + \frac{\tau \Omega_r}{T} \quad , \tag{16}$$

where, for simplicity, pressure is taken as zero. From Eq. 1, the net strain rate is

$$\dot{\epsilon} = \dot{\epsilon}_c e^{-\Delta F_o^o / k T_o} \left[\frac{\Delta H_r - \Delta H_f}{k} \left(\frac{1}{T} - \frac{1}{T_o} \right) + \frac{\tau(\Omega_f + \Omega_r)}{kT} \right] \quad . \tag{17}$$

For the same strain rate, if the flow stress at T_o is τ_o, the flow stress at T is

$$\tau = \tau_o + \frac{\Delta H_r - \Delta H_f}{\Omega_f + \Omega_r} \left(1 - \frac{T_o}{T} \right) \quad . \tag{18}$$

In other words, for an exothermic reaction, or $\Delta H_r > \Delta H_f$, the flow stress should increase with increasing temperature, contrary to observed facts.

Since we have ruled out the near-equilibrium possibility, the stress-temperature relation should be sought from Eqs. 2 and 4:

$$\left(\frac{\partial \tau}{\partial T}\right)_{\dot{\epsilon},P} = -\frac{\Delta H}{T\Omega_\tau} , \tag{19}$$

which provides a way of measuring the activation enthalpy from the temperature dependence of flow stress. The activation enthalpy for AC1220 high-molecular weight PE is found[15] to be about 1.9 ev at room temperature, 1 atm, and a strain rate of 3.5×10^{-3}/sec in tension. It varies with temperature and shows a transition at 265 K. This is discussed later.

5 STRESS AND STRAIN-RATE EFFECTS

As indicated earlier, the nonlinear relation between stress and strain rate is evidence that plastic deformation is far from equilibrium. From Eqs. 1 and 6, even if Ω_τ is a constant, the strain rate would increase exponentially with stress.

Since it is well known that polymers can be deformed at any temperature, it implies that the activation free energy can be supplied by stress alone. But since stress cannot do work without strain, it follows that one of the reaction coordinates can be the strain. Along the strain coordinate, the free energy of the mobile defect must start from a minimum in the normal state and pass over a maximum in the activated state. It follows that the stress must start from zero in the normal state, pass over a maximum, and return to zero in the activated state. Hence, it is likely that the shear-strain-volume decreases with increasing stress and becomes zero at the maximum stress.

In a region in which the shear-strain-volume can be approximated to be inversely proportional to stress, a power relation is seen between strain rate and stress from Eqs. 1 and 6:

$$\dot{\epsilon} = B\tau^m . \tag{20}$$

Similarly, if the shear-strain-volume is inversely proportional to the square of stress, an exponential relation results:

$$\dot{\epsilon} = \dot{\epsilon}_\infty \exp(-D/\tau) . \tag{21}$$

These relations have been found valid in many crystalline materials but not yet in polymers.

For most materials investigated, there seems to be a general correlation[16] between the activation shear-strain-volume (also known simply as activation volume or activation area) and stress as a decreasing function. This general correlation seems to work also for polymers, as shown in Fig. 6. The range of values is due to varying strain. The reason for the shear stress being considered the same for all strains is explained later. The torsion data are those of Wu.[17]

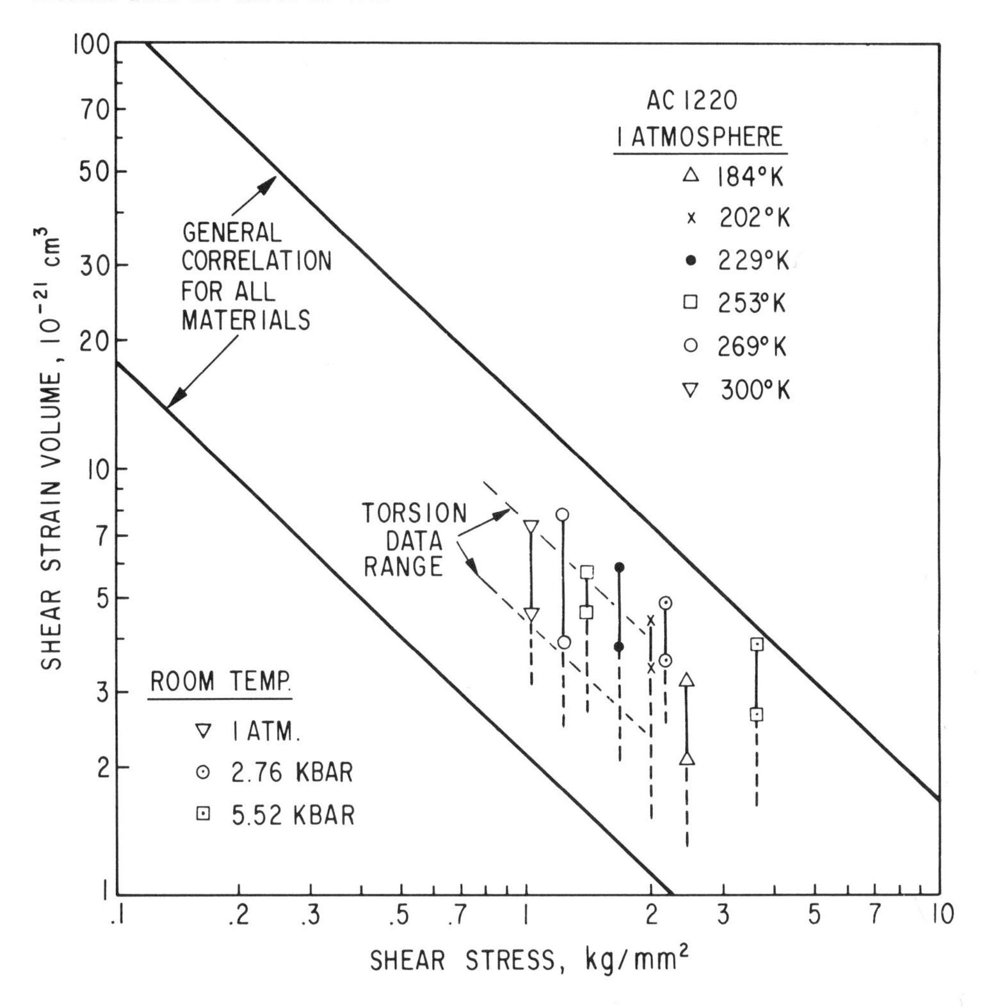

Fig. 6. Shear strain volume — stress relation for polymers.

6 SIMULTANEOUS PROCESSES

The rate-theory formulism represented by Eqs. 1 through 7 is meaningful only for a single process. In reality, there may be two or more processes operating simultaneously. The macroscopic behavior and the nature of these simultaneous processes are discussed elsewhere. Briefly, the single process formulism still can be used when the range of variables is large enough to show a transition between processes. Then it is possible to identify activation parameters for each process.

These techniques have been used to analyze the data for AC1220 PE, and the results are summarized in Fig. 7. Two transitions are found in the variation of flow stress with temperature: one at about 175 K and the other at about 265 K. The first is probably related to the glass transition. From the way the flow stress changes with temperature it can be deduced that the 265 K transition involves two processes operating in parallel.

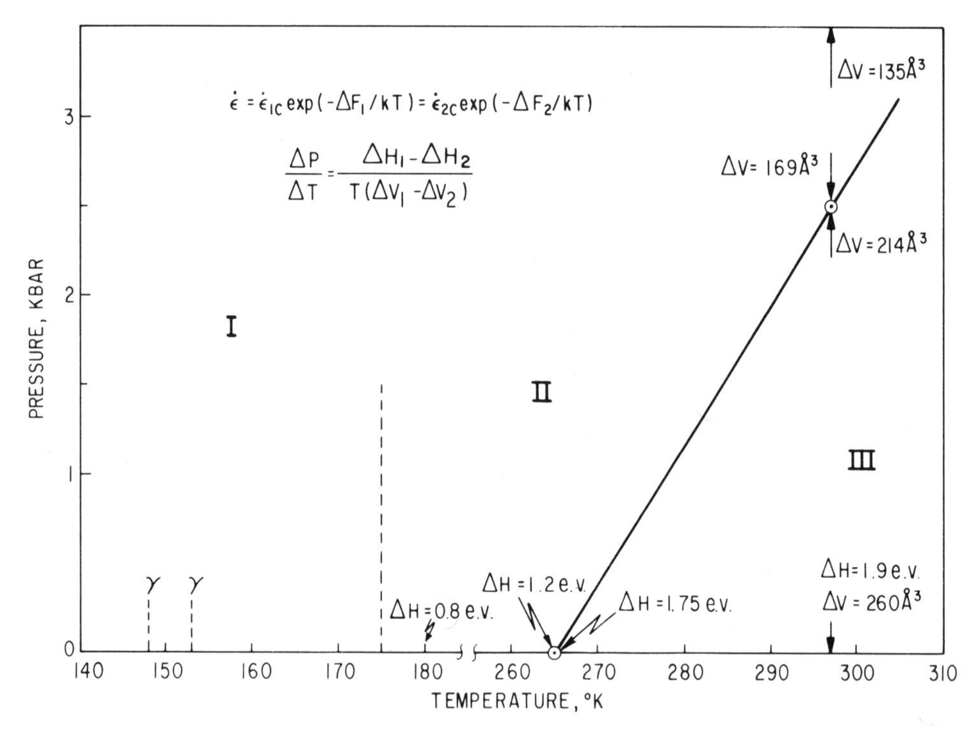

Fig. 7. Simultaneous processes in the deformation of AC1220 polyethylene.

Three processes are then assigned: I, below 175 K; II, between 175 K and 265 K; and III, above 265 K. Consistent with the flow stress-temperature relation, the activation enthalpy-temperature relation shows an increase of about 0.55 ev at about 265 K. This is an expected behavior for parallel processes. As mentioned earlier, the activation volume-pressure relation also shows an increase of about 45 A^3 at about 2.5 kbar. This is also the behavior for parallel processes. These two observations prompted a suggestion that perhaps the 265 K, 1-atm transition and the 297 K, 2.5-kbar transition are for the same two parallel processes. To verify this suggestion, a Clapeyron-type equation, as shown in Fig. 7, was used which relates $\Delta P/\Delta T$ to the ratio of the changes of enthalpies and volumes based on the fact that at the transition, the rates of the two processes are equal. This relation was found obeyed within experimental uncertainties.

Consistent with Fig. 7, the ductility[15] at atmospheric pressure has a transition at about 265 K. Similarly, the ductility at room temperature has a transition at about 2.5 kbar. From these two transitions, it seems that the more ductile process is the rate-controlling one for parallel processes.

7 WORK HARDENING IN TENSION

The increase of stress with strain is known as work hardening. The work hardening in tension is considerably more rapid than in torsion. This prompted a suggestion that perhaps the lining up of chain directions with the tensile axis is the major reason for work hardening in tension. On the basis of this conjecture, a work hardening theory is developed as follows.

Plastic deformation in the shear mode is thought to take place by sliding between molecules in the chain direction. The driving force for the process is taken as the shear stress operating in a plane containing the chain direction. If the chain direction makes an angle θ with the tensile axis, the plane which receives the most shear stress is that whose normal makes an angle $(\pi/2) - \theta$ with the tensile axis.

Hence the driving force or the shear stress for chain sliding is related to the tensile stress through the Schmid law:[19]

$$\tau = \sigma \sin\theta \cos\theta \quad . \tag{22}$$

In a region in which all molecules are oriented at an angle θ with respect to the tensile axis and the elongation of this region takes place by sliding of these molecules relative to each other in the plane of maximum shear stress, the elongation ℓ is directly related to θ as follows:

$$\ell \sin\theta = \ell_o \sin\theta_o \quad . \tag{23}$$

Hence, the local tensile strain relative to the original elongation ℓ_o is:

$$\epsilon^* = \ln(\ell/\ell_o) = \ln(\sin\theta_o/\sin\theta) \quad . \tag{24}$$

If the resolved shear stress is a constant, the relation between the tensile stress, σ, and the average tensile strain ϵ is:

$$\frac{\sigma}{\sigma_o} = \frac{\cos\theta_o \exp(\epsilon/f_\epsilon)}{[1 - \sin^2\theta_o \exp(-2\epsilon/f_\epsilon)]^{1/2}} \quad , \tag{25}$$

where ϵ^* is replaced by ϵ/f_ϵ, with f_ϵ being the fraction of deformed material.

Equation 25 suggests that at large strains, a plot of log σ with ϵ should be linear. This is shown in Fig. 8 for AC1220. The linear part determines the deformed fraction f_ϵ which is seen to be between 0.555 and 0.685, a reasonable range. It also seems consistent with an estimated crystallinity of about 60 percent from density measurements. The small strain range of the stress-strain curve can be fitted to Eq. 25, with the θ_o values shown in Fig. 8. These values are reasonable, being 45 ± 8, which indicates random distribution at yielding.

The success of the foregoing simple work-hardening theory suggests the possibility that the shear stress which causes sliding of molecules relative to each other is probably independent of strain. This is the basis for Fig. 6 when the shear-strain volume as a function of strain is plotted against shear stress. The shear-strain volume is found linear with reciprocal exponential strain and hence is limited within a range of values as shown in Fig. 6. The dotted lines indicate extrapolated values to very large strains. The decrease of shear-strain volume with strain seems to suggest an increase of obstacles with strain. But this must be compensated by an increase in the number of mobile units so that the shear stress remains unchanged.

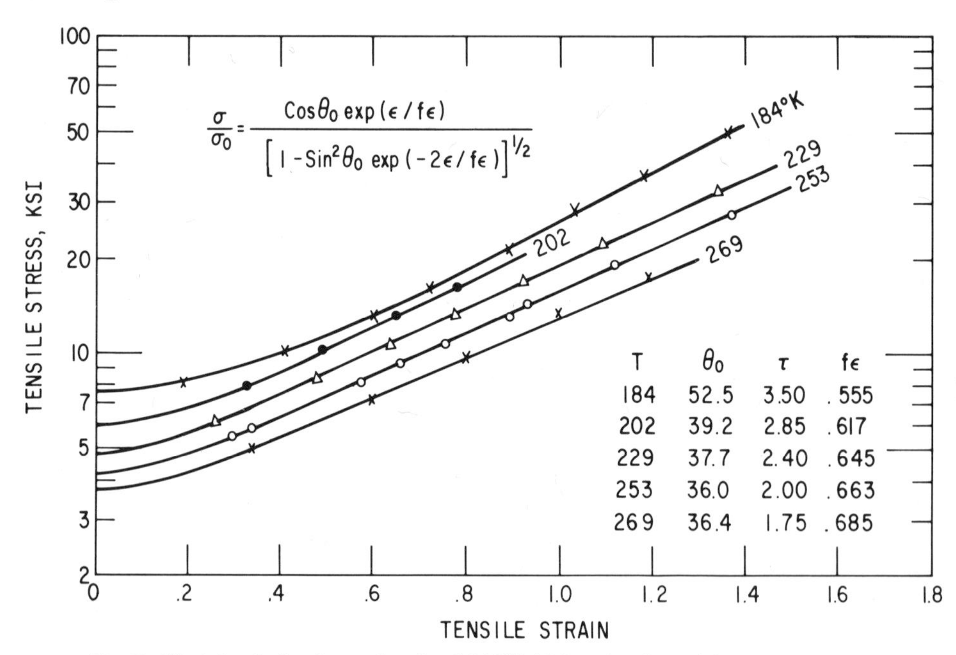

Fig. 8. Work hardening in tension for AC1220 high-molecular-weight polyethylene.

Accepting the stress-strain relation of Eq. 25, it is possible to estimate the Lüders strain. First of all, the rate of work hardening is:

$$\frac{d\ln\sigma}{d\epsilon} = \frac{1}{f_\epsilon} \cdot \frac{1 - 2\sin^2\theta_o \exp(-2\epsilon/f_\epsilon)}{1 - \sin^2\theta_o \exp(-2\epsilon/f_\epsilon)} \quad . \tag{26}$$

Hence, yielding will be unstable or a Lüders band will appear as long as:

$$\cos^2\theta_o < 1/(2 - f_\epsilon) \quad . \tag{27}$$

When a tensile specimen necks, the stress-strain relation inside the neck is approximately $\sigma = \sigma_o \exp(\epsilon)$ by assuming constant volume. The neck is stabilized when this stress becomes lower than the flow stress of Eq. 25. Thus, the Lüders strain is given by:

$$\epsilon_L = \frac{f_\epsilon}{2(1 - f_\epsilon)} \ln \frac{1 - \sin^2\theta_o \exp(-2\epsilon_L/f_\epsilon)}{\cos^2\theta_o} \quad . \tag{28}$$

For typical values of $f_\epsilon \sim 0.6$ and $\theta_o \sim 45$ degrees, a typical Lüders strain is 0.42. This seems to explain the much larger Lüders strain in polymeric materials as compared with that in metals.

8 SOME POSSIBLE MOLECULAR PROCESSES

On the basis of the theory of work hardening just discussed, the primary process is probably the sliding between molecules in the chain direction for the shear mode of deformation. However, since the deformation is inhomogeneous, the sliding process probably takes place by a defect mechanism such as the motion of dislocations.

The possible existence of dislocations in amorphous materials was first suggested by Gilman.[18] It can be demonstrated as follows: when polystyrene is compressed slightly, slip lines are developed on a polished surface as shown in Fig. 9. Note that some of the slip lines terminate inside the material. When the slip lines cut through each other, slip steps are observed as shown in Fig. 10. These two observations confirm the existence of dislocations, because, when a slipped area expands into the material, the boundary of the area separates the slipped and unslipped regions and therefore by definition is a dislocation.

As expected from the foregoing considerations, the slipped area should contain dislocations and may be chemically inhomogeneous and more reactive than the rest of material. This is demonstrated in Fig. 11, which shows a slightly deformed polystyrene sample after etching with sulfochromic acid at ~70 C for about 1 hour. The etch pits formed on the slip lines may or may not be associated with dislocations but are certainly indications of inhomogeneity within the deformed area.

Fig. 9. Terminating slip lines in polystyrene.

Fig. 10. Intersection of slip lines in polystyrene showing slip steps.

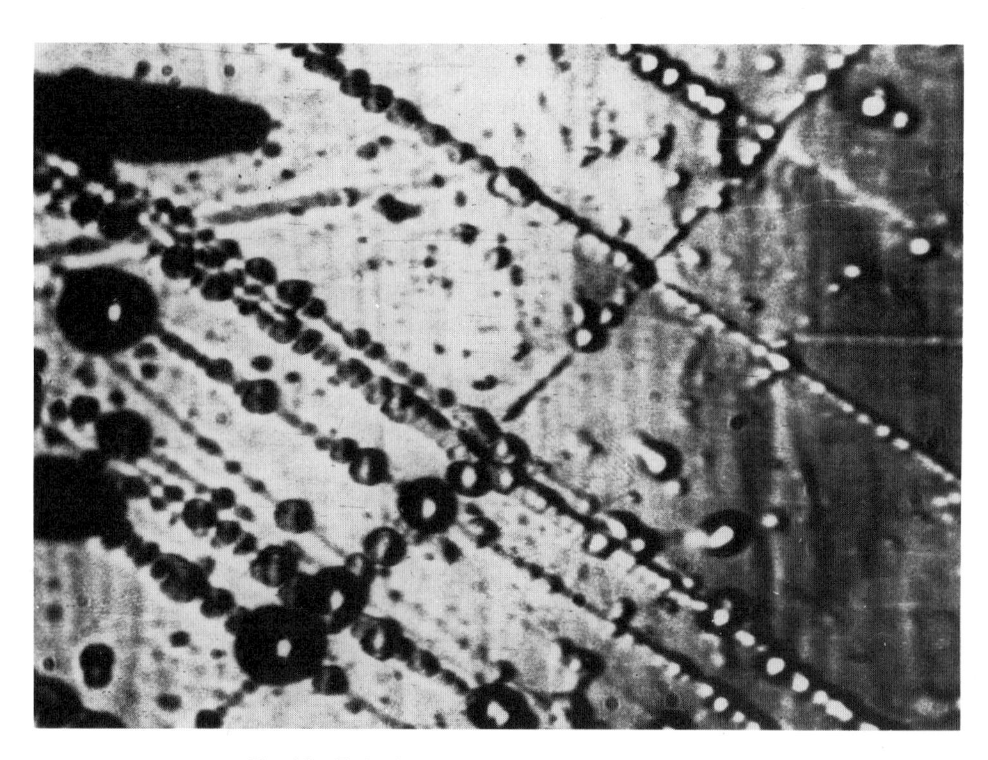

Fig. 11. Etch pits along the slip lines in polystyrene.

The dislocation mechanism is consistent with Fig. 6 in which the activation strain volume for polyethylene obeys a general correlation for all materials. However, such correlation was understood in terms of random distribution of obstacles along the dislocation. What could be the obstacles for dislocation motion in polymers?

Tie molecules[19] are chosen as possible obstacles at this time because they are capable of blocking the motion of dislocations in the amorphous region. They also offer two possible parallel mechanisms for the rate-controlling processes II and III. Process II may be the scission of tie molecules and process III may be the pulling out (a short segment of the length of Burgers vector) of tie molecules from the crystalline region. The activation enthalpy for process II extrapolated to zero stress is about 1.7 ev. This energy is about the energy required to break a hydrogen abstracted molecule (1.0 ev) plus that required for the diffusion of one of the chain ends (0.7 ev) to avoid recombination.

9 SUMMARY AND CONCLUSIONS

1. Several volumetric changes take place during the deformation of a polymer (see Fig. 1), namely, changes in the instantaneous volume, the elastic volume, the plastic volume, and the residual volume. None of their effects is found to correlate with the effect of pressure on the plastic deformation of polymers.

2. The plastic deformation of polymers under ordinary conditions is not a near-equilibrium process. The activation volume is the only correct quantity for correlation with the pressure effects on deformation.

3. The shear-strain-volume is found within 10^3-10^4 A^3 and decreases with shear stress in a manner consistent with the behavior of many other materials.

4. At least three processes are found to control deformation in AC1220 high-molecular-weight polyethylene (see Fig. 7). Process I is the low-temperature process which seems to associate with the glassy state of the amorphous region. Process II is the low-temperature, high-pressure process with low activation energy and low activation volume. Process III is the high-temperature, low-pressure process with high activation energy and high activation volume. The latter two processes operate in parallel.

5. Work hardening in tension can be understood by assuming a certain shear stress needed to slide molecules relative to each other and the gradual alignment of molecules with the tensile axis. The stress-strain relation can be fitted with two parameters: the initial average orientation of the molecules and the deformable volume fraction. Reasonable values are found for these parameters in all the cases. On the basis of this work-hardening theory, the large Lüders strain can be easily understood.

6. Some possible molecular processes are suggested.

REFERENCES

1. Einstein, A., *Ann. Phys., Lpz.* (4), **17**, 549 (1905).
2. See Dorn, J. E., *Creep and Recovery,* American Society for Metals (1957), p 255.
3. See Li, J.C.M., in *Dislocation Dynamics,* A. R. Rosenfield, et al. (Eds.), McGraw-Hill, New York, N.Y. (1968), p 87.
4. Becker, R., *Z. Physik.,* **26**, 919 (1925); *Z. Techn. Phys.,* 7, 547 (1926).

5. Sherby, O. D., and Dorn, J. E., *J. Mech. Phys. Solids,* **6**, 145 (1958).
6. Li, J.C.M., *J. Appl. Phys.*, **42**, 4543 (1971).
7. Pampillo, C. A., and Davis, L. A., *J. Appl. Phys.*, **42**, 4674 (1971).
8. Alksne, K. I., Ainbinder, S. B., and Slonimskii, G. L., *Mekhan. Polimerov Akad. Nauk Latv. SSR,* **2**, 355 (1966).
9. Ainbinder, S. B., Laka, M. G., and Maidrs, I. Yu., *Mekhan. Polimerov Akad. Nauk. Latv. SSR,* **1**, 65 (1965).
10. Litt, M. H., and Tobolsky, A. V., *J. Macromol. Sci. (Phys.)* **1**, 433 (1967).
11. Litt, M. H., Koch, P. J., and Tobolsky, A. V., *J. Macromol. Sci. (Phys.),* **1**, 587 (1967).
12. Sardar, D., Radcliffe, S. V., and Baer, E., *Polymer Eng. Sci.,* 8, 290 (1968).
13. Sternstein, S. S., Ongchin, L., and Silverman, A., *Applied Polymer Symposia,* No. 7, 175 (1968).
14. Davis, L. A., and Pampillo, C. A., *J. Appl. Phys.*, **42**, 4659 (1971).
15. Pampillo, C. A., and Davis, L. A., *J. Appl. Phys.*, **43** (1972).
16. Balasubramanian, N., and Li, J.C.M., *J. Materials Sci.*, **5**, 434 (1970).
17. Wu, Wen-Li, "Plastic Deformation of Polymers", MIT Thesis (1972).
18. Gilman, J. J., in *Dislocation Dynamics,* A. R. Rosenfield, et al. (Eds.), McGraw-Hill, New York, N.Y. (1968), p 3.
19. Brown, N., Duckett, R. A., and Ward, I. M., *J. Phys. D: Appl. Phys.*, **1**, 1369 (1968).

DISCUSSION on Paper by J.C.M. Li, C. A. Pampillo, and L. A. Davis

LEE: You used the relation

$$\dot{\epsilon} = \dot{\epsilon}(T,P,\tau)$$

and mentioned the avoidance of the difficulty, raised in the discussion yesterday, of deducing, for example, creep behavior from relaxation behavior. It seems to me that such a law implies very special properties, for example, that the creep function is linear ($\dot{\epsilon}$ constant for constant stress τ), and that the relaxation "function" yields constant stress ($\dot{\epsilon} = 0, \tau = $ constant). Moreover, you showed a stress-strain relation which would be quite arbitrary with such a law. In view of these difficulties, perhaps you would explain the range of applicability of such considerations.

LI: Perhaps I did not make myself clear; the relation is for the plastic strain rate and also for an unchanged microstructure. The creep rate may vary with strain owing to a change in microstructure (transient creep behavior). The plastic strain rate varies during stress relaxation by replacing the elastic strain rate. These viewpoints have all been treated extensively in the case of crystalline materials.

The stress-strain relation for polymers is based on the change of average molecular orientation with strain. Since this involves a change in microstructure, it cannot be derived from the rate equation.

ANDREWS: You explain your data in terms of molecular fracture of tie molecules, but this seems to be an unnecessarily drastic mechanism. In polyethylene, above -30 C at least, the amorphous phase is rubberlike and can extend by hundreds of percent without network fracture, while the crystalline phase has several easy slip modes not involving covalent bond fracture. If large forces are applied to tie molecules during deformation, I would expect the crystal regions to which they are tied to yield plastically before the tie molecules break.

LI: Tie molecules are chosen as random obstacles for dislocation motion in the amorphous region, in our attempt to understand the correlation between the activation shear-strain-volume and shear stress. The molecules are proposed to break in the temperature range between glass transition and 265 K. Above 265 K, they are proposed to slip out from the crystalline region a segment long enough to allow the dislocation to pass. The activation energy obtained for the chain-scission process seems reasonable.

HULL: Does your analysis allow you to calculate the number of breakages or the rate of breaking? If so, can the results be used to resolve the difficulty mentioned by Professor Andrews?

LI: Yes. From the magnitude of shear-strain volume, 10^3-10^4 A^3, it can be estimated using dislocation parameters that there is a tie molecule for every few hundred molecules. This is a reasonable number for the tie-molecule density. Depending on how many free radicals are generated per 1000 tie molecules broken, the number of free radicals produced during deformation may escape detection.

HALPIN: In your presentation you exhibited Lüders bands in an amorphous polymer. The appearance of these bands is not to be interpreted as providing a priori proof of dislocation systems in amorphous bodies. The capability of a solid to undergo a bulk instability termed yielding is in itself sufficient to produce shear lines. The consequences of a solid exhibiting Lüders bands may find a correspondence with phenomena observed in metal technology, but for different physical reasons. The physics of the problem resides in the processes which permit a solid to undergo yield or plastic flow.

LI: It may be a question of semantics. But when a slipped area, as shown by one of the slip lines, extends into the material, the boundary which separates the slipped and unslipped areas is by definition a dislocation. The macroscopic understanding of bulk instability or yielding does not provide an answer on the molecular level. The physics of the problem may well be the nucleation and motion of dislocations which cause yielding and plastic flow.

ANTHONY: Did you already measure the magnitude of the Burgers vector associated with your slip lines? In polymers, the plastic deformation may be due to dislocations as well as to disclinations.

LI: No, we have not measured the magnitude of Burgers vectors associated with the slip lines. It is rather difficult. However, the slip step does indicate the average direction of the Burgers vectors. There may be disclinations in the slip lines but on the average, the slip boundary is a dislocation.

PECHHOLD: Professor Li, you have discussed the pressure activation volume. Has anyone to your knowledge tried to measure this directly by dilatometry? If you assume that no cracks or crazes are being produced, then, in principle, it should be possible to measure these intermediate volumes during strain by dilatometry.

LI: That is an interesting question. The time that the volume is occupied by the activated state is very short, so the number of activated-state species existing at a given time is very small. I think it would be impossible to measure this experimentally. I think it is immeasurably small.

DEFORMATION OF CROSS-LINKED POLYMERS BELOW Tg

E. H. Andrews and P. E. Reed

Materials Department
Queen Mary College
London, England

ABSTRACT

Tensile tests were carried out on oriented elastomers in the glassy condition, and the deformation process was monitored by ESR. Both the nature and quantity of radicals formed are discussed. It is found that the radical concentration varies with the rate of deformation, the crosslink density of the sample used, the degree of deformation and the purity of the sample. Tests to explore the effect of crystalline morphology on radical formation are also briefly discussed.

Different modes of deformation occur, depending on the preorientation the sample receives prior to testing at temperatures below T_g. The three possible modes are found to be brittle failure, ductile failure associated with crazing, and ductile deformation with no observable crazing. When visible crazing occurs, it is always associated with the absorption of environmental gases, and where oxygen is present, this leads to the formation of peroxy radicals.

1 INTRODUCTION

An interesting phenomenon associated with the deformation of elastomers at low temperatures has been previously reported.[1] Samples of sulphur-cured natural rubber (cis-polyisoprene) preoriented above T_g were subsequently subjected to tensile testing below T_g. The samples deformed by a further 100 to 200 percent during the tensile test and were found to foam on subsequent warming to room temperature (Fig. 1). In subsequent studies[2], it was shown by ESR investigations that the deformation at low temperatures was associated with the production of free radicals, and that the subsequent foaming was due to the expansion both of environmental gases absorbed during the test and of hydrogen released from the polymer. The phenomenon was also shown to result in an increase in the crosslink density of the material.

The production of ESR signals during mechanical deformation has been shown to be a common occurrence in many polymeric materials[3-6], particularly highly oriented fibrillar materials. The ESR signals in these cases result from molecular rupture and thus provide insight into the mechanism of deformation and fracture in polymers.

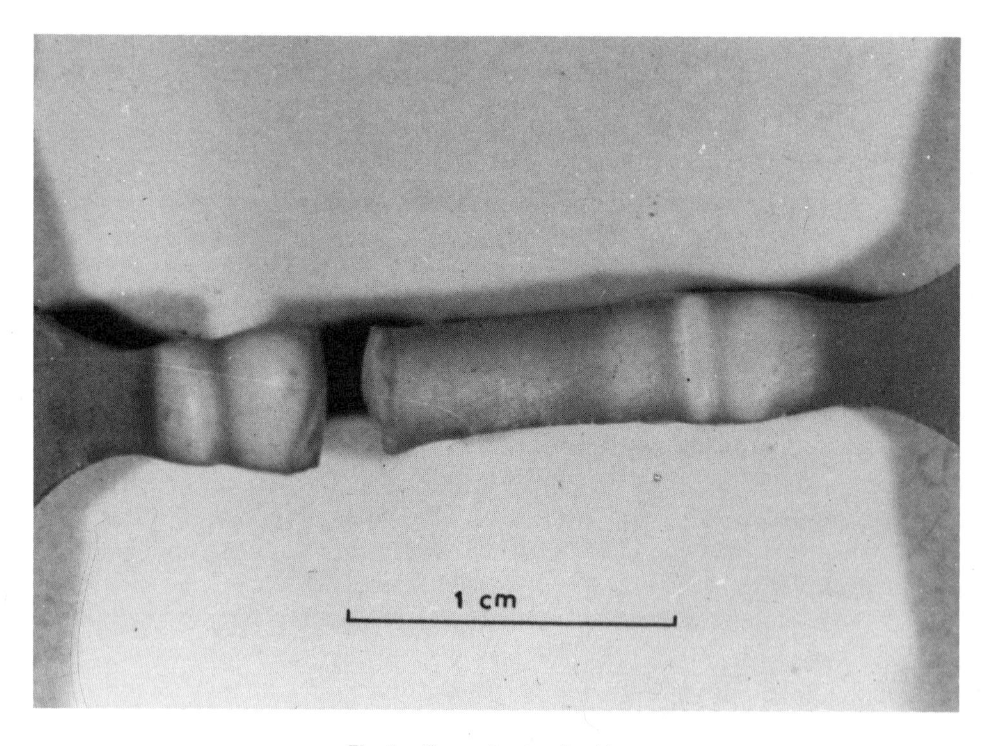

Fig. 1. Foamed natural rubber.

In a previous paper[2] on the deformation of slightly oriented (100 percent) sulphur-cured natural rubber, it was shown that three kinds of deformation are possible below T_g, depending on the temperature and strain rate at which the tensile test is performed. At temperatures immediately below T_g, the material deforms by shear yielding or cold drawing. At temperatures below -110 C and above -130 C, brittle behavior obtains. Further decrease in the test temperature results in ductile deformation due to microvoiding in narrow bands. Fig. 2 shows these three regions of different behavior for a strain rate of 0.02 min^{-1}.

The ESR and related studies of the deformation of oriented elastomers at temperatures below T_g have continued, and further parameters have been studied. Investigations have been extended to both polychloroprene and, more recently, polybutadiene. The effects of changes in initial crosslink density, strain rate, and the introduction of various crystalline morphologies on the deformation process have been investigated.

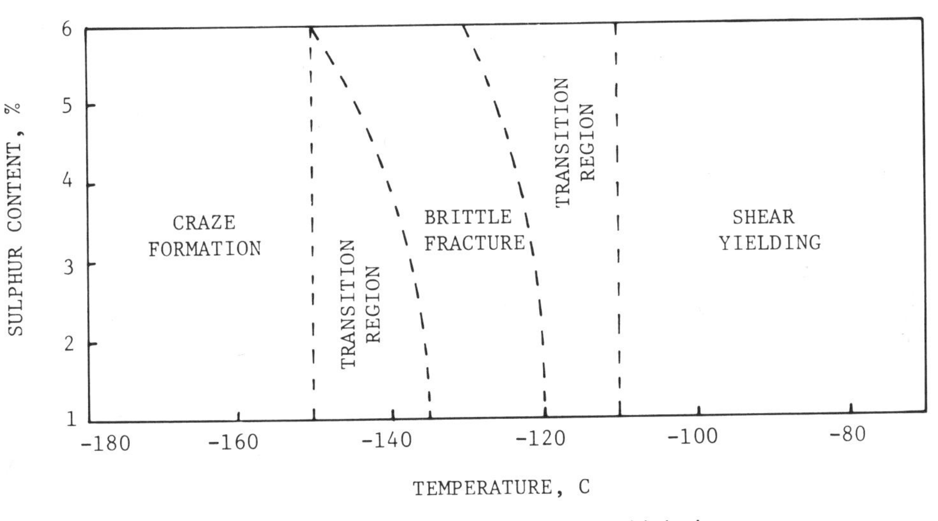

Fig. 2. Three regions of different mechanical behavior.

2 EXPERIMENTAL

The manner in which the experiments are performed is briefly as follows. Tensile specimens are prestretched to the required amount at room temperature before quenching to the test temperature in the extended condition. The tensile tests are then performed on the oriented glassy material in the usual way at the required extension rate. The tensile test is terminated either at fracture or at an earlier stage, and a section of deformed specimen is taken for ESR or gas-analysis investigations. In either case, the section is maintained at -196 C prior to subsequent testing.

3 NATURE OF RADICALS FORMED

The section taken from the specimen after tensile testing is transferred to the ESR equipment at -196 C and placed in the cavity precooled to -170 C. The temperature of the specimen is then adjusted incrementally, spectra and spin concentrations being recorded at each temperature increment.

Fig. 3 shows the spectra of sulphur-cured polyisoprene, dicumyl peroxide-cured polyisoprene, and polychloroprene taken at -160 C from specimens tensile tested at -180 C. The spectra of the peroxide-cured polyisoprene and polychloroprene are attributed to a peroxy radical, $RO_2^{\cdot}$. This radical is frequently found[6,7] in polymers degraded when oxygen is present. The presence of further lines in the spectrum of the sulphur-cured polyisoprene is attributable to sulphur radicals, RS· due to further scission either of the sulphur cross links or of sulphur hexacycles in the main chain. Peroxy radicals are secondary radicals formed by the combination of the primary radicals with oxygen dissolved in the samples. The primary radicals are thought to be allyl radicals produced by scission of the main chain between the α-methylene groups, i.e., at the weakest link.

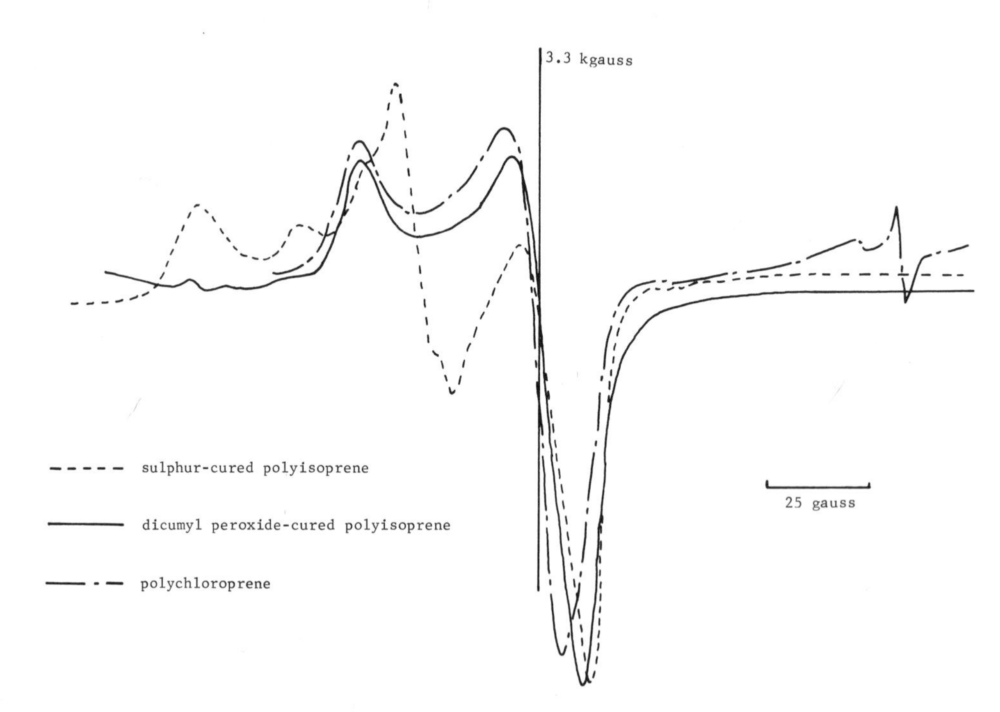

Fig. 3. Spectra of sulphur-cured polyisoprene, dicumyl peroxide-cured polyisoprene, and polychloroprene taken at -160 C.

The peroxy radicals observed at -160 C are very stable. On warming the specimens through T_g, the form of the spectrum changes to a broad singlet (Fig. 4) and the radical concentration decreases rapidly (Fig. 5).

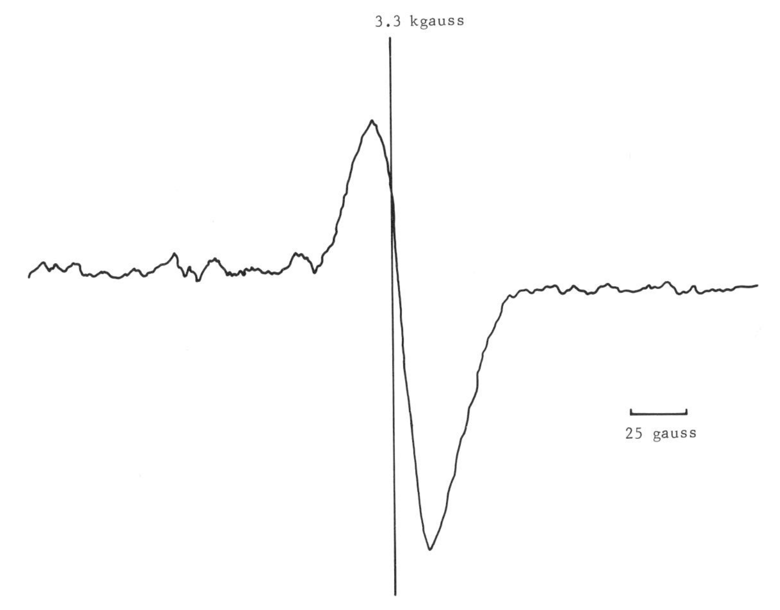

Fig. 4. Spectrum of sulphur-cured polyisoprene above T_g

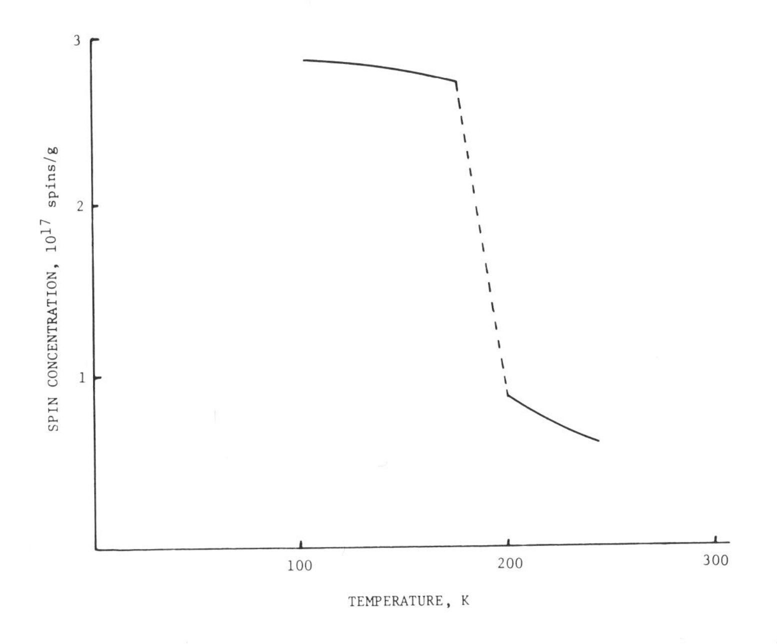

Fig. 5. Variation in radical concentration with temperature.

4 FACTORS AFFECTING THE RADICAL CONCENTRATION

So far, the effects of five variables on the formation of radicals have been investigated, viz., the initial crosslink density, impurity content, strain rate, initial orientation of the sample, and the extent of deformation during tensile testing.

The effect of varying the initial crosslink density of a sulphur-cured natural rubber specimen is shown in Fig. 6. All the spin concentrations shown in the figure are recorded on specimens which had been preoriented by 100 percent and subsequently strained by the same amount during the tensile test. It can be seen that the spin concentration increases with increase in the initial crosslink density (sulphur content). The more highly crosslinked material will have shorter network chains. Since random crosslinking occurs during vulcanization, more of the network chains will be fully extended and liable to rupture at any given extension in a more highly crosslinked sample.

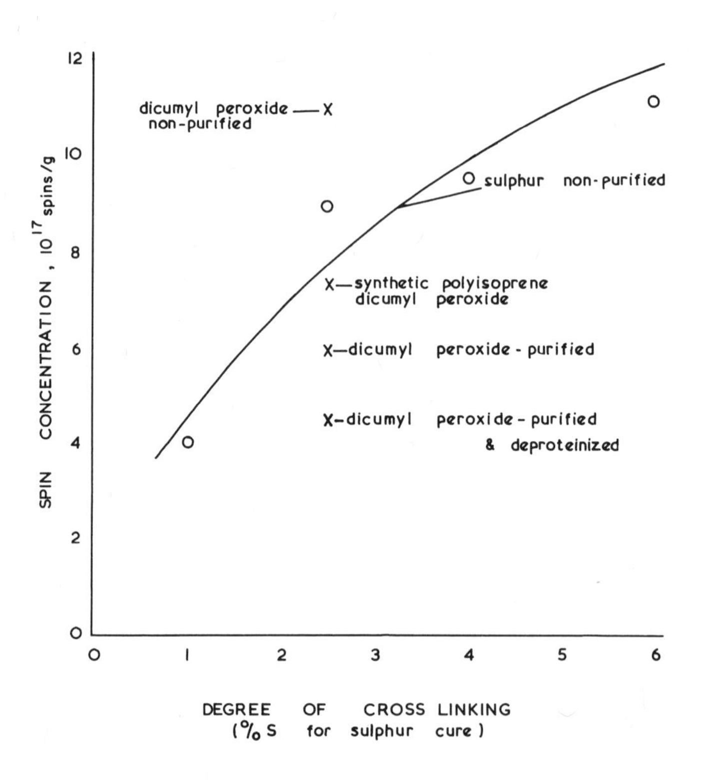

Fig. 6. Variation in radical concentration with crosslink density and purity of sample.

Fig. 6 also shows the effect on spin concentration of purifying a dicumyl peroxide-cured sample with crosslink density equal to that of a 2.5 percent sulphur-cured sample. Increasing purification, first by acetone extraction and then by deproteinizing, steadily decreases the spin concentration. The reason for this is not fully understood. Impurities appear to increase the probability of chain rupture at a given strain. This may be due to the impurities causing localized stress or strain intensification.

The spin concentration increases monotonically with strain once yielding is established. No evidence of radical formation has been found prior to the upper yield point of the load-extension curve.[2,8] This increase in the number of free radicals with deformation is accompanied by an observed increase in either the number or size of the striations (craze bands), as shown on a polychloroprene sample in Fig. 7. The increase in the size of the striations may, in some instances, be due to the coalescence of adjacent craze bands and thus to an increase in size of the crazed region rather than of any individual band.

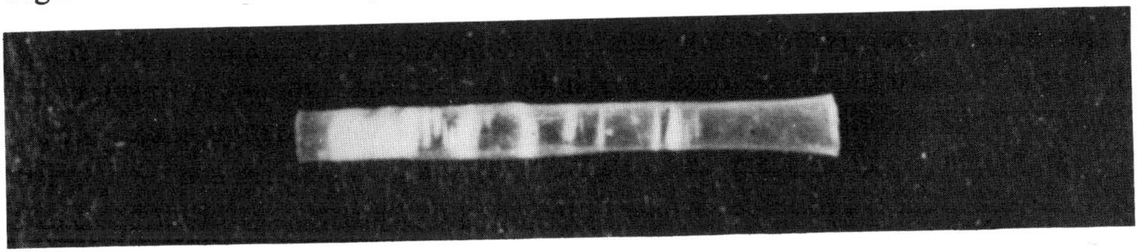

Fig. 7. Striations in polychloroprene sample.

Strain rate also affects the spin concentration.[9] Fig. 8 shows the variation of spin concentration at a given overall strain for polychloroprene, for extension rates varying over two decades. The graph shows a maximum at an extension rate of 6 cm/min. Since the number of radicals formed is related to the number of craze bands formed, Fig. 8 suggests that at extension rates lower and higher than the optimum, the concentration of craze bands decreases. At lower extension rates, there will be adequate time for viscous flow within a few craze bands to account for the overall extension of the specimen. If the strain rate is sufficiently great, brittle fracture obtains which may be considered as rapid fracture passing through a single propagating craze band. The reduction in spin concentration at the higher strain rates would suggest that the situation of a single, propagating, brittle fracture is approached gradually from a ductile region involving crazing, over a range of strain rates.

Table I shows the variation of spin concentration at rupture with change in preorientation prior to tensile testing for polychloroprene.

The table shows the averages of results from three types of specimens: (a) amorphous samples, (b) samples containing row-nucleated crystallites (type I), and (c) a second semicrystalline form consisting of deformed spherulitic material (type II). The results in Table I show no

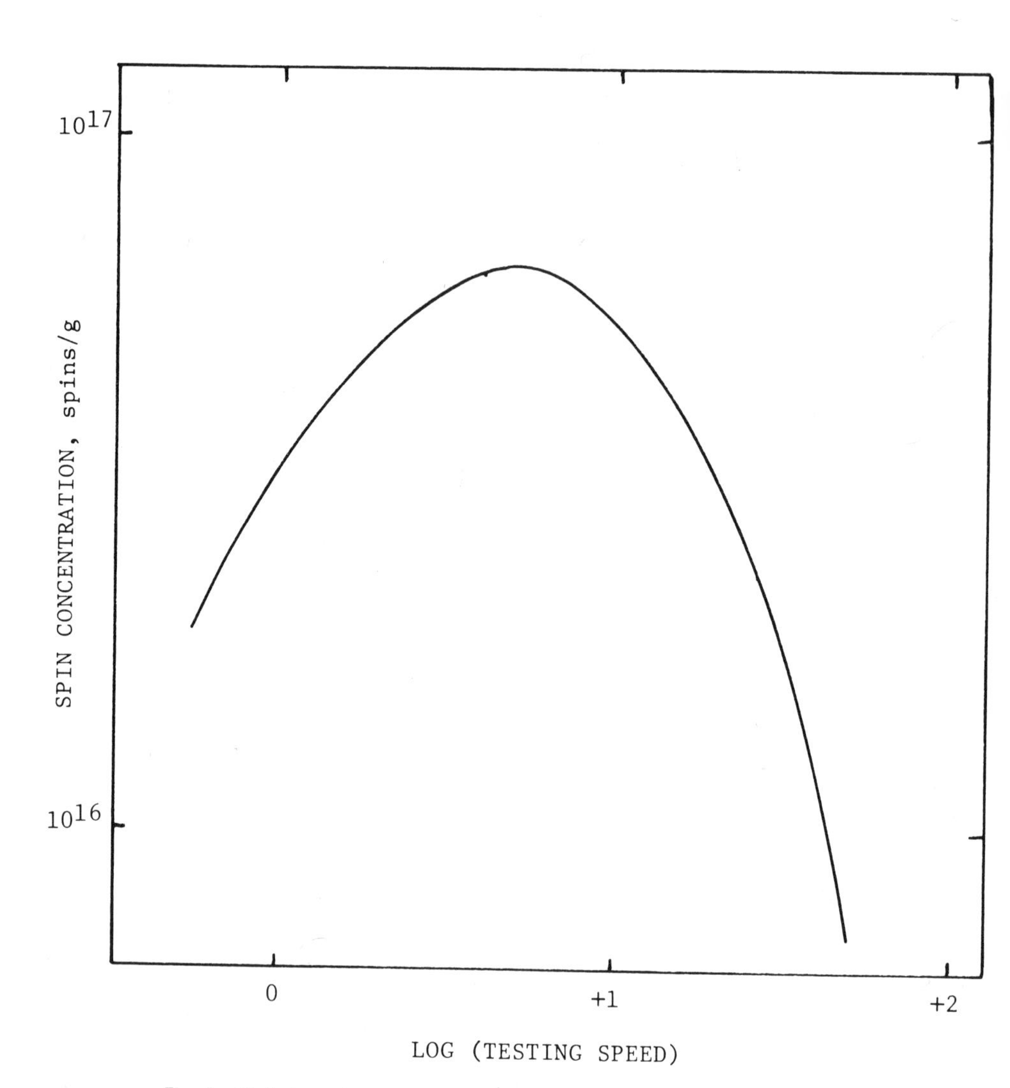

Fig. 8. Variation of spin concentration with strain rate for polychloroprene.

Table I. Variation of Polychloroprene Spin Concentration at Rupture With Change in Preorientation

Preorientation, %	Spin Concentration, 10^{16} spins/g		
	Amorphous	Crystalline (Type I)	Crystalline (Type II)
100	2.4	3.0	1.62
200	2.5	1.55	3.68
300	–	5.6	7.10

clear trend. However, detailed study of the stress-strain curves for these specimens shows that the strain to rupture varies in a manner similar to that of the spin concentration. Hence, the spin concentration appears to be a function of the extensibility of the material, and probably of the extensibility of the amorphous regions alone. The crystalline morphology does not appear to play any significant role in radical formation other than that of modifying the strain in the amorphous regions. Fig. 9 shows the variation in spin concentration with increasing postyield strain in polyisoprene.

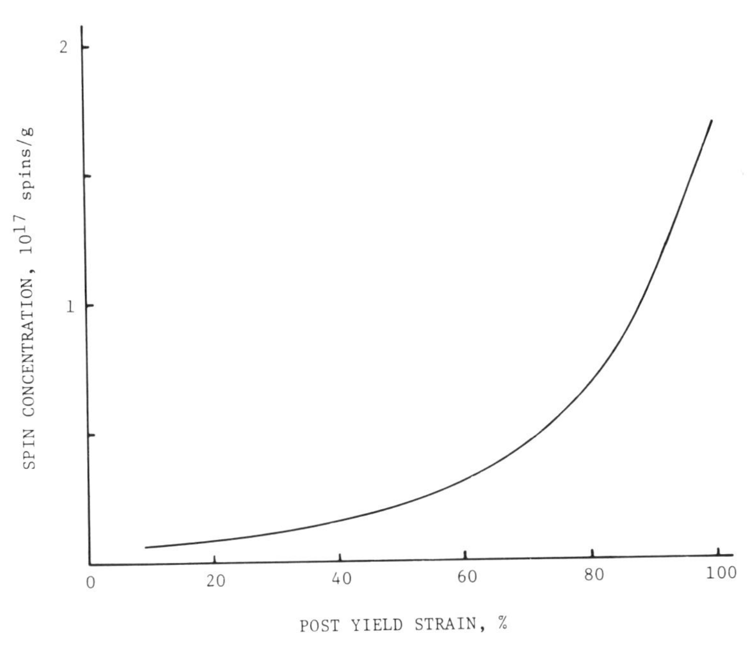

Fig. 9. Variation in spin concentration with deformation for polysioprene.

5 GAS EVOLUTION

It has been previously stated that the production of ESR signals is associated with the formation of numerous white striations which foam on subsequent warming of the specimens to room temperature. The foaming is due to the expansion of gases trapped in the specimen after tensile testing is complete. Analysis of the released gas shows it to be largely a mixture of environmental gases (nitrogen and argon) but including up to 15 percent of volume of hydrogen.* Recent investigations on polybutadiene[10] indicate that hydrogen is not present at the tensile test temperature

*Recent results suggest that the hydrogen, observed in mass spectrographic analysis, *may* come from breakdown of a low-molecular-weight hydrocarbon rather than hydrogen gas.

but is subsequently released on warming the specimen through T_g. The hydrogen release is thus associated with the decay in the ESR signal. The nitrogen and argon are basorbed environmental gases from the atmosphere in which the tensile tests are performed. The total quantity of gas released follows the same trends as the spin concentration[2], increasing both with increase in strain and with increase in crosslink density.

Striations, ESR signals, subsequent foaming, and gas evolution are obtained with both polychloroprene and natural-rubber specimens when the degree of preorientation is small, i.e., 100 to 200 percent. Tests on polychloroprene having 300 percent preorientation showed a different mode of deformation when tensile tested at -180 C. These specimens deformed in a ductile manner but did not exhibit white striations, nor foam or release environmental gases, on warming to room temperature. They gave a slightly different ESR spectrum to specimens preoriented by lesser amounts. Preextending by 300 percent thus suppresses craze formation and hence the absorption of the environmental gases, including oxygen, into the voids in the craze bands. The exclusion of oxygen reduces the peroxy radical concentration and causes the change in ESR spectrum. However, the production of the same intensity of radicals as when craze bands are formed suggests that the broken chains are still present but more evenly dispersed throughout the specimen and not concentrated in the craze bands. It is suggested that, with specimens preoriented by 300 percent, the microvoiding still exists but that the voids are smaller and more widely distributed in the deformed specimen. This is in accordance with the work of Zhurkov et al.[11] on microcavitation in several oriented polymers.

Considerable quantities of environmental gases are absorbed during the tensile test.[2] Values measured have been as high as 60 ml of gas for each 1 ml of material, the gas volume being measured at standard temperature and pressure. Such a volume cannot be contained in gaseous form in the specimen at the test temperature even after allowing for the contraction of the gas on cooling and the volumetric change in the specimen during deformation. The gas must thus be stored in the specimen in either liquid form or as an adsorbed layer on the inside surface of the voids. Calculations show the latter suggestion to be possible and that the number of voids, assumed spherical, required to contain the observed volume of gas approximates the spin concentration observed if the gas is present as a monolayer.

6 CONCLUSION

The phenomenon being investigated is an interesting example of mechano-chemistry. Molecular rupture is obtained by purely mechanical means which gives rise to the observed ESR spectra. Owing to the presence of oxygen, the radicals observed are predominantly peroxy radicals formed by the combination of the primary radicals with oxygen. The prerequisite to obtaining *observable* radical formation appears to be any means of pinning the molecules such that they become effectively network chains unable to flow past one another. Crosslinking has been used in the elastomers tested in this paper to provide this molecular anchorage, since it is, to some extent, quantitative, but crystallization can clearly serve the same function.

The degree of preorientation of the amorphous phase determines the deformation mode of the elastomers tested. Zero preorientation results in brittle behavior. Small amounts of orientation, 100 to 200 percent, result in free-radical formation, and the related absorption of environmental gases. Further preorientation gives rise to ESR spectra alone, without observable craze formation, probably owing to a more even dispersion of much smaller microvoids such as were observed using small-angle X-ray scattering by Zhurkov et al.[11]

It appears that microcavitation is a universal phenomenon in the tensile deformation and fracture of polymers. In uncrosslinked amorphous materials, localized crazing is known to be a precursor to brittle fracture. The effect of crosslinking is to increase, progressively, the likelihood of molecular fracture, and the consequent microcavitation becomes delocalized, giving rise to a macroscopic yielding. Molecular orientation and/or crystallinity further delocalizes the cavitation, eventually rendering it undetectable by visual inspection or by gas absorption analysis. The persistence of strong ESR signals and low-angle X-ray scattering, nevertheless, points to the continued formation of microcavities, both as a precursor to macroscopic fracture and as a plastic deformation mechanism.

ACKNOWLEDGMENTS

The authors wish to thank Dr. R. Natarajan and Dr. B. Reeve for performing the experimental work and the Scientific Research Council for a Research Grant.

REFERENCES

1. Andrews, E. H., and Reed, P. E., *J. Polym. Sci.*, Part B, **5**, 317 (1967).
2. Natarajan, R., and Reed, P. E., *J. Polym. Sci.*, Part A-2, **10**, 585 (1972).
3. Kausch, H. H., *Reviews in Macromol. Chem.*, **5**, 97 (1970).
4. Roylance, D. K., DeVries, K. L., and Williams, M. L., *Fracture 1969*, Chapman & Hall (1969), p 551.
5. Becht, J., and Fischer, H., *Kolloid. Z. u. Z. Polymere*, **229**, 167 (1969).

6. Zhurkov, S. N., and Tomashevsky, E. E., *Physical Basis of Yield ana Fracture,* Institute of Physics, London (1966).
7. Carstersen, P., *Macromol. Chem.,* **142**, 131 (1971).
8. Brown, R., DeVries, K. L., and Williams, M. L., Paper presented at American Chemical Society meeting, Chicago, September, 1970.
9. Reeve, B., "Morphology and Tensile Properties of Polychloroprene", Ph.D. Thesis (University of London) (1972).
10. Mead, W., and Reed, P. E., to be published.
11. Zhurkov, S. N., Kuksenko, V. S., and Slutsker, A. I., *Fracture 1969,* Chapman & Hall (1969), p 519.

DISCUSSION on Paper by E. H. Andrews

MÜLLER: The decrease in spin concentration for high-speed deformation could be a consequence of increase in temperature with such high testing speeds. An increase in temperature may cause more rapid recombination processes.

ANDREWS: This is a possible explanation. The testing rate was not very high, being limited to the speeds available on an Instron testing machine. However, the *local* strain rate in small regions of the specimen could be much higher.

THOMAS: Some work carried out at N.R.P.R.A. has indicated that at room temperature and at high stresses (a significant fraction of the breaking stress), rupture of chemical bonds takes place in a crosslinked network. A good proportion of these broken bonds re-form in the strained state, producing permanent set. Changes in equilibrium swelling also occur, and in many cases a reduction in swelling is found, apparently indicating a net increase in crosslinking. However, this interpretation is not generally correct, as a crosslink formed in the strained state will, on the two-network theory, produce a larger reduction in swelling than one introduced in the unstrained state, a point which may be relevant to Dr. Andrews' measurements. It also appears that if polysulphide crosslinks are present, rupture takes place preferentially at these crosslinks rather than in the main chain.

ANDREWS: We have wondered whether crosslink breakage may be an alternative explanation to main-chain fracture. However, the sulphur and peroxide cures give similar behavior as regards crosslink density increase after deformation. This suggests that the nature of the crosslink may not be of primary importance. The point is worth studying, however.

KNAPPE: In crosslinking rubbers by sulphur, some cyclization occurs which will result in a stiffening of the chain and a shift of transition temperature to higher values, more than in the case of crosslinking by peroxides or radiation. Has the effect of cyclization some influence on your results?

ANDREWS: Yes, it is possible, but we do not know. What we do believe is that the difference between the ESR spectra of sulphur-cured and peroxide-cured cis-polyisoprene is due to the presence of sulphur radicals in the former case. The spin concentration is higher in peroxide cures, but since this concentration can be reduced below that for the sulphur cure by purification, it is impossible to comment on the more detailed role of molecular microstructure.

KNAPPE: The development of hydrogen is well known in the case of radiation crosslinking of PE. Can these results be correlated to your experiments?

ANDREWS: Possibly yes, but would have to be a different mechanism from irradiation of PE. We have irradiated our materials and looked at the ESR spectrum produced. It is different from the spectra produced by stretching, however.

LANDEL: Were the samples purified before or after crosslinking?

ANDREWS: Deproteinization was carried out before crosslinking, but acetone extraction after crosslinking.

LANDEL: You showed a decrease in the number of radicals (spins) with purification – if there is a change in the type of radical it would show up in the nature of the signal or in the course of the reactions leading to the hydrogen production. Therefore, I would like to ask if there was a change in the amount of hydrogen evolution in the purified samples, and was there a change in the nature of the SV signal in these samples?

ANDREWS: The spectrum of purified samples was the same as for unpurified material. Hydrogen volume has not yet been studied as a function of purity.

KAUSCH: The group of people investigating fracture phenomena by ESR is particularly thankful for the extension of this work to elastomers. Thus, our basis is broadened from which the effect of cohesive energy density and orientation on radical formation is judged. In this context, I have two questions:

1. Does the number of radicals observed correspond to the increase

in degree of crosslinking?

2. What significance do you attach to the fact that cavitation is observed only simultaneously with the breakage of chains?

ANDREWS:

1. Yes, approximately. About two new crosslinks are formed for each molecular fracture event.

2. We think that cavity formation is nucleated by molecular fracture.

TIME-TEMPERATURE SUPERPOSITION IN HETEROPHASE BLOCK COPOLYMERS

A. Kaya, G. Choi, and M. C. Shen*

Department of Chemical Engineering
University of California
Berkeley, California

ABSTRACT

The viscoelastic behavior of a styrene-butadiene-styrene block copolymer cast from several different solvents was studied in order to examine the validity of the time-temperature superposition principle in heterophase polymers. It is found that stress-relaxation isotherms over a wide temperature range can be easily superposed into smooth master curves. The master curves compare favorably with other mechanical data determined at long times and at high frequencies. The shift factors used in superposition can be rationalized by an additive model. However, it is pointed out that this apparent thermorheological simplicity is strictly valid only in the region where one of the relaxation mechanisms dominates. Consequently, only those portions of the master curve that meet this criterion are expected to be useful. Dynamic mechanical data illustrating the effect of multiple relaxations are presented.

*Present address: Department of Chemical Engineering, University of Michigan, Ann Arbor, Michigan.

1 INTRODUCTION

The time-temperature-superposition principle is an extremely powerful tool in the study of viscoelastic properties of polymers. On the basis of this principle, viscoelastic master curves covering many logarithmic decades of time can be obtained by simple horizontal shifting of modulus-time isotherms (over short periods of time) along the time axis in a log-log plot. Materials obeying this principle are said to be thermorheologically simple. That most homogeneous amorphous polymers are thermorheologically simple has been amply demonstrated.[1,2]

It is now generally accepted, on the strength of evidence provided by electron microscopy, small-angle X-ray diffraction, light scattering, and other techniques, that most of the block and graft copolymers and polyblends show microphase separation. In fact, many of the most desirable characteristics of these polymers, such as impact resistance and thermoplasticity, are directly attributable to their nonhomogeneity. A number of workers have applied the time-temperature superposition principle to the study of the viscoelastic properties of these heterophase polymers. However, it has recently been pointed out by Fesko and Tschoegl[3] that heterophase polymers should, in general, be considered thermorheologically complex. The purpose of this paper is to examine to what extent the simple time-temperature superposition may still be useful for such materials, and the caution one must exercise in the interpretation of the resulting viscoelastic master curves.

2 EXPERIMENTAL

A styrene-butadiene-styrene (SBS) triblock copolymer was obtained from the Shell Chemical Company as Kraton 1101. This sample contains 28 weight percent polystyrene. The number average molecular weight of the block copolymer is 76,000. The polybutadiene portion consists of 51 percent trans-1,4, 41 percent cis-1,4, and 8 percent vinyl structure.[4] Pellets were dissolved in benzene, carbon tetrachloride, and tetrahydrofuran (THF) containing 10 percent methyl ethyl ketone (MEK) by volume. The solution was poured on a mercury surface and the solvent was permitted to evaporate slowly. Sheets 0.15 cm thick were cast for stress-relaxation experiments, and 0.01 cm thick for measurements in the Rheovibron. After 2 weeks, the sheet was removed and dried *in vacuo* at 45 C until it had reached constant weight.

Stress-relaxation data were taken on an Instron Model TM-SM universal testing machine. Temperature control was provided by a Missimers Model PITC temperature controller to ±0.5 C. The tensile mode was used up to a modulus of 10^9 dynes/cm^2. Above this value, the flexural mode was found to be more accurate. Samples were cut into rectangular strips by high-speed rotating blades. They were maintained at the temperature of the experiment for 2 hours prior to the application of strain. In all measurements, strains were kept below 3 percent in order to avoid yielding.

Dynamic mechanical measurements were made on a Rheovibron DDV-II-B with low-frequency sweep accessories (Toyo Measuring Instrument Company). An IMASS temperature chamber was used to achieve low temperatures.

3 RESULTS

Figure 1 shows the stress-relaxation isotherms for the THF-MEK-cast Kraton 1101. The tensile moduli shown in the isotherms were reduced to a reference temperature of 23 C according to the formula[1,2]

$$E_r(t) = 296\ E(t)\ \rho_0/T\rho \quad , \tag{1}$$

where ρ_0 and ρ are densities at reference and current temperatures. Equation 1 is strictly applicable to polymers near or above their glass transition temperatures. In the case of heterophase polymers, most of the isotherms were determined above the T_g of one component but below

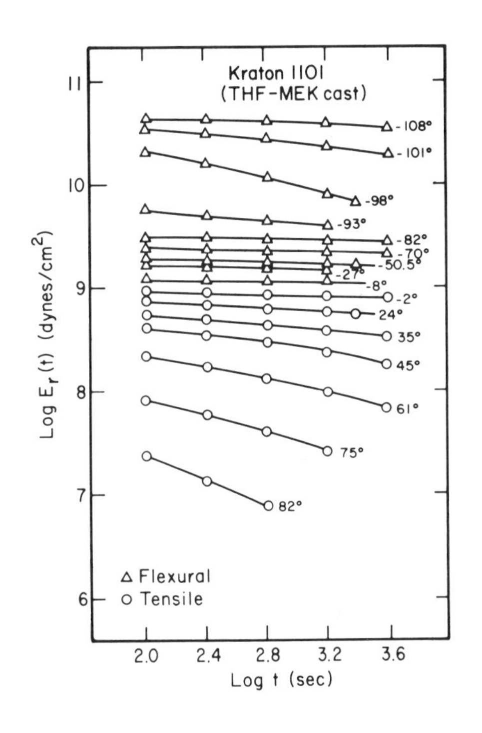

Fig. 1. Modulus-time isotherms for Kraton 1101 cast from the solution in tetrahydrofuran and methyl ethyl ketone. Moduli were reduced to 23 C by Eq. 1.

that of the other. It has been demonstrated by McCrum and Morris[5] that for polymeric glasses, the vertical shift should be determined separately from high-frequency data. There is thus some question, as yet unresolved, as to whether the vertical shift provided by Eq. 1 is entirely satisfactory for our block copolymer.

It is of interest to examine the modulus-temperature behavior of this heterophase polymer by cross-plotting data from Fig. 1 for a fixed reference time of 100 seconds. The modulus-temperature curve for the THF-MEK-cast Kraton 1101 is shown in Fig. 2. The transition at -90 C is attributable to the glass transition of polybutadiene phase, while that above 80 C is related to the T_g of the polystyrene phase. Between these transitions there is a rather extended plateau region. Also shown in Fig. 2 is the modulus-temperature curve for the CCl_4-cast Kraton 1101. Moduli in the plateau region for the latter are much lower than those for the THF-MEK-cast sample. This is to be expected since it has been demonstrated by electron microscopy[4] that styrene domains show much more interconnectivity if the sample is cast from THF-MEK rather than from CCl_4.

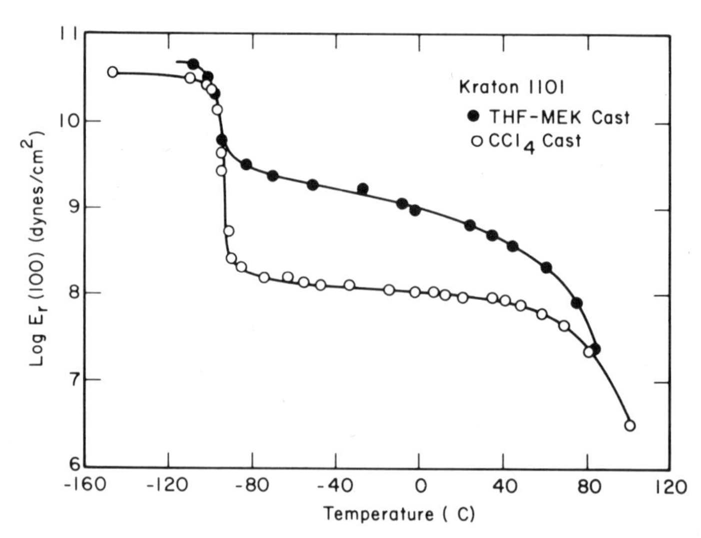

Fig. 2. 100-second moduli versus temperature for Kraton 1101 cast from the solution of tetrahydrofuran and methyl ethyl ketone, and from carbon tetrachloride.

The modulus-time isotherms for the THF-MEK-cast Kraton 1101 presented in Fig. 1 can be readily superposed into a smooth master curve by simple horizontal shifts along the log time axis. The same can also be accomplished for the CCl_4-cast sample, the isotherms for which have been published previously.[6] Master curves for these two samples are shown in Figs. 3 and 4. Shift factors used in obtaining these superpositions are given in Figs. 5 and 6.

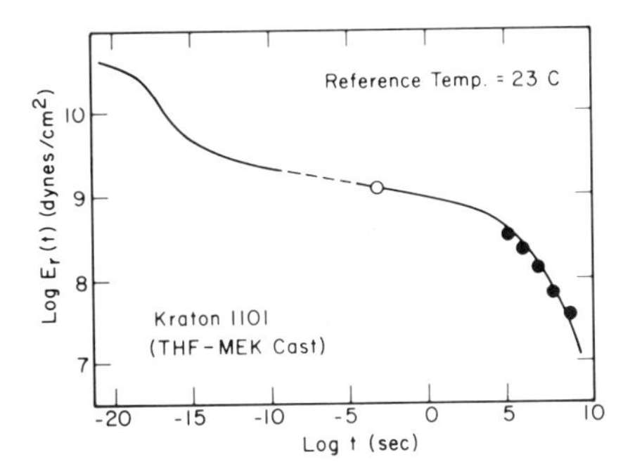

Fig. 3. Viscoelastic master curve for Kraton 1101 cast from the solution in tetrahydrofuran and methylethyl ketone (solid curve). Filled circles: long-time stress-relaxation data determined at 59 C and shifted to 23 C by shift factors given in Fig. 5; open circles: dynamic storage modulus determined at 2 kc at 23 C[8].

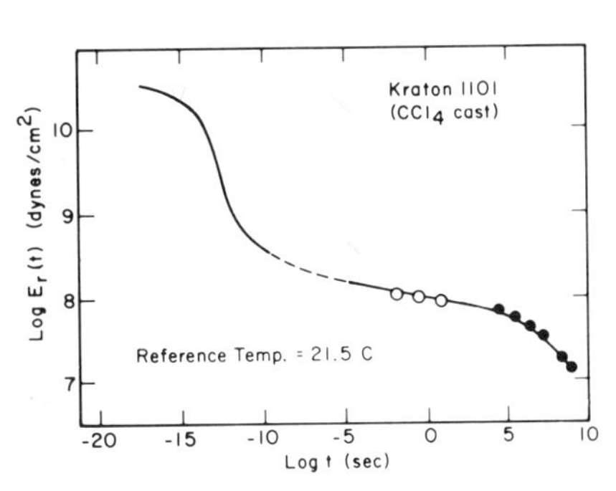

Fig. 4. Viscoelastic master curve for Kraton 1101 cast from carbon tetrachloride (solid curve). Filled circles: stress-relaxation data determined at 56 C and shifted to 23 C by shift factors given in Fig. 6; open circles: dynamic storage moduli determined at 0.1, 3.5, and 110 cps in the Rheovibron.

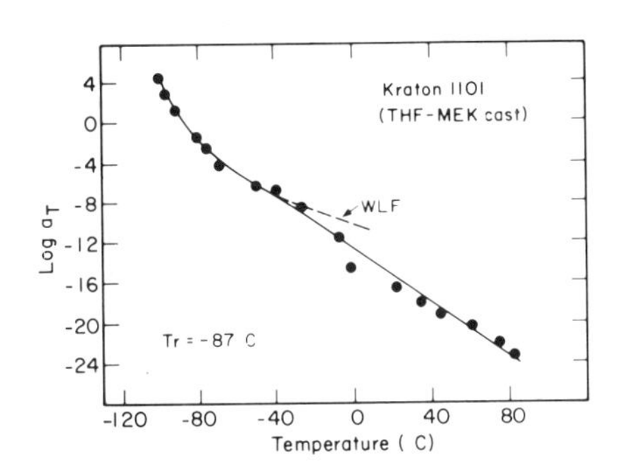

Fig. 5. Shift factor-temperature curve for Kraton 1101 cast from the solution of tetrahydrofuran and methylethyl ketone.

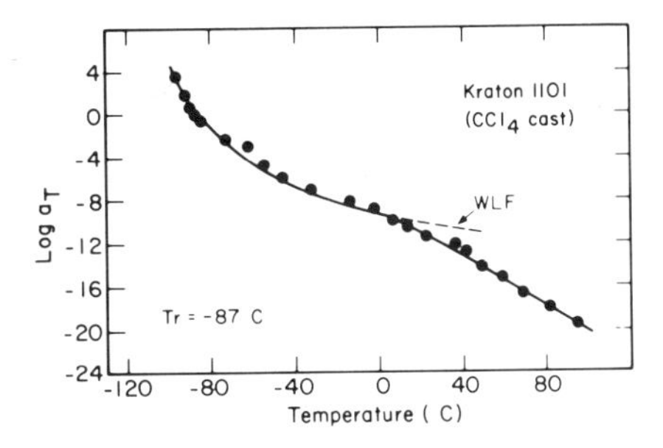

Fig. 6. Shift factor-temperature curve for Kraton 1101 cast from carbon tetrachloride.

4 DISCUSSION

Even though these heterophase polymers are thermorheologically complex, it would be of interest to determine whether the master curves obtained through the simple time-temperature superposition are at all useful, and if so, what the limitations are. First, we note that the shift factors do not follow the classical Williams-Landel-Ferry equation.[1,2] It is, however, possible to fit these data to a composite model, taking into account the contributions from the two phases. Alternative models describing the shift factors for block copolymers have been proposed previously.[11,12]

Shift factors for homogeneous polymers are defined as the ratio of the relaxation time for a given temperature to that for the reference temperature.[1,2] We assume that for heterophase polymers the shift factor can be defined by the following equation:

$$a_T = \left[\frac{(\tau_{ir})_T}{(\tau_{ir})_{T_r}}\right]\left[\frac{(\tau_{jg})_T}{(\tau_{jg})_{T_r}}\right] \quad . \tag{2}$$

In the above equation, τ_{ir} is the ith relaxation time of the rubbery phase, τ_{jg} the jth relaxation time of the glassy phase, and T_r the reference temperature. What the model suggests is that the relaxational behavior of the individual phases is the same as that of the pure polymers and, thus, any shift that affects one of the phases affects the shift behavior of the whole system. The cumulative effect of the temperature dependence of the shift factor on the whole system follows, as the equation above suggests, a multiplicative rule. Since the constituent homopolymers themselves are thermorheologically simple, the symbols i and j in Eq. 2 are, in reality, "dummy" symbols – meaning that it makes no difference which relaxation time of a given phase is chosen, provided that the same relaxation time is considered at the reference temperature.

Our model is essentially that for heterophase polymers at low temperatures near the T_g of the rubbery phase; the relaxation behavior of the system is controlled by this phase, whereas, at temperatures near the T_g of the glassy phase, the behavior is closely associated with the latter phase. At very low temperatures, the glassy phase acts essentially as a filler, the τ_{jg} become very large, and the term $(\tau_{jg})_T/(\tau_{jg})_{T_r}$ is unaffected by a change in temperature, thus acting essentially as a constant. In this case, the temperature dependence of the shift factor is affected only by the $(\tau_{ir})_T/(\tau_{ir})_{T_r}$ term. The reverse situation takes place at the other end of the temperature range of interest, or near the T_g of the glassy phase. In between the two T_g's, the relaxation behavior of the system is affected by both phases.

Taking logarithms of both sides of Eq. 2, we get:

$$\log a_T = \log \frac{(\tau_{ir})_T}{(\tau_{ir})_{T_r}} + \log \frac{(\tau_{jg})_T}{(\tau_{jg})_{T_r}} \tag{3}$$

or

$$\log a_T = \log a_T^r + \log a_T^g \quad , \tag{4}$$

where a_T^r is the shift factor of the rubbery phase and a_T^g is that of the glassy phase. Equation 4 shows that if we consider the curve for log a_T versus the temperature, the shift factor of the heterophase system is simply obtained by adding the shift-factor curves of the individual homopolymers. A similar model proposed by Fesko and Tschoegl[3] requires weighting factors. Since the latter are difficult to determine, we shall in this work assume that the contributions were weighted automatically by the temperature dependence of the respective phases.

The shift factor in the rubbery phase can be represented by the Williams-Landel-Ferry equation[1,2]

$$\log a_T^r = -\frac{C_1 (T-T_r)}{C_2 + (T-T_r)} \quad . \tag{5}$$

For polybutadiene[2,6], C_1 = 11.2, C_2 = 60.5.

The reference temperature is taken to be -87 C. The shift factor in the glassy phase is given by the modified WLF equation proposed by Rusch[7]:

$$\log a_T^g = -\frac{C_1 (T_e-T_r)}{C_2 + (T_e-T_r)} \quad . \tag{6}$$

In Eq. 6, T_e, the "effective temperature", is defined as

$$\begin{aligned} T_e &= T + w_f(T)/\Delta\alpha_f \qquad \text{at } T \geqslant T_2 \\ T_e &= T_2 + w_f(T)/\Delta\alpha_f \qquad \text{at } T \leqslant T_2 \\ T_e &= T \qquad \text{at } T \qquad \text{at } T > T_g \quad , \end{aligned} \tag{7}$$

where $w_f(T)$ is the fraction of nonequilibrium free volume and $\Delta\alpha_f$ is the thermal-expansion coefficient of the free volume. For polystyrene, Rusch found C_1 = 12.4, C_2 = 41, T_2 = 59 C, T_r = 100 C, and $\Delta\alpha_f$ = 4.5 x 10^{-4}/C. Values of $w_f(T)$ are taken from the work of Rusch. The solid lines in Figs. 5 and 6 are the calculated curves. In order to obtain a good fit between the calculated and experimental curves, the polystyrene phase was assumed to begin contributing to the relaxation at 12 C for the CCl_4-cast sample, and at -26 C for the THF-MEK-cast sample.

In order to test the validity of the master curve, we performed stress-relaxation measurements over a period of time that is long (>5 decades) compared with the span covered by the isotherms (~1.5 decades) at a temperature (59 C) different from the reference temperature (room temperature). These long-time relaxation moduli are then shifted to the reference temperature by the shift-factor curve of Fig. 5. These are shown in Fig. 3 as filled circles for THF-MEK-cast Kraton 1101. The agreement with master curves is excellent. The same can be said of the CCl_4-cast sample (Fig. 4).

It is well known that the dynamic storage moduli (E′) determined at a comparable time (or frequency) scale have values negligibly different from the relaxation moduli.[1] In Fig. 3 we show the room-temperature modulus of THF-MEK-cast Kraton 1101 determined at ~2 kc (open circle) by an acoustic spectrometer.[8] In Fig. 4, the room-temperature moduli of the CCl_4-cast sample, determined by means of the Rheovibron at 0.1, 3.5 and 110 cps, are compared with the master curve. The agreements are very good in both cases.

However, despite the apparent utility of time-temperature superposition principle for our heterophase block copolymer, the latter should not be considered a truly thermorheologically simple material. The master curve for a heterophase polymer is supposed to cover at least two transition regions. Now the relaxation time for any of the relaxation processes is governed, at least over a narrow temperature range, by its own activation energy:

$$\tau = \tau_r \exp(\Delta H_a/RT) \quad . \tag{8}$$

Unless the activation energies of the two transitions are the same, their temperature dependence will not be the same. The activation energies for polystyrene and polybutadiene can be estimated from the WLF equations[2] by

$$\Delta H_a = 2.303 \, d \log a_T / d(1/T) \quad . \tag{9}$$

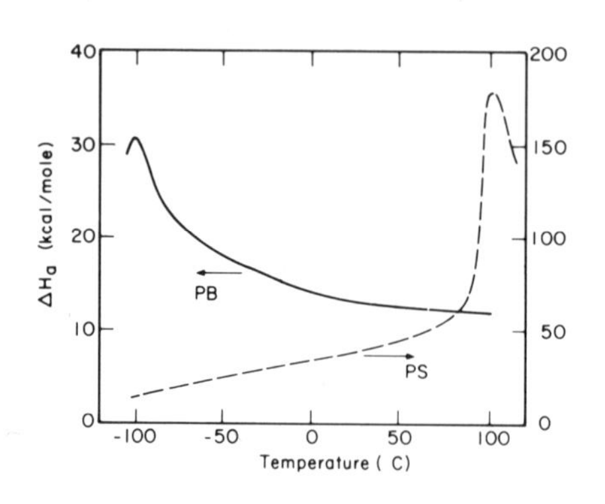

Fig. 7. Activation energies of polystyrene (PS) and polybutadiene (PB) estimated by differentiating Eqs. 5 and 6 according to Eq. 9.

As seen in Fig. 7, the quantity is very different for polystyrene and polybutadiene throughout the entire temperature region. Thus, if the master curve shown in Figs. 3 and 4 is valid for room temperature, it will not remain valid at another temperature. The shapes of the master curves would be similar at all temperatures, but the two transitions will not be separated by the same logarithmic time scale at different temperatures because of the difference in activation energies.

The above considerations indicate that the time-temperature superposition principle is strictly applicable only in the region where one of the relaxations dominates. This is the reason that the long-time end of the master curve agrees well with experimental data. The effect of multiple relaxations has been discussed previosuly by Ferry[9], who demonstrated that the presence of secondary relaxations would render superposition of the isotherms difficult unless the latter's contributions were removed. In a recent paper[10] we have found that the shift factors necessary in obtaining smooth master curves and the resulting master curves themselves are different for the same block-copolymer sample, depending on the time region covered by the isotherms. One set of isotherms was obtained by mechanical stress-relaxation experiments within the log time region of 2.0 and 3.5, while the other set of data were determined dielectrically in the frequency range of 10^2 to 10^5 (or approximately equivalent to the log time region of -2.0 to -5.0). These observations are consonant with the conclusions reached by Fesko and Tschoegl[3] in their model calculations.

One manifestation of this effect can be seen in Fig. 8. Here, the loss tangent of a benzene-cast Kraton 1101 was determined as a function of temperature at three different frequencies. As the frequency is increased, the loss peaks are shifted to higher temperatures. However, because of the difference in activation energies, the low-temperature peak (associated with the glass transition of polybutadiene) moved much less with changes in frequency than did the high-temperature peak (polystyrene glass transition). Thus, the shapes of these curves are similar, but the separation in temperature between the peaks is different for each frequency. For this reason we have used broken lines in Figs. 3 and 4 to indicate that the time scale may change with temperature for these master curves.

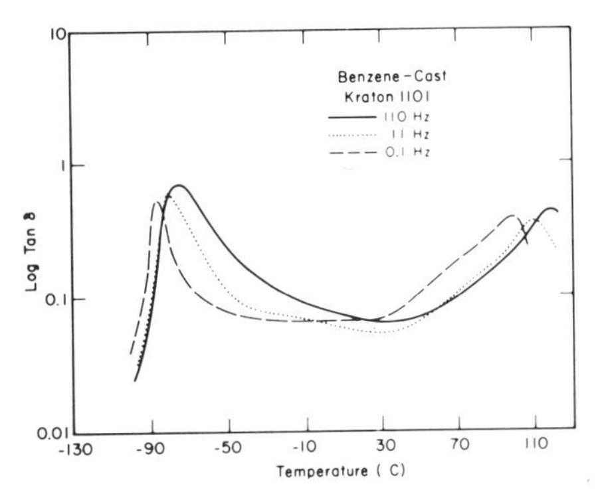

Fig. 8. Loss tangent-temperature data of Kraton 1101 cast from a benzene solution.

Another factor that must be considered in studying the viscoelasticity of heterophase polymers is the role of the morphological structure. It is well known from electron-microscope investigations that the same block copolymer may have different morphological features, depending upon the solvent from which it was cast. In this work we found that by assuming that the polymer in microphases behaves in the same way it does in bulk, shift-factor curves may be satisfactorily explained (Figs. 5 and 6). However, these figures also show that shift-factor curves for the same heterophase polymer may be drastically different if cast from a different solvent. Thus, a detailed account of the supramolecular structure may be required in attempting a complete description of the viscoelastic behavior of heterophase polymers.

ACKNOWLEDGMENT

We are indebted to Professor N. W. Tschoegl for very helpful discussions. This work was supported by the Petroleum Research Fund, administered by the American Chemical Society, and by the Advanced Research Projects Agency of the Department of Defense, monitored by the Office of Naval Research under Contract No. N00014-69-A-0200-1053.

REFERENCES

1. Tobolsky, A. V., *Properties and Structure of Polymers,* Wiley, New York (1960).
2. Ferry, J. D., *Viscoelastic Properties of Polymers,* 2nd Ed., Wiley, New York (1970).
3. Fesko, D. G., and Tschoegl, N. W., *J. Polymer Sci.,* Part C, **35**, 51 (1971); *Int. J. Polym. Mat.,* in press.
4. Beecher, J. F., Marker, L., Bradford, R. D., and Aggarwal, S. L., *J. Polymer Sci.,* Part C, **26**, 163 (1969).
5. McCrum, N. G., and Morris, E. L., *Proc. Roy. Soc.,* **A281**, 258 (1964).
6. Jamieson, R. T., Kaniskin, V. A., Ouano, A. C., and Shen, M., *Advances in Polymer Science and Engineering,* K. D. Pae, D. R. Morrow, and Y. Chen (Eds.), Plenum, New York (1972), p. 163.
7. Rusch, K. C., *J. Macromol. Sci.,* **B2**, 179 (1968).
8. Shen, M., Circlin, E. H., and Kaelble, D. H.,in *Colloidal and Morphological Behavior of Block and Graft Copolymers,* G. E. Molau (Ed.), Plenum, New York (1971), p 307.
9. Ferry, J. D., Child, W. C., Zand, R., Stern, D. M., Williams, W. L., and Landel, R. F., *J. Colloid Sci.,* **12**, 53 (1957).
10. Kaniskin, V. A., Kaya, A., Long, A., and Shen, M., *J. Appl. Polymer Sci.,* in press.
11. Shen, M., and Kaelble, D. H., *Polymer Letters,* **8**, 149 (1970).
12. Lim, C. K., Cohen, R. E., and Tschoegl, N. W., *Advan. Chem. Series,* **99**, 397 (1971); Cohen, R., and Tschoegl, N. W., *Int. J. Polym. Mat.,* **2**, 49 (1972).

DISCUSSION on Paper by A. Kaya, G. Choi, and M. Shen

KNAUSS: In your closing remarks you indicate that you have difficulty in interpreting your data in a conclusive way. You have tried to predict the thermorheological behavior of heterogeneous polymers from the properties of the individual phases which later are taken as thermorheologically simple. In this attempt, you have not at all referred to the mechanics of deformation in multiphase solids. In fact, you are dealing with a composite material made up of thermorheologically simple constituents which adhere to each other. It seems to me that if one calculates the thermorheological response of such a composite, then the mutual, mechanical interaction of the two viscoelastic phases is automatically accounted for. It would be surprising, it seems to me, if the response of the composite were thermorheologically simple except near those limits on temperature when both phases are simultaneously in either the glassy or in the rubbery state.

SHEN: We have made no attempt to calculate the thermorheological response of our block copolymers from the point of view of composite mechanics. It may very well be a fruitful approach. However, it is important to note that the relaxation behavior of the same polymer (Kraton 1101) cast from different solvents (THF-MEK and CCl_4) is not the same because of the difference in morphology. There is no evidence at present to indicate that relaxations of these microphases (near molecular dimensions) are the same as these phases in macroscopic dimensions.

TSCHOEGL: The treatment of your data as you presented it here follows the treatment developed by us in a paper published last year [Lim, Cohen, and Tschoegl, *Advan. Chem. Ser.*, **99**, 397 (1971)] which we now know to be wrong. The criteria you gave for thermorheological simplicity are based on the appearance of the data. This is totally misleading in two-phase materials in which both phases are viscoelastic. The proper criterion for thermorheological simplicity is given by the requirement that all relaxation or retardation times should be equally affected by a change in temperature. This requirement is inherently *not* fulfilled in two-phase materials, since the transitions of the two phases wander across the logarithmic time scale at different rates. In a later paper [Fesko and Tschoegl, *J. Polymer Sci.*, Part C, **35**, 51 (1971)] we have made a thorough examination of the problem of time-temperature superposition in two-phase materials. In such materials, the shift factors become functions of time as well as temperature, except in certain regions of low and high temperatures, where one of the phases dominates. In these regions, empirical shifting is possible. In the intervening region, shifting must be carried out point by point. I hope to have an opportunity later this evening to elaborate on this with the aid of a few slides.

SHEN: I wish to reemphasize that we do not claim that hererophase polymers are thermorheologically simple, as your paper with Fesko has already pointed out (see Ref. 3 of our paper). What we intended to investigate is the extent to which simple time-temperature superposition may still be useful. As we stated in the paper, the latter is only applicable where the relaxation is dominated by only one phase, namely, either at very long or very short time regions (or high- or low-temperature regions). Because the relaxation times of the respective phases do not, in general, have the same temperature dependence, master curves are not expected to have the same time scale at different temperatures. As evidence of this type of phenomenon, Fig. 8 of this paper shows that the shapes of the loss tangent – temperature curves are similar at different frequencies, but the separations between the loss peaks are not the same. Our composite model for shift factors is not at all the same as the one put forth by Lim, Cohen, and Tschoegl (Ref. 12 above), which attributes that portion of the shift not accounted for by the WLF equation to an activation-energy term.

TSCHOEGL: Were your measurements made on annealed samples?

SHEN: Yes.

LANDEL: Was the morphology in your materials thermally stable, so that after doing a high-temperature test, one could take the same specimen, repeat a low-temperature run, and obtain the same relaxation curve?

SHEN: The current available evidence indicates that the morphology of a styrene-butadiene-styrene block copolymer is stable thermally to the highest testing temperature (80 C) used in this work. However, it might be of interest to mention that a phase transition seems to occur around 120 C, and the heterophase structure disappears (becomes homogeneous) above 170 C. It has been shown that the morphological structure is affected by mechanical deformation. For this reason, we used only fresh samples in our measurements. Thus, the type of testing you mentioned might yield different results, unless the deformation is very small.

KOVACS: I think that the conditions of homogeneity are not realized in heterophase block copolymers since the size of the phase domains is comparable to the radius of gyration of the blocks. This results in a rather high value for the interface/volume ratio, which means that an important fraction of the chain segments does not have the same surrounding near the interface as do those in the center of the domains. This probably provides for a gradient in the segmental mobility inside of each of the domains, rather than for a constant value, which is generally assumed to be equal to that of the respective bulk homopolymer.

SHEN: There is considerable indirect evidence that the interfacial region between the microphases may be very substantial. Unfortunately, there is, at present, no means for studying the relaxation behavior of these mixed domains.

LOAD-EXTENSION CURVES AND FRACTURE TOUGHNESS

P. I. Vincent

Imperial Chemical Industries Limited
Plastics Division
Welwyn Garden City
Herts, England

ABSTRACT

The shape of the load-extension curve, observed when stretching an unnotched specimen at constant speed, provides an indication of the rate of hardening beyond yield. Evidence is provided that this rate has an important influence on propensity to craze, on the plastic-zone shape, and on the fracture toughness of notched specimens. This has relevance to the effects of chemical constitution, molecular orientation, molecular weight, and some additives.

1 INTRODUCTION

It is now well known that even in apparently brittle fractures of thermoplastics, it is possible to demonstrate that there has been deformation beyond the yield point at the tip of the notch, crack, or natural defect which initiated fracture. In extreme cases, such as the fracture of notched specimens of isotropic polystyrene at room temperature, the zone of plastic deformation may be a single craze whose dimensions are small compared with those of the test specimen; then it is permissible to calculate stress distributions and to derive stress-intensity factors assuming Hooke's law. On the other hand, in many practical cases, the size of the zone of plastic deformation is not negligible and theories based on the assumption of Hooke's law can be in significant disagreement with experiment.[1]

It can be of practical importance to consider stress-strain behavior beyond the yield point, because it throws light on discrepancies between observed stress distributions and those calculated from linear elastic theory. Also, it provides information on the phenomena in the plastic zone around a notch. When trying to relate fracture toughness to polymer structure, it is not desirable to ignore the plastic zone because this is the part of the specimen in which fracture actually takes place.

The main body of this paper is divided into two parts. In the first part, it is shown that a large difference in stress-strain behavior beyond yield can be the cause of qualitative differences in several of the phenomena associated with fracture. The samples are taken as elasto-plastic continua because structural considerations are not necessary to establish the essential points. In contrast, in the second part, the samples are considered as aggregates and examples of the effects of structural changes on stress-strain behavior beyond yield are presented. It is suggested that these behavioral differences should not be ignored in theoretical fracture mechanics.

2 HARDENING RATE AND LOCALIZED PLASTIC DEFORMATION

In order to emphasize the importance of postyield behavior, it is helpful to consider two very different shapes of load-extension curve and their consequences. As one example, consider a specimen of a biaxially drawn and heat-set poly(ethylene terephthalate) film (PET) whose conventional tensile stress-strain curve at 20 C and 5 percent per minute has a bend at the yield point and no drop in load before fracture (Fig. 1). As the other example, consider an approximately isotropic sample of the polycarbonate of bisphenol A (PC). Because this material necks beyond the yield point, its conventional stress-strain curve cannot be determined by simply stretching a parallel-sided strip. Figure 2 shows a tensile load-extension curve obtained on a dumbbell-shaped specimen at 20 C and

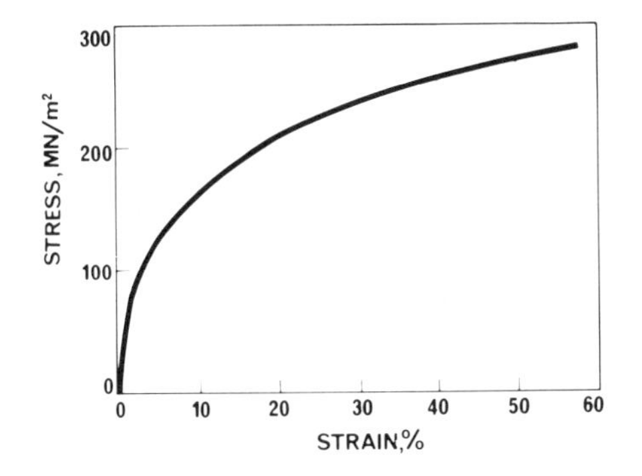

Fig. 1. Stress-strain curve at 20 C (5%/min) for a specimen of biaxially drawn and heat-set poly(ethylene terephthalate) film.

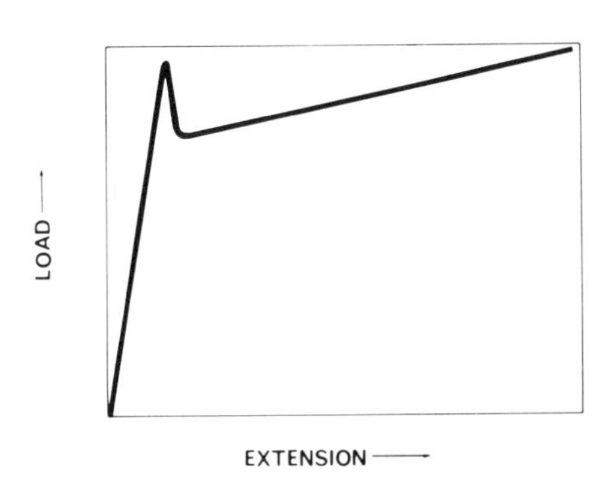

Fig. 2. Load-extension curve at 20 C (5 mm/min) for a dumbbell specimen of a polycarbonate of bisphenol A.

Fig. 3. Shear bands and crazes in a specimen of polycarbonate film at yield. Crossed polars and monochromatic light. Crazes are the short black lines lying vertically in the photograph.

5 mm per minute. There is a drop in load after the yield point, associated with the formation of the neck, and then a subsequent rise in load up to the point of fracture.

A decline in the conventional stress beyond the yield point causes plastic instability and therefore gives rise to localized plastic deformation. Figure 3 illustrates this by showing shear bands in PC specimens. PET specimens, with the greater hardening rate exemplified in Fig. 1, do not suffer plastic instability and thus do not form necks and shear bands.

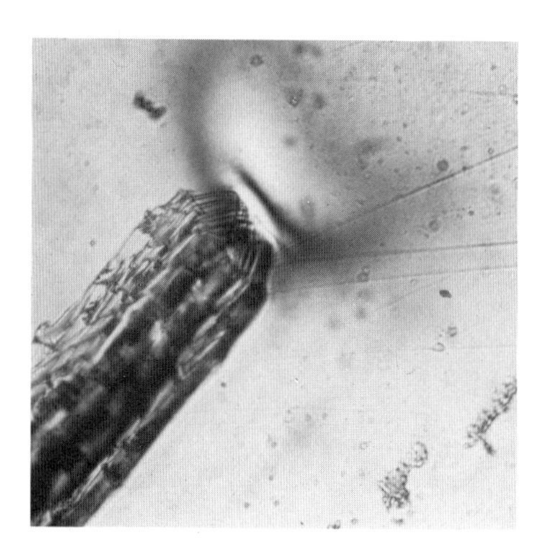

Fig. 4. Kidney-shaped zone of high deformation near the tip of a razor notch in a strained specimen of biaxially drawn and heat-set poly(ethylene terphthalate) film.

Fig. 5. Kidney-shaped zone of deformation near the tip of a razor notch in a strained specimen of polycarbonate film.

Figure 4 shows the shape of the plastic zone observed around the tip of a razor notch in PET. Although this shape has not been calculated in quantitative detail from the stress-strain relation in Fig. 1, it is clear that it has a certain similarity to the contours of constant stress derived from linear elastic theory. A kidney-shaped zone of this type is observed in PET at all stages of plastic-zone development and crack growth.[1]

In contrast, the development of the plastic zone in PC is somewhat more complex. In the earliest stages, when the local strains are below the yield strain and the slope of the load-extension curve is positive, the zone is kidney shaped as illustrated in Fig. 5. In the second stage, when the local strains are above the yield strain and the slope of the load-extension curve is negative, the zone becomes wedge shaped, as illustrated in Fig. 6. Finally, at higher strains, when the slope of the load-extension curve becomes positive again, there is a kidney-shaped zone within the wedge (Fig. 7). These observations suggest the generalization that a positive slope causes a kidney-shaped zone, whereas a negative slope causes a wedge-shaped zone.

When an unnotched specimen of PC is stressed in tension, adventitious surface defects lead to the development of crazes – wedge-shaped plastic zones which have been reduced in density by plastic constraint. Examples are visible as short black lines perpendicular to the applied tensile stress in Fig. 3. It seems plausible to suggest that craze formation is also a consequence of the plastic instability caused by the negative slope of the conventional load-extension curve.

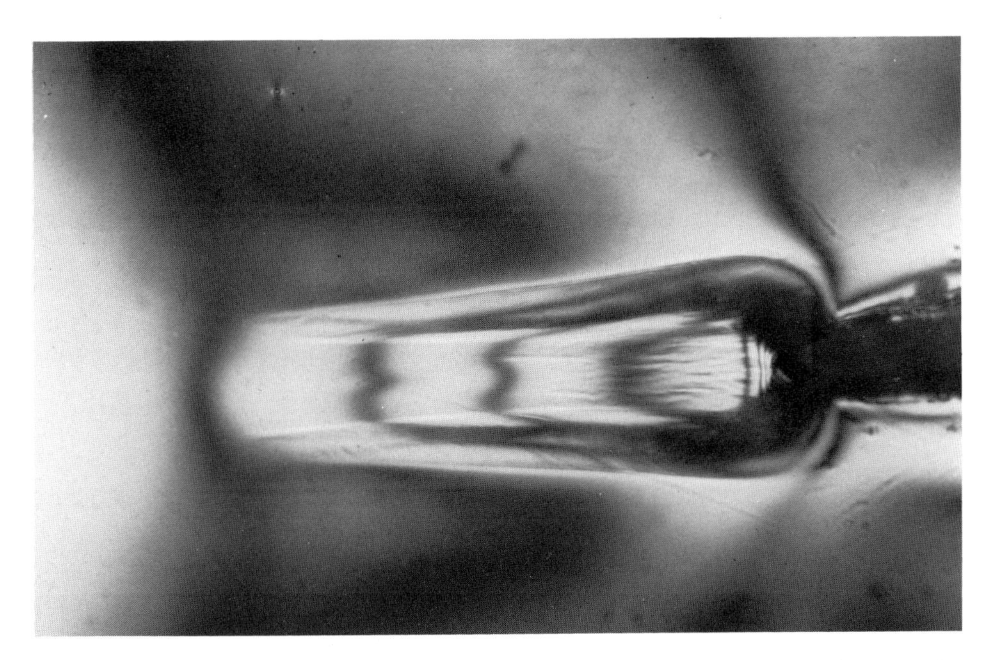

Fig. 6. Photograph of same specimen as in Fig. 5, revealing a wedge-shaped zone at higher extension.

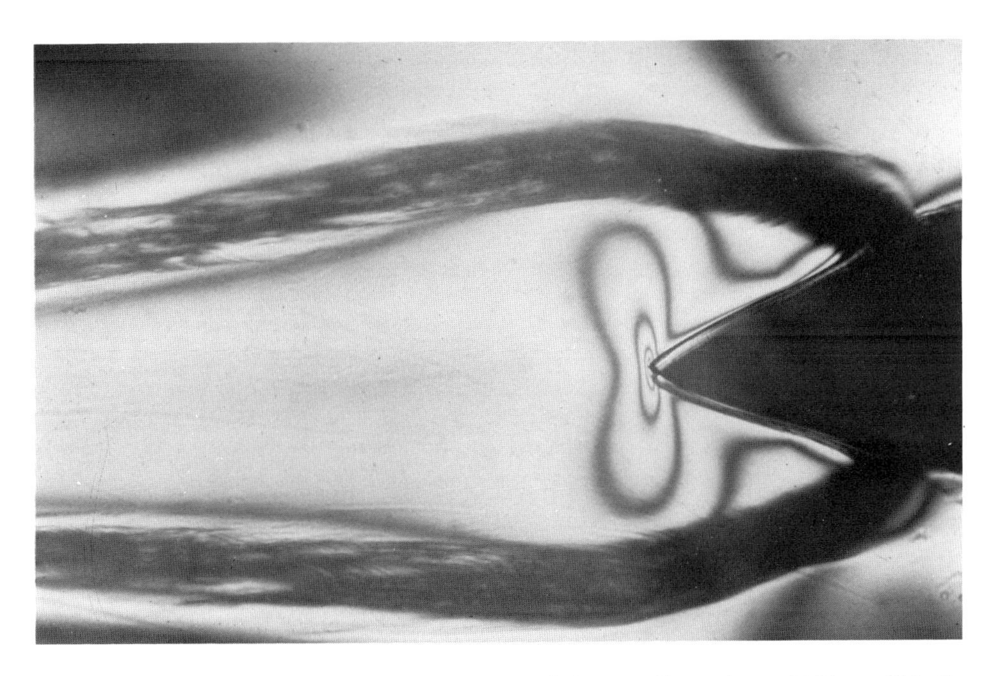

Fig. 7. Photograph of same specimen as in Figs. 5 and 6, but revealing an internal kidney within the wedge at still higher extension.

These suggested generalizations are supported by consideration of the behavior of several other thermoplastics at room temperature. Normal samples of low-density polyethylene and polytetrafluoroethylene extend uniformly with a rising load beyond yield and do not form crazes or wedge-shaped plastic zones. Ductile amorphous polymers which are below their glass transitions at room temperature, such as unplasticized polyvinyl chloride, isotropic poly(ethylene terephthalate), and poly(2,6 dimethyl-p-phenylene oxide), invariably have a region of negative slope associated with plastic instability and neck formation; they all craze at adventitious defects and have wedge-shaped plastic zones at notches. This is also true of some partially crystalline polymers such as high density polyethylene, polypropylene, polyoxymethylene, and poly 4-methyl pentene-1. (The "crazes" in some of these crystalline polymers are fatter than those typical of amorphous polymers and may perhaps be preferably called "normal deformation bands", depending on the definition of a craze; however, this does not alter the fact that they are wedge-shaped plastic zones.) Consideration of brittle amorphous plastics, such as polystyrene, leads to a slight complication, viz., their behavior beyond yield in tension cannot be determined directly because brittle fracture supervenes. However, the compressive load-deformation curve of polystyrene shows a pronounced drop in load after yield, and it may, therefore, be safely assumed that there is a region of plastic instability in tension to account for its pronounced craze formation.[2]

To summarize, there seems to be no obvious exception to the rule that the formation of crazes and other wedge-shaped plastic zones is correlated with the plastic instability which follows from a region of negative slope in the tensile load-extension curve.

3 HARDENING RATE AND TOUGHNESS

Having established a connection between the rate of hardening beyond yield and various phenomena associated with fracture, it is pertinent to consider whether a knowledge of postyield behavior can help in the interpretation of trends in fracture toughness. Presumably, the ideal way to attack this problem would be to use a reliable theory to calculate stress and strain distributions around notches from stress-strain relations measured on unnotched specimens. However, with the combination of nonlinearity, anelasticity, anisotropy, inhomogeneity, time and temperature dependence, and large strains, any theory of three-dimensional distributions would be highly complex, and certainly no such theory is available at present. In these circumstances, it is potentially more effective to propose a hypothesis and to test it by searching for appropriate empirical correlations.

From impact tests on notched specimens, for example, fracture toughness is known to be affected by changes in molecular orientation and molecular weight and by various types of additives. Shortage of space prohibits a thorough discussion of each of these factors, with full experimental details, but it can be demonstrated that each of them affects the rate of hardening beyond yield in the same sense as they affect fracture toughness.

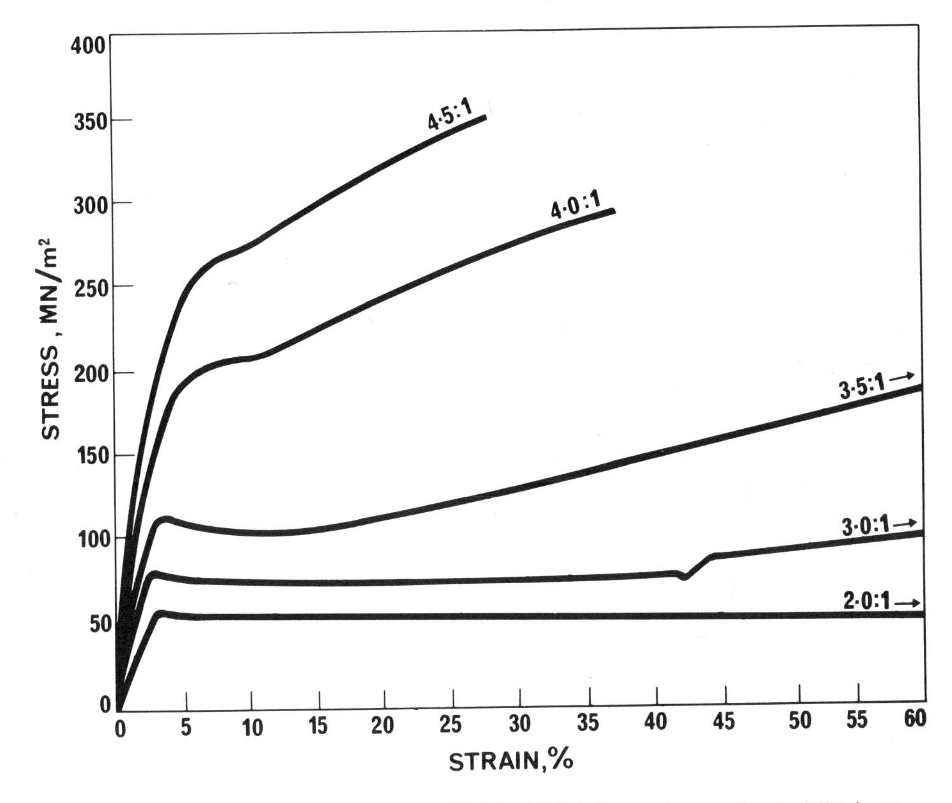

Fig. 8. Conventional stress-strain curves at 20 C (5%/min) for specimens of uniaxially drawn poly(ethylene terephthalate) film, stretched in the predraw direction. Five different draw ratios.

Figure 8 shows conventional tensile stress-strain curves on a series of poly(ethylene terephthalate) films which had been uniaxially predrawn to draw ratios up to 4.5; the applied stress in the tensile test was parallel to the draw direction. As the prior draw ratio increases, the drop in load after yield becomes smaller and the rate of hardening beyond yield becomes higher.

It is, therefore, not surprising that, whereas isotropic samples craze and neck, samples with sufficient prior molecular orientation do neither when stressed along the orientation direction. The photographs in Fig. 9 show plastic-zone shapes around sharp notches for films at various initial

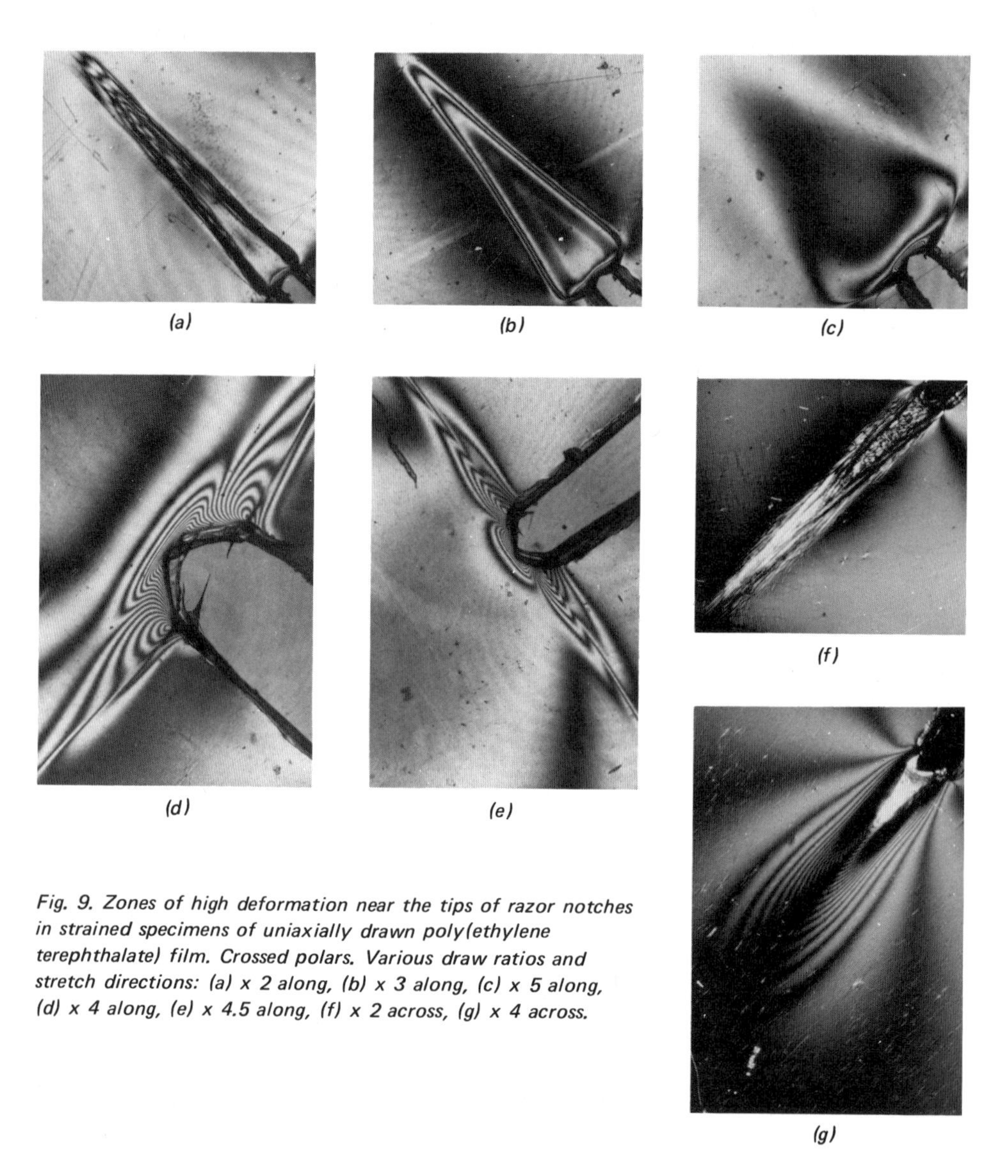

Fig. 9. Zones of high deformation near the tips of razor notches in strained specimens of uniaxially drawn poly(ethylene terephthalate) film. Crossed polars. Various draw ratios and stretch directions: (a) x 2 along, (b) x 3 along, (c) x 5 along, (d) x 4 along, (e) x 4.5 along, (f) x 2 across, (g) x 4 across.

draw ratios with the tensile stress applied either parallel or perpendicular to the preorientation direction. When the stress is parallel to the draw, the plastic zone becomes shorter and wider with increasing draw ratio. It is more sharply wedge shaped when the stress is perpendicular to the draw. When unnotched specimens are tested perpendicular to the preorientation direction, the load drops after yield and crazing is pronounced. These results support the general rule suggested in the previous section. Additionally, they support the hypothesis that the greater fracture toughness observed, when oriented samples are stressed along rather than across the orientation direction, is associated with more rapid hardening beyond yield and a wider plastic zone. Presumably, a wider plastic zone corresponds to a larger amount of energy needed for a given amount of crack development.

Figure 10 shows true stress-strain curves beyond yield for three samples of polypropylene. The rate of hardening increases with decreasing melt flow index, which is taken as a measure of increasing molecular weight. Similar results have been obtained with samples of branched and linear polyethylene, polytetrafluoroethylene and poly(methyl methacrylate) under suitable test conditions. In general, increasing molecular weight increases both fracture toughness and the rate of hardening beyond yield.

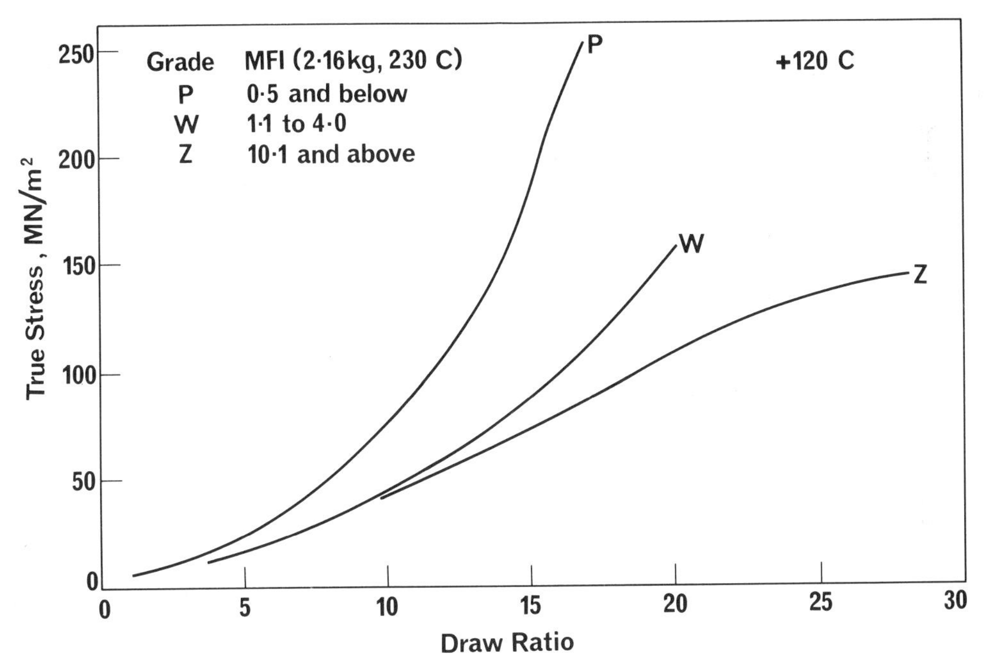

Fig. 10. Curves of true stress versus draw ratio at 120 C for three samples of polypropylene of different molecular weights.

Figure 11 illustrates the effect of the addition of rubbery chlorinated polyethylene particles on the shape of a tensile load-extension curve of PVC. Figure 12 shows the effect on the brittle point of very sharply notched specimens in Charpy-type impact tests. Again, the absence of a yield drop and the greater hardening beyond yield are associated with an improvement in fracture toughness.

Figure 13 shows the effect on a true stress-strain curve of PVC produced by the addition of finely divided titanium dioxide powder, and Figure 14 shows the effect on Charpy-type-impact test results. Again the improvement in impact behavior is associated with more rapid hardening.

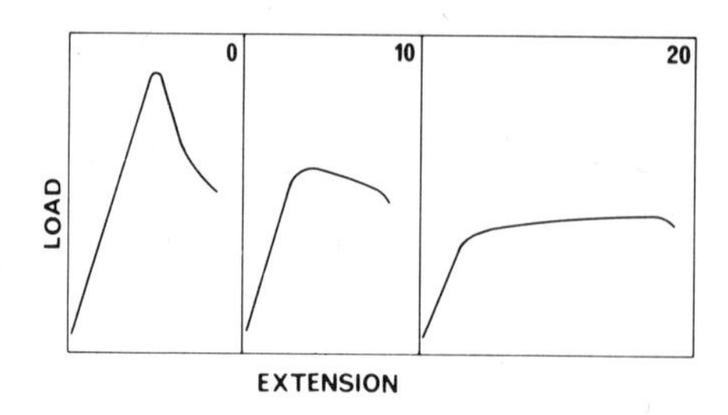

Fig. 11. Load-extension curves at 20 C (5 mm/min) of dumbbell specimens of PVC with 0, 10, and 20 percent additions of chlorinated polyethylene.

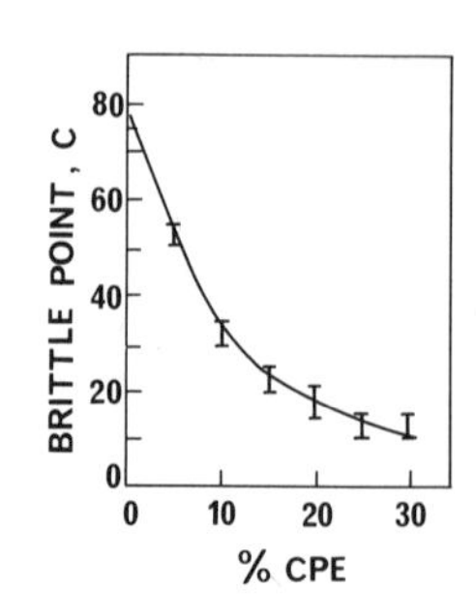

Fig. 12. Brittle point of very sharply notched specimens of PVC in a Charpy-type impact test, as a function of the concentration of a chlorinated polyethylene addition.

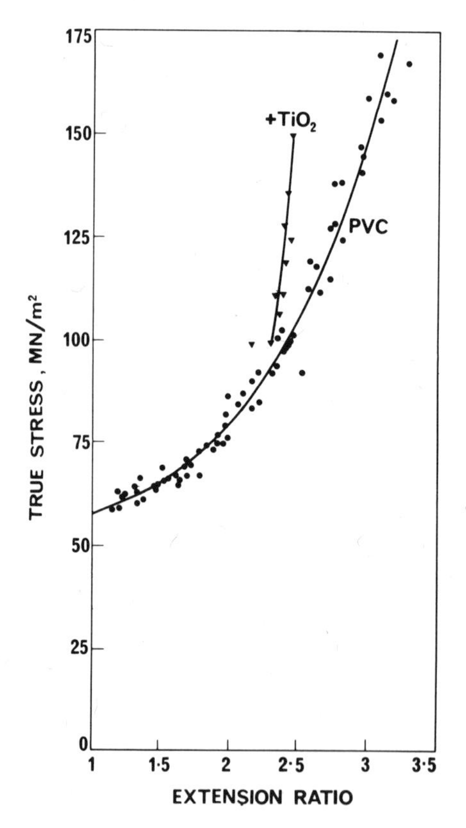

Fig. 13. Curves of true stress versus draw ratio at 20 C for PVC unfilled and with a 13 percent addition of titanium dioxide powder.

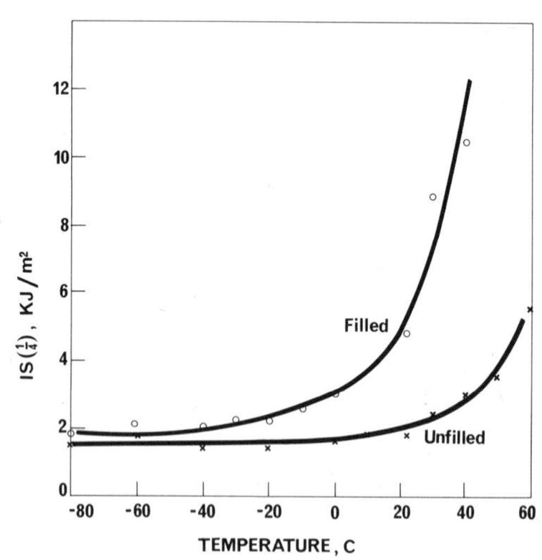

Fig. 14. Charpy-type impact strength (1/4-mm notch-tip radius) versus temperature for the same samples as in Fig. 13.

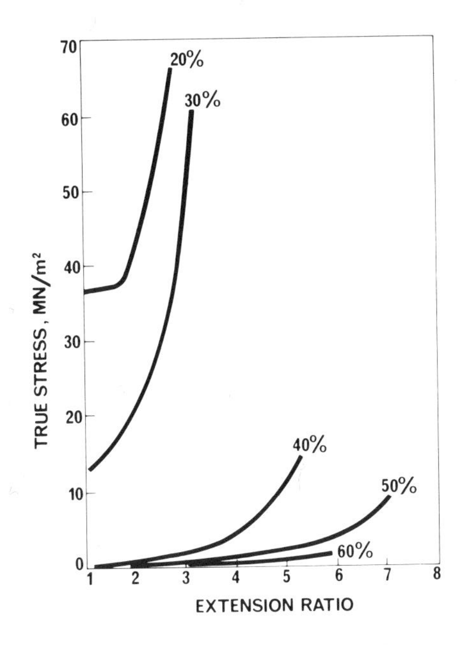

Fig. 15. Curves of true stress versus draw ratio at 25 C for poly(methyl methacrylate) with the addition of various concentrations of dibutyl phthalate.

Figure 15 shows true stress-strain curves at 25 C for samples of poly(methyl methacrylate) with a 20 to 60 percent addition of dibutyl phthalate. It is plain that the toughness of the polymer has been increased by adding the plasticizer since, for example, the specimen with 50 percent DBP stretched uniformly to a strain of around 600 percent before breaking. It had no yield point, but its proportionate rate of hardening was very high.

In the absence of a satisfactory basic theory, it is not possible to say that all these observed trends in fracture toughness can be entirely accounted for by the parallel trends in hardening rate. However, the experimental evidence strongly suggests that it is an important contributory factor.

REFERENCES

1. Vincent, P. I., "Ductile Crack Growth in Poly(Ethylene Terephthalate) Film, *Polymer,* **12,** 534-536 (1971).
2. Haward, R. N., "Effect of an Elastomeric Phase on the Properties of Rigid Thermoplastics", *Brit. Polymer J.,* **2,** 209-216 (1970).

DISCUSSION of Paper by P. I. Vincent

HULL: I do not wish to dispute the general approach which you have introduced in your paper. However, I think we could well have a "chicken and egg" situation in that you could present your observations on the interrelation between the yield drop and the occurrence of crazing and shear bands in a completely different way. Thus, it would not be unreasonable to suggest that the yield drop arises because deformation occurs by crazing and/or shear banding. There is clear evidence for this approach since crazing or shear banding can be detected before the maximum load (upper yield stress), and it is as a result of the multiplication and propagation of these zones of large shear that instability occurs. The strain rate in the specimen is greater than the imposed strain rate.

The same point can be made about the correlation between these processes and fracture. Brittle fracture is preferred because the shearing or crazing concentrates the strain and produces high stress concentration.

ANDREWS: I agree with Professor Hull that the load drop is probably a consequence rather than a cause of crazing and slip-band formation. I think Dr. Vincent is on safer ground in connecting these effects with the strain-hardening propensities of the material, since this will affect both shear-band formation (via the natural draw ratio) and craze-growth characteristics.

VINCENT: This is an important question and needs careful consideration. The picture of the mechanism of crazing, which I have suggested, involves four steps:

1. The yield stress is reached at a point of stress concentration.
2. Plastic instability follows because of the load drop after yield, observed in the uniaxial load-extension curve; this gives a wedge-shaped plastic zone perpendicular to the applied tension.
3. Void formation is a consequence of constraint from the surrounding unyielded material.
4. Final stability of the craze is a consequence of molecular orientation which causes a rapid increase in the stress.

If there were no load drop after yield, then step 2 would not occur, there would be no plastic instability, and the craze would not grow. Thus, I believe that the load drop causes craze growth rather than that the crazing causes the load drop.

There are some exceptional cases where an enormous amount of craze development causes premature yielding. One example of this can be seen in moderate-speed tensile tests at room temperature on certain types of rubber-modified polystyrene. Here, yielding occurs unusually abruptly, at an unusually low strain and over an unusual length of the specimen. There is no necking and little or no drop in load after yield.

It is particularly pertinent to note that in this "false yielding" or "craze formation yielding", the yield drop is suppressed by the crazing rather than being caused by it. The crazing is explained by the yield drop observed in compression on unmodified polystyrene, mentioned in the paper.

Professor Andrews mentions the natural draw ratio; this cannot account for the formation of the craze (step 2) but only for its final stability (step 4). It may be relevant to mention that, in glassy polymers with sufficiently low molecular weight, crazes do form but do not stabilize because there is insufficient strain hardening.

I agree with Professor Hull that brittle fracture is preferred because of the concentration of the strain, but I am suggesting that one needs the yield drop to provide the localized plastic instability which concentrates the strain.

KOBAYASHI: Previous experiments with double-edge-notched, mild-steel tension specimens showed that in addition to a Dugdale-type yield zone, a typical butterfly type of yield zone grew out of the crack tip. Thus, for loading beyond net section yielding of the notched specimen, the overall plastic yield zone assumed a conventional butterfly shape. I suspect that the same is true for notched polycarbonate tension specimens if the specimen thickness is increased sufficiently.

VINCENT: It may be significant that you say that the butterfly (kidney) shape occurs after net section yielding. In polycarbonate I observed kidney-shaped isochromatic fringe patterns both before and after the formation of the wedge-shaped zone. We have only tested notched polycarbonate up to 3 mm thick, but there are still wedge-shaped zones in that case.

BRINSON: In response to the author's point on the relationship of the shape of the stress-strain diagram and the shape of plastic-zone formation in front of notches or cracks, it has been shown by Rice and Drucker that the shape of the plastic zone depends on whether the yield surface of the material is of the von Mises or Tresca type (the latter giving rise to Dugdale-type zones). Other factors such as local anisotropy, and whether the Dugdale boundary conditions are met, may also be quite important. I wonder if you would comment on which factor you feel is the most important in determining the shape of plastic zones?

VINCENT: I think that the rate of hardening beyond the yield point is the dominant factor. I cannot feel that the relatively subtle difference between a Tresca and a von Mises yield criterion could be the cause of the large differences illustrated in the paper.

MENGES: It is difficult to understand why so much excellent research work is done on tensile tests at strain rates where the materials' behavior is very time dependent. It is suggested that in research work, one should begin with low rates and then use higher ones. Thus, for example, the production of voids (the first step of crazing) is often not visible in the tensile test, but only the second step – formation of shear bands. In investigations started at low rates, it was actually found that the first step is always the production of voids.

VINCENT: If one is concerned with the practically important problem of notch brittleness in impact, as I am, it is necessary to perform tests at high strain rates, where yield stresses are high and deformation processes are adiabatic.

ROSENFIELD: The author's hypothesis that fracture toughness (K_{Ic}) is dependent on the work-hardening rate (n) is in accord with a body of literature on high-strength metallic alloys. The original suggestion was made by Krafft a decade ago. In 1968, Hahn and I suggested* that the connection between the two quantities was through the effect of n on plastic-zone shape. Both Krafft and ourselves developed simple equations, the essence of which was $K_{Ic} \propto n$. Subsequent investigations conclude that this approach contains elements of the correct relations between precracked and smooth properties but that the actual situation is somewhat more complex. It would appear from the present paper that the situation in brittle plastics may not be too different from that in metals of low toughness.

VINCENT: It is encouraging that this relation has some truth in metals as well as plastics. It could be quite general, irrespective of material structure, since it depends only on strain distributions around notches in work-hardening solids.

---

*Hahn, G. T., and Rosenfield, A. R., *ASTM-STP* No. 432, 5 (1968).

DEFORMATION OF POLYMERS AS EXPLAINED BY THE MEANDER MODEL

W. Pechhold

Abteilung für Experimentelle Physik II,
Universität Ulm,
Stuttgart, West Germany

ABSTRACT

After a brief discussion, the meander model of condensed polymers is used to explain the relaxation properties as well as the deformation behavior of polymer materials. The latter is considered from the geometrical and the stress-strain points of view for different materials: muscle fibrils, rubbers, and semicrystalline polymers, including drawn fibers.

1 INTRODUCTION

Deformation processes in any material are, in general, highly nonlinear phenomena regardless of whether paraelastic (reversible) or plastic (irreversible) deformation is concerned. A quantitative description containing only a few meaningful parameters which is useful for engineering purposes is often difficult or impossible to formulate. Therefore, the molecular approach, which for linear phenomena is mainly of academic interest, seems to be a competitive method to the phenomenological description, considering the theoretical effort to be made.

The dislocation theory for metals and low-molecular-weight crystals has already been developed into a powerful tool for tackling problems of work hardening and fracture in such materials. In these crystalline materials, the mobile and interacting defects as well as the deformation geometry can be readily defined on the basis of the well-known ideal state. This is why important questions about inhomogeneous strain distribution or stress concentrations leading to cracks can be hopefully approached.

For polymers in their condensed state, however, we do not yet know enough about microstructure, mobile defects, or micromorphology of deformation. This is because condensed macromolecules do not like the perfectly ordered state.

But what state do they prefer instead? Since, until now, neither computer calculations nor electron micrographs give sufficient details of microstructure and superstructure, one has to assume models (Fig. 1):

(1) The random-coil model – established for dilute solutions – in which one has to envisage dense-packed and mutually penetrating coiled molecules. To achieve a dense package, however, a considerable amount of chain segments have to be parallel, a condition which possibly violates this model.
(2) The bead-string model – qualitatively proposed by Schoon – in which molecules pass from one folded region to another.
(3) The meander model – in which bundles of molecules fold to give an isotropic and dense-packed polymer material.

Fig. 1. Models of microstructure (superstructure) in amorphous polymers: random-coil model, bead-string model (Schoon), meander model.

2 THE MEANDER MODEL[1-9]

Starting with a nondefective bundle of macromolecules (the ideal crystal) one gets a low-defective bundle (the real crystal) by introduction of stable defects (kinks, torsional defects, jogs, and folds) which are compatible with the intra- and intermolecular potential. The cooperative statistical treatment of this bundle reveals – under a certain condition – a first-order transition, connected with a strong increase in defect concentration.

In the case of polyethylene, the bundle model has been shown to explain quantitatively the transition data (T_m, ΔH_m, ΔV_m) as well as the expansion coefficient and the compressibility of the melt, together with their dependence on static pressure. The calculated short-range order also is in accordance with the X-ray and electron diffraction data.

By introduction of cooperatively arranged gauche areas (for planar molecules), the meander model is established which guarantees the isotropy of the melt. Assuming that the bundle diameter (calculated to be about 50 A, which is in the range of the observed superstructure in amorphous polymers) is constant during crystallization, a lamellae structure results, in which the meander thickness is determined by the crystallization temperature.

In a recent paper[8], the meander model has been applied also to solution-grown crystal lamellae (Fig. 2) in which the single molecules fold by adjacent reentry as well as by meander curvatures, thus forming favorable gauche areas but also an amorphous fold surface. The most-probable ratio of meander radius, r, over interchain distance, d, has been shown to be about 3 for PE.

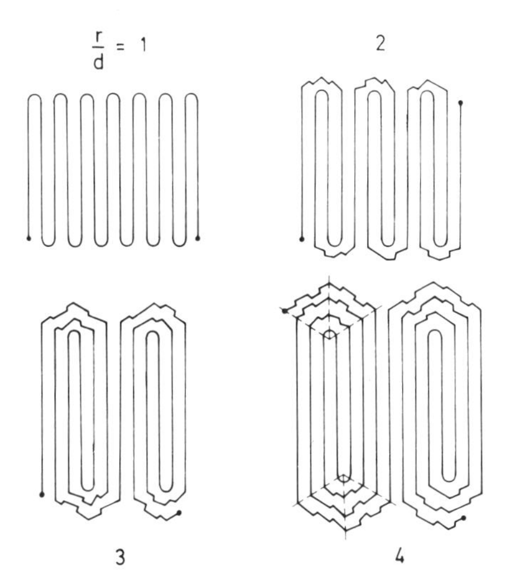

Fig. 2. Models of crystal lamellae grown from dilute solution. It has been shown [8] that $\frac{r}{d}$ = 3 is most probable.

Another refinement of the meander model was achieved by introduction of a kinetic condition which allows definition of the defect state of the crystalline regions: during annealing, on the average, two-segment layers in a compressed molten state should exist on one of the two crystal boundaries (Fig. 3a, middle), the inner layer becoming dissociated into single crystal defects on cooling (Fig. 3a, right-hand side). With this model, X-ray and thermal transition data of PE[8] and PET[9] can be quantitatively explained. This two-phase meander model, moreover, implies the key to the understanding of molecular motion in polymers.

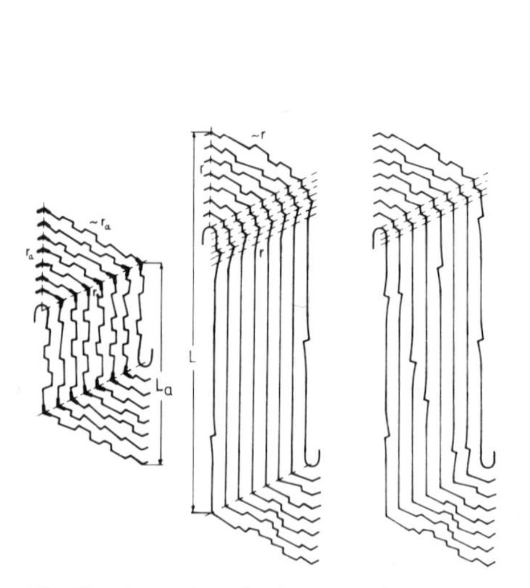

Fig. 3a. Amorphous half-mender (left-hand side). Semicrystalline half-meander with two defect layers (Middle). At lower temperature (some 10 degrees below the melting point), the inner layer will dissociate into single kinks or small kink blocks (right-hand side).

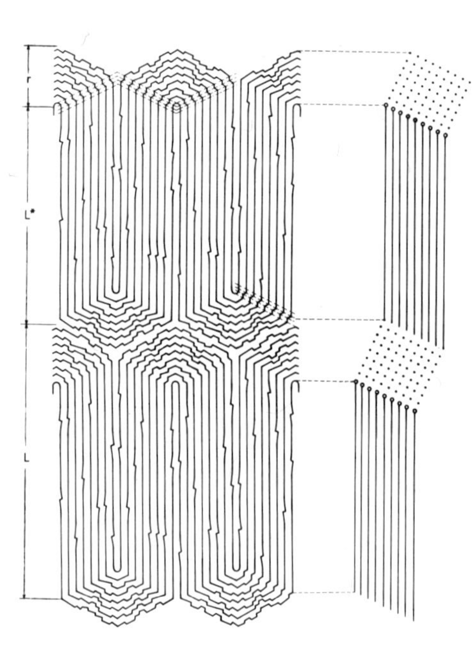

Fig. 3b. Total view of two meander layers. On the right, a side view of the meander piles.

3 MOLECULAR MOTION

The study of the relaxation processes at small amplitudes and their quantitative interpretation in terms of thermally activated molecular motions have essentially contributed to the development and verification of the meander model:

(1) The secondary dispersion (γ process) due to the main chain is attributed to the rearrangement of small kink blocks[5,6] concentrated in the amorphous regions (kink concentration $x_{1a} \approx 0.2$ per backbone atom) and of the few kinks in the crystalline

regions ($x_{1c} \approx 0.004$). These kinks can switch to different positions, which leads to a paraelastic shear deformation perpendicular to the chains. The corresponding relaxation strength, ΔJ, is in the range of 10^{-10} cm²/dyne.

(2) The relaxation strength of the main dispersion – usually in the range of 10^{-7} cm²/dyne for amorphous polymers – is the high-frequency rubberelastic compliance. It is due to glide motion of layers of molecules enabled by overall kink mobility which, in turn, is achieved by a sufficient amount of "free volume" (112 vacancies[4]). This short-time glide motion perpendicular and parallel to chain direction is restricted to one interchain distance, d, by the nonplanar meander geometry. A derivation of the two microscopic compliances for an uncrosslinked rubber is presented in Fig. 4. Assuming the meander geometry $\frac{r}{d} \approx 10$ to be constant for different polymers, the calculated macroscopic shear compliances J are in accordance with experimental results.[12] Figure 5 shows measurements of five anisotropic compliances for NR. The theoretically derived microscopic compliances (Fig. 4) are in good agreement with those determined from an analysis of Fig. 5.

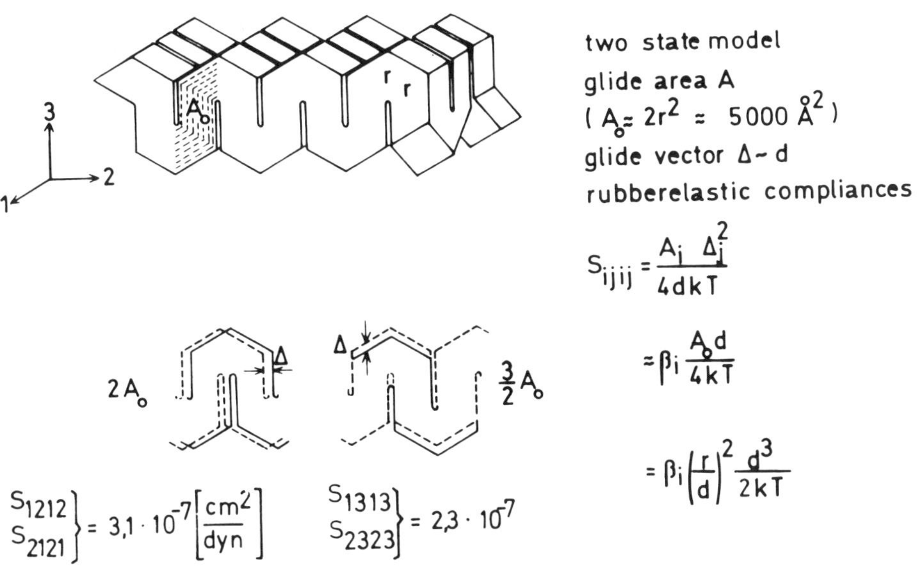

Fig. 4. High-frequency rubberelastic compliances enabled by glide of layers of molecules perpendicular to or in chain direction by one interchain distance. This motion takes place above T_g

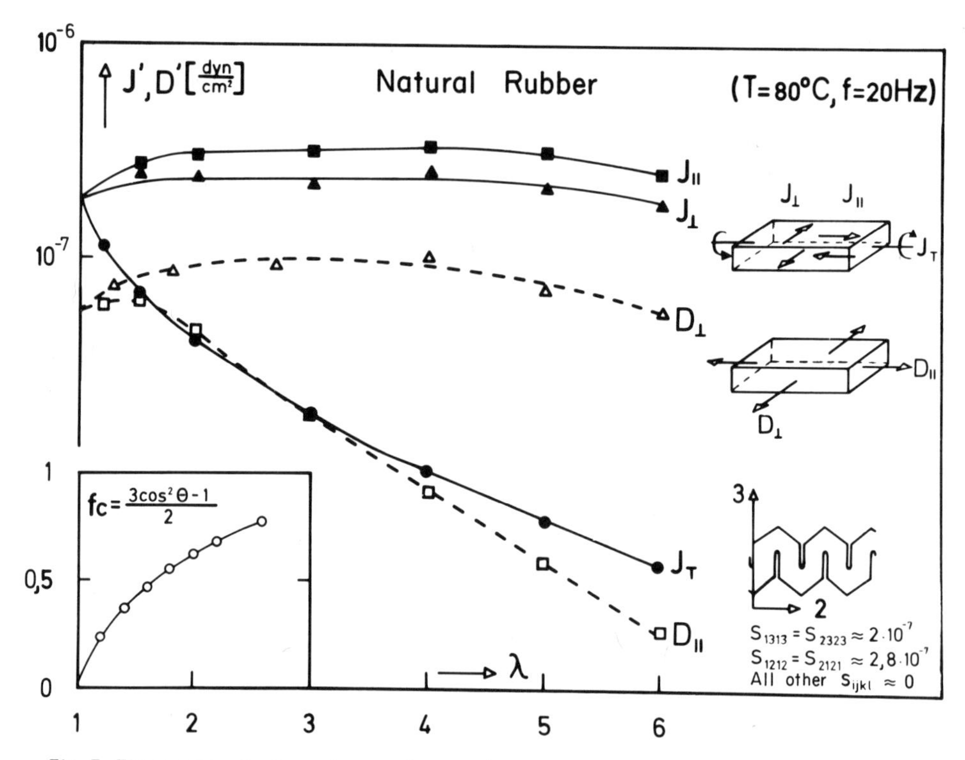

Fig. 5. Five anisotropic dynamic compliances of one rubber sample measured as function of its extension ratio. For λ less than 3, the macroscopic compliances can be explained by the two microscopic compliances derived from the meander model (Fig. 4).

(3) For a quantitative interpretation of the two mechanically observed α processes which occur in the crystalline regions, e.g., of PE, a motion of screw dislocations is reasonable. As already pointed out[5,6], jog bands will act as the cores of such screw dislocations running across the lamella with a Burgers vector of $\bar{c}$. The mean distances between such glide dislocations have been taken to be equal to the mosaic-block dimensions determined by Hosemann and co-workers.[10,11] In a recent paper[8], we proposed the mosaic-block structure to be the result of a minimum energy arrangement of the already described defect layers instead of the previously assumed dislocation network. In this new picture the jog blocks need not be permanent screw dislocations but can as well be spontaneously generated at the mosaic-block boundaries.

4 GEOMETRY OF LARGE-SCALE DEFORMATION

In the second part of my paper, I will shortly describe what kind of information – to our present knowledge – can be gained from the meander model concerning large-scale deformation of solid polymers, rubbers, and muscle fibrils. I will restrict myself to deformations with negligible amount of chain scission as well as negligible influence of chain ends, i.e., fracture and viscous phenomena are to be excluded here.

Figure 6 shows two possibilities for the conversion from lamellar to fibrillar semicrystalline structure. During this conversion, the matrix connectively is maintained. The fibrils can further glide along each other until they are stopped by tie molecules or tie-bundle action. In the most simple and, therefore, most probable type of forming ultrafibrils (Fig. 6, left-hand side), the instantaneous-fibril long period is drastically reduced, but may be increased by annealing, dependent on drawing temperature.

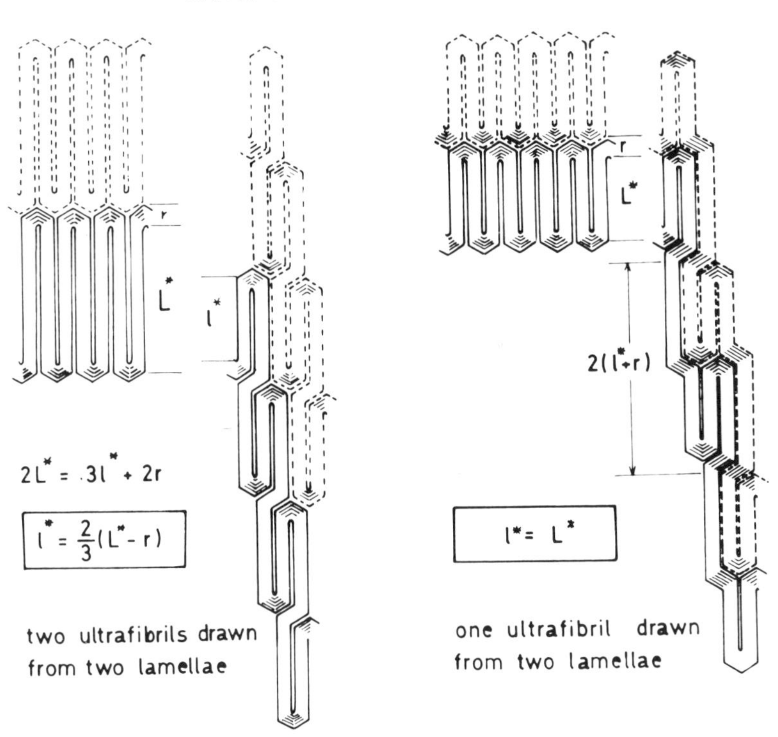

Fig. 6. Two models for the conversion of lamellar to fibrillar structure, maintaining matrix connectivity and allowing further plastic shear.

Figure 7 indicates a possible rubberelastic deformation (low-frequency mode)[12] consisting of shear parallel to interconnected meander lines and shear perpendicular to meander layers.

A proposal for a bundle or meander model of a myofibril from striated muscle is presented in Fig. 8 (on the right-hand side). The shadowed meanders represent the less contractile myosin filaments, be-

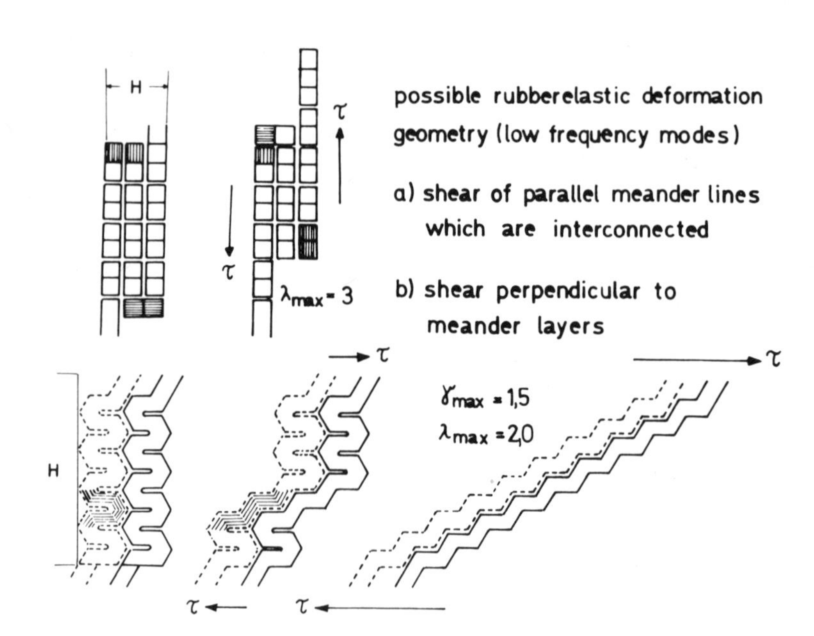

Fig. 7. A model possibly describing the large-scale (low-frequency) deformation of rubbers.

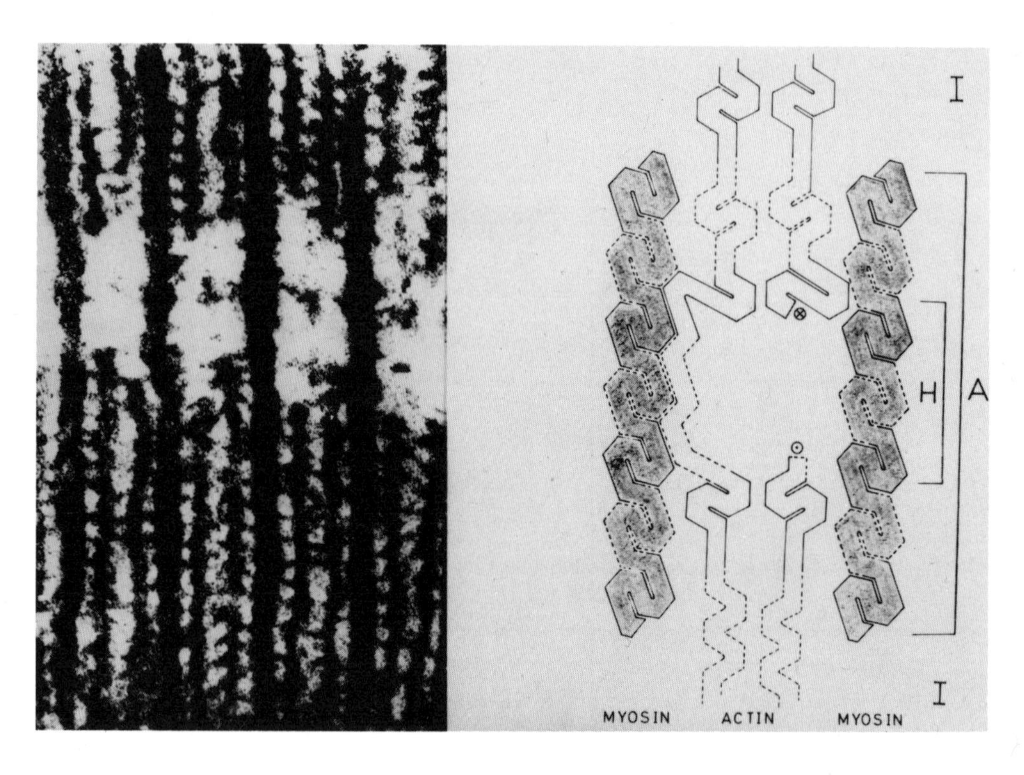

Fig. 8. Electron micrograph of a longitudinal section of myofibril (Huxley 1957, from Ref 16); bundle and meander model of myofibril.

tween which one or two action filaments are situated, clamped outside at the Z-zones (not shown). This model is in accordance with all electron-microscopic evidence, including the H-zone and the observed cross bridges (compare the micrograph on the left-hand side). In contradiction to Huxley's sliding-filament model[13], the actin filaments are not interrupted in the H-zone but are attached there to the myosin rods. They can maintain, therefore, a stretching force not only in the active but also in the passive state of the muscle, which property is lost in Huxley's model.

5 STRESS-STRAIN BEHAVIOR

From a molecular model containing only well-defined parameters which can be determined by atomistic potential calculations[1-3], one should, in principle, be able to derive all macroscopic properties, including the stress-strain behavior. For the meander model, I will shortly describe some first steps towards this aim. Figure 9 shows a section of an actin

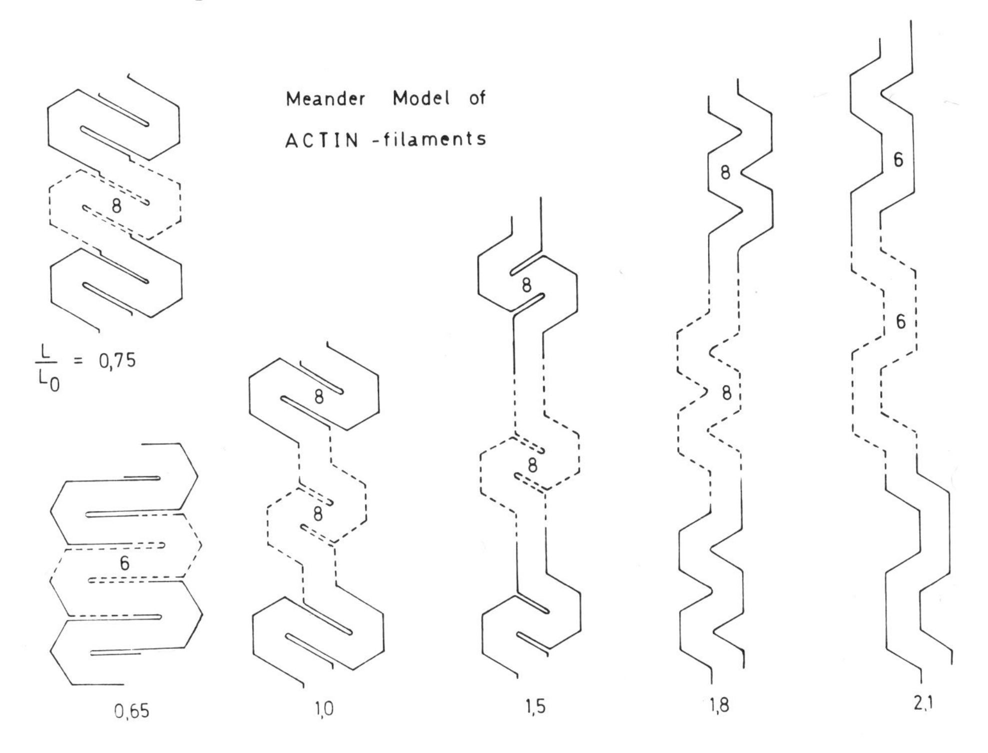

Fig. 9. Models showing an actin filament at different lengths. The unit of protein content was chosen to be one full meander in the contracted state ($\frac{L}{L_o}$ = 0.65). The change in length is accomplished by reduction of fold length and unfolding of meanders (decrease in the number of gauche areas).

filament at different lengths. The unit of protein content was chosen to be one full meander in the contracted state $\left(\frac{L}{L_o} = 0.65\right)$. The increase in length is accomplished by reduction of fold length, followed by unfolding of meanders (decrease in the number of gauche areas). In order to calculate the passive stress-strain relation to an actin filament, the indicated discrete shapes $\left(\frac{l_i}{l_o} = \Lambda_i = 0.65, 0.75, 1.5, 1.8, 2.1\right)$ and the related Gibbs free energies ($-g_i = -9$, $1.8 + RT \ln 5$, $1.2 + RT \ln 10$, $RT \ln 5$, $RT \ln 10$) were assumed for each of the N units, together with their rest length $l_o = 150$ Å, the cross section of actin bundle A - 2500 Å^2 and the temperature T = 300 K. The concentration u_i of each shape depends, via g_i, on the numbers of gauche areas, folds, and fold lengths of closed meanders. It also depends on chosen total length $\frac{L}{L_o}$ of the filament. The stretching force $\sigma \cdot A$ as a function of $\frac{L}{L_o}$ is derived from ordinary equilibrium statistics:

$$\text{Mean Biggs free energy per unit} \quad \frac{G}{N} = \sum_i u_i\, g_i + kt \sum_i u_i \ln u_i$$

$$\text{Secondary conditions} \quad \sum_i u_i = 1 \quad \sum_i u_i \Lambda_i = \frac{L}{L_o}$$

$$\text{Strain-stress relation} \quad L_o = \frac{\sum_i \Lambda_i \cdot \exp\left[\frac{\partial A \ell_o \Lambda_i - g_i}{kT}\right]}{\sum_i \exp\left[\frac{\partial A \ell_o \Lambda_i - g_i}{kT}\right]}$$

The result shown in Fig. 10 comes close to experimental data. The stress-strain relation for a rubber sample has not yet been calculated from the respective large-scale-deformation model (Fig. 7). Fig. 11 shows clearly the discrepancy between the high-frequency modulus, E′, and the differential modulus, $\frac{\partial \sigma}{\partial \lambda} \cdot \lambda$, derived from the measured stress-strain curve. This indicates that the model proposed for large-scale deformations should contribute an additional compliance at long times[12] to the high-frequency compliance.

True paraelasticity can also be observed in drawn polyamid fibers (Fig. 12). It has been shown[5,6] that the meander model can explain the stress-strain curve of such a fibre by superposition of elastic ($E\infty$) and

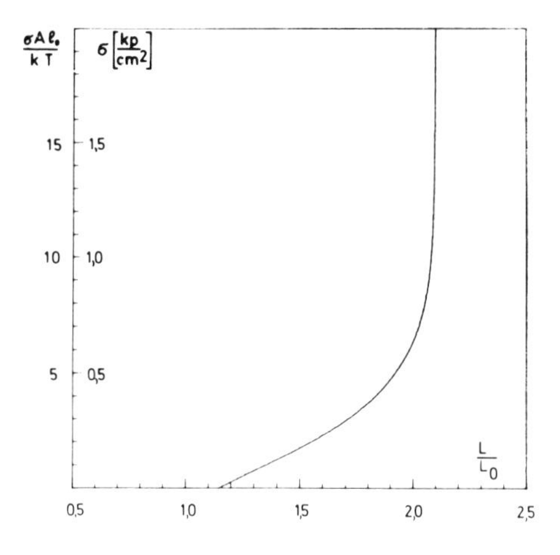

Fig. 10. Calculated stress-strain relation for an actin filament according to the model of Fig. 9, assuming discrete shapes (Λ_i) and Gibbs free energies (g_i) of filament units.

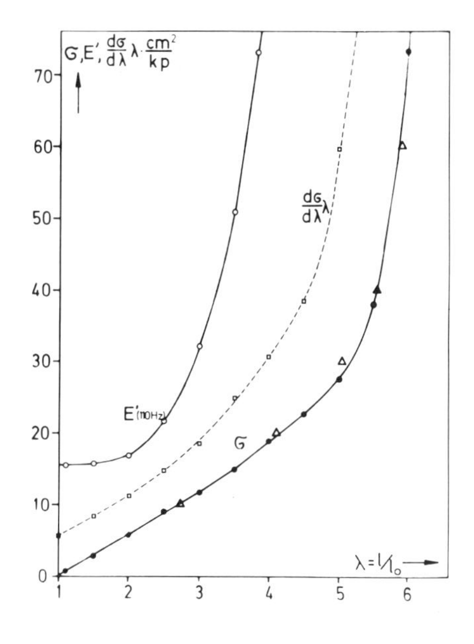

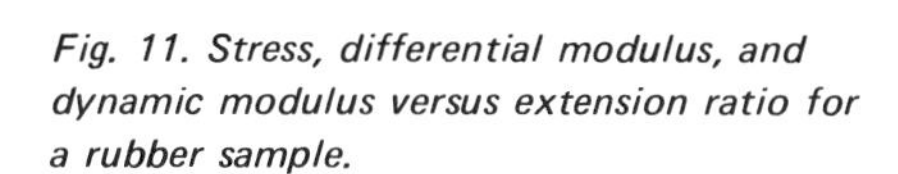

Fig. 11. Stress, differential modulus, and dynamic modulus versus extension ratio for a rubber sample.

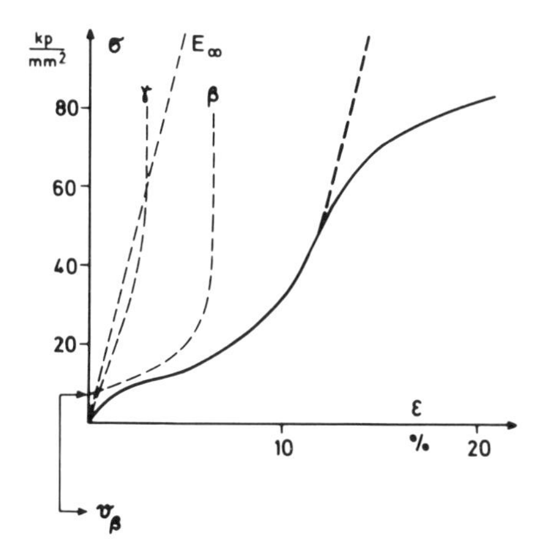

Fig. 12. Stress-strain curve and its analysis for a drawn polyamide fiber.

paraelastic deformation due to fast rearrangements of kinks (γ) and rubberelastic deformation of the meander curvatures (β), respectively. This explanation fits the experiments up to an elongation of 12 percent. At higher extensions, scission of tie molecules or – if we believe in our meander model – of tie bundles occurs. The value of 12 percent as that strain where chain scission begins is in accordance with results from EPR.[14]

In order to explain the other limiting case, that of true plasticity, which, e.g., is observed in polyethylene at least at slight deformations, one takes advantage of the knowledge of screw-dislocation movement which has been tested in the quantitative interpretation of the mechanical α-processes. From this, a thermally activated component σ_s of the yield stress $\sigma(T)$ of linear polyethylene can be derived which varies with temperature according to the formulae in Fig. 13. The remaining part of the yield stress, its athermal component σ_μ, depends on temperature only via the shear modulus of the crystal. Figure 13 shows the model which has been used to explain σ_μ and σ_s quantitatively: whereas a short-range dislocation motion implies a thermally activated overcoming of the activation energy of the α-process, a long-range dislocation motion – necessary for plastic deformation – causes larger voids to be formed for a short time. The latter process therefore is athermal in nature and can be brought about only by an external stress.

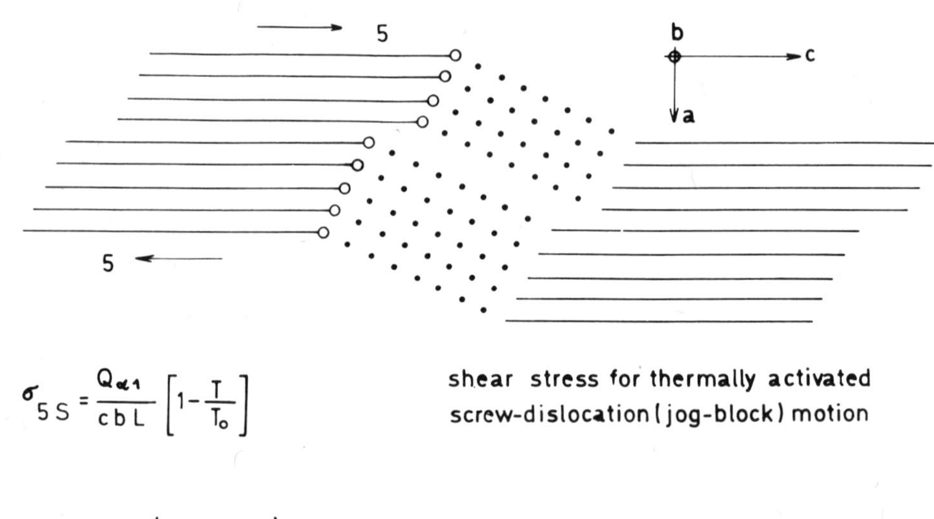

Fig. 13. Model for calculation of the thermal σ_s and a thermal σ_μ component of the yield stress σ in linear PE.

In Fig. 14, the measured yield stress $\sigma(T)$ of linear PE[15] has been decomposed into σ_s and σ_μ, which come close to their theoretically

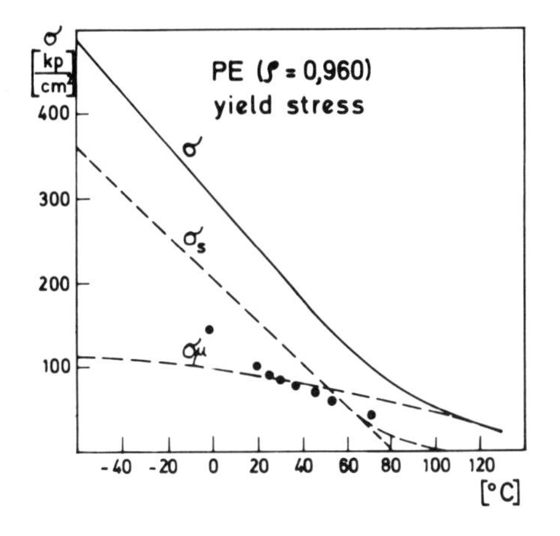

Fig. 14. Yield stress $\sigma(T)$ of linear PE[17] *and its decomposition into σ_s and σ_μ. The points represent dynamically measured critical stress amplitudes (Fig. 15).*

calculated values. The points represent dynamically measured critical stress amplitudes, σ_c which are an apparently constant fraction ($\sigma_c/\sigma \approx 0.43$) of the yield stress. This indicates that crystallites of preferred orientation probably start to deform plastically.

An example of such an amplitude dependence of the dynamic compliance is presented in Fig. 15. The critical amplitude, σ_c, is defined by the steep increase in J. Investigations of the amplitude dependence of metal single crystals have been widely used in our group for testing dislocation models and for nondestructive determination of flow stress.

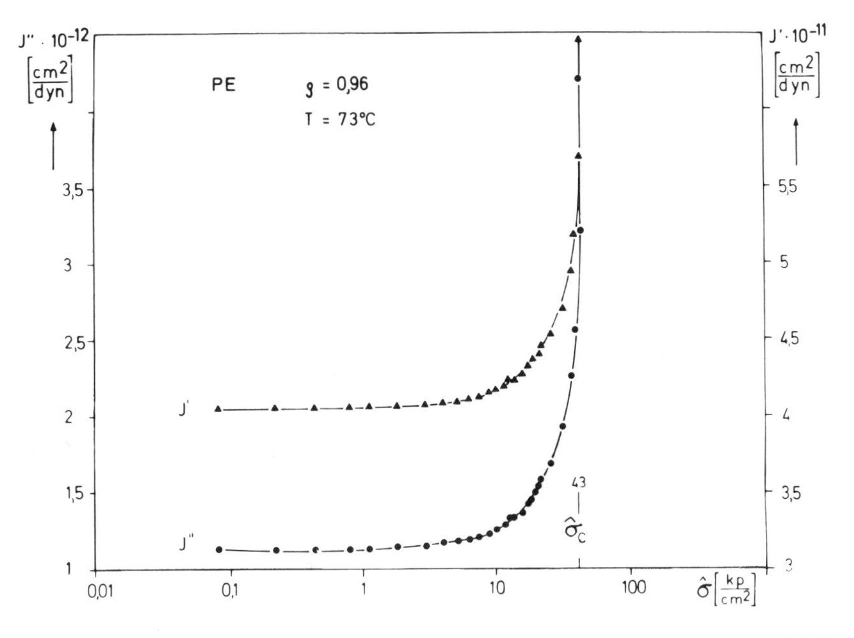

Fig. 15. Amplitude dependence of the dynamic compliance measured at 2.5 kHz and 73 C for linear PE. The critical amplitude, σ_c, is defined by the steep increase in J.

REFERENCES

1. Hägele, P. C., Wobser, G., Pechhold, W., and Blasenbrey, S., IUPAC-Symposium, Boston, Massachusetts (1971).
2. Wobser, G., Thesis, Stuttgart, West Germany (1971).
3. Hägele, P. C., Thesis, Ulm, West Germany (1972).
4. Blasenbrey, S., and Pechhold, W., *Ber. Bunsenges. Phys. Chem.*, **74**, 784 (1970).
5. Pechhold, W., and Blasenbrey, S., *Kolloid-Z. u. Z. Polymere,* **241**, 955 (1970).
6. Pechhold, W., Polymer Conference Series, University of Utah, July, 1971.
7. Pechhold, W., *Plenarvortrage Physik,* DPG (1971).
8. Pechhold, W., Liska, E., and Baumgartner, A., *Kolloid-Z. u. Z. Polymere,* in press.
9. Baumgärtner, A., Blasenbrey, S., Dollhopf, W., Liska, E., and Pechhold, W., *Kolloid-Z. u. Z. Polymere,* in press.
10. Čačković, H., Hosemann, R., and Wilke, W., *Kolloid-Z. u. Z. Polymere,* **234**, 1000 (1969).
11. Wilke, W., Vogel, W., and Hosemann, R., *Kolloid-Z. u. Z. Polymere,* **237**, 317 (1970).
12. Ferry, J. D., *Vicsocleastic Properties of Polymers,* Wiley (1970), p 337.
13. Huxley, H. E., *Science,* **164**, 1356 (1969).
14. Kausch, H. H., and Becht, J., *Rheol. Acta,* **9**, 137 (1970).
15. Andrews, J. M., and Ward, I. M., *J. Mater. Sci.,* **5**, 411 (1970).

DISCUSSION on Paper by W. R. Pechhold

THOMAS: I would like to enquire what is the relation of the theory you describe when applied to rubberlike deformation to the established kinetic theory of rubber elasticity.

PECHHOLD: At first sight, the only relation seems to me to be the application of the same well-known principles of statistics to a different model of microstructure. As pointed out in my paper, the meander model of the amorphous state contains anisotropic domains of some hundred A in dimension, the average deformation of which determines the macroscopic strain. In this model, the high-frequency rubber-elasticity is due to a shear between stratified molecules parallel and perpendicular to their axis, a deformation γ which is probably restricted to less than 2 because of the nonplanar geometry of the meander (compare Fig. 4 in my paper). This type of rubberelasticity explains the anisotropic dynamic compliances as measured on rubber samples with different extension ratios (Fig. 5). The low-frequency mode of rubber-elasticity can be visualized as a large-scale shear deformation parallel or perpendicular to meander layers involving a partial defolding of meanders and leading to a fibrillar type of structure (Figs. 7 and 6).

Looking at a single molecule, which can be shown in the meander model to have about the same radius of gyration in two dimensions compared with the 3-dimensional random coil, the large-scale mode of deformation will probably cause some geometrical relation to the extended randomly coiled molecule.

LANDEL: I want to ask a question on the calculation from the meander model in regard to muscle stress-strain. For muscle you showed a calculation of the force elongation. Has the calculation for an unloading curve been made?

PECHHOLD: I assumed the stress-strain relation to be reversible. This, of course, should not be and in fact is not true in reality [e.g., Wilkie, D. R., "Facts and Theories about Muscle", *Progr. Biophys.*, **4**, 288 (1954)] because of the interactions of actin bundles with the myosin rods which compare with those within the actin bundles. The interactions between neighboring bundles have not yet been taken into account in the calculation. By doing this, I would expect the observed stress-strain hystereses to be explainable.

I think, all polymer scientists will gratefully acknowledge the application of non-Euclidean geometries to achieve a continuous-function representation of superstructures in polymers and of their deformation behavior. Taking Gibbs free energies into account, which are intramolecularly involved in structure curvature, I hope Dr. Anthony will succeed in describing many details of micromorphology and its changes during deformation.

THE ROLE OF INTRA- AND INTERMOLECULAR COHESION IN FRACTURE INITIATION OF HIGH POLYMERS

H. H. Kausch and J. Becht

Battelle-Institut e.V.
Frankfurt (Main), Germany

ABSTRACT

The ultimate behavior of solid high polymers depends critically on the elastic energy stored within molecular chains. In this paper, the mechanism of the axial loading of chains through secondary bonds is studied in detail. For polyethylene and 6-polyamide, the maximum stresses which can be exerted by a defect-free crystallite in static equilibrium onto a chain are calculated. They are 0.75×10^{11} dynes/cm^2 (PE) and 2.24×10^{11} dynes/cm^2 (6-PA), respectively. The thermal excitation of stressed chains causes dislocations to migrate into the crystallite. The energy of dislocations and that of their activation are calculated as a function of molecular stresses. The effect of dislocation motion on chain breakage and macroscopic strength is discussed.

1 INTRODUCTION

The molecular description of fracture initiation is largely a molecular description of the deformation of a polymer under load. A loaded polymer forms a complicated, reacting system, the reaction path of which seems to be well determined but – at present – not well enough understood. In recent years, attention has been focused on the role of chain breakage in high-polymer fracture. There is strong evidence suggesting that the breakage of chain molecules in certain loaded polymeric samples actively influences the further development of macroscopic failure.[1] The breaking of a chain molecule may affect neighboring molecules in several ways: through reactions of the free radicals formed by homolytic chain scission, by the redistribution of load or increased stress concentration, and by release of elastically stored energy from the highly stressed breaking segments.

In this paper, we will briefly discuss these effects. The major part of the paper will be concerned, however, with an investigation of the role of the molecular superstructure in mechanical excitation and breakage of chains.

As the breakage of a chain leads to the formation of two free radicals (the two unpaired electrons of a broken valence bond), the electron-spin-resonance method (ESR) may be expected to provide the desired information on the development of fracture on a molecular level.

In general, free radicals are highly reactive and have a very short lifetime, so that stationary radical concentrations usually are not high enough – in relation to spectrometer sensitivity – to be detected. In the case of solid polymers, however, radicals – at least secondary ones – are sometimes quite stable, and the ESR signals can be readily detected in straining experiments. (For instance, in the case of nylon 6, the primary chain-end radicals produced in straining experiments at room temperature are not detectable, only the more stable secondary chain radicals. Checks suggest that there is very probably a 1:1 conversion of primary to secondary radicals formed.[2])

The number and species of free radicals can be determined by the ESR method. In structurally similar highly oriented fibers of different polymers, vastly different numbers of chain breakages are observed: from more than 5×10^{17} cm^{-3} in 6-polyamide (PA) to less than 10^{15} cm^{-3} in polyethylene terephthalate (PET) or polypropylene (PP). In the review paper[1], a few of the difficulties of relating the number of *observed* radicals to the number of chain breakages have been discussed. Chain reactions may lead to a multiplication of the number of broken bonds, and the reactions with oxygen or other impurities to a reduction of free radicals.

The chain breakages occur at high strains ($\epsilon > 10$ percent) and – at least in the case of PA – exclusively in the amorphous regions of the stressed samples.[3] It is generally assumed[1], that the breaking chains are tie molecules subjected to high stress concentrations. This is an extremely important point. The breakage of chains, i.e., of primary valence bonds, is induced by the axial loading of these chains through the action of secondary (van der Waals) forces. The study of chain breakages provides a new means of comparing primary and secondary bonds and their roles in fracture initiation.

2 TIE MOLECULES AND CRYSTALLITES

2.1 Effect of Concentrated Stress on Folded-Chain Crystallites

The highly oriented tie chains in drawn fibers are a very suitable object for studying the mechanism of load application on chain segments. Tie chains are partly embedded in crystallites through which the axial forces are transmitted. The large axial forces acting on the tie molecules, on the other hand, also affect the crystallites – but to different extents in different samples.[3] It is this effect which we wish to investigate further to derive the strength of a crystal for axial loading of a chain.

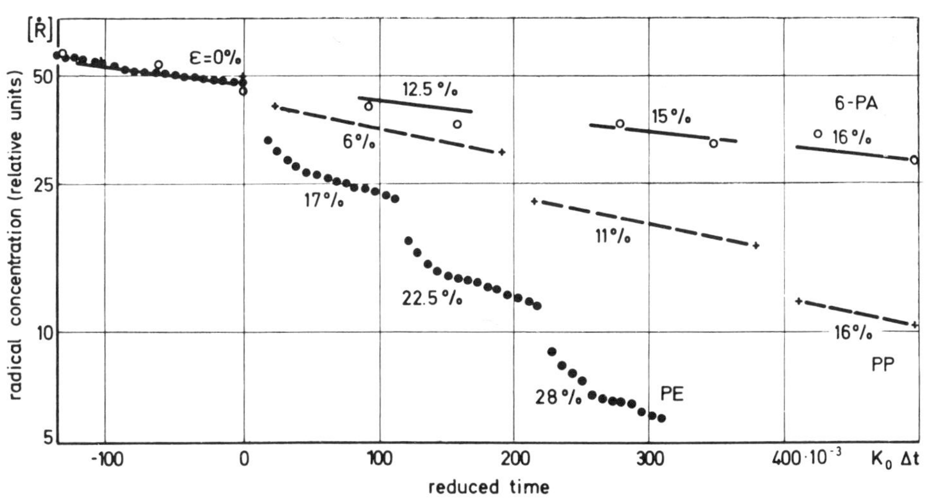

Fig. 1. Radical decay in irradiated fibers strained in air (6-PA, PP) or N_2 (PE).

For illustration, we show (Fig. 1) the effect of sample strain on the rate of decay of chain radicals. Samples of highly oriented 6-PA, PP, and polyethylene (PE) had been irradiated by 1-Mev electrons at liquid-nitrogen temperature. Subsequently, all samples were heated to their glass-transition-temperature (or above) for at least 5 minutes; thus, all radicals in the amorphous phase had disappeared, and radicals remained within the crystallites only. These samples were then strained within the cavity of an ESR spectrometer at room temperature.

The rate-determining steps of the radical-decay reactions in these cases are transfer reactions and oxygen diffusion. They are both very strongly correlated with the mobility of the radical-carrying molecular segments with respect to their surroundings, i.e., the crystal lattice. An increase in these rate constants, therefore, is a measure of an increase in mobility, which may be due to a decrease in intermolecular attraction and/or to the introduction of new crystal defects.

In Fig. 1, the concentration of free radicals is plotted versus a reduced time $K_o \Delta t$, where K_o is the rate of radical decay at zero strain and Δt is the time elapsed from the inception of sample straining. It should be noted that in PE and PP, the application of sample strain has a twofold result: an immediate and irreversible decrease of the (crystalline) radical population and a stress-dependent reversible increase in the rate of decay. In 6-PA, neither of these two effects can be observed. It was concluded from this experiment that in strained 6-PA, the tie molecules fail before crystal disintegration sets in, whereas in PP and PE, crystal distortion occurs before major numbers of chains are broken. This result is worth noting in view of the size and structure of PE and PA crystallites, and necessitates a further investigation of the different modes of micro-structure deformation at near-fracture level. We have calculated, therefore, how an axially stressed elastic chain is elongated and displaced within a rigid crystalline lattice represented by a periodic potential (Fig. 2). We assume that within a regular (folded-chain) crystal, the chain segments are bound with an average (lateral) cohesive energy U_{coh}. Upon axial displacement in the z-direction (chain axis), the chain will rotate and translate along a temperature-dependent minimum energy path. For PE, Lindenmeyer[4] has given the maximum potential energy to be overcome by a CH_2 group in twisted translation as $\Delta U_{tt} = 500$ cal/mole at room temperature. The potential energy of a CH_2 group within the amorphous phase can be determined from the difference of the enthalpies of crystalline and amorphous PE at absolute zero temperature, $H_{am,o} - H_{cr,o} = 661$ cal/mole.[5] We also take into account that the internal energy of a partly oriented PE may be smaller than that of an amorphous one. The energy difference will be comparable to the enthalpy difference ($H_{am} - H_{def}$), which was estimated by Fischer and Hinrichsen[6] to be 140 cal/mole.

For calculating the chain displacements within periodic potentials, we use the scheme set up earlier[7], but a potential as shown in Fig. 2. We

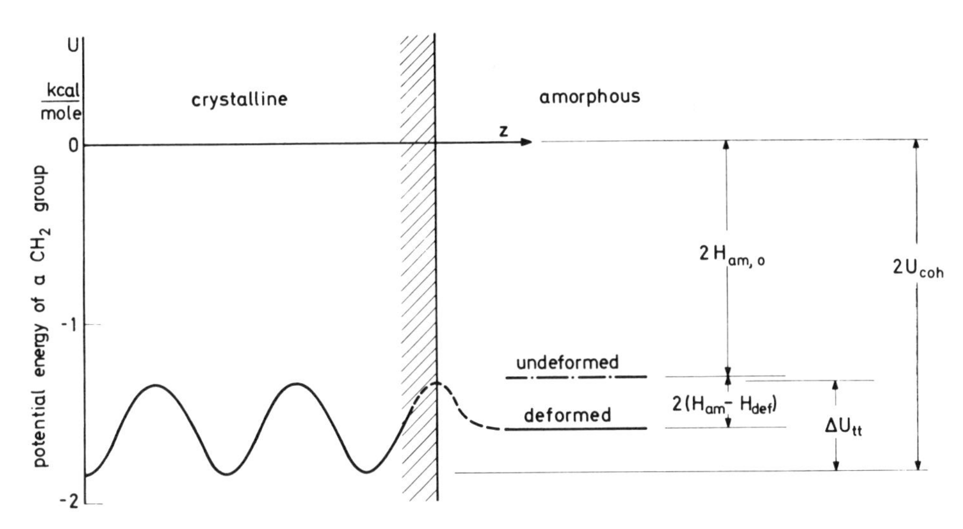

Fig. 2. Cohesive energy of crystalline and amorphous PE.

consider the case of a large folded-chain crystal with its fold surface perpendicular to the z-direction, bounded by z = 0. One tie molecule extends through this surface in the z-direction. If a tension s_a in axial direction is applied to this tie molecule and if the (large) crystal is kept in place by (arbitrary) boundary forces, then the atoms of the tie molecule will be displaced against the crystal lattice. The counterforces thus exerted on the chain atoms just balance the tension s_a. In static mechanical equilibrium, the local tensions, s, and displacements, u, of the tie-chain atoms within the crystallite are a unique function of s_a.

In Fig. 3, tension and displacement of a PE chain within a PE crystallite are shown for maximum axial chain tension s_o equal to 1.372 x 10^{-4} dyne. For any force larger than s_o, no static equilibrium can be established within a defect-free crystallite. It should be noted that this force which pulls out of the crystal a chain of full crystal length is 15.8 times larger than the force $K_1 v_1 d_1$ which is necessary to pull out a single monomer unit. If the tension s_o is divided by the chain cross section (18.24 A^2), we arrive at the maximum axial stress ψ_o which a perfect PE crystallite is able to exert on a tie molecule:

$$\psi_o = 0.75 \times 10^{11} \text{ dynes/cm}^2 \ .$$

As can be seen from Fig. 3, the maximum mechanical excitation of a chain penetrates about 50 A into the crystallite. At a distance of 60 A from the crystal boundary, it is smaller than the average thermal amplitude at room temperature, which is about 0.08 A.

Similar calculations have been attempted for 6-polyamide chains in PA crystallites. Here it was assumed that the difference between the

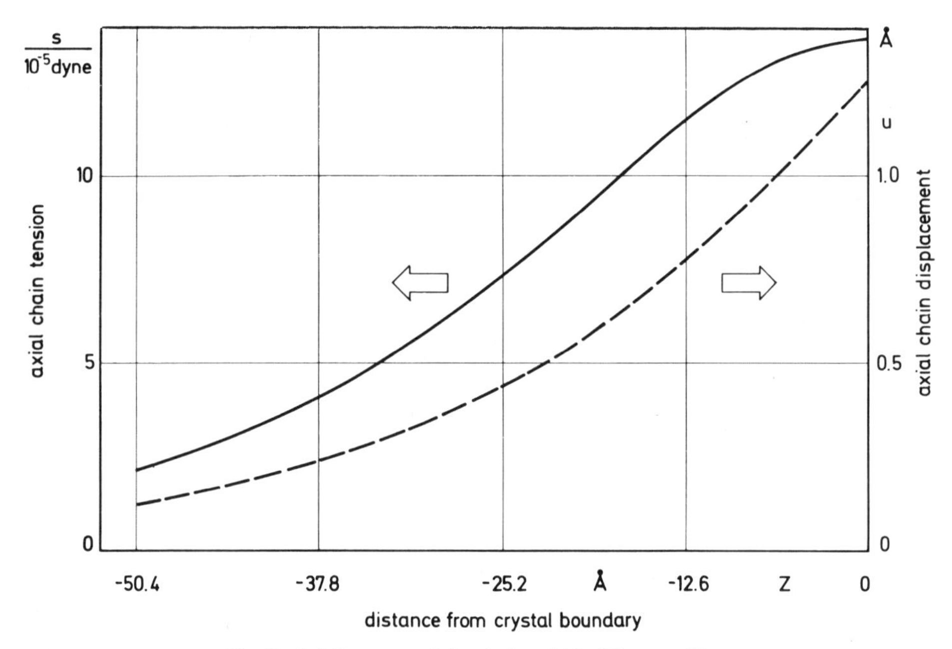

Fig. 3. Axially stressed tie chain within PE crystallite.

cohesive energies U_{coh} of PA (18.4 kcal/mole) and PE (6.3 kcal/mole) is caused by the presence of hydrogen bonds between the carbonamide groups of neighboring molecules within the grid planes of the PA crystallites. This excess energy contributes to the intermolecular potential v_2 acting between carbonamide groups. The potential v_2 thus is (12.1 + 0.50) kcal/mole, which is equivalent to 36.9 x 10^{-6} erg/cm.

Chain tension and displacement of a 6-PA tie molecule within a 6-PA crystal are shown in Fig. 4. The strong attractive effect of the hydrogen bonds results in a rapid drop in chain tension at the position of the carbonamide groups. The decrease in both tension and displacement is much more rapid than in the case of PE. At a distance of 21 A from the crystal boundary, the displacement has already dropped to the average level of thermal vibration at room temperature.

With the crystal boundary positioned – as shown in Fig. 4 – at the end of a $(-CH_2-)_5$ segment, the maximum chain tension is s_o = 3.94 x 10^{-4} dyne. This tension is only 1.7 times the force necessary to remove one carbonamide group from the crystal, i.e., 59 percent of the maximum chain tension is expended to break the hydrogen bonds of the first CONH group.

It should be noted that the value of s_o depends upon the geometric arrangement of the atoms at the crystal boundary. With a carbonamide group at the crystal boundary, the maximum tension s_o is about 20 percent smaller than in the case where the crystal boundary runs through

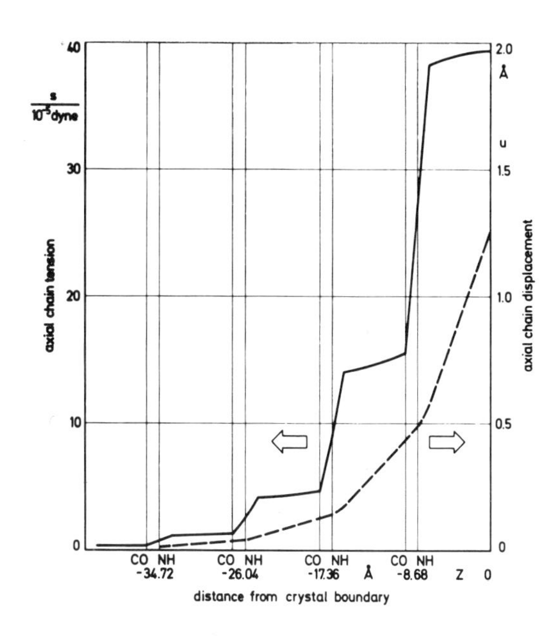

Fig. 4. Stressed tie chain within 6-PA crystallite.

the middle of the $(-CH_2-)_5$ segment.[7]

If s_o is divided by the molecular cross section (17.6 A^2), a stress of

$$\psi_o = 2.24 \times 10^{11} \text{ dynes/cm}^2$$

is obtained, which may be termed the static strength of a perfect PA crystal.

The effect of thermal vibrations on the strength of a crystal will be considered in the following.

2.2 Chain Translation Through Thermomechanical Activation

Figure 2 shows the potential energy of a CH_2 group in static mechanical equilibrium with the crystal lattice and with its neighbors within the chain.

The same potential is effective in quasi-static translation of a (tie) chain. The crystal strengths calculated from that potential therefore refer also to the same state, and do not take into account thermal vibrations of the chain. We have to study, however, the effect of thermal vibrations on chain translation in order to obtain – as desired – crystal strengths as a function of loading times or rates.

We use the same terminology as before.[7] The forces acting on an atom or a group of atoms within a folded-chain crystal are given by the displacements u of the chain atoms with respect to their position within the crystal lattice at zero axial stress. In particular, we obtain the potential energy W_p of a group with respect to the crystal lattice as:

$$W_p(u_i,x_i) = \frac{d_i v_i}{2} [1 - \cos K_i (u_i + x_i)] \quad , \tag{1}$$

where d_i is the distance between alternative atoms, $K_i = 2\pi/d_i$, and x_i is the amplitude of thermal vibration of atom i. The axial force s_i between atoms i+1 and i is given by

$$s_i = \lambda (u_{i+1} - u_i + x_{i+1} - x_i) \quad . \tag{2}$$

Here, λ is the constant of chain elasticity.

The energy W_s necessary to displace the atom group i by x_i is:

$$W_s(x_i) = W_p + W_t - \lambda \int_0^{x_i} (x'_{i+1} - 2x'_i + x'_{i-1})\, dx'_i \quad , \tag{3}$$

where W_t stands for $\lambda(u_{i+1} - 2u_i + u_{i-1})\, x_i$.

We would like to characterize, in the following, the shape of $W_s(x_i)$ for a number of boundary conditions.

(a) If x_{i-1} and x_{i+1} are zero, i.e., if only the central group is displaced, we have

$$W_s = W_p + W_t + \lambda\ x_i^2 \quad . \tag{4}$$

(b) If $x_{i-1} = x_i = x_{i+1}$, we have a translation of the chain segment and

$$W_s = W_p + W_t \quad . \tag{5}$$

(c) If the chain is uniformly stretched or if the stress is released with thermal amplitudes varying as $x_i - x_{i-1} = bx_i = x_{i+1} - x_i$, then

$$W_s = W_p + W_t \quad . \tag{6}$$

(d) Neither of the above cases is of general interest, as the random thermal excitation of atom groups within a chain leads to more irregular boundary conditions.

As a first approximation we assume that W_s contains x_i in linear and quadratic form, i.e., that

$$W_s = W_p + W_t + \frac{1}{2}\lambda (u_{i+1} - u_{i-1})\, A\, x_i + \frac{1}{2}\lambda A^2 x_i^2 \quad . \tag{7}$$

Here, A is a parameter determined by the phase differences between vibrational amplitudes of neighboring atom groups, which depend on the frequency distribution of the longitudinally polarized phonons exciting the atom group i. The term $\frac{1}{2}\, A(u_{i+1} - u_{i-1})$ is the average deviation of $(u_{i+1} - u_{i-1})$ from its equilibrium value.

In Figs. 5, 6, and 7, we have depicted the vibrational potentials for different boundary conditions and different levels of chain tensions. As shown in Fig. 5, a displacement of an atom group according to boundary condition (a) meets with very strong energetic hindrances and will only lead to small amplitudes. Any translation of a chain segment of neighboring potential wells is impossible per definitionem.

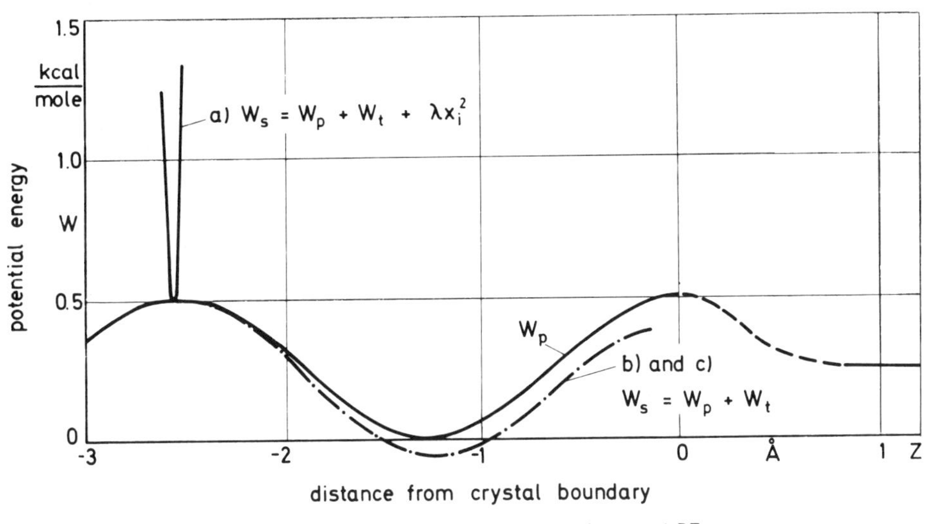

Fig. 5. Potential for cooperative displacement of stressed PE segment.

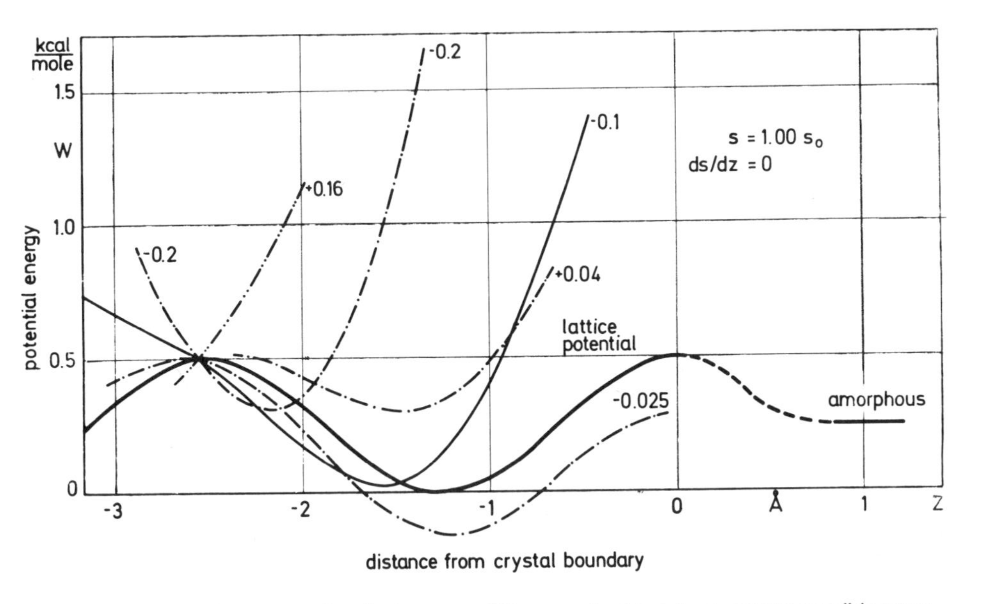

Fig. 6. Potential for cooperative displacement, PE segment subject to maximum possible stress.

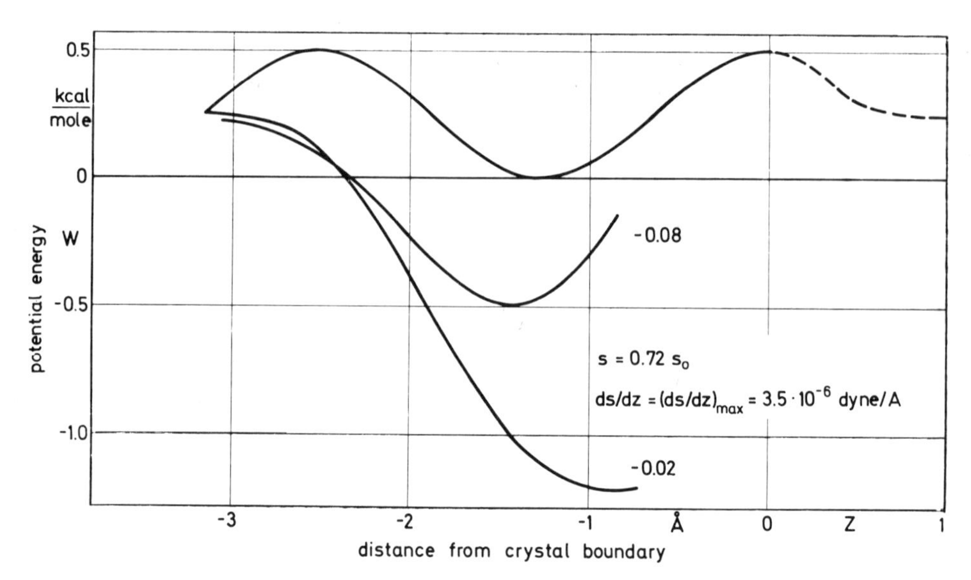

Fig. 7. Potential for cooperative displacement of stressed PE segment.

The potential for simple translation and uniform extension or compression of the stressed chain is also shown in Fig. 5 for a PE segment ($s = s_0$) subjected to the maximum possible stress.

The more probable potentials for cooperative displacement following from boundary condition (d) are depicted in Fig. 6 for a PE chain subjected to the maximum possible stress, with A as a parameter. A comparison with Fig. 7 reveals that the depth of the stress-biased potential minima depend on the stress gradient ds/dy, rather than on the level of stress s. From the shape of the curves in Figs. 5 through 7, we draw a number of conclusions:

1. It is evident that chain translation from one potential well to the next involves the cooperative motion of several atom groups within the chain. Their number should be of the order of 1/A.
2. For small negative A, the potential minima are deepest. Only if A lies between -0.2 and +0.004, does a translation of a chain segment seem possible.
3. An atom group subjected to a maximum positive stress gradient requires no activation energy for chain translation.
4. Atoms at less than maximum stress gradient ($u_i < d_i/4$) require an activation energy for translation.
5. A chain subjected to maximum axial tension s_0 is basically not resistant to translation [boundary condition (b)] or uniform chain elongation or compression [boundary condition (c)].
6. At maximum tension s_0 we have to expect a partial translation of the chain. This introduces a defect (dislocation) into the crys-

tallite. Stability and motion of these dislocations determine the stresses which can be exerted by crystallites onto tie molecules.

2.3 Energy of Defects

In Section 2.2 we have discussed that a chain subjected to maximum tension s_o is not resistant to translation. Even the smallest thermal vibration will cause further displacement of the chain and a decrease in tension s_a. The point of maximum tension will start to propagate into the crystallite (Fig. 8). The dislocation is fully developed once u_a has reached the value of d. The energy W_d of the defect introduced into the crystal by displacement of the atoms of one chain is easily calculated as the sum of all potential energies W_p (u_i) and of the elastic chain energy. Both are a function of external tension. In Fig. 9, the defect energy has been plotted as a function of displacement u_a and tension s_a at the crystal boundary. The energy of a completed defect (displacement of the chain by d) in PE amounts to 32.85 kcal/mole. At tensions smaller than s_o, the displacements u_a are smaller than d/2, the defect energy is smaller than $W_d(s_o)$, and the defect has a stable position. At tensions larger than s_o, the system becomes unstable as the chain tension decreases with increasing displacement u_a.

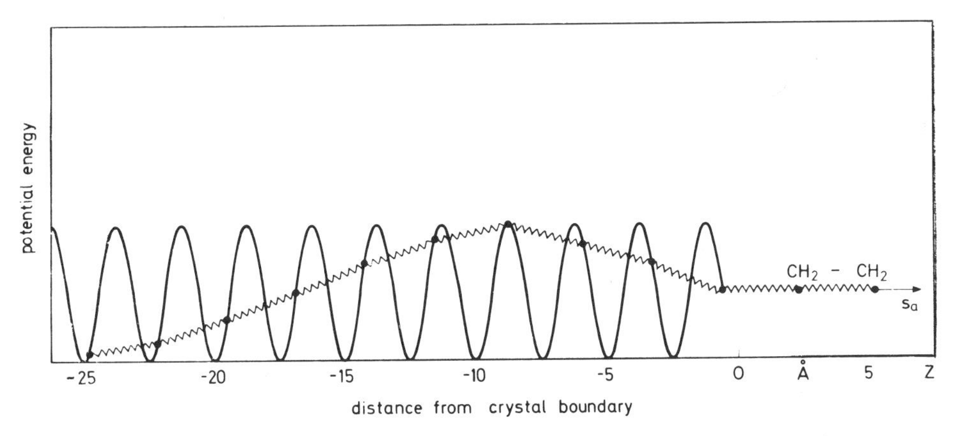

Fig. 8. Schematic representation of stress-induced dislocation within crystal.

As shown in Fig. 9, transitions between the stable and the unstable state can be caused by thermal activation. The required activation energy U_d is given by

$$U_d(s_a) = 2\left[W_d(s_o) - W_d\ (s_a)\right] . \qquad (8)$$

The activation energies required for translation of CH_2-CH_2 segments for various displacements u_a and tensions s_a are listed in Table I.

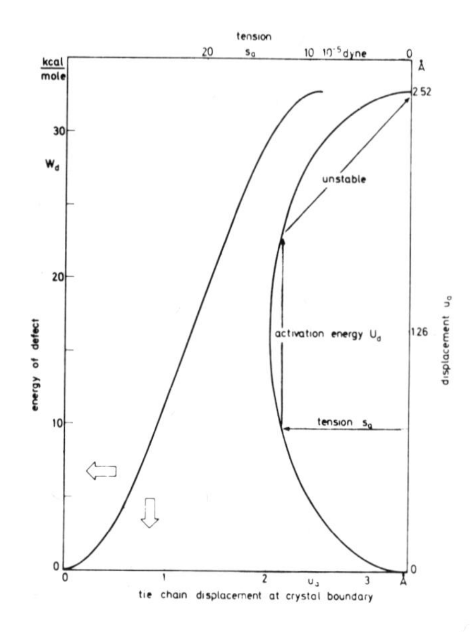

Fig. 9. Tension and energy of defect.

Table I. Activation Energies for Chain Translation in PE

u_a, A	s_a, 10^{-5} dyne	U_d	
		10^{-5} dyne/A	kcal/mole
0	0	23.08	33.0
0.104	1.79	22.89	32.7
0.195	3.38	22.38	32.0
0.303	5.18	21.40	30.6
0.412	6.91	20.00	28.6
0.494	8.12	18.69	26.6
0.589	9.41	16.90	24.2
0.699	10.71	14.60	20.8
0.824	11.93	11.68	16.6
0.914	12.63	9.39	13.40
1.009	13.18	6.88	9.82
1.108	13.56	4.19	5.98
1.158	13.67	2.80	4.12
1.209	13.72	1.40	2.06
1.260	13.72	0	0

Then, the probability of translation of a PE chain by one monomeric distance is proportional to that fraction of time during which the cooperatively moving groups of the chain are excited to a translational energy $E^* > U_d$ (s_a). We would like to treat this problem in terms of the statistical mechanics of polyatomic molecules.[8] We consider the cooperatively moving chain atoms as a group of r coupled (harmonic) oscillators. In thermal equilibrium, the average energy of these r oscillators is:

$$\bar{E} = \frac{h\,\nu\, r \exp(-h\nu/kT)}{1 - r \exp(-h\nu/kT)} \quad , \tag{9}$$

where ν is the oscillator frequency, and h and k are Planck's and Boltzmann's constants, respectively. For the case under consideration, the C-C bond stretching vibration at room temperature,

$$\bar{E} = 0.0075 \cdot h\,\nu \cdot r \quad ,$$

which corresponds to 2.14 r cal/mole. The C-C bond stretching vibration is largely unexcited. Nevertheless, there is a finite probablity that the r oscillators have a combined energy $E^* > U_d$:

$$p\,(E^* > U_d) = r^{(U_d/h\nu \cdot N_L)} \exp(-U_d/RT) \quad , \tag{10}$$

where N_L is Avogadro's number.

The total number of energy states per unit time accessible to the group of oscillators is equal to ν times a constant c_ν. Thus, the time $t_t(s_a)$ which, on the average, elapses before the stressed tie-chain atoms acquire enough energy for a translation in the +z direction is:

$$t_t(s_a) = 2 \cdot \nu^{-1}\; c_\nu^{-1}\; r^{(U_d/h\nu N_L)} \exp(U_d/RT) \quad . \tag{11}$$

The stress dependency of t_t enters through $U_d(s_a)$. The inverse function, $s_a(t_t)$, is only given implicitly; it corresponds to the time-dependent crystal strength, since it is the largest force not leading to crystal disintegration that may be transmitted onto a crystal for a period of time τ equal to t_t. Obviously, it is also the largest force that can be transmitted onto a tie molecule – if the tie molecule is strong enough to bear such a load. If not, chain scission will occur prior to crystal disintegration.

3 CRYSTAL STRENGTH, CHAIN SCISSION, AND FRACTURE DEVELOPMENT

In this paper, we have discussed the special case of the mechanical excitation of tie molecules through crystallites. We have studied the effect of intermolecular attraction (v) and chain elasticity (λ) on the maximum axial tensions s_o which can be transmitted from a crystallite onto a chain (and vice versa). The results of this study are of far-reaching consequences.

It is evident that the fraction of deformation energy stored elastically is large if large axial tensions can be transmitted onto chain segments. Large axial chain stresses are a necessary condition for the occurrence of brittle fracture. In this respect, the molecular stress distribution in amorphous regions is of considerable interest. The potential shown in Fig. 2 for unoriented amorphous PE is just an average potential. It has nearly the same value as that of two lattice CH_2 groups in opposition. The value of the average potential of oriented amorphous material indicates that about one half of the CH_2 groups is in opposition and one half is in latticelike arrangement. The largest gradients of this potential will, therefore, be of the same magnitude as those of the lattice potential, but irregularly spaced. If this is really so, the same maximum stress ψ_o may be expected – but, much larger distances than 60 A are required to transmit stresses of this order from the amorphous matrix onto extended chain segments.

The magnitude of ψ_o also determines the mode of failure of the system crystallite – stressed tie molecule. If ψ_o is equal to or larger than the (time dependent) chain strength, chain scission will occur and free radicals will be formed. For the two molecular species, 6-polyamide and polyethylene, we have compared chain strength with crystal strength. According to the reaction-rate theory of strength, the chain strength ψ_b is given[9] by:

$$\psi_b = \frac{1}{\beta}(U_o - RT \ln \nu\tau) \quad , \tag{12}$$

where β is the so-called activation volume of chain scission; it has been determined to be 1.32×10^{-10} dyne/cm^2 (kcal/mole) for PA[9] and 2.00×10^{-10} dyne/cm^2 (kcal/mole) for PE[10]. We are further using a frequency of C-C bond vibration of 3.10^{13} sec^{-1} ($\bar{\nu}$ = 1000 cm^{-1}), and the values of U_d as given in Tables I and II. The values of r and c may still be subject to discussion; we have chosen r = 3 as the number of modes of vibration contributing towards chain translation and c_ν as unity. Crystal strength $\psi_o = s_a(t_t)/q$ and chain strength ψ_b have been calculated from Eqs. 11 and 12, respectively, and are plotted in Fig. 10.

From Fig. 10 it becomes apparent that crystal strength and chain strength depend quite differently upon time. For 6-PA, ψ_o is larger than ψ_b for all excitation times larger than exp(-5) min, i.e., in most experimental time regimes. For PE we have shown the crystal strength calculated from Eq. 11. The chain strength was derived for two different values of U_o, namely 80 kcal/mole and the quite recently proposed low value of 25 kcal/mole.[10] It is obvious that in the first case, no chain scission should occur, whereas in the second case, chain scission and crystal disintegration have comparable probabilities. It should be mentioned, however, that the low activation energy is probably the result of a reaction involving other

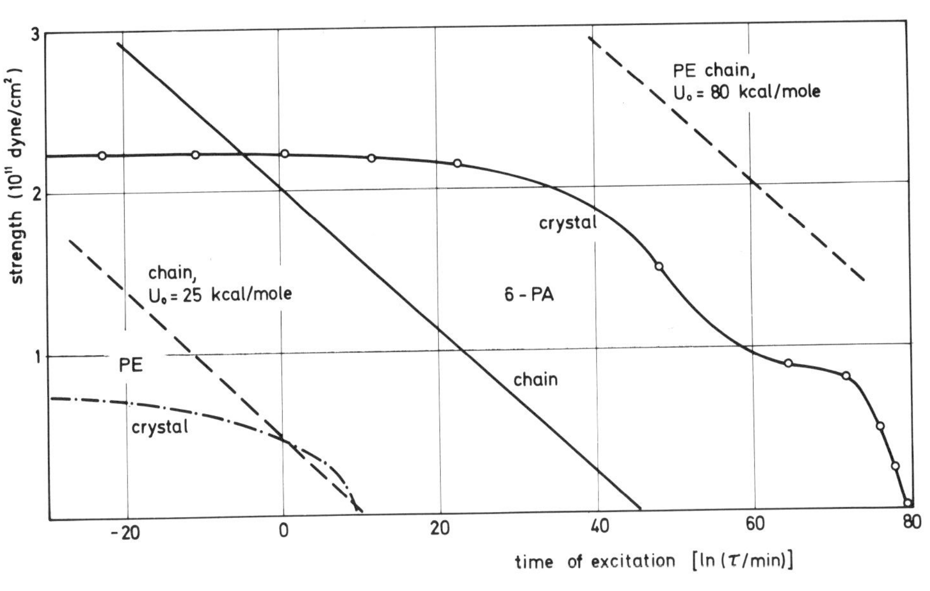

Fig. 10. Strengths of chains and of defect-free crystalline regions.

Table II. Activation Energies for Chain Translation in 6 PA

u_a, A	ψ, 10^{11} dynes/cm^2	U_d, kcal/mole
0	0	85.15
0.125	0.26	84.20
0.14	0.50	82.94
0.22	0.81	79.60
0.43	0.88	74.34
0.48	1.51	61.92
0.58	2.17	42.68
0.71	2.19	34.60
0.85	2.22	26.20
0.98	2.23	17.58
1.12	2.24	8.80
1.26	2.24	0

atoms than C or H. So Eq. 12 may not give the proper time dependency of ψ_b.

If ψ_o is smaller than the chain strength, dislocations of the kind discussed in this paper will be formed. As we have seen, the chain displacements penetrate into PE crystallites a sizable distance, even if the external stresses are smaller than the maximum one. It is fair to assume that the stress-induced displacement of a chain increases the mobility of neighboring segments and thus the rate of recombination of any free radicals attached to the chain or its surroundings.

It should be noted that in many cases, the conformation of the tie molecules within the crystalline phase deviates from a pure trans-conformation. Thus, the defect energy and the apparent crystal strength may be considerably amplified. In the absence of such defects, stress-induced dislocations enter the crystallite and migrate through it, permitting the tie molecule to the pulled out of the crystallite. It should be noted that the formation of one such defect in PA releases the axial stress on a 50 A tie molecule by $2.52 \times 2 \times 10^{12}$ dynes/50 cm^2 = 10^{11} dynes/cm^2, so that it assumes a subcritical level.

The formation of free radicals may in many ways affect the initiation and development of macroscopic fracture. This has been discussed in detail elsewhere.[1] We would only like to mention that the strength of microfibrils in oriented fibers or of filaments in highly drawn material is determined by chain-scission kinetics. The statement that the necking of glassy polymers is also initiated and controlled by radical reactions has not yet been confirmed, however.

ACKNOWLEDGMENT

The authors would like to thank Professors K. L. DeVries, Salt Lake City, and D. Langbein, Frankfurt, for fruitful discussions.

REFERENCES

1. Kausch, H. H., and Becht, J., *Kolloid-Z. u. Z. Polymere,* **250**, 1048 (1972).
2. DeVries, K. L., private communication.
3. Becht, J., Dissertation, Technische Hochschule, Darmstadt, 1970.
4. Lindenmeyer, P., *Mechanical Behavior of Materials,* Soc. Mat. Sci., Japan (1972), Special Vol., p 74.
5. Baur, H., and Wunderlich, B., *Fortschr. d. Hochpolymeren Forschung,* **7**, 388 (1970).
6. Fischer, E. W., and Hinrichsen, G., *Kolloid-Z. u. Z. Polymere,* **213**, 28 (1966).
7. Kausch, H. H., and Langbein, D., *J. Polymer Sci.,* Polymer Physics Edition, Vol 11 (1973).
8. Fowler, R., and Guggenheim, E. A., *Statistical Thermodynamics,* University Press, Cambridge (1952), p 495.
9. Kausch, H. H., and Becht, J., *Rheol. Acta,* **9**, 137 (1970).
10. Tomashevskii, E. E., *Soviet Physics – Solid State,* **12**, 2588 (1971).

DISCUSSION on Paper by H. H. Kausch and J. Becht

HULL: What is the Burgers vector of your dislocations?

KAUSCH: The Burgers vector is parallel to the chain axis.

HULL: So they are screw dislocations?

KAUSCH: No, they have been introduced by Reneker [*J. Polymer Sci.*, **59**, S39 (1962)] as point dislocations.

HULL: So they are not really dislocations as we normally understand them?

KAUSCH: Point dislocations carry characteristic properties of a dislocation: they can be generated, they can move, they give rise to a displacement field; there is however, no dislocation line but rather a point. Wherever this difference is important, we should talk of Reneker defects instead of dislocations.

AGENDA DISCUSSION: MOLECULAR DESCRIPTIONS OF DEFORMATION

*F. H. Müller**

University of Marburg

*T. O'Neill***

Battelle-Geneva

1 INTRODUCTION

The papers in this session considered phenomena associated with large deformation (nonlinear behavior) as well as fracture. The Meander Model should be discussed a little more intensively in order to see how the well-known random-coil explanation of rubber elasticity can be replaced by this model. The Meander Model also assists us in understanding the dense ordering of chains, which is necessary to explain the high density of the rubberlike and glassy states.

The discussions in Session II have shown that, in principle, a

*Chairman.
**Secretary.

phenomenological formulation of nonlinear behavior is possible. But, I do not believe that we are as yet able to describe, in a general way, all of the complex processes which may occur. Further, it seems to be nearly impossible to take into account the prehistory of a sample. Moreover, other factors often influence the deformation of the sample via processes which develop during the deformation itself, e.g., chain scission, stress-induced radical reactions, cavitation, and crystallization.

In addition, the deformation process influences the deformation behavior. High-speed deformation leads to behavior different from that produced by slow speed deformation, when crystallization or orientation, for example, can occur. An interesting point of discussion would be deformation with and without necking. It seems that the necking process depends on the test conditions, in a manner somewhat analogous to that of the laminar and turbulent flow of fluids. Finally, the influence of local heat production during deformation, along with other temperature effects including local cooling, and the influence of local fluctuations in the packing of molecules (short-range order) should be considered.

2 CRYSTALLIZATION AND CRYSTALLINE POLYMERS

The most sensitive method for the investigation of crystallization in the stretching processes, e.g., of natural rubber, seems to be a calorimetric technique: where one follows simultaneously the work done on a sample and the heat generated during and after elongation. An analysis of both the stress-strain curves and the heating curves provides information on the degree of crystallinity as a function of elongation. In this case, both temperature, measured as a function of time, and maximal final elongation are parameters. (See Figs. 1 and 2.) It is clear from the figures that

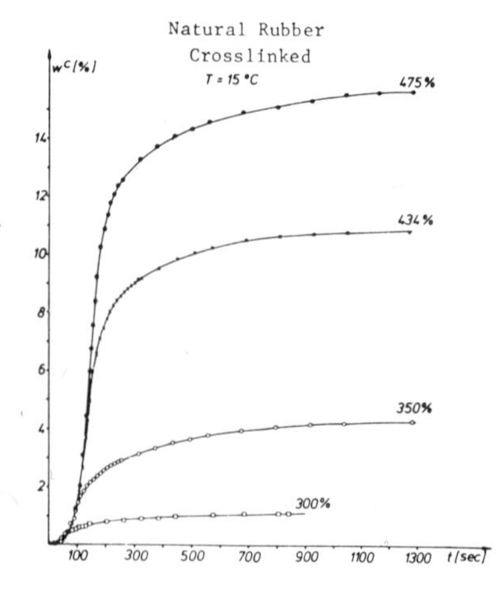

Fig. 1. Crystallinity as function of time.
Parameter: maximal elongation of the sample in percent.

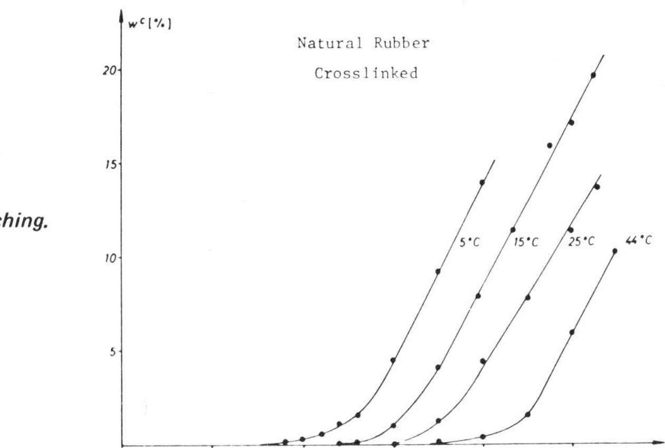

Fig. 2. Degree of crystallinity as function of elongation. Parameter: temperature of stretching.

calorimetry is superior to all other methods in regard to determining the total amounts of crystalline regions. Using this method, the kinetics of crystallization was studied in detail with much success.

Moreover, this technique, stretching calorimetry, provides a tool to investigate many stress conditions. In certain cases, conclusions have been drawn on chain breakage by stepwise loading and unloading samples of vulcanized natural rubber. From the overall balance of deformation and heat, it was concluded that chain breakage was occurring. With knowledge of the bond energies of the polymer chain, an estimate of the number of broken chains is possible [see *Koll.-Z.*, **172**, 1 (1960) and, the Dissertation by Gerth, Marburg University, 1972].

Andrews studied the crystallization of natural rubber from the melt and was able to interpret the results only in terms of crystallization of the single molecules. In other words, in order to explain the results he must assume that the rubber molecules are transferred from the melt to the crystal as single molecules. The critical secondary nucleus which has to be formed must be formed from one molecule, otherwise the results which he obtained could not possibly be explained. If such a single-molecule transfer does occur, this would suggest that the Meander Model is not physically possible. If there were parallel groups of molecules, then it would be easy for these parallel groups to transfer as a whole to the crystal, and it would be much easier to make a secondary nucleus by this means. However, this would not agree with the experimental results on the kinetics of lamellar growth.

The actual velocity of growth of a single-crystal lamella was measured by electron microscopy. This velocity had been measured as a function of temperature and gives the familiar crystallization curve which relates to single-folded-chain lamellae. A 1000-fold decrease in velocity is obtained by putting 10 percent trans units into the cis-polymer. Both the dependence and the magnitude of the decrease (can be explained) by simply assuming that the molecule has to form a secondary nucleus from the melt, involving three folds of the platelet growth surface. All one has to do is to calculate the probability of finding enough monomer units to give three folds of the homopolymer without any foreign units intervening. It seems that this concept only is capable of explaining an effect of this magnitude on the growth velocity. This implies, however, that only a single molecule can join the crystal.

If a "packet" of molecules exists in the melt, then the whole packet could attach itself to the growing surface of the lamellar crystals. In the case of "packet growth" of molecules on the crystal surface, a large effect of imperfections would be inconceivable. Andrews stated that unless his interpretation of these crystallization results is completely erroneous, he feels that the melt simply cannot contain crystallizable "packets" of molecules [Andrews, Owen, and Singh, *Proc. Roy. Soc. London, Series A*, **324**, 79-97 (1971)].

With regard to the proposed Meander Model, Fischer had two comments concerning the deformation of crystalline polymers and the configuration of chain molecules in the amorphous state:

1. According to the proposed model for the conversion from the lamellar to the fibrillar structure, the long spacing should remain approximately constant. It was shown, however, that the long spacings in drawn linear and branched polyethylene depend strongly on drawing temperature [E. W. Fischer and G. F. Schmidt, *Angew. Chem., Int. Ed.*, **1**, 488 (1962)], i.e., the long period of the starting material does not determine the morphological structure of the drawn fibrillar sample [A. Peterlin, *J. Mater. Sci.*, **6**, 490 (1971)]. These findings are in complete agreement with the observations of Gleiter, reported during this conference.
2. In a recent study by Kirste, et al. (to be published in *Makromol. Chem.*), the radius of gyration of a polymethyl methacrylate chain, in the glassy state, was measured by means of neutron small-angle scattering. For that purpose undeuterated chains were embedded in deuterated chains with the same molecular weight of M_w = 250,000. The evaluation of the scattering curves revealed $\bar{r}$ = 126 A, in close agreement with the value found in a θ solvent. This result yields strong support for Flory's conclusion about the coiled configuration of a chain in the bulk amorphous state [P. J. Flory, *Statistical Mechanics of Chain Molecules*, Interscience Publishers (1969)].

To these comments Pechhold added that there is no single phenomenon in physics which cannot be explained by more than one model. He had calculated roughly what radius of gyration of the PMMA just mentioned would fit his model and found that the result more or less coincides with Kirste's. In the Meander Model, two-dimensional statistics are used in which the x and y directions of the molecular bundle are defined, but the z direction is not.

According to the analyses by Peterlin and by Hosemann, there is some doubt about the exclusive temperature dependence of the long period. Pechhold cannot disprove Fischer's contention, but would suggest doing experiments on a range of samples of differing initial state, to see if any small break or discontinuity is perceptible. For in this model, where a fibril is made from one lamella, the long period is split into two, and for material with a short period, the crystal thickness remains nearly constant. This question has to be investigated further. A crystallization theory from the Meander Model has not yet been worked out. In dilute solution, of course, one does have single molecules which fold. Thus, in Krimm's work on crystallization of PE single crystals from solution, a big difference is observed in the IR spectra of a mixture of deuterated and undeuterated samples. In the single crystal, a folding exists which does not depend on a precise adjacent reentry folding or Meander folding. It is the folding of a single molecule, and thus if the sample is melted, the rocking-bond phenomenon disappears. Consequently the solution-grown lamella structure is destroyed if passed into the molten state. The shape of the crystal remains for a short time, but the IR bands change markedly. So as far as crystallization is concerned, I would say that in our case the molecular bundle being attached to the growth surface corresponds to the removal of kinks from the bundle. This process would also be strongly influenced by the introduction of foreign units or imperfections in the molecular chain.

Halpin raised a number of questions concerning Pechhold's model in which the anisotropic bulk state consists of packets of bundles of polymer chains. Similar models were proposed several years ago on the basis that the density of the glassy state was not consistent with the spatial requirements of a randomly coiled interpenetrating chain model. Halpin asks whether this model was evaluated in light of specific volume, thermal expansion, and other phenomena characteristic of the rubberlike and the glasslike state of amorphous polymeric solids. Furthermore, did one develop an exclusion principle to define what are permissible and what are forbidden modes of dislocation motions in polymeric crystals? Can an amorphous body have a dislocation?

Pechhold took up these questions and indicated that he had proposed the model of screw dislocations sliding along the chain. This should be possible in all crystal planes perpendicular to the lamellae. Other types of dislocation, according to electromicroscopy, are not very numerous, and probably negligible. Surface dislocations in these specially arranged lamellae are distinct from this, of course.

3 CRAZES AND GAS EVOLUTION

J. G. Williams reported some observations he had made on slowly growing crazes in PMMA in methanol. It appears that for times greater than 5 hours or temperatures above 20 C significant amounts of gas are evolved. There is evidence that it has a significant hydrogen content. This is not in contrast to Andrews' work. Andrews' results showed that the number of free radicals decreased to "zero" as the crosslink density decreased to zero, but it should be clear on what this zero means. His ESR equipment can measure only radical concentrations above a limit of about 10^{13}/g. So "zero radical concentration" in this ESR work means only that the number of radicals has decreased to below the detection limit. It is therefore possible that bonds are being broken in the uncross-linked case and that hydrogen is formed by recombination of H atoms..

Andrews pointed out the possibility of gases being actually dissolved in the PMMA in, for example, molecular voids. The quantity of gas evolved is considerable that one would be surprised if it came out of solution in the bulk polymer. Low pressure does exist in crazes because of cavitation effects, so it is possible that the gas is coming out of solution in the liquid. However, J. G. Williams claims that this is unlikely for it should go back into solution when pressure goes up to atmospheric. The most likely explanation is that the gas is H_2 generated by a molecular fracture process, since the molecular weight of PMMA is very high and one would expect a large concentration of "entanglement" crosslinks. It is possible that if the concentration of crazes in the specimen were increased an ESR signal would be detectable.

Andrews mentioned that he can get 60 ml of gas from 1 ml rubber (only 15 percent of that is hydrogen, and the rest is dissolved gas). Hydrogen is evolved when the secondary peroxy radical recombines with a molecule. It is only when the peroxy-radical signal begins to decay on heating through the glass transition that hydrogen evolution begins. Perhaps tertiary carbons are formed rather than double bonds. Roughly, he gets one molecule of hydrogen for each molecular fracture.

To the remark of Knappe that evolution of H_2 is well known in studies of radiation effects on PE, Andrews answered that although the ESR spectrum of the irradiated material is quite different from the secondary spectrum of his test specimen, it may well be that the primary spectra are the same.

4 MOLECULAR DEFORMATION ANALYSIS

As an illustration of a fundamental method of attacking the ultimate problem of predicting the stress-strain behavior in a bulk polymer, Williams presented some initial results obtained by R. H. Bond, R. E. Stephenson, and S. M. Breitling, University of Utah, in calculating the atomic force-displacement relation of a single, isolated, polymer chain composed of approximately 100 atoms. This development, from conformational analysis, shows an isolated n-octane molecule elongated from its local minimum-energy conformation. The energy required for such an elongation was calculated by using valence-force potentials to simulate the interactions of bond stretching, bond bending, bond rotation, and nonbonded interactions. For each increment of elongation, the entire molecule was allowed to relax to a conformation of minimum energy, with the restriction of a fixed end-to-end distance. Thus, such an energy-elongation curve will represent a minimum energy path corresponding to the elongation for a given initial conformation.

It was highly instructive to see the digital computer output in the form of computer graphics displayed in a motion picture of two molecules: one the rotation about the carbon-carbon bond in ethane, and the other a short segment of polyoxymethylene in a minimum-energy helical state.

5 STRESS-STRAIN CURVE AND THE NATURE OF DAMAGE

In addition to the methods discussed so far or described in the literature, Retting used a mechanical method which is based on the existence of a brittle-tough transition [*Europ. Polymer J.*, **6**, 853 (1970)] to examine the continuous damage of polymers during deformation. He prestrained PVC specimens to certain small amounts with a high tensile speed (at which the ultimate extensibility is low) and continued the straining with a low tensile speed (at which the extensibility is high). The loss of ultimate extensibility caused by prestraining is due to the early (beyond the yield point) initiation of microcracks and similar irregularities. From this and from other experiments like those of the authors mentioned before, Retting concludes:

1. Fracture of polymers can be understood only by taking into consideration the stress-strain, strain-rate time, and temperature history of the specimen during formation from its very beginning.
2. Every gross deformation of polymers consists of a competition

between local deformation and fracture. Which of the above will be dominant at a certain elongation depends on several parameters (e.g., time and temperature).

Lee commented that the influence of various histories of straining may need interpretation through the associated stress variation. This is particularly true if the stress variation leads to local necking of the specimen, a phenomenon which can be most directly assessed in terms of stress history; for then the strain is not uniform along the specimen and average strain does not provide information about the material in the neck, which is currently deforming. However, if inertia forces are negligible, which is the case for many tests other than high-velocity impact tests, the tensile force is uniform along the specimen and so provides information concerning behavior in the necked region where the deformation is currently taking place. The Chairman stressed the importance of taking into account the influence of necking in interpreting test results. Necking in a tensile test is a macroscopic instability phenomenon, which can perhaps be most lucidly discussed in terms of the true and nominal stress-strain relations. The variation of true stress reveals the material properties as determined by the microstructure and loading or straining history. If, for example, internal crazing of the material causes a reduction in the rise of true stress, this might then be dominated by the reduction of area and lead to necking which would not have occurred without crazing. Crazing itself is an instability of a similar nature but on a smaller scale. The initiation of voids leads to an effective reduction of area which will lead to a local instability, with increase in the void volume and corresponding reduction in average stress. Taking the "true" stress to be averaged over the total area, this will reduce the increase of stress and hence increase the tendency to neck on the macroscopic specimen scale. The theory also applies if the true stress passes through a maximum as, i.e., the upper yield point in some metals. Then instability is certain to occur and take the form of Lüders band. This approach also carries through for compression, where the area increases, but a sharp peak in the true-stress curve will still lead to instability. It is important to know whether the testing machine is subjected to prescribed elongation or prescribed force. However, in the latter case, true stress and hence precise material behavior is still not determined since change in section, and hence necking, can cause large variations of true stress in the specimen.

Lee emphasized the main point, viz., that necking is caused by a combination of material properties expressed in the true stress-strain curve, and the geometrical factor of reduction of area. The former can be influenced by structure, anisotropy, rate effects, etc., and hence necking is in part determined by these influences. It seems that such influences will also arise in Retting's measurements, and it would be helpful to know how the stress at fracture varies for the different straining histories in his experiments.

However, in his experiment, Retting deformed a solid at the rate R_1 and found a rate of damage accumulation which eventually induced the final fracture of the sample. A similar experiment was then performed at R_2 to obtain a new rate of damage accumulation and different ultimate properties. An additional set of experiments was performed by combining the two stress histories R_1 for t_i and R_2 until fracture. The fracture was intermediate between the basic experiments R_1 and R_2. Retting pointing out that the summation of the expected damage accumulated during the initial stage $R_1 t_i$ and the final stage $R_2 (t_b - t_i)$ will lead to this result and is in support of Kausch and Becht. Halpin commented that Retting is performing a linear summation of damage, a summation which is analogous to a Miner's Rule fatigue calculation. In fact, one should expect to see a damage acceleration adjacent to the discontinuity in the stress or strain history analogous to the results of Halpin and Polly, *J. Comp. Mater.*, **IX** (1968), and the thesis work of Knauss at California Institute of Technology.

According to Lee, in a tensile test the stress decrease at the yield point may be a consequence of the necking of the sample. Bauwens pointed out that some polymers (PVC, PC, PMMA) exhibit the same behavior in a compression test where there is an increase of the cross section. In this case, the instability at the yield point seems to be a consequence of the intrinsic properties of the polymer.

Radcliffe added that reliable experimental measurements have established for several polymers that a true yield drop does occur in uniaxial compression. In such cases, a yield drop also occurs in tension and is *intrinsic,* i.e., it relates to the characteristic flow mechanism involved and appears in the true stress-strain curve, as well as in the engineering load curve. In contrast, other polymers exhibit a yield-stress drop in tension but not in compression. In such cases, derivation of the tensile true-stress versus true-strain curve shows the drop to be absent, i.e., a geometrical effect unrelated to the yield mechanism.

Andrews found that true-stress versus true-strain curves for highly irradiated HD polyethylene exhibit a maximum at yield (in the region of homogeneous deformation), whereas no such maximum occurs for unirradiated material. Parallel work on the electron microscopy of deformation in single crystals shows that irradiated crystals are subject to extensive microfracture which enables them to deform without the operation of crystallographic mechanisms. The strain softening after yield may thus be associated with the microvoiding in the crystalline phase which reduces the load-bearing cross section in this phase.

Kausch drew attention to the fact that the mechanical excitation, and the rate of breakage, of extended chains is a function of the rate of straining of their environment and suggested that the "undetected damage" in Retting's experiments might be caused by chain scissions. Retting's experiments were supported by Menges who referred to his paper and also reported that bars with crazes of the same length break instantly without new plastic deformations, if they are tension tested at 10 mm/min. When tested at 1 mm/min, the material begins to shear, Lüders bands form, and finally necking occurs. This is viewed by Müller as another argument in support of the fact that the occurrence of chain breakage is often determined by how rigidly the molecule is held at its ends. This matter is dealt with in Kausch's paper. M. L. Williams also endorsed the type of calculation presented by Kausch. Within the context of the discussion of Kausch's paper and the relevence of his model to a realistic fracture of the tie chains, his calculation using a constant load on a single chain could be viewed as the limit of a real situation in which all the tie chains except one were slack, thus leaving the one chain to carry all the load. He suggests that Kausch should extend his calculations, if desired, to include a statistical distribution of tie-chain lengths (or loading) to represent more realistically the intercrystalline situation.

Other questions were raised by Halpin concerning ESR: has the experimental experience in studies of free radical generations provided evidence to support or criticize the assumption of fracture by flaw development? Specifically, has sufficient experimental evidence been obtained about the rate of free-radical generation to evaluate the possible form of a fracture-propagation law from molecular phenomena?

Kausch replied that the radical production curve follows an exhaustion process. That is, one has a certain, obviously finite, number of chains which can undergo chain scission because of their conformation and degree of attachment to the surroundings. The various aspects of how chain scission contributes to crack initiation have been discussed extensively (*Kolloid-Z. u. Z. Polymere,* **250**, 1048 (1972)] but not conclusively. It can be stated, however, that present experimental techniques are insufficient to effectively count the number of free radicals formed by *one* growing crack. So no evidence is available from ESR right now as to how localized chain scission occurs.

Andrews added that although the rate of radical production decreases with time at constant load, this effect will not be significant in the case of a crack propagating at constant velocity. In this case, the high stress region propagates continuously into new regions, thus replenishing the population of highly stressed bonds. This corresponds to a tensile experiment carried out at constant rate of strain in which the radical concentration, C, obeys the equation

$$dC/dt = A \exp(-\alpha\, \sigma/RT) \quad , \tag{1}$$

where A and α are constants, t is the time, and σ is the stress.

6 TIME-TEMPERATURE SUPERPOSITION IN TWO-PHASE MATERIALS

Tschoegl made some further comments on the complex temperature relations in the materials which Shen (in his paper) called "polyalloys" i.e., the block copolymers, graft copolymers, and polyblends. These materials contain two or more phases. They thus exhibit at least two transitions and are, therefore, necessarily complex thermorheologically. The criterion that all relaxation or retardation times should be equally affected by a change in temperature, which leads to thermorheological simplicity, is not fulfilled in these materials. D. G. Fesko has clarified the basic issues in his doctoral work at Caltech.

Figure 3 shows two isothermal segments of the creep compliance as a function of temperature in the customary doubly logarithmic plot. The

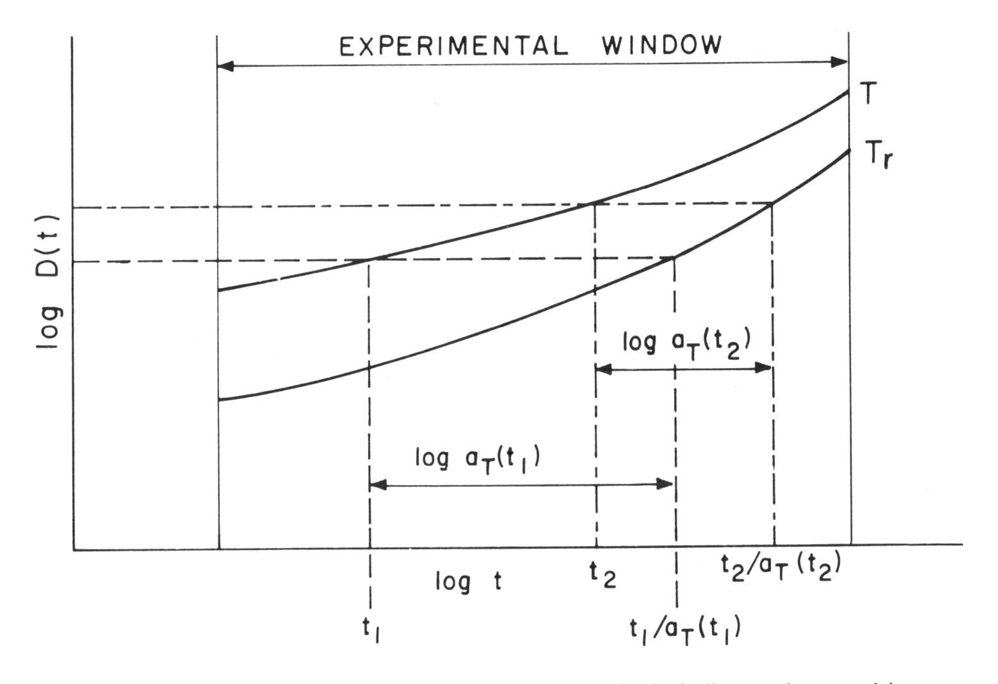

Fig. 3. Superposition along the log t-axis for a thermorheologically complex material.

curves represent the behavior of a hypothetical thermorheologically complex material. They cannot be superposed by a simple shift along the log t-axis. As shown in the figure, the shift factors become functions of time as well as temperature if the material is thermorheologically complex.

A somewhat similar situation exists if one records the creep compliance of a thermorheologically simple material as a function of temperature at two different times as shown in Fig. 4. The curves are not superposable by a shift along the T-axis. However, if the WLF equation

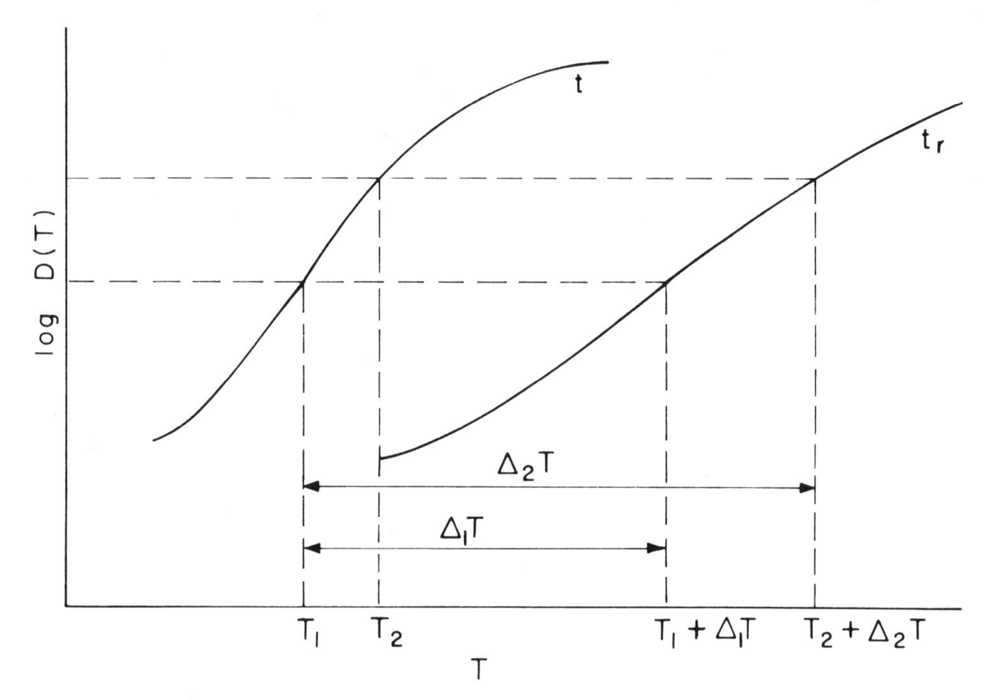

Fig. 4. Superposition along the T-axis for a thermorheologically simple material.

for the material is known, the amount can be calculated by which each point recorded at time t_1 has to be shifted to fall onto the curve obtained at t_2. Fesko and Tschoegl considered, therefore, the circumstances under which such a point-by-point shift could be made for two-phase materials along the log t-axis.

By Taylor expansion[1] of the compliance in the variables t and T, one obtains

$$\left(\frac{\partial \log a_T(t)}{\partial T}\right)_t = -\left(\frac{\partial D(T)}{\partial T}\right)_t \left(\frac{\partial D[t/a_t(t)]}{\partial \log t/a_T(t)}\right)_{T_r}^{-1} \tag{2}$$

This equation shows that the slope of the (time-dependent) shift distance with respect to temperature at constant time is related to the slope of the compliance as a function of temperature at the constant time, as well as to the slope of the compliance as a function of time at the reference temperature. For a thermorheologically simple material, the partial derivative on the left of the equation becomes an ordinary derivative and the equation reduces to the one used by Ferry in his discussion of the temperature-axis shift for simple materials.[2]

A similar equation can be derived for the change of the (time-dependent) shift distance with logarithmic time:

$$\left(\frac{\partial \log D(t)}{\partial \log t}\right)_T = \left(\frac{\partial \log D[t/a_T(t)]}{\partial \log t/a_T(t)}\right)_{T_r} \left[1 - \left(\frac{\partial \log a_T(t)}{\partial \log t}\right)_T\right] . \quad (3)$$

Unfortunately neither of these equations can be integrated.

Earlier experiments had led Tschoegl to believe[3] that in his material, in which roughly spherical glassy domains are embedded in a rubbery matrix, the compliances of the two phases are additive, i.e.,

$$D = w_B\ D_B + w_S\ D_S \ , \quad (4)$$

where D stands for compliance, w is the weight fraction, and B and S refer to the two phases, respectively. In a material in which the rubbery phase is embedded in a glassy phase, presumably an additive modulus model would be appropriate. Intermediate cases could be covered by the more general Takayanagi model.

Application of the additive compliance model to Eq. (2) leads eventually to the relations

$$\left[\frac{\partial \log a_T(t)}{\partial T}\right] = N_B(t)\ \frac{d \log a_{TB}}{dT} + N_S(t)\ \frac{d \log a_{TS}}{dT} \quad (5)$$

$$N_B(t) = \frac{W_B \left[\dfrac{\partial D_B(t)}{\partial \ln t}\right]_T}{W_B \left[\dfrac{\partial D_B[t/a_T\ (t)]}{\partial \ln t/a_T\ (t)}\right]_{T_r} + W_S \left[\dfrac{\partial D_S[t/a_T\ (t)]}{\partial \ln t/a_T\ (t)}\right]_{T_r}} \quad (6)$$

and

$$N_S(t) = \frac{W_S \left[\dfrac{\partial D_S(t)}{\partial \ln t}\right]_{TT}}{W_B \left[\dfrac{\partial D_B[t/a_T\ (t)]}{\partial \ln t/a_T\ (t)}\right]_{T_r} + W_S \left[\dfrac{\partial D_S[t/a_T\ (t)]}{\partial \ln t/a_T\ (t)}\right]_{T_r}} \ . \quad (7)$$

The contributions of the individual phases are weighted by time-dependent weighting factors. These relations are intuitively gratifying because one would expect that the contribution of each phase to the total shift would be proportional to the amount of that phase, its temperature function, and some rate of change in its mechanical response with time. Figure 5 shows a schematic representation of the weighting factors. At low temperatures the behavior is dominated by the polybutadiene phase, and at high temperatures by the polystyrene phase.

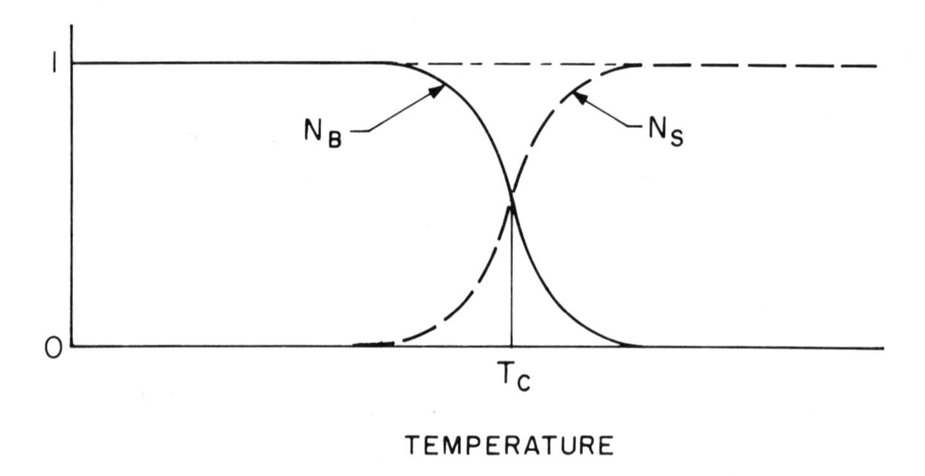

Fig. 5. Weighting factors, $N_B(t)$ and $N_S(t)$, as function of temperature.

These equations still cannot be integrated. One may, however, construct the appropriate shift factors if the compliances and the temperature dependence of the two phases are known. In that case one can construct a reference master curve of the compliance at the reference temperature according to the additive compliance model. Similar short response curves can then be generated for the temperatures and times of interest. Then, by comparing the time, t, at which a given compliance appears at temperature, T, to the time, $t/a_T(t)$, at which the same compliance appears on the reference master curve and taking the difference, one can obtain the shift, $\log a_T(t)$, for that time and temperature.

Figure 6 shows the temperature dependences which were assumed for the polybutadiene and the polystyrene phases. The shift factors generated at the reference temperature of 85 C with these temperature dependences are shown in Fig. 7 for the frequencies $\log \omega = -6$ and $\log \omega = 5$. The general shape of these shift-factor curves is very similar to those shown earlier today by Shen. However, in contrast to the curve that Shen showed, it is clear that the shift factors are functions of the frequency. Empirical shifting leads to the shift-factor curves shown by the dotted lines in Fig. 7. Empirical shifting of data at log $\omega = -6$ in the temperature region over which the curve is flat would produce "polybutadiene behavior" tacked onto the polystyrene behavior, as indicated by the dotted line marked B. The empirical shift function, therefore, would never converge to the behavior required when the polybutadiene phase dominates. Similarly, at log $\omega = 5$, empirical shifting would produce the behavior indicated by the dotted line marked S. In master curves constructed with such shift factors, therefore, the two transitions would appear displaced from their true locations.

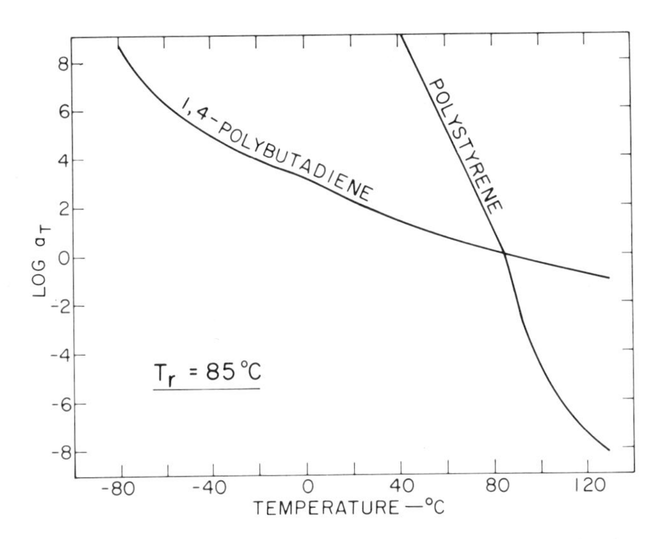

Fig. 6. Temperature functions, log a_T, for 1,4-polybutadiene and polystyrene.

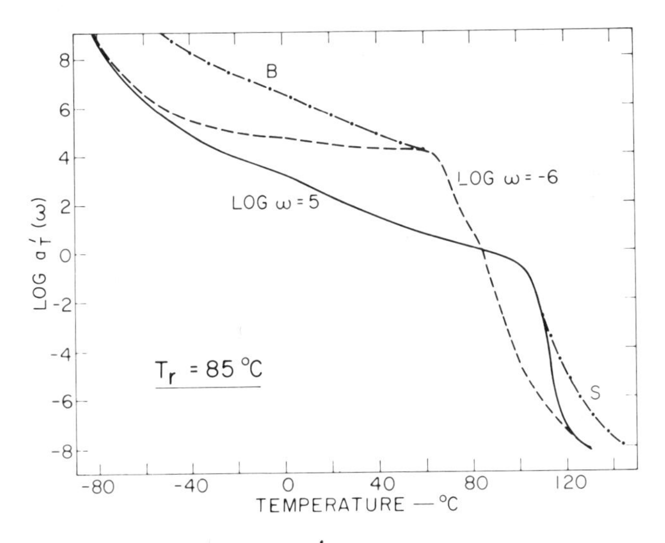

Fig. 7. Temperature function, log $a_T'(\omega)$, at log $\omega = -6$ and 5 for composite.

In conclusion, Tschoegl pointed out that Fesko and he had shown[1] that it is possible to construct master curves for two-phase materials by the point-by-point shifting of isothermal temperature segments, using time-dependent shift factors which can be obtained on the basis of a suitably chosen model if the compliances or moduli and temperature dependences of the two phases are known. This treatment assumes, of course, that the two phases do not interact. This assumption must be handled with some caution.

Shen supports Tschoegl's analysis for thermorheologically complex materials. It is clear that, in general, heterophase polymers can be expected to be thermorheologically complex. The simple time-temperature superposition procedure can be applied for these materials only under the following limited conditions:

1. If relaxation times of all phases have the same temperature dependence
2. If one limits the superposition to the region where the relaxation is dominated by one mechanism.

In the region where both relaxations contribute significantly, the simple superposition cannot in general be used, at least not without modification. Thus, if one could obtain master curves at different temperatures by some direct means, they would be similar in shape but not in time scale. An indirect demonstration of this phenomenon is given in Fig. 6 of Shen's paper (loss tangent as a function of temperature at different frequencies).

In addition to the elegant but rather involved calculations Tschoegl has shown, Kovacs called attention to some further complications which were not taken into account. In fact, in a glassy polymer the shift factor depends not only on temperature but also upon the thermal history of the sample [A. J. Kovacs, R. E. Stratton, and J. D. Ferry, *J. Chem. Phys.* (1973)]. Moreover, the relaxation and retardation spectra show some modification below Tg – which excludes "thermorheological simplicity". Both of these phenomena perturb considerably the straightforward applicability of the stress-temperature superposition in pure homopolymers, and even more so – because of the lack of homogeneity – its applicability in heterophase block copolymer systems.

REFERENCES

1. Fesko, D. G., and Tschoegl, N. W., *J. Polymer Sci.*, Part C, No. 35, p 51 (1971).
2. Ferry, J. D., *Viscoelastic Properties of Polymers*, 2nd ed., Wiley, New York (1970), p 343.
3. Lim, C. K., Cohen, R. E., and Tschoegl, N. W., *Advances in Chemistry Series*, ACS Monograph No. 99, p 397 (1971).

Part Four

CONTINUUM DESCRIPTIONS OF DEFORMATION

PLASTIC DEFORMATION OF CRYSTALLINE POLYMERS IN SOLID-STATE EXTRUSION THROUGH A TAPERED DIE

M. Takayanagi

Faculty of Engineering
Kyushu University
Fukuoka, Japan

ABSTRACT

A theoretical analysis based on the free-body and upper-bound approaches for the solid-state extrusion of crystalline polymers is developed by taking into account their remarkable strain-hardening behavior. The strain-hardening process in uniaxial extension is formulated in a generalized form. The calculations are found to compare favorably with observed extrusion pressures as a function of area reduction of die for linear polyethylene, polypropylene, and nylon 6. The critical area reduction for smooth extrusion is predictable in terms of the unfoldability parameter of folded molecules in lamellar crystals, the strain-hardening parameter, the initial yield stress, and the tenacity of fibers.

1 INTRODUCTION

The extrudate obtained with a high reduction in area from extrusion of high-density polyethylene (PE) in a solid state through a tapered die under high pressure exhibits outstanding transparency and high dimensional stability nearly up to the melting temperature.[1] This paper presents the results of an analysis of such an extrusion process for crystalline polymeric solids. The analysis is based on continuum theory, using methods employed in the field of metals. The results are compared with our experimental results.

The continuum approaches used previously neglect the difference between the plastic deformation behavior of metals and that of crystalline polymers. For example, Sachs[2] presented an equation for evaluating the extrusion pressure with the assumption that metal is perfectly plastic – that is, the strain-hardening effect is negligible. However, crystalline polymers are usually composed of spherulites that are changed during deformation into a fiber structure with very high modulus. Such a structural change is accompanied by a remarkable strain-hardening effect. This phenomenon must be taken into account if continuum methods are to be applied to crystalline polymers.

2 THEORY

2.1 Equation for Strain Hardening of Polymeric Solids During Plastic Deformation

As mentioned in the introduction, the superstructure of solid crystalline polymers changes from a spherulitic structure to a fiber structure during extrusion through a tapered die.[1] Almost the same structural change can be seen during uniaxial deformation of crystalline polymers in a simple extension test. The main difference between the solid-state extrusion process and uniaxial deformation lies in the stress structure – hydrostatic pressure is the controlling factor in the former, while atmospheric pressure controls in the latter. If the effect of hydrostatic pressure on the deformation process is assumed to be negligible, the data from uniaxial extension tests can be used for evaluating the extrusion pressure.

In order to describe the process of uniaxial extension, the use of true stress (σ) and true strain (ϵ) is preferable to the use of engineering stress or nominal stress (s) and nominal strain (e), because polymeric materials change their structure drastically during deformation, and, hence, the quantities associated with every stage of structural change better represent the changing structure.

Engineering stress (s) is given by

$$s = F/A_0 \tag{1}$$

and engineering strain (e) by

$$e = \Delta l/l_0 \quad , \tag{2}$$

where l_0 and A_0 are the original length and cross-sectional area of the undeformed specimen, respectively. If the specimen changes its dimensions to l and A at force F, true stress (σ) is

$$\sigma = F/A \quad , \tag{3}$$

while true strain (ϵ) is

$$\epsilon = \int_{l_0}^{l} \frac{dl}{l} = \ln \frac{l}{l_0} \quad . \tag{4}$$

Because of constant volume ($A_0 l_0 = Al$), Eqs. 3 and 4 can be rewritten thus:

$$\sigma = Fl/A_0 l_0 = sl/l_0 \tag{3'}$$

and

$$\epsilon = \ln(A_0/A) = \ln(1 + e) \quad . \tag{4'}$$

It should be noted that, if necking takes place, engineering strain (e) is the average of the large strain in the necking region and the small strain in the undeformed region. On the other hand, true strain (ϵ) represents the strain at the necking region that is associated with the drastic structural change.

By using σ and ϵ, we derive the following general equation, which is applicable to the uniaxial drawing process of crystalline polymers such as PE[3], nylon 6[3], isotactic polypropylene (PP)[4], polyoxymethylene (POM)[5], and polytetrafluoroethylene (PTFE)[5]:

$$\log(\sigma/\sigma^*) \cdot \log(\epsilon/\epsilon^*) = -c \quad , \tag{5}$$

where c is the constant characteristic of chemical species of polymers and σ^* and ϵ^* are the normalizing factors of σ and ϵ, which are determined by shifting the σ–ϵ curve along the logarithmic axes of σ and ϵ to fit the curve represented by Eq. 5.

Figure 1 shows the composite curves for PE[3], nylon 6[3], and PP[4]. Figure 2 shows those for POM[5] and PTFE[5]. Table I lists the drawing conditions and the values of c, σ^*, and ϵ^* for these polymers.

From these data it was concluded that c is a constant characteristic only of chemical species of polymers, being independent of molecular weight, drawing temperature, and elongation rate. ϵ^* and σ^* are depen-

Table I. Drawing Conditions and Some Parameters Characterizing the True Stress – True Strain Curves for PE, Nylon 6, POM, PTFE, and PP

	Drawing	Elongation Rate[a],		kg/mm^2	
Sample	Temperature, C	mm/min	c,	σ^*,	ϵ^*
PE-1[b]	30	5	0.384	1.28	5.75
	50	5	0.384	0.84	5.33
	70	5	0.384	0.63	5.38
	90	5	0.384	0.46	5.55
	110	5	0.384	0.27	5.58
	90	20	0.384	0.50	5.75
	50	50	0.384	1.26	6.31
	70	50	0.384	0.82	5.91
	90	50	0.384	0.56	5.93
PE-2[c]	90	5	0.384	0.41	4.82
PE-3[d]	50	5	0.384	0.80	5.55
Nylon 6[e]	90	5	0.175	1.37	1.70
	120	5	0.175	1.26	1.78
	160	5	0.175	0.89	1.76
POM[f]	130	5	0.30	1.23	3.98
PTFE[g]	50	5	0.384	0.43	3.40
	75	5	0.384	0.36	3.33
	100	5	0.384	0.32	3.27
	120	5	0.384	0.28	3.23
	140	5	0.384	0.24	3.32
	160	5	0.384	0.20	3.26
PP-1[h]	40	5	0.23	1.82	3.45
	70	5	0.23	1.20	3.60
	100	5	0.23	0.65	3.45
	130	5	0.23	0.34	3.65
PP-2[i]	40	5	0.23	1.55	3.10
	70	5	0.23	1.00	3.50
	100	5	0.23	0.55	3.55
	130	5	0.23	0.30	3.55

(a) Original length of the specimen is 20 mm.
(b) Hizex 1200J (Mitsui Petrochemical Co.).
(c) Sholex 4002B (Nihon Olefin Co.).
(d) Novatec JV040 (Mitsubishi Chemical Co.).
(e) Amilan CM1030 (Toray Co.).
(f) Delrin (Du Pont Co.).
(g) Commercial product.
(h) Noblen W101 (Sumitomo Chemical Co.).
(i) Noblen D501 (Sumitomo Chemical Co.).

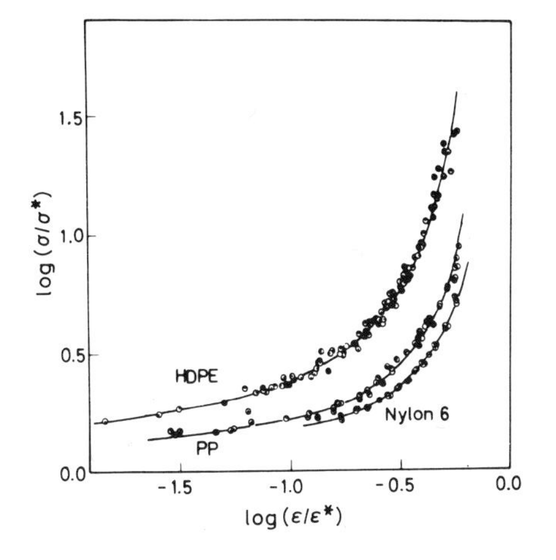

Fig. 1. Composite curves of log (σ/σ) versus log (ε/ε*) for linear polyethylene (PE), isotact-polypropylene (PP), and nylon 6.*

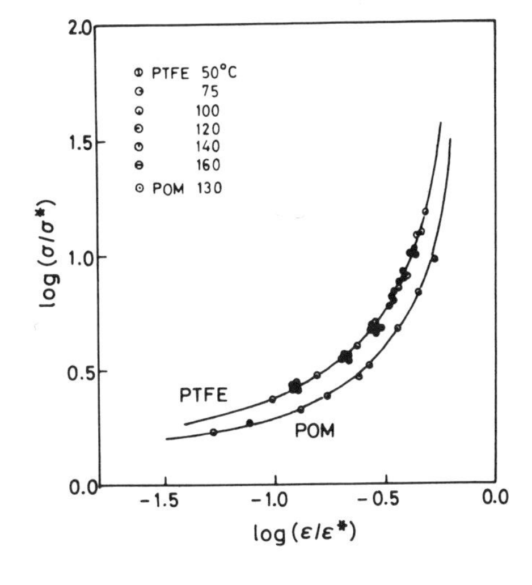

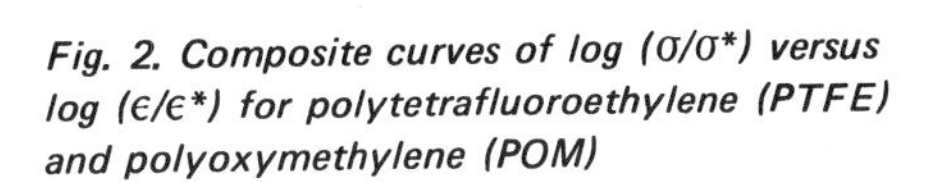

Fig. 2. Composite curves of log (σ/σ) versus log (ε/ε*) for polytetrafluoroethylene (PTFE) and polyoxymethylene (POM)*

dent on drawing conditions and characterizing parameters of crystallized state. Especially, σ^* is very sensitive to drawing temperature and steadily decreases to zero at the melting temperature of the polymer, whereas ϵ^* does not change so remarkably with drawing temperature. Figure 3 is a schematic representation of Eq. 5. Figure 3(a) represents the case of $c = 0$, which corresponds to idealized deformation of crystalline polymers. The plastic deformation starts at $\sigma > \sigma^*$, with σ increasing to infinity discontinuously at $\epsilon = \epsilon^*$. Figure 3(b) represents the actual case, according to which the strain-hardening phenomenon becomes more remarkable with increasing values of c. That is, the true stress increases more rapidly with increasing true strain. In this context, the constant c may be appropriately called the "strain-hardening parameter" of the individual polymer. Figure 4(a) and 4(b) show some examples of σ–ϵ data (circles) and the calculated curves according to Eq. 5 (solid lines) for PE and PP, respectively. Equation 5 effectively describes the process of solid-state extrusion of crystalline polymers, as mentioned later.

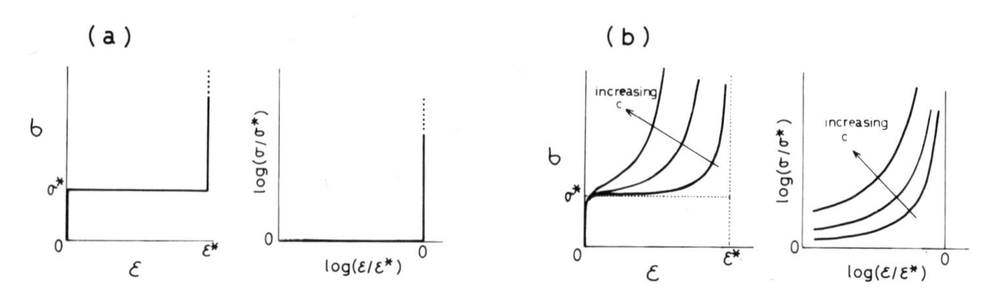

Fig. 3. Schematic representation of true stress — true strain relationship according to the general equation for the (a) idealized case of c = 0 and (b) actual systems of crystalline polymers.

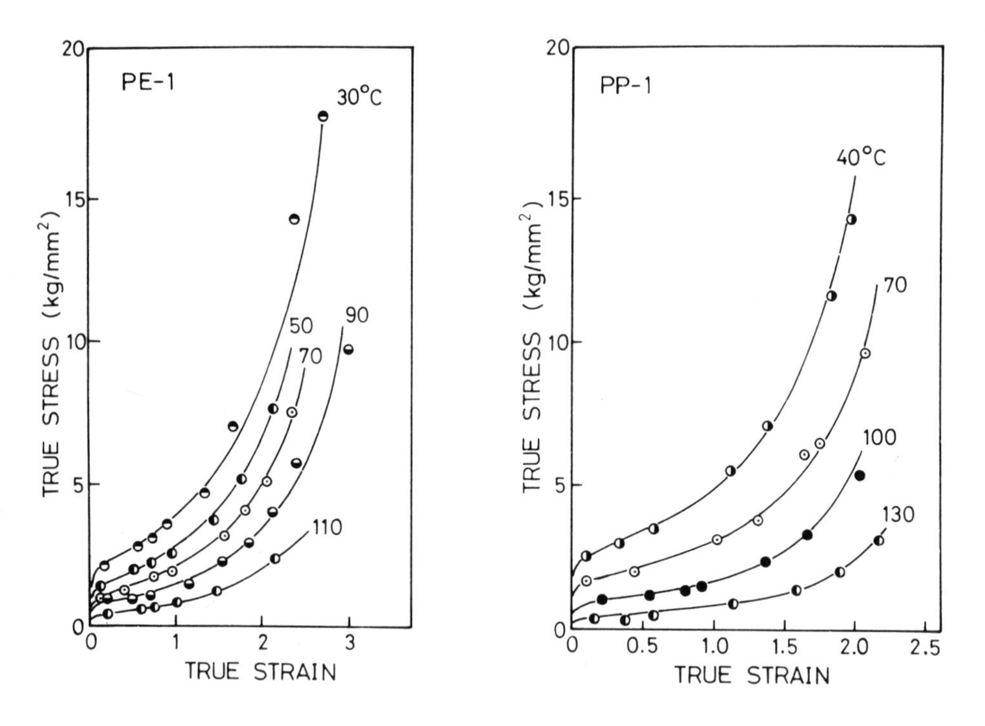

Fig. 4. The true stress (σ) versus true strain (ϵ) at various drawing temperatures for linear polyethynene (on the left) and polypropylene (on the right). Circles denote experimental data and solid curves are the values calculated with the general equation.

It will be instructive to explain in molecular terms the physical meanings of the various parameters in Eq. 5. These enable us to better visualize the plastic deformation process. Figure 5 illustrates the molecular

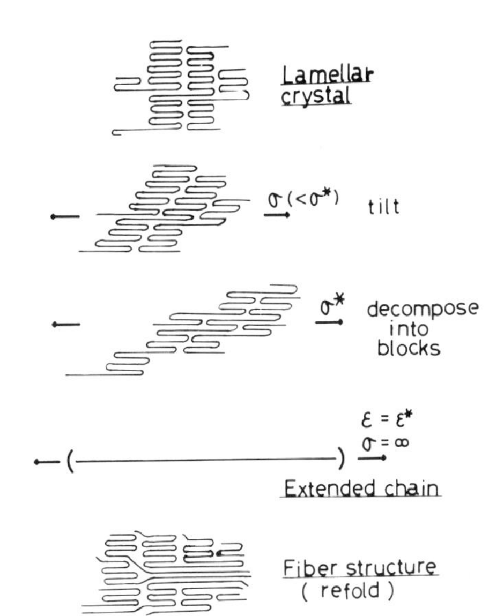

Fig. 5. Schematic representation of molecular process of plastic deformation of lamellar crystals.

mechanisms occurring during the deformation of lamellar crystals in the bulk crystallized sample. Under shear, the lamellar crystals become thin by tilting of their molecular axes. They then break into blocks by slip of mosaic-block crystals several hundred angstroms wide; this slip occurs at their interfacial region at $\sigma = \sigma^*$. In the idealized case of $c = 0$, the unfolding of molecular chains proceeds at constant stress σ^* and, when the molecular chains are fully extended at $\epsilon = \epsilon^*$, the stress increases to the infinite value. Such a behavior agrees well with that of a purely plastic body although the latter cannot achieve an infinite stress at the final stage. Rather, its yield stress remains constant. In the actual case, the molecules cannot be fully extended because the forces sustaining both ends of the molecules are intermolecular and very weak compared with those of the covalent bonds in the main chain. Thus, some of the unfolding molecules becoming free from restraint are refolded midway in the process to $\epsilon = \epsilon^*$, and then reform as folded crystals, which bury the still extending molecules. Such molecules will form tie molecules or crystal bridges between neighboring lamellar crystals. The superstructure thus formed is essentially the same as predicted for the fiber structure. This is shown at the bottom of Fig. 5. According to the molecular models discussed, the strain-hardening parameter (c), reflects the quantity of crystal bridges or tie links formed during plastic deformation. The reason is that molecules thus oriented display a very high modulus and bond breakage strength and resist further unfolding of molecules from the lamellar crystals. The fact that, according to our results, PE and PTFE have a comparatively large c value of 0.384 and POM has a c value of 0.30 indicates that the crystal-bridge formation is easier in these polymers. The values of $c = 0.23$ for PP and 0.175 for nylon 6 are rather small. Polymers of the first group are known to be highly crystalline and form morphologically well-defined lamellar crystals, whereas those of the second group show a rather low degree of crystallinity (about 60 percent) and it is comparatively difficult to obtain extensively developed lamellar crystals. Such crystallization tendencies of the various polymers might be related to the strain hardening as represented by the c value. Concurrently, it is noted that the average values of ϵ^* over the temperature range covered here for various polymers have tendencies similar to those found for c. The values of $\bar{\epsilon}^*$ for PE (5.5) and POM (4.0) are larger than those for PP and nylon 6 (3.4 and 1.8, respectively). According to the molecular models shown in Fig. 5, $\bar{\epsilon}^*$ is a measure of the unfoldability of molecular chains from lamellar crystals. It may be reasonable to assume that the polymers exhibiting well-developed lamellar crystals will show a relatively high value of ϵ^*. Hereinafter, in this sense, we will call ϵ^* the "unfoldability parameter". The relationship of σ^* to the polymer structure is, at present, difficult to determine owing to the large temperature sensitivity of σ^*, although it is certain that the values of σ^* decrease with increasing temperature and become zero at the melting temperature of the individual polymer.

2.2 Evaluation of the Extrusion Pressure of Solid Polymers With the Free-Body Approach[11]

Figure 6 shows the extrusion of a solid-state polymer through a tapered die with half-die angle α. Referring to Sach's analysis for wire drawing and extrusion of a completely plastic body[2], a slice of infinitesimal thickness (dx) is considered to be in equilibrium under the external force. Cylindrical symmetry is assumed, with the cylindrical coordinates (r, θ, x) as the principal axes of stress. The stress is uniformly distributed over the surface normal to the x axis and, obeying Coulomb's law of friction, friction acts over the die wall. Equilibrium of the free slice is expressed by

$$(\sigma_x + d\sigma_x)\,\pi\,(r + dr)^2 = \sigma_x \pi r^2 + \mu\sigma_r \cos\alpha \frac{dr}{\sin\alpha} 2\pi r$$

$$+ \sigma_r \sin\alpha \frac{dr}{\sin\alpha} 2\pi r \;, \qquad (6)$$

where σ_x is assumed to be the principal stress acting along the x axis, σ_r is the stress normal to the die wall, and μ is the Coulomb coefficient of friction. Equation 4′, with $A = \pi r^2$ and $A_0 = \pi R_0{}^2$, becomes

$$\epsilon = 2\ln\left(\frac{R_0}{r}\right)$$

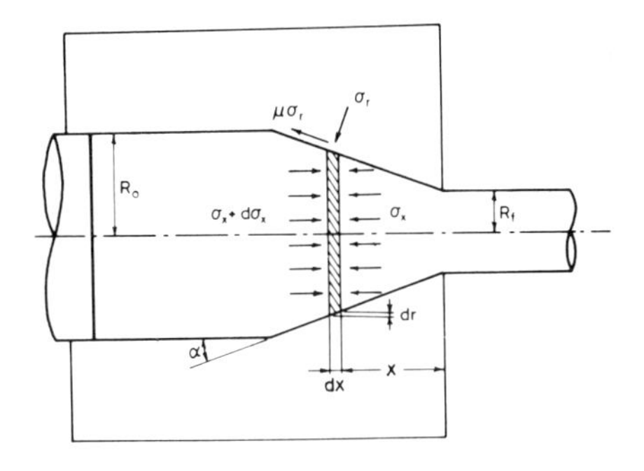

Fig. 6. Free-body equilibrium in solid-state extrusion through a tapered die.

or

$$d\epsilon = -2\,\frac{dr}{r} \quad , \tag{7}$$

where r is the radius of the free slice at x. Substituting Eq. 7 into Eq. 6 and setting $B = \mu \cdot \cot \alpha$, we obtain the differential equation

$$\frac{d\sigma_x}{d\epsilon} = \sigma_x - \sigma_r\,(1 + B) \quad . \tag{8}$$

A relation between σ_x and σ_r is associated with the yield criterion. The yield criteria of both von Mises and Tresca reduce to the same form as Eq. 9:

$$Y(\epsilon) = \sigma_x - \sigma_r$$

for

$$\sigma_r < \sigma_x < 0 \quad . \tag{9}$$

In solid-state extrusion, $Y(\epsilon)$ is assumed to be the yield stress of the strain-hardening materials deformed by ϵ. The functional form of $Y(\epsilon)$ is assumed here to be represented by Eq. 5, which is empirically found at atmospheric pressure. Sachs's treatment is based on the constant value of yield stress. Here, we have introduced a nonlinear function of the yield stress, which is largely different from the strain-hardening function in metals as given by the power law.[6] The latter shows a steadily increasing curve, with decreasing slope with increasing elongation, whereas the strain-hardening curve of crystalline polymers shows increasing slope. Of course, it is preferable to use the strain-hardening curve under the corresponding hydrostatic pressure. At present, however, we assume that $Y(\epsilon)$, as given approximately by Eq. 5 obtained at atmospheric pressure, is sufficient.

Substituting Eq. 9 into Eq. 8 results in

$$\frac{d\sigma_x(\epsilon)}{d\epsilon} = (1 + B)\, Y(\epsilon) - B\sigma_x(\epsilon) \quad , \tag{10}$$

where $\sigma_x(\epsilon)$ is the normal stress to the slice surface when true strain is ϵ. Solving the differential (Eq. 10),

$$\sigma_{x0} = -(1 + B)\int_0^{\epsilon_f} Y(\epsilon)\exp(B\epsilon)\, d\epsilon + \sigma_{xf}\exp(B\epsilon_f) \quad , \tag{11}$$

where σ_{x0} and σ_{xf} are the principal stresses along the x axis at the entrance and the exit of the tapered part of the die, respectively. σ_{x0} in Eq. 11 can be replaced by extrusion pressure P_0 after changing its sign, and ϵ_f is represented by $2 \ln (R_0/R_f)$. Thus, Eq. 11 is rewritten as:

$$P_0 = (1 + B)\int_0^{2\ln\left(\frac{R_0}{R_f}\right)} Y(\epsilon)\exp(B\epsilon)\, d\epsilon + P_f\left(\frac{R_0}{R_f}\right)^{2B} \quad . \tag{12}$$

As there is no drawing stress in our case, it is reasonable to assume that the pressure at the exit of the die (P_f) is zero.

Körber and Eichinger[7] proposed that the correction term of $(2/3\sqrt{3})\, Y(\epsilon_0)\cdot \tan\alpha + (2/3\sqrt{3})\, Y(\epsilon_f)\cdot \tan\alpha$ should be added to the extrusion pressure evaluated by Sachs's method when ϵ_0 and ϵ_f are the true strains at the die entrance and the die exit, respectively. This term was introduced by taking into account the shear stress necessary to change the direction of movement of the material in the die. Detailed inspection of Eq. 12 reveals that the integrand in the first term of the right side of Eq. 12, especially the term $\exp(B\epsilon)$, is remarkably increased by increasing ϵ for solid crystalline polymers; then, the contribution of the Körber-Eichinger term can no longer be expected to have any meaning. For numerical calculation, $Y(\epsilon_0)$ can be substituted by σ^* in Eq. 5. Thus, μ (or $B = \mu \cot \alpha$) is the only term to be determined by the trial-and-error method to achieve agreement between the experimental data and the calculated values.

2.3 Evaluation of Extrusion Pressure by the Upper-Bound Approach[12]

The analytical methods for the treatment of the metal-forming process are known to include the free-body approach and limit analysis. The free-body approach was introduced by von Karman[8] as early as 1925 and has been pursued extensively since then. The limit analysis of metal forming is relatively new, and the upper-bound solutions were developed only recently.[9] In such situations, we have tried to apply the latter method to the crystalline solid polymers; then it again becomes important to take into account the remarkable strain-hardening phenomenon of crystalline polymers, which is the main difference between solid polymers and metals. The analytical method employed in metals must be modified to fit the situation found in crystalline polymers.

Prager and Hodge[10] generalized the upper-bound theorem. Among the possible conceivable types of velocity field, the actual one minimizes the externally supplied power (J*) given by Eq. 13*:

$$J^* = \dot{W}_i + \dot{W}_s \quad , \tag{13}$$

where $\dot{W}_i$ is the power for internal deformation of the material and $\dot{W}_s$ is the power for energy consumption at the interfaces of discontinuity in velocity fields. The actual externally supplied power (J*) is never higher than that computed by using Eq. 13.

Figure 7 shows three zones – I, II, and III – in which the dead zone has been neglected because the tapered die with low angle is used (the effect of the dead zone caused by the high angle has also been calculated). In Zones I and III, the velocity is uniform and has only an axial component. In Zone II, the velocity vector is directed toward the apex (0) of a cone of an included angle 2θ, e.g., along the line LM in Fig. 7. Velocity components normal to the surfaces are continuous across the boundaries Γ_1, Γ_2, and Γ_3, while velocity discontinuities exist parallel with these surfaces.

First, the power for internal deformation of the material in Zone II of Fig. 7 is obtained, after several stages of calculation, by

$$\dot{W}_i = \pi v_f R_f^{\,2} f(\alpha) \int_0^{\epsilon_f} Y(\epsilon)\, d\epsilon \quad , \tag{14}$$

where v_f is the velocity at interface Γ_1 and $f(\alpha)$ is the function of α only:

$$f(\alpha) = \frac{1}{\sin^2\alpha}\left[1 - \cos\alpha\sqrt{1-\frac{11}{12}\sin^2\alpha} + \frac{1}{\sqrt{11\cdot 12}}\ln\frac{1+\sqrt{\frac{11}{12}}}{\sqrt{\frac{11}{12}}\cos\alpha + \sqrt{1-\frac{11}{12}\sin^2\alpha}}\right] \tag{15}$$

The introduction of a nonlinear function of $Y(\epsilon)$, as employed in the previous section, into the integrand of Eq. 14 is characteristic of our derivation.[12]

The more exact expression of J is given in References 9 and 10.

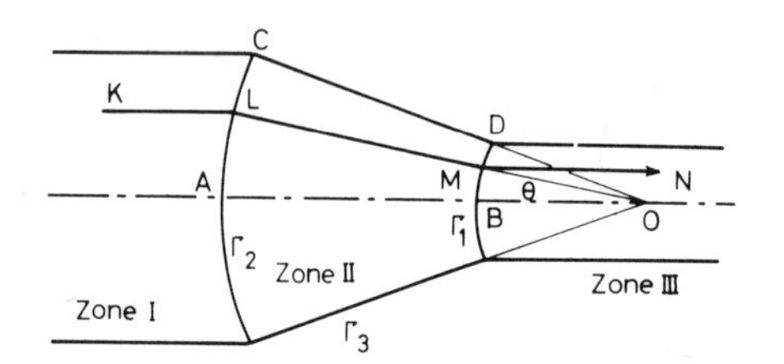

Fig. 7. Spherical velocity field in the upper-bound approach. O: the apex of die cone; ABO: symmetry axis; CD: die wall; KLMN: assumed stream line; Zones I and III: uniform velocity zones; Zone II: zone with heterogeneous velocity field; Γ_1, Γ_2 and Γ_3: the boundaries of velocity discontinuities perpendicular to these planes.

Next, we calculate the powers at interfaces Γ_1, Γ_2, and Γ_3. The power at surface Γ_1 is given by taking into account von Mises's criterion:

$$\dot{W}_{s1} = \frac{1}{\sqrt{3}} Y(\epsilon_f)\pi v_f R_f^2 \left(\frac{\alpha}{\sin^2 \alpha} - \cot \alpha \right), \tag{16}$$

where the suffix f denotes the values at Γ_1. Similarly, at Γ_2,

$$\dot{W}_{s2} = \frac{1}{\sqrt{3}} Y(\epsilon_0)\pi v_f R_f^2 \left(\frac{\alpha}{\sin^2 \alpha} - \cot \alpha \right). \tag{17}$$

At Γ_3, the Coulomb friction acts on the material. The normal stress on interface Γ_3 is not known, but the normal stress for zero frictional force is given by Sachs. If we assume that his representation is valid, then $\dot{W}_{s3}$ at interface Γ_3 is given approximately by Eq. 18:

$$\dot{W}_{s3} = \mu \pi v_f R_f^2 \cot \alpha \int_0^{\epsilon_f} Y(\epsilon)\,(1 + \epsilon_f - \epsilon) d\epsilon \,. \tag{18}$$

The sum of the powers given by Eq. 14, 16, 17, and 18 is equal to J*, which in turn is given by Eq. 19 because of the constant-volume assumption:

$$J^* = \pi v_f Rf^2 P_0 = \pi v_0 R_0^2 P_0 \,. \tag{19}$$

Thus,

$$P_0 = f(\alpha) \int_0^{\epsilon_f} Y(\epsilon) d\epsilon + \frac{1}{\sqrt{3}} [Y(\epsilon_0) + Y(\epsilon_f)] \left(\frac{\alpha}{\sin^2 \alpha} - \cot \alpha \right) + \mu \cdot \cot \alpha \int_0^{\epsilon_f} Y(\epsilon)\,(1 + \epsilon_f - \epsilon) d\epsilon \,, \tag{20}$$

where $Y(\epsilon_0)$ is assumed to be equal to σ^* in Eq. 5, which represents the initial yield stress of crystalline polymers.

3 EXPERIMENTAL PROCEDURES

The samples used are the commercial products listed in Table I. Figure 8(a) shows the piston-cylinder type of apparatus used to measure the extruding process. Figure 8(b) shows the vertical section of the tapered die. Billets 10 mm in diameter are cut out of the melt-crystallized block materials with a lathe and inserted into the cylinder. The length of the extrudate is recorded as a function of time, to evaluate the extrusion rate. By extrapolating the straight line of the relation between extrusion rate and pressure to zero rate, the "extrusion pressure" is determined.

The conical converging die with half-die angle α is used. The standard half-die angle is 20 degrees, but angles different from that are also used to evaluate the effect of α. The diameter at the die entrance ($2R_0$) is 10 mm, while that at the die exit varies from 2.0 to 7.0 mm.

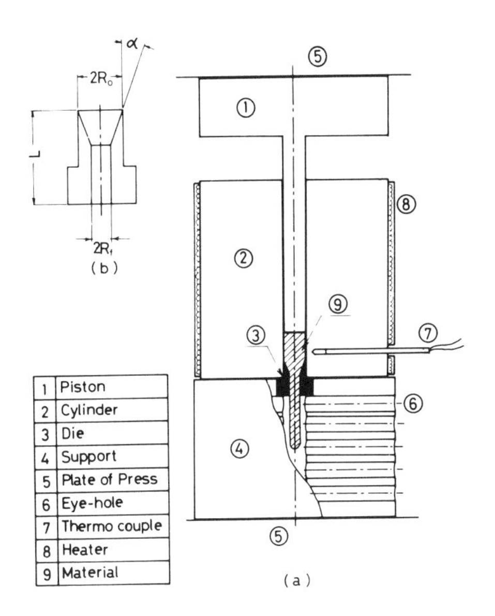

Fig. 8. Extrusion device: (a) piston, cylinder, and die assembly; (b) vertical section of die.

4 RESULTS AND DISCUSSION

4.1 Effect of Area Reduction on Extrusion Pressure[4,11,12]

Figures 9 and 10 are examples of measurements of the extrusion process, under the conditions indicated in the figures, for PE and nylon 6, respectively. Following the transient rapid extrusion, the stationary extrusion takes place. The extrusion rate is evaluated from its slope. Figures 11, 12, and 13 show the relationships between extrusion rate and pressure. As

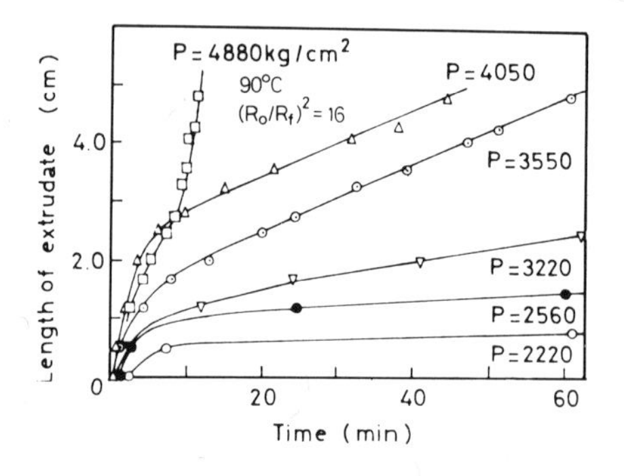

Fig. 9. Example of extrusion of high-density polyethylene, Hizex 1200 J. Area reduction 16.0.

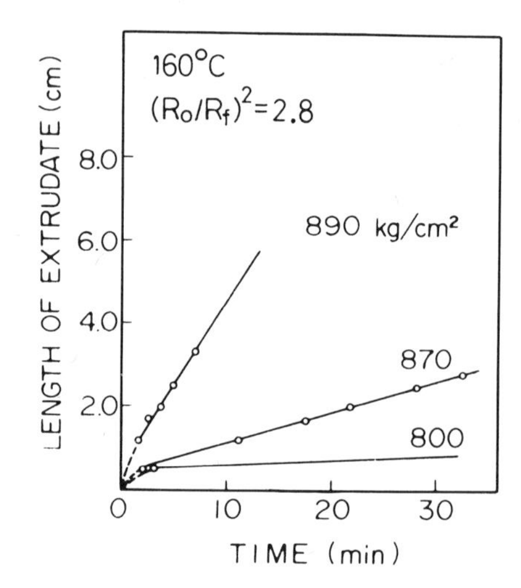

Fig. 10. Example of extrusion of nylon 6. Area reduction 2.8.

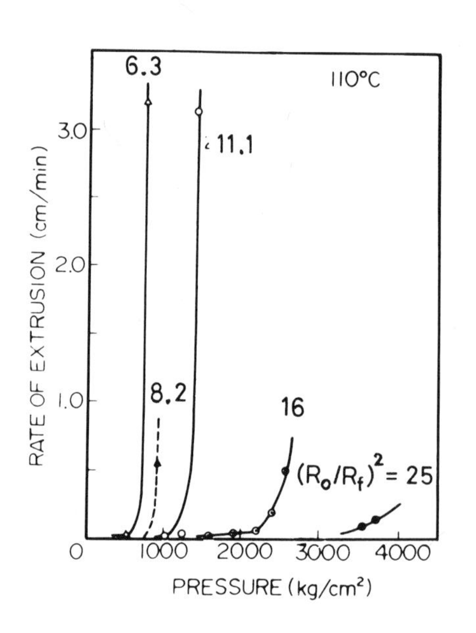

Fig. 11. Rate of extrusion versus extrusion pressure for linear polyethylene.

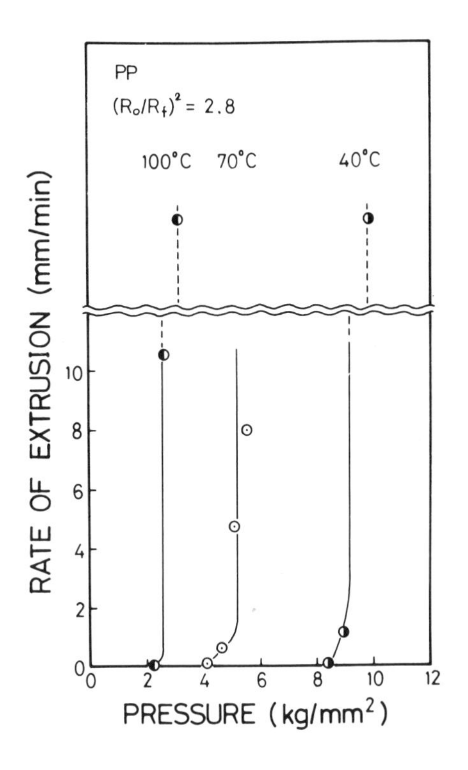

Fig. 12. Rate of extrusion versus extrusion pressure for polypropylene.

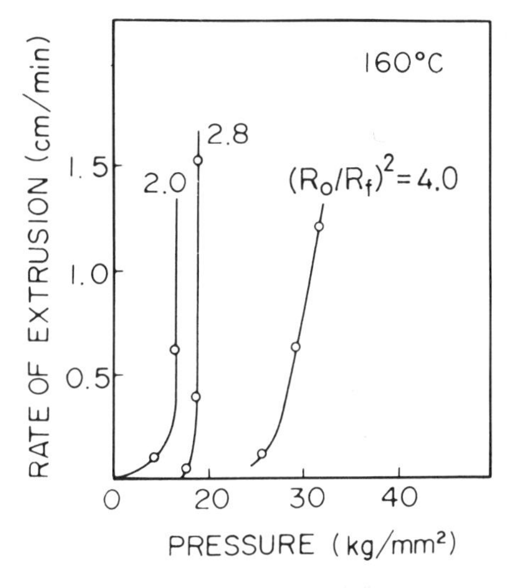

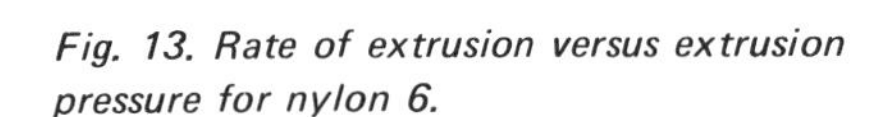

Fig. 13. Rate of extrusion versus extrusion pressure for nylon 6.

seen in these examples, while remarkably fast extrusion takes place above the critical pressure, below the critical pressure, the extrusion rate can scarcely be observed. By extrapolating the straight-line part to the zero rate, as mentioned above, we can determine the "extrusion pressure". Rapid increase of extrusion rate at the extrusion pressure supports the existence of a yield phenomenon for crystalline polymers, as discussed in the theoretical section of this paper.

Figures 14, 15, and 16 show the relationships between the extrusion pressure and the degree of processing, 2 ln (R_0/R_f), which corresponds with the true strain of the material at the die exit according to Eq. 7. The solid lines in the figure are calculated according to Eq. 12, the derivation of which is based on the free-body equilibrium FBE, taking into account the strain-hardening phenomenon as described by Eq. 5. The values of the frictional coefficient μ assumed in these calculations are summarized in Table II. The broken lines in Figs. 14, 15, and 16 are calculated according to Eq. 20, the derivation of which is based on the upper-bound theorem (UBT), using the same strain-hardening equation as in the case of FBE – Eq. 5. The μ values in this case are also included in Table II.

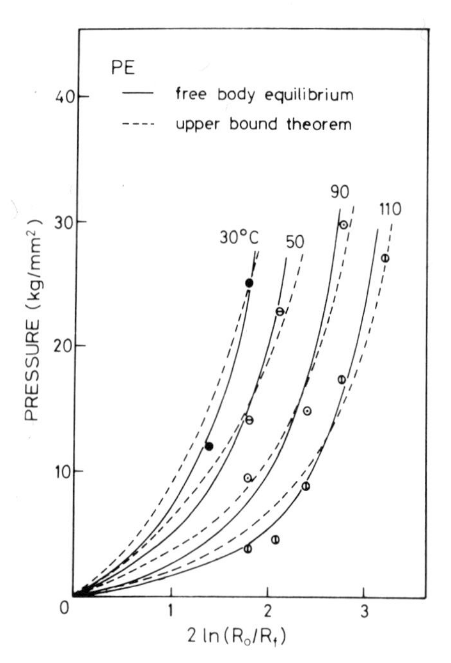

Fig. 14. Extrusion pressure versus 2 ln (R_0/R_f) or true strain at die exit at various temperatures for linear polyethylene. Solid lines represent the free-body equilibrium (FBE) approach. Broken lines represent the upper-bound-theorem (UBT) approach. Circles are experimental data.

Generally speaking, the FBE approach data (solid lines) represent the experimental data rather well, especially in the case of PE. In the case of PP, shown in Fig. 15, the deviation becomes somewhat greater for a higher degree of processing. In the case of nylon 6 (Fig. 16), the calculated curves represent the tendency of the relationship. Solid curves are obtained at fixed values of μ for each extrusion conducted at the various temperatures listed in Table II.

The UBT approach data, represented by broken lines, seem to deviate largely from the experimental data at the high degree of processing. The UBT approach is not much different from the FBE approach in nylon 6.

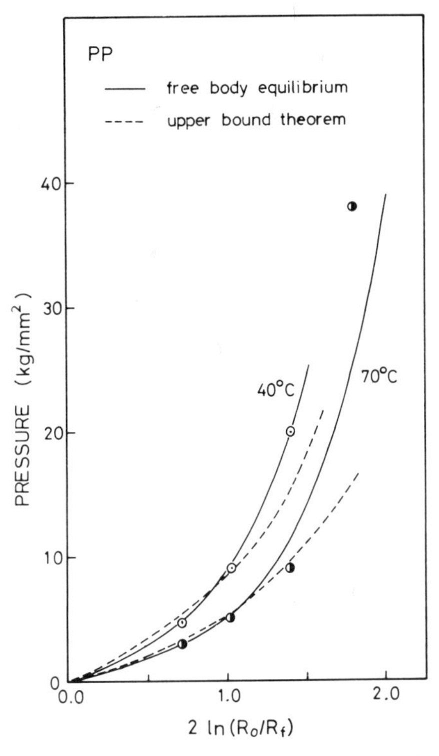

Fig. 15. Extrusion pressure versus 2 ln (R_o/R_f) for polypropylene. Solid lines represent the FBE approach and broken lines the UBT approach.

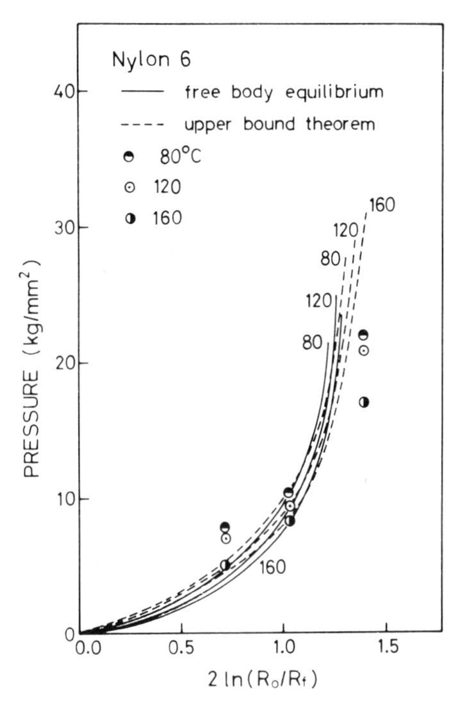

Fig. 16. Extrusion pressure versus 2 ln (R_o/R_f) for nylon 6. Solid lines represent the FBE approach and broken lines the UBT approach.

Table II. Extrusion Temperatures and Frictional Coefficients for Various Crystalline Polymers

Polymer Materials	Extrusion Temperature, C	Coefficient of Friction: From Free-Body Equilibrium	Coefficient of Friction: From Upper-Bound Theorem
PE-1	30	0.230	0.230
	90	0.182	0.230
	110	0.167	0.230
PE-2	90	0.189	0.230
PE-3	50	0.220	0.230
PP-2	40	0.30	0.15
	70	0.30	0.17
Nylon 6	80	0.28	0.15
	120	0.30	0.15
	160	0.36	0.40

The μ values in the UBT approach for PE are set equal with each other at different extrusion temperatures. The reason for such a deviation in the values obtained by the two approaches is explained by comparing Eq. 12 of FBE with Eq. 20 of UBT. The integrand of the first term in the right side of Eq. 12 includes the function $\exp(B\epsilon)$ which makes P_0 increase rapidly with increasing ϵ. The absence of this term in Eq. 20 of UBT precluded generation of the higher value of extrusion pressure for the high area reduction. Concerning this point, the FBE approach seems to be more preferable. However, the UBT approach could be improved by improving the approximation made in deriving Eq. 20.

The μ value of PE in UBT is constant from 30 to 110 C, whereas the μ values in FBE vary with extrusion temperatures, e.g., 0.230 at 30 C and 0.167 at 110 C. These values are very near those for PE given in the literature: 0.2 to 0.3.[13] The μ value of PP is 0.30 by FBE and 0.15 – 0.17 by UBT, while the literature values are 0.3 to 0.4.[14] The μ values of nylon 6 are 0.28 to 0.30 by FBE and 0.15 by UBT, while the literature value is 0.15.[13] On the average, the values of μ evaluated by the extrusion experiment are reasonable. The high values of μ of crystalline polymers are distinctive compared with those of metals. Solid n-paraffin included in the crystalline texture of PE does not essentially change the extrusion behavior, which is analyzed by the same method as that conducted in pure PE to give sufficiently good agreement between the calculated and observed values.[12] The μ values for the systems including 10 to 33 percent of n-paraffin in PE are 0.204 to 0.175 at 50 C by FBE, which indicates that n-paraffin plays the role of a solid lubricant[12] in these systems.

4.2 Die Angle and Extrusion Pressure

The effect of half-die angle (α) on the extrusion pressure is shown in Fig. 17 for PE. The case with an area reduction of 6.9 is shown on the left, and that for a reduction of 9.8 on the right. The extrusion pressure increases with increasing half-die angle. The effect of half-die angle on extrusion pressure is more remarkable in the case of higher reduction in area. In spite of the success in predicting the effect of area reduction on the extrusion pressure mentioned in the last section, both the FBE and UBT approaches failed in predicting the relationships between extrusion pressure and half-die angle (α). Figure 17 shows the comparisons between the experimental data (solid lines) and the values calculated by the UBT approach (broken lines) and the FBE approach (dotted lines). Insofar as we rely on the FBE approach, it seems to be necessary to assume that the frictional coefficient should vary with the half-die angle, although we have no experimental support at present. The UBT approach is preferable to the FBE approach because of the former's tendency to increase the pressure value with half-die angle after passing through the minimum. The introduction of a dead zone into the UBT solution for a higher half-die angle predicts a decreasing extrusion pressure, which makes the situation more undesirable. Even by employing the UBT approach, a quantitative explanation of the relationship between extrusion pressure and half-die angle is far from realistic.

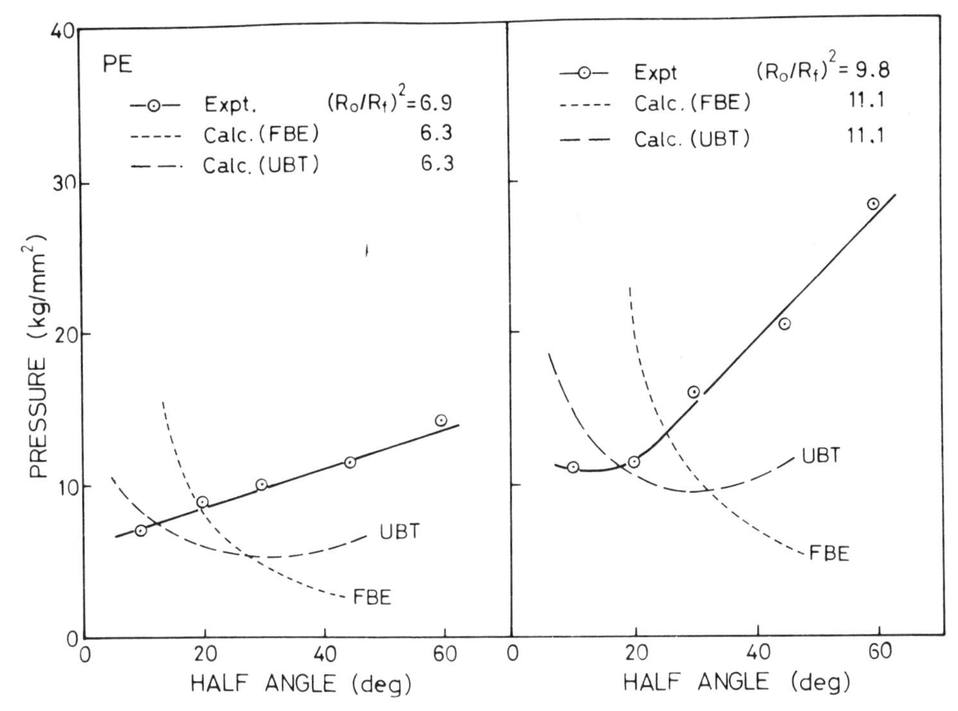

Fig. 17. Extrusion pressure versus half-die angle for linear polyethylene. Bold lines represent experimental data. Broken lines represent values calculated by the UBT approach, dotted lines represent values calculated by the FBE approach. All calculated results are based on a constant μ value.

4.3 Critical Degree of Processing

Critical degree of processing (CDP) is defined as the maximum degree of processing or logarithmic area reduction, $\ln\,(R_0/R_f)^2{}_{lim}$. Below this value, smooth extrusion is realized, while above this value, the extrusion is accompanied by occurrences of fracture and generation of cracks or irregular surfaces caused by the local destruction of superstructure in the extrudate. Extrudate with the conditions above CDP therefore is impractical. Figure 18 shows the transparent appearance of the solid-state extrudate of PP, with indications of area reduction and extrusion temperature. The extrudate obtained with an area reduction of 6.3 shows irregular surfaces, and with greater reduction, the extrudate was destroyed. Thus,

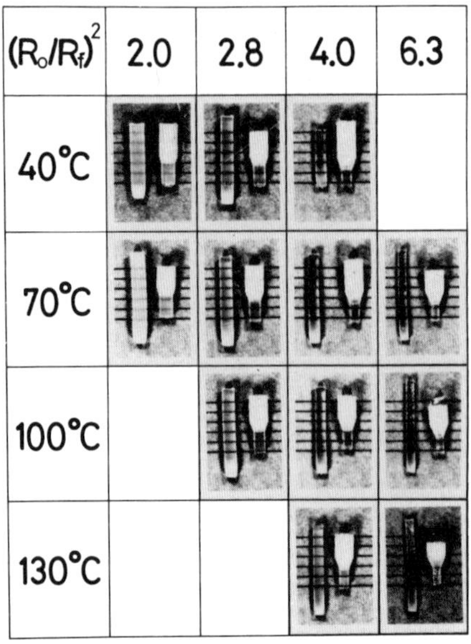

$(R_0/R_f)^2$	2.0	2.8	4.0	6.3
40°C				
70°C				
100°C				
130°C				

Fig. 18. The appearance of extrudates at various extrusion conditions for polypropylene.

the CDP of PP was determined to be 1.8. An extrusion temperature of from about 70 to 100 C seems to give the most transparent extrudate of PP.

The CDP value denoted by 2 ln $(R_0/R_f)_{lim}$ varied with polymer species, e.g., 2.1 to 2.4 for PE, 1.8 for PP, and 1.4 for nylon 6. There is a close correlation between CDP and ϵ^* in Eq. 5, as shown in the left in Fig. 19. Here we have taken the average value of ϵ^* over the range of drawing temperatures as indicated in Table I. In Section 2.1, we called ϵ^* the "unfoldability parameter", a term relating to the condition in which the folded molecular chains in the lamellar crystals are unfolded by the shear stress, and the polymers crystallizing in well-developed lamellar crystals, such as PE and POM, show comparatively large values of ϵ^*. According to Eqs. 4 and 7, 2 ln $(R_0/R_f)_{lim}$ corresponds to the critical true strain (ϵ_c), which represents the limiting strain for smooth deformation of materials during solid-state extrusion. Thus, the polymers with higher values for the unfoldability parameter, ϵ^*, are expected to allow higher area reduction or degree of processing for smooth extrusion.

The broken line on the left in Fig. 19 is drawn with a unit slope. If the unfoldability parameter ($\overline{\epsilon^*}$) is a unique factor determining CDP, the open circles should agree with the broken line. But the deviation of the data from the unit slope line increases systematically with increasing ϵ^*. Therefore, it is necessary to introduce another variable to make the critical true strain, $\epsilon_c = 2 \ln (R_0/R_f)_{lim}$, in solid-state extrusion consistent with the true strain in uniaxial extension. This can be realized by simply assuming that

$$\epsilon_c = \overline{\epsilon}^* \exp(-2.303\ c/\kappa) \tag{21}$$

and

$$\kappa = \log(\sigma_b/\sigma^*) \quad , \tag{22}$$

where σ_b is the true stress at which the extrudate is broken. The right-hand side of Fig. 19 shows plots of ϵ_c versus $\overline{\epsilon}^* \exp(-2.303\ c/\kappa)$ for PE-1, -2, and -3, PP, and nylon 6, in which the data points represented by the filled circles are calculated according to Eq. 21 by using an a priori assumption of $\kappa = 1.1$. The values of $\overline{\epsilon}^*$ and c are taken from Table I. By adopting the provisional assumption of $\kappa = 1.1$, we are able to predict the CDP or the ratio of the limiting logarithmic area reduction in solid-state extrusion by using the data obtained from uniaxial extension tests.

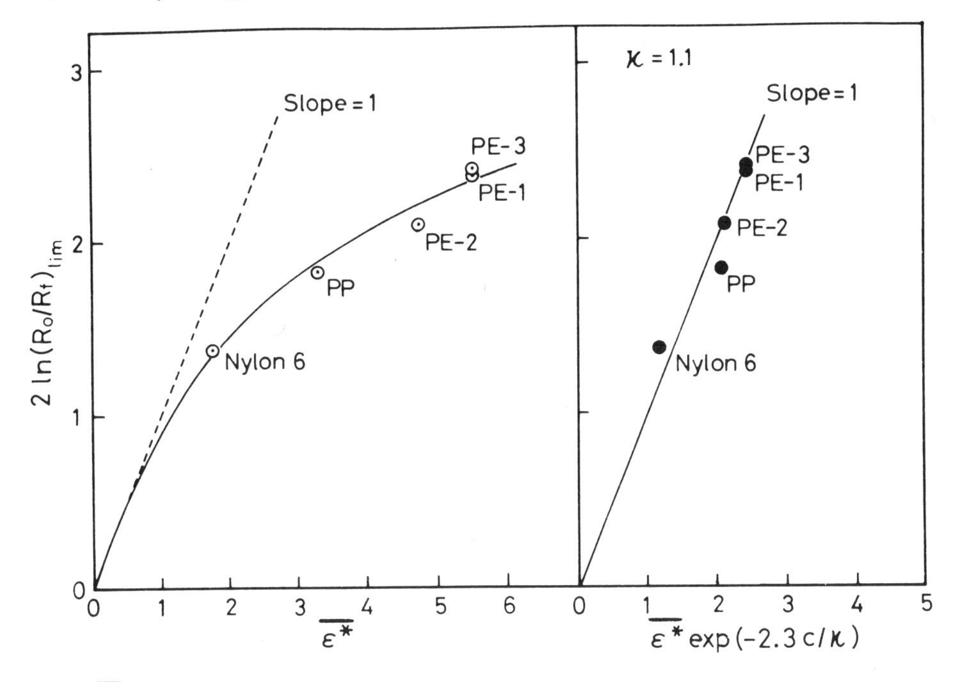

Fig. 19. Limiting true strain, $2 \ln (R_0/R_f)_{lim}$, versus ϵ^* (on the left) and $\epsilon^* \exp(-2.3\ c/\kappa)$ (on the right) for various polymers. ϵ^* is the unfoldability parameter and c is the strain-hardening parameter in the general equation for the σ–ϵ relationship. κ represents $\log(\sigma_b/\sigma^*)$ and is provisionally assumed to be 1.1.

Before discussing the provisional assumption of $\kappa = 1.1$, the effect of the newly introduced parameter c on $\epsilon_c = 2 \ln (R_0/R_f)_{lim}$ will be considered. The parameter c is a constant characteristic only of polymer species and is called the strain-hardening parameter in connection with the general equation (Eq. 5). The larger the c value, the higher is the buildup of stress with the progression of deformation. As seen in Table I, the polymers with larger values of unfoldability parameter ϵ^* tend to have larger values of strain-hardening parameter (c). Thus, a large ϵ^* value of some polymer increases the CDP or ϵ_c in extrusion, but the same polymer usually has a higher value of c, which acts to reduce the solid processability of the polymer. Such a compensative interrelationship between ϵ^* and c remains unexplained at present. A hypothetical polymer with high ϵ^* and low c, if it exists, would be most desirable for solid-state extrusion.

Next, we will discuss the provisional assumption of $\kappa = 1.1$ in Eq. 22. Detailed inspection of Eq. 21 reveals that the selection of the values of σ_b has little effect on the calculated value of ϵ_c. In other words, even if the value of σ_b varies to some extent, Eq. 21 apparently still holds. For example, in the case of nylon 6, if we assume that the error of ϵ_c is neglected and the deflection of the value of ϵ_c, calculated from Eq. 21, from the straight line with unit slope is due to the selection of σ_b value, then $\sigma_b = 12.6\sigma^*$ for the $\kappa = 1.1$ assumption, while $\sigma_b = 51.2\sigma^*$ or $\kappa = 1.71$ for the actual data. Therefore, the justification for assuming $\kappa = 1.1$ is intrinsically questionable. However, the experimental errors in determining the $\epsilon_c = 2 \ln (R_0/R_f)_{lim}$ value and $\overline{\epsilon}^*$ should be taken into account. The value of ϵ^* covers the range from 5.3 to 5.8 for PE (referring to Table I). In spite of these uncertainties, it is convenient to evaluate ϵ_c from the uniaxial extension data to adopt the provisional assumption of $\kappa = 1.1$.

It is interesting to relate the σ_b values determined by Eq. 21 and 22 to the tenacity of the fibers of the corresponding polymers. Smooth extrusion of solid crystalline polymers becomes impossible for the stress condition at which the structural elements of extrudate (such as the macro- or microfibrils) are fractured during the extrusion process. The superstructure of well-oriented solid-state extrudate[1] is considered to be almost the same as that of commercial fibers prepared by cold-drawing melt-spun undrawn fibers. The engineering stress-strain curves of standard fibers are given in the *Rheology Handbook*[16]. For these data, the stress unit is represented by g/d, which was converted to kg/mm^2 by multiplying by 9ρ (density). Further, the nominal stress thus obtained is converted to true stress by using Eq. 3. Table III shows the comparison between the true stress at fracture evaluated from the data obtained through solid-state extrusion by using Eqs. 21 and 22 and the true stress converted from the tenacity data for commercial fibers in the literature.[16]

Table III. Comparison of True Stress at Fracture Determined by Solid-State Extrusion With That of Fibers Converted From the Tenacity Data Given in Literature[a]

		σ_b, kg/mm^2	
Polymers	Temperature	Calculated by Eqs. 21 and 22	Converted From the Tenacity Data for Fibers[b]
PE	30	15.3	46
	50	10.0	41
	70	7.6	25
	90	5.5	21
	110	3.2	18
PP	40	12.5	70
	70	8.0	54
	100	4.4	42
	130	2.4	19
Nylon 6	80	71	57[c]
	120	66	48[c]
	160	46	40[c]

(a) *Rheology Handbook*[16].
(b) The unit of g/d given in the literature is converted to the unit of kg/mm^2 by multiplying by 9ρ (ρ = density). Nominal stress is converted to the true stress by Eq. 3$'$.
(c) Fiber fineness is 70 d.

Comparison between these data, obtained from quite different sources, gives an impression of large deviations: σ_b from fiber tenactity is three or four times greater than σ_b evaluated from solid-state extrusion data for PE, six to eight times for PP, and comparable for nylon 6. But, the fact that the fracture strength is remarkably dependent on the specimen size must be taken into account. For example, Anderegg[17] reported that the tensile strength of fused quartz fiber was increased by a factor of 100 with a 100-fold increase in diameter. Hirakawa et al.[18] showed the relationship between tenacity and fineness for PTFE fibers, according to which there is a sixfold increase in tenacity with an increase of fineness from 10^4d to 10d. In light of these data, the disparity between the σ_b values given in Table III is not surprising. Due to the random selection of σ_b to ϵ_c values, evaluation of ϵ_c can be approximated by substituting σ_b values for fiber tenacity reported in the literature into Eqs. 21 and 22. Thus, it may be concluded that the ultimate strength of highly oriented polymers is one of the factors limiting the degree of processing or logarithmic area reduction in solid-state extrusion. The other factors are the unfoldability parameter (ϵ^*), initial yield stress (σ^*), and the strain-hardening parameter (c). All these parameters are derived from the polymer characteristics and have a complex interrelationship. Clarification of these relationships in molecular terms will be interesting future work.

Finally, it should be noted that our analyses of the process of solid-state extrusion were based on the σ-ϵ relationship at atmospheric pressure. This method was especially successful for extrusion of PE. For PP, the agreement between the experimental and calculated values is restricted to a low degree of processing. For nylon 6, experimental data are in the range of calculated values, but the agreement cannot be obtained throughout the whole curve. It seems necessary to consider the effect of hydrostatic pressure on the σ-ϵ curve, especially for crystalline polymers such as PP and nylon 6 which have a more amorphous fraction in the crystalline texture. An amorphous structure is more sensitive to hydrostatic pressure than is a crystalline structure.[15] This is one possible reason for the difficulty in predicting the deformation behavior during solid-state extrusion of the last two polymers discussed. The heterogeneous pressure field in a tapered die makes analysis more difficult, even when the hydrostatic-pressure dependence of yield stress can be determined.

The effect of the half-die angle on the extrusion pressure remains an unsolved problem. It will be necessary to conduct a more detailed study of the effect of normal stress on the frictional coefficient, μ. The approximation employed in the UBT approach should also be examined.

Despite these unknowns, the remarkable strain-hardening effect observed in crystalline polymers is expected to introduce into the field of metals new methods patterned after those used in the analysis.

ACKNOWLEDGMENTS

The author wishes to express his heartfelt thanks to Messrs. K. Imada, S. Maruyama, and K. Nakamura for their cooperation in this work. He also thanks Sumitomo Chemical Ind. Co., Ltd., Mitsubishi Chemical Ind. Co., Ltd., and Asahi Glass Association for Promotion of Industrial Engineering for their partial financial support to this work.

REFERENCES

1. Imada, K., Yamamoto, T., Shigematsu, K., and Takayanagi, M., *J. Mater. Sci.,* **6,** 537 (1971).
2. Sachs, G., *Z. angew. Math. u. Mech.,* **7,** 235 (1927).
3. Maruyama, S., Imada, K., and Takayanagi, M., *Int. J. Polymeric Materials,* **1,** 211 (1972).
4. Nakamura, K., Imada, K., and Takayanagi, M., *Int. J. Polymeric Materials,* **2,** 71 (1972).
5. Takayanagi, M., Imada, K., Maruyama, S., and Nakamura, K., paper presented at the 6th International Congress on Rheology, Lyon, France, September, 1972.
6. Refer to G. Dieter, *Mechanical Metallurgy,* McGraw-Hill, New York (1961), p. 247.
7. Körber, F., and Eichinger, A., *Mitt. Kaiser-Wilhelm-Institut Eisenforschung,* **22,** 57 (1940).
8. Karman, Th. von: *Z. Angew. Math. Mechanik,* 1925.
9. Avitzur, B., *Metal Forming Processes and Analysis,* McGraw-Hill, 1968.

10. Prager, W., and Hodge, P. G., *Theory of Perfectly Plastic Solids,* John Wiley, New York (1951).
11. Imada, K., Yamamoto, T., Kanekiyo, K., and Takayanagi, M., *Zairyo,* **20,** 606 (1971); *Rep. Progr. Polym. Phys. Japan,* **14,** 393 (1971); *Proceedings of the 1971 International Conference on Mechanical Behavior of Materials,* (1971), Vol 3, p 476; *Int. J. Polymeric Materials,* **2** 89 (1973).
12. Maruyama, S., Imada, K., and Takayanagi, M., *Int. J. Polymeric Materials,* **2,** 105 (1973).
13. *Rheology Handbook,* Society of Polymer Science Japan (Ed.), Maruzen Co., Tokyo (1962), p 276.
14. Gregory, R. B., *SPE J.,* **25,** 55 (1969).
15. Hirakawa, S., and Takemura, T., *Japan. J. Appl. Phys.,* **7,** 814 (1968).
16. *Rheology Handbook,* Society of Polymer Science, Japan (Ed.), Maruzen, Tokyo (1965), p 353.
17. Anderegg, F. O., *Ind. Eng. Chem.,* **31,** 290 (1939).
18. Hirakawa, S., Matsushige, K., and Takemura, T., *Tech. Rep. Kyushu University,* **44,** 712 (1971).

DISCUSSION on Paper by M. Takayanagi

PECHHOLD: Did you investigate the distribution of orientations and superstructures (long period) across the radius of the extruded sample?

TAKAYANAGI: We have enough data concerning the orientation and superstructure of the solid-state extrudate, a part of which has already been published in the *Journal of Materials Science,* **6,** 537 (1971) (Orientation Issue).

MAYER: Concerning the geometrical conditions of your extrusion process, I would think that problems of redundant work and homogeneity of deformation might come into play here on the mechanical properties and structures of the polymers thus produced. For example, difficulties have been encountered in this regard (in the case of metals) when the die angle becomes large for a given small reduction.*

TAKAYANAGI: Our method of analysis could not explain quantitatively the effect of die angle on extrusion pressure as shown in Fig. 17. The tendency of increasing extrusion pressure with increasing half-die angle might be explained by the redundant work. This tendency is more remarkable for larger area reduction. At present we have developed no explanation of this phenomenon. The FBE approach clearly failed to explain these tendencies. Strict X-ray analyses revealed that the structure of extrudate of PE showed some heterogeneity along radius direction, that is, the small-angle X-ray scattering (SAXS) in the central part of extrudate showed symmetrical meridional spots, but the SAXS

*See, e.g., *Deformation Processing,* W. A. Backofen, Addison-Wesley (1972), Chapters V–VIII.

pattern at the wall surface showed an unsymmetrical pattern, which means that alignments of crystallites or lamellar directions tilt to the extrusion axis. The a-axis orientation measured by WAXS along the radius was more remarkable near the wall than in the central part. We did not encounter the difficulties for extrusion insofar as the half-die angle of 20 degrees widely used in our study was concerned. The surface roughness could be explained on the basis of ultimate strength of uniaxial elongation of the same sample as shown in Fig. 19 or Table III. This means that the redundant work as indicated by Dr. Mayer is not so remarkable with this geometrical condition. But extrusion with larger half-die angle than 20 degrees might give rise to different types of difficulties owing to redundant work. We have conducted no experiments on it as yet.

MENGES: In the laboratories of IKV in Aachen we constructed an extrusion machine with four high-pressure cylinders with pressures up to 8 kbars. Under this pressure, the raw-material particles are compressed and forced through small orifices of 0.05 mm. There the material is melted and pushed into a central mixer, where the four streams of melt are mixed and guided to the outlet. The output of this machine is 200 to 300 kg/hr.

We calculated the pressures and the heating in the same way as Professor Takayanagi, with formulas of metal plastic transformation developed by Sachs and later by Siebel. Our measurements agree with those by Takayanagi.

TAKAYANAGI: Thank you very much for your information.

WILLIAMS: Does the agreement between the simple theory, which assumes that plane sections remain plane, and the experimental results imply that there is very little redundant shearing with polymers?

TAKAYANAGI: Observation of the lattice distortion in the longitudinal cross section of the extruding sample does not conform with the plane-section assumption of the free-body approach. Our treatment is, therefore, an approximate one. It is necessary to develop a more elaborate theory to explain the actual deformation pattern. At present, the upper-bound approach seems to give an excessively distorted lattice pattern and, therefore, the slip-line analysis reproduces rather well the actual pattern. The applicability of the latter method is confined to the plane-stress state.

LANDEL: How were the samples prepared for introduction into the chamber of the extruder? According to your picture, σ^* and ϵ^* are related to lamellae deformation and unfolding. Therefore, these parameters should be sensitive to prior working of the sample, to irradiation, etc. Have you made any investigations on the dependence of these parameters on such morphology changes or crystallite "damage"?

TAKAYANAGI: We have prepared the billet by molding the polymer melts and cutting the bulk crystallized sample in a rod shape to fit the cylinder. We are planning to inspect the effect of lamellar thickness on σ^* and ϵ^* by using the samples with different long-period values. The irradiated samples, as you suggest, can be used effectively to clarify the characteristics of initial yield stress σ^* and unfoldability parameter ϵ^*.

VINCENT: We have also measured true stress – strain curves in tension for several polymers over wide temperature ranges. We find that the rate of strain hardening is not a constant for a polymer species but depends on molecular weight and degree of crystallinity.

TAKAYANAGI: If you will replot your true stress – true strain curves in both logarithmic curves as we have proposed and shift them along both axes, you will find that your data fit the hyperbolic curve represented by our general equation. We have already verified the validity of our equation on several crystalline polymers with different molecular weights and degrees of crystallinity.

LEE: A question was asked concerning the meaning of the term "lost work" in the extrusion process. Friction at the die surface causes shearing, since the surfaces are held back in the extrusion motion, as illustrated in the attached figure. For contrast, the deformation of an initially rectangular grid by stretching in simple tension is shown. The shearing deformation which causes the cross-scribed lines to become curved, does not contribute to the section reduction and yet absorbs additional plastic work, known as lost work.

The influence of the coefficient of friction may be limited by the fact that the frictional shear stress on the surface cannot exceed the yield stress in shear, since slip then occurs within the material before the full surface friction can be developed. At high die pressures, this circumstance is likely to occur, and I am wondering whether you have observed this in your assessment of the effect of friction coefficient. If extrusion causes anisotropy with a larger tensile yield stress, this inability to generate the maximum possible friction will arise more easily since, because of anisotropy, the yield in shear parallel to the cylindrical surface will be correspondingly smaller compared with the tensile yield.

TAKAYANAGI: Professor Lee's indication on the friction coefficient seems to be acceptable in molecular terms. But we have found no polymeric materials at the die surface after extrusion and, therefore, slip within materials is not necessary to assume. Energy calculations based on the upper bound theorem prefers the die surface slip to the yielding within materials.

NONLINEAR BEHAVIOR OF POLYISOBUTYLENE SOLUTIONS

L. J. Zapas

National Bureau of Standards
Washington, D.C.

ABSTRACT

Experiments on various histories of simple shear were obtained with a polyisobutylene solution. A possible inadequacy of the BKZ theory is presented, together with a modified theory whose predictions are in agreement with the experimental data.

1 INTRODUCTION

The elegant theory of simple materials[1,2] has been formulated for the description of nonlinear viscoelastic behavior of isotropic materials. In this theory, the stress is given by a general functional of the history of the relative strain of a material point, but because of this generality it is sometimes a very cumbersome theory to handle. Simpler specialized theories have been formulated and numerous constitutive equations have been published which have been used mostly in engineering calculations and in preparing numerous manuscripts. In 1963, Bernstein, Kearsley, and Zapas[3] proposed a specialized theory which has come to be called the BKZ elastic fluid. This theory was designed to be as simple as possible consistent with the then existing data on polymers. Experimental data obtained on uncrosslinked amorphous polymers[4] and on lightly cross-linked rubbers[5] were consistent with the theory. Subsequent work on polymer melts and polymer solutions[6,7] were also in good agreement with the theory. In this paper I intend to show a possible inadequacy of the BKZ theory and to propose a modification to the theory.

2 EXPERIMENTAL

The data reported in this paper were obtained on a 19.3 percent solution of polyisobutylene (Vistanex L-100, Enjay Chemical Co.)* in cetane. A Weissenberg rheogoniometer was used to shear the sample between a flat plate (7.5-cm diameter) and a cone with a gap angle of 0.0268 radian. The cone was at the bottom and was connected to the driving shaft. The plate was at the top and was connected to a torsion bar which was used to measure the torque. For most of the experiments, a torsion bar 0.35 in. in diameter was used. The stress-time measurements were recorded on an oscillograph.

3 RESULTS AND DISCUSSION

This work will be concerned with simple shearing histories only. The BKZ theory for isothermal simple shearing histories of incompressible materials gives the value of the shearing stress at time t by the following relation:

$$\sigma(t) = -\int_{-\infty}^{t} \mathcal{K}_{*}\,(\gamma(t) - \gamma(\tau),\, t-\tau)\, d\tau \qquad (1)$$

where $\gamma(\tau)$ is the amount of shear at time τ and $\mathcal{K}_*$ is a function of strain and time, obeying $\mathcal{K}_*(0,\tau) = 0$. If $\mathcal{K}_*$ is known, the shearing stress $\sigma(t)$ for any simple shearing history can be calculated through Eq. 1. Of course, the form of $\mathcal{K}_*$ is not specified by the theory and in general the function

*Certain commercial equipment, instruments, or materials are identified in this paper in order to adequately specify the experimental procedure. In no case does such identification imply recommendation or endorsement by the National Bureau of Standards, nor does it imply that the material or equipment identified is necessarily the best available for the purpose.

$\mathcal{K}_*$ must be determined through experiments. If the material had been at rest up to time $t = 0$ and at that time we introduce a single step in shear of magnitude γ, we can show from Eq. 1 that

$$\mathcal{K}_*(\gamma,t) = \frac{\partial \mathcal{K}(\gamma,t)}{\partial t} \quad , \tag{2}$$

where $\mathcal{K}(\gamma,t) = \sigma(t)$, $\mathcal{K}(\gamma,\infty) = 0$ and $\mathcal{K}(0,t) = 0$.

Bernstein[8] has shown that, in a suddenly applied steady rate of shear where the material has been at rest up to time $t = 0$, the stress $\sigma(t)$ is given by

$$\sigma(t) = \kappa \int_0^t \mathcal{K}'(\kappa\xi,\xi)d\xi \quad , \tag{3}$$

where $\mathcal{K}'(\gamma,t) = \dfrac{\partial \mathcal{K}(\gamma,t)}{\partial \gamma}$ and κ is the rate of shear.

Differentiation of Eq. 3 with respect to time yields

$$\frac{\dot{\sigma}(t)}{\kappa} = \mathcal{K}'(\kappa t,t) \quad . \tag{4}$$

Thus, one can obtain $\mathcal{K}'(\gamma,t)$ from experiments on suddenly applied steady rate of shear at various rates of strain. $\mathcal{K}(\gamma,t)$ can then be calculated by integrating isochronal values of $\mathcal{K}'(\gamma,t)$, i.e.,

$$\mathcal{K}(\gamma,t) = \int_0^\gamma \mathcal{K}'(\xi,t)d\xi \quad . \tag{5}$$

We attempted to do that by taking data on suddenly applied steady-shear experiments on a 19.3 percent solution of Vistanex L-100. Some of the results are shown in Fig. 1. The maximum of the stress overshoot occurred at a constant strain, whose value is 2.5. This is in agreement with the value of 2.44 reported by Zapas and Phillips[7] for the same type of measurements on a 10 percent polyisobutylene solution. In Fig. 2 we show a plot of $\gamma\mathcal{K}'(\gamma,t)$ versus $\log \gamma$ for $t = 0.6$ second. The area under the curve up to γ, according to Eq. 5, is the value of $\mathcal{K}(\gamma,t)$. It is evident from Fig. 2 that for high values of γ, this area is negative, and therefore $\mathcal{K}(\gamma,t)$ is negative for some positive values of γ. This would imply that in a single-step stress-relaxation experiment, the initially positive stress will decay to a negative value, which is unacceptable and suggests an inconsistency in the BKZ theory. There is the possibility that at the high rates of shear, we did not have a simple shearing flow or that some other

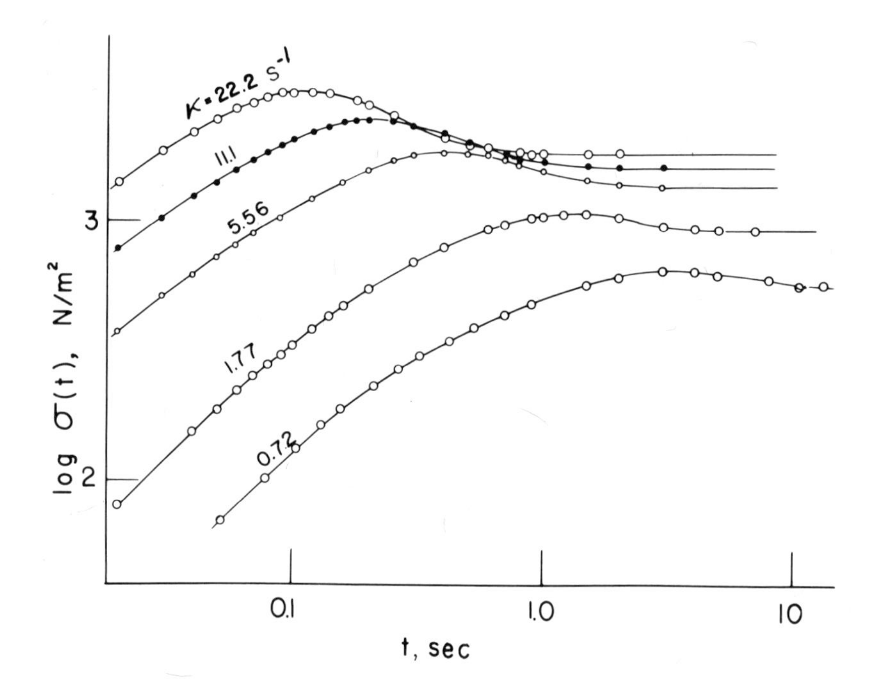

Fig. 1. Suddenly applied constant rate of shear of a 19.3 percent solution of Vistanex L-100, at 25 C.

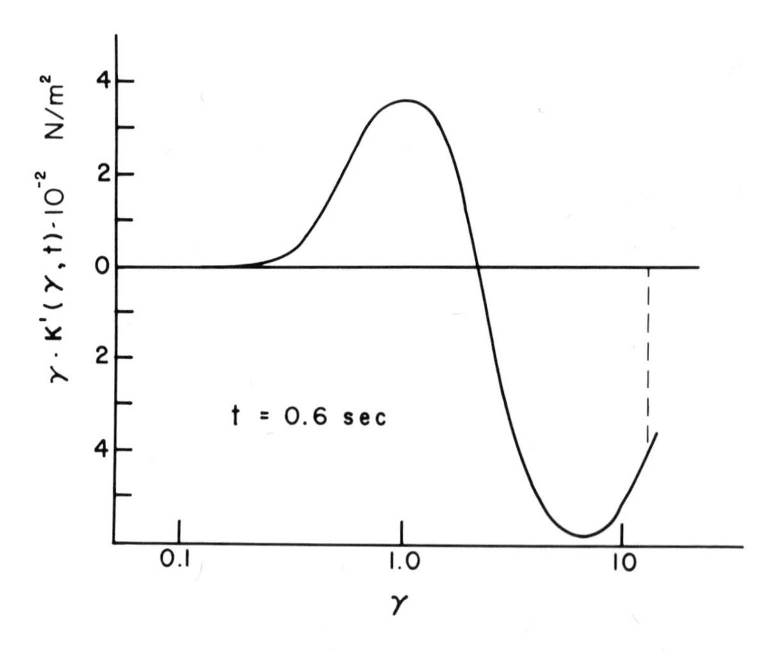

Fig. 2. A plot of γ K$'(\gamma,t)$ versus log γ for t = 0.6 second.

unknown error may have occurred, but we shall proceed by taking the data at face value.

By examining Eq. 1 we see that the stress is given by the sum of contributions between the present strain at time t and all past strains at time τ, weighted properly with the elapsed time $t-\tau$. At this point we introduce the concept of a material clock whose rate depends on the strain history in some way. Then, what is to an observer an isochronal set of data could be at different times according to the material clock. The material clock is not a new idea, since in a sense it was introduced by Leaderman[9] for his time-temperature superposition principle, and by Moreland and Lee[10] for the thermorheologically simple material.

We shall proceed to write the modified BKZ theory for isothermal conditions assuming incompressibility. We use a Cartesian tensor notation, letting the point $\mathbf{x}$ be given by Cartesian coordinates x_i, $i = 1,2,3$. For a given motion, the position of a particle $\mathbf{X}$ at time t, $\mathbf{x}(t)$, can be expressed in terms of its position at time τ, $x(\tau)$, by a function

$$x_i(t) = x_i\ [\mathbf{x}(t), t, \tau] \quad , \tag{1a}$$

from which we may define the relative deformation gradients

$$x_{ik}\ (t,\tau) = \frac{\partial x_i(t)}{\partial x_k(\tau)} \quad . \tag{1b}$$

The Cauchy-Green relative strain tensor $C_{ij}(t,\tau)$ is defined by

$$C_{ij}(t,\tau) \equiv x_{k\,i}(t,\tau)\ x_{k\,j}(t,\tau) \tag{1c}$$

and the Green-St. Venant relative strain tensor is

$$E_{ij}(t,\tau) \equiv \frac{1}{2}\left[C_{ij}(t,\tau) - \delta_{ij}\right] \quad . \tag{1d}$$

For the matrix of $E_{ij}(t,\tau)$ and $C_{ij}(t,\tau)$, we write $\mathbf{E}(t,\tau)$ and $\mathbf{C}(t,\tau)$, respectively. Now we assume the existence of a scalar-valued function

$$S\ [\mathbf{E}(\xi,\tau),t] \quad ,$$

which is defined for all $t \geqslant 0$ and depends on a strain $\mathbf{E}(\xi,\tau)$ only through

$$I_1(\xi,\tau) = \lambda_1^2 + \lambda_2^2 + \lambda_3^2$$

$$I_2(\xi,\tau) = \lambda_1^2\lambda_2^2 + \lambda_2^2\lambda_3^2 + \lambda_3^2\lambda_1^2$$

$$I_3(\xi,\tau) = \lambda_1^2\lambda_2^2\lambda_3^2 = 1 \quad ,$$

where $\lambda_i^2 = \lambda_i^2(\xi,\tau)$, i = 1, 2, 3, are the principal values of $\mathbf{C}(\xi,\tau)$.

We write

$$S_{kl}\,[\mathbf{E}(\xi,\tau),t] \equiv \frac{\partial S}{\partial E_{kl}} \quad . \tag{1e}$$

Furthermore, we shall assume the existence of a material property $\psi > 0$ which scales time intervals in the manner detailed below. The quantity ψ depends on the previous strain history. We can write then the stress $\sigma_{ij}(t)$ at time t as

$$\sigma_{ij}(t) = -p\delta_{ij} + \int_{-\infty}^{t} x_{ik}(t,\tau)x_{jl}(t,\tau)S_{kl}\left[\mathbf{E}(t,\tau), \int_{\tau}^{t} \Psi(\cdot,\xi)d\xi\right] \Psi(\cdot,\tau)d\tau \ . \tag{6}$$

When the function $\Psi(\cdot,\xi)$ is unity, Eq. 6 gives the incompressible elastic fluid given by Bernstein, Kearsley, and Zapas.[3]

For simple shear-strain histories we shall write $\Psi(\cdot,\xi)$ as $\dot{\phi}(|\gamma(t)|-|\gamma(\xi)|,t-\xi)$. Furthermore, hereinafter we shall omit the bars indicating the absolute values. We also take the function $\phi(\gamma,t)$ to be the function whose partial derivative with respect to the second argument gives $\dot{\phi}(\gamma,t)$.

The motion

$$x_1(t) = x_1(\tau) + (\gamma(t) - \gamma(\tau))x_2(\tau)$$
$$x_2(t) = x_2(\tau)$$
$$x_3(t) = x_3(\tau)$$

represents a simple shear. The relative deformation matrix is

$$\left\| x_{ik}(t,\tau) \right\| = \left\| \begin{matrix} 1 & \gamma(t) - \gamma(\tau) & 0 \\ 0 & 1 & 0 \\ 0 & 0 & 1 \end{matrix} \right\| \quad \text{and } \mathbf{E}(t,\tau) =$$

$$\frac{1}{2} \left\| \begin{matrix} 0 & \gamma(t) - \gamma(\tau) & 0 \\ \gamma(t) - \gamma(\tau) & [\gamma(t) - \gamma(\tau)]^2 & 0 \\ 0 & 0 & 0 \end{matrix} \right\|$$

Thus, we can write for simple shear

$$x_{ik}(t,\tau)x_{jl}(t,\tau)S_{kl}\left[\mathbf{E}(t,\tau), \int_{\tau}^{t} \Psi(\cdot,\xi)d\xi\right] = \mathcal{K}_{*_{ij}}\left(\gamma(t)-\gamma(\tau), \int_{\tau}^{t} \dot{\phi}(\gamma(t)-\gamma(\tau), t-\xi)d\xi\right) \tag{6a}$$

for an appropriately defined function $\mathcal{K}_{*_{ij}}$, i,j = 1, 2, 3. Thus the shearing stress $\sigma_{12}(t) \equiv \sigma(t)$ and $\mathcal{K}_{*_{12}} \equiv \mathcal{K}_*$ is given by

$$\sigma(t) = -\int_{-\infty}^{t} \mathcal{K}_*\left(\gamma(t)-\gamma(\tau), \int_{\tau}^{t} \dot{\phi}(\gamma(t)-\gamma(\xi), t-\xi)d\xi\right)\dot{\phi}(\gamma(t) - \gamma(\tau), t-\tau)d\tau \tag{7}$$

In a single-step stress-relaxation experiment, the shearing history is given by:

$$\begin{aligned}\gamma(\tau) &= 0 \text{ for } \tau < 0\\ \gamma(\tau) &= \gamma \text{ for } \tau > 0\end{aligned} \quad . \tag{8}$$

Using this history in Eq. 7 we obtain

$$\sigma(t) = -\int_{-\infty}^{o} \mathcal{K}_*\left(\gamma, \int_{\tau}^{o} \dot{\phi}(\gamma,t-\xi)d\xi + \int_{o}^{t} \dot{\phi}(o,t-\xi)d\xi\right)\dot{\phi}(\gamma,t-\tau)d\tau \quad , \tag{9}$$

where $\mathcal{K}_*(o,\xi) = 0$.

By substitution of $t-\xi = \theta$ and integration within the kernel we obtain

$$\sigma(t) = -\int_{\infty}^{o} \mathcal{K}_*(\gamma,\phi(\gamma,t-\tau)-\phi(\gamma,t)+t)\dot{\phi}(\gamma,t-\tau)d\tau \quad , \tag{10}$$

and by further substitution in Eq. 10 where $\phi(\gamma,t-\tau)-\phi(\gamma,t)+t = \zeta$ we get

$$\sigma(t) = -\int_{t}^{\infty} \mathcal{K}_*(\gamma,\zeta)d\zeta = \mathcal{K}(\gamma,t) \quad . \tag{11}$$

Equation 11 is identical with the prediction of a single-step stress-relaxation experiment given by the BKZ theory; furthermore, the rates of the laboratory clock and the material clock coincide. The difference in the clock rates can be shown in multistep stress-relaxation experiments. For example, for a two-step stress relaxation where the strain history is

$$\gamma(\tau) = 0 \text{ for } -\infty < \tau < -t_1$$

$$\gamma(\tau) = \gamma_1 \text{ for } -t_1 < \tau < 0$$

$$\gamma(\tau) = \gamma_2 \text{ for } 0 < \tau < t \quad ,$$

we obtain

$$\sigma(t) = \mathcal{K}\left(\gamma_2, \phi(\gamma_2-\gamma_1, t+t_1) - \phi(\gamma_2-\gamma_1, t)+t\right) + \mathcal{K}(\gamma_2-\gamma_1, t) - \mathcal{K}\left(\gamma_2-\gamma_1, \phi(\gamma_2-\gamma_1, t+t_1) - \phi(\gamma_2-\gamma_1, t)+t\right) \quad ; \quad (12)$$

for comparison, we show the prediction of the BKZ theory for the same history

$$\sigma(t) = \mathcal{K}(\gamma_2, t+t_1) + \mathcal{K}(\gamma_2-\gamma_1, t) - \mathcal{K}(\gamma_2-\gamma_1\, t+t_1) \quad . \qquad (13)$$

For an experiment on a suddenly applied steady rate of shear, the strain history is given by

$$\gamma(t) = 0 \text{ for } -\infty < \tau 0$$

$$\gamma(\tau) = \kappa\tau \text{ for } 0 < \tau \leqslant t \quad ,$$

where κ is the rate of deformation. With this strain history and Eq. 7, we obtain

$$\sigma(t) = \kappa \int_0^t \mathcal{K}'\left(\kappa\xi, \int_0^\xi \dot{\phi}(\kappa\zeta, \zeta)d\zeta\right) \quad . \qquad (14)$$

Taking the derivative with respect to time we obtain

$$\frac{\dot{\sigma}(t)}{\kappa} = \mathcal{K}'\left(\kappa t, \int_0^t \dot{\phi}(\kappa\zeta, \zeta)d\zeta\right) \quad . \qquad (15)$$

Comparing Eq. 15 with Eq. 4 we can see the influence of the material clock. The time derivative of the stress at time t divided by the rate of shear gives again the function $\mathcal{K}'$, but at a time which may be different from t. Since our experiments were not designed to test the validity of this theory, I shall proceed with certain simplifications on the function $\mathcal{K}(\gamma,t)$ to show from our experiments some of the behavior of the material clock. In Fig. 3 we show data obtained from single-step stress-relaxation experiments, where $W(\gamma,t)$ is the stress-relaxation function given by

$$W(\gamma,t) = \frac{\mathcal{K}(\gamma,t)}{\gamma}$$

$$\lim_{\gamma \to 0} W(\gamma,t) = G(t) \quad , \tag{16}$$

where $G(t)$ is the shear modulus. The experimental values of $W(\gamma,t)$ at small strains agree very well with the calculated values of $G(t)$, which we obtained from $G'(\omega)$ and $\eta'(\omega)$ (the dynamic modulus and dynamic viscosity as a function of frequency). The isochrones of $W(\gamma,t)$ in the log-log plot are parallel. So we can simplify the function $\mathcal{K}$ by writing it as a product of two functions, one involving the time and the other the strain.

We can write, then,

$$\mathcal{K}\left(\gamma, \int_0^t \dot{\phi}\,(\kappa\zeta,\zeta)d\zeta\right) = G(\xi) \quad \left(\int_0^t \dot{\phi}\,(\kappa\zeta,\zeta)d\zeta\right) \; f(\gamma) \quad , \tag{17}$$

where $f(\gamma)$ is an odd function in γ, and $G(\xi)$ is the shear modulus at time ξ. Moreover, since up to strains of 2.5 the experimental values of $f(\gamma)$ are identical with those obtained by Zapas and Phillips, we shall use their form. The function $f(\gamma)$ is plotted and shown in Fig. 4. We can determine then the integral $\int_0^t \dot{\phi}(\kappa\zeta,\zeta)d\zeta$, using the $G(t)$ curve shown in Fig. 5. By following this procedure we found that $\phi(\gamma,t)$ could be very well represented by the relation

$$\phi(\gamma,t) = t\text{-}g(\gamma)t^{\alpha} \quad , \tag{18}$$

where $\alpha = 0.886$ and $g(\gamma)$ is as shown in Fig. 6. From Eq. 18 and Fig. 6, we see that the clock slows down when, for example, we go from a small step to a bigger step. The opposite happens when we introduce first a big step and then follow with a step down to a final strain smaller than the first step.

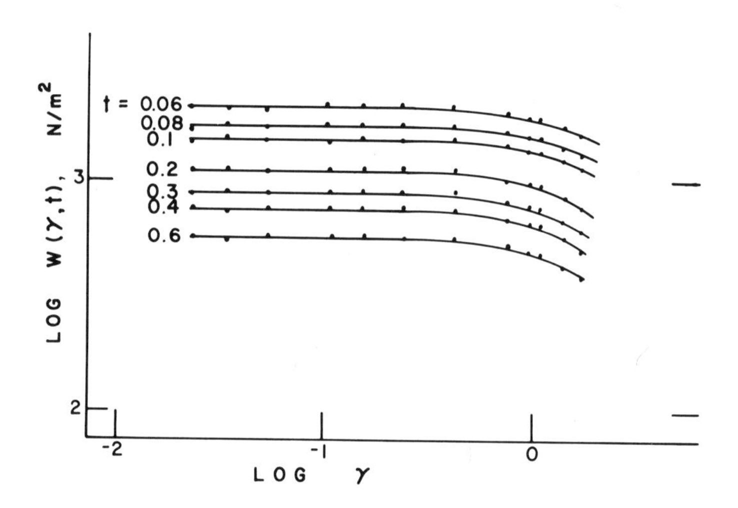

Fig. 3. Isochronal data of the stress-relaxation function $W(\gamma,t)$.

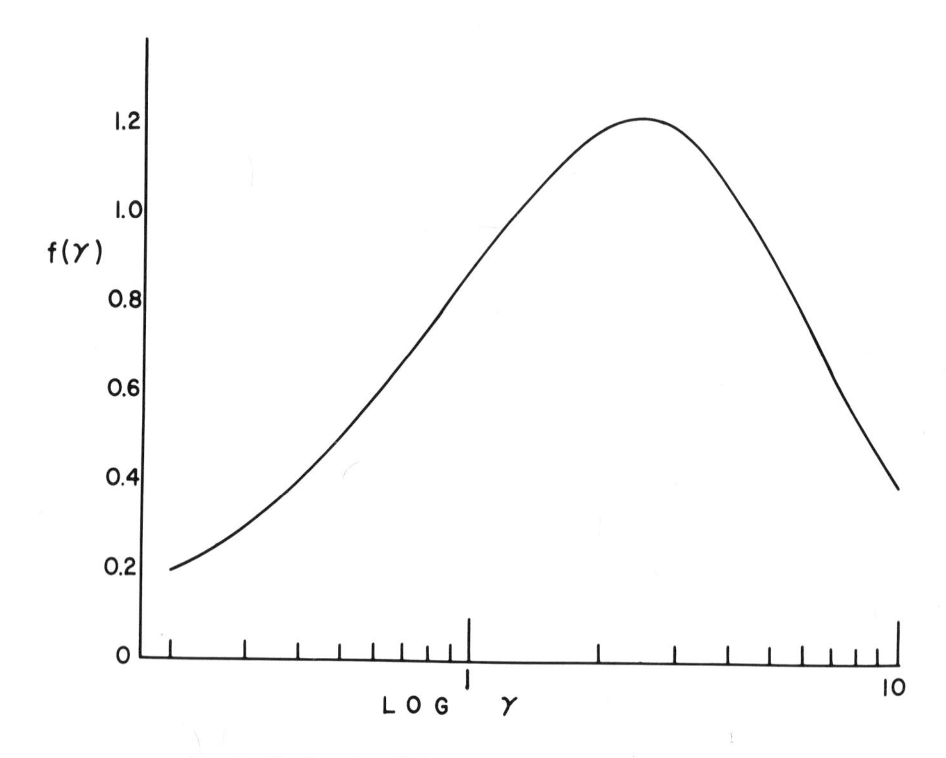

Fig. 4. The function $f(\gamma)$ given in Eq. 17 plotted versus γ.

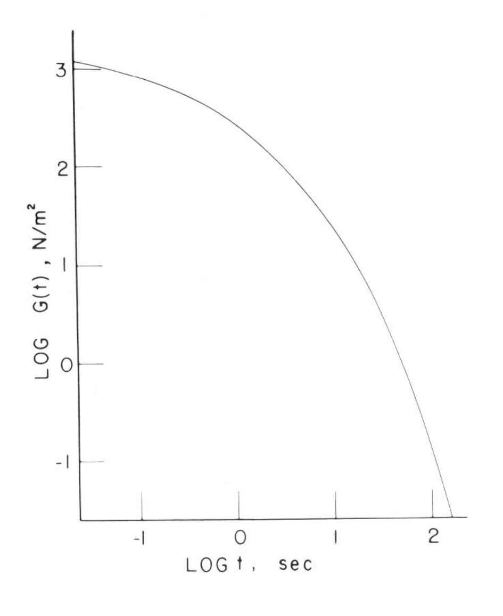

Fig. 5. The shear-relaxation modulus, G(t), of a 19.3 percent solution of Vistanex L-100, at 25 C.

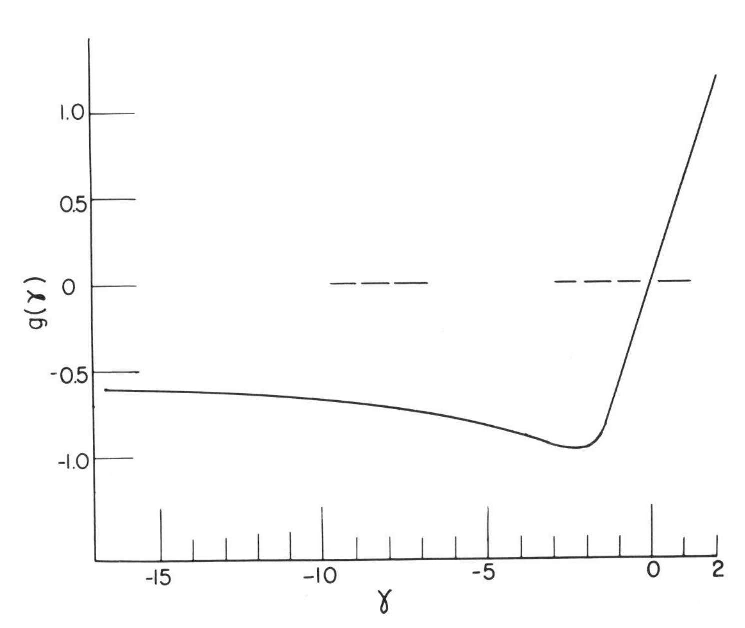

Fig. 6. The calculated behavior of g(γ).

If $\sigma_1(t)$ is the stress given by the BKZ elastic fluid, then for a two-step stress relaxation we obtain from Eqs. 12 and 13:

$$\sigma(t)-\sigma_1(t) = \mathcal{K}\left(\gamma_2,\beta(t+t_1)\right) + \mathcal{K}\left(\gamma_2-\gamma_1,\beta(t+t_1)\right) \quad -$$

$$-\mathcal{K}(\gamma_2 t+t_1) - \mathcal{K}(\gamma_2-\gamma_1, t+t_1) \quad , \tag{19}$$

where, for convenience, we use $\beta(t+t_1)$ to represent

$$\phi(\gamma_2-\gamma_1, t+t_1) - \phi(\gamma_2-\gamma_1, t)+t \quad .$$

By combining Eqs. 17 and 19, we get

$$\sigma(t) - \sigma_1(t) = \left\{ G[(\beta(t+t_1)] - G(t+t_1) \right\} \left\{ f(\gamma_2) - f(\gamma_2 - \gamma_1) \right\} \quad , \tag{20}$$

where γ_1 and γ_2 are the strains of the first and second step. For the case where $\beta \neq 1$, it is possible for the difference in the stresses to be zero, since the function $f(\gamma)$ can have the same value for different values of γ (see Fig. 4). If, for example, $\gamma_1 = 1.1$ and $\gamma_2 = 3$, we get $f(\gamma_2) = f(\gamma_2\text{-}\gamma_1) = 1.186$ and $\sigma(t)\text{-}\sigma_1(t) = 0$.

Equation 21 can be used to select those two-step stress-relaxation experiments which emphasize the difference between the theories. Since $f(\gamma)$ is an odd function of γ, when $\gamma_2 > 0$ and $\gamma_2\text{-}\gamma_1 < 0$, Eq. 20 can be written as

$$\sigma(t) - \sigma(t_1) = \left\{ G[(\beta(t+t_1)] - G(t+t_1) \right\} \left\{ f(\gamma)+f(\gamma_1-\gamma_2) \right\} , \tag{21}$$

which can be zero only when $\beta = 1$. In a two-step stress relaxation for which the first step is bigger than the second, the difference between the two theories will always be pronounced. Several such experiments were attempted and the results are shown in Figs. 7 and 8. The open circles show the prediction of the BKZ theory, the solid circles are the values calculated from the modified theory, and the line is the experimental data.

From Eq. 6 we can write the first normal stress difference as follows:

$$\sigma_{11}-\sigma_{22} = \kappa \int_0^\infty \mathcal{K}_* \left(\kappa\xi, \int_0^\xi \dot{\phi}\,(\kappa\zeta,\zeta)d\zeta \right) \xi\dot{\phi}(\kappa\xi,\xi)d\xi \quad , \tag{22}$$

where κ is the rate of shear. Using $\mathcal{K}_*(\gamma,t) = G(t)f(\gamma)$, as before, we calculated the first normal stress difference. The results are shown in Fig. 9, where again the solid circles are the values calculated using the modified theory. The experimental data were obtained from the total thrust measurements by means of a Weissenberg rheogoniometer.

The BKZ elastic-fluid theory needed only a complete set of single-step stress-relaxation data for the complete description of a material. This modified theory, being more general, requires more data to describe a material. From experiments on single-step stress relaxation, we can obtain

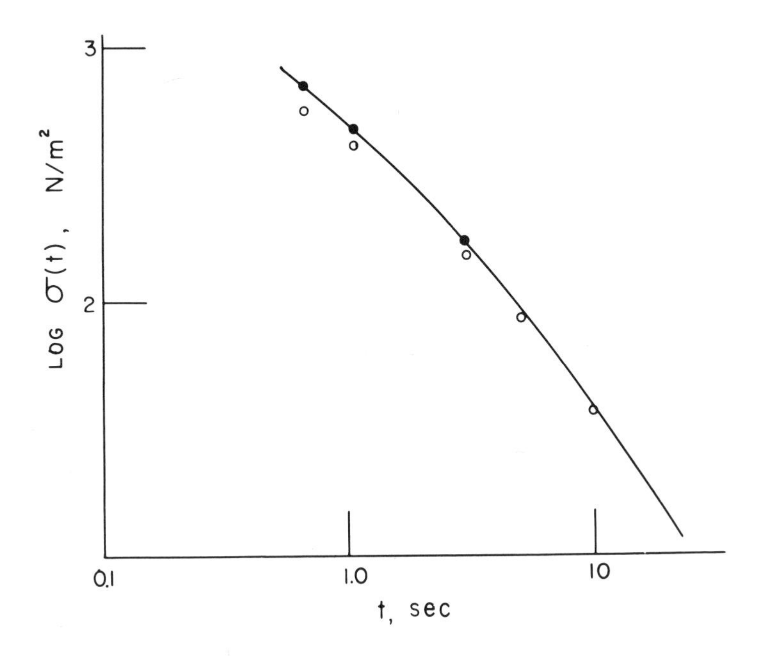

Fig. 7. The stress of the second step, in a two-step stress relaxation, where t_1 = 2.8 seconds, γ_1 = 0.532, γ_2 = 1.77. Solid line is the experimental data, and open and solid circles are the values calculated from BKZ and the modified theory, respectively.

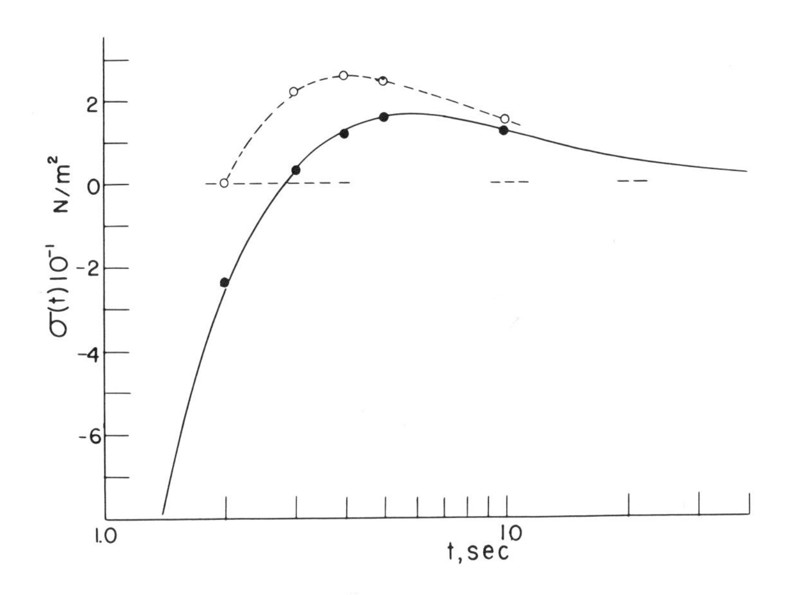

Fig. 8. The stress of the second step, in a two-step stress relaxation, where t_1 = 1.62 seconds, γ_1 = 1.77, γ_2 = 0.85. Solid and open circles are the values calculated from BKZ and the modified theory, respectively.

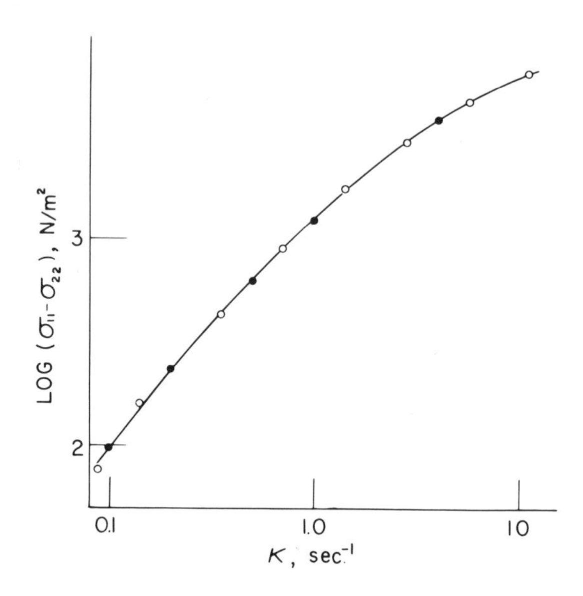

Fig. 9. The first normal stress as a function of rate of shear. Open circles are experimental data, and the solid circles are values calculated using the modified theory.

the function $\mathcal{K}(\gamma,t)$ and from a series of two-step stress-relaxation experiments, knowing $\mathcal{K}(\gamma,t)$, we can obtain the material clock behavior of $\phi(\gamma,t)$. I tried this modified theory successfully, with torsion data on pure aluminum obtained by Dr. Penn[11] in our laboratories.

The special form of $\mathcal{K}(\gamma,t)$ which was used in the calculations, is by no means unique, and it was used to facilitate numerical computations. This is also true for $\phi(\gamma,t)$, which represents the material clock behavior. The form of $\phi(\gamma,t)$ depended highly on the functional form of $\mathcal{K}(\gamma,t)$.

We expect this modified theory to be useful in cases of high nonlinearity coupled with strong time dependence.

REFERENCES

1. Green, A. E., and Rivlin, R. S., *Arch. Rational Mech. Anal.*, **1**, 1-21 (1957).
2. Coleman, B. D., and Noll, W., *Rev. Mod. Phys.*, **33**, 239 (1961).
3. Bernstein, B., Kearsley, E. A., and Zapas, L. J., *Trans. Soc. Rheol.*, 7, 391 (1963).
4. Zapas, L. J., and Craft, T. J., *J. Res. Nat. Bur. Stand.*, **69A**, 541 (1965).
5. Zapas, L. J., *J. Res. Nat. Bur. Stand.*, **70A**, 525 (1966).
6. Tanner, R. I., and Williams, G., *Trans. Soc. Rheol.*, **14**, 19 (1970). Chen, I-Jen, and Bogue, D. C., *Trans. Soc. Rheol.*, **16**, 59 (1972). Bernstein, B., and Huilgol, R. R., *Trans. Soc. Rheol.*, **15**, 731 (1971).
7. Zapas, L. J., and Phillips, J. C., *J. Res. Nat. Bur. Stand.*, **75A**, 33 (1971).
8. Bernstein, B., *Int. J. Non-linear Mech.*, **4**, 183 (1969).

9. Leaderman, H., *Elastic and Creep Properties of Filamentous Materials,* Textile Foundation, Washington (1943).
10. Morland, L. W., and Lee, E. H., *Trans. Soc. Rheol.*, **4**, 233 (1960).
11. Penn, R. W., Paper presented at the VIth International Congress on Rheology, Lyon, France (1972).

DISCUSSION on Paper by L. J. Zapas

RIVLIN: In view of the recently discovered difficulties in making meaningful normal stress measurements, would Dr. Zapas be kind enough to tell us how his normal stress measurements were made?

ZAPAS: Thank you, Dr. Rivlin, for your question. Because of the lack of time I did not go into the details of the experimental methods. The first normal stress measurements were obtained from total thrust measurements.

RIVLIN: In the BKZ theory, the assumption is made that the contribution to the stress at time t made by the strain at a previous time τ depends only on the lapsed time t-τ and not on the strain at any other time. Would Dr. Zapas be kind enough to state the assumptions of the modified theory in a corresponding form?

ZAPAS: In the modified theory the assumption is made that the contribution to the stress at time t made by the strain at a previous time τ depends on a time interval whose value depends on the previous strain history.

BECKER: You have presented the factorization of $\mathcal{K}(\gamma,t)$ into the functions G(t) and f(γ). Is the function f(γ) strictly empirical or is there any other reason for this special formula? And furthermore: would you use this expression also for other materials or solutions, respectively?

ZAPAS: The function f(γ) is empirical, obtained from curve fitting the data. Its form was restricted so that it can be consistent with the second normal stress difference. For the second part of your question, I can say that for polyoxymethylene solutions in water we could not factorize the function $\mathcal{K}_*(\gamma,t)$ into a product of G(t) and a function of the strain.

THE VISCOELASTIC-PLASTIC BEHAVIOR OF A DUCTILE POLYMER

H. F. Brinson

Department of Engineering Science and Mechanics
Virginia Polytechnic Institute and State University
Blacksburg, Virginia

ABSTRACT

A brief discussion of the use of polymers in general and polycarbonate in particular in experimental mechanics is presented both from the viewpoint of materials for use in photomechanics and as materials for use in the validation of continuum constitutive equations. Experimental data are presented showing polycarbonate to be a ductile material with elastic, viscoelastic, and plastic behavior regimes. Using constant-head-rate tests (0.002 in./min to 20 in./min), the rate-dependent yield behavior of polycarbonate is described and is compared with that proposed for metals by Ludwik. By means of creep tests, the time-dependent yield behavior of polycarbonate is also described and compared with viscoelastic-plastic theories of Nagdi and Murch and of Crochet. Finally, possible application of the results to photoplasticity and ductile fracture are discussed.

1 INTRODUCTION

Natural and synthetic polymers have long been used in experimental mechanics. As early as 1850, Maxwell used an isinglass gel for his investigations leading to the now famous Maxwell-Neumann stress-optic law of photoelasticity.[1,2] Cellulose nitrates, phenolics, epoxies, etc., have been used by photoelasticians since shortly after discovery and introduction by polymer scientists. These and other polymers have not only been used in photomechanics, but have also been used as materials for the experimental validation of continuum theories of viscoelasticity, finite deformations, and various fracture-related phenomena.

Almost since the beginning of modern photoelasticity, efforts have been under way to utilize the principles of birefringence* in the study of plasticity.[2,3] In the last decade, a number of investigators have used polycarbonate as a photoelastic-plastic model material. Indeed, polycarbonate appears to be an ideal material for such studies. It is very ductile and flows under high stress, has a well-defined yield point ($\sigma_{y.p.} \simeq$ 9000 psi, $\epsilon_{y.p.} \simeq .06$ in./in.), exhibits a tensile-instability phenomenon with accompanying Lüder's bands similarly to some ductile metals, and, of course, is birefringent under load. Brill[4], Gurtman et al.[5], Whitfield and Smith[6], and Dally[7] have performed experiments to characterize the mechanical and optical behavior of polycarbonate. By means of a limited number of constant-strain-rate tests, creep tests, and relaxation tests, they have shown that polycarbonate is essentially elastic at stresses below $\sigma \simeq$ 4000 psi and that the material is viscoelastic between $\sigma \simeq$ 4000 psi and $\sigma_{y.p.} \simeq$ 9000 psi. That is, below the yield point, nearly all mechanical and optical effects are recovered within a relatively short time after unloading. Above the yield point, large irrecoverable deformations and birefringence are obtained.

While Gurtman et al.[7] reported the stress at localized yield in polycarbonate to be rate dependent, all the authors cited above, including Gurtman, elected to ignore such rate dependency in their photoplasticity efforts. Attempts were made to minimize viscoelastic effects as well as rate effects by performing characterization tests and model testing at the same strain rate[7] by using rates below which differences in the rate of testing were negligible[4], or by using a common time to yield in characterization and model testing[6]. With the exception of Dally[7], all investigators limited

*In photoelastic investigations, birefringence is understood to mean the division of the light vector into two mutually orthogonal components which propagate through a stressed transparent medium with different velocities. This leads to a relative retardation between the components and appears as isochromatic and isoclinic fringes when viewed with polarized light. The relation between such fringes and the internal principal stresses is given by the Maxwell-Neumann stress-optic law. Birefringence in regions of plasticity generally has the same physical meaning but the Maxwell-Neumann law may be no longer rigorously valid.

strains to less than yield-point strains in the actual application of photoplasticity to the solution of boundary-value problems. As a result, much of their work likely could be interpreted in terms of viscoelasticity theories rather than plasticity theories. Both Dally[7] and Brinson[8] explored the interpretation of the birefringence of polycarbonate when stressed beyond the yield point, but these efforts did not include rate- or time-dependent effects.

In addition, polycarbonate was used as a material to validate and extend Dugdale's strip-type yielding models for ductile fracture.[9-11] In these endeavors, it was shown that the postyield birefringence of polycarbonate could be utilized in plastic-zone-size measurements, and that such measurements are quite compatible with those in ductile metals. Again, rate-dependent effects were minimized by performing all tests at the same strain rate.

It also was shown by Sauer et al.[12] that the stress at localized yield in polycarbonate is dependent upon hydrostatic pressure.

The foregoing comments provide a portrait of a material (polycarbonate) which has elastic, viscoelastic, and plastic regions of behavior and which has yield properties that are both rate and pressure dependent. Such yield properties, as are other properties of polymers, should be strongly temperature sensitive. In order to develop viable theories of plasticity, viscoplasticity, and photoplasticity for this material, it is necessary to have detailed knowledge of the properties of the material in all behavior regimes as a function of time or strain rate, pressure, and temperature and to assess such properties in terms of continuum theories of viscoelasticity and viscoplasticity. The present endeavor is directed toward quantifying the rate- and time-dependent yield behavior of polycarbonate.

2 EXPERIMENTAL PROGRAM

There are inherent difficulties in obtaining accurate mechanical and optical properties of polycarbonate. Electrical strain gages are, of course, the most accurate method for measuring mechanical properties. However, reinforcement effects of the strain gage for thin sheets or films, heat generation due to gage current, and large deformations tend to make electrical-strain-gage measurements unreliable.[13] The isochromatic fringe density for polycarbonate is so great that fringe measurements are unreliable for stresses and strains near or past the yield point. For these reasons, all previous investigators have used grid techniques or extensometers for the measurement of strain in polycarbonate, and those who have also performed birefringence measurements have limited their investigations to very thin sheet material (0.020 to 0.60 inch).

2.1 Equipment and Procedures

For the present investigation, commercially available 1/8-inch-thick sheets of polycarbonate were used. All uniaxial tensile specimens were machined from a single sheet of material to the geometry shown in Fig. 1(a). High-elongation, foil strain gages capable of 10 to 15 percent maximum strain were mounted both in the direction of the load and transverse to the load. Strains were monitored with the aid of a modified amplification and recording system capable of indicating strains up to 10 percent. Voltages were intentionally kept low to minimize heating of the material due to gage current. Preliminary tests were performed which indicated that use of the gages did not appreciably change the properties of the material, due to either heating or reinforcement effects. Some nonlinearities are inherent in the instrumentation system used at high strains. However, these were believed to be small as results were quite comparable to those of other investigators.

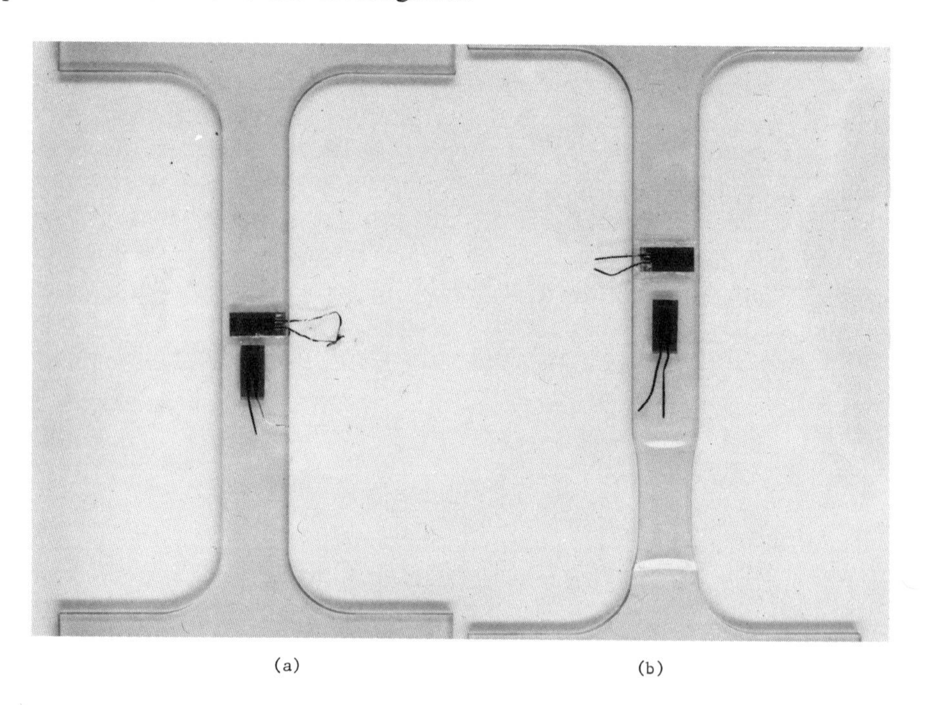

Fig. 1. Specimen geometry: (a) geometry prior to loading and (b) geometry after slip-band formation.

All testing was performed on three separate testing machines. Two were of the controlled-head-rate type, one with rates accurate from 0.002 to 2 in./min and the other with rates accurate from 0.5 to 20 in./min. The third was a pneumatic testing machine. The first two were used for constant-strain-rate and relaxation tests and the third for creep tests.

For some of the tests, a polariscope utilizing a laser light source was mounted around the test specimen to observe and photograph birefringence.

All tests were performed under controlled environmental conditions of approximately 70 F and 70 percent relative humidity.

2.2 Results

With the instrumentation briefly described above, constant-head-rate tests were performed at various rates ranging from 0.002 to 20 in./min. Longitudinal and transverse strains at the strain-gage site were monitored through localized yield and, in several cases, to fracture. Figure 2 shows the stress-strain results obtained for the head rate of 0.002 in./min. All other constant-rate tests displayed behavior similar to that shown in Fig. 2, except for rate- and time-dependent effects which will be discussed subsequently. As a result, the stress-strain behavior of polycarbonate described below (see Fig. 2) is typical of all results obtained regardless of strain rate.

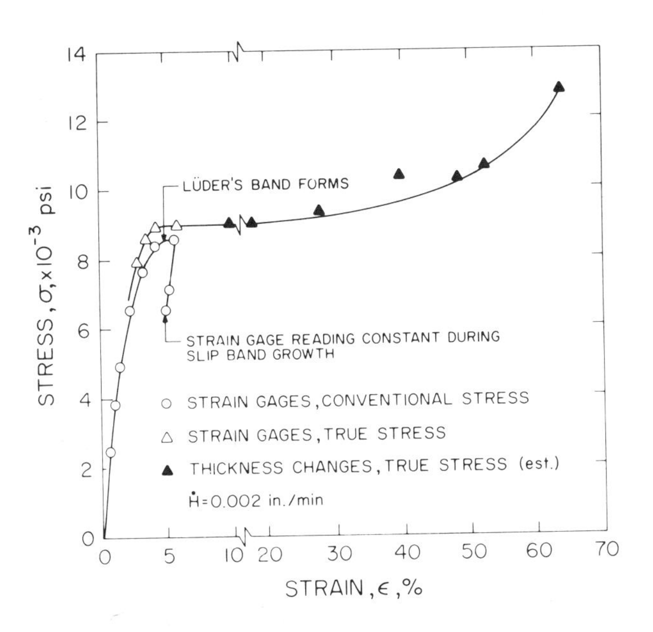

Fig. 2. Stress-strain behavior of polycarbonate.

Conventional stress and true stress versus conventional strain from strain-gage measurements are plotted in Fig. 2. As may be observed, stress and strain increased until a point of tensile instability was reached, at which time, V-shaped Lüder's bands formed in accordance with the observations of Nádai[14] and Hill[15]. These slip bands always formed in the same way for all strain rates and never occurred at the strain-gage locations. Because of the localized yield, both stress and strain decreased nearly elastically at the location of the strain gages until equilibrium was obtained with stresses within the yielded region. After equilibrium was obtained, with a constant growth of slip band occurring in the direction of the load, the strains outside the slip-band region remained constant for the duration of the test or until the slip band impinged upon the strain gage. Figure 1(b) is a photograph of a fully developed slip band during growth.

For this one test, thickness and width measurements were made at the point of localized yield during slip-band initiation and subsequent growth, with the aid of a precision dial indicator and micrometer. From these measurements, strain-gage readings, and estimates of Poisson's ratio prior to localized yield, an estimate was obtained for the stress-strain behavior to actual fracture. As indicated in Fig. 2, while unloading may occur in unyielded regions or in regions loaded only to incipient yield, stresses and strains in yielded regions generally increase until fracture occurs, with significant strain hardening being encountered near fracture.

Figure 3 shows several isochromatic photographs which indicate: a typical pattern at incipient yield [Fig. 3(a)], the unloaded pattern immediately following localized yield [Fig. 3(b)], and the unloaded pattern after subsequent flow [Fig. 3(c)]. These photographs illustrate the above comments on yield initiation and subsequent flow, and also illustrate that by using a highly monochromatic light source, birefringence can be observed up to and beyond the yield point even for specimens as thick as those used (1/8 inch).

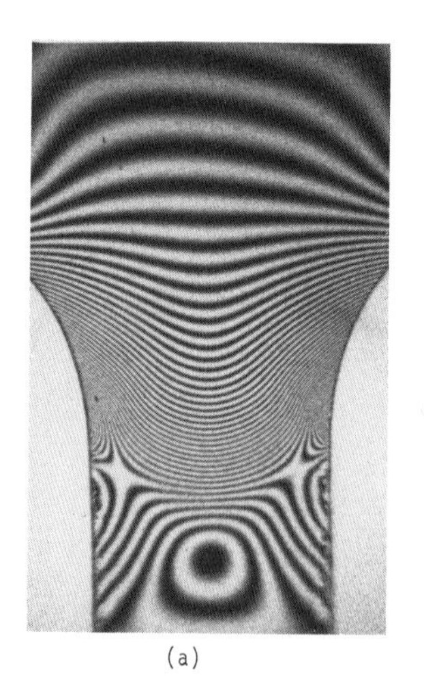
(a)

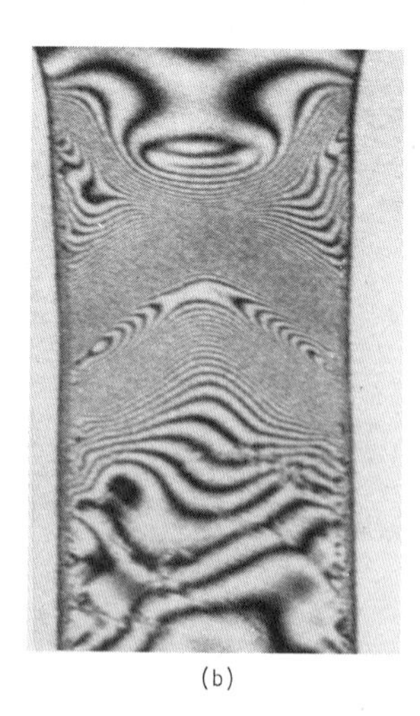
(b)

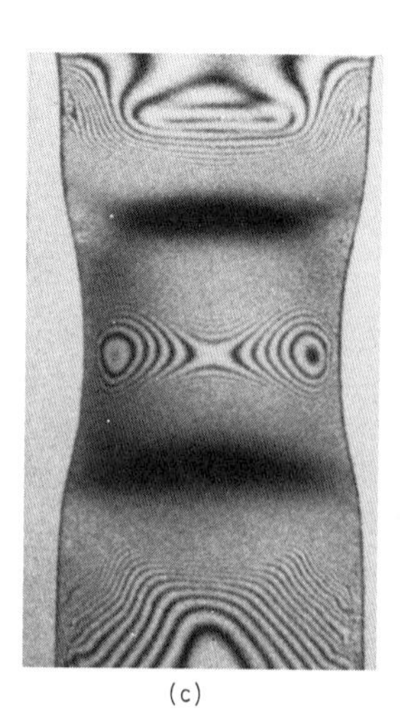
(c)

Fig. 3. Typical isochromatic patterns: (a) at insipient yield, (b) enlargement of localized yield region after unloading, and (c) enlargement of subsequent flow region after unloading.

Birefringence was also recorded during the test for the head rate of 0.002 in./min. Figure 4 shows the results and confirms those reported earlier for polycarbonate.[4-6]

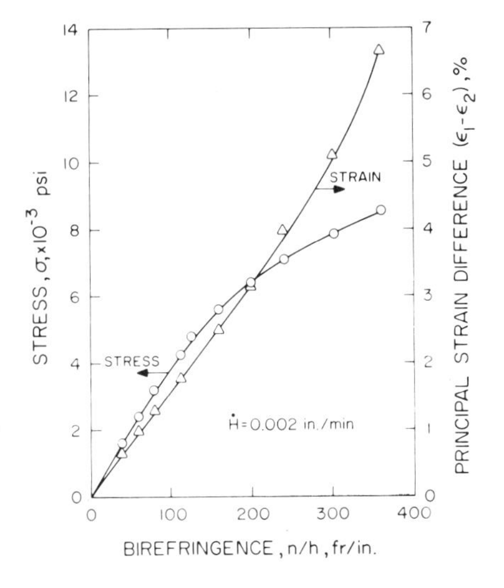

Fig. 4. Stress- and strain-optic behavior of polycarbonate.

Stress-strain data obtained for polycarbonate are shown in Fig. 5. It is apparent from this figure that polycarbonate possesses a region of nearly linear elastic behavior, a region of viscoelastic and/or nonlinear behavior, and a region of plastic behavior. The data show that polycarbonate

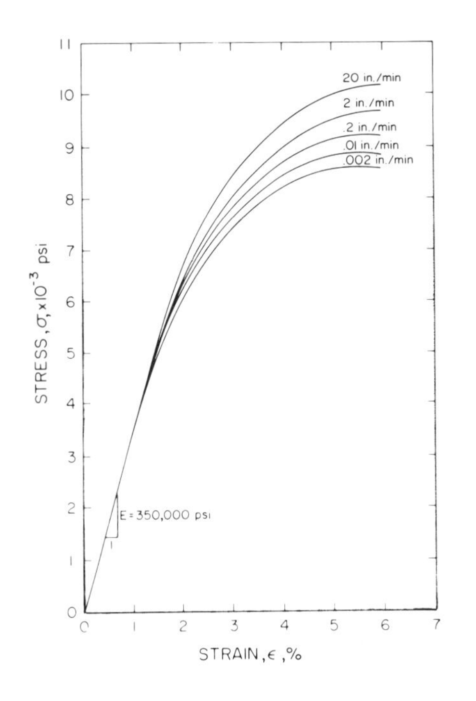

Fig. 5. Stress – strain – strain rate behavior.

possesses a rate-dependent yield point, identified as the point of tensile instability or the formation of Lüder's bands, which occurred in all tests at the point of maximum stress and strain as determined from strain-gage measurements. Figure 5, together with Fig. 2, can be used as an indication of the rate-dependent response of polycarbonate from zero load to fracture.

During these constant-head-rate tests, Poisson's ratios were also determined. In general, Poisson's ratio increased from a value of $\nu \simeq 0.35$ in the elastic range to a value of $\nu \simeq 0.40$ at incipient yield, and this ratio tended to decrease with increasing strain rate. Because the material at the location of the transverse gage never actually yielded, measurements of Poisson's ratio in the yielded region were not possible. However, Poisson's ratios were somewhat less than those reported earlier, $\nu \simeq 0.5$ at or near yield.[4,5] The observation was made that the time of maximum strain significantly lagged behind the time of maximum stress in all tests, further indicating the presence of viscoelastic effects.

The yield stresses for the various rates as defined above are plotted in Fig. 6 against machine head rate, initial elastic strain rate, and initial

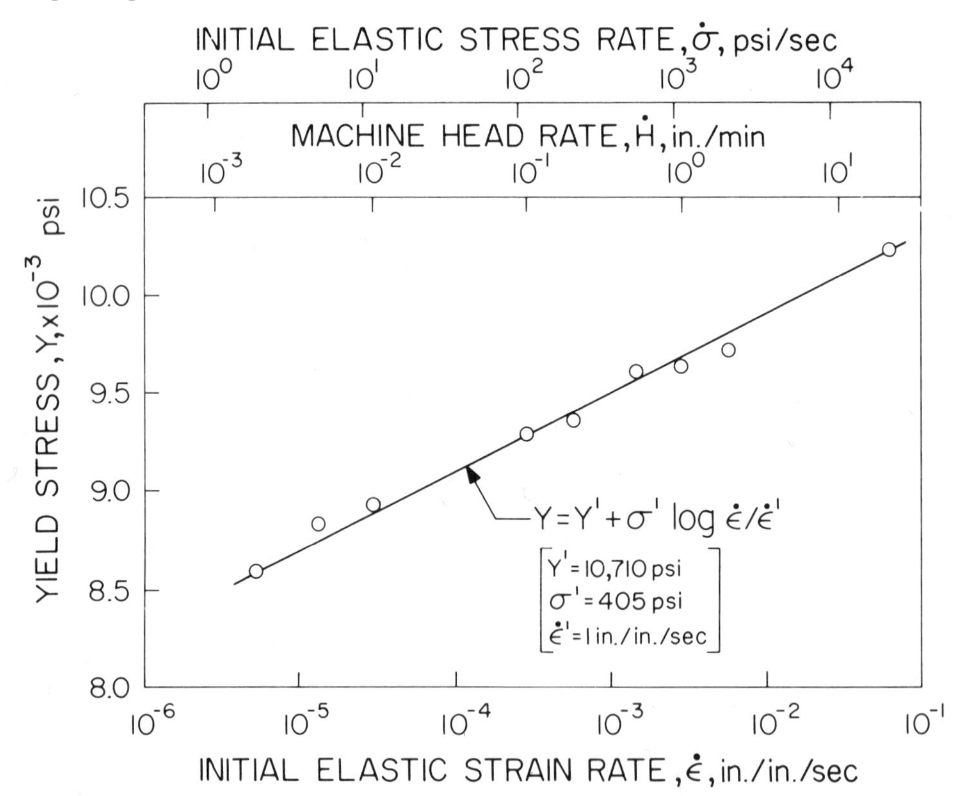

Fig. 6. Rate-dependent variation of yield stress.

elastic stress rate. At this point it should be noted that, in general, stress initially increased at a constant rate and subsequently at a decreasing rate (i.e., at stresses above the elastic limit), whereas strain initially increased at a constant rate, followed by increasing rates and finally by decreasing rates near yield. As a result, it was not possible to accurately predict either stress or strain rates at incipient yield. However, initial strain rates were approximately the same as average strain rates for the duration of each test.

While the yield-stress variation with strain rate is not large (at the rates used), it is extremely predictable. The data are fitted well by an equation proposed for metals by Ludwik (as reported by Thorkildsen[16]) of the form

$$Y = Y' + \sigma' \log \frac{\dot{\epsilon}}{\dot{\epsilon}'} \quad , \tag{1}$$

where Y', σ', and $\dot{\epsilon}'$ are constants for a given material.

To evaluate the nature of the response between the apparent elastic limit and the yield point, shown in Fig. 5, relaxation and creep tests were performed at various stress levels. These results, shown in Figs. 7 and 8, indicate that at low stress levels, little viscoelastic behavior occurred, whereas pronounced short-time viscoelastic behavior occurred at higher stress levels, especially at levels near the yield point of the material.

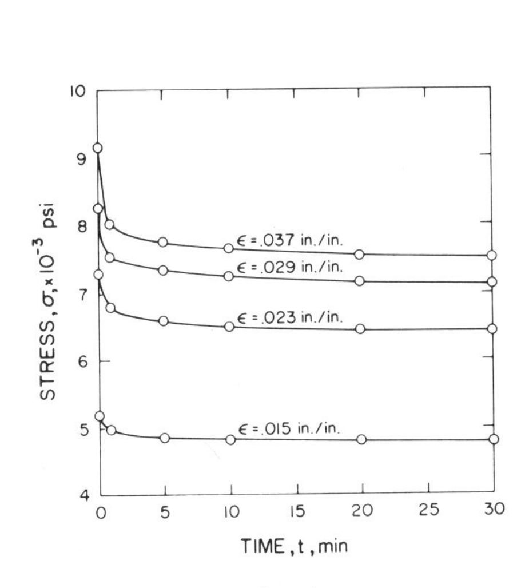

Fig. 7. Stress-relaxation response.

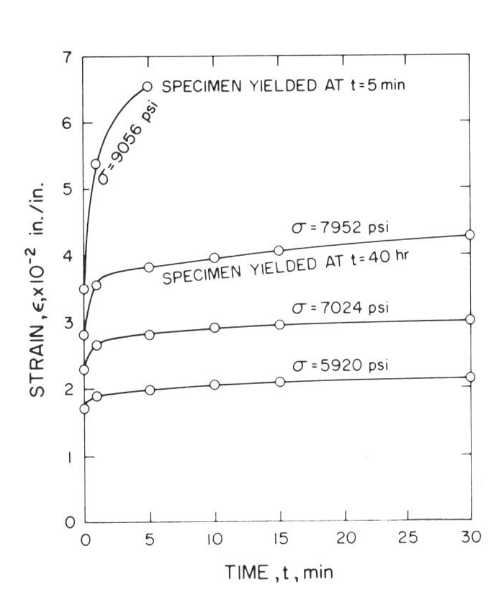

Fig. 8. Creep response.

The relaxation tests were performed at the fastest rate possible (20 in./min). These were not truly relaxation tests because significant amounts of stress relaxation occurred during the loading time. Thus, while the relaxation behavior appears to be nonlinear (Fig. 7), this cannot be ascertained without appropriate analysis.

The creep tests were performed with a pneumatic test machine, and the initial load was applied at the fastest possible rate. This rate, as determined from stress-time and strain-time records, was approximately 20 in./min (the stress rate was approximately 13×10^3 psi/sec). Again, the behavior appears nonlinear (Fig. 8) but, because of short-time behavior during the loading period, this cannot be evaluated without further analysis. An analysis is presented in the next section.

It may be noted in Fig. 8 that at stresses above approximately 8000 psi, a creep-to-yield behavior was observed. The time intervals necessary for creep to yield were observed to decrease rapidly above this level. This leads to the conclusion that the yield process in polycarbonate is not only rate dependent but also time dependent. A rational basis for such behavior is presented in the next section.

In both creep and relaxation tests, attempts were made to photograph the birefringence. However, initial relaxation and retardation times were too short and exposure times too long for photographs of isochromatic time response to be possible.

3 ANALYTICAL CONSIDERATIONS

Obviously, rate-independent instantaneous theories of plasticity[15] are not adequate for quantifying the rate-dependent yield behavior of polycarbonate. Equation 1 can accurately predict the rate-dependent yield phenomena shown in Fig. 6 but cannot predict the time-dependent phenomena illustrated in Figs. 7 and 8.

Efforts to develop a viable theory of viscoelastic-plastic behavior were reported by Nagdi and Murch.[17] These efforts and those of others were summarized by Cristescu.[18] In the theory of Nagdi and Murch, strains in plastic regions are represented by the relation

$$\epsilon_{ij} = \epsilon_{ij}^{V} + \epsilon_{ij}^{P} \quad , \tag{2}$$

where the symbols identify total strain, viscoelastic strain, and plastic strain, respectively. Also, the yield surface is assumed to have the following functional form:

$$f(\sigma_{ij}, \epsilon_{ij}^{P}, \chi_{ij}, \kappa_{ij}) = 0 \quad , \tag{3}$$

where χ_{ij} is a tensor depending on time history and κ_{ij} is a measure of work hardening. The constitutive equations are written as

$$\dot{\epsilon}_{ij}^{P} = \Lambda \frac{\partial f}{\partial \sigma_{ij}}$$

$$e_{ij}^{V} = \int_{-\infty}^{t} J_1(t - \tau) \frac{\partial s_{ij}(\tau)}{\partial \tau} d\tau \tag{4}$$

$$e^{V} = \int_{-\infty}^{t} J_2(t - \tau) \frac{\partial s(\tau)}{\partial \tau} d\tau \quad .$$

In Eq. 4, Λ is a scaler and stress and strain are separated into dilitation (e_{ij}^{V}, s_{ij}) and spherical (e^{V}, s) components. Also, $J_1(t)$ and $J_2(t)$ are shear and bulk creep compliances.

Further, the functional form of χ_{ij} in Eq. 3 is given as $\chi_{ij} = \chi_{ij}(\epsilon_{kl}^{V} - \epsilon_{kl}^{E})$, with ϵ_{ij}^{E} being representative of elastic strains. Nagdi and Murch[17] and Crochet[14] suggest that Eq. 3 be specialized to a Mises yield criterion,

$$f = \frac{1}{2} s_{ij} s_{ij} - Y^2 = 0, \quad Y = Y(\chi, \kappa) \quad . \tag{5}$$

Crochet further proposes that χ have the specific form

$$\chi = \left[\left(\epsilon_{ij}^{V} - \epsilon_{ij}^{E} \right) \left(\epsilon_{ij}^{V} - \epsilon_{ij}^{E} \right) \right]^{1/2} \quad . \tag{6}$$

With this formulation, Crochet shows that one-dimensional stress-strain curves for different rates of loading should appear as is shown schematically in Fig. 9. That is, for instantaneous loading to yield, usual perfectly elastic-plastic behavior should be observed. For slower rates of loading or for instantaneous loading to stresses lower than the instantaneous yield stress, yield stress should decrease exponentially with either rate or time. Thus, Crochet proposes a time-dependent yield phenomenon for one-dimensional behavior to have the form

$$Y(t) = A + B \exp(-C\chi) \quad , \tag{7}$$

where A, B, and C are material parameters. This form appears to be particularly suitable for polycarbonate since the strain-hardening function κ is neglected, and it is therefore representative of the behavior shown in Figs. 2 and 5.

Using arguments similar to those of Crochet[19] and Wnuk and Knauss[20], χ can be written in the following form for creep testing:

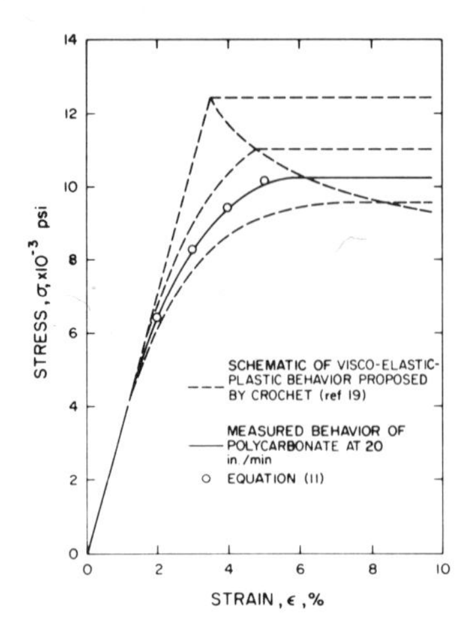

Fig. 9. Comparison of ideal viscoelastic-plastic behavior to that of polycarbonate.

$$\chi = \sigma_0 D(t) \left\{1 - 2[\nu(t)]^2\right\}^{1/2} \quad . \tag{8}$$

The evaluation of Eq. 7 depends on being able to experimentally determine the uniaxial creep compliance, D(t), and Poisson's ratio in creep, $\nu(t)$.

The simplest way of incorporating D(t) and $\nu(t)$ into Eq. 7 is by using mechanical models, as was done by both Crochet and Wnuk, et al. The question then is: what kind of model could be formulated to be representative of the material in question, e.g., polycarbonate? Reiner[21], Alfrey and Gurne[22], and Kachanov[23] have proposed various modifications to the Bingham or Schwedoff model. One particular variation of this model which seems to be appropriate for polycarbonate is shown in Fig. 10. In such a model, stresses would be elastic for $\sigma < \theta$, viscoelastic for $\theta \leqslant \sigma < Y$, and plastic for $\sigma \geqslant Y$. Figures 2, 5, 7, and 8 show that the above conditions are also appropriate for polycarbonate. The model in Fig. 10 and polycarbonate have an elastic limit with subsequent flow proportional to an "overstress" (the stress above the elastic limit), similar to suggestions of Prager.[24] However, the process of yielding and, hence, the nature of the plastic flow region are somewhat different. The rheological equations for the model shown in Fig. 10 can be written as

$$\begin{aligned} \epsilon &= \frac{\sigma}{E}, \quad \sigma < \theta \\ \dot{\epsilon} &= \frac{\dot{\sigma}}{E} + \frac{\sigma - \theta}{\mu}, \quad \theta \leqslant \sigma < Y \ . \end{aligned} \tag{9}$$

For a constant-strain-rate test, i.e., $\epsilon = Rt$ and $\dot{\epsilon} = R$, the solution of Eq. 9 becomes

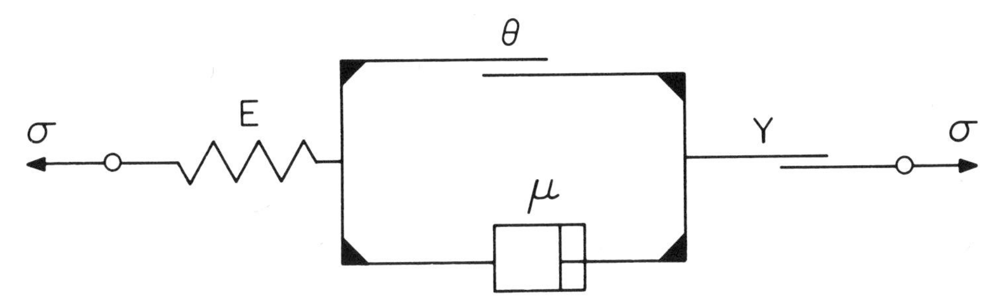

Fig. 10. Modified Bingham model.

$$\sigma(t) = \theta + \mu R \left[1 - e^{-\lambda(t-t_0)}\right] \tag{10}$$

or in terms of strain

$$\sigma(\epsilon) = \theta + \mu R \left[1 - e^{-\lambda(\epsilon-\epsilon_0)/R}\right], \tag{11}$$

where $\lambda = \frac{E}{\mu}$ is the inverse of the relaxation time and t_0 and ϵ_0 are the time and strain at the elastic limit $\sigma = \theta$. The constants were determined to be θ = 4000 psi, E = 350,000 psi, and λ = 3.32 sec^{-1} from the experimental data given in Fig. 4 for a head rate of $\dot{H}$ = 20 in./min.

Both the experimental data for $\dot{H}$ = 20 in./min and the values of stress and strain from Eq. 11 have been superimposed on the schematic of the behavior proposed by Crochet in Fig. 9. Obviously, the model shown in Fig. 10 and its equivalent, Eq. 11, provide an excellent representation of observed behavior for this particular strain rate. Other rate-dependent curves could be fitted by altering the constants of Eq. 11. However, to obtain a single equation representative of all strain data would require the use of multiple elements to provide for a spectrum of relaxation times.

Equation 9 can now be solved for the condition of creep to provide the functions necessary for the solution of Eq. 7. Such a procedure results in the following equation for creep-to-yield behavior:

$$Y(t) = A + B \exp\left\{-C\left[\frac{t}{\mu}\left[Y(t) - \theta\right] + \frac{Y(t)}{E}\right]\left[1 - 2\nu^2(t)\right]^{1/2}\right\}. \tag{12}$$

Using the constants previously determined from the rate-dependent behavior of $\dot{H}$ = 20 in./min and experimental data for creep-to-yield times, and assuming Poisson's ratio to be ν = 0.4 = constant (which is a reasonable assumption based on the experimental evidence previously discussed), the constants A, B, and C were determined to be 8000 psi, 2250 psi, and 0.0771, respectively. With these results, Eq. 12 has been plotted in Fig. 11, together with the observed experimental data. For short times, $t < 10$ min, agreement is excellent. Again, to predict both short- and long-time behavior, a distribution of relaxation times would be necessary.

Whitfield and Smith[6] determined a yield surface for polycarbonate, using a biaxial testing program at a single strain rate, and found that a Mises yield condition was a reasonable representation of their results. Combining the present uniaxial results with those of Whitfield and Smith infers that polycarbonate can be represented quite well with the viscoelastic-plastic theory of Nagdi and Murch. These comments are represented by the illustration in Fig. 12, which shows the yield surface to be a function of stress (including pressure), time, and rate of loading.

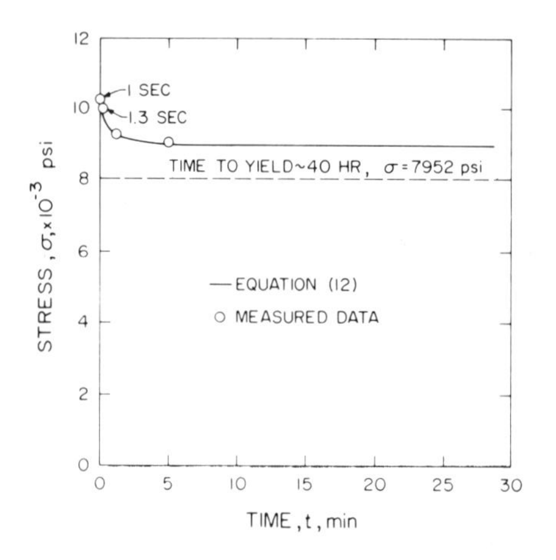

Fig. 11. Comparison of creep-to-yield data and theory.

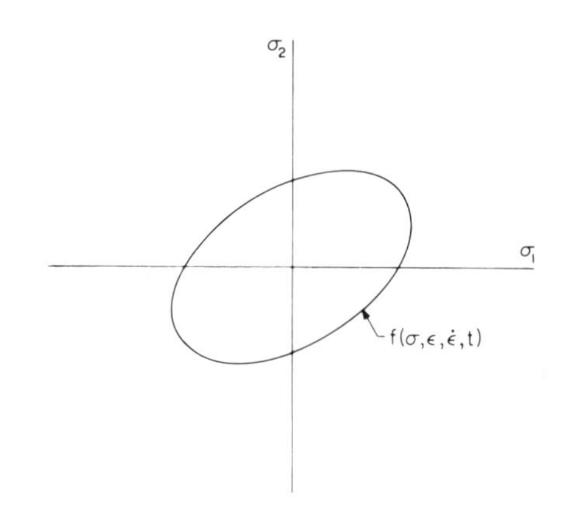

Fig. 12. Yield locus for polycarbonate.

4 DISCUSSION

It was demonstrated that yielding in polycarbonate is both time and rate dependent. Rate-dependent variation of yielding was shown to be in good agreement with a model proposed for metals by Ludwik, whereas time-dependent variation of yielding was shown to be representable by viscoelastic-plastic theories due to Nagdi and Murch and to Crochet. In the latter case, a simple modification of a Bingham model was shown to give a reasonable representation of experimental data for stress-strain rate behavior for a head rate of 20 in./min. The model was fitted to that particular rate since it was the rate used in establishing creep-to-yield times. The model could be fitted to other individual rates and could be used for all rates simultaneously, provided that the model were altered by the inclusion of more Maxwell elements to give a spectrum of relaxation times.

Perhaps it should be mentioned that the mechanical model used is not the only one which could be used, nor does it adequately represent all experimental observations. For example, it was observed during the experimental program that nearly all viscoelastic deformation which occurred during either the constant-head-rate tests or creep tests was recovered with time after unloading if stresses were kept below the yield stress of the material. Such behavior would imply that a mechanical model of the solid type would be more appropriate for polycarbonate. An attempt was made to use a model of the solid type to fit the constant-strain-rate behavior, but the representation was poor. Further, the modified Bingham model used predicts a linear creep behavior contrary to that shown in Fig. 8.

The behavior described herein represents behavior prior to, and including, initial yield. It would, of course, be valuable to ascertain creep or relaxation behavior within the yielded region to assess the nature of subsequent yield. Two relaxation tests were conducted at, or slightly after, the occurrence of localized yield. These results were not presented because in the present study there was no way to separate the effects of unloading due to the reduction of area in slip-band formation (as shown in Fig. 2) and unloading due to stress relaxation (as shown in Fig. 7). Separation of these two effects would be essential to the study of the time behavior of subsequent yield.

With the interpretation of the behavior of polycarbonate depicted herein, it is anticipated that new efforts can be made to quantify the photo-viscoelastic-plastic response of polycarbonate and to use the results in the solution of meaningful boundary-value problems. In so doing it will be necessary to decrease photographic exposure times to the point that birefringence-time effects can be recorded accurately. Modification of the laser-light-source intensity might be possible in order to achieve this result. Mechanical measurements and birefringence measurements of polycarbonate could be used to study the fracture mechanisms of ductile polymers similar to the efforts of Wnuk and Knauss.[20] Efforts of this type would be valuable in assessing the nature of rate-dependent fracture mechanisms which might also be applicable to ductile metals. For example, it has already been ascertained that strip-type-yielding models are equally suitable for polycarbonate and mild steel.[9-11] Since the rate-dependent yield behavior described in the preceding pages is not unlike that reported for metals by Kendall[25], it is expected that rate-dependent-fracture models that could be developed for polycarbonate would have application to metals as well.

In conclusion, it is noted that the behavior described has been limited to uniaxial constant-temperature and constant-humidity testing. Future programs should certainly include temperature effects, biaxial or triaxial effects, bulk-viscoelastic-property effects, etc. It is unlikely that a single experimental study could delineate the response of polycarbonate or other ductile polymers to all possible variables. However, over a period of time, it is hoped that rational mathematical models of the yielding process can be established to accurately predict pressure, rate, time, temperature, and multiaxial initial and subsequent yield processes in ductile or quasi-ductile polymers.

ACKNOWLEDGMENTS

The financial support provided by the Department of Defense, U.S. Army Contract No. DAA-F07-69-C-0444, through Watervliet Arsenal, Watervliet, New York, is gratefully acknowledged. Further appreciation is extended for the many helpful suggestions supplied by colleagues at both Virginia Polytechnic Institute and State University and Watervliet Arsenal and for the technical assistance of V. K. Wright and G. K. McCauly of VPI&SU.

REFERENCES

1. Drucker, D. C., in *Handbook of Experimental Stress Analysis,* M. Hetényi (Ed.), John Wiley & Sons, New York (1950), p 933.
2. Hetényi, M., in *Structural Mechanics,* N. J. Hoff (Ed.), Pergamon Press, New York (1960), p 483.
3. Hetényi, M., *Proceedings of the First National Congress of Applied Mechanics,* 1952, p 499.
4. Brill, W. A., "Basic Studies in Photoplasticity", Ph.D. Thesis, Stanford University, 1965.
5. Gurtman, G. A., Jenkins, W. C., and Tung, T. K., *Characterization of a Birefringent Material for Use in Photoelasto-Plasticity,* Douglas Report SM 4T196 (1965).
6. Whitfield, J. K., and Smith, C. W., *Experimental Mechanics,* **12**, 72 (1972).
7. Dally, J. W., and Mulc, A., "Polycarbonate as a Material for Three-Dimensional Photoplasticity" (to be published in *J. Appl. Mech.*).
8. Brinson, H. F., *Experimental Mechanics,* **11**, 467 (1971).
9. Brinson, H. F., *Experimental Mechanics,* **10**, 72 (1970).
10. Brinson, H. F., and Gonzalez, H., *Experimental Mechanics,* **12**, 130 (1972).
11. Brinson, H. F., Underwood, J. H., and Rosenfield, A. R., VPI-E-72-7, April, 1972 (to be published, *Int. J. Frac. Mech.*).
12. Sauer, J. A., Mears, D. R., and Pu, K. D., *European Polymer J.,* **6**, 1015 (1970).
13. Muller, R. K., "The Influence of Measuring Current and Preheating in Measurements with Strain Gages on Resins", presented at Second International Congress on Experimental Mechanics, Washington, D.C., 1965.
14. Nádai, A., *Theory of Flow and Fracture of Solids,* McGraw-Hill, New York, (1950), Vol 1, p 316.
15. Hill, R., *Plasticity,* Oxford University Press, London (1950), p 324.
16. Thorkildsen, R. L., in *Engineering Design for Plastics,* E. Baer (Ed.), Reinholt Book Co., New York (1964), p 295.
17. Nagdi, P. M., and Murch, S. A., *J. Appl. Mech.,* **30**, 321 (1963).
18. Cristescu, N., *Dynamic Plasticity,* North-Holland Publishing Company, Amsterdam (1967), p 559-579.
19. Crochet, M. J., *J. Appl. Mech.,* **33**, 327 (1966).

20. Wnuk, M. P., and Knauss, W. G., *Int. J. Solids Struc.,* **6**, 995 (1970).
21. Reiner, M., *Advanced Rheology,* H. K. Lewis and Co. Ltd., London (1971), p 208.
22. Alfrey, T., and Gurne, E. F., *Organic Polymers,* Prentice-Hall, Inc., New Jersey (1967), p 80.
23. Kachanov, L. M., *Foundations of the Theory of Plasticity,* North-Holland Publishing Co., Amsterdam (1971), p 459.
24. Prager, W., *Introduction to Mechanics of Continua,* Ginn and Co., New York (1961), p 136.
25. Kendall, D. P., *J. Basic Engr.,* **94**, 207 (1972).

DISCUSSION on Paper by H. F. Brinson

KAUSCH: It might be mentioned that polycarbonate is – to my knowledge – the only one of the amorphous thermoplastics in which free radicals are formed in the necking process [cf. Zaks, Yu. B., Lebedinskaya, M. L., and Chalidze, V. N., *Polymer Sci. USSR,* 12, 3025 (1971)].

ONOGI: I would like to make just short comments on Dr. Brinson's paper. We are measuring infrared dichroism and birefringence during stress-strain and stress-relaxation measurements. These rheo-optical techniques serve as powerful methods to correlate stress-strain behavior with structural change, particularly of amorphous and crystal orientations, in high polymers. [See, for example, *J. Polymer Sci.,* Part C, **16**, 1445 (1967); *Zairyo,* **14**, 322 (1965); ibid., **16**, 746 (1967)].

VINCENT: Biaxially drawn poly(ethylene terephthalate) can have certain advantages as a photoplastic material. Because of the absence of inhomogeneous plastic deformation, high strains beyond yield are readily measurable. Constant-strain and strain-recovery experiments demonstrate that the relation between stress and birefringence is nonlinear and multivalued but that the relation between strain and birefringence is linear and single valued. In addition, varying the drawing and heat-treatment conditions, and the test direction, can provide materials with a wide range of different shapes of stress-strain curves.

BRINSON: Your suggestion for the use of biaxially drawn poly(ethylene terephthalate) is a good one and should be followed up by someone to ascertain its suitability for photo-viscoelastic-plastic investigations. This material might be particularly applicable to orthotropic or anisotropic birefringence analyses which are currently being developed by photoelasticians. However, the mere fact that the strain-optic response is linear would not necessarily make it a suitable photoplastic material, especially if time effects are important. It has been shown by many people that stress- or strain-optic laws which are correctly formulated for viscoelastic materials are extremely complicated. [For references, see Brinson, H. F., *Experimental Mechanics* (December, 1968)].

KOBAYASHI: You mention birefringence measurement up to and including the yield point. Have you characterized the optical response of polycarbonate in the viscoplastic region with changing principal directions of the strain tensor?

BRINSON: No, I have not attempted to relate the principal axis of birefringence with either the principal axis of stress or of strain. Others have attempted to relate the isoclinic directions with the principal axis of stress or strain in polycarbonate (see for example Refs. 4 through 6 of my paper). However, this was done assuming that time effects were negligible, and the results, in my opinion, are inconclusive. The general opinion seems to be that the principal axis of birefringence is probably more closely related to that of strain rather than stress. I was able to show, at least to my satisfaction, that isochromatics tend to conform to lines of constant thickness change in yielded regions (Ref. 8). This fact would tend to support the strain-optic view.

In the present effort I have attempted to show that the yield process in polycarbonate is rate- and time-dependent, which indicates that the optic laws may be viscoelastic-plastic in nature rather than elastic-plastic. As you know, the optic law for a viscoelastic material is quite complicated and must be written as an integral equation involving both isochromatics and isoclinic parameters [for references see Brinson, H. F., *Experimental Mechanics* (December, 1968)]. Because of these facts, I feel additional work should be performed to establish simple mechanical-optical relationships which would include time- and rate-dependent effects.

ALFREY: Glassy polystyrene, in the elastic regime, exhibits a positive stress-optical coefficient. On the other hand, plastic and viscoelastic extension of polystyrene is characterized by a negative coefficient. In the case of polycarbonate, I would not expect a sign reversal, but I would expect a different value of the mechanical-optical coefficient for elastic deformation and plastic deformation. If true, this would complicate any quantitative elastic-plastic-viscoelastic optical analysis.

BRINSON: As far as I know, there is no sign reversal of birefringence response in polycarbonate. However, you are quite correct in expecting different stress- or strain-optic coefficients in the elastic and plastic regions (see Refs. 4 and 8 of my paper). This certainly would present a complication, and I have attempted to show today that an additional complication may be time- and rate-dependent yield effects.

KNAPPE: You find at the beginning of the stress-strain diagram ideal elastic behavior, which is not dependent on strain rate. Later on, viscoelasticity will arise (see Fig. 2 of the paper). Can this shift be attributed to structural changes? In polycarbonate we must suppose some short-range order which will perhaps be dissolved by a certain strain.

BRINSON: You may be quite correct. I suspect that the polymer chemist and polymer physicists present could answer this question on micromolecular structure better than I, and I, for one, would be very interested in learning more about this aspect of polycarbonate.

HASSELL: Dr. Brinson, you have mentioned and emphasized the influence of strain rate, time, temperature, and strain on the deformation properties of polycarbonate. The influence you have not mentioned is that of frozen stress due to the method of sample preparation. Have you determined the frozen orientation of your samples?

As Dr. Vincent pointed out, orientation can have a large effect on the stress-strain behavior. Would you not expect similar effects? And how does the frozen orientation affect the three regions you have observed, namely, the elastic, viscoelastic, and plastic or yield regions.

BRINSON: I believe you are referring to the residual stress-strain birefringence and anisotropy of the material prior to loading rather than the standard load-induced "frozen stress" technique of photoelasticity. We know that there is some residual birefringence and anisotropy in the material as received from commercial sources. This birefringence was in the order of one half of a fringe, but because we have fringe orders of about 30 to 55 in the viscoelastic-plastic region, we feel the residual effects to be quite small. We do subtract these residual fringe effects from our measurements, as is obvious from an examination of Fig. 4.

To avoid orientation effects, we cut all specimens from the same direction in the sheet and we used only a single sheet of material to avoid reducibility from batch to batch.

We did not measure anisotropic properties in the present study, but we have done this in the past with thinner sheets (see Ref. 9). For the thinner sheets, E_x/E_y was found to be about 1.1. A 10 percent orthotropic effect has not been found to be serious in connection with our previous ductile fracture work.

MENGES: How did you attach the strain gages to the testing bars? We and other researchers used glues and found crazing and cracking under the strain gages, influenced by glue diffusion.

When you did not find such faults, maybe you had a reinforcing effect due to the wires of the strain gages. It is very surprising that you had no time dependence in the elastic range of stress-strain curves, which contradicts many results of research on the long-time, stress-strain behavior of polycarbonate.

BRINSON: The foil gages used were attached with a strain-gage cement called "Eastman 910". We found no evidence that the glue caused crazing or cracking of the polycarbonate beneath the strain gage. However, we inspected the gages and the specimens only with the unaided eye. Microscopic examination might reveal some evidence of the damage you speak of, but I don't believe that this was a serious factor in the interpretation of our results.

There is undoubtedly a reinforcing effect of the strain gages. This is discussed in the text of the paper, but for the thickness of material used, I do not believe a substantial effect existed. Our measurements of strain are in agreement with the measurements of others using grid or extensometer techniques. Also, yield stresses were obviously unaffected by the strain gages or the glue because yielding always occurred at a location different from that of the strain-gage location.

I did not mean to imply that there were no time effects at low stress and strain levels. I meant only that they were small, and that within the accuracy of our measurements, polycarbonate could be considered as essentially elastic below a stress of about 4000 psi and for relatively short times, i.e., 1 to 3 hours.

THE EFFECT OF DIFFERENT BEHAVIOR IN TENSION THAN IN COMPRESSION ON THE MECHANICAL RESPONSE OF POLYMERIC MATERIALS

E. F. Rybicki and M. F. Kanninen

BATTELLE
Columbus Laboratories
Columbus, Ohio

ABSTRACT

High polymers exhibit different stress-strain behavior in tension than in compression. The effect of this characteristic is examined in terms of three-point bending tests on polyethylene and polypropylene. A model to predict the experimentally obtained force-deflection curves is presented and used to illustrate the importance of including both tension and compression stress-strain data when these differ.

1 INTRODUCTION

High polymers exhibit many macroscopic characteristics that complicate the analysis of their deformation and fracture. This paper is concerned with one of these: behavior different in tension than in compression. Different behavior in tension than in compression, the property that the stress-strain curve obtained by elongation is not the same as that obtained by compression, can be described as stress-sign dependent (SSD). Example data for polypropylene are shown in Fig. 1.

While the most likely source of SSD behavior is in oriented materials, Ward[1] reports that SSD behavior occurs in isotropic polymers as well as in oriented polymers. For example, Fig. 2, which is based on data obtained by Rabinowitz, shows that the yield stress for drawn poly(ethylene terephthalate) is different in tension than in compression. The data obtained by Argon et al.[2] show SSD behavior for polystyrene in biaxial states of stress (see Fig. 3). In addition, Ward[1] presents yield-stress data for isotropic polyethylene (obtained by Duckett) in which SSD behavior appears over a range of strain rates.

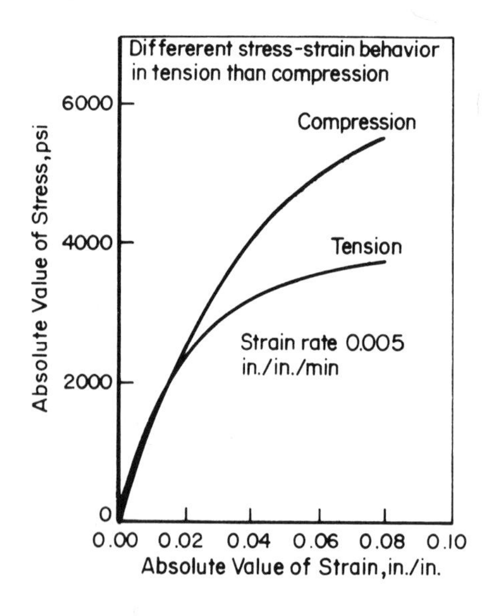

Fig. 1. Stress-strain curves for polypropylene.

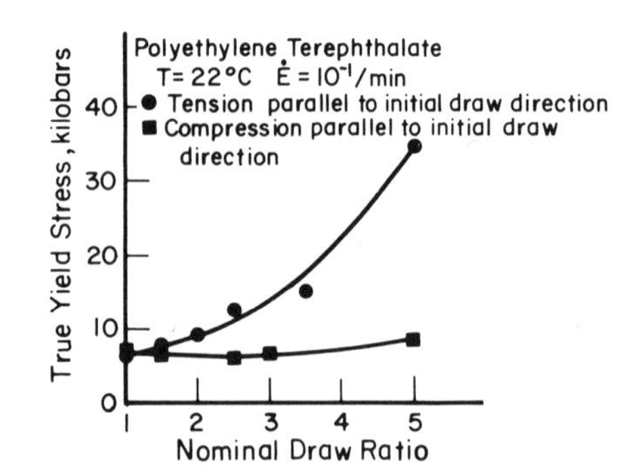

Fig. 2. Different yielding behavior in tension than in compression. Graph of yield stress in tension and compression vs. nominal draw ratio (Data obtained by Dr. S. Rabinowitz). Fig. 29 from Ward, I. M., Molecular Order-Molecular Motion: Their Response to Microscopic Stresses, edited by H. H. Kausch, Interscience, New York (1971), p 195.

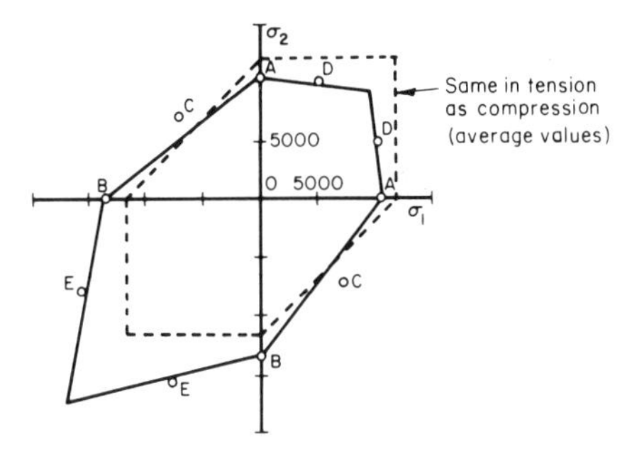

Fig. 3. Different 'yielding' behavior in tension than in compression. Yield locus of polystyrene showing the pressure dependence of yielding. Figure 4a from Ward, I. M., Molecular Order-Molecular Motion: Their Response to Microscopic Stresses, edited by H. H. Kausch, Interscience, New York (1971), p 195.

In most load-carrying applications, a uniform state of stress is not expected. Consequently, an analysis that can take into account the effect of different mechanical behavior in tension than in compression may be necessary, both for the analysis of laboratory experiments with nonuniform stress (e.g., fracture-test specimens) and for structural applications of polymers and other materials that exhibit such behavior. With this in mind, the objectives of this investigation are: (1) to present a continuum mechanics analysis that can take account of SSD material behavior, and (2) to investigate the significance of SSD material behavior in terms of the overall mechanical response of structural components to applied loads.

2 MODEL FOR THREE-POINT BEND TESTS OF A MATERIAL WITH SSD BEHAVIOR

The model presented in this paper is an extension of work reported previously.[3,4] It might be noted that the models described in Refs. 3 and 4 successfully demonstrated a capability to predict experimental results.

The three-point bending model incorporating SSD behavior is based on assumed deformations that are consistent with beam theory. That is, while the variation of strain along the length of the beam is unrestricted, the variation through the depth is linear. For these conditions, the three-point bending test is a statically determinant configuration independent of the type of stress-strain behavior. The bending moment has a maximum value at midspan and decreases linearly to zero at both ends. The net force in the length direction is zero everywhere. Thus, there are equilibrium equations on the bending moment and the force to be satisfied along the length of the beam.

Let x be the coordinate along the length of the beam and z be the coordinate through the depth of the beam. Consider a section of the beam at x where the bending moment is M(x) and the net force along the length is zero. The two equilibrium equations can be expressed as

$$F = 0 = \int_{-h}^{h} \sigma_x(x,z)\,dz \tag{1}$$

and

$$M = M(x) = \int_{-h}^{h} \sigma_x(x,z)\,z dz \quad , \tag{2}$$

where $\sigma_x(x,z)$ denotes the stress distribution through the depth of the beam at x and the depth of the beam is 2h. The assumed strain in the x-direction has the form

$$\epsilon_x(x,z) = A(x) + B(x)z \quad , \tag{3}$$

where A(x) and B(x) are unknown and dependent only on x. The stress-strain relation can be represented by the dependence

$$\sigma_x = \sigma_x(\epsilon_x) \quad . \tag{4}$$

The procedure for evaluating A(x) and B(x) is to substitute Eq. 4 into Eq. 3, thus obtaining an expression for σ_x as a function of A(x) and B(x). This relationship can be denoted by

$$\sigma_x = \sigma_x[A(x), B(x)] \quad . \tag{5}$$

Substituting Eq. 5 into Eqs. 1 and 2 gives two equations in terms of the two unknowns A(x) and B(x). The resulting equations will, in general, be nonlinear in A(x) and B(x). Notice that the different stress-strain behavior in tension than in compression will, in general, produce a neutral axis that is not at the center of the beam, i.e., at $z = 0$. For such cases, A(x) is not zero.

After the values for A(x) and B(x) have been determined at a set of points along the length, the deflection w(x) of the beam can be obtained. An expression involving w(x) is obtained by assuming that the shear strain γ_{xz} is zero. This corresponds to neglecting shear deformations. The result of this assumption is

$$\frac{dw(x)}{dx} = -\frac{\partial u(x,z)}{\partial z} \quad . \tag{6}$$

The expression for u(x,z) comes from integrating the following strain-displacement equation:

$$\frac{\partial u(x,z)}{\partial x} = \epsilon_x(x,z) = A(x) + B(x)z \quad . \tag{7}$$

The result is

$$u(x,z) = \bar{A}(x) + \bar{B}(x)z \quad , \tag{8}$$

where $\bar{A}$ and $\bar{B}$ are integrals of A and B. Note that the integration constant is zero because the origin of the coordinate system is at the center of the beam. Substituting Eq. 8 into Eq. 6 and integrating gives

$$w(x) = -\int_o^x \bar{B}(\eta)d\eta + w_o \quad , \tag{9}$$

where w_o is the center deflection (at x = 0). If the length of the beam is L, then w_o is determined from the zero-deflection boundary condition

$$w\left(\frac{L}{2}\right) = \int_o^{L/2} \bar{B}(\eta)d\eta + w_o = 0 \quad . \tag{10}$$

In the analysis, loadings for the three-point bending tests were given and a center deflection, w_o, was predicted using Eq. 10. All integrations in the x-direction were carried out using numerical integration techniques.

Having presented a general description of the analysis, some of the details for representing the stress-strain curves and solving for A(x) and B(x) are now given. The stress-strain curves were represented by a series of straight-line segments. Up to 15 were used in these computations. Concerning the solution for A(x) and B(x), recall that substituting Eq. 5 into Eqs. 1 and 2 gave two simultaneous nonlinear algebraic equations in A(x) and B(x). Solving these equations directly was avoided by noting that values for A and B need not be computed at a predetermined set of points. The procedure was to first assign a value to B. The corresponding value of A was determined by satisfying Eqs. 5 and 1, which is equivalent to solving a single nonlinear algebraic equation. This was accomplished by an iterative technique using the Newton-Raphson method[5] to start the iteration and the Regula Falsi method[5] when applicable. With values for A and B, the moment was computed using Eqs. 5 and 2. Knowing the moment, the value of x was determined from the moment distribution along the beam. The procedure was repeated with a new value of B until values of A(x) and B(x) were obtained for a set of points along the beam. Several sets of points were considered to be certain that the numerical integrations had converged.

3 COMPARISON OF PREDICTED AND EXPERIMENTAL RESULTS

Test specimens of polypropylene and polyethylene were used to obtain tension and compression stress-strain curves. The curves are shown in Figs. 4 and 5. Next, several beams were placed in three-point bending tests to obtain curves of applied force versus center deflection. The experimental data were found to be quite reproducible in that data from distinct tests fell on the same curve. Small strain rates were used to reduce strain-rate-dependent effects.

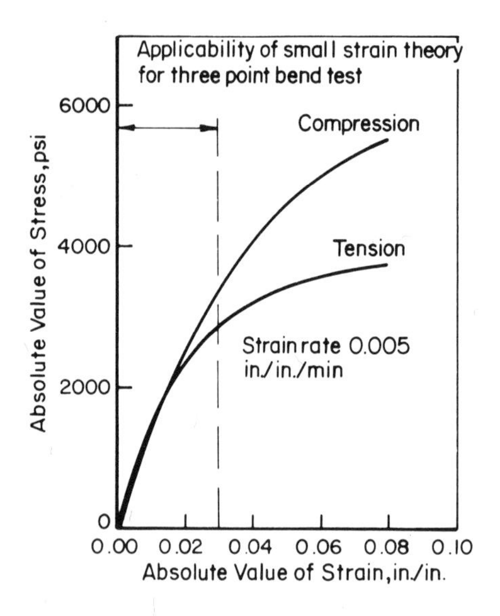

Fig. 4. Stress-strain curves for polypropylene.

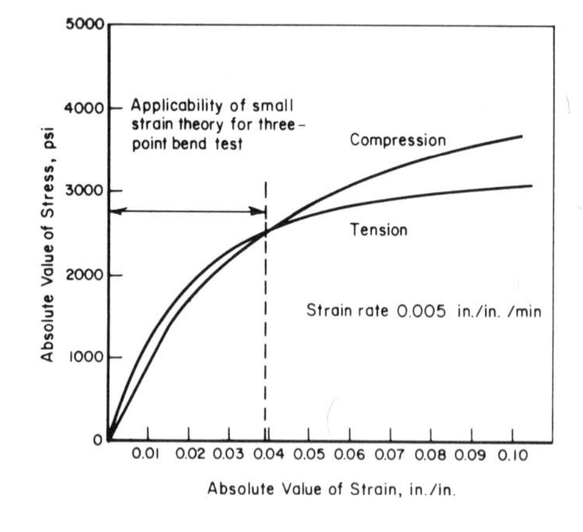

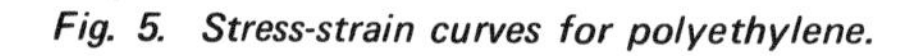

Fig. 5. Stress-strain curves for polyethylene.

For comparative purposes, the curves were predicted using the stress-strain data in the following three ways:

(1) Assuming same behavior in tension as in compression – using tension stress-strain data

(2) Assuming same behavior in tension as compression – using compression stress-strain data
(3) Assuming different behavior in tension than in compression – using both tension and compression stress-strain data.

The results of the three calculations and the experimental curves are shown in Figs. 6 and 7.

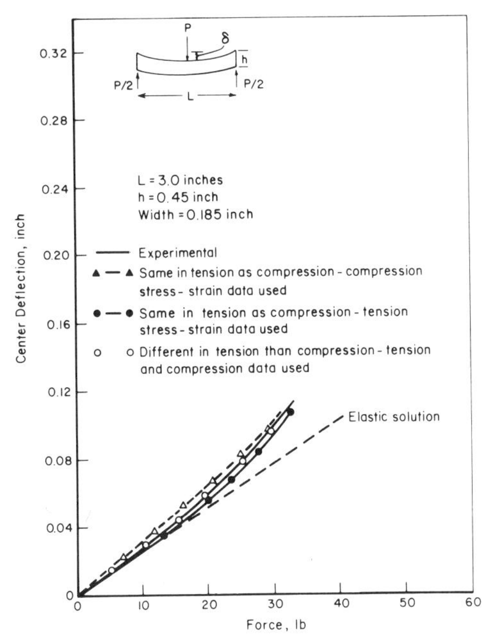

Fig. 6. Effect of different behavior in tension than in compression for polypropylene in a three-point bending test.

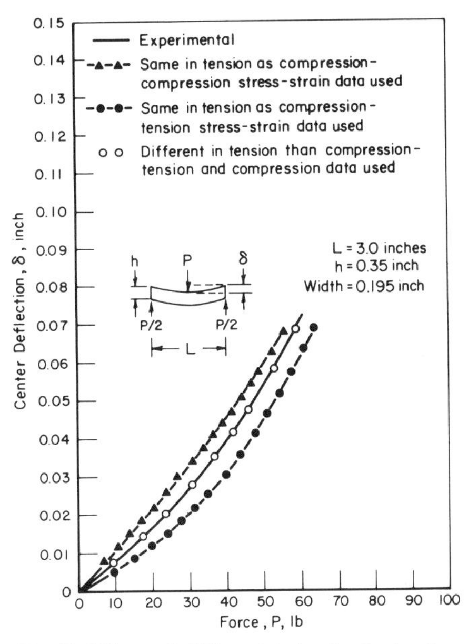

Fig. 7. Effect of different behavior in tension than in compression for polyethylene in a three-point bending test.

While the force-deflection response is a convenient way of evaluating the analysis, a more important computational result is the amount of work required to obtain a given displacement of the beam. This is of crucial importance, for example, in fracture mechanics, where the strain-energy-release-rate approach plays a dominant role. The effect of SSD behavior on the amount of energy to deform the beam, for each of the three cases given above, is presented in Table I. It can be seen from these results that the predicted energy can vary as much as ±35 percent, depending on the type of stress-strain behavior assumed.

It is pointed out that the model is not applicable to cases where large deflections are involved, because equilibrium conditions were based on the undeformed beam configuration and large deflections distort this geometry. The allowable limits for deflections were set so that the ratio of center deflection to beam depth was less than one sixth. The range of strains corresponding to this deflection limit is shown in Figs. 4 and 5.

Table I. Comparison of Energy Values Obtained From Force-Deflection Curves for Three-Point Bending of Polyethylene

Center Deflection, inch	E_D = Energy From Data Curve[a], lb-in.	$E_C^{(b)}/E_D$, %	$E_T^{(c)}/E_D$, %
0.01	0.0625	73.6	144.2
0.02	0.2425	70.4	134.4
0.03	0.5240	65.8	127.8
0.04	0.8890	73.2	123.8
0.05	1.3275	79.0	120.4
0.06	1.8330	82.6	117.8

(a) Energy = $\int_0^\delta$ force X D (deflection).

(b) Energy calculated assuming same behavior in tension as in compression — compression stress-strain data used.

(c) Energy calculated assuming same behavior in tension as in compression — tension stress-strain data used.

As a further application of the theory, a material that is markedly different in tension than in compression can be considered. Ward[1] has shown that the yield stresses in tension and compression can be made to vary by drawing the material. For illustrative purposes, a cantilever beam of such a highly drawn material loaded by a bending moment is considered (Fig. 8). The stress-strain behavior is taken to be piecewise linear but different in tension than in compression (Fig. 8b). The influence of various assumed ratios of the tension yield stress and compression yield stress on the end deflection of the beam can then be computed using the technique described above.

For example, the results presented in Fig. 8c show that for an applied bending moment of 30 in-lb, a deflection of 0.475 inch would be obtained if the behavior in tension and compression were the same. In contrast, a deflection of about 1.625 inches would be obtained if, as suggested by Ward[1], $\sigma_Y^T = 4\sigma_Y^c$. These values differ by a factor of 3.4, which indicates that the assumption that the tension-compression yield stresses are the same can be grossly in error.

It can also be seen that the geometry of the beam cross section can affect the overall response for a material with SSD behavior. For example, bending a T-shaped cross section could require only a small part of the tension stress-strain curve and a much larger part of the compression curve. Thus, the bending-deflection curve for such a beam would not only

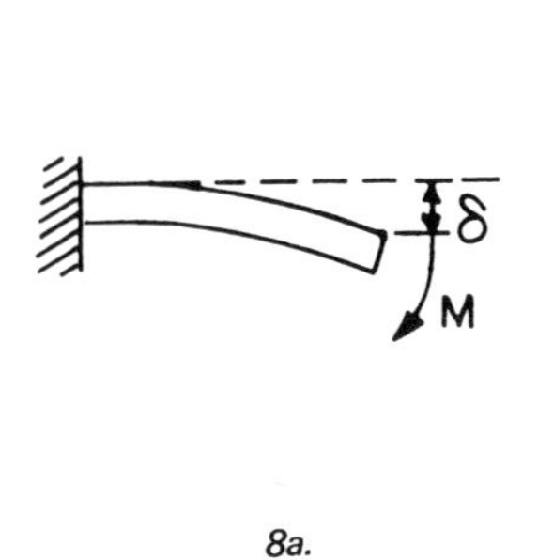

8a.

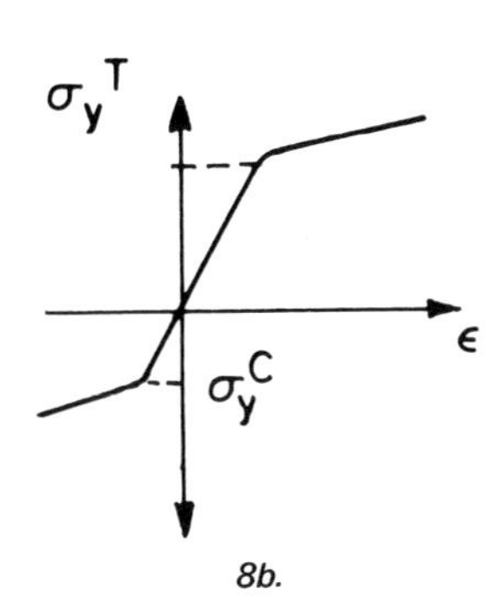

8b.

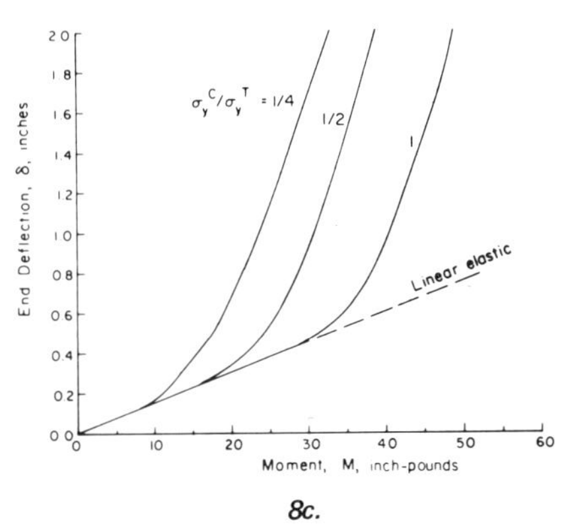

8c.

Fig. 8. Cantilever beam with bending moment (a); stress-strain curve different in tension than compression (b); moment deflection curves (c).

be dependent on the nature of SSD material behavior, but also on the sign of the bending moment.

4 CONCLUSIONS

Three main conclusions can be drawn from the work described in this paper. First, the continuum analysis gives good agreement with experiments. Second, even for the least sensitive material examined (polyethylene), the SSD behavior can significantly influence the overall response. Third, it is not sufficient to examine only the force-deflection response, as this may be a poor indication of the strain-energy content.

Finally, the analytical treatment of SSD behavior given in this paper can be extended to include other complications arising in the deformation and fracture of polymers, time-dependent viscoelasticity and large deformations, in particular. Rate-dependent behavior can be included in a straightforward manner by allowing different stress-strain response at each point along the axis of the beam. The analysis of large deformations is more difficult, but can be handled by using incremental loading and taking into account the deformed geometry of the beam after every load increment.

ACKNOWLEDGMENTS

The authors wish to acknowledge Mr. Paul Mincer and Dr. Alan Rosenfield of Battelle's Columbus Laboratories for their discussions and contributions to the experimental portion of this study.

REFERENCES

1. Ward, I. M., *Molecular Order – Molecular Motion: Their Response to Microscopic Stresses,* H. H. Kausch (Ed.), Interscience, New York (1971), p 195.
2. Argon, A. S., Andrews, R. D., Godrick, J. A., and Whitney, W., *J. Appl. Phys.,* **39,** 1899 (1968).
3. Iwamura, Y., and Rybicki, E. F., "A Transient Elastic-Plastic Thermal Stress Analysis of Flame Forming", Presented at 27th ASME Petroleum Division Conference, New Orleans, Louisiana, September 17-21, 1972 (Paper No. 72-PVP-20).
4. Rybicki, E. F., and Schmit, L. A., Jr., "An Incremental Complementary Energy Method of Nonlinear Stress Analysis", *AIAA Journal,* 8 (10), 1805 (1970).
5. Scarborough, J. B., *Numerical Mathematical Analysis,* Fourth Edition, The Johns Hopkins Press, Baltimore, Maryland, pp 185-204.

DISCUSSION on Paper by E. F. Rybicki and M. F. Kanninen

KANNINEN: In the calculations you described, it was sufficient to use uniaxial stress-strain data. The method of analysis is general enough, however, that two-dimensional problems can also be handled. As examples, the computational techniques can be applied to cylinders, plates with holes, and bodies containing cracks.

RYBICKI: Reference 4 of the paper describes an application of the method to the two-dimensional analysis of a cylinder exhibiting stress-sign-dependent behavior.

TSCHOEGL: I am not sure that I understand what you mean when you refer to differences in behavior in tension and compression. Compression is really equivalent to uniform biaxial tension. I see no a priori reason why the same stress should produce the same strain in different types of deformation. If you knew the constitutive equation for the material, you could, of course, predict the behavior in any deformation.

RYBICKI: I do not understand how compression is equivalent to uniform biaxial tension. One motivation for doing this study is that there has been very little attention given to developing capabilities for predicting the deformations of bodies that behave differently in tension than in compression. Granted many studies to characterize this behavior for small pieces of material (under uniform stress conditions) have been conducted, but very few studies have been done on the response of large bodies with stress varying from point to point.

BECKER: I would like to make a remark to the comment of Dr. Tschoegl. I think it depends on the shape of the specimen, the type of material, and the amount of strain, whether you have differences in tension and compression or not. For example, bars with a ratio of length to thickness large enough under small strains and of materials like metals or polymers in the glassy state, you will not get any difference in the responses in tension or compression. You may easily prove this by an experiment with acoustical vibrations with a compression and a tension phase and a sensitive electronical device for controlling the linearity. Differences arise, of course, if the deformations become larger.

LEE: It seems to me that the comparison between behavior in tension and compression does not fall within the scope of isotropy or anisotropy. Anisotropy is associated with response to stressing when the body in its reference state is subjected to rigid-body rotation. For a body in tension in the direction χ_1, with axes of anisotropy parallel to the coordinate axes, stretch ratios in the directions n_1, n_2, and n_3 will be λ_1, λ_2, and λ_3, respectively, with $\lambda_1 > 1$ and, in general, $\lambda_2, \lambda_3 < 1$. For compression along the same axis, λ_1 will be < 1, and $\lambda_2, \lambda_3 > 1$. These are independent deformations, not related simply by a rigid-body rotation. Thus symmetry, as normally understood, does not call for similar response in tension and compression. There seems to be no basic reason, according to continuum theory, to expect such behavior.

A DEFORMATION THEORY OF POLYMERS

K. H. Anthony and E. Kröner

Universität Stuttgart
Institut für Theoretische und Angewandte Physik
Stuttgart, Germany

ABSTRACT

For the bundle model of polymers, a quantitative nonlinear deformation theory is established using the methods of non-Euclidean geometries. The physical basis is discussed. Similarities between polymers and atomic crystals are shown.

The structure defect "disclination" is introduced into the bundle model. By means of this defect, the meander model is reduced to a particular arrangement of disclinations.

The properties of polymers are assumed to depend on the existence, the properties, and the interaction effects of structure defects. Continuum theory may thus be a powerful tool for quantitative investigations. The deformation theory established in this paper is the basis of such a continuum theory.

1 INTRODUCTION

Continuum theory was developed originally to describe the macroscopic behavior of materials the mass distribution of which could be assumed to be continuous. Nevertheless, continuum theory is as well a very good tool for investigating microscopic behavior of materials for which the atomic structure really becomes important. For instance, in atomic crystals the lattice strain in the neighborhood of a lattice defect may be calculated with sufficient accuracy by means of a continuum theory even at a distance of about one or two lattice constants from the defect. Thus, it is quite obvious that a continuous mass distribution is really not essential for the usefulness of continuum theory. Continuum theory merely is a theory using continuous functions, which mathematically may easily be handled. Its usefulness especially to investigate atomic, i.e., noncontinuous problems is limited only by our demand for accuracy.

Continuum theory of the mechanical behavior is based on a deformation theory of the material under consideration. The methods of non-Euclidean geometries have proved to be powerful tools for describing the deformation of atomic crystals. The chief properties of non-Euclidean geometries (which are continuum theories in the sense of using continuous functions) are exactly analogous to characteristic properties of discrete lattice geometry. For instance, the torsion of non-Euclidean geometry corresponds to the concept of Burgers vector which is associated with dislocations in atomic crystals. The Riemann-Christoffel curvature of non-Euclidean geometry corresponds to the rotation failure of crystal disclinations. This means that the continuous geometry to be used is always selected on the basis that it best fits the discrete geometry of the material.

From the same points of view, we try to establish a continuum theory of polymers which is based on the bundle model. We adopt the methods used for atomic crystals. In order to take into account the large chain deformations occurring in the bundle model, the intended theory has to be nonlinear from the very beginning. Preparatory to the discussion of corresponding investigations in polymers, some properties of the lattice geometry of atomic crystals* are briefly reviewed below.

The real state of an atomic crystal lattice is compared with the ideal state, the lattice of which is an ideal and unstrained one. With the introduction of lattice defects, we get from the ideal to the real state. Thus, the properties of a real-state crystal may be considered to be imparted by the existence, the properties, and the interaction effects of lattice defects.

---

*For detailed presentations, see Refs. 1 through 6.

A disclination is one type of lattice defect. Figure 1 shows a so-called wedge disclination in a flat, hexagonal, triangle lattice. In a lattice with sixfold symmetry, this defect produces a structure with fivefold symmetry. The center of the pentagon is the disclination. The lattice is topologically disturbed only in the disclination center. Elsewhere, it is continuous. It suffers a very strong strain. The disclination shown in Fig. 1 has actually been observed in the flux-line lattice of a lead-indium crystal (Fig. 2), which is a superconductor of the second kind.

A disclination may geometrically be characterized as follows: on a closed circuit we drag along a lattice vector in quite a natural way in the sense of the lattice (Fig. 1). After returning to the starting point, we find that the final vector differs from the initial vector. On the circuit, the vector suffers a rotation which equals a symmetry angle of the lattice. In our example, this angle equals 60 degrees. This "rotation failure" is the main characteristic of a disclination. It is analogous to the Burgers vector of a dislocation and does not depend on the choice of the circuit.

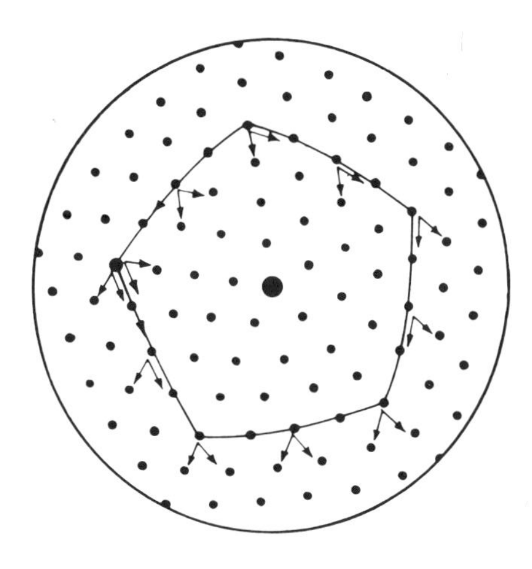

Fig. 1. Wedge disclination in a hexagonal, triangle lattice. Definition of the rotation failure.[2]

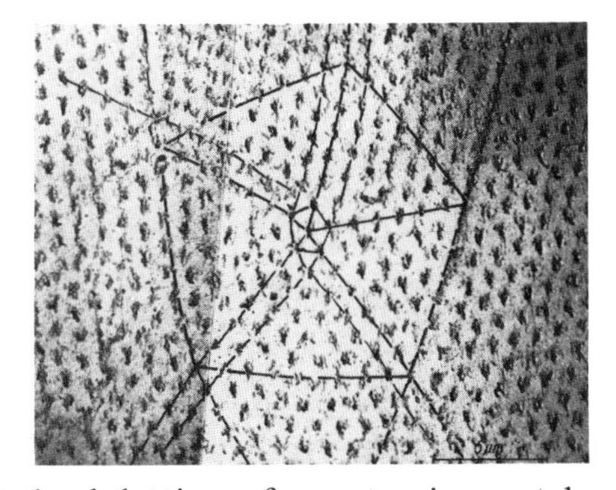

Fig. 2. Wedge disclination observed in the flux-line lattice of lead-indium. Photograph taken from Träuble and Essmann.[7]

To describe quantitatively the strained lattice of an atomic crystal, we use the idea of parallel displacement, which is associated with an affine connection. As shown in Fig. 3, we may define in the lattice, in quite a natural way, an infinite number of lattice vector fields. Going on from point x to point x + dx, such a vector field $\vec{v}$ (x) = $[v^k(x)]$ changes according to the formula

$$dv^k = - \Gamma_{ij}{}^k(x)\, v^j(x)\, dx^i \quad . \tag{1)*$$

(All quantities are referred to a coordinate system x^k, k=1,2,3.) This formula is well known in geometry. It describes a parallel displacement of the vector $\vec{v}$ along $\vec{dx}$ in the sense of the affine connection Γ. We therefore consider equivalent lattice vectors as being parallelly displaced in the sense of the lattice parallelism. This parallelism is completely determined by the field Γ(x) = $\Gamma_{ij}{}^k(x)$ of the affine lattice connection (27 components). In Fig. 3, each of the fields, 1, 2, 3, 4, are parallelly displaced vector fields. In particular, the fields 1 and 4 are parallel to each other. We see, that the lattice parallelism in general differs from the Euclidean parallelism which everyone knows from daily life (Fig. 3, field 5). Only in the special case of an unstrained lattice, do the lattice parallelism and Euclidean parallelism coincide: equivalent lattice vectors of the ideal state are parallelly displaced in the Euclidean sense. In each case, the lattice variation is completely determined by the lattice connection Γ(x).

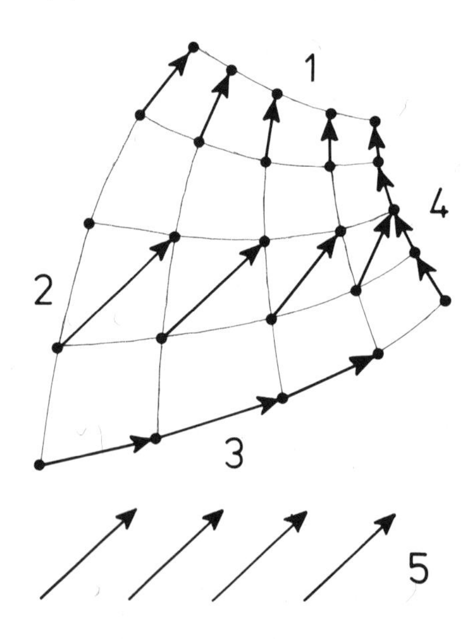

Fig. 3. Lattice parallelism.

In geometry, two different parallelisms are considered: In the case of an absolute or distant parallelism, the result of a parallel displacement

*Einstein's summation convention is implied throughout the paper.

from Point P to point Q does not depend on the choice of the path (Fig. 4a). As is well known, this case is realized in Euclidean geometry. A nonabsolute parallelism gives rise to different results of parallel displacement on different paths (Fig. 4b). Absolute and nonabsolute parallelism are distinguished by a vanishing and by a nonvanishing Riemann-Christoffel tensor field

$$R_{ijk}{}^{l}(x) = 2\left(\partial_i \Gamma_{jk}{}^{l}(x) + \Gamma_{ip}{}^{l}(x)\Gamma_{jk}{}^{p}(x)\right)_{[ij]} \quad , \tag{2)*}$$

respectively. Γ(x) is the basic connection defining the parallelism under consideration.

A disclination obviously produces a nonabsolute lattice parallelism (Fig. 1). The rotation failure is due to the nonunique parallel displacement depending on the path. Whereas the rotation failure is a global characterization of the disclination, we get a local characterization by means of the Riemann-Christoffel tensor associated with the lattice connection. We get

$$R \sim \delta(x) = \begin{cases} \infty \text{ at the disclination center} \\ \text{o outside the disclination center.} \end{cases} \tag{3}$$

Using Fig. 1, we easily can show that the parallel displacement is unique if we do not enclose the disclination. Only if the disclination center is enclosed does the nonabsolute lattice parallelism appear, i.e., the disclination is a center of nonabsolute parallelism, which is expressed in Eq. 3 by means of Diracs δ-function.

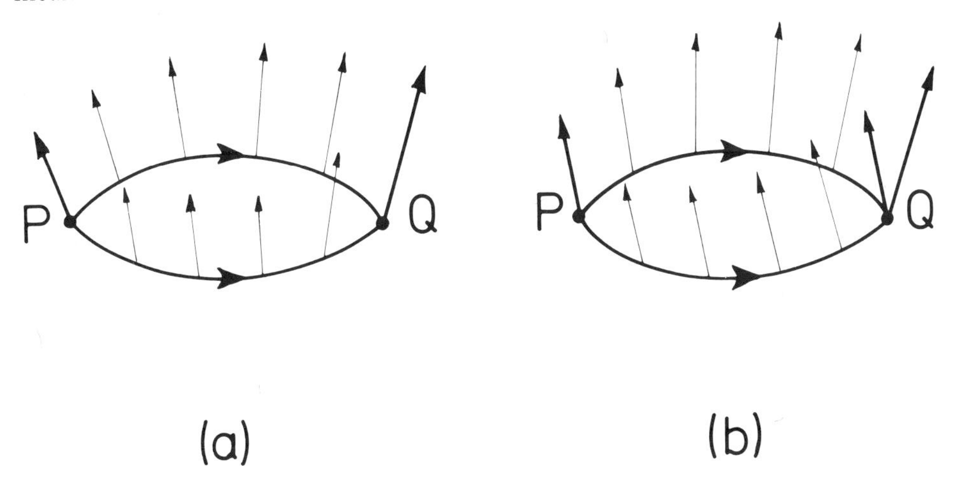

Fig. 4. Absolute (a) and nonabsolute (b) parallelism.

*$\partial_i = \partial/\partial x_i$, (ij) and [ij] denote the symmetric and the asymmetric parts of the tensor.

On the basis of these and other considerations, a complete, nonlinear elasticity theory of disclinations in atomic crystals has been developed[2,3] by means of which the lattice strains around the disclination may be completely calculated.

2 REMARKS ON THE PHYSICAL BASIS OF A CONTINUUM THEORY OF POLYMERS

We now transfer to polymers the above-mentioned methods. We refer to a model discussed by Blasenbrey and Pechhold[8,9]. Polymers are built up by molecular chains of a great length. It is suggested that a number of parallel molecular chains join together to form a bundle (*bundle model*). These bundles are assumed to arrange themselves in a meander-like (meander model, Fig. 5a) and a honeycomb-like (honeycomb model, Fig. 6a) fashion. The zigzag structure on the chains indicates thermally activated kinks. For our purpose we disregard this effect and keep in mind only the quite obvious directional structure due to the molecular chains.

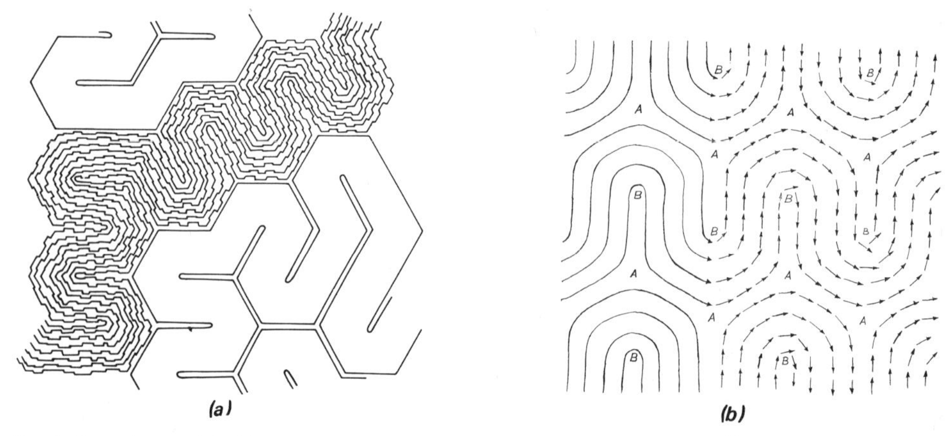

Fig. 5. Meander model: (a) photograph taken from Pechhold and Blasenbrey,[9] (b) chain parallelism, meander model as described by means of disclinations.

To each point we attach a unit vector $\vec{\xi}$ which defines the direction of the molecular chain at this particular point (Fig. 5b). Thus, we may reduce the meander model to a particular disclination arrangement. Let us go clockwise around the points A and B of Fig. 5b and observe on the circuit the variation of the vector field $\vec{\xi}(x)$ (Figs. 7a and b). We find a rotation of -180 degrees at A and +180 degrees at B. (The sign of the angle is chosen positive if the vector $\vec{\xi}$ rotates in the same sense as the circuit.) These properties are exactly the same as those of the disclination in an atomic crystal (Fig. 1). We therefore denote the centers A and B in Fig. 5b

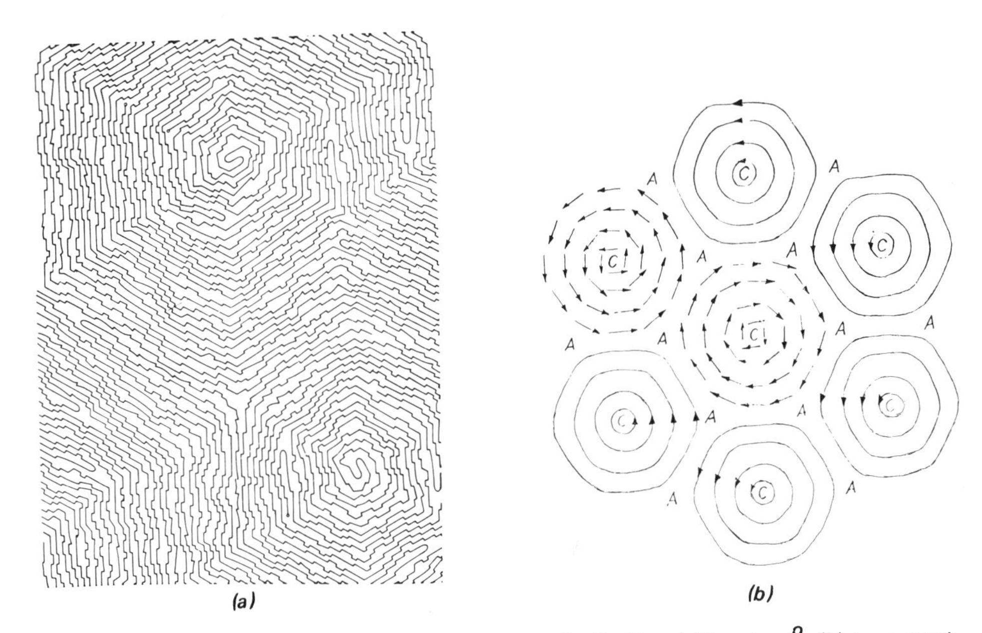

Fig. 6. Honeycomb model: (a) photograph taken from Pechhold and Blasenbrey[9]; (b) honeycomb model as described by disclinations.

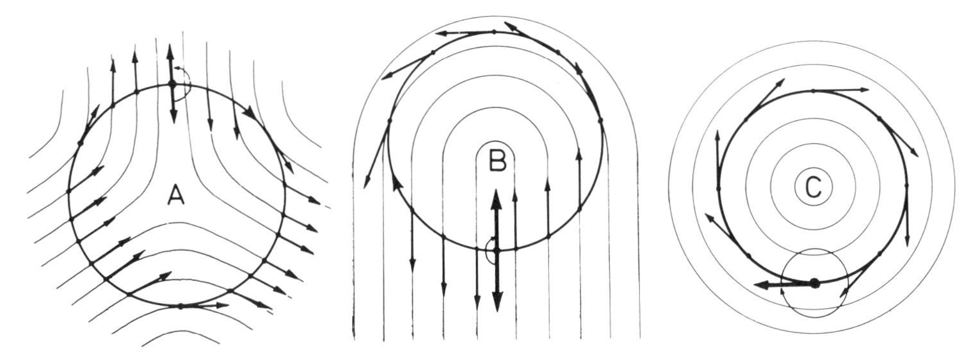

Fig. 7. Polymer disclinations: (a) minus 180-degree disclination; (b) plus 180-degree disclination; (c) plus 360-degree disclination.

as -180-degree polymer disclinations and +180-degree polymer disclinations.*

Thus, the meander model may be considered as a series of disclinations, which arrange themselves in a sequence of layers, the sign of the disclinations changing from one layer to the other (Fig. 5b).

In the honeycomb structure (Fig. 6b), we find another type of polymer disclination located at points labelled C. From Fig. 7c we see that it is a +360-degree disclination. Furthermore, we find -180-degree disclinations at points labelled A.

*See Refs. 2 and 10.

Disclinations in polymers are centers of very strong chain curvature, which decreases with increasing distance from the center. They extend along a line, which cannot end within the probe. In the case of Figs. 5 and 6, the disclination lines are located at A,B,C. They extend perpendicular to the drawing plane.

If besides the meander and the honeycomb structure an ideal state of the polymer is defined, we may apply exactly the same ideas as in the case of an atomic crystal. In the ideal state, all molecular chains are straight and parallel. Then the real state (= meander and honeycomb structure) is produced from the ideal state by introducing a number of structure defects. The properties of real-state polymers may thus be explained by the existence of structure defects, by their properties, and by their interaction effects. Polymer disclinations obviously are important structure defects. Other defects are also discussed.

Through the foregoing considerations we intended to make it clear that within the models discussed there is no methodical difference between polymers and atomic crystals. Thus, the time is near at hand when the quantitative methods of the continuum theory of atomic crystals can be adopted for studies of polymers. In this paper we confine ourselves to the basic deformation theory.

3 MATHEMATICAL FOUNDATION OF DEFORMATION THEORY OF POLYMERS

All quantities are referred to a coordinate system x^k, k=1,2,3 of Euclidean space. $\vec{e}_k$, k=1,2,3 are the basis vectors of the system x^k.

Similarly to the case of atomic crystals, we consider the vector field $\vec{\xi}(x)$, defined before as a parallelly displaced vector field. Then the polymer disclinations are sources of an nonabsolute chain parallelism. But, if we look for the corresponding affine connection $\Lambda(x) = \left(\Lambda_{ij}{}^k(x)\right)$, which describes the variation of the vector field $\vec{\xi}(x)$ by an equation analogous to Eq. 1, we are stopped by a fundamental difficulty: a connection Λ may be defined if and only if there exist at least three linearly independent vector fields, all of which are parallelly displaced with respect to the same parallelism. In an atomic crystal these vector fields are quite obvious, whereas in polymers, only one field $\vec{\xi}(x)$ is given immediately. But in the following way we may associate to the field $\vec{\xi}(x)$ three linearly independent vector fields $\vec{n}_\lambda(x)$, $\lambda = 1,2,3$:* $\Gamma(x) = \left(\Gamma_{ij}{}^k(x)\right)$ is the affine

*The reader is assumed to be familiar with Ricci calculus.[11]

connection corresponding to Euclidean parallelism. We define from the vector field

$$\underset{1}{\xi}^k(x) = \xi^k(x) \tag{4}$$

two vector fields

$$\underset{2}{\xi}^k(x) = \underset{1}{\xi}^i(x) \overset{(\Gamma)}{\nabla_i} \underset{1}{\xi}^k(x) \tag{5}$$

and

$$\underset{3}{\xi}^k(x) = \underset{1}{\xi}^i(x) \overset{(\Gamma)}{\nabla_i} \underset{2}{\xi}^k(x) \quad , \tag{6}$$

where $\overset{(\Gamma)}{\nabla_i}$ denotes the covariant derivative with respect to Γ:

$$\overset{(\Gamma)}{\nabla_i}\xi^k(x) = \partial_i\xi^k(x) + \Gamma_{ij}{}^k(x)\,\xi^j(x) \quad . \tag{7}$$

The field $\underset{2}{\vec{\xi}}(x)$ describes the Euclidean absolute variation of the vector field $\underset{1}{\vec{\xi}}(x)$ referred to the direction $\underset{1}{\vec{\xi}}(x)$. In the same way, $\underset{3}{\vec{\xi}}(x)$ describes the Euclidean absolute variation of $\underset{2}{\vec{\xi}}(x)$ referred to the direction $\underset{1}{\vec{\xi}}(x)$. Now we have to consider three cases: either the three vector fields $\underset{\lambda}{\vec{\xi}}(x)$, λ=1,2,3 are linearly independent (main case), or $\underset{3}{\vec{\xi}}(x)$ depends linearly on $\underset{1}{\vec{\xi}}(x)$ *and* $\underset{2}{\vec{\xi}}(x)$ [partial flattening of the field $\vec{\xi}(x)$], or $\underset{3}{\vec{\xi}}(x)$ and $\underset{2}{\vec{\xi}}(x)$ depend linearly on $\underset{1}{\vec{\xi}}(x)$ [total flattening of the field $\underset{3}{\vec{\xi}}(x)$]. It can be shown that no other case exists. The partially flattened vector field $\vec{\xi}(x)$ is a plane field, i.e., there exists a manifold of parallel planes, so that the vector $\vec{\xi}$ at each point x lies in one of the planes. The totally flattened vector field $\vec{\xi}(x)$ is parallelly displaced in the Euclidean sense. For example, the director field $\vec{\xi}(x)$ of the ideal state of a polymer is totally flattened.

In general, the vector triplet $\underset{\lambda}{\vec{\xi}}(x)$, λ = 1,2,3 is not orthonormalized in the Euclidean sense. Therefore, we construct another vector triplet $\underset{\lambda}{\vec{n}}(x)$, λ = 1,2,3, which is orthonormalized, i.e.,

$$|\underset{\lambda}{\vec{n}}| = 1,\ \lambda = 1,2,3;\ \underset{\lambda}{\vec{n}} \perp \underset{\kappa}{\vec{n}},\ \lambda \neq \kappa \quad , \tag{8}$$

and which is defined from $\vec{\xi}_{\lambda}(x)$ as follows:

$$\vec{n}_1(x) = \vec{\xi}_1(x) = \vec{\xi}(x) \quad . \tag{9}$$

The vector $\vec{n}_2$ lies in the plane spanned by $\vec{\xi}_1$ and $\vec{\xi}_2$:

$$\vec{n}_2(x) = \alpha^1(x) \cdot \vec{\xi}_1(x) + \alpha^2(x) \cdot \vec{\xi}_2(x) \quad . \tag{10}$$

Finally, $\vec{n}_3(x)$ is spanned by all vectors $\vec{\xi}_{\lambda}$:

$$\vec{n}_3(x) = \beta^1(x) \cdot \vec{\xi}_1(x) + \beta^2(x) \cdot \vec{\xi}_2(x) + \beta^3(x) \cdot \vec{\xi}_3(x) \quad . \tag{11}$$

The basis system $\vec{n}_{\lambda}(x)$ shall be a right-hand system. Then in the main case, the coefficients α and β in Eqs. 10 and 11 are uniquely determined. In the case of partial flattening we take into account only Eqs. 9 and 10. In the case of total flattening, only Eq. 9 makes sense. In this case, $\vec{n}_2(x)$ and $\vec{n}_3(x)$ are assumed to be parallelly displaced in the Euclidean sense.

Thus, in each case we may uniquely associate with the director field $\vec{\xi}(x)$ an orthonormalized vector triplet $\vec{n}_{\lambda}(x)$, $\lambda = 1,2,3$, which we call the accompanying triplet. If the director field $\vec{\xi}(x)$ is sufficiently continuous, the vectors $\vec{\xi}$ may be interpreted as unit tangent vectors of a manifold of curves, physically defined by the molecular chains. Then $\vec{n}_2(x)$ is the first normal vector of the curve going through x. $\vec{n}_2(x)$ lies in the winding plane of the curve at point x. $\vec{n}_3(x)$ is the second normal vector of the curve. It is the unit normal of the winding plane.

Now we can define the chain parallelism associated with the director field $\vec{\xi}(x)$: the three vector fields $\vec{n}_{\lambda}(x)$, $\lambda = 1,2,3$, are assumed to be parallelly displaced with respect to the affine chain connection Λ. Because of Eq. 9, the field $\vec{\xi}(x)$ is parallelly displaced, too. The connection Λ is uniquely determined. We get

$$\Lambda_{ij}{}^{k}(x) = A^{k}_{\lambda}(x)\,\partial_i A^{\lambda}_{j}(x) \tag{12}$$

if all the fields $\underset{\lambda}{\vec{n}}(x)$ are continuously differentiable. This assumption is supposed to be broken only on isolated surfaces, curves, and points. In Eq. 12, A^k_λ are the components of the vectors $\underset{\lambda}{\vec{n}}$ with respect to the basis vectors $\underset{k}{\vec{e}}$ of the coordinate system x^k:

$$\underset{\lambda}{\vec{n}} = A^k_\lambda \underset{k}{\vec{e}} \quad . \tag{13}$$

A^λ_j is the inverse of the nonsingular matrix A^k_λ:

$$A^\lambda_j A^k_\lambda = \delta^k_j, \; A^\lambda_j A^j_\kappa = \delta^\lambda_\kappa \quad . \tag{14}$$

In our polymeric piece of material we have now defined two different parallelisms: the Euclidean parallelism Γ and the chain parallelism Λ. We take the difference

$$H_{ijk}(x) = \left(\Gamma_{ij}{}^{l}(x) - \Lambda_{ij}{}^{l}(x) \right) g_{lk}(x) \tag{15}$$

and obtain a tensor, which measures the deviation of the real state of the polymer from its ideal state. In Eq. 15, $g(x) = \left(g_{lk}(x)\right)$ is the Euclidean metric tensor. Going on from point x to point x + dx, the vector triplet $\underset{\lambda}{\vec{n}}(x)$ suffers only an infinitesimal rotation which is described by an infinitesimal tensor $d\omega_{jk}$. We get

$$d\omega_{jk} = H_{i[jk]}(x)\,dx^i \quad . \tag{16}$$

Because of the pure rotation we also get

$$H_{i(jk)}(x) = 0 \quad . \tag{17}$$

This equation may also be proved by means of the equations

$$\overset{(\Gamma)}{\nabla}_i g_{jk} = 0 \tag{18a}$$

and

$$\overset{(\Lambda)}{\nabla}_i g_{jk} = 0 \quad . \tag{18b}$$

Equation 18b means that Λ is a metric connection with respect to the

Euclidean metric.[11] It may be established by means of "naturalization processes".[1]

We call

$$\kappa_{ijk}(x) = H_{i[jk]}(x) \tag{19}$$

the structure curvature tensor of the polymer. It describes the rotation of the accompanying triplet $\vec{n}_{\lambda}$ if we pass through the material, i.e., κ involves all information concerning the arrangement of the molecular chains of the polymer.

In order to describe disclination arrangements, we use the Riemann-Christoffel tensor $K_{ijk}{}^{l}(x)$ defined by the connection Λ (see Eq. 2). Lowering the upper index by g and inserting the structure curvature we get

$$K_{ijkl}(x) = -2\left(\overset{(\Gamma)}{\nabla}_i \kappa_{jkl}(x) + \kappa_{ilp}(x)\,\kappa_{jkq}(x)\,g_{pq}(x)\right)_{[ij]} \quad . \tag{20}$$

As in the case of atomic crystals (see Eqs. 2 and 3), this quantity equals the sources of nonabsolute chain parallelism, i.e.,

$$K \sim \delta(x) \quad , \tag{21}$$

where the δ-peaks are located at the disclination lines.

So far we have established the framework of a deformation theory of polymers. Adding statics and constitutive equations, we get a complete continuum theory of polymers.[12]

REFERENCES

1. Anthony, K. H., *Arch. Rat. Mech. Anal.*, **37**, 161 (1970).
2. Anthony, K. H., *Arch. Rat. Mech. Anal.*, **39**, 43 (1970).
3. Anthony, K. H., *Arch. Rat. Mech. Anal.*, **40**, 50 (1971).
4. Kröner, E., *Arch. Rat. Mech. Anal.*, **4**, 273 (1960).
5. Kröner, E., and Seeger, A., *Arch. Rat. Mech. Anal.*, **3**, 97 (1959).
6. Teodosiu, C., *Rev. Roumaine Sci. Tech., Sér. Mécan. Appl.*, **12**, 961 (1967).
7. Träuble, H., and Essmann, U., *Phys. Status Solid.*, **25**, 373 (1968).
8. Blasenbrey, S., and Pechhold, W., *Ber. Bunsenges. Physik. Chem.*, **74**, 784 (1970).
9. Pechhold, W., and Blasenbrey, S., *Kolloid-Z. u. Z. Polymere*, **241**, 955 (1970).
10. Nabarro, F.R.N., *Theory of Crystal Dislocations*, Clarendon Press, Oxford (1967).
11. Schouten, J. A., *Ricci-Calculus*, Springer, Berlin, Göttingen, Heidelberg (1954).
12. Anthony, K. H., to be published.

DISCUSSION on Paper by K. H. Anthony and E. Kröner

SCHWARZL: As far as I understood, you have prepared a scheme to describe the geometry of the meander model in terms of the theory of atomic crystals. The essential step, however, is still to be made, viz., to set up relations between the geometrical order in the model and the external stresses which will occur. Did you calculate the stress-strain curves for meanderlike structures?

ANTHONY: The theory given here is similar to continuum theories of atomic crystals. I presented these similarities only for the sake of a better illustration of some complicated topics.

I described the meander model in terms of non-Euclidean geometry. On the basis of this concept, a continuum theory of the mechanical behavior of polymers will be established. This theory includes interactions between the external stresses and the internal geometrical structure of molecular chains. Stress-strain curves are closely related to these interactions. Theory is in progress.

RADCLIFFE: What modifications, if any, are needed to apply these concepts to the disordered regions of semicrystalline polymers and to amorphous polymers?

ANTHONY: In terms of the meander model, the disordered regions of semicrystalline polymers as well as the amorphous polymers are due to a very high density of kinks. Nevertheless, in these regions the chain director field is preserved in the sense of a continuously varying averaged chain direction.

Kinks are a special sort of polymeric point defects which in terms of continuum theory may be described by a "nonmetric connection"* or by a "quasi-disclination and quasi-dislocation density"*. Adding these concepts to the concepts of disclinations and chain parallelism, the amorphous regions may be included into a continuum theory of polymers.

HULL: I was very interested in your attempt to describe polymer deformation in terms of the existence, properties, and interactions of disclinations. Do you have any views on the properties of disclinations and the way they will interact with each other or with other obstacles?

ANTHONY: In my talk I have been concerned with the geometrical properties of polymer disclinations only. Adding constitutive assump-

*For these concepts, see Anthony, K. H., *Arch. Rat. Mech. Anal.,* **40,** 50-78 (1971).

tions and equilibrium equations, we get the mechanical properties of disclinations, i.e., stress and strain fields and chain curvatures due to disclinations. These investigations are in progress.

Interaction effects of disclinations with each other or with other obstacles are due to strains and chain curvatures produced by the defects under consideration. Thus, interaction effects cannot be studied before the properties of the single defects are well understood.

HULL: Is there any physical justification for the existence of disclinations which have invariant Burgers vectors along their length?

ANTHONY: Disclinations are characterized by the so-called Frank's vector – we had better say Frank's rotation – which is quite analogous to Burgers vector of dislocations. Frank's rotation is constant along the disclination line. It is defined by means of a circuit around the disclination and does not depend on the choice of the circuit. Besides Frank's rotation, we may also introduce the concept of Burger's vector in the case of disclinations. But this quantity is no invariant of the disclination. It depends on the circuit by which it is defined.

Disclinations are observed for instance in the flux-line lattice of superconductors, in the director field of liquid crystals, and in magnetic structures. As to disclinations in high polymers, they are inherent obstacles of the meander model, which have not as yet been directly observed. But if the concept of a single crystal in polymers is accepted, the disclination seems to be a quite natural polymer defect.

AGENDA DISCUSSION: CONTINUUM DESCRIPTION OF DEFORMATION

E. H. Lee*

Stanford University
Stanford, California

L. E. Hulbert**

Battelle
Columbus Laboratories
Columbus, Ohio

1 INTRODUCTION

The suggested overall plan for this Agenda Discussion was to see whether, utilizing the broad range of expertise in various aspects of the study of polymers represented at the Colloquium, the participants could jointly reach some conclusions on what kind of constitutive relations are satisfactory for what types of materials under what testing conditions.

---

*Chairman.
**Secretary.

This objective clearly involved some overlap with the earlier session on Phenomenology, and the Chairman suggested that the discussion continue from that session with an attempt to focus on the various representations of nonlinear viscoelasticity with the hope of assessing which are most satisfactory in practice. Dr. J. G. Williams and Professor R. S. Rivlin had agreed to summarize experimental and analytical investigations which bear on this question.

Nonlinear Viscoelasticity

For later comparison, Williams first mentioned the linear viscoelastic laws in the form of superposition integrals:

$$\sigma(t) = \int_{-\infty}^{t} E(t - \tau) \frac{d\epsilon}{d\tau} d\tau \quad , \tag{1}$$

which gives the stress variation $\sigma(t)$ for simple tension in terms of the strain variation $\epsilon(t)$ and the relaxation modulus E(t). The equivalent reciprocal relation in terms of the creep compliance D(t) takes the form

$$\epsilon(t) = \int_{-\infty}^{t} D(t - \tau) \frac{d\sigma}{d\tau} d\tau \quad . \tag{2}$$

He mentioned that, for glassy polymers, these linear laws are usually satisfactory for strains less than 1/2 percent.

Williams stated that nonlinear theories may be needed in order to account for deviations from linear behavior. Such theories fall into two categories, single- and multiple-integral theories. Examples of the former are:

Single-Integral Nonlinear Theory

(a) Separable creep and relaxation behavior, Leaderman[1]
(b) Modified superposition, a generalization of (a)[2]
(c) Schapery's thermodynamic approach[3]
(d) The BKZ method[4], discussed in Session IV.

Multiple-Integral Theory

For small strains, this can take the form[5]:

$$\sigma(t) = \int_{-\infty}^{t} E_0(t-\tau) \frac{d\epsilon}{d\tau} d\tau + \int_{-\infty}^{t}\int_{-\infty}^{t} E_1(t-\tau_1, t-\tau_2) \frac{d\epsilon}{d\tau_1} \frac{d\epsilon}{d\tau_2} d\tau_1 d\tau_2 + \ldots \tag{3}$$

The first group consists of generalizations of Boltzmann's superposition principle, making it nonlinear in various ways.

Leaderman's theory is the simplest nonlinear single integral theory, and it can be applied only to materials which exhibit separable relaxation or creep behavior, that is, materials for which the stress is given by:

$$\sigma(t) = f(\epsilon)\, E(t) \quad , \tag{4}$$

in a relaxation test in which the constant strain ϵ is suddenly applied at time $t = 0$ and maintained. The function f is nonlinear, and for transient straining, a Boltzmann integral in $d[f(\epsilon)]$ applies:

$$\sigma(t) = \int_{-\infty}^{t} E(t-\tau) \frac{df}{d\epsilon} \frac{d\epsilon}{d\tau} d\tau \quad . \tag{5}$$

An analogous result applies for separable creep. These relations are similar to the Boltzmann superposition integrals (Eqs. 1 and 2), but a nonlinear function of stress or strain replaces stress or strain, respectively.

The modified superposition theory applies to materials which are not separable in the above sense. If

$$\epsilon(t) = g(\sigma,t) \tag{6}$$

gives the creep function for suddenly applied constant stress σ, then for steps of stress from σ_{i-1} to σ_i at $t = t_i$, $i = 0, 1, .. N$; for $t > t_N$, the strain is given by[2]

$$\epsilon(t) = \sum_{i=0}^{N} [g(\sigma_i, t-t_i) - g(\sigma_{i-1}, t-t_i)] \quad , \tag{7}$$

and for continuous variation of stress

$$\epsilon(t) = \int_{-\infty}^{t} \frac{\partial g[\sigma(\tau), t-\tau]}{\partial \sigma(\tau)} \frac{d\sigma(\tau)}{d\tau} d\tau \quad . \tag{8}$$

Schapery[6] originally developed a theory of linear viscoelasticity including thermo-mechanical coupling by means of the theory of the thermodynamics of irreversible processes. He extended this to nonlinear materials[3], and expressed it in single-integral form. His development included a log (time) - temperature shift hypothesis which modified the time variable of integration. The BKZ theory due to Bernstein, Kearsley, and Zapas, and described by Dr. Zapas at the Colloquium, takes on a form somewhat similar to Schapery's representation.

Williams reported that these various theories with parameters selected to fit a particular set of data would give approximately the same predictions for continued loading which were in satisfactory agreement with experiment.

Williams continued his discussion: "The other form of nonlinear theory is the multiple integral theory and this produces its nonlinearity not by tinkering with this function or that function, but by series expansion (Eq. 3). The first term is identical with linear theory and higher terms introduce multiple integrals.

"Multiple integral theory written with strain on the right hand side – the strain history approach (Eq. 3) – evaluated for a simple relaxation test gives:

$$\sigma(t) = E_0(t)\epsilon + E_1(t,t)\epsilon^2 + \ldots \quad (9)$$

The first term corresponds to linear theory and the double integral gives the squared term and so on. Experimental data can be so represented by fitting polynomial terms. The point that Professor Rivlin made earlier today that different responses can occur in tension and compression is apparent here in that the even and odd powers behave differently with change of sign of strain.

"A multiple integral theory can also be written with stress on the right hand side and a similar form results with creep given by:

$$\epsilon(t) = D_0(t)\sigma + D_1(t,t)\sigma^2 + \ldots\text{''} \quad (10)$$

Williams pointed out that interconversion between the sets of kernel functions E_i and D_i, and also for the nonlinear single-integral theories, is largely intractable.

He finally commented on a limitation of all the theories discussed. If a ramp stress-time input as shown on Fig. 1a is applied, the strain response is as shown in Figs. 1b and 1c. For appropriate choice of material functions, all the nonlinear theories provide a satisfactory representation for the loading part of the cycle. Linear theories may be in error by 10 or 15 percent, but usually one or two terms of the multiple-integral theory will

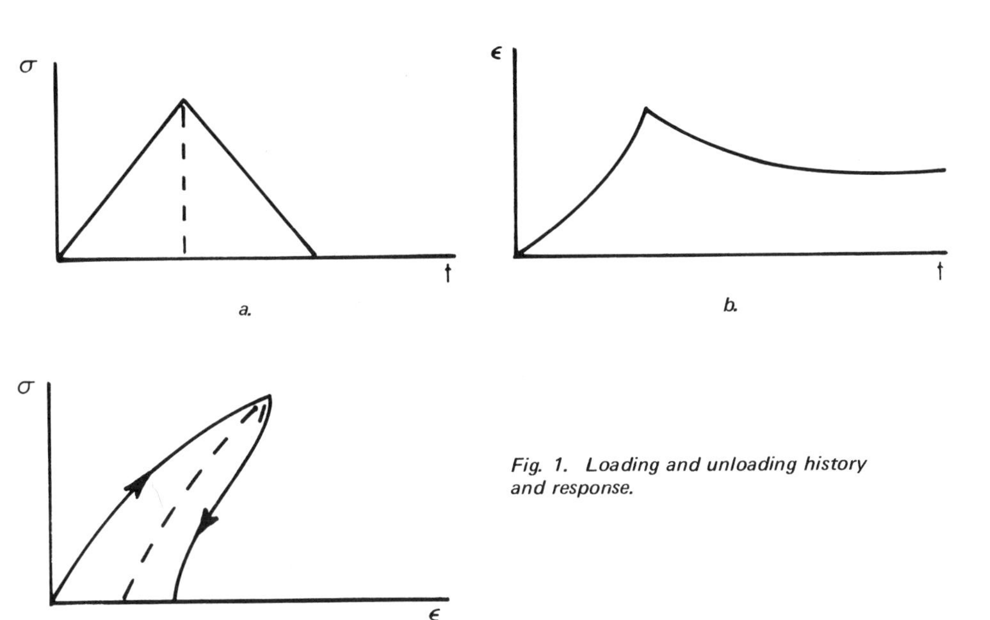

Fig. 1. Loading and unloading history and response.

eliminate this error. However, unloading gives a σ-ϵ response, as shown by the broken line in Fig. 1c, which differs markedly from the full line predicted by the theories. Moreover, the prediction of the unloading line tends to be very sensitive to the choice of material functions. He stated that good prediction was achieved for general loading functions, but that a major problem remains, viz., to make some modification of these theories to permit satisfactory prediction of unloading histories. It is interesting to observe that Leaderman[1] did achieve accurate prediction of creep *and* recovery. The Chairman mentioned that the difficulty in accurately predicting unloading response conformed with the comments made by Zapas in the morning session concerning the difficulty of dealing with multiple steps of stress which included steps down in stress magnitude.

With regard to functional forms which determine the stress in terms of strain history, or vice versa, in the range of finite strain, the Chairman pointed out the appearance of finite rotations which appear in the former in addition to strain history. Such terms can be considered as influences of geometrical nonlinearity. It was agreed that such terms lead to difficulties if attempts are made to represent strain in terms of stress history. Commenting on this question, Rivlin said that he did not think that for a one-dimensional theory, there are major technical difficulties in inverting between stress versus strain and strain versus stress relations. For three dimensional theory, the problem is clear from the structure of the relations. Taking equations from his introductory lecture at the Colloquium, for deformation given by $x = x(\mathbf{X},\tau)$, the Cauchy stress, σ_{ij}, can be expressed as

$$\sigma_{ij}(t) = x_{i,A}\ x_{j,B}\ \mathop{\mathcal{F}}_{-\infty}^{t}{}_{AB}[E_{PQ}(\tau)] \quad , \tag{11}$$

where $\mathcal{F}$ is a functional of the Lagrange strain E_{PQ}, that is, it depends on the history, $E_{PQ}(\tau)$, of the strain. Current instantaneous values $E_{PQ}(t)$ and its derivatives can appear in $\mathcal{F}$. The inverse theory has to be written in the form:

$$E_{AB} = \mathcal{D}_{AB}[P_{PQ}(\tau)] \quad , \tag{12}$$

where P_{PQ}, the Piola stress, is defined as

$$P_{PQ} = \frac{\partial X_P}{\partial x_i(\tau)} \frac{\partial X_Q}{\partial x_j(\tau)} \sigma_{ij}(\tau) \quad . \tag{13}$$

Thus, in the inverse relation (Eq. 12), deformation derivatives appear on the right-hand side in combination with the Cauchy stress σ_{ij}. Rivlin pointed out that there is no simple physical interpretation of the Piola stress in terms of the applied forces and the current configuration of the body. Clearly, Eq. 11 can be written in the form:

$$P_{AB} = \mathop{\mathcal{F}}_{-\infty}^{t}{}_{AB}[E_{PQ}(\tau)] \quad , \tag{14}$$

but this simply conceals the difficulty of isolating the stress and deformation variables, respectively.

Rivlin then turned to another aspect of viscoelastic constitutive relations. Eq. 1 is written in terms of the strain derivative, which by integration by parts can be modified to the form:

$$\sigma(t) = [E(t-\tau)\ \epsilon(\tau)]_{-\infty}^{t} + \int_{-\infty}^{t} E'(t-\tau)\ \epsilon(\tau)\ d\tau$$

or

$$\sigma(t) = \int_{-\infty}^{t} f(t-\tau)\ \epsilon(t)\ d\tau + K\ \epsilon(t) \quad . \tag{15}$$

Equation 1 implies instantaneous elastic response if E(0) is not zero, and then K is not zero. The multiple integral theory can be modified in a similar manner. In order to develop the multiple-integral expression corresponding to Eq. 15 to represent a functional relation, it is necessary to assume that the functional relation in terms of $\epsilon(t)$ is continuous, which does not imply continuity as a functional of $\dot{\epsilon}(\tau)$. In writing the form Eq. 1, continuity as a functional of $\dot{\epsilon}(\tau)$ is assumed, and this also implies the validity of the form Eq. 15. Thus, there is a subtle difference in the physical conditions which each can represent. This difference takes on greater significance for nonlinear, multiple-integral theories.

Rivlin went on to discuss the related topic of the accuracy of prediction of material response. In the development of the multiple-integral theory[7], the extension history was approximated by a cosine Fourier series, and the nonlinear functional law was expressed as a function of the Fourier coefficients. In the case of a material for which the nonlinearity increases with rate of deformation, the coefficients for the higher harmonics will have greater influence. Consider the ramp-function straining history shown in Fig. 2a. This is continuous in terms of strain, which provides a limitation on the magnitudes of the Fourier harmonics. Analysis in terms of strain rate, Fig. 2b, which is discontinuous, generates more dominant higher harmonics, and hence slower convergence of the multiple-integral series. Even in cases as shown in Fig. 2a, the sudden change from loading to unloading will cause lack of convergence of the multiple-integral series, in conformity with the prediction difficulties for unloading mentioned by Williams. Rivlin suggested that such sudden changes in stress history can lead to prediction difficulties, and that the constitutive equations with a few integrals may be satisfactory only for

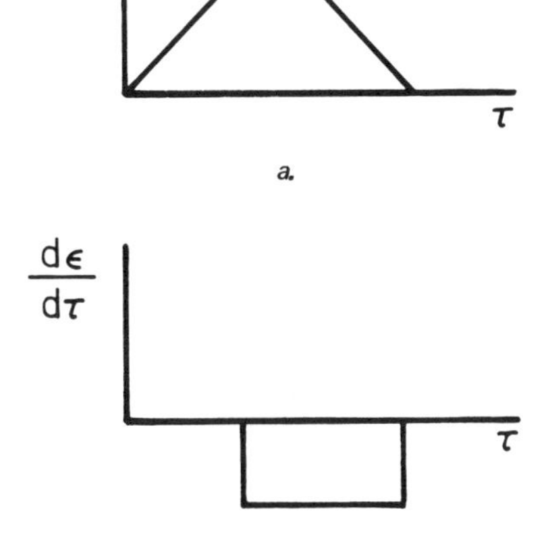

Fig. 2. Straining history.

smooth histories of straining or loading. If sudden changes occur, it may be necessary to wait until their influence has been attenuated through fading memory to the point where satisfactory prediction is again achieved.

Andrews asked whether other participants in the discussion had experimental confirmation of the ability to predict the response to slowly varying inputs using multiple-integral laws involving only a few terms. He suggested that most experiments in fact exhibited sinusoidal loading, and Max Williams reiterated the difficulty in experimentally producing discontinuities in strain rate. Gordon Williams confirmed that his experiments involved sharp changes in loading rate, and that satisfactory agreement was achieved only some time after the change. When pressed by Andrews as to whether the limitation to smooth responses eliminated the difficulty of not being able to deal with unloading which Gordon Williams had mentioned earlier, Rivlin observed: "There is absolutely no guarantee that any theory will be able to fit any particular material. What I'm saying is that if one deals with relatively small, smooth and slowly varying changes, then it will be possible to find polymeric nonlinear materials for which a two or three integral type theory really works." In answer to queries about the type of material under study, Gordon Williams listed rigid PVC, polyethylene, and polypropylene. Tests ranged over strains up to about 3 percent, so that geometrical nonlinearity was not significant. The amount of nonlinearity in the PVC was quite modest, but that in polypropylene was appreciable. In all these cases, the difficulty of reproducing unloading behavior arose.

Zapas commented that the reason that he encountered problems in analyzing steps down in stress was the difference in behavior for loading in tension and loading in compression. He observed that the effects of such differences occur at long times. However, Rivlin pointed out that the approximations arising in the BKZ theory, which comprises the sum of many integrals of a particular form, differ from the assessment of a multiple-integral theory containing only a few terms.

In pursuing an earlier question concerning the particular materials under discussion, Landel remarked: "One of the reasons for raising my question was the fact that whether or not one applies a linear theory or a nonlinear theory to a particular material, the assumption is made that what one started with remains the same material throughout the experiment. If the material is changed by the process of straining, then one runs into difficulties. For instance, if a material cavitates, then unloading involves quite a different process from the loading process. And in particular, in crystalline materials, straining the material is going to change its morphology, and then a second loading curve with this changed morphology will be closer to the unloading curve than to the first loading relation. This relates to a point that was made earlier, to the effect that we don't have enough information on the effect of the strain on the changes in properties of materials. For example, the material described by Dr. Schwarzl looks linear on a loading curve in terms of true stress and true strain over quite a wide range. One would expect to be able to handle the unloading curve quite nicely, but for unloading it is quite a different material."

Rivlin observed that it is not true to state that nonlinear theory cannot account for changes in properties of the material, although it cannot account for all types of changes of structure. The introduction of strain history into the constitutive relation does, in principle, introduce the influence of material-properties changes, for how the material behaves at a given instant then depends on what has happened to it before. However, some effects, such as the development of Lüder's bands when the deformation ceases to be homogeneous, are not included in the types of theories we have discussed.

The Chairman suggested that having discussed various aspects and difficulties of nonlinear theory, it would be useful to consider under what circumstances the simple linear theory can provide a satisfactory basis for analyzing material response.

Linear Viscoelasticity

The Chairman commented that the analysis of stress and deformation distributions for viscoelastic materials can play an important role in the technology of the application of polymers and also in the correct evaluation of mechanical tests on polymers. The latter is particularly pertinent for dynamic tests in which inertia forces have a significant influence. In general, stress and strain distributions are much more easily analyzed on the basis of linear viscoelastic laws, and so it is extremely important to know for what materials and under what circumstances linear viscoelasticity provides a satisfactory representation of material response. The answer clearly depends on the accuracy required of the analysis. In view of the variability of material properties in this field, I suggest that an accuracy of 5 or 10 percent in stress and strain magnitudes be considered satisfactory. Is it possible to state ranges of material type and loading characteristics which permit linear analysis to this order of accuracy? For example, how do crystalline and amorphous polymers compare in this connection? Is there an essential difference from this standpoint between the glassy and rubbery states, and what can we say about the strain and temperature ranges for these various material conditions?

Landel opened the discussion with the comment that for materials in the rubbery region, linearity thus defined extends to about 5 percent strain. He pointed out that in Session II, Becker, for example, had observed stress-strain response for both uniaxial and multiaxial tests which followed classical theory up to about this order of magnitude. In answer to a query, Landel stated that by "rubbery region" he referred to rubbery amorphous materials at least, probably excluding PVC and possibly others which exhibit strong hydrogen bonds. In reply to a question from Ferry, he confirmed that he was referring to unfilled polymers, normally cross-linked to a shear modulus of up to about 10^7 dynes/cm^2, beyond which the material becomes leathery and falls outside the linear range.

Andrews made the general point that as soon as one goes beyond a simple rubber to either a filled system or a semicrystalline material in which the amorphous phase is rubberlike, the strain becomes concentrated in the matrix or amorphous region respectively, where it therefore exceeds the average strain. One would thus expect linearity to fail for a smaller average strain since this will be governed by the response of the highly strained regions.

Retting commented that for the glassy state of amorphous polymers and for semicrystalline polymers, he has found that the linear region holds only for a few tenths of a percent strain. For PVC and polyethylene, the modulus was found to be a function of both time and strain above this range, the maximum range being 1/2 percent. Landel responded that for a rubbery material the change in modulus from tension to compression occurs smoothly, and linearity strictly applies only to the tangent to the stress-strain curve at the origin. However, the curvature is slight, and for the accuracy suggested by the Chairman in posing the question, he felt that a 5 percent strain was a reasonable bound. Gordon Williams stayed with his earlier statement of a 1/2 percent strain limit, and stated that he was satisfied with an accuracy of 10 to 15 percent for stresses. Retting stated that the 1/2 percent limit applied for all thermoplastic materials in the glassy state. Since the glassy state is quite rigid, this would of course correspond to an appreciable stress.

Kovacs commented: "I am glad Dr. Retting has raised this question, since I have the feeling that Dr. Williams overestimated the strain limit below which the viscoelastic behavior of polymeric glasses may be described by the linear approximation.

"In this respect I would like to mention the results reported by G. Goldback and G. Rehage[8] on volume creep and recovery under hydrostatic pressure. These authors have shown that the practical limit within which linearity seems to hold is of the order of $\pm 1 \times 10^{-4}$. The corresponding limit in uniaxial tension or compression would be (assuming a Poisson ratio of 0.3) 2.5×10^{-4} which is about 50 times less than that given by Dr. Williams. This rough calculation is based on a general rule given by Prof. Ferry[9] according to which the applicability of linear approximation is essentially limited by the volume strain involved rather than by shear.

"In addition to this type of limitation I would like to call your attention to another source of nonlinearity caused by the structural instability of semicrystalline and glassy polymers, a general feature which is completely overlooked by most of the authors who have investigated such materials. This instability results in a spontaneous variation of the magnitude of the viscoelastic parameters which characterize the stress-strain behavior of these materials at constant temperature and pressure and also that of their temperature and pressure dependence.[10]

"These variations depend essentially on the thermal history of the sample for small deformations, but they may be amplified by the strain history also when the deformation becomes large. If the sample is carefully annealed prior to mechanical investigations, these variations of the parameters may be quite small – at least during the duration of relatively short testing periods. However, without annealing, a twofold change in the storage moduli and a 10 to 100-fold change in the loss factor may be easily observed.[10] Even in the first case when the variations are small, one must realize that measured parameters characterize only the specimen with its particular thermal history.

"I think that important progress in the science of glassy and semicrystalline polymers could be achieved if these phenomena were taken into account and investigated more thoroughly."

Gordon Williams responded: "My statement was made in the context of the engineering applications of polymers and the practical utility of the theory. I fully admit that if you make precise measurements of the modulus, you will find nonlinear behavior down to any level of strain. However, if you are interested in engineering results and engineering accuracy, you will find that within these limits of strains, you can get useful results from the applications of the theory."

Kovacs reiterated that change in temperatures would change the material characteristics of, for example, polystyrene, but agreed that relaxation measurements at the new temperature could form the basis for satisfactory application of linear viscoelastic theory.

The Chairman emphasized the significance of Kovac's comments, that in stating a strain limit for linear behavior it is extremely important to prescribe whether dilatational or shear strain is being specified. Five percent volume strain would correspond to a very high pressure. He suggested that since viscoelasticity is commonly dominated by shear deformation, it is this component that is commonly referred to as the strain limit for linearity.

Schwarzl commented: "I am quite convinced that if one looks only at the limits of linearity in pure shear, then linear behavior occurs over an appreciable strain range, while shear strain with superimposed hydrostatic pressure gives nonlinear response. With linear behavior, hydrostatic effects and shear effects are independent. In fact there may very well be a strong interdependence between these two, which we don't generally realize because we simply don't test for it. They are in fact interrelated through free volume. The hydrostatic pressure changes the free volume and the shear properties are extremely sensitive to this." Schwarzl stated that he considered these effects to be the basis for Kovac's remarks. Similarly, temperature variation changes the free volume which affects the shear properties.

Gordon Williams mentioned that the assumption of constant bulk modulus, rather than constant Poisson's ratio, provided a usable basis for linear viscoelastic stress analysis of polymers. Schwarzl agreed that the bulk modulus changes only by a factor of two or three, while the shear modulus can change by a factor of 10,000 or more, thus supporting Williams' suggestion.

Takayanagi mentioned his measurements on single crystals, for which he carried out fast relaxation tests and dynamic modulus measurements. These were related according to linear viscoelastic theory below .01 percent strain, but not at 0.1 to 0.3 percent strain. The crystal response was linear only below 80 C. Thus, single crystals appear to be particularly sensitive for the onset of nonlinearity.

Rivlin mentioned work he had been associated with[11,12] on relaxation of PVC with various fillers. He found the modulus to be quite constant for unfilled PVC, but to become both strain and time dependent with the addition of fillers. Although the response was nonlinear, the time factor was separable. Menges corroborated such an effect in all amorphous materials such as PVC, polycarbonate, and other cellulosic materials which contain elastic filler particles larger than the grain size. He had found that the strain at which craze-cracking begins drops from about 1 percent in the unfilled material to about 0.5 percent. Andrews commented that this could be caused by a stress-concentration effect around the filler particles without the need to assume crazing, as he had suggested earlier in the discussion.

Thermo-Mechanical Coupling

The Chairman introduced this topic by commenting: "Viscoelasticity is an irreversible process during which mechanical energy is dissipated into heat. However, most polymers are poor conductors of heat, so that heat generated is not easily conducted away without appreciable temperature changes, to which the viscoelastic properties tend to be extremely sensitive. Thus thermal and thermo-mechanical coupling effects can play a dominant role in stress and strain analysis, and can introduce a significant complication in interpreting test results, for example, in steady-state oscillatory tests in which mechanical energy is continuously dissipated.

"In the mechanics literature, the thermodynamics of viscoelastic materials has been treated in terms of hidden variables and irreversible thermodynamics, or by replacing the thermodynamic functions of classical theory by corresponding functionals, i.e., by replacing quantities (e.g., internal energy, entropy, etc.) which depend on the current values of the state variables, by corresponding quantities which depend on the history of variation of the state variables. I would like us to consider whether either of these approaches comprises the most appropriate form for representing the thermodynamics of polymers, and what new information is needed to establish satisfactorily applicable theories in this area.

"To present an introductory example of what has been accomplished in this area, I will describe briefly the theory presented by R. A. Schapery[6] which is based on the hidden variable approach, and an application of this theory to a problem of wave propagation or oscillation.

"In Dr. Schapery's treatment, the state of a viscoelastic material is considered to be a function of n variables q_i, i = i, 2, . . . n; q_1 to q_6 being the strain components ϵ_{ij}. The remainder of the variables are termed "hidden", in the sense that the corresponding generalized forces Q_i, i = 7, 8 . . . n, are zero, the forces corresponding to ϵ_{ij} being the stress components σ_{ij}. Following the thermodynamics of irreversible processes, entropy is defined to be a function of q_i (i = i . . . n) and the internal energy U. Heating with q_i constant is considered to be reversible. Reversible components of generalized forces are defined from the entropy expression, and this leads to a relation for the increase in entropy while the coordinates are changing. Onsager's principle is introduced, and the coefficients of the linear form in the velocities which expresses entropy increase are chosen to be temperature dependent so that the time-temperature shift concept will apply. If a quadratic form in the generalized coordinates and temperature is chosen for the Helmholtz free energy, elimination of the hidden variables yields linear viscoelastic operators for the stress-strain relations and for the specific heat and thermal expansion.

"Application of this theory to forced oscillation of a rod or slab of material has been presented by Huang and Lee.[13] An oscillatory stress is applied at one end of the rod or one face of the slab. The resulting oscillatory wave causes heating due to the dissipation of mechanical energy, and this leads to softening of the material according to the time-temperature shift concept. The wave speed and hence wavelength are thus reduced, the attenuation and energy dissipation increased, so that localized high temperatures can occur. Such effects have been shown by J.

F. Tormey and S. C. Britton[14] to lead to overheating in oscillated solid propellant rocket grains.

"Schapery's approach has been generalized both by himself and by other authors to include nonlinear and finite deformation effects, and to remove the dependence on Onsager's relations. Some of its characteristics have been verified for certain materials. For example, the reduced time variable associated with the time-temperature shift concept has been shown by T. A. Johnson, C. W. Fowles, and E. H. Dill[15] to be valid for an epoxy resin in a varying temperature cycle. What about the operators corresponding to specific heat and thermal expansion? Is there sufficient information to specify these by hereditary integral kernels or by the constants appearing in differential operators?

"As mentioned previously, an alternative approach is to represent thermodynamic quantities directly as functionals of, say, temperature and strain history (see, for example, B. D. Coleman[16]). This formulation proceeds directly to the operator structure as developed from the thermodynamics of irreversible processes, but without certain connections arising from that theory. Is this a valid approach to the thermodynamics of polymers, and how can it be developed for practical application? Are there other ways of analyzing the thermo-mechanical problem for polymers?"

Rivlin opened the discussion: "As far as I can see, all that is needed in order to produce a viscoelastic theory that takes into account temperature effects is simply to make the kernels in the integrals to be functions of temperature, whether it is linear viscoelastic theory or nonlinear theory. I would like to repeat a remark that I made earlier in the Colloquium in connection with the nonlinear theories. Since we couldn't find the kernels of one expansion even at one temperature, I think that it is premature to try to find them at all temperatures. I don't think that the question of temperature dependence in viscoelasticity is a thermodynamic question at all. I think it is a question really of constitutive equations."

The Chairman suggested that one needed to go beyond simply taking the constitutive relation to depend on temperature, since one must face the problem of thermo-mechanical coupling and how the mechanical work done influences the temperature and subsequent behavior of the material. This influence, in the case of viscoelasticity, is certainly not so simple as for elasticity which is reversible. When asked about Newtonian fluids, the Chairman suggested that the assumption that the mechanical work dissipated by viscosity appears immediately as an equivalent heat source seems to be satisfactory. However, with viscoelasticity, relaxation effects arise, and such a simple assumption would not be valid.

In order to define the problem, Rivlin presented a discussion on the foundations of thermodynamics including irreversibility and functional relations: "With regard to the second item discussed by Dr. Lee, the attempt to apply thermodynamic principles to materials with memory, it might be well to remind ourselves of what classical Gibbsian thermodynamics actually says. The first law, expressing the conservation of energy, can be equally well applied in the case of irreversible deformation and temperature change processes, the internal energy being taken, where this is appropriate, as a functional of the deformation and temperature histories. The situation with regard to the second law is less happy.

"In Gibbsian thermodynamics, the entropy difference between two equilibrium states connected by a reversible path is defined as the integrated reduced heat flux (i.e., heat flux/absolute temperature). This is independent of the actual reversible path taken. If the same equilibrium states are connected by an irreversible path, then the integrated reduced heat flux is different and the second law of thermodynamics states that it is less than the entropy difference between the two equilibrium states.

"In application to an elastic material with internal friction, we can define the entropy difference for an isothermal quasi-static path connecting two equilibrium states and the second law then tells us that, for a dynamic path connecting these states, the integrated reduced heat flux is less than this entropy change. However, for a rate-independent material such as a plastically deforming metal, we cannot connect two states of deformation by a reversible path and, accordingly, we cannot define entropy change within the framework of Gibbsian thermodynamics. This is also true for rate-dependent materials in which changes between situations of thermodynamic nonequilibrium are considered. This statement may perhaps be qualified by the recognition that in certain situations it may be useful to define the entropy of a body in a nonequilibrium state as the entropy associated with an equilibrium state for which certain of the variables defining the situation are the same. As an example, we may take the case of a statically deformed elastic metal in which the temperature is nonuniform. Strictly, due to the heat flux, the situation is not one of thermodynamic equilibrium. However, if we are concerned only with the forces necessary to support the deformation, we may define the entropy at each point as that associated with the same state of deformation and temperature in a body in which the temperature is uniform. Alternatively and, I think, advantageously, we may make the constitutive assumption that the stress is independent of the temperature gradient and thus avoid the necessity of applying the second law to the nonequilibrium situation under discussion.

"Coleman's[16] much propagandized application of the second law to irreversible situations is, insofar as it is correct, a heavily disguised application to thermodynamically reversible situations. The materials to which his theory is applicable show instantaneous elasticity and, for sufficiently rapid deformations, we can talk about entropy change in a meaningful way. His theory is thus, insofar as it is correct, essentially a restatement of the Gibbsian thermodynamics of elastic materials.

"The second law of thermodynamics is usually extended to irreversible processes in the form of the Clausius-Duhem inequality

$$\dot{S} > \nabla \cdot \mathbf{Q}/\theta \quad , \tag{16}$$

where $\dot{S}$ is the rate of change of entropy, S, θ is the absolute temperature, and $\nabla \cdot \mathbf{Q}$ is the divergence of the heat flux vector **Q**. The difficulty in giving meaning to this statement resides in that of giving meaning to S and θ in irreversible situations. We may attempt to take refuge in the approach of statistical mechanics and define the entropy at a point on an irreversible path as the ensemble average of the logarithm of the reciprocal distribution function in phase space corresponding to the microscopic variables (or the deformation and temperature histories) defining the point on the path considered. The definition of θ presents even greater problems. However we define θ, it is not clear that it can have the relationship to the entropy and heat flux implied by the Clausius-Duhem inequality. To see this, we reflect that in a perfect gas in thermodynamic equilibrium, the internal energy and absolute temperature define completely the Maxwellian velocity distribution in the gas. On the other hand, if the gas is not in thermodynamic equilibrium, the velocity distribution will, in general, not be Maxwellian. In the absence of further knowledge of what is this distribution, we must assume that the internal energy and moments of the distribution of all orders must be known in order to define it with the same completeness as do internal energy and absolute temperature in the equilibrium case.

"It is for this reason that I feel it to be dangerous, in the absence of any clear definition of entropy and temperature for irreversible situations, to build too much superstructure on the Clausius-Duhem inequality. Of course, there may be situations of thermodynamic nonequilibrium in which the Clausius-Duhem inequality may be made physically meaningful and valid. These are likely to be situations in which the departure from thermodynamic equilibrium is not too great."

Halpin mentioned that in order to apply the theory of irreversible thermodynamics one has to make the initial assumption of Onsager's reciprocity principle. Rivlin agreed, but commented that irreversible thermodynamics has validity only in the limit of the linear case, and in his remarks he was not considering such a restriction. The Chairman observed that Schapery's work was initially carried out for linear relations[6], but had later been extended to nonlinear laws and finite strain[3], and that the need to assume Onsager's reciprocal relations had been removed by Valanis[17]. Also, the time-temperature shift concept had been introduced into the development. Pechhold felt that Rivlin's comments had been somewhat skeptical since one can construct many models which must be formulated in terms of statistical mechanics. If they are simple enough to be treated in this way, they can then also be treated by irreversible thermodynamics. Only plasticity cannot be modelled in this way. In replying, Rivlin again reiterated that irreversible thermodynamics can represent only situations close to equilibrium.

The Chairman interjected that he would like the discussion to turn more directly to the topic of viscoelasticity, and he posed certain problems. If the temperature of a polymer were to change suddenly, how would the volume change respond? Would it also increase suddenly, or grow like a creep function? What about the analagous problem of adding heat? Can one take the specific heat as a constant, or following heat input does the temperature rise over a period of time? Can it be assumed that the operator nature of thermal expansion and specific heat have natural times of such short duration that it would usually be permissible to consider the coefficient of thermal expansion and the specific heat simply to be constants?

Halpin commented that such questions can play an important role in the theory of polymer processing and the design of equipment, since in many situations appreciable amounts of heat are transferred from or to the material. He said that the assumption of constant values was common, but that these phenomena are not well documented. The Chairman mentioned that even for a constant coefficient of thermal expansion, the resulting thermal stress would relax with time because of the operator nature of the stress-strain relation. Halpin suggested that Dr. Kovacs be invited to comment since he has been active in this area. Kovacs responded: "Professor Lee's question may be answered in a general manner rather simply. Since temperature changes produce volume variations similar to the effect of pressure (P), temperature (T) may be conceived as a stress related to the strain through the expansion coefficient and the compliance.

"Considering only the case of homogeneous distribution of temperature in the sample, viscoelastic effects may be observed if the volume recovery at constant T has a smaller rate than the rate of temperature changes, controlled by the heat diffusion in the specimen. The glass transition phenomenon of super-cooled liquids, as revealed by dilatometry, is one of the most common aspects of bulk viscoelasticity.

"In dynamic tests, such as mentioned by Prof. Pechhold, the expansion coefficient may be imagined as a complex factor, just as the shear compliance. Its real part has two limiting values, the upper one corresponding to the liquid-like (equilibrium) and the lower one to the glassy expansion, to which only the vibrational motions contribute.

"The main difference between shear and volume viscoelasticity is that the latter, as I already mentioned, has a much smaller practical strain limit within which linear approximation can be safely applied. This is due to the fact that in condensed systems the configuration mobility of the molecules is extremely sensitive to their packing density. In fact, according to the free-volume concept, nonlinear behavior prevails for any finite change in volume.

"Similar considerations may be applied to the simultaneous changes of enthalpy, the heat capacity being the compliance. According to our present state of knowledge, the time dependences of volume and enthalpy changes, produced by temperature variations, are loosely related."

Shen asked whether the question concerned the influence of deformation on the heat capacity. The Chairman replied: "The question I was asking was whether the relationship between added heat and the resulting temperature rise is a time operator relation or a function relation, whether the increase in temperature is proportional to the heat added or whether it is a viscoelastic-type operator relation. A direct effect of deformation on the heat capacity would be a coupling effect which would transfer the relations beyond the scope of linear analysis and I was thinking in terms of a linear system."

Halpin closed the discussion on this topic with the comments: "I think that, in general, transport properties, specific heat being one of them, depend upon the instantaneous free volume of the system. If one studied the molecular volume, it would be a continuously varying parameter of the system, and depend upon the previous history of the body. For engineering calculations, it might be permissible to assume some of them to be constant, but in fact they are all variables just like the modulus."

The Yield Phenomenon in Polymers

The Chairman opened the discussion: "In this morning's session, Prof. Brinson discussed the yield behavior of polycarbonate. I would like to bring up the question of yield in polymers for general discussion, particularly since it can play an important role in processing techniques, as illustrated by Prof. Takayanagi also in this morning's session.

"As brought out by Dr. Vincent yesterday, yield comprises flow at approximately constant stress, which is less rate dependent than the deformation prior to yield. I would like comments on the macroscopic characteristics of the yield phenomenon, and also on the molecular motions which yield such behavior."

Vincent opened the discussion: "I would like to try to clarify the two basic problems of yield behavior. It doesn't really matter as far as basic behavior is concerned, whether tension, compression, shear, or any other suitable stress system is applied, nor whether one loads the material at a constant strain rate or by doing the experiments in an isochronous manner. And it is not of basic importance whether the material crazes or not. The essential phenomenon is that the true stress – true strain curve is smoothly turning and generally levels out; the yield is not a special phenomenon, it is merely part of the curvature of the stress-strain curve. The reason why it is normally thought to be a special phenomenon is that it is a very simple constitutive measure and thus, experimentally, many yield surfaces are measured, and also it is of great practical importance. The stress-strain curves for large strains might go gently upward, might level out flat, or could even go downwards slightly, even if we are considering true stress. From an analytical point of view, therefore, yield is only one part of the general nonlinear viscoelastic problem, and it is no more or no less solved than the whole of nonlinear viscoelastic theory. From a molecular point of view, however, in a general qualitative way it is clear what is happening. The shear component of the applied strain is increasing the segmental molecular mobility in much the same way as would increasing the temperature. It is of course complicated by the effects of the hydrostatic component. This gives the differences between the application of tension or compression, but the basic behavior in the curvature of the stress-strain curve is related to the shear strain. Thus the problem one has to face is either to extend nonlinear viscoelasticity or to know in more detail what is happening to the molecular segments as the shear strain is increased."

Andrews asked if all polymers without exception are covered by the description, and Vincent replied that he was talking about thermoplastics, that is, amorphous polymers below the glass transition temperature or crystalline polymers below the melting point.

Andrews commented: "I have found that very often in semicrystalline polymers a sharp transition occurs. I agree that one always gets curvature but I believe that there is such a thing in physical terms as the yield phenomenon in semicrystalline polymers."

Vincent replied: "I know of two special cases, although there may be others. For instance, in some high-impact polystyrenes, one does get an abrupt craze phenomenon, and Kambour and Bernier reported on a phenomenon in PPO submerged in acetone where an abrupt change takes place, but these are rare, not the general problem which I was addressing myself to."

Whether yield is a separate phenomenon or simply an aspect of nonlinear viscoelasticity appears to depend on just how it is defined. Gordon Williams pointed out that in plasticity, yield is determined by the onset of permanent deformation. With viscoelastic strain, recovery after unloading prevents such a precise definition, and yield has come to mean an abrupt change in the stress-strain curve. This definition is not directly related to irreversibility. Hull mentioned that for polystyrene which has been quenched and subsequently aged, the stress-strain curve in compression exhibits a maximum which leads to macroscopic instability as for necking in tension. Inhomogeneous deformation occurs and then it is difficult to interpret the measured true strain. Williams pointed out that this definition of yield is completely divorced from irreversibility. Menges observed that yield of the type described by Hull depends on the geometry of the test specimen, and he added that for a material such as polycarbonate, the mechanical behavior can depend greatly on the conditions of manufacture.

Schwarzl commented: "If stress-strain curves are investigated at different temperatures, at low temperatures very sharp yielding occurs and the curves change more or less gradually over to the rubbery curves. This whole phenomenon occurs over a range of 30 to 40 C below or around the T_g of the material. The molecular process of yielding over this range is the same as that occurring along the stress-strain curve for the rubber. The orientation of the chain molecules is finally the same. The processes occurring right at the yield point may be different, but the ultimate state is the oriented state of the strained molecules. If one waits long enough or if the temperature is increased, all of the yielded samples return to their original state, just as the rubber samples do when they are released.

Vincent expressed surprise at the sharpness of the yield described, and Schwarzl commented that a very sharp peak occurs at a certain temperature. He found that if a normal rubber is drawn at a low enough temperature, a sharp peak and necking phenomenon results.

Pechhold commented on yield from the molecular standpoint. As mentioned by Schwarzl, it is connected with the glass transition temperature and the compliance of rubbers at high frequencies. Pechhold's comments referred to materials exhibiting smooth stress-strain curves without maxima and minima. Then, at low temperatures, the bend in the stress-strain curve can be expected to be sharp because the glass transition is sharp. The activation spectrum at a fixed temperature is sharp at low temperatures, but broadens out as the temperature increases. Above T_g, normal rubberlike response occurs. Referring to the first slide of his paper in Session III, he assumes for an elementary process in high-frequency shear, gliding of a half meander. Knowing the dimensions of meanders from electron microscopy, the rubberelastic compliance can be calculated and is found to agree with measurement. He found that the same model could be used to reproduce the results Brinson had shown in the morning for the relation between stress and the logarithm of the deformation rate. The activation energy relation for loading with shear stress σ gives:

$$\dot{\epsilon} \sim e^{\sigma\Omega} \quad , \tag{17}$$

where

$$\Omega = kT\ \partial\sigma/\partial(\ln \dot{\epsilon}) \quad . \tag{18}$$

Brinson's results of a linear relation between σ and $\ln \dot{\epsilon}$ give Ω to be about 1000 A^3, which compares in a general way with the size of the meanders. Pechhold also suggested that brittleness might arise because the stress cannot raise the molecular mobility to high enough levels to create sufficiently large strain rates. The glass transition temperature is mainly due to lack of free volume. The activation curve has a free-volume component and also an internal elastic-potential component. If the temperature is too low for stress to activate the thermal processes, then brittleness may result.

Referring to Schwarzl's comments, Knappe remarked that he did not think that the deformation in cold drawing is rubberlike. Very high anisotropy is observed in the cold-drawn regions which cannot be explained by the segmental model. Schwarzl replied that the mechanism was stress induced, but that the final effect was approximately the same. He mentioned specimens drawn into sheets above T_g, and then cooled below T_g and tested. Their properties were similar to those of cold-drawn sheets, which may be an indication that the final state is about the same.

With regard to Pechhold's comment, Kovacs mentioned work by Dr. Petrie of the Kodak Laboratory on polyethylene-terephtalate (PET), which shows that temperature may not be the most important factor in the yielding and fracture behavior of glassy polymers. "A freshly quenched amorphous PET film can show at room temperature (T_g - 65 C) a typical yielding behavior and it can be drawn to about 5 or 6 times its original length before fracture occurs. The same sample at the same temperature shows, however, a brittle fracture behavior, at a few percent strain, if kept at room temperature 1 day prior to the tensile test. This effect is well known in every plant where PET sheets are manufactured, and probably Dr. Vincent is familiar with it." Schwarzl suggested that this behavior can be attributed to the decrease of free volume during the 1-day storage period. Kovacs replied that this was exactly what he wished to suggest.

From the discussion of yield in polymers, it is clear that much work is needed to define it precisely for different materials and conditions of testing. Brinson suggested that yield is much more complicated in polymers than in metals, and that it is important to delineate the phenomenon for further study.

REFERENCES

1. Leaderman, H., *Trans. Soc. Rheol.*, **6**, 361-382 (1962).
2. Lai, J.S.Y., and Findley, W. N., *Trans. Soc. Rheol.*, **17**, 129-152 (1973).
3. Schapery, R. A., *Proc. 5th U.S. Natl. Cong. Appl. Mech.*, ASME (1966), pp 511-530.
4. Berstien, B., Kearsley, E. A., and Zapas, L. J., *J. Research Nat. Bur. Std.*, Ser. B, **68**, 103-113 (1964).
5. Pipkin, A. C., *Rev. Mod. Phys.*, **36**, 1034-1041 (1964).
6. Schapery, R. A., *J. Appl. Phys.*, **35**, 1451-1465 (1964).
7. Green, A. E., and Rivlin, R. S., *Arch. Ratl. Mech. Anal.*, **1**, 1-21 (1957).
8. Goldbach, G., and Rehage, G., *J. Polymer Sci.*, Part C, **16**, 2289 (1967).
9. Ferry, J. D., *Viscoelastic Properties of Polymers*, 2nd Ed., J. Wiley & Sons, New York (1970).
10. Kovacs, A. J., Stratton, R. S., and Ferry, J. D., *J. Phys. Chem.*, **67**, 152 (1963).
11. Rivlin, R. S., *Viscoelasticity*, J. T. Bergen (Ed.), Academic Press (1960), pp 93-108.
12. Bergen, J. T., *ibid.*, pp 109-132.
13. Huang, N. C., and Lee, E. H., *J. Appl. Mech.*, **34**, 127-132 (1967).
14. Tormey, J. F., and Britton, S. C., *AIAA J.*, **1**, 1763 (1963).
15. Johnson, T. A., Fowles, C. W., and Dill, E. H., *Proc. 5th Int. Congr. on Rheol.* (1970), Vol 3, pp 349-355.
16. Coleman, B. D., *Arch. Ratl. Mech. Anal.*, **17**, 1-46 (1964).
17. Valanis, K. C., *J. Math. Phys.*, **45**, 197-212 (1966).

Part Five

FRACTURE

CRACK GROWTH AND FAILURE IN RUBBERS

A. G. Thomas

The Natural Rubber Producers' Research Association
Welwyn Garden City
Hertfordshire, England

This paper is in part a review of the work by the author and his associates on the crack growth and associated failure processes of rubbers.

The approach is essentially that of fracture mechanics, the relevant parameter being the energy available for crack propagation, and which has been termed the tearing energy, denoted by T. Numerous experiments have been carried out under conditions of both steady and repeated stressing[1-4], using a variety of test pieces to show that when the crack growth, or tear behavior, is expressed in terms of T the results are independent of the shape of the test piece.

Under repeated stressing in which the test piece is returned to the unstressed state during each cycle the crack grows from an initial razor cut first at a high rate with the tip remaining sharp, but after a few hundred cycles, the tip becomes rough and, for a similar T value, the rate of growth diminishes by about a factor of 10. This effect is important when considering the relation between fatigue failure under many repeated stressings and tensile failure in one stressing. The theory that has been developed for fatigue failure of rubbers[3,5] is based on the concept of crack growth from small flaws present in the material. This has been successful in explaining the general dependence of fatigue life on strain, suggesting a flaw size of the order of 2×10^{-3} cm. The origin of these flaws can be particulate dirt, or cutting or molding imperfections, but there are indications that the rubber itself has inhomogeneities of a not much smaller scale which may be associated with the tip roughening referred to above. At high fatiguing strains, greater than about 250 percent, the life becomes shorter than the simple fatigue theory predicts, and this may be ascribed to the relatively rapid growth of a smooth tip for the first few hundred cycles. When still higher strains and shorter lives are considered, the situation is complicated by the fact that the growth of a crack during a stressing cycle is comparable with the size of the crack at the start of the cycle. Allowance has therefore to be made for this effect. Carrying out this approach enables a theory of fatigue failure to be developed which embodies tensile failure in one cycle, that is, failure as in a normal tensile strength test, viz., as essentially a one-cycle fatigue failure.[6]

Fatigue experiments at high strains have been carried out to test the theory. The size of the flaw which initiates failure is an important parameter, and it was found convenient to vary this by exposing the slightly stretched natural rubber samples to attack by ozone. The agreement with theory was, in general, found to be quite good.

The variation of tensile strength with exposure to ozone suggests that measurement of this property may provide a suitable means of assessing the susceptibility of materials to ozone attack. The reduction in ozone crack growth due to nonrubbers present in the vulcanizate is readily shown by the smaller effect on the tensile strength.

Under sufficiently low stresses, either steady or repeatedly applied, it has been found that a T value exists below which, in the absence of ozone, no crack growth occurs. This limiting value, which has been termed the crack-growth limit, is denoted by T_0. There is evidence that this quantity, which is of the order of 5×10^4 ergs/cm^2, is determined by the molecular constitution of the vulcanizate and, in particular, the mechanical strength of the chemical bonds.[7] At higher T values, the crack grows either in a time-dependent manner, in the case of noncrystallizing elastomers, or only while the load, and hence the T value, is increasing in the case of strain-crystallizing elastomers such as natural rubber. The crack-growth behavior in this region is strongly influenced by the hysteresial behavior at the high strains existing at the tip of the crack.[7]

The influence of oxygen and antioxidants is well established. In particular, the value of T_0 is substantially reduced in the presence of oxygen, the effect being mitigated by the addition of antioxidant. Corresponding effects occur at T values somewhat above T_0, the value of T for a given crack-growth rate being proportional to T_0. This would be expected on the basis that the fundamental strength property of the vulcanizate is T_0 and that the T value required to produce a given crack-growth rate r is proportional to T_0 and another factor involving the hysteresis. This would parallel the behavior found for the failure of soft adhesives by Gent and Schultz[8] where interfacial energy plays a similar role to T_0. However, at high T values, this simple approach appears not to hold, since in this region the effects of oxygen and antioxidants diminish.[5] It is now suggested that this is because at the higher crack-growth rates, when the rubber at the crack tip is stretched to rupture in a very short time, oxygen has insufficient time to produce a deleterious effect. If tests at the high tearing energies are carried out sufficiently slowly, then the effects reappear.[9] Tensile-strength measurements in air are not usually sensitive to the presence of antioxidants, which, on the basis of the above theory, would be due to the relatively rapid average rate of crack growth during the loading process. On reducing the stretching rate so that rupture occurs after some hours, the expected influence of antioxidant does appear.

The crack-growth behavior of natural rubber under repeated stressing is substantially altered if the stress is not relaxed to zero during each cycle. That is, the crack-growth rate is reduced considerably, and in fact appears to become zero at T values much higher than the T_0 value obtained when the test piece is completely relaxed during each cycle. This is believed to be associated with the ability of natural rubber to crystallize under strain, and the tendency of the tip to remain crystalline if the stress is not completely removed. A semiquantitative theory[10] has been developed to explain this effect, based on an approach by Andrews[11]. This treats the crystallization as introducing some set at the tip and hence effectively increasing the tip diameter.

REFERENCES

1. Rivlin, R. S., and Thomas, A. G., *J. Polymer Sci.,* **10**, 291 (1953).
2. Thomas, A. G., *J. Polymer Sci.,* **31**, 467 (1958).
3. Gent, A. N., Lindley, P. B., and Thomas, A. G., *J. Appl. Polymer Sci.,* **8**, 455 (1964).
4. Lake, G. J., Lindley, P. B., and Thomas, A. G., Proceedings of the Second International Fracture Conference (1969), p 493.
5. Lindley, P. B., and Thomas, A. G., Proceedings of the Fourth Rubber Technology Conference (1962), p 428.
6. Thomas, A. G., Proceedings of the Conference on Physical Basis of Yield and Fracture, Institute of Physics and the Physical Society, Oxford (1966), p 134.
7. Lake, G. J., and Thomas, A. G., *Proc. Roy. Soc.* (London), **A20**, 211 (1967).
8. Gent, A. N., and Schultz, J., Proceedings of the International Rubber Conference, Brighton (1972).

9. Lake, G. J., to be published.
10. Lindley, P. B., to be published.
11. Andrews, E. H., *J. Mech. Phys. Solids,* **11,** 231 (1963).

DISCUSSION on Paper by A. G. Thomas

ANDREWS: Our studies of fatigue fracture in polyethylene show that the approach presented by Dr. Thomas is applicable and valuable for thermoplastics as well as rubbers. There are several important differences, of course. Firstly, it is necessary to modify the analysis to take account of the bulk inelasticity of the polymer. Secondly, the curve of crack growth per cycle versus the energy release rate (on a log-log basis) shows two separate and noncontinuous regions corresponding to brittle and microductile crack propagation, respectively. Thirdly, the intrinsic flaw size, c_0, correlates closely with the spherulite size in the semicrystalline polymer. We think c_0 may be associated with the interspherulite boundaries.

FAST FRACTURE IN PMMA

M. F. Kanninen, A. R. Rosenfield, and R. G. Hoagland

Battelle
Columbus Laboratories
Columbus, Ohio

ABSTRACT

A combined experimental and analytical study of rapid crack propagation in PMMA is described. The experiments employed wedge-loaded, double-cantilever-beam test specimens with blunt starting notches. For the analysis, a finite-difference procedure was used to solve the equations of motion of a beam-on-elastic-foundation model of the test specimen. In both experiment and analysis, crack propagation was found to occur at an essentially constant speed which was maintained until shortly before crack arrest. The magnitude of the observed speeds depended upon the initial bluntness (but only slightly on other dimensions of the specimen) and bore the same relation to the elastic-bar-wave speed C_0 as that observed in similar experiments on high-strength steel. The frequency of parabolic markings on the fracture surface was found to be related to the dynamic fracture toughness.

1 INTRODUCTION

Unlike metals and other essentially rate-independent materials, the behavior of polymers under an applied stress depends on time. The consequent analytical difficulties have generally prevented appropriate treatments of experimental results to be developed. A way of circumventing some of the problems is to devise an integrated program of analysis and experiment centered on a test specimen in which many of the complications arising in the general case can be safely ignored. In this way, quantitative descriptions of the deformation and fracture process in polymers can be obtained. An approach employing the double-cantilever-beam (DCB) specimen is described in this paper, with specific application being made to the problem of rapid, unstable, crack propagation and crack arrest in PMMA.

Previous experiments in this laboratory[1-5] have demonstrated that when a crack is propagated from a blunt notch in the DCB specimen, a region of approximately constant speed is observed. Provided the starting notch is not too blunt, deceleration and crack arrest also occur within the sample, thus encouraging its use as a vehicle for characterizing crack arrest. A beam-on-elastic-foundation augmentation of the simple cantilever-beam models of the DCB specimen used by Benbow and Roessler[6], Gilman[7], and Berry[8] was devised to provide quantitative interpretation of these experiments. While this model was found to be quite accurate in the static situation[9], a quasi-static extension to treat unstable crack propagation was unsuccessful. Accordingly, a fully dynamic treatment was undertaken in which the inertial forces were incorporated into the beam-deflection equation. The model response was then in accord with many of the essential features exhibited by the experimental results on steel.[4,5] In this paper, propagation and arrest measurements on PMMA are described and compared with the predictions of an improved version of the dynamic beam-on-elastic-foundation model.

2 EXPERIMENTAL PROCEDURE AND RESULTS

The data appearing in the literature on the dynamic toughness (or, equivalently, the dynamic fracture energy) of PMMA for crack speeds comparable to those observed in the experiments reported in this paper support two main generalizations:

(1) Dynamic toughness increases sharply with increasing crack speed[10-16]

(2) The fracture surface contains parabolic markings whose density increases with speed.[15,19]

These observations are consistent if, as is often assumed, the formation of parabolic markings is a major energy-absorbing process in rapid fracture of PMMA.

Upon closer examination, however, discrepancies can be found among the results reported by different investigators. These arise not only because of differences in interpretation, but because of the significant physical variations in the properties of PMMA. The variations in properties arise not only among different batches but also within a single batch. Consequently, wide scatter bands accompany the presentation of many of the dynamic-fracture data on PMMA appearing in the literature.

The experimental techniques used in this study have been described extensively in previous papers.[1,3,4] Of greatest interest is the fact that the wedge-loaded DCB specimen allows rapid fracture to extend from a blunt starting notch and to propagate under conditions in which no external work is done on the specimen. In this way, a curve for crack length versus time was generated using a resistance grid whose output is fed into a storage oscilloscope. With the exception of cracks arrested in or just beyond the test grid, the crack length – time data were almost always in direct proportion to one another. Hence, the corresponding crack speed was easily and accurately determined.

A typical crack length – time test result for PMMA is shown in Fig. 1. A photograph of the fracture surface is attached to the diagram. It can be seen that deceleration (i.e., decreasing slope of the crack-length – time curve) is accompanied by a sharp drop in the frequency of the parabolic markings. It might also be noted that fairly consistent behavior was always obtained between the new and old batches of specimens even though difficulties have been reported by others.[20]

Figure 2 shows a collection of steady-state crack-speed results (typified by the result shown in Fig. 1) as a function of the applied stress-intensity factor K_q associated with the blunted notch. These results demonstrate that the experimental points lie within the limits of the scatter band of the results reported previously.[3] In addition, the results demonstrate that the crack speed for a given bluntness is not particularly affected by the specimen dimensions – a key point which the analytical model also reflects, as discussed below.

Specifying crack arrest in PMMA presents a problem in that propagation continues at an ever-decreasing rate for several days.[2] However, there is often a point where an extremely rapid deceleration occurs, reducing the velocity by several orders of magnitude over a short distance and producing "halt lines" on the fracture surface.[21,22] The positions of these halt lines were thus used with the static beam-on-elastic-foundation model to provide an ostensible value of K_a, the stress intensity at arrest. More precisely, in PMMA, K_a defines the end of the rapid-propagation phase

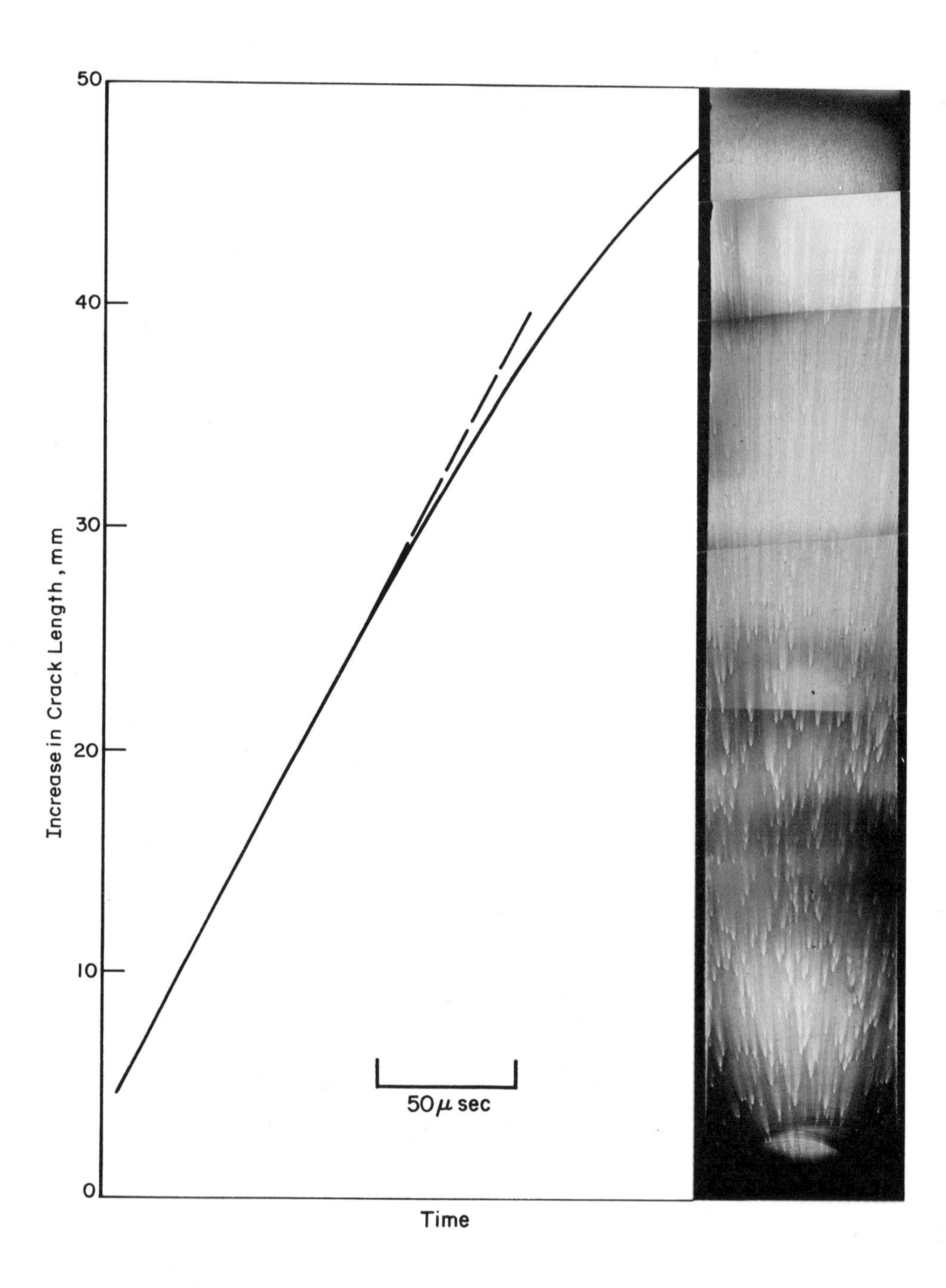

Fig. 1. Steady-state crack propagation and eventual deceleration in a wedge-loaded DCB specimen of PMMA. The crack length vs. time record is accompanied by a photograph of the fracture surface. For this specimen, K_q = 4 $MNm^{-3/2}$ and V = 240 m/s.

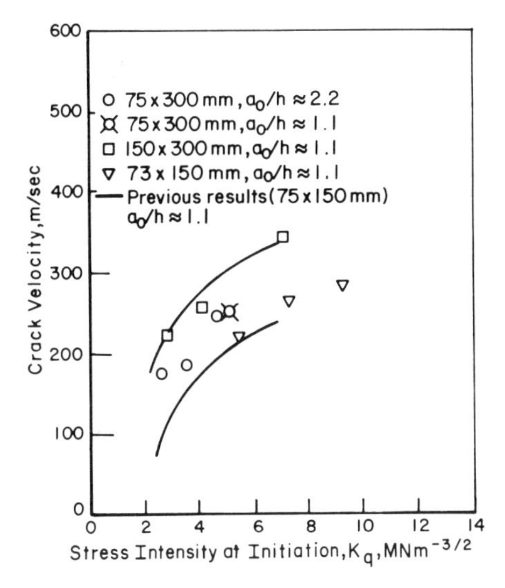

Fig. 2. Relation between steady-state crack speed and the stress intensity at the onset of crack extension in wedge-loaded DCB specimens of PMMA.

and the onset of very slow crack growth.

Since the parameter K_q reflects the stress distribution in the notch-tip region only at the onset of fracture, additional measurements are needed to characterize the situation throughout the propagation event. We use the symbol K_d to represent dynamic toughness. An estimate of the average value of K_d during crack propagation can be calculated from crack-arrest data by assuming that the change in potential energy during crack propagation is all converted to fracture energy. Then, it can be shown that[1]

$$K_d = \sqrt{K_q K_a} \tag{1}$$

As shown in Fig. 3, our experiments indicate that K_a is substantially independent of K_q between initiation values of 1.3 and ~3 $MNm^{-3/2}$. The constancy may extend even higher, as suggested by the one point at 3.5 $MNm^{-3/2}$. But, starting stress intensities at this level and above resulted in crack propagation splitting the sample in two with no arrest. The value of K_{Ic} for this batch of material was determined by reinitiating arrested cracks. It was found that $K_{Ic} = K_a = K_d(V = 0) = 1.3\ MNm^{-3/2}$.

With the aid of Eq. 1, the average dynamic toughness can be calculated for values of K_q below 3.5 $MNm^{-3/2}$. For higher values of K_q, K_a is not known. In this range, K_d was estimated by assuming that the constancy of K_a extends to all of the experimental conditions used here. An approximate check of this assumption was also made from measurements of the frequency of parabolic markings on the fracture surface (see Fig. 1), which Cotterell[19] has suggested is directly proportional to dynamic fracture energy. Our data at the low values of K_q (where K_a is known) display a distinct intercept. This is shown in Fig. 4. The best straight line through these data has the equation:

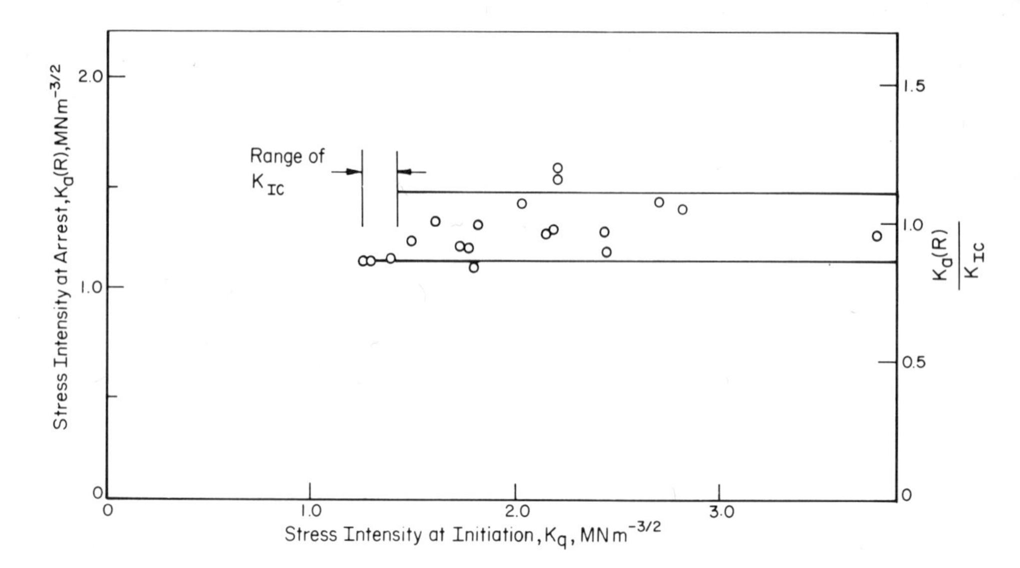

Fig. 3. Effect of stress intensity at the onset of crack extension on stress intensity at crack arrest in PMMA.

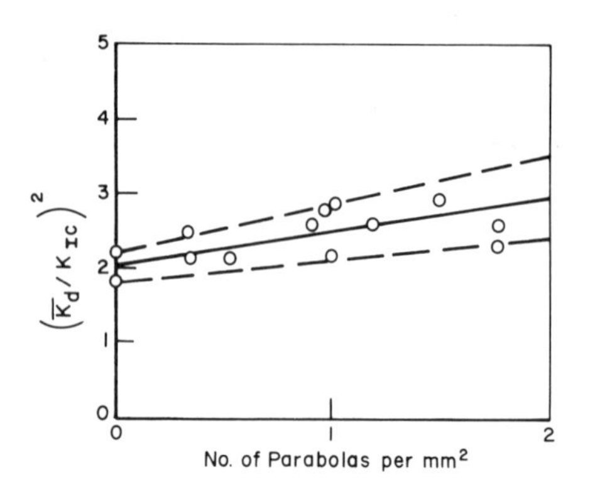

Fig. 4. Relation between dynamic fracture toughness and the frequency of parabolic markings on the fracture surface in PMMA.

$$\frac{\overline{K}_d}{K_{Ic}} = \left(2 + \frac{N}{2}\right)^{1/2} , \tag{2}$$

where N is the number of parabolas per mm^2. When Eq. 2 was extrapolated to the higher values of N associated with higher values of K_q, almost all of the resulting values of $\overline{K}_d$ fell between 0.9 and 1.0 times the values calculated from Eq. 1. This result is encouraging considering the scatter in the data used in defining Eq. 2 and the large extrapolations required to define the highest $\overline{K}_d$ values. The result is also consistent with the idea that there are two additive contributions to fracture energy: formation of craze material and of parabolas.

3 THEORETICAL ANALYSIS

The one-dimensional beam model is very attractive for dynamic crack propagation because the additional variable of time can be accommodated without an inordinate increase in mathematical complexity. The simple built-in cantilever-beam models[6-8] cannot, however, be easily employed in a dynamic treatment. Consequently, an extension of the simple model was developed by taking the upper half-specimen as a beam which is partly free and partly supported by an elastic foundation, the foundation modulus representing the transverse elasticity of the specimen in the uncracked region. As shown in Ref. 9, the stress-intensity factors obtained from this model are then in excellent agreement with well-established experimental and computational results for the DCB specimen. Moreover, the presence of the foundation allows a direct calculation of crack growth without the necessity of postulating an additional arbitrary crack-extension criterion such as that used by Burns and Bilek[23].

The computational procedure to be followed here is illustrated in Fig. 5. In essence, the dynamic model employs the classical Euler-Bernoulli beam and the Winkler foundation. Hence, only the lateral inertial forces enter into the equation of motion. The model is therefore the same as that used previously[4], with the exception of the manner in which the initial blunting of the crack tip is simulated. As shown in Fig. 5(a), the latter effect is accomplished by considering that a shearing force Q acts at the crack tip. The magnitude of Q is not arbitrary. Instead, the force acts to keep the beam deflection at the crack tip from exceeding w_c, the crack-tip deflection that must be achieved for a perfectly sharp crack to extend. In this way, δ, the displacement of beam ends, can be made arbitrarily greater than the level that would cause crack extension at a fracture toughness K_d of the material. This allows an "overstress" corresponding to the stress-intensity factor K_q to be applied, just as is actually accomplished by blunting the crack.

The view that a critical-strain-energy release rate exists is equivalent to considering that crack extension occurs when the deflection at the crack tip exceeds a critical value. While the distinction is unimportant in the static case, it is very useful in the dynamic situation because it provides a quantitative (and consistent) crack-extension criterion. Specifically, the "springs" contained in a small interval of length Δa ahead of the crack tip can be considered to break when they are stretched a critical amount w_c. For linear springs, the energy lost from the system is then $\frac{1}{2}kw_c{}^2\Delta a$, where k is the foundation modulus. This must be just equal to $\frac{1}{2}Rb\Delta a$, where R is the energy absorbed per unit area of crack

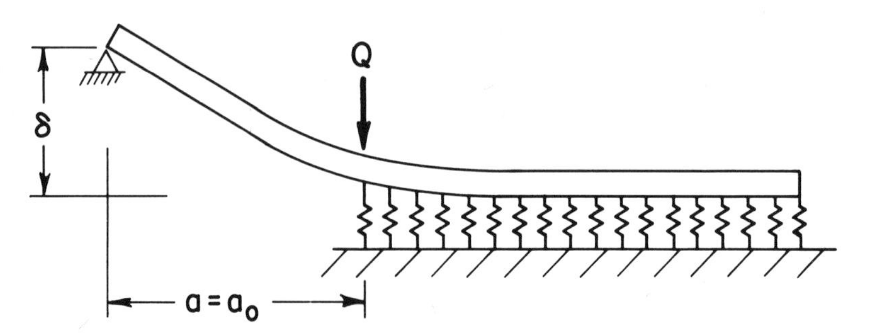

(a) Equilibrium Configuration Prior to Onset of Unstable Crack Propagation

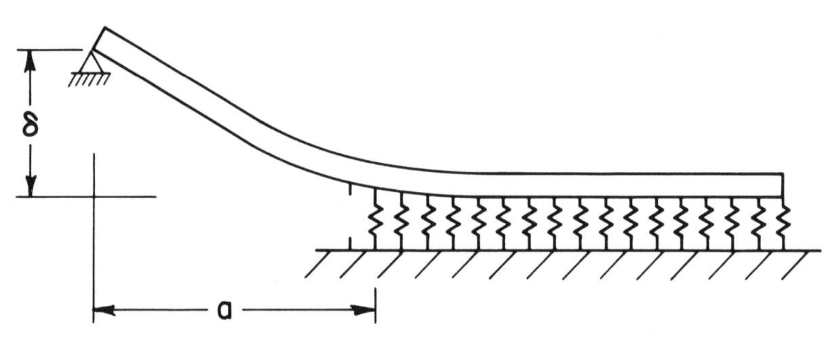

(b) Configuration Just Following the Onset of Unstable Crack Propagation

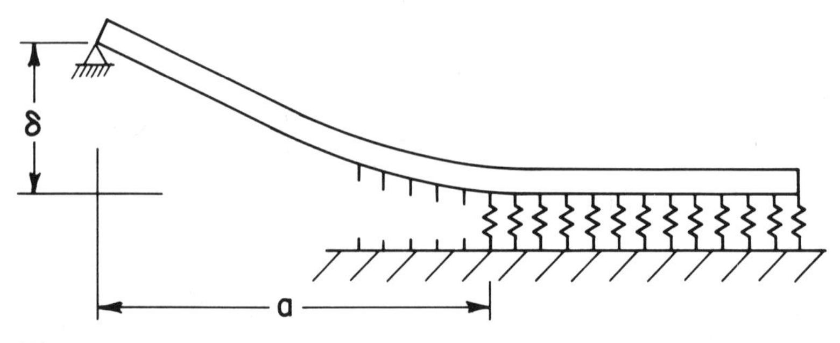

(c) Configuration During Unstable Crack Propagation

Fig. 5. Calculation of unstable crack propagation using the beam-on-elastic-foundation model.

extension and b is the specimen thickness. Equating these two expressions then gives

$$w_c^2 = \frac{Rb}{k} \tag{3}$$

In analogy with the static situation, an equivalence between the critical-strain-energy release rate and the fracture toughness can be postulated. Hence, using the relation $k = 2Eb/h$, given in Ref. 9, allows Eq. 3 to be rewritten in the alternative form

$$w_c = \left(\frac{h}{2}\frac{R}{E}\right)^{\frac{1}{2}} = \left(\frac{h}{2}\right)^{\frac{1}{2}} \frac{K_d}{E} \quad , \tag{4}$$

where h is the beam height, E is the elastic modulus, and, as before, K_d denotes the dynamic fracture toughness.

For the case where K_d, E, and h are constants, Eq. 4 shows that w_c is similarly a constant, independent of time and position. In this circumstance, the equation of motion for the beam deflection $w = w(x,t)$ in the problem shown in Fig. 5 is

$$EI \frac{\partial^4 w}{\partial x^4} + \rho A \frac{\partial^2 w}{\partial t^2} + kH^*(w_c - w)\, w = 0 \quad , \tag{5}$$

where I is the moment of inertia of the beam, ρ is its density, and A is the cross-sectional area. Also, $H^*(x)$ denotes the ordinary Heaviside step function

$$H(x) = \begin{cases} 1, & x > 0 \\ 0, & x < 0 \end{cases}$$

modified such that a spring once broken remains broken.

The boundary conditions for the problem shown in Fig. 5 are those for a simply supported left-hand end and a stress-free right-hand end. Hence,

$$w(o,t) = \delta, \; w''(o,t) = 0$$

and

$$w''(L,t) = w'''(L,t) = 0 \quad , \tag{6}$$

where the primes denote differentiation with respect to x. The initial conditions are obtained by finding the solution to the static counterpart of Eq. 5, which includes a shearing force Q such that $w(a_o,o) = w_c$. This is most easily done by separating the region of interest into the two parts where the differential equation has constant coefficients. General solutions are then readily obtained, and the arbitrary constants are used to satisfy Eq. 6 and to match the deflection and its first three derivatives at the interface between regions, it being recognized that the third derivative has a jump discontinuity proportional to Q. The resulting equation for the beam deflection is similar to that given in Ref. 9 and will not be written out explicitly here. Suffice to say that the initial beam deflection is known in terms of the specimen geometry and applied loads. The auxiliary equations are then completely specified by considering that the initial deflection velocity is zero.

The solution to Eq. 5 was obtained using a finite-difference procedure appropriate for a parabolic partial-differential equation.[24] To facilitate the computation, the specific relations for a rectangular cross section, $I = bh^3/12$, $A = bh$, and $k = 2Eb/h$, were introduced into Eq. 5, whereupon convenient dimensionless variables were found to be $W = w/w_c$, $\xi = x/h$, and $\tau = C_o t/\sqrt{12}\,h$. Here, $C_o = \sqrt{E/\rho}$ is the elastic-bar-wave speed. As a consequence of these transformations, the computational results depend only upon the ratios L/h, a_o/h, and either δ/w_c or, equivalently, K_q/K_d. The resulting crack speeds will similarly be determined in ratio form; specifically, as V/C_o.

An important use of the analysis in connection with rapid crack propagation and arrest experimentation is that it provides a way of determining the partition of energy during the crack-extension process. Specifically, for the Euler-Bernoulli beam and the Winkler layer, the strain energy is given by*

$$U = \int_0^L EI \left(\frac{\partial^2 w}{\partial x^2}\right)^2 dx + \int_0^L kH^*(w_c - w)\, w^2\, dx \ , \tag{7}$$

while the kinetic energy is given by

$$T = \int_0^L \rho A \left(\frac{\partial w}{\partial t}\right)^2 dx \ . \tag{8}$$

These can be routinely evaluated as a part of the computation. A typical apportionment of energy for the geometric configuration used in the PMMA experiments is shown in Fig. 6. Most importantly, it can be seen that the kinetic energy is negligibly small at crack arrest. This supports Eq. 1, the approximate relation for K_d, which is based upon the difference between the static-strain-energy values at initiation and arrest.

---

*Note that the usual factor $\frac{1}{2}$ does not appear because U and T each represent a total for the two halves of the specimen.

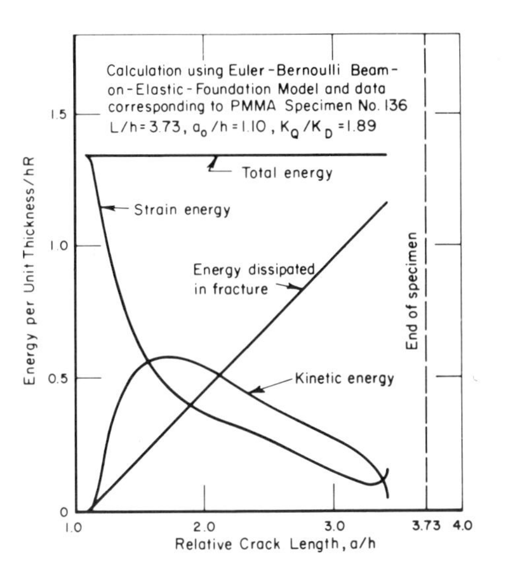

Fig. 6. Components of energy as a function of crack length during unstable crack propagation.

Finally, for completeness, it must be shown that energy is completely accounted for in the formulation described here. This can be done by showing that the change in the sum total of the energy with respect to crack length is zero. Equivalently, the sum of the strain energy and the kinetic energy can be shown to be equal to the rate of change in the fracture energy. First, note that

$$\frac{d}{da}(U+T) = \frac{d}{dt}(U+T)\frac{dt}{da} = \frac{1}{V}\frac{d}{dt}(U+T) \quad , \tag{9}$$

where V = V(a) is the crack speed. Upon substituting from Eqs. 7 and 8, Eq. 9 becomes

$$\frac{d}{da}(U+T) = \frac{1}{V}\frac{d}{dt}\int_{o}^{L}\left\{EI\left(\frac{\partial^2 w}{\partial x^2}\right)^2 + kH(x-a(t))\,w^2 + \rho A\left(\frac{\partial w}{\partial t}\right)^2\right\}dx \;. \tag{10}$$

In Eq. 10, the step function $H^*(w_c\text{-}w)$ has been replaced by H(x-a), in accord with the computational results which show that the foundation ruptures in a continuous fashion. This simplifies the proof.

Interchanging integration and differentiation in Eq. 10, integrating by parts, and using Eq. 6 then gives

$$\frac{d}{da}(U+T) = \frac{2}{V}\int_0^L \frac{\partial w}{\partial t}\left\{EI\frac{\partial^4 w}{\partial x^4} + \rho A\frac{\partial^2 w}{\partial t^2} + kH(x-a)w\right\}dx +$$

$$+\frac{k}{V}\int_0^L w^2 \frac{\partial}{\partial t} H(x-a(t))\,dx \quad .$$

The bracketed term is the same as the left-hand side of Eq. 5 under the condition that $H^*(w_c-w) = H(x-a)$. Hence, the first integral vanishes identically. Recognizing that $\frac{dH}{dt}$ is the Dirac delta function, the remaining term can be integrated to give

$$\frac{d}{da}(U+T) = \frac{k}{V}\left[-w^2\frac{da}{dt}\right]_{x=a} = -kw_c^2 \quad . \tag{11}$$

Finally, by using Eq. 3, Eq. 11 can be rewritten to show that

$$\frac{dU}{da} + \frac{dT}{da} + Rb = 0 \quad , \tag{12}$$

as required for energy conservation when no external work is done on the specimen.

4 DISCUSSION

Figure 7 compares propagation data for PMMA and high-strength-steel[5] samples both fracturing at the same ratio of K_q/K_d. Note that on this graph the time scales have been adjusted to the ratios of the elastic-bar-wave speed C_o where a "fast" modulus of 3.46 GN/M^2 was used.* Although it is likely that viscoelastic effects are important during the initial loading process, it is unlikely that they are important while the crack is propagating rapidly. Consequently, it is not surprising that the crack speeds in two very different materials – high-strength steel and PMMA – display a similar dependence relative to their elastic wave speeds. It can also be seen that there is only a minor dependence of specimen dimensions on the resulting crack speeds.

*All K values in this paper have been calculated using this modulus value.

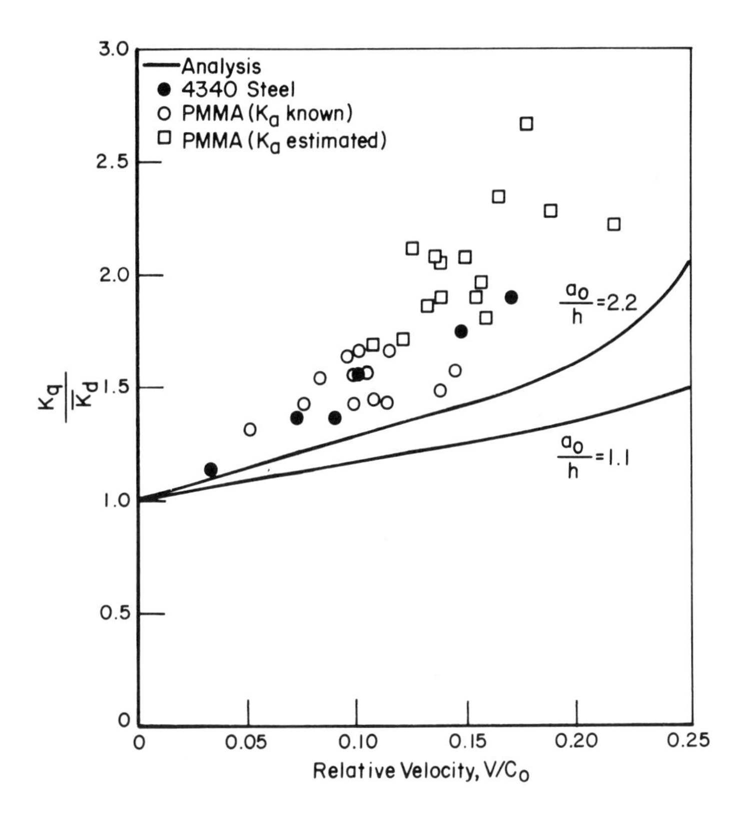

Fig. 7. Comparison between theoretical predictions and experimental results on PMMA and steel.

There are two main conclusions that can be drawn from these results:

(1) The time dependence of the elastic modulus plays little role in these experiments. Consequently, an arbitrary choice of modulus is not a concern provided the same value is relevant for both K_q and $\overline{K}_d$. This further implies that memory effects associated with the loading rates used in the experiments are small.

(2) The inelastic aspects of crack propagation are all contained in the term $\overline{K}_d$. This does not imply that the dependence of K_d on V/C_o is the same for both materials, but merely that the same relative overload results in the same relative steady-state velocity.

Some tentative comments may also be made regarding crack arrest. Without further refinement, the current analysis does not provide a good prediction of the arrest length, since the change in toughness due to the lower speeds accompanying deceleration is not explicitly included. For this reason, and because of the arbitrariness necessary to define the arrest point in PMMA, it was not felt useful to attempt to define a K_q value which would produce arrest. However, the calculations do suggest that the arrest/no arrest transition lies somewhere between $K_q/\bar{K}_d$ values of 2 and 3. In comparison, for steel, when $K_q/\bar{K}_d$ was less than 2, arrest always occurred. For the polymer, arrest was observed in one specimen where $K_q/\bar{K}_d$ was 1.7. On the other hand, the crack passed entirely through specimens where $K_q/\bar{K}_d$ was as low as 1.3. It is felt that this behavior is within the accuracy of the model as it currently stands and has to do with the details of the deceleration and arrest process and not with the phenomenon of rapid steady-state crack propagation, a point which needs more extensive study.

Finally, some remarks concerning the applicability of the analysis presented in this paper are appropriate. As can be seen in Fig. 7, the predicted crack speeds are about a factor of two in excess of the observed values. Nevertheless, the results given by the dynamic model represent a significant improvement over those of the previous quasi-static models[6-9]. In the quasi-static calculations, the crack speeds are roughly 5 to 10 times higher (often exceeding C_0) and, in addition, do not exhibit the constant-speed behavior observed when using the dynamic model.

It should also be pointed out that the model presented here is perhaps the simplest dynamic model that could be employed. The more realistic theories that could be devised would certainly tend to reduce the present discrepancies between theory and experiment. In fact, some preliminary calculations employing Timoshenko beam theory (i.e., including shear deformation and rotary inertia) and a generalized elastic foundation (including rotational stiffness) have also given constant-speed propagation, but at a level roughly half that given by the Euler-Bernoulli beam-Winkler layer model used in this paper. It is expected that the details will be published subsequently.

5 CONCLUSIONS

(1) The dynamic beam-on-elastic foundation analysis with the further refinements discussed in this paper provides a good description of rapid crack propagation in the wedge-loaded, double-cantilever-beam specimen.

(2) Formation of parabolic markings on the fracture surface of PMMA contributes markedly to the dynamic fracture toughness and therefore must be considered in a microstructural formulation.

ACKNOWLEDGMENTS

The experimental research was sponsored by Battelle's Columbus Laboratories and the analysis by the Ship Structure Committee. We are grateful to G. T. Hahn for discussions and to P. N. Mincer for experimental assistance.

REFERENCES

1. Hoagland, R. G., Rosenfield, A. R., and Hahn, G. T., *Met. Trans.*, **3**, 123 (1972).
2. Rosenfield, A. R., and Mincer, P. M., *J. Polymer Sci.*, Part C, **32**, 283 (1971).
3. Hoagland, R. G., Kanninen, M. F., Markworth, A. J., Mincer, P. N., and Rosenfield, A. R., *Intern. J. Fracture Mech.*, **8**, 337 (1972).
4. Hahn, G. T., Hoagland, R. G., Kanninen, M. F., and Rosenfield, A. R., *Proceedings of the International Conference on Dynamic Crack Propagation*, Lehigh University, July 10-12, 1972, to be published.
5. Hahn, G. T., Hoagland, R. G., Kanninen, M. F., and Rosenfield, A. R., paper presented at Third International Conference on Fracture, Munich, Germany, April, 1973.
6. Benbow, J. J., and Roesler, F. C., *Proc. Phys. Soc. (London)*, **B70**, 201 (1957).
7. Gilman, J. J., *Fracture*, Averbach, B. L., et al. (Ed.), MIT Press, Cambridge (1959), p 193.
8. Berry, J. P., *J. Mech. Phys. Solids*, **8**, 194 (1960).
9. Kanninen, M. F., *Intern. J. Fracture Mech.*, **9**, 83 (1973).
10. Cotterell, B., *Appl. Mater. Res.*, **4**, 227 (1965).
11. Manogg, P., *Physics of Non-Crystalline Solids*, Prins, J. A. (Ed.), North-Holland Press, Amsterdam (1965), p 481.
12. Williams, J. G., Williams, J. C., and Turner, C. E., *Polymer Eng. Sci.*, **8**, 130 (1968).
13. Radon, J. C., and Fitzgerald, N. P., *Eng. Mater. Design*, **13**, 1125 (1970).
14. Chubb, J. P., and Congleton, J., *Intern. J. Fracture Mech.*, **8**, 227 (1972).
15. Green, A. K., and Pratt, P. L., Imperial College, London, private communication (1972).
16. Carlsson, J., Dahlberg, L., and Nilsson, F., *Proceedings of the International Conference on Dynamic Crack Propagation*, Lehigh University, July 10-12, 1972, to be published.
17. Paxson, T. L., and Lucas, R. A., ibid.
18. Broutman, L. J., and Kobayashi, T., ibid.
19. Cotterell, B., *Intern. J. Fracture Mech.*, **4**, 209 (1965).
20. Chubb, J. P., and Congleton, J., *Proceedings of the International Conference on Dynamic Crack Propagation*, Lehigh University, July 10-12, 1972, to be published.
21. Clark, A.B.J., and Irwin, G. R., *Exp. Mech.*, **6**, 323 (1966).
22. Bradley, W. B., and Kobayashi, A. S., *Eng. Fracture Mech.*, **3**, 317 (1971).
23. Burns, S. J., and Bilek, Z. J., *Proceedings of the International Conference on Dynamic Crack Propagation*, Lehigh University, July 10-12, 1972, to be published.
24. Forsythe, G. E., and Wasow, W. R., *Finite Difference Methods for Partial Differential Equations*, Wiley, New York (1960), p 131.

DISCUSSION on Paper by M. F. Kanninen

KOBAYASHI: One of the most intriguing features of the authors' double cantilever specimen is the fact that side grooving is not needed to make the crack run straight. I hope that this advantage is shown more forcefully to the ASTM E-24 committee that is responsible for setting fracture toughness testing standards. As a question, how much does the compressive load acting at the loading pin parallel to the crack contribute to the stress-intensity factor?

KANNINEN: The beam-on-elastic-foundation analysis can incorporate an axial loading without too much difficulty. We have included this effect in some calculations and have found that, provided the wedge angle is not too great, the contribution of the axial force to the stress-intensity factor is small. The available experimental evidence is in agreement with this finding. This is fortunate because the accurate determination of the force components acting on the load pins would require the solution of a difficult contact-stress problem. In connection with this, we have noticed that a large error in measurement of stress intensity can arise if the pin diameter is too small. Then, large local displacements can arise (from distortion of the material around the holes) which produce spurious shifts in the clip-gage reading.

KNAUSS: It should be stated for the record that as early as 1933, Ludwig Prandtl used the model of an elastic beam on an elastic foundation to describe the fracture process, although no dynamic effects were considered. [See Prandtl, L., "Ein Gedankenmodell für den Zerreissvorgang spröder Körper", *ZAMM*, 13, 129 (1933); also Prandtl, L., *Gesammelte Abhandlungen,* Erster Teil, Springer (1961).].

KNAPPE: The parabolas in a fracture surface result from inhomogeneities. Since the density of the parabolas strongly influences fracture energy, it should be possible to influence fracture energy by inhomogeneities in the material, which you introduce in processing. What fracture behavior can be predicted?

KANNINEN: The effect of inhomogeneities is what we would expect. In fact, deliberate introduction of heterogeneities (e.g., to make high-impact polystyrene) is a practical application of this phenomenon.

FRACTURE DYNAMICS OF HOMOLITE-100

A. S. Kobayashi and B. G. Wade

University of Washington
Seattle, Washington

W. B. Bradley

Shell Development Company
Houston, Texas

ABSTRACT

The fracture dynamics of Homalite-100 plates is discussed in the context of linear theory of fracture mechanics. Dynamic photoelasticity is used for dynamic analysis and finite-element method is used for static analysis. Static and dynamic strain-energy-release rates and kinetic-energy-release rates for constant-velocity, accelerating, and decelerating cracks are determined in fracturing tension plates and in pretensioned plates impacted by a projectile.

1 INTRODUCTION

The first international meeting on fracture dynamics[1], which attracted over forty papers, is an indication of the recent attention given to this specific subject. A brief but up-to-date survey on the theoretical as well as experimental aspects of fracture dynamics was presented by Hahn et al.[2] at this meeting. As described by Hahn et al., research in fracture dynamics, unlike static fracture mechanics, has been hampered by the lack of a single elastodynamic solution for a realistic crack propagating with variable velocity.

In the absence of suitable theoretical analysis in fracture dynamics, the authors have used, during the past 4 years, dynamic photoelasticity to investigate the transient stress fields surrounding a running crack. Although some ambiguity exists in scaling a dynamic fracture criterion for a photoelastic material to that for a structural material, photoelasticity provided a whole field picture and enabled one to visualize the influence of some of the structural parameters which govern fracture dynamics.

In the following, fracture dynamic results obtained in the past by the writers[3-7] are reviewed. These results are then evaluated from the standpoint of the rate of energy change instead of the concept of dynamic stress-intensity factor used previously.

2 EXPERIMENTAL AND ANALYTICAL PROCEDURES

The experimental setup used in these series of investigations consisted of a modified Cranz-Schardin 16 spark-gap camera and associated dynamic polariscope. Time intervals between each of the 16 frames were measured by a Lite-Mike. The test specimens were 3/8-inch-thick Homalite-100 plates with 10 x 10-inch test sections. Prescribed boundary conditions included uniform, linearly increasing, or linearly decreasing displacements along the gripped edges of the specimen. For a statically loaded specimen, the crack propagated from a single, edge-notched starter crack which was saw cut and chiseled.[3-6] Dynamic loading was accomplished by impacting a starter crack in a subcritically loaded specimen by a projectile.[7] Figure 1 shows typical dynamic photoelastic results for a 3/8-inch-thick plate loaded in uniform tension. In this test, the crack reached a constant velocity of approximately 15,000 ips after approximately 2-1/2 inches of propagation. The last frame in this photograph shows the characteristic butterfly pattern of isochromatics being reflected at the far edge of the plate after complete crack penetration. Other typical isochromatic patterns of tension plates with arresting holes, branch cracks, and pretensioned plates impacted by a projectile are shown in the references cited above.

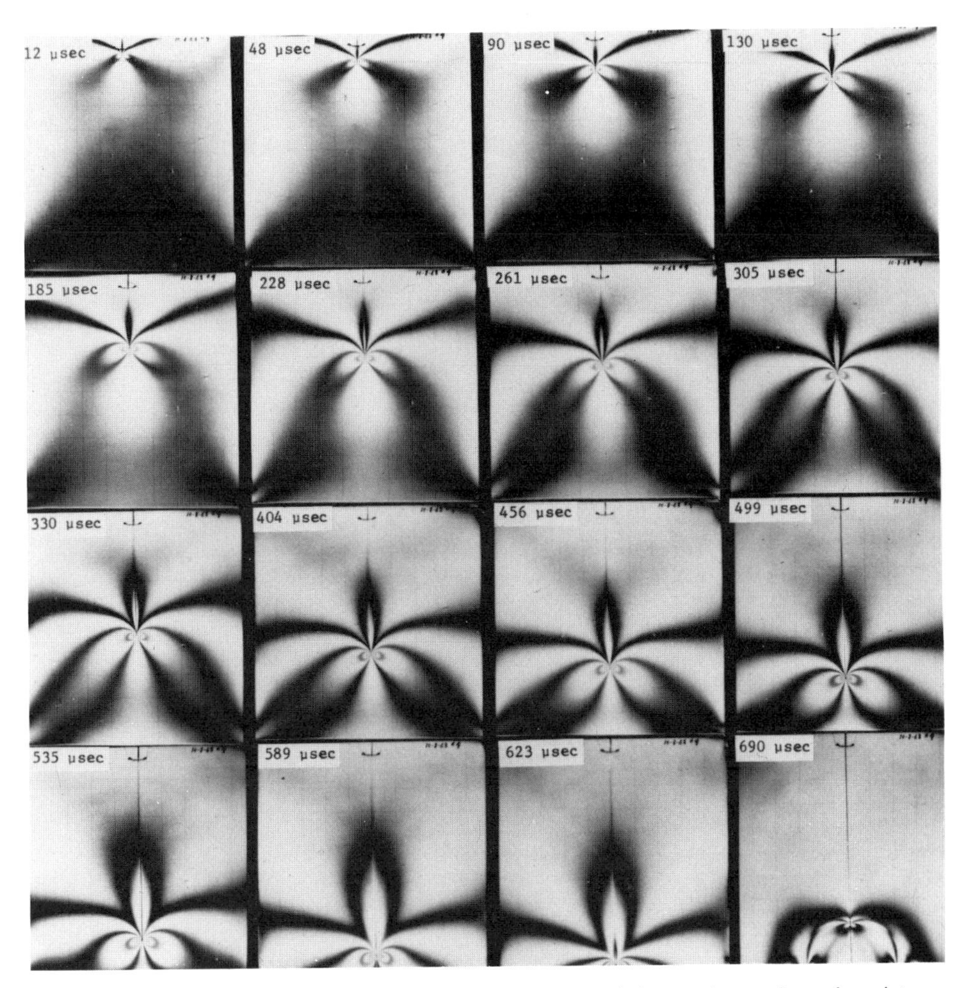

Fig. 1. Dynamic photoelastic patterns of a propagating crack in an edge crack tension plate.

Most of the Homalite-100 sheets used in these experiments were calibrated by Bradley[3]. An average dynamic modulus of elasticity, Poisson's ratio, stress-optic coefficient, and static fracture toughness of 675 ksi, 0.345, 155 psi-in./fringe, and 579 psi $\sqrt{\text{in.}}$, respectively, obtained by Bradley were used in all data evaluation.

In previous papers, dynamic stress-intensity factors were determined from the near-field dynamic isochromatic lobes using Bradley's modified static procedure[3] which is a variation of the original static analysis suggested by Irwin[8]. In these investigations as well as those of C. W. Smith[9,10], distortions of the static isochromatics due to either crack propagation or geometric effects were accounted for by an appropriate variation of the unknown remote stress component σ_{ox}. A possible error introduced by such matching of static and dynamic isochromatic lobes is discussed below.

A near-field isochromatic lobe can be represented as[7,8]

$$\tau^2_{max} = \frac{K^2_s}{8\pi r}\left[\sin^2\theta + 2\delta\sqrt{\frac{2r}{a}}\sin\theta\sin\frac{3\theta}{2} + \frac{2r\delta^2}{a}\right], \qquad (1)$$

where r and θ are polar coordinates with origin at the crack tip

K_s is the static, opening-mode stress-intensity factor

a is the half crack length of a totally embedded crack in an infinite solid

$\delta = \frac{\sigma_{ox}}{\sigma}$ is the remote stress factor

σ is the applied stress.

In applying Eq. 1 to a single, edge-notched plate, a is assumed to be the total crack length. Variations in K_s and δ then account for the geometric effects imposed on the near-field stress distributions. In particular, δ serves to tilt the static isochromatic lobe forward or backward.[7,9] The matching technique thus consisted of determining an appropriate δ in addition to K_s. For fast crack propagation where the crack tends to lean backwards[11], appropriate values of δ ranged between 1 and -0.4. Fortunately, the dynamic stress-intensity factor was relatively insensitive to slight variations in δ when Bradley's procedure of calculating the differences of τ_{max} for two adjacent isochromatic lobes was used. As a result, δ was set to 1 in previous computations and the dynamic stress-intensity factor, K_D, which was assumed to be equal to K_s with $\delta = 1$, was determined by Eq. 1.

The accuracy of the dynamic stress-intensity factor determined by the above procedure can be evaluated by comparing theoretical static and dynamic isochromatic lobes. The latter was derived from Sih's near-field solution[12,13] associated with a crack propagating with constant velocity in an infinite plate. This isochromatic lobe can be expressed as

$$\tau^2_{max} = \left[\frac{K_s F_1(s_1,s_2)}{\sqrt{2\pi r}}\right]^2 \left\{\left[(1+s_2^2)(1+s_1^2)f(s_1) - 4s_1s_2 f(s_2)\right]^2 + 4s_1^2(1+s_2^2)^2[g(s_1) - g(s_2)]^2\right\}, \qquad (2)$$

where

$$s_1 = \left[1 - \left(\frac{c}{c_1}\right)^2\right]^{1/2}$$

$$s_2 = \left[1 - \left(\frac{c}{c_2}\right)^2\right]^{1/2}$$

$$f^2(s_j) + g^2(s_j) = \frac{1}{\sqrt{\cos^2\theta + s_j^2 \sin^2\theta_1}}, \quad j = 1,2$$

$$f^2(s_j) - g^2(s_j) = \frac{1}{\cos\theta + s_j^2 \sin\theta \tan\theta}, \quad j = 1,2$$

c is the crack velocity

$c_1 = \sqrt{\dfrac{E(1-\nu)}{\rho(1+\nu)(1-2\nu)}}$ compressional wave velocity

$c_2 = \sqrt{\dfrac{G}{\rho}}$ shear-wave velocity

E and G are the modulus of elasticity and shear modulus, respectively, and ν is Poisson ratio

ρ is the mass density

K_s is the static stress-intensity factor

$F_1(s_1,s_2)$ is the dynamic correction factor which varies with the boundary and initial conditions.

The normalized isochromatic lobe of Eq. 2 varies with the crack velocity, c, and is plotted in Fig. 2 for typical crack velocities observed in fracture tests of Homalite-100 plates for $K_s F_1/[\sqrt{2\pi}\,\tau_{max}] = 1.0$. Also plotted in Fig. 2 are the best-matched static isochromatic lobes represented by δ = -0.2 in Eq. 1. Peak distances of the static isochromatics are matched with corresponding dynamic isochromatics for comparison purposes. As shown in this figure, little difference exists between the shapes of static and dynamic isochromatics. The dynamic stress-intensity factor of Eq. 1 thus cannot be significantly different from the more correct dynamic stress-intensity factors determined by Eq. 2 which incorporates the distortion of the isochromatics associated with a constant-velocity crack.

Using Eq. 2, the unknown function of $F_1(s_1,s_2)$ and hence the dynamic stress-intensity factor, which varies slightly with angular orientation θ, can be determined from the dynamic isochromatics. The dependence of K_D on θ indicates that a dynamic stress-intensity factor cannot be defined as an equilvalent static stress-intensity factor, K_s, with $K_D = \lim_{r\to 0} \sqrt{2\pi r}\,\sigma_{yy}$, where σ_{yy} is the near-field normal stress vector which is oriented in the direction normal to the crack surface. In practice, this discrepancy does not affect K_D significantly, as shown in Fig. 2, and thus the above static equivalence of K_D will not introduce errors larger than the scatter of experimental data. A more logical criterion for fracture dynamics, however, would be energy-release rate, which is independent of

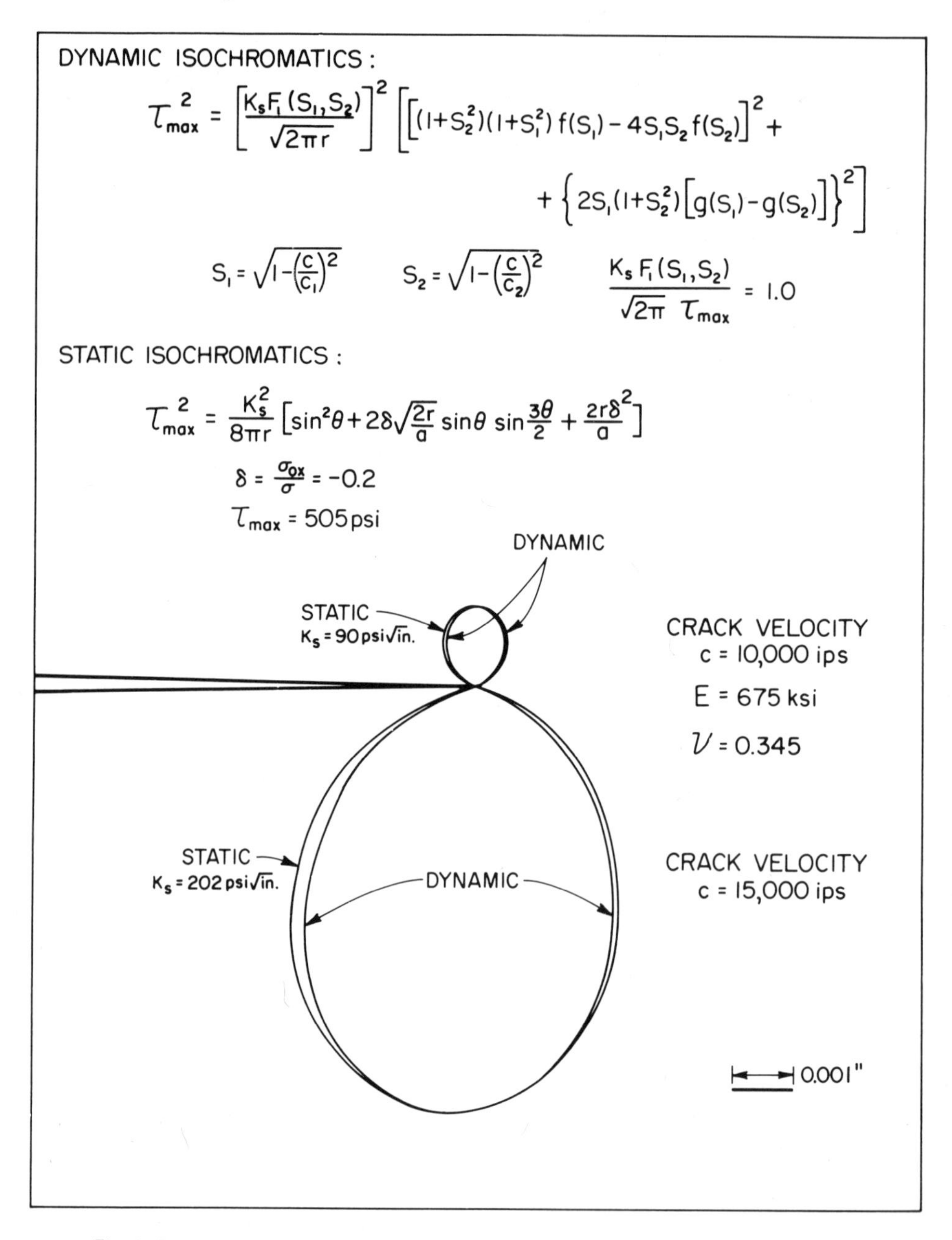

Fig. 2. Dynamic and static isochromatic lobes at the crack tip for constant τ_{max}/K_s.

such angular orientation.

Sih has derived the dynamic strain-energy-release rate for a constant-velocity crack as [12,13]

$$\frac{G_D}{G} = (1+\nu)\, s_1 \left(1-s_2^2\right) \left[4\, s_1 s_2 - \left(1+s_2^2\right)^2\right] F_1^2\,(s_1, s_2) \quad , \qquad (3)$$

where

ν is the Poisson ratio

G_D is the dynamic strain energy released and is considered as the dynamic resistance to fracture.

G is the elastic strain energy released by the propagating crack.

Furthermore, from total energy consideration

$$G_D - G + \frac{dT}{da} = 0 \ , \tag{4}$$

where $\frac{dT}{da}$ is the kinetic-energy-release rate.

Equation 4 enables one to determine the kinetic-energy-release rate from the measured dynamic strain-energy-release rate of G_D and the static strain-energy-release rate of G calculated by finite-element analysis. The value of kinetic-energy-release rate is of particular interest in view of the recent discussion involving the relative significance of this quantity in the presence of a propagating crack.[1]

It should, however, be noted that Sih's solution does not account for the dynamic effects due to reflecting stress waves at the various boundaries of the finite-size test specimen or the stress waves due to projectile impact. Thus, the kinetic-energy-release rate in Eq. 4, which encompasses the entire test specimen, cannot be computed from G_D which ignores the stress-wave effects. In practice, however, the stress-wave effects in simple tension specimens, such as that shown in Fig. 1, are small and thus the kinetic-energy-release rate can be estimated through the use of Eq. 4.

3 RESULTS

Selected test results reported in Refs. 3 through 7, as well as heretofore unreported test results, were evaluated using Sih's expression of Eqs. 2 and 3 for a constant-velocity crack in an infinite plate. The kinetic-energy-release rate was computed for selected tests by using Eq. 4. Figures 3, 4, and 5 show the rates of energy change in fracturing single, edge-notched plates loaded with two types of displacement boundary conditions. In both specimens with almost uniform edge displacements of boundary conditions (Figs. 3 and 5), the crack ran steadily at a maximum velocity of approximately 15,000 ips. For the specimen with decreasing edge displacement (Fig. 4), crack arrest was achieved at approximately 70 percent of the plate width. The relatively large variations in dynamic strain-energy-release rate G_D, which appears to be related to small variations in crack velocities, implies that the dynamic strain-energy-release rate

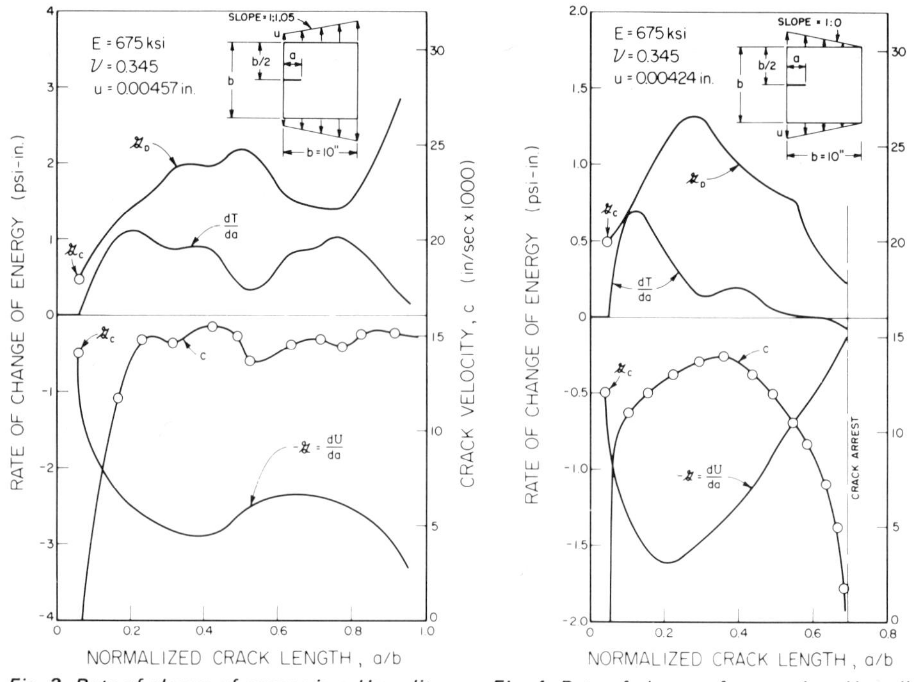

Fig. 3. Rate of change of energy in a Homalite-100 tension plate shown in Fig. 1.

Fig. 4. Rate of change of energy in a Homalite-100 tension plate, showing point of crack arrest.

and hence the dynamic resistance of a running crack is not a constant value. This variation cannot be attributed to strain-rate effect alone, as evidenced by the sharp rise in dynamic strain-energy-release rate, G_D, at the end of crack propagation in Figs. 3 and 5. Nor could such large changes be attributed to errors introduced by the use of Eq. 1 and 2, derived for an infinite plate, in regions close to the free edge of the test specimen. Thus, the dynamic strain-energy-release rate appears to be a function of the crack velocity as well as the boundary and initial conditions imposed on the test specimen.

The results of the three tests plotted in Figs. 3, 4, and 5 reveal a sharp increase in the estimated kinetic-energy-release rate with rapid crack acceleration immediately after the onset of fracture. Variations in the estimated kinetic-energy-release rates, $\frac{dT}{da}$, despite the relatively constant crack velocity (Figs. 3 and 5) illustrate also the effect of multiple reflected stress waves at the test-specimen boundaries. This effect is less noticeable in Fig. 4 where the crack is propagating at a lower velocity most of the time, indicating the lower magnitude of the stress waves generated by such low-velocity cracks. As expected then, the kinetic-energy-release rate, $\frac{dT}{da}$, in much of Fig. 4 is, at the most, about one-third of those in Figs. 3 and 5.

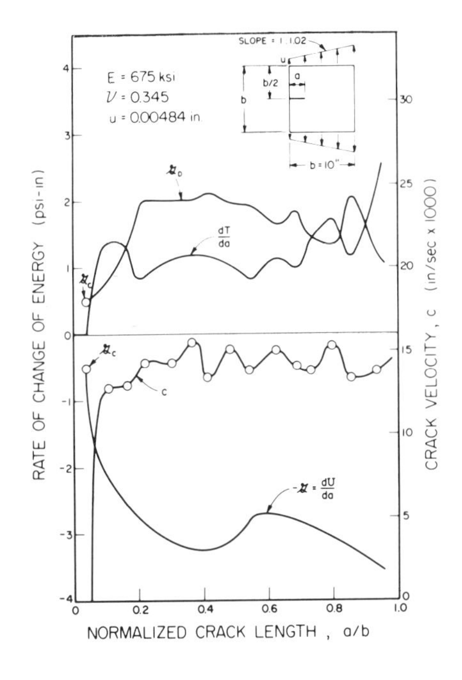

Fig. 5. Rate of change of energy in a Homalite-100 tension plate.

In the case where the crack was arrested by propagating into a decreasing strain field (Fig. 4), the static and dynamic strain-energy-release rates for crack arrest differ slightly. This difference accounts for the excess kinetic energy which the propagating crack must reabsorb before coming to complete arrest.

Figure 6 shows similar rates of energy change for cracks which ran through a 0.150-inch-diameter hole, and reappeared on the other side to continue their propagation. The large value for dynamic strain-energy-release rate, which exceeds the available static strain-energy-release rate, is due to negative value of kinetic energy released by the multiple reflecting stress waves at the arresting hole boundary. This excess kinetic energy contributes to crack acceleration.

Rapid oscillation in the dynamic strain-energy-release rate due to fluctuation in kinetic-energy-release rates is also observed after the crack restarts from the hole. These fluctations are due to multiple-stress-wave reflection caused by sudden increase in strain energy released when the running crack impacted the hole. It is interesting to note that the static maximum stress at the hole, as shown in Fig. 6, exceeded the ultimate strength of Homalite-100 at crack penetration.[5] Assuming an overload of 20 to 30 percent due to dynamic effects, the maximum stress at this point not only exceeded the static ultimate strength but also the dynamic ultimate strength at crack penetration, causing a secondary crack to form at the far end of the hole.

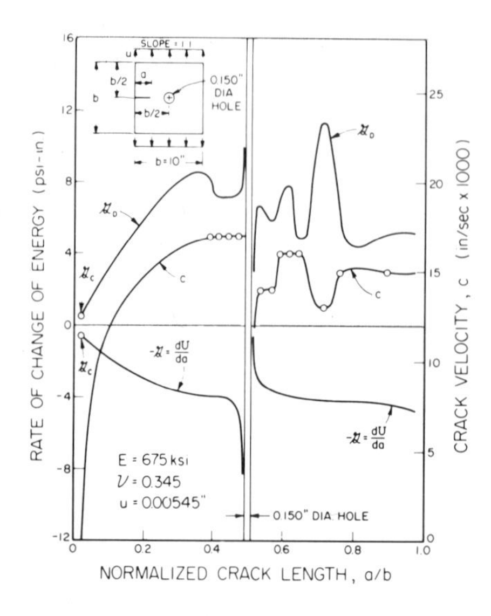

Fig. 6. Rate of energy change in a Homalite-100 tension plate, for cracks running through a 0.150-inch-diameter hole.

Figures 7(a) and 7(b) show the rates of energy change for pretensioned plates impacted by a projectile. The rapid fluctuations in dynamic strain-energy-release rate as well as in the crack velocity evident in Fig. 7(a) illustrate the effect of transient stress waves caused by the impact. The high excess-kinetic-energy-release rate due to the impact of

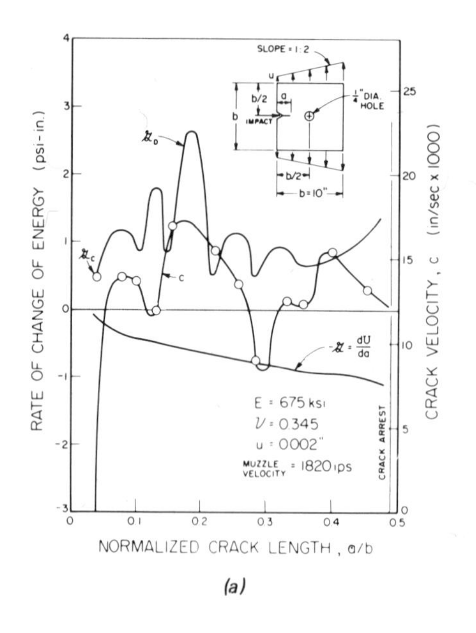

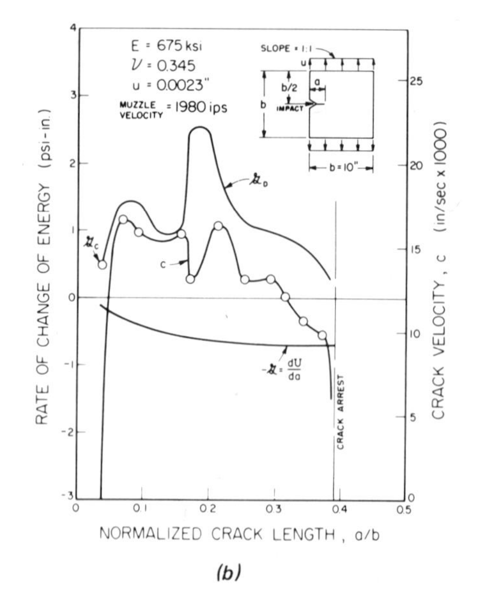

Fig. 7. Rate of energy change in a pretensioned Homalite-100 plates fractured by impact.

the projectile, which undoubtedly exceeds the available static strain-energy-release rate in parts of the test specimen, generated the large dynamic strain-energy-release rate. In contrast to that for the test specimen of Fig. 6, the static maximum stress at the hole boundary for the test specimen of Fig. 7(a) was only 56 percent of the ultimate strength at crack penetration. The running crack thus was arrested at the instant of crack penetration. In all tests involving impacted plates with an arresting hole, the elastic strain energy released at the instant of the crack penetration was not large enough to cause the crack to restart at the other end of the hole.[7]

The dynamic strain-energy-release rate at arrest, as shown in Fig. 7(b), is considerably smaller than the static strain-energy-release rate, which indicates that the fluctuating kinetic-energy-release rate can act to arrest the crack temporarily while being used to accelerate other portions of the test specimen surrounding the crack. After this temporary arrest, the crack apparently restarted to complete the fracture of the specimen.

Figures 8(a) and 8(b) show the rates of energy change in a branching crack. A notable feature here is the lack of an apparent dynamic strain-energy-release rate that corresponds to the static strain-energy-release rate for branching. Thus, crack branching cannot be explained in terms of a branching dynamic strain-energy-release rate since the crack velocity is decreasing with decreasing dynamic resistance to crack propagation despite increasing static strain energy released. The kinetic-energy-release rate at crack branching computed from Eq. 4, obviously will reach a maximum value. As a result, the crack must branch in order to absorb such excess

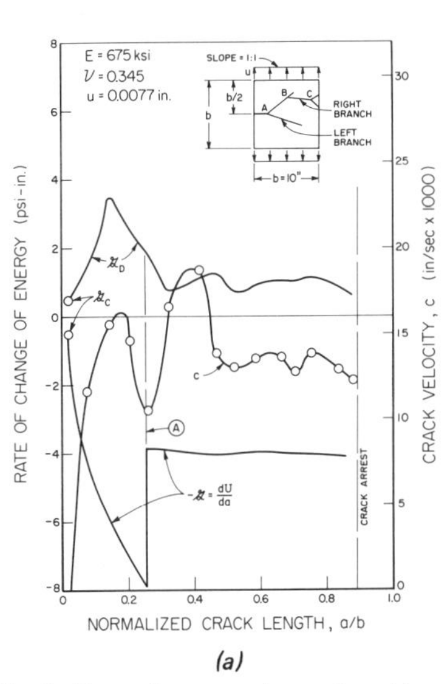

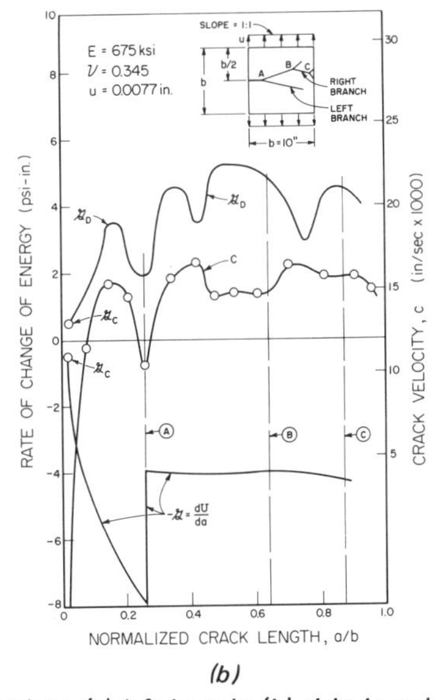

Fig. 8. Rate of energy change in a Homalite-100 tension plate: (a) left branch; (b) right branch.

energy. Obviously, this condition persists when the static strain-energy-release rate becomes excessively large, and thus the general conclusion of the previous paper is still correct. However, the exact sequence of events leading to crack branching is not very clear since a complete elastodynamic solution is needed to assess qualitatively the exact amount of kinetic energy absorbed prior to and immediately after crack branching.

Crack branching can also be duplicated in the presence of high impact energy, which increases the release rate of kinetic energy. Such crack branches in impacted plates will not continue to propagate since they are generated by the transient passage of stress waves as shown in Fig. 9. This dynamic photoelastic picture was taken from the fifth frame of a pretensioned plate impacted by a high-velocity projectile.

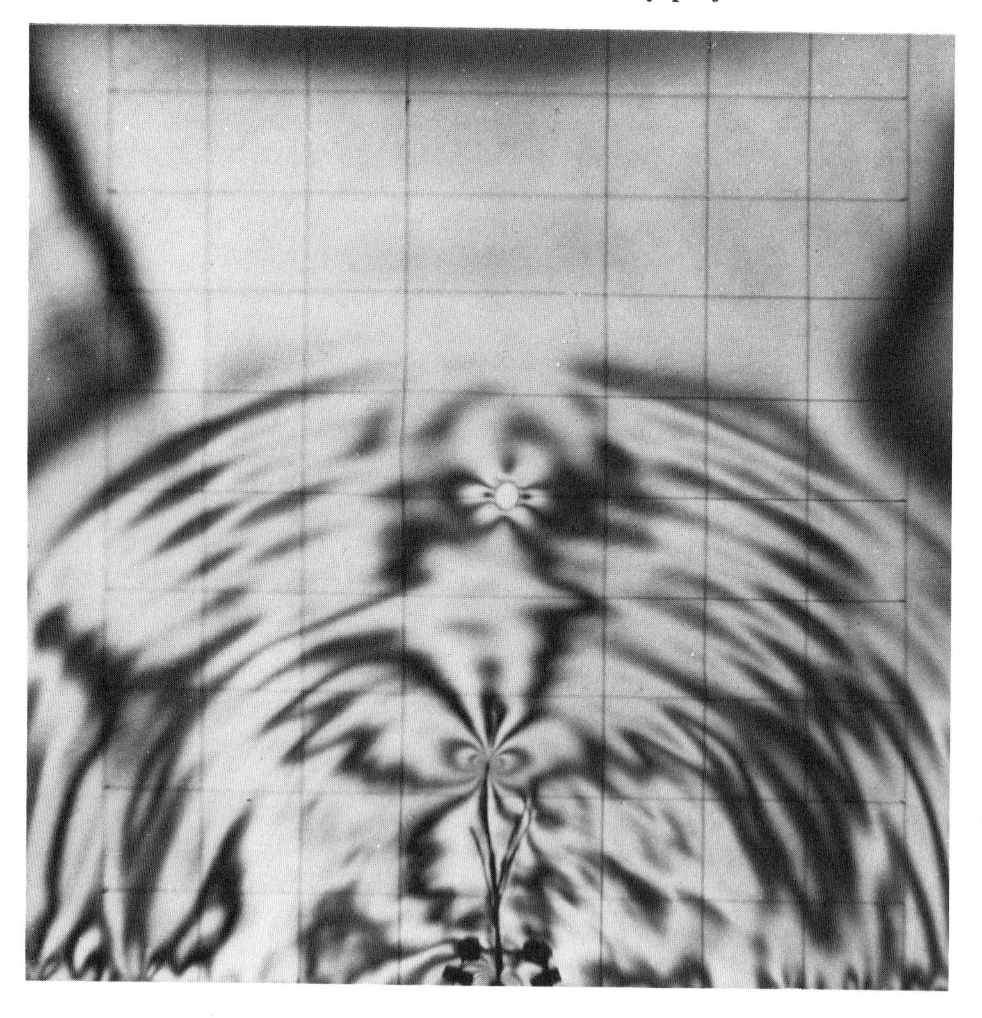

Fig. 9. Dynamic photoelastic pattern of a propagating crack in a pretensioned plate subjected to high-velocity impact.

Figure 10 shows a plot of the dynamic strain-energy-release rates or dynamic resistance to crack propagation at various crack velocities taken from various tests. At first glance, no correlation is observed between these two quantities. A broad correlation between the dynamic strain-energy-release rate and the crack velocities is observed, however, in plates fractured by static loading alone. This preliminary evidence indicates that the dynamic fracture resistance of this material is related to crack velocities for simple loading condition.

Another notable feature is the apparent dependence of dynamic strain-energy-release rate at crack arrest on test conditions. The current arrest strain-energy-release rates, G_a, for the test specimens of Figs. 4, 7b, and 8(a) were 0.23, 0.28, and 0.40 psi-in., respectively. The coincidence of the two arrest energy-release rates may be fortuitous since the variety of transient stress conditions existing in these tests would have probably damped out at the time of crack arrest, just providing an impression that arrest strain-energy-release rate (or an arrest fracture stress-intensity factor) is a constant material property. The above three arrest energy-release rates, G_a, are substantially lower than the critical strain-energy-release rates, G_c, reflecting the influence of the dynamic condition.

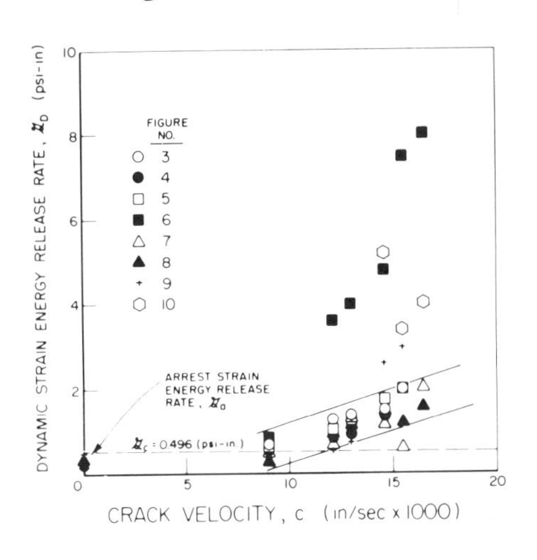

Fig. 10. Dynamic strain-energy release rates at various crack velocities.

4 CONCLUSIONS

The above summary of dynamic photoelastic results evaluated from the viewpoint of energy-release rate led to the following conclusions.

1. For simple loading conditions, the dynamic strain-energy-release rate is a function of the crack velocity.
2. The strain-energy-release rate at arrest is governed by the boundary and initial conditions and does not exceed the critical strain-energy-release rate.

3. Branching is caused by excess-kinetic-energy-release rate which is a function of the boundary and initial conditions.

ACKNOWLEDGMENT

This research project was sponsored by the Office of Naval Research, Contract No. N00014-67-0103-0018 NR 064 478. The writers wish to thank Mr. J. H. Crowley, now retired, Dr. N. Basdagas and Dr. N. Perrone of ONR for their patience and encouragement during the course of this investigation.

REFERENCES

1. *Proceedings of the International Conference on Dynamic Crack Propagation,* Lehigh University, July 10-12, 1972, to be published.
2. Hahn, G. T., Hoagland, R. G., Kanninen, M. F., and Rosenfield, A. R., to be published in Ref. 1.
3. Bradley, W. B., and Kobayashi, A. S., *Exp. Mech.,* **10e,** 106-113 (March, 1970).
4. Bradley, W. B., and Kobayashi, A. S., *Eng. Fracture Mech.,* **3,** 317-332 (1971).
5. Kobayashi, A. S., Wade, B. G., and Maiden, D. E., *Exp. Mech.,* **12,** (1), 32-37 (1972).
6. Kobayashi, A. S., Wade, B. G., Bradley, W. B., and Chiu, S. T., to be presented at the Third International Conference on Fracture, Munich, April, 1973.
7. Kobayashi, A. S., and Wade, B. G., to be published in Ref. 1.
8. Irwin, G. R., *Proc. of SESA,* **XVI** (1), 93-96 (1958).
9. Marrs, G. R., and Smith, C. W., to be published in *ASTM STP 513.*
10. Schroedl, M. A., McGowan, J. J., and Smith, C. W., to be published in the *J. Eng. Fracture Mech.*
11. Cotterell, B., "On Brittle Fracture Path", *Intern. J. Fracture Mech.,* **2,** 96-103 (1965).
12. Sih, G. C., "Some Elastodynamic Problems of Cracks", *Intern. J. Fracture Mech.,* **5**, 51-68 (1968).
13. Sih, G. C., *Inelastic Behavior of Solids,* M. F. Kanninen, W. F. Adler, A. R. Rosenfield, and R. I. Jaffee, (Eds.), McGraw-Hill (1970), pp 607-639.

DISCUSSION on Paper by A. S. Kobayashi, B. G. Wade and W. B. Bradley

HULL: Have you been able to correlate changes in the appearance of the fracture surface with changes in the dynamic stress-intensity factor or with changes in crack velocity?

KOBAYASHI: Yes we have. It has been reported in Ref. 4. Further studies on crack surfaces and crack branching will be reported at the Third International Conference on Fracture, to be held in Munich, Germany, in April, 1973.

ON THE STEADY PROPAGATION OF A CRACK IN A VISCOELASTIC SHEET: EXPERIMENTS AND ANALYSIS

W. G. Knauss

California Institute of Technology
Pasadena, California

1 INTRODUCTION

In the following pages we present a continuum mechanical description of steady crack propagation in a linearly viscoelastic solid. The analysis is exact within the realm of linear viscoelasticity theory. Inasmuch as this colloquium deals with the deformation and fracture of high polymers from the molecular, microscopic, and macroscopic viewpoint, it is appropriate to recognize the complementary nature of the microscopic and the macro-continuum approach to polymer fracture. This remark seems essential because the later development contains assumptions which must appear rather gross to those who concern themselves with the microscopic or even atomistic aspect of the fracture process.

If one is interested in the fracture analysis of an engineering structure, it is natural to be concerned with the way in which macroscopic cracks propagate. For, if cracks propagate with varying velocity in dependence on the loading applied to the structure as a function of time, then it is, in principle, necessary only to determine the crack size history in order to predict the failure time for the structure. Moreover, from the engineer's point of view it is desirable to characterize the crack-growth process *quantitatively* with a minimum number of material parameters or functions. How large this minimum number is depends, of course, on how carefully the experimental data have been or can be gathered and in how much detail one describes the fracture process analytically. In the following description we have attempted to illuminate the relation between the crack-propagation rate, the rheological material properties, and the load acting on a structure.

It is a truism that the propagation of a crack is fundamentally resolved in terms of interatomic and intermolecular forces. But it is also clear that the quantitative synthesis of the macroscopic-crack-propagation process from atomistic considerations invites staggering difficulties; the number of degrees of motion available to atoms at the tip of a crack in an amorphous, semicrystalline, or polycrystalline bulk polymer is too large to be tractable except possibly in cases when the molecular structure at the crack tip is spatially highly ordered. As a consequence, one must consider the summary response of many atoms or molecules and attribute to them average properties or average response characteristics. The question of how large the domains should be to which this averaging process applies can be answered only by the application one has in mind. For example, for a crazed material which connects the craze walls, or one may be interested only in the average forces which are transmitted by many of these elements across the craze. If one *assumes* that such an average representation of molecular properties is sufficient for the analytical description of a process, then one must also require an experimental examination of whether the average representation of the molecular properties is justified. We remind ourselves in this context that the continuum mechanical description of constitutive behavior is a representation of the average properties of an assemblage of molecules or atoms, and that this average representation is justified only because it has proven many times to be a powerful implement for the solution of many practical problems.

In the following we shall consider the fracture of a crosslinked polymer above the glass transition temperature (elastomer). This restriction is made for two reasons. First, uncrosslinked polymers tend to flow like a liquid at sufficiently high temperatures or low deformation rates, thus inducing phenomena which are described more aptly in fluid than in solid mechanical terms. Second, at temperatures close to or below the glass transition temperature, many polymers exhibit nonlinear viscoelastic response to deformation which may be much more pronounced than that encountered above the glass transition. We think here of behavior that resembles the yield phenomenon in metals. In view of the fact that we shall later on employ linear viscoelasticity theory – the only practical viscoelasticity theory available to date – it seems advisable to exclude from consideration materials with such visco-plastic behavior. It is natural to question the wisdom of employing *linear* viscoelasticity theory when large deformations are potentially involved at the tip of a crack. In partial answer to this question we note that our later development generates only bounded stresses and strains; this is in contrast to the classical Inglis-Griffith solution and in keeping with the work of Barenblatt. Thus, the assumptions underlying the linear theory are not violated per se; it is only when the analysis is applied to the fracture response of particular materials that one may infringe upon these assumptions. At the same time, we must confess that we know of no elastomer which permits strict adherence to the suppositions of the linear theory. In applying the linear theory of viscoelasticity to the fracture of polymeric materials, we are, therefore, guided by some faith that the endeavor is not without merit. This faith draws on experience in the fracture of rate-insensitive solids where the linear theory of elasticity has been a cornerstone of fracture mechanics. Moreover, we are motivated by the belief that, at least for some materials, a linearly viscoelastic analysis of fracture is more descriptive than none at all; a linear analysis may thus further the understanding of polymer fracture and raise new questions for a more complete characterization of the fracture process.

Our considerations here are limited to constant crack-propagation rates. Although we are ultimately interested in arbitrary crack-propagation histories such as those resulting from, say, monotonically or cyclicly varying loads, it is prudent first to investigate thoroughly the simpler problem of steady crack propagation. As a consequence of this restriction, we shall not discuss the problem of how a crack is generated, nor how it begins to move under some transient initial loading if it exists already.

The application of mechanics to the fracture of polymers involves several disciplines: one draws first on the mechanics of continua and on a certain, if restricted, amount of applied mathematics; further, one involves experimental techniques, concepts of polymer physics, and even (potentially) polymer chemistry. It is clearly impossible to do justice to the subject with satisfaction for those concerned primarily with these various fields of study. In addition, there are fewer concepts formulated and explored in connection with polymeric solids than with the usual rate-insensitive engineering metals. Both considerations should be an inducement for a detailed and explicit treatment of the subject. However, editorial restriction on the length of this paper – bound to be broken – precludes such a thorough account. But rather than present a particular aspect of the overall problem in appropriate detail, a summary, if not thoroughly detailed, presentation is given; a more exhaustive report is left for a future publication. In an attempted sensitivity towards the viewpoint of the polymer physicist, the physical aspects of the problem are emphasized.

Several individuals or groups of researchers have contributed towards the understanding of crack-propagation analysis. Of these, we note particularly Barenblatt, Entov and Salganik[1-6], Greensmith, Lake, Rivlin and Thomas[7-16] Knauss and Mueller[17-21], Kostrov and Nikitin[22], Culver, Marshall, and J. G. Williams[23,24] Prandtl[25], M. L. Williams[26-28], and Wnuk[29-31]. The approaches to the problem of time-dependent crack propagation offered by these authors are varied; to show how they differ from each other and from the present work* is not possible in a paper of this length. In view of the fact that a recent review[32] presents this comparison more comprehensively than would be reasonable in this place, we dispense here with an abbreviated repetition.

But we take the author's prerogative to comment briefly on the motivation for the present work and its relation to the earlier investigations[17-21]. In that earlier work, comparison of analysis with experimental data yielded a relatively small but definitely systematic discrepancy. This discrepancy could be made small by the choice of the size of the crack-tip disintegration zone.** However, on the basis of physical reasoning, the size of this zone turned out so small that even an accounting for the approximate nature of the analysis could not dispel the doubt that this past description of the crack-propagation process was incomplete. The following work thus constitutes a reexamination of a problem considered earlier.

The presentation consists of seven sections. Following this Introduction we describe first the mechanical characterization of the polyurethane material whose crack-propagation behavior is to be studied. Within the space limitations, the consequences of material variability are considered and their effect upon a comparison with later calculations is documented. In addition, we record the measured dependence of the (constant) crack-propagation speed on the load or deformation imposed upon a cracked strip geometry cut from a sheet.

In the third section we present some background on the mechanical degradation of material at the microscopic size scale, which is needed to make later approximations plausible. These approximations pertain primarily to the summary treatment of the disintegrating but still cohesive

---

*Of the above-mentioned authors, the work of Kostrov and Nikitin is most similar to that presented here. With the exception of some experimental work, the present analysis was already completed when Wnuk called this reference to the writer's attention. While there are definite similarities with the work of Kostrov and Nikitin, the main difference – beside the fact that Kostrov and Nikitin deal with the idealized case of a Maxwell material and do not present experimental results – lies in this author's avoidance of some a priori assumptions common to the fracture analysis in rate-insensitive solids, and to which Kostrov and Nikitin adhere.

**The physical meaning of this term will be explained fully in Section 3.

material at the crack tip, and we discuss the implicit (physical) restrictions which they place on the application of the subsequent calculations. The fourth section deals with two possible criteria for crack propagation. Both are predicated on a knowledge of the velocity-dependent displacements in the crack-tip region. These displacements are documented in the fifth section on the viscoelastic-boundary-value problem, and some rather general guidelines are presented concerning how they were obtained.

The sixth section deals with a comparison of the two hypothetical failure criteria with each other and with the experimental data documented in the second section. Besides a brief summary, the final section enumerates several topics whose exploration would be necessary adjuncts or logical extensions of this work, the detailed discussion of which is beyond the limited scope of this presentation.

2 EXPERIMENTAL WORK

In this section we record first those material properties of a polyurethane elastomer which are necessary for later computations; second, some additional useful information on material behavior which aids in developing and checking the crack propagation model; third, the dependence of the (steady) crack-propagation speed on the loads applied to a test specimen.

In work with elastomers, a major problem arises out of the lack of material reproducibility. This observation is true whether one is concerned with different production batches, or, though to a lesser degree, with respect to samples taken from the same production batch. The result of this variation in material properties is that the effect of some phenomena which occur at the tip of a moving crack may be masked by the scatter in the experimental results. To gain insight into the experimental limits set by this fact, we include some information on variability of material properties.

The material employed is a polyurethane elastomer traded by the Thiokol Chemical Corporation under the name Solithane 113. It is obtained by mixing the liquid "resin", a trifunctional isocyanate*, and the liquid "catalyst", which is essentially castor oil; the mixture used in this study consisted of equal parts by volume of resin and catalyst. The liquid mixture is cast into sheet molds at 60 C in a special apparatus where it is in contact only with dry nitrogen; thereafter, the material is allowed to cure for 2 hours at 140 C. The molds were constructed of two aluminum plates 0.5 in. thick against which ferrotype plates** were held for flatness

*The product of a reaction between castor oil and tolylenediisocyanate (TDI).

**Used normally for the production of glossy photographic prints.

by a low-grade vacuum and which were separated by a spacer. The finished sheet measured 12 x 12 x 1/32 inch and exhibited very smooth surfaces, apart from small, gentle irregularities resulting from the application of a spray-on silicone-mold release agent. After removal from the mold, the sheets were stored over dry silica gel. Earlier experiments had indicated that some "postcuring" took place during this storage, with the rate of postcuring decreasing continuously. Because crack-propagation and material-characterization work usually extends over several weeks, the material was to be stored for a sufficiently long time that postcure produced only insignificant changes during the course of the measurements; therefore, the material was stored for 1 year before it was used in tests. Each sheet produced was marked with an identification number and one half of each sheet was set aside for work on nonsteady crack propagation, not described here. From the remainder, strip specimens were prepared in accordance with Fig. 1. The brass stock served as a connector to the testing apparatus and was glued to the elastomer sheet with Eastman 910 adhesive. The flowing adhesive provided an imperfect bond in that it protruded from under the brass stock, in some places forming a wet radius while failing to wet the edge of the brass in other places; in spite of great care, the gage width of the strip specimen thus varied, in an unspecified manner, approximately by ±3 percent about the nominal value of 1.387 inch. However, this glue-joint method was superior to any clamping arrangement, which is potentially more precise; for it was found that the local clamp pressure had to be so great to prevent the sheet from slipping out of the clamps that fracture occurred almost invariably under the clamps instead of at the crack tip.

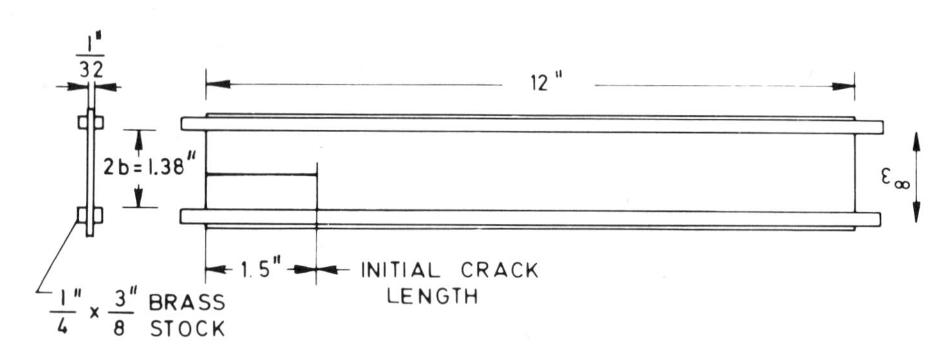

Fig. 1. Crack propagation specimen (pure shear specimen).

The easy fracture under the clamp pressure is an indication that, in contrast to most commercially useful elastomers, the present material fractures under relatively small strains. Since we shall be interested in comparing a fracture model based on the small-deformation theory of linearly viscoelastic solids, this low strain capability is a desirable property. Furthermore, for any crack-propagation velocity experienced, the fracture surface is entirely featureless, even when examined under a scanning electron microscope[33], and is as smooth as a surface produced by brittle fracture in a silica glass. One would argue, therefore, that microscopic, structural irregularities arising in the tearing of the crack-tip material should be small compared with the wavelength of visible light. We shall return to this observation at the end of the paper when we compare these experimental findings with the computations of the crack-propagation model.

The polyurethane possesses a strong photoelastic response (approximately 400 fringes per unit thickness and unit strain). This property was exploited in determining the stress-free condition of the test strip: the vanishing of photoelastic response at the crack tip was taken as an indicator that the specimen was stress free, compression not being permitted to occur at all. The straining was accomplished by moving the brass-stock rails shown in Fig. 1 apart, parallel to each other, in a straining device, the relative displacement being monitored by a dial gage having 1/1000 inch as its smallest division.

After the crack had severed the specimen, a strip 1/2 x 8 inches was cut from one half of the old specimen for the determination of the tensile creep properties. This was done once only for each of the cast sheets, so that each sheet was individually characterized. There was no evidence that the material from which the tensile specimen was cut had sustained any damage while it was under stain in the crack-propagation test. The latter tests and the creep-properties determination on any one sheet were carried out within at least 2 weeks in order to minimize possible aging effects during these tests, although a time period two to three times that length would have probably sufficed.

The creep compliance was determined under dead-load conditions, the elongation being determined with a linearly variable differential transformer (LVDT) possessing a resolution of better than 1/2000 inch; the gage length of the specimen was nominally 6 inches; its uniform cross section, 1/2 x 1/32 inch, was determined with 2 percent accuracy. No account was taken of end effects since the latter are small for the quoted dimensions. The tensile stress was based on the undeformed cross section, since most measurements were made at strains below 3 percent. These measurements were carried out in a temperature-conditioning chamber, the temperature being monitored by a mercury thermometer and a copper-constantan thermocouple. To eliminate the sensitivity of the LVDT to the different test temperatures, it was placed outside of the temperature chamber along with the load pan. The structure connecting the specimen inside the chamber with the LVDT and the load pan outside deformed only a small amount under the larger loads used (at cold test temperatures), but this deformation was taken into account through calibration.

The results of the creep measurements are shown in Fig. 2 for five different sheets; of these five sheets, only two (sheets 1 and 5) had identical properties within plotting accuracy. Moreover, these differences exist with respect to both the shape of the curves and the time scale of the relaxation and/or retardation times. For later calculations, the curves 1 and 5 were employed. The discrepancy along the logarithmic time axis – but not the shape discrepancy – of the remaining curves, was used to adjust the crack-propagation-velocity data obtained from sheets 2 and 4*, so that they were normalized approximately to the properties of sheets 1 and 5. The construction of the master curves shown in Fig. 2 was accomplished with the time-temperature shift factors[34] plotted in Fig. 3. Unless indicated otherwise, T denotes the absolute temperature in all figures.

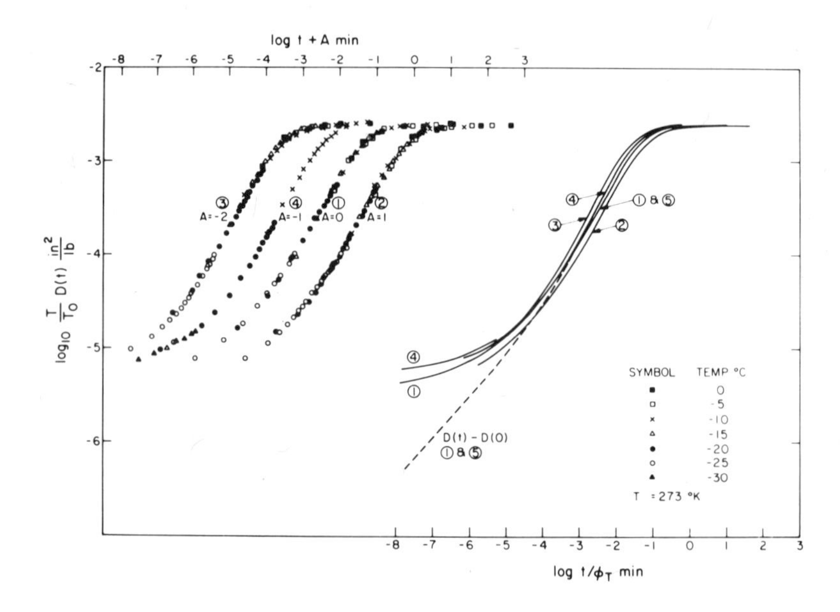

Fig. 2. Master curves of the creep compliances for different sheets of Solithane 113 (equivoluminal composition).

With three exceptions, crack velocities were kept below 1 ipm in order to avoid heat generation at the crack tip, which could affect the rheological properties of the material at that point. Since this phenomenon

*Crack-propagation data of sheet 3 were erroneous because of equipment malfunctioning. The creep properties are shown, nevertheless, to afford an appraisal of material variability.

is very difficult to account for[32], we can only surmise that these crack speeds were sufficiently low to reduce crack-tip heating to insignificant proportions. Slow crack propagation, on the other hand, invites environmental effects to enter the experiments; in this regard, we can only tender the opinion that these effects were not important when cracks propagated in the unpurified laboratory air.

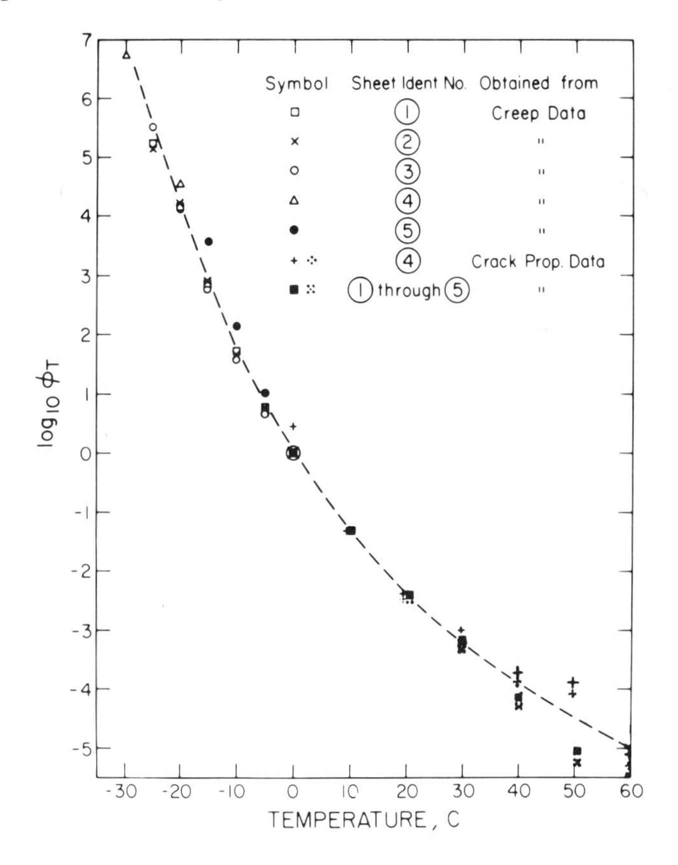

Fig. 3. Time-temperature shift factors for Figs. 4, 5, and 6.

The crack speeds are shown in Fig. 4 as a function of the strain ϵ_∞ across the strip height 2b = 1.38 inches. The set of data on the left of Fig. 4 was derived from different sheets (1 through 5), while the set on the right was obtained entirely from sheet 4. The temperature ratio T/T_0 in the abscissa arises from the normalization of the temperature-dependent rubbery modulus in accordance with the statistical theory of rubber elasticity.

The choice of 0 C as the base line for the data reduction was completely arbitrary. In fact, initially, the coldest test temperature was

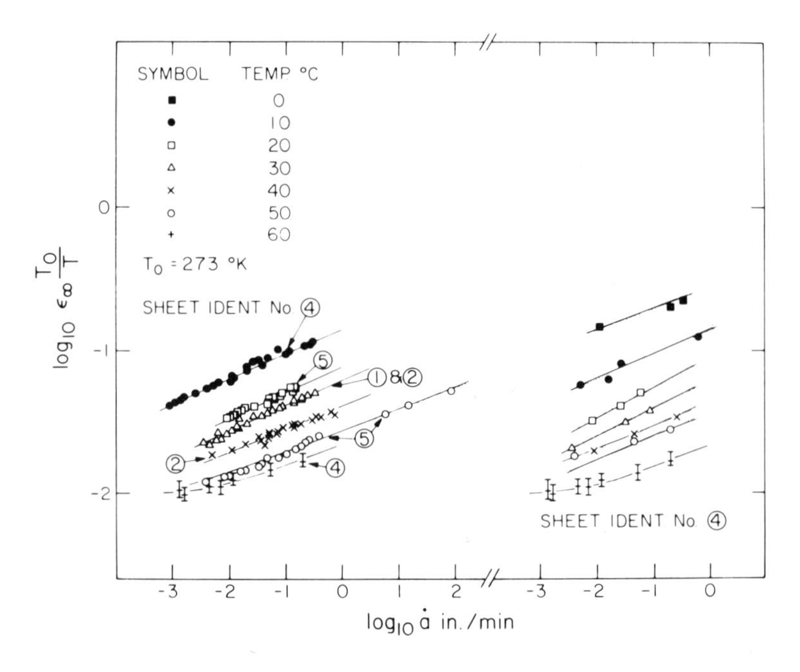

Fig. 4. Crack-propagation rates at different strains ϵ_∞ and different temperatures. Left set of data from various sheets; right set from one sheet only to assess the effect of sheet-to-sheet variability in properties.

10 C; the additional tests at 0 C were later used as a base line by time-temperature shifting all the data through the WLF equation (dashed curve in Fig. 3, Ref. 34). This procedure accounts for the fact that not all shift data cover the point ($\log\phi_T = 0$, T = 0 C) in Fig. 3.

In accordance with the developments in Section 6, the crack velocity – temperature reduction may require a factor $(T_0/T)^2$ in addition to the usual shift factor for thermorheologically simple materials. All data sets were shifted independently of each other, with the raw shift factors recorded as dotted symbols; when the additional correction with $(T_0/T)^2$ was taken into account, the data were recorded as the solid square and + symbols. Master curves of crack-tip speed as a function of the strip strain ϵ_∞ are shown in Figs. 5 and 6.

3 CONSIDERATIONS PERTAINING TO THE MATERIAL BEHAVIOR AT THE TIP OF A STEADILY MOVING CRACK

It was already stated in the Introduction that fracture can be understood ultimately in terms of force interactions at the atomic level, but that, for a macroscopic description of fracture, a less detailed account of

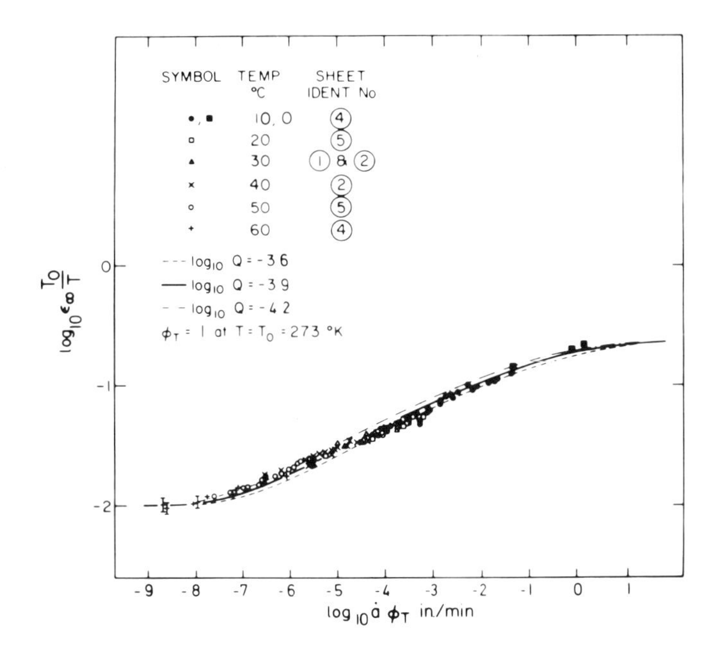

Fig. 5. Master curve of crack-propagation rates obtained on various sheets.

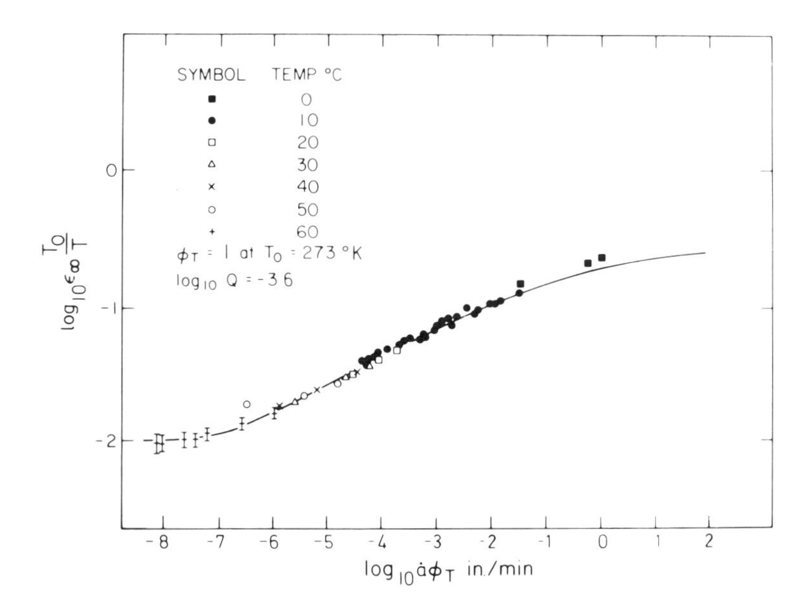

Fig. 6. Master curve of crack-propagation rates obtained on specimens cut from a single sheet.

the failure process appears necessary. We admit from the outset that we have in mind the application of certain analytical procedures to the formulation of the crack-propagation problem, which depends, however, on the rational development of an appropriate fracture model. We search, thus, for a description of the crack-tip-material behavior which contains the processes occurring at the microscopic and molecular level in some average sense, but which is still simple enough so that it can be incorporated into an analytical framework yielding numerical results for comparison with physical measurements.

We start by noting that, for some materials, direct microscopic observation of material behavior at the crack tip is possible. For instance, the role played by the formation and determination of void-rich craze material from the continuous phase in the fracture of rigid, uncrosslinked polymers is well understood: the continuous phase deteriorates thus at the crack tip into craze material which then decays further as the crack-tip advances.[35-37]. These resulting surfaces are not necessarily smooth. A related phenomenon occurs during peel in relatively low-molecular-weight polymers used for pressure-sensitive adhesives. Here the viscoelastic adhesive forms voids under the locally high stress at the peel front.[38] The voids or cavities then grow and coalesce, producing a stringy substance with temporary cohesive strength. As the peel process continues, these strips elongate and become thinner, thus losing their ability to transmit a force. Also, in some metals, the formation, plastic growth, and coalescence of cavities under the high crack-tip stresses are held accountable for the propagation of cracks.[39,40] In unfilled crosslinked polymers above the glass transition (elastomers), the generation of a fibrous structure at the crack tip with cohesive properties seems to play a role.[18] The fibers are apparently the result of coalesced voids; the same observation is made with regard to the disintegration of carbon-black-reinforced rubber as documented photographically in ref. 15.

While the phenomenon of cavitation at an advancing crack tip can thus be observed directly or deduced for some materials as just described, there are others for which such observation is either extremely difficult or impossible. A case in point is the polyurethane elastomer described and characterized in the previous section. It should be expected from the mirrorlike appearance of the fracture surface that the irregularities associated with possible cavitation are small compared with the wavelength of light and should, therefore, be difficult to observe. Microfractures are enlarged in the strain field at the crack tip while the crack propagates; but it becomes clearly difficult to attempt direct observation when the phenomenon is not static and on the order of, or smaller than, the wavelength of light. In postfracture examinations, voids formed in the crack-propagation process probably close to invisible dimensions unless the material can sustain permanent deformations. There seems to be no compelling reason, however, why the cavitation and ligamentation phenomena observed on a relatively large scale as described above should not occur on a smaller scale also, unless the size scale of consideration becomes so small that one must deal with the discrete structure of molecules and atoms rather than with the continuous concept of void geometries. If direct experimental observation of such phenomena is not possible, one might nevertheless hope to deduce indirectly, from an appropriate analysis and experiments, whether cavitation takes place or whether an atomistic description or interpretation is more appropriate. Anticipating our later results, we find, however, that this hope was not realized to the point where this indeterminacy was resolved clearly. The fundamental reason for this (mild) shortcoming of this work is that it is a hopeless undertaking to follow individual cavities through their stages of development and coalescence. One must therefore deal with the average effect which the (potential) voids have upon the surrounding material at the expense of detailed information which would have been obtained with a more refined description of the fracture process.

With the representation of this average effect of the growing cavities in mind, we digress briefly and consider a small cube which contains sources from which microscopic voids or cracklike flaws may grow. Let a tensile stress be applied to two opposing faces of the cube, which increases from zero; since we are here interested in a qualitative exposition, we need not be concerned with the stress history. If we now imagine that at some value of the stress, voids or cracks appear in the cube, then their effect is to decrease the rigidity of the cube. As the defects grow under continued straining, a stress-deformation dependence develops as shown qualitatively in Fig. 7. We may associate the decrease in stress past the strain ϵ^* with pronounced void growth and coalescence, in other words, with mechanical disintegration.

Let us return now to our original intention of incorporating the development of the voids in an average sense into a crack-propagation formulation. We consider, therefore, a small element of material under

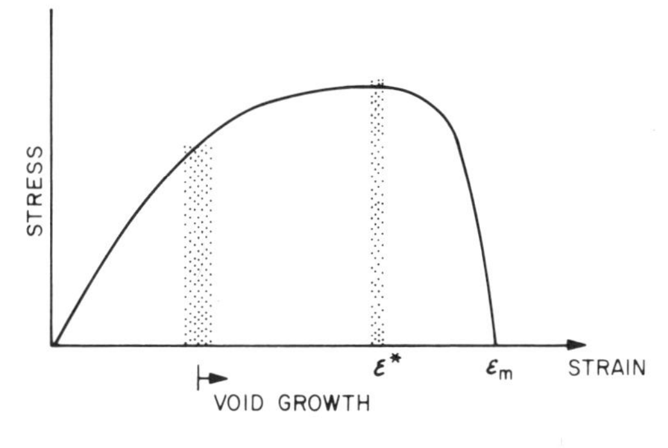

Fig. 7. Qualitative stress-strain characteristic of a strained rod containing growing voids.

tension ahead of the crack tip as shown in Fig. 8a. Deformations are assumed symmetric with respect to the crack axis. Although a finite material element near the crack tip experiences a stress state more complex than the uniaxial one, it suffices for the present purposes to consider the element in this simple stress state; the reason for this simplification will be understood along with later and additional assumptions. As this material element is approached by the advancing crack tip, it develops microscopic voids much as the cube discussed above. In the process of passing from an undamaged state ahead of the crack through the crack-tip region to a "mechanically disintegrated state", the element has provided cohesive stresses which pass through a maximum as in Fig. 8b and then drop to zero on the crack surfaces; the corresponding stress distribution holding the crack tip together is shown qualitatively in Fig. 8b, but it may differ considerably with respect to detail.

The material providing these cohesive stresses spans the crack axis over a – possibly thin – layer, as shown by the dots in Fig. 8c, although this region would possess rather diffuse boundaries. In the formulation of the viscoelastic-boundary-value problem for a moving crack tip, we would have to choose a boundary of the type shown in Fig. 8d, for which the indentation moves to the right; the question of how one would have to account for the damaged material along the crack flanks would remain open. However, if we choose to eliminate the cohesive material while

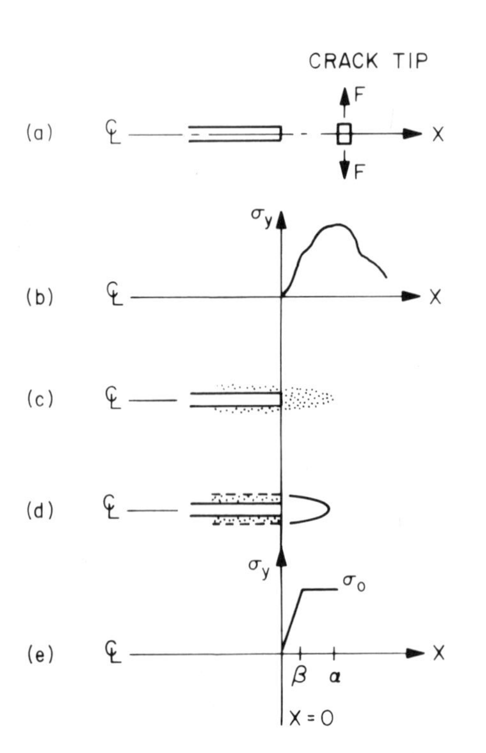

Fig. 8. Development of crack-tip parameters.

retaining the forces which it exerts on the upper and lower segments of the crack tip, we arrive – tentatively – at a boundary which in the undeformed state is straight. The surface of the boundary is traction free over that portion representing the crack surface ($X < 0$ in Fig. 8e), an adjacent section of which ($0 \leqslant x < \alpha$ in Fig. 8e) carries prescribed tractions representing the cohesive forces. The remainder ($x \geqslant \alpha$ in Fig. 8e) is restrained from motion normal to the crack axis, which represents the uncut portion of the solid.

It should be stated specifically that with the elimination of the disintegrating material, the model introduces considerable analytical simplification at the expense of completeness in the description of the fracture process. By reducing the volume of the cohesive material, we have assumed implicitly that the cohesive material properties do not produce an interaction between the deformations and stresses in the cohesive zone and its surroundings. This interaction has been reduced to the prescription of a cohesive stress; in keeping with the viscoelastic properties of the cohesive material, this stress may, however, be a function of the rate of crack propagation. A second implicit assumption is that the energy dissipated against viscous forces in the cohesive material is small compared with that dissipated in the surrounding material. For one fracture criterion explored subsequently, we assume that the only energy expended in the cohesive zone is that needed to rupture (chemical) bonds. Nevertheless, we can readily conceive of materials, as for instance some glassy or crystalline polymers, for which such an assumption may be improper, because the major time dependence in crack propagation may derive from the time-dependent behavior of the disintegrating material itself.[1-6] The following development is thus further restricted beyond the use of linear viscoelasticity theory, because for some glassy polymers, a more careful analysis of the interaction between the disintegrating material and the bulk solid around the crack may have to be taken into account.

We now simplify the distribution of the cohesive forces further to a bilinear distribution, as shown in Fig. 8e, leaving the magnitude σ_0 and the lengths α, β arbitrary. In contrast to usual practice for the fracture of rate-independent material, where $\beta = 0$, the linear portion of this distribution ($0 \leqslant x \leqslant \beta$) is retained for two reasons. First, the extra parameter β may be used to investigate whether the change in stress distribution, resulting from a variation in β, is of any consequence for the crack-propagation behavior, or whether the latter is insensitive to such simple variations. Second, it seems reasonable from a physical viewpoint that the stress distribution should be at least continuous for a continuously moving crack tip. Moreover, the occurrence of step functions, such as afforded by a distribution for which $\beta = 0$, engenders mathematical inconvenience in the application of the theory of viscoelasticity which is circumvented by allowing β to be different from zero. The case of a constant cohesive stress may be obtained by letting $\beta \to 0$, if desired.

Finally, we remark that the distribution of cohesive forces for which $\beta = 0$ in Fig. 8e was used by Dugdale[41] to describe the fracture of thin steel sheets, for which σ_0 denotes the more or less constant yield stress in the plastically deformed crack extension region. In that work, the size of the plastically deformed zone ($= \alpha$ in Fig. 8e, $\beta = 0$) was large compared to the sheet thickness; that mode of plastic defomation is not observed for specimens which are thick compared to the plastic zone in a sheet and, therefore, Dugdale's analysis is usually not applicable to the fracture of thick sections favoring conditions of plane strain at the crack tip. Although we have performed our experiments with sheets 1/32 inch thick, we shall find that the disintegration zone size α is much smaller than 1/32 inch. This observation should not be construed, however, as an invalidation of our results in view of the just-mentioned inconsistency between Dugdale's calculations for a steel sheet and the fracture behavior of thick sections. The coincidence of the special stress distribution used here (for $\beta = 0$) with that employed by Dugdale is more or less accidental and a matter of convenience; the physical processes which underlie them are quite different.

Let us summarize the important points of this section. We have deduced from qualitative experimental observations, a physically plausible model of material deterioration at the tip of a moving crack. The decohering material provides the traction boundary conditions for a simply posed boundary-value problem which is amenable to an exact solution within the realm of linear viscoelasticity theory. Implicit in the formulation of this simplified boundary-value problem is the assumption that the time dependence of the crack-propagation process is derived primarily from the bulk solid surrounding the moving crack tip and not from the rate sensitivity of the disintegrating material. This assumption needs to be checked experimentally. Finally, we see no a priori limitation of the crack-propagation model to thin sheets, but argue that the physical processes described can occur at crack tips in both thick and thin sections.

4 CRITERIA FOR CRACK PROPAGATION

In the last section we gave an outline of a physically plausible process leading to crack propagation. We shall retain in our model only the average stresses exerted by the decomposing material for the purpose of satisfying the equilibrium of the quasi-static stress field; since we do not include the details of the material decomposition process we must supplant the latter by an additional hypothesis. This hypothesis is referred to as a criterion of fracture or of crack propagation. We shall consider here two such criteria.

We call the first criterion, a statement of the law of conservation of energy, the "energy criterion". In view of the assumptions of the last section on the effect of the cohesive material, we demand that the work done against the cohesive forces be a constant, i.e., independent of the velocity of crack propagation.

We call the second criterion the "maximum strain criterion". This criterion assumes that the disintegrating material loses cohesiveness at a finite strain indicated as ϵ_m in Fig. 7. Since we have allowed the weakened material to occupy a vanishingly thin layer or line along the crack axis, it is consistent with this approximation to require as an equivalent criterion that the crack-opening displacement be a specified value at the point where the cohesive forces decay to zero, i.e., at $X = 0$ in Fig. 8e. Moreover, we assume that this strain or displacement be a constant, i.e., independent of the crack-tip velocity.

To give these criteria substance, we first state the boundary-value problem which is to simulate the propagating crack and then state the maximum strain (crack-opening) and energy criteria. These statements connect the (steady) crack-propagation velocity with the applied far-field stresses inducing the crack propagation.

Consider, then, the tip of a crack in a region R of a plane possessing isotropic and linearly viscoelastic properties as shown in Fig. 9, with tractions T^A on the boundary A of the region producing deformations which are symmetric with respect to the crack axis. In a two-dimensional reference frame X, y, affixed to the crack tip moving to the right with velocity $\dot{a}$, let τ_{xy}, σ_X, and σ_y be, respectively, the shear stress, normal components of the stress, and normal components of the stress tensor, and let u_X and u_y be the displacement components in the X- and y-direction. For reasons of symmetry in loading and geometry, the shear traction vanishes on $y = 0$. In accordance with the developments of the last section and with reference to Fig. 8e, we prescribe the remaining boundary conditions as

$$\sigma_y(X, 0) = \begin{cases} 0 & X \leqslant 0 \\ \dfrac{\sigma_o}{\beta}\, X & 0 < X \leqslant \beta \\ \sigma_o & \beta < X \leqslant \alpha \end{cases} \tag{1}$$

$$u_y(X, 0, \dot{a}) = 0 \quad . \qquad X > \alpha$$

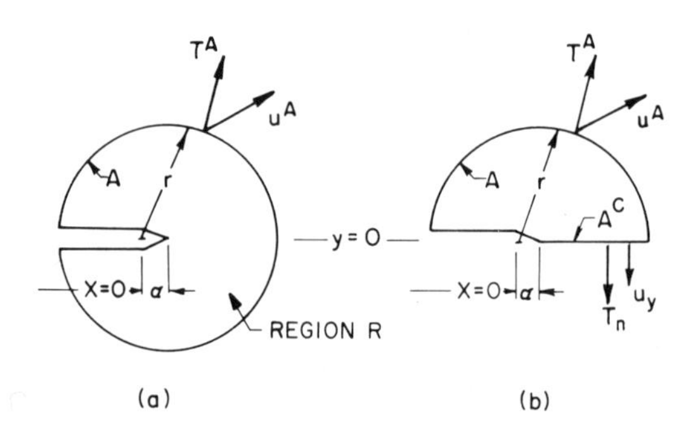

Fig. 9. Crack-tip domain.

The question whether σ_o, α, and β are material constants or not remains open; its answer will be discussed in connection with the evaluation of the two criteria via the experimental evidence. The velocity $\dot{a}$ is supposed to be so small that neither inertial effects nor crack-tip heating due to viscous energy dissipation are important. The question that now motivates us further is: How are the tractions T^A in Fig. 9 related quantitatively to the crack-tip velocity $\dot{a}$?

(a) *Crack Opening Displacement or Ultimate, Constant Strain Criterion.*

Suppose that the displacement at the crack tip is always constrained by

$$u_y\ (0, 0, \dot{a}) = u_o \quad , \tag{2}$$

with u_o an as yet unknown constant, i.e., independent of the applied loads and independent of the crack-tip velocity. To evaluate this criterion it is necessary to determine the crack surface displacement u_y normal to the crack axis as a function of the traction applied to R and as a function of the crack-tip velocity $\dot{a}$. We then need to examine whether experimental evidence permits a coordination of these tractions with the crack-tip velocity $\dot{a}$ when $u_y(0, 0, \dot{a})$ remains a constant.

(b) *Energy Criterion.* For rate-insensitive solids Barenblatt[42] has shown how the unloading tractions at the crack tip do work which is equal to the surface-energy requirement postulated by Griffith. The question now arises whether a similar argument is sufficient to establish not only the onset of fracture propagation in rate-insensitive, brittle solids, but also the condition for continuous crack propagation. If we denote by

$$A = \int_A \vec{T}^A \cdot \dot{\vec{u}}^A \, ds \tag{3}$$

the rate at which the tractions T_i^A do work on the surface A of the region R in Fig. 9a and if $\dot{D}$, $\dot{W}$, $\dot{S}$, and $\dot{Q}$ denote, respectively, the rate of energy dissipation in R, the rate at which the stored energy changes, the rate of change of the surface energy, and the rate at which heat is added to R, then the first law of thermodynamics requires that

$$\dot{Q} + \dot{A} = \dot{D} + \dot{W} + 2\dot{S} \quad , \tag{4}$$

provided we include in our consideration *only* that part of the crack surface $X \leqslant 0$ which is free of tractions. Now remove the lower half of the region R and replace it by the equipollent traction T_n (cf. Fig. 9b) which it exerts on the upper half. We apply now the first law of thermodynamics to this (half) body and note that it is only a body in equilibrium under surface tractions all around without any semblance of a crack. Therefore, the surface energy does not now play a role, but we have a term representing the work done on the half-body over the surface A^c as defined in Fig. 9b. By taking into account the directions of the traction and displacement vectors, one obtains

$$\frac{1}{2}\dot{Q} + \frac{1}{2}\dot{A} - \int_{A^c} T_n(X) \cdot \dot{u}_y(X, 0, \dot{a})dX = \frac{1}{2}\dot{D} + \frac{1}{2}\dot{W} \quad . \tag{5}$$

Upon comparing Eq. 4 and 5 and recalling Eq. 1, one finds for any instant in time

$$\int_0^\alpha T_n(X)\, \dot{u}_y(X, 0, \dot{a})dX = \dot{S} \quad . \tag{6}$$

If the energy required to form a unit of new surface is independent of the loading and of the rate of crack growth, then $\dot{S} = \Gamma\dot{a}$ where Γ is a constant which we call the intrinsic fracture energy. Equation 6 now becomes

$$\int_0^\alpha T_n\,(X)\dot{u}_y)(X, 0, \dot{a})dX = \Gamma\dot{a} \quad .$$

Having already chosen the traction $T_n(X)$ in accordance with Fig. 8e, we need to determine the displacement profile of the "crack surface" in the cohesion zone $0 \leqslant X \leqslant \alpha$ as a function of the tip velocity $\dot{a}$. For the steadily moving crack, we can transform the time derivative in Eq. 6 into a spacial derivative to obtain

$$-\int_0^\alpha T_n(X) \cdot \frac{\partial u_y}{\partial X}(X, 0, \dot{a})dX = \Gamma \quad . \tag{7}$$

Now recall again Eq. 1 and in particular that $T_n(0) = 0$, $u_y(\alpha, 0, \dot{a}) = 0$, and $T_n = \sigma_o = \text{const.}$ for $\beta < X < \alpha$; these facts and integration of Eq. 7 by parts yields

$$\int_o^\beta u_y(X, 0, \dot{a})dX = \frac{\beta\Gamma}{\sigma_o} \quad . \tag{8}$$

In order to evaluate both the displacement criterion and the energy criterion, we need to calculate, thus, only the velocity-dependent displacement profile in that part of the material disintegration zone given by $0 \leqslant X \leqslant \beta$.

It is evident from the energy criterion (Eq. 8) that if Γ/σ_o is constant, then for $\beta \rightarrow 0$, the two criteria (Eqs. 8 and 2) are equivalent, and independently so, of the crack speed $\dot{a}$ (u_y continuous at $X = 0$ for $\beta \rightarrow 0$).

5 SOLUTION TO THE VISCOELASTIC-BOUNDARY-VALUE PROBLEM REPRESENTING THE MOVING CRACK

We remind ourselves once more that the length α of the cohesive zone is small compared to any other dimension of the solid. This observation permits working with only the asymptotic solution valid in the close proximity of the crack tip. Bearing in mind the boundary conditions (Eq. 1), we wish to calculate the displacement $u_y(X, 0, \dot{a})$ along $0 \leqslant X \leqslant \beta$ while the crack tip translates to the right with constant velocity $\dot{a}$.

The solution to this boundary-value problem with moving boundary tractions is not particularly difficult, but is lengthy in detail. Because of the limit on the length of this paper we sketch, therefore, only its development here, and defer a more comprehensive treatment to a future publication on this subject.

We consider, first, the relation between the solution resulting from linearly elastic and viscoelastic properties for the isotropic and homogeneous solid and observe that the distribution of stresses in plane, quasi-static problems is the same for elastic and viscoelastic solids, provided: (a) only tractions are prescribed on the boundary, including the crack surfaces; (b) the tractions on the crack surfaces are self-equilibrating; and (c) no body forces are present. This observation follows from the fact that under this set of restrictions, the material properties can be factored out of the linearized field equations, thus leaving the latter independent of the material properties.[22] The components of the strain tensor and of the displacement vector are, however, dependent upon the material properties and hence are, functions of the stress or load history. That this result should hold also for a crack enlarging quasi-statically and rectilinearly along the x-axis can be shown in a straightforward manner. In the interest of brevity we describe only the proof here. The latter may be supplied by resorting to the Kolosoff-Muskhelishvili method of complex potentials for the linearly elastic plane under the provisions (a)-(c) mentioned above. In the potential functions describing the elastic stress states, replace the coordinate x by x-a(t), where the crack-tip position a(t) need not necessarily grow linearly with time. In place of a multiplication with the elastic material constants, one forms Riemann convolutions with the material functions, say, with the relaxation or creep properties in shear and bulk deformation. The displacements derived from these – so far tentatively constructed – potentials may be substituted into the linearized strain-displacement relations and the strains into the stress-strain equations for this viscoelastic solid. One obtains, then, stresses which are independent of the material properties. Moreover, they satisfy the equations of equilibrium and meet the appropriate boundary conditions on the crack axis if the proper potentials were chosen. We can thus construct the viscoelastic solution from the elastic one if we adhere to the provisions (a)-(c) stated earlier in this section. This close and simple relation between the elastic and viscoelastic solution for the enlarging crack is a consequence of the facts that the order of differentiation and integration with respect to time and space are interchangeable and that only tractions are prescribed on the surfaces.*

With our application of these calculations to an elastomer in mind, we introduce one further simplification by assuming an incompressible solid $\left(\text{Poisson's ratio } \nu = \frac{1}{2}\right)$. Let J(t) be the creep compliance in shear and define for plane strain the constant $\kappa = 3-4\nu = 1$ and for generalized plane stress $\kappa = 3-\nu)/(1+\nu) = 5/3$. In accordance with the fleeting remarks above on the construction of the displacement components, we write the latter in terms of a single** function ϕ as a result of the symmetry of the problem with self-equilibrating tractions on the crack surfaces. Define $X = x-\dot{a}t$, $z = X+iy$, $\mathring{z} = x+iu$ and the displacement components which depend on the crack speed $\dot{a}$ by

$$u_x(X,y,\dot{a}) + i\,u_y(X,y,\dot{a}) = 1/2\,J(t)\left\{\kappa\phi(\mathring{z}) - \phi(\overline{\mathring{z}}) - (\mathring{z}-\overline{\mathring{z}})\overline{\phi'(\mathring{z})}\right\}$$

$$+\frac{1}{2}\int_{o}^{t} J(t-\xi)\,\frac{\partial}{\partial\xi}\left\{\kappa\phi(z) - \phi(\overline{z}) - (z-\overline{z})\,\overline{\phi'(z)}\right\}d\xi \quad . \qquad (9)$$

*In particular, only stresses are prescribed on the newly generated surfaces of the enlarging crack. For the *closing* crack, the displacements must be prescribed along the crack surfaces and therefore the above method of calculating the viscoelastic deformations breaks down.

**For an exposition of this method see, e.g., Refs. 43 or 44; an exhaustive treatment is given in Ref. 45.

The first term on the right of Eq. 9 represents the displacement contribution due to an initial step-load application. We are interested only in the motion of the crack long after the starting transients have subsided and the crack has enlarged from its initial geometry. Under these conditions, the first term contributes nothing to the *crack opening* displacement* and we may neglect it in our further development. As stated earlier, the stresses derived from Eq. 9 via the stress-strain equations for a homogeneous, isotropic, and linearly viscoelastic solid $\left(\nu = \frac{1}{2}\right)$ yield a stress distribution which moves with the crack tip and which satisfies the boundary conditions Eq. 1 *on the crack,* provided (see, e.g., Refs. 43-45)

$$\phi'(z) = \frac{1}{2\pi i\sqrt{z-\alpha}} \int_0^\alpha \frac{\sigma_y(\tau)\sqrt{\tau-\alpha}}{\tau - z}\, d\tau \ , \tag{10}$$

where $\sigma_y(\tau)$ is the normal traction on the crack surface (Fig. 8e). In addition, $\phi(z)$ must also satisfy the far-field boundary conditions on the tractions T^A (cf. Fig. 9) which drive the crack. We have already indicated that we wish to retain only the dominant part of the stress field near the crack tip which results from these far-field boundary conditions. This is accomplished by including in $\phi(z)$ an additive function

$$\phi_f(z) = \frac{K}{\sqrt{2\pi}}\sqrt{z-\alpha} \tag{11}$$

which leaves the crack surface completely traction free for $X < \alpha$. The factor K is termed the stress-intensity factor; it contains the far-field boundary conditions** as well as the geometry if the solid has finite dimensions in the plane, such as other cracks or holes, etc. Because of the assumed symmetry in our problem, such boundaries must fall into the same symmetric pattern. The function $\phi(z)$ in Eq. 9 is thus the sum of $\phi_f(z)$, Eq. 11, and the function, say, $\phi_c(z)$, obtained by integrating Eq. 10, and which satisfies the boundary conditions on the crack surfaces; one has thus

$$\phi(z) = \phi_f(z) + \phi_c(z) \ . \tag{12}$$

The function $\phi_c(z)$ may be evaluated readily from Eq. 10 by employing Eq. 1, but the calculations are tedious. We therefore record only the final result

*In fact, for any point $x > \alpha$ to the right of the initial crack-tip position, the first term vanishes on the crack axis.

**Strictly speaking these must be stress boundary conditions according to provision (a) at the beginning of this section.

$$\phi_c(z) = i\,\frac{\sigma_o \alpha}{\pi}\left\{\frac{\beta-z}{2\alpha}\,\ell n\left|\frac{C+C_o}{C-C_o}\right| - C_o C\right.$$

$$+\frac{\alpha}{2\beta}\left[\frac{1}{2}\,\frac{\beta^2-z^2}{\alpha^2}\,\ell n\left|\frac{C-C_o}{C+C_o}\right| - \frac{1}{2}\,\frac{z^2}{\alpha^2}\,\ell n\left|\frac{C+1}{C-1}\right|\right. \tag{13}$$

$$\left.\left. - \left|(1-C_o)\left(1+\frac{z}{\alpha}\right)C + \frac{1}{3}\,(1-C_o^3)\right|C\right]\right\} \quad ,$$

where $C = [1\text{-}z/\alpha]^{1/2}$ and $C_o = [1\text{-}\beta/\alpha]^{1/2}$. Now Eqs. 11 and 13 render $\phi(z)$ through Eq. 12. Both Eq. 11 and Eq. 13 provide stresses which diverge as the inverse square root of the distance from the point ($X = \alpha$, $y = 0$). The combination of Eqs. 11 and 13 allows the elimination of such singular stresses by a proper choice of σ_o, α, and β in terms of the stress-intensity factor K. This is accomplished by letting the coefficient, multiplying the (combined) singularity and resulting from $\phi(z)$ in Eq. 12, vanish. This coefficient, say K_o, is given by (see, e.g., Ref. 44)

$$K_o = 2\sqrt{2\pi}\lim_{z\to\alpha}\sqrt{z-\alpha}\,\phi'(z) \quad . \tag{14}$$

The vanishing of K_o gives a relation, subsequently termed the finiteness condition, which links K, σ_o, α, and β by

$$\sigma_o = \frac{1}{2}\sqrt{\frac{\pi}{2}}\frac{K}{\sqrt{\alpha}}\left\{\frac{\alpha}{\beta}\left[1-C_o - \frac{1}{3}\,(1-C_o^3)\right] + C_o\right\}^{-1} \quad . \tag{15}$$

We emphasize that although the stress-intensity factor K appears here as well as subsequently, the stresses are now finite everywhere in the crack-tip region; therefore, K is only a parameter representing the far-field loading conditions and the geometry of the body apart from the crack tip.

Note that for $\beta \to 0$, Eq. 15 reduces to the well-known result for a constant cohesive (yield) stress (see, e.g., Ref. 44):

$$\sigma_o = \frac{1}{2}\sqrt{\frac{\pi}{2}}\frac{K}{\sqrt{\alpha}} \quad , \tag{16}$$

while for $\beta = \alpha$, representing a cohesive force distribution which rises linearly from $X = 0$ to $X = \alpha$, one obtains for the maximum stress

$$\sigma_o = \frac{3}{4}\sqrt{\frac{\pi}{2}}\frac{K}{\sqrt{\alpha}} \quad . \tag{17}$$

Upon defining

$$A = \left\{ \frac{\alpha}{\beta} \left[1 - C_o - \frac{1}{3}(1-C_o^3) \right] + C_o \right\}^{-1} \tag{18}$$

and taking account of Eqs. 11, 12, 13, and 15, we obtain

$$\phi(z) = i \frac{K\sqrt{\alpha}}{\sqrt{2\pi}} \left\{ C + \frac{A}{2} \left[\frac{\beta - z}{2\alpha} \ell n \left| \frac{C+C_o}{C-C_o} \right| - C_o C \right] \right.$$

$$+ \frac{A}{4} \frac{\alpha}{\beta} \left[\frac{\beta^2 - z^2}{2\alpha^2} \ell n \left| \frac{C-C_o}{C+C_o} \right| - \frac{1}{2} \frac{z^2}{\alpha^2} \ell n \left| \frac{C+1}{C-1} \right| \right.$$

$$\left. \left. -(1-C_o)\left(1 + \frac{z}{\alpha}\right) C + \frac{1}{3}(1-C_o^3) C \right] \right\} . \tag{19}$$

For the determination of the crack surface displacement over $0 \leqslant X \leqslant \beta$, we evaluate Eq. 9 on the crack axis $y = 0$ through the use of Eq. 19; it can be shown that on $y = 0$, $(z-\overline{z})\ \overline{\phi'(z)}$ in the integrand on Eq. 9 vanishes and that $-\phi(\overline{z}) = \phi(z)$ there. We observe, furthermore, that for all $X \geqslant \alpha$, the normal displacement u_y is zero. This fact can be used to change the lower limit of the integral in Eq. 9. To see this, we consider a point x_o on the crack axis ($y = 0+$) and ahead of the crack; we suppose that the crack tip is stationary and allow the point x_o to move to the left with the crack-tip velocity $\dot{a}$, and are then interested in the displacement u_y which the point x_o experiences normal to the crack axis. Evidently, the point experiences no displacement at all until the time t_o when it reaches the tip at $X = \alpha$. Moreover, any displacement subsequent to this event at t_o is a function only of the time which has elapsed since the point passed $X = \alpha$; by virtue of the constancy of $\dot{a}$ it is thus also only a function of the distance from $X = \alpha$. We now use this latter fact to transform the time integral for the displacement u_y in Eq. 9 to a spatial integral. To that end, denote by *Im* the imaginary part of a complex function, so that

$$u_y(x, 0, \dot{a}) \underset{y \to 0+}{=} \frac{\kappa+1}{2} \, Im \int_{t_o}^{t} J(t-\xi) \frac{\partial}{\partial \xi} \phi(z) d\xi \ . \tag{20}$$

Define

$$\underset{y \to 0+}{Im\ \phi(z)} = \frac{\kappa \sqrt{\alpha}}{\sqrt{2\pi}} f(X) \ , \tag{21}$$

where

$$f(X) = C + \frac{A}{2} \left[\frac{\beta - X}{2\alpha} \ell n \left| \frac{C+C_o}{C-C_o} \right| - C_o C \right]$$

$$+\frac{A\alpha}{4\beta}\left[\frac{\beta^2-X^2}{2\alpha^2}\,\ell n\left|\frac{C-C_o}{C+C_o}\right| - \frac{X^2}{2\alpha^2}\,\ell n\left|\frac{C+1}{C-1}\right|\right. \tag{22}$$

$$\left.-(1-C_o)\left(1+\frac{X}{\alpha}\right)C+\frac{1}{3}(1-C_o^3)C\right]$$

$$C=\left[1-\frac{X}{\alpha}\right]^{1/2}$$

$$C_o=\left[1-\frac{\beta}{\alpha}\right]^{1/2} \quad .$$

Furthermore, let

$$J(t) = J_o + \Delta J(t) \quad , \tag{23}$$

where $\Delta J(t)$ represents the transient creep response such that $\Delta J(0) = 0$, and J_o is the initial or glassy compliance. Upon remembering that $X = x-\dot{a}t$, Eq. 20 may then be written as

$$u_y(X,0,\dot{a}) = \frac{\kappa+1}{2}\,\frac{K\sqrt{\alpha}}{\sqrt{2\pi}}\left\{J_o f(x-\dot{a}t) + \int_{t_o}^{t}\Delta J(t-\xi)\,\frac{\partial}{\partial\xi}\,f(x-\dot{a}\,\xi)d\xi\right\}. \tag{24}$$

We may, without consequences, let $t_o = 0$. With reference to Fig. 10, define the intermediate variables ζ and η by

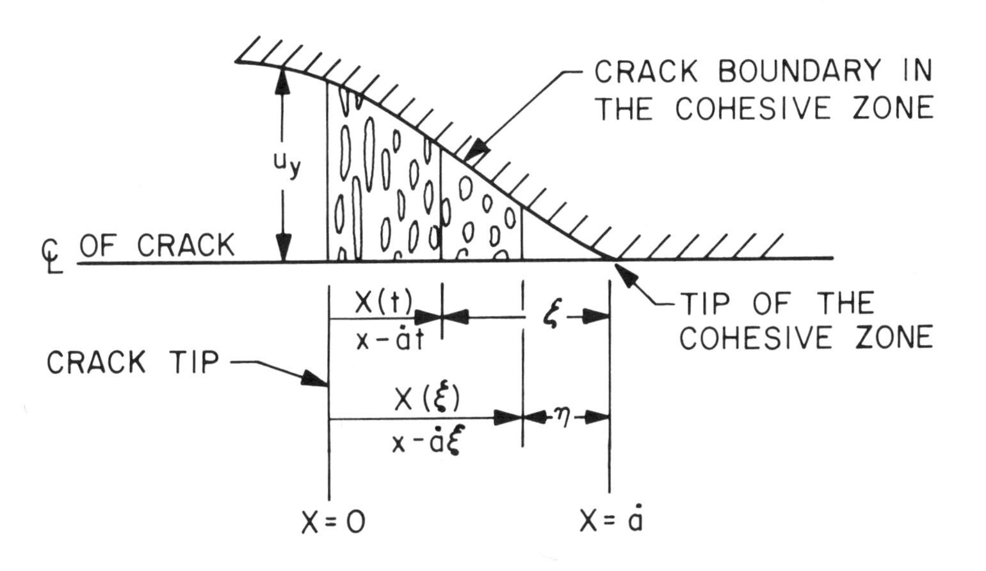

Fig. 10. Definition and transformation of crack-tip parameters and geometry.

$$X(t) = x-\dot{a}t = \alpha-\zeta \tag{25}$$

$$X(\xi) = x-\dot{a}\ \xi = \alpha-\eta$$

Then

$$t-\xi = \frac{\zeta-\eta}{\dot{a}} = \frac{X(\xi)-X(t)}{\dot{a}} \quad . \tag{26}$$

If, for the purpose of nondimensionalization, we furthermore let

$$\rho = \frac{X(t)}{\alpha}; \; r = \frac{X(\xi)}{\alpha}$$

$$F(r) = f(\alpha r) \qquad F'(r) = \alpha f'(\alpha r) \quad . \tag{27}$$

We can rewrite Eq. 24 as

$$u_y(\rho,0,\dot{a}) = \frac{\kappa+1}{2}\frac{K\sqrt{\alpha}}{\sqrt{2\pi}}\left\{J_o F(\rho) - \int_\rho^1 \Delta J\left[\frac{\alpha}{\dot{a}}(r-\rho)\right] F'(r)dr\right\}, \tag{28*}$$

with

$$R = \beta/\alpha$$

$$A = A(R)$$

$$F(r) = C + \frac{A}{2}\left[\frac{R-r}{2}\ell n\left|\frac{C+C_o}{C-C_o}\right| - C_o C\right]$$

$$+ \frac{A\alpha}{4\beta}\left[\frac{R^2-r^2}{2}\ \ell n\left|\frac{C-C_o}{C+C_o}\right| - \frac{r^2}{r}\ell n\left|\frac{C+1}{C-1}\right|\right.$$

$$\left. - \left|(1-C_o)(1+r)C + \frac{1}{3}(1-C_o^3)\right| C\right]$$

$$C = [1 - r]^{1/2}$$

$$C_o = [1 - R]^{1/2} \quad .$$

We note that for the special case of a rate-insensitive, elasto-plastic solid ($\Delta J(t) = 0$) with a constant yield stress in the cohesive zone ($R = \beta/\alpha = 0$), Eq. 28 reduces in view of Eq. 16 to

---

*In explanation of the minus sign in front of the integral, note that $F'(r)$ is negative so that the contribution of the integral term is positive.

$$u_y(r, 0) = \frac{\kappa+1}{2\mu} \frac{\sigma_o \alpha}{\pi} \left[C - \frac{r}{2} \ln \left| \frac{C+1}{C-1} \right| \right]. \tag{29}$$

This result agrees with that given in Ref. 46 if one takes into account that the latter expresses the total crack opening rather than the displacement of one crack surface from the crack axis. It remains to write Eq. 28, in terms of the tensile creep compliance determined in section 2 for the numerical evaluation of Eq. 28. Because of the assumed incompressibility of the material, we have

$$J(t) = 3D(t) \tag{30}$$

and separate D(t) into its glassy and transient components D_o and $\Delta D(t)$, $\Delta D(0) = 0$.

We shall see later that the disintegration zone α is much smaller than the sheet thickness employed in the experiments. Consequently we assume that for the local, crack-tip-dominated solution, conditions of plane strain prevail. However, we point out that subsequently we use a stress-intensity factor which is determined from conditions of plane stress far away from the crack tip; this condition is indicative of the thin-sheet geometry which governs the stress state throughout the major portion of the strip geometry used in the experiments. The present assumption of *local* plane strain conditions should thus be viewed as an improvement on the crack-tip stress state as afforded by the plane stress field. We write thus for Eq. 28 with $\kappa = 1$

$$u_y(\rho, 0, \dot{a}) = \frac{3K\sqrt{\alpha}}{\sqrt{2\pi}} \left\{ D_o F(\rho) - \int_\rho^1 \Delta D \left[\frac{\alpha}{\dot{a}} (r-\rho) \right] F'(r)\, dr \right\}. \tag{31}$$

6 EVALUATION OF THE CRACK PROPAGATION CRITERIA

We may now substitute the displacement Eq. 31 into the maximum strain criterion Eq. 2 and the energy criterion Eq. 8. The further evaluation has to be accomplished numerically. We compare the two criteria first with each other and then with the experimental results of section 2. As a consequence of the comparison with experimental measurements, some of the hitherto undetermined parameters will be specified.

6.1 Comparison of the Two Criteria.

We discuss first the implications of the finiteness condition Eq. 15 in connection with its effect on the displacement (Eq. 31). The latter contains the load parameter K and the size of the cohesive zone α in the argument of the creep compliance; but K, α β, and σ_o are connected by Eq. 15 and are therefore not independent of each other. In general, one must expect, therefore, that with changing K – usually implying a change in the crack-propagation speed* – all three parameters σ_o, α, and β change and become, thus, functions of the crack velocity. We choose for consideration *two special cases.* First, we consider that α remain constant and that σ_o and β vary for different values of the stress intensity factor K in such a way that Eq. 15 is satisfied. One then finds that a change in β alone cannot satisfy Eq. 15 for all values of K of interest, but that for sufficiently large K, an increase in the latter requires an increase in the maximum cohesive stress σ_o. Since an increase in K results usually in an increased crack velocity, it would follow that an increase in crack velocity is associated also with an increase in the cohesive stress; such a finding would be consistent with other rate-sensitive properties of viscoelastic solids. Second, we consider that σ_o instead of α remains constant and represents some ultimate stress of the solid, similar to but not necessarily equal to the molecular strength of the material. In that case α and β depend on the magnitude of the load parameter (stress intensity factor) K.

With regard to the choice of β, we anticipate here some numerical results. On the basis of computations to be discussed presently, we have examined the effect of varying β between the limits zero and α. Within plotting accuracy, the results were, practically speaking, independent of β. Since β was, in part, incorporated into the analysis to effect a change in the cohesive force *distribution,* we conclude that this *distribution* is relatively unimportant for the crack-propagation process.** For rate-insensitive solids, this conclusion had been reached by Barenblatt[48] also, for the case when the cohesive zone is small compared with any other dimensions of the solid. That this result should be true for rate-sensitive solids is not immediately obvious. In our numerical work we have arbitrarily set $R = \frac{1}{2}$.

For brevity of notation, define with $E_\infty = 1/D(\infty)$

$$\vartheta\left(\frac{\alpha}{\dot{a}}, \rho\right) = E_\infty\left\{D_o F(\rho) - \int_\rho^1 \Delta D\left[\frac{\alpha}{\dot{a}}(r-\rho)\right] F'(r)dr\right\} \qquad (32)^{***}$$

*It is also possible, in general, that a change in K does not increase or decrease the crack velocity from a finite value.

**At least for the material examined here.

***$\vartheta \frac{\alpha}{\dot{a}}, \rho$ is a function of R also by virtue of F(r), Eqs. 27 and 28.

$$\Theta_R\left(\frac{\alpha}{\dot{a}}\right) = \frac{A(R)}{R}\int_o^R \vartheta\left(\frac{\alpha}{\dot{a}}, \rho\right) d\rho \quad . \tag{33}$$

By virtue of Eq. 32, Eq. 31 becomes

$$u_y(\rho, 0, \dot{a}) = \frac{3K\sqrt{\alpha}}{\sqrt{2\pi}\, E_\infty}\, \vartheta\left(\frac{\alpha}{\dot{a}}, \rho\right) \quad . \tag{34}$$

The crack opening criterion (ultimate, constant strain criterion) of Eq. 2 requires then that the crack velocity $\dot{a}$ be related to the load parameter K through

$$\frac{3K\sqrt{\alpha}}{\sqrt{2\pi}\, E_\infty}\, \vartheta\left(\frac{\alpha}{\dot{a}}, 0\right) = u_o \quad . \tag{35}$$

If $\alpha = \alpha_o$ = constant, this relation reads

$$\frac{3K\sqrt{\alpha_o}}{\sqrt{2\pi}\, E_\infty}\, \vartheta\left(\frac{\alpha_o}{\dot{a}}, 0\right) = u_o \quad , \tag{36}$$

while for σ_o = constant, Eq. 15, in the form

$$\sqrt{\alpha} = \sqrt{\frac{\pi}{2}}\, \frac{K}{\sigma_o}\, \frac{A(R)}{2} \quad , \tag{37}$$

demands

$$\frac{3}{4}\, \frac{K^2}{\sigma_o}\, \frac{A(R)}{E_\infty}\, \vartheta\left(\frac{\pi K^2 A^2(R)}{8\sigma_o^2\, \dot{a}}, 0\right) = u_o \quad . \tag{38}$$

The same considerations apply to the energy criterion Eq. 8. In view of the normalization preceding Eq. 27 and of the definition Eq. 33, this criterion may be written as

$$\frac{3}{4} K^2\, \Theta_R\left(\frac{\alpha_o}{\dot{a}}\right) = E_\infty \Gamma \tag{39}$$

for $\alpha = \alpha_o$ = constant, and for constant σ_o instead of constant α,

$$\frac{3}{4} K^2\, \Theta_R\left(\frac{\pi K^2 A(R)}{8\sigma_o^2\, \dot{a}}\right) = E_\infty \Gamma \quad . \tag{40}$$

We now particularize these results for the geometry of a long strip (cf. Fig. 1) employed in the crack-propagation measurements in section 2.

Denote the strain across the strip by ϵ_∞. If, as we have stated earlier, we restrict our interest to times when the crack moves into the long-time equilibrium stress field, then the stress-intensity factor for a quasi-statically moving crack is $\left(\text{for generalized plane stress and } \nu = \frac{1}{2}\right)$

$$K^2 = \frac{4}{3} E_\infty^2 \epsilon_\infty^2 b \quad . \tag{41}$$

Insertion of Eq. 41 into Eq. 36 and Eqs. 38 through 40 relates the steady rate of crack propagation $\dot{a}$ to the applied strain ϵ_∞ through the two variants of the two hypothetical fracture criteria. If we use the further notational definition

$$Q = \frac{\pi b \, E_\infty^2 \, A^2(R)}{6\sigma_o^2} \quad , \tag{42}$$

we find that the criteria read now as follows:

From Eq. 36, the *crack opening displacement criterion* follows as

$$\epsilon_\infty \, \vartheta \left(\frac{\alpha_o}{\dot{a}}, 0 \right) = \sqrt{\frac{\pi}{6}} \frac{u_o}{\sqrt{b\alpha_o}} \qquad \alpha_o = \text{constant}; \tag{43}$$

and from Eq. 37,

$$\epsilon_\infty^2 \, \vartheta \left(Q \frac{\epsilon_\infty^2}{\dot{a}}, 0 \right) = \frac{\sigma_o \, u_o}{E_\infty A(R) b} \qquad \sigma_o = \text{constant}; \tag{44}$$

while the *energy criterion* Eqs. 39 and 40 render, respectively,

$$\epsilon_\infty^2 \, \Theta_R \left(\frac{\alpha_o}{\dot{a}} \right) = \frac{\Gamma}{E_\infty b}, \qquad \alpha_o = \text{constant}; \tag{45}$$

and

$$\epsilon_\infty^2 \, \Theta_R \left(Q \frac{\epsilon_\infty^2}{\dot{a}} \right) = \frac{\Gamma}{E_\infty b} \qquad \sigma_o = \text{constant}. \tag{46*}$$

For the numerical evaluation of the relations in Eqs. 43-46, the creep compliance marked 1 and 5 in Fig. 2 was used. In order to keep these computations economical and in view of the intrinsic uncertainty of the experimental data, which is on the order of ±5 percent, an error of 0.1

---

*For the limit of $\beta \rightarrow 0$, Eq. 46 reduces to the result given in Ref. 22, while Eq. 46 – as well as Eqs. 43-45 – applies strictly only to *steady* crack propagation owing to assumptions incorporated into Eq. 31; the result in Ref. 22 is derived with the *intent* to treat a *general* crack-speed history.

percent in the integration was deemed sufficient for the present purpose. We need not report much more about this integration than that a routine based on Simpson's rule was used and that care was exercised in evaluating the limits of the products involving the logarithmic functions in the functions F(r) and F′(r).

The two criteria Eqs. 43 and 44 and Eqs. 45 and 46 each involve three undetermined parameters, if $R = \frac{1}{2}$ as indicated earlier in this section; for the crack opening displacement criterion we lack the values of u_o and either α_o or σ_o, while for the energy criterion we need to determine Γ and α_o or σ_o. Since both u_o and Γ were assumed to be independent of the crack-propagation speed, they may be evaluated from conditions prevailing as $\dot{a} \to 0$. The remaining parameters are then determined by matching the velocity scale to experimental data, if such matching is at all possible. For an initial comparison, u_o or Γ were chosen such that $\epsilon_\infty = 0.01$ for $\dot{a} \to 0$. The graphical representation of Eqs. 43 - 46 is given in Fig. 11. The crack opening displacement criterion for constant α deviates most from the others, primarily because it involves the strain ϵ_∞ to the first power, while the others demand the square of the strain.

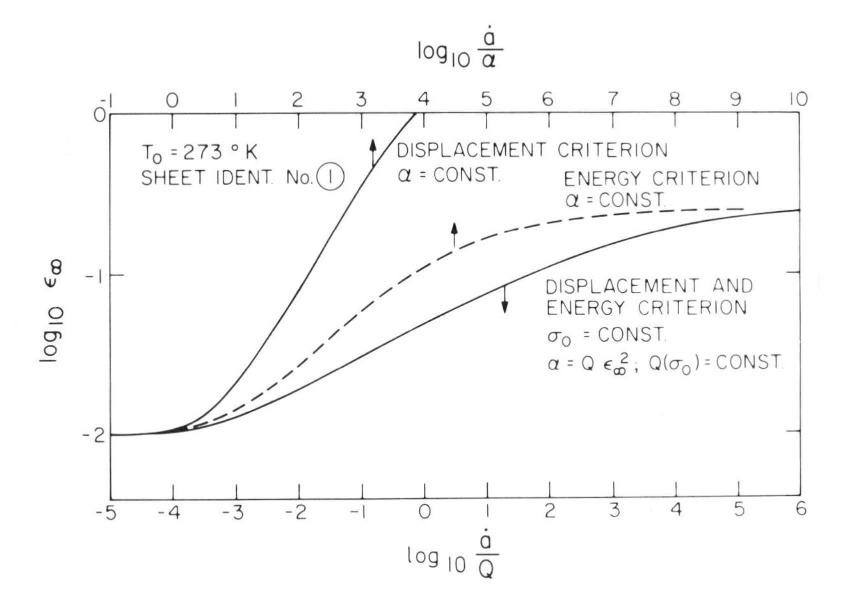

Fig. 11. Comparison of the ultimate strain and energy criteria for constant α and σ_o; creep properties same as those shown in Fig. 2. Parameters u_o and Γ adjusted so that $\dot{a} \to 0$ for $\epsilon_\infty = 0.01$.

The result that both the energy and the crack opening displacement criteria are coincident may be, superficially speaking, surprising. It is true, though, that for rate-insensitive elastic-plastic solids, the Dugdale model for sheet fracture [for R = 0, A(R) = 1] equates the two criteria for predicting initial crack instability if one sets $u_o \sigma_o = \Gamma$.[31] Furthermore, we have already seen at the end of the previous section that the two criteria are identical if Γ and σ_o are constant (Γ/σ_o = const) and if $\beta \rightarrow 0$. The thought is therefore close that for different values of β, the two criteria should also give the same crack-propagation behavior in agreement with the stated insensitivity of the results in Fig. 11 to variations in β. However, the close agreement between the two criteria Eqs. 44 and 46 – both for constant σ_o – is a consequence of the creep properties displayed in Fig. 2. Some reflection, too lengthy to be detailed here, shows that the two criteria should predict, closely, the same crack-propagation behavior if the maximum slope of log D(t) vs log t is not great* (about 0.7 in Fig. 2). For a significantly greater slope, the criteria should not agree, while for slopes smaller than or equal to 0.7, they should agree as well as for the properties employed here. To check this assertion, the two criteria Eqs. 44 and 46 were evaluated for a standard linear solid having the same short and long time limits as the polyurethane elastomer characterized in Fig. 2. The result of this separate calculation was a different crack-speed prediction** from the two criteria. Even when they were brought to agreement for $\dot{a} \rightarrow 0$, the predicted velocity differed by approximately 25 percent over a range of velocities spanning about seven decades; the energy criterion predicted the slower crack propagation. The agreement for the two criteria is thus, in principle, fortuitous and the result of the material properties. From a practical viewpoint, we must bear in mind, however, that a standard linear solid, for example, is hardly a good representative of a polymer, and that the polyurethane elastomer employed in this study is much more representative of some polymeric solids. Equation 46 may thus well be the rule, rather than the exception, in the fracture of some viscoelastic solids. Before proceeding to the exploitation of the experiments, we make two deductions from the criteria as stated so far.

If the material is thermorheologically simple, one sees with the aid of Eq. 31 that a change in temperature affects the time scale through a shift factor ϕ_T which multiplies the velocity $\dot{a}$. Moreover, on the basis of the

*In a recent report ("A Theory of Crack Growth in Viscoelastic Media", Mechanics and Materials Research Center, Texas A & M University, MM 2764-73-1, March 1973), R. A. Schapery develops an approximate theory of crack propagation in linearly viscoelastic solids. One approximation – valid if the second derivative of the logarithm of the material function, e.g., creep compliance, with respect to the logarithm of time is "small" – replaces the material function by a power law. This substitution allows the approximate evaluation of convolution integrals such as Eq. 24. The reason that the two criteria discussed in this work render nearly the same result is intimately connected with the justification for the Schapery approximation.

**σ_o was the same and $\beta = \alpha/2$ for both.

classical theory of rubber elasticity, the equilibrium modulus E_∞ is proportional to the absolute temperature T in Eqs. 44 and 46; E_∞^2 enters the argument of ϑ or Θ_R through Q (see Eq. 42). If we write thus the temperature-dependent equivalent of Eq. 46, for example, we obtain, with T_o as a reference (here 273 K) and with E_* the equilibrium modulus at T_o,

$$\epsilon_\infty^2 \Theta_R \left[\frac{\pi b E_*^2 A^2(R)}{6\sigma_o^2} \frac{\epsilon_\infty^2}{\dot{a}} \frac{T^2}{T_o^2 \phi_T} \right] = \frac{\Gamma T_o}{E_* T b} \quad . \tag{47}$$

Thus, the crack-velocity data need not necessarily time-temperature shift according to the shift law for thermorheologically simple solids, but *may* involve *further* temperature corrections. Moreover, these possible corrections, as evidenced in Eq. 47, are occasioned by the *test method,* since they result from the prescription of the load in the form of displacements (strain) rather than tractions. In practical terms, this correction is small compared to the shift factor ϕ_T.

We add here a second brief observation on the relation of the rate-dependent fracture energy which plays a prominent part in Refs. 8, 16 and 28. Let this energy be denoted by $S(\dot{a})$. For the strip geometry in Fig. 1, this quantity can be expressed for constant temperature by[19]

$$S(\dot{a}) = \frac{4}{3} E_\infty \epsilon_\infty^2 b \quad . \tag{48}$$

Substituting, for example, Eq. 48 into Eq. 45 yields

$$S(\dot{a}) = \frac{4\,\Gamma}{3\Theta_R\left(\frac{\alpha_o}{\dot{a}}\right)} \quad \left(= \frac{8}{3}\Gamma;\ \dot{a} \to 0 \right) \quad . \tag{49}$$

Thus, the rate-dependent fracture energy[9-12] is essentially the product of the intrinsic fracture energy Γ, presumably of molecular origin, and a nondimensional function $\Theta_R(\alpha_o/\dot{a})^{-1}$ which embodies the rheology of the material surrounding the crack tip. This result contradicts the suggestion in Ref. 28 that the rate-dependent fracture energy be the *sum* of the intrinsic fracture energy (surface energy) and of the viscous contribution. Finally, if Eq. 48 is combined with Eq. 46 instead of Eq. 45, a less simple relation results, namely,

$$\frac{3}{4} S(\dot{a}) \cdot \Theta_R \left[\frac{\pi}{8} \frac{E_\infty A(R)}{\sigma_o^2} \frac{S(\dot{a})}{\dot{a}} \right] = \Gamma \quad . \tag{50}$$

In concluding this discussion on the equivalence of the crack opening displacement and the energy criteria, we emphasize once more, with a view towards Fig. 11, that this equivalence has been shown so far to hold only if the maximum cohesive stress σ_0 is a constant. The same is not necessarily true if σ_0 is a function of the crack-propagation speed; it is obviously not true if $\alpha = \alpha_0$, independent of the crack-propagation speed. Henceforth, we shall refer only to the energy criterion if σ_0 = constant, and not to the crack opening displacement criterion.

6.2 Deductions from Experiments

In section 3, we have presented a detailed argument for the plausibility of a crack-propagation model. In essence, we have questioned whether the *realistic* dependence of crack speed on the loading can be described in terms of such a simple model. So far we have evaluated this model only analytically in preparation for the answer to this question. In applying these calculations to measurements, we search for two kinds of information. First, we are interested in whether one or another of the crack-propagation equations (Eqs. 43-46) can be made to agree with the experiment by a suitable choice of the unknown parameters. Second, if a coincidence between the calculations and the experimental results can be achieved, it would be appealing if the parameters so determined were physically reasonable. In this way, one might "make a case" for the model, although one would not necessarily prove that the supposed crack-propagation process delineated in section 3 is uniquely true.

In Figs. 5 and 6, the energy criterion for σ_0 = constant is shown as the solid curve through the experimental points. It is clear, upon consideration of Fig. 11, that the crack-propagation criteria Eqs. 43 and 45 for $\alpha = \alpha_0$ will fit the data to a distinctly lesser degree* than Eqs. 49 and 46 for σ_0 = constant. The dashed curves in Fig. 5 represent the same criterion as the solid line, but shifted along the log $\dot{a}$-axis.

We observe, thus, that the proposed model does not only fit reasonably well to the experimental data, but specifies further that the maximal cohesive stress at the crack tip remains – at least for this polyurethane elastomer – independent of the applied load or strain ϵ_∞ and independent of the crack-growth speed. We turn next to the evaluation of the parameters σ_0, Γ, and u_0, as well as the range of variation for α. To determine Γ we evaluate Eq. 46 for $\dot{a} \to 0$, where, on account of Figs. 5 and 6, ϵ_∞ = 0.01, and by noting that $\Theta_R(\infty) = \frac{1}{2}$, which latter relation follows from

*We note in passing that the energy criterion for $\alpha = \alpha_0$ corresponds closely to the results obtained in Ref. 19.

the numerical calculations since Eq. 33 was not integrated in closed form for $\dot{a} \to 0$.

By matching the velocity scale of Fig. 11 to the velocity scales in Figs. 5 and 6, we obtain values of $\log_{10} Q$ = -3.9 and -3.6 inch, from which we can determine σ_o by virtue of Eq. 42.

Upon evaluating Eqs. 44 and 32 for $\dot{a} \to 0$, we obtain u_o by letting $\epsilon_\infty = 0.01$. Using b = 0.69 inch, E_∞ = 398 psi, ν = 0.5, and R - 0.5, one obtains

Γ = 0.014 lb/in.
σ_o = 1.7 to 2.5 $\cdot$ 10^4 psi
u_o = 1.0 $\cdot$ 10^{-6} in. (225 A)
α = 2.5 to 1.3 $\cdot$ 10^{-4} ϵ_∞^2 in.
= 6.3 to 3.1 $\cdot$ 10^4 ϵ_∞^2 A.

The values of these parameters invite several comments. In earlier experiments on nominally the same material, with a different pretest history, the intrinsic fracture energy Γ was found to be of the same order of magnitude ($\Gamma \sim$ 0.09 lb/in.) and in reasonable agreement with calculations based on the molecular origin of this quantity.[46] A value of 0.014 lb/in. appears, therefore, not unreasonable.* Whether the microstructurally oriented parameters u_o and α have realistic values is difficult to say since they depend strongly on the use of the linearly viscoelastic theory. The most pertinent experimental information is the fact that this polyurethane elastomer breaks with a mirrorlike fracture surface like inorganic glass at any test strain and velocity measured in these experiments. Irregularities of the fracture topology which may be an indication of how large the cohesive zone is, are therefore less than about a quarter of the wavelength of visible light, i.e., about 1500 A. For a strain of ϵ_∞ = 0.1, one obtains a value of α = 600(300) A, while for the minimum strain of ϵ_∞ = 0.01 where $\dot{a} \to 0$, this value would be only 6(3) A. On the average, the value of α is acceptable in the light of the upper bound provided by the smoothness of the crack surface, but for $\dot{a} \to 0$, the zone represents clearly molecular rather than continuum mechanical dimensions. In looking at these small values of α critically we must not forget that these data (and the value of σ_o) are the result of matching the theoretical velocity scale to the experimental one, which was in turn assembled by a none-too-accurate time-temperature correspondence. In addition, we recall that we have not accounted for the nonlinearly viscoelastic response of the highly deformed material at the crack tip. If the large strains** at the crack tip have the

*Our past experience with this polyurethane elastomer has shown that different batches of material can exhibit significantly different properties. It was for this reason that the experimental work for this paper was repeated with a new set of material to eliminate possible aging effects, rather than use the results of earlier investigations.[17-21]

**We deal here with strains potentially so large that they induce crystallization in natural rubber.

effect of shifting the relaxation spectrum to longer times, then this would result in a value of α which is erroneously too small.

With regard to the crack opening displacement, a value of 225 A is not unreasonable in connection with the just cited values of α. We bear in mind, however, that u_0 is, physically speaking, a less meaningful quantity than, for instance, the cohesive stress σ_0 or the disintegration-zone size. The picture that presents itself of the *deformed* disintegration zone near the limit $\dot{a} \rightarrow 0$ is thus one of a short zone highly elongated normal to the crack axis for $\epsilon_\infty = 0.01$, while at higher crack velocities ($\epsilon_\infty = 0.1$), the length of the zone is about as long as its dimension across the crack axis ($2u_0$). This calculated shape of the crack-tip zone indicates large, if finite, deformation gradients which violate the precepts of linear viscoelasticity theory; obviously, this observation cannot be left out of consideration when one evaluates the overall merit of the crack-propagation model presented here.

Finally we remark that to our knowledge there exists no good estimate for the magnitude of σ_0. If the material were a crystal, one would expect a value on the order of 10 percent of the elastic modulus. The fact that the above range of σ_0 is between 7 and 10 percent of the glassy modulus (estimated to be about 2.5×10^5 psi from Fig. 2) may be significant, but could be equally well fictitious if one considers the amorphous molecular structure of the material and the large deformations which this structure experiences at the crack tip.

7 SUMMARY AND CONCLUDING REMARKS

In the foregoing pages we have attempted to develop a plausible model of the fracture process in a crosslinked elastomer. That development was constrained by the availability – or lack – of analytical tools for a more general description of the process. In view of the uncertainties arising out of the mathematical modelling as well as from the lack of definite knowledge concerning the physical mechanism by which the material at the crack tip gives way to crack propagation, it was essential to draw on experiments for an evaluation. This interaction between experiment and analysis has shown that either of two criteria can describe the crack propagation of at least one material: The criteria are the ultimate, constant strain (crack opening displacement) criterion on the one hand and the energy criterion on the other. Either criterion is applicable, provided the maximum cohesive stress at the crack tip is a constant.

The model explains why the time-temperature superposition principle is reasonable for reducing crack-propagation data of thermorheologically simple materials. Moreover, it explains, as did an earlier model proposed in

Ref. 19, that the experimentally conceived rate-dependent fracture energy[11,12] is constituted of a product, one factor of which is the intrinsic fracture energy Γ required to rupture the elastomer in the absence of viscous energy dissipation ($\dot{a} \rightarrow 0$) and the other factor of which represents essentially the rheological properties of the material. It was found also that for thermorheologically simple solids, the time-temperature shift of crack-propagation data need not necessarily be precisely the same as that obtained from the reduction of relaxation or creep data.

In spite of the good agreement between this thought model and the experimental facts, one cannot claim that the conjectures on the fracture process have been proven; nor have they been disproven. The reason for this remaining uncertainty is the fact that several phenomena have been neglected in the analysis. While these phenomena have been deemed to be of secondary importance, their influence may not be negligible; some of these have already been discussed and we state a further one below, but without discussion. Furthermore we mention, primarily with a comprehensive presentation of related topics in mind, some ideas which offer themselves as natural further pursuits but for which the limited writing space does not permit a more thorough discussion.

1. We must be aware of the possibility that the parameter α does not literally correspond to the size of the zone in which fracture occurs actually in the experiments. It could possibly also stand for the size in which the molecules reach their ultimate extensibility. If this were so, then β would have to be considered as the length over which the material breaks down. However, α would probably still determine the local *size scale* in the crack-tip region which measures the *domain* in which the most significant portion of viscous energy dissipation occurs.

2. Statistical crack-speed variation about the mean, steady value: for relatively short times a crack may slow down or even come to a stop. The source of this intermittently hesitant response is not quite clear; an uneven advance of the crack front through the sheet was not alone responsible. Consequently, it is uncertain whether such temporary crack retardation is to be included in calculating an average speed or whether the stopping of the crack should be excluded from the speed determination. This effect was not taken into account here, but the velocity was taken as the ratio of the distance travelled by the crack to the time required therefore.

3. The temporarily nonsteady crack propagation is closely linked with the problem of crack propagation under nonsteady loads. While the crack-propagation Eqs. 36 and 38-40 may, as approximations, be applicable at any instant of time, it is not certain alone from the considerations presented here under which conditions such an approximation is permissible. Consider the following as an illustrative example of the type of

question to be answered: If a crack has propagated under a high load with a correspondingly large cohesive zone, how will it propagate when the load is reduced and requires only a smaller cohesive zone for the equilibrium of the crack-tip stresses? It is quite possible that the answer to this question cannot be supplied through so simple a model as the one presented here. For accelerating cracks, it is probably sufficient that $\alpha/\dot{a} \leqslant K/\dot{K}$ for Eqs. 36 and 38-40 to be good approximations for nonsteady crack growth (see, for example, Ref. 21).

4. From the numerical calculations, the function $\Theta_R(t)/E_\infty$ is very closely equal to one-half of the creep compliance.* This simple relation offers a way to approximate crack-propagation analyses of test data; it is possibly one explanation for the success of the less refined crack-growth models of Bueche and Halpin[47] and of Mueller and Knauss[19,21].

5. As a final note, we mention an experience with a different elastomer than the one studied here, namely, carboxy-terminated poly-butylacrylic acid. It was observed that at low strains and low velocities, the crack tip exhibited a foamlike or fibrillar disintegration zone on the order of 1 or 2 mm which produced a very rough fracture surface.[18] In contrast to the calculations presented here, this macroscopically evident decomposition zone *decreased in size* with *increasing* load and crack speed to produce, above a certain velocity, a glassy surface like the polyurethane Solithane studied here. We cite this example – which contradicts our present findings – as an indication of our awareness that the work presented on Solithane here deals with a class of materials which is a subset of all elastomeric solids. However, we do not know at this time how to define the characteristics of this material subset clearly.

REFERENCES

1. Barenblatt, G. I., Entov, V. M., and Salganik, R. L., "Kinetics of Crack Propagation, General Considerations: Cracks Approaching Equilibrium Cracks", *Mech. Solids (Mekhan. Tverdogo Tela)* **1** (5), English version p 82 (1966).
2. Barenblatt, G. I., Entov, V. M., and Salganik, R. L., "Kinetics of Crack Propagation: Conditions of Fracture and Long Term Strength", *ibid.*, p 76.
3. Barenblatt, G. I., Entov, V. M., and Salganik, R. L., "Kinetics of Crack Propagation: A Note on the Rule of Summation of Damageabilities", *ibid.*, **2** (2), English version p 148 (1967).
4. Barenblatt, G. I., Entov, V. M., and Salganik, R. L., "Kinetics of Crack Propagation: Fluctuation Fracture", *ibid.*, **2**, (1), English version p 122 (1967).
5. Barenblatt, G. I., Entov, V. M., and Salganik, R. L., "On the Influence of Vibrational Heating on the Fracture Propagation in Polymeric Materials", *Proceedings of the IUTAM Symposium in East Kilbride,* June 25-28, 1968.

*The reasons for this close resemblance are the same as those that led Schapery (see footnote regarding Schapery's work in Section 6.1) to employ the power-law approximation for the viscoelastic-material properties.

6. Barenblatt, G. I., Entov, V. M., and Salganik, R. L., "Some Problems of the Kinetics of Crack Propagation", in *Inelastic Behavior of Solids,* Materials Science and Engineering Series, Kanninen, M. F., Adler, W. F., Rosenfield, A. R., and Jaffee, R. I. (Eds.), McGraw-Hill (1970).
7. Rivlin, R. S., and Thomas, A. G., "Rupture of Rubber, I. Characteristic Energy for Tearing", *J. Polymer Sci.,* **10** (3), 291 (1953).
8. Thomas, A. G., "Rupture of Rubber, II. The Strain Concentration at an Incision", *J. Polymer Sci.,* **18,** 177 (1955).
9. Greensmith, H. W., and Thomas, A. G., "Rupture of Rubber, III. Determination of Tear Properties", *J. Polymer Sci.,* **18,** 189 (1955).
10. Greensmith, H. W., "Rupture of Rubber, IV. Tear Properties of Vulcanizates Containing Carbon Black", *J. Polymer Sci.,* **21,** 175 (1956).
11. Greensmith, H. W., Mullins, L., and Thomas, A. G., "Rupture of Rubber", *Trans. Soc. Rheology,* **4,** 179 (1960).
12. Thomas, A. G., "Rupture of Rubber, VI. Further Experiments on the Tear Criterion", *J. Appl. Polymer Sci.,* **3** (8), 168 (1960).
13. Greensmith, H. W., "Rupture of Rubber, VII. Effect of Rate Extension in Tensile Tests", *J. Appl. Polymer Sci.,* **3** (8), p 175 (1960).
14. Mullins, L., "Rupture of Rubber, Part IX. Role of Hysteresis in the Tearing of Rubber", *Trans. Rubber Ind.,* **35,** 213 (1959).
15. Greensmith, H. W., "Rupture of Rubber, VIII. Comparison of Tear and Tensile Rupture Measurements", *J. Appl. Polymer Sci.,* **3,** 183 (1960).
16. Greensmith, H. W., "Rupture of Rubber, XI. Tensile Rupture and Crack Growth in a Noncrystallizing Rubber", *J. Appl. Polymer Sci.,* **8,** 1113 (1964).
17. Knauss, W. G., "The Time Dependent Fracture of Viscoelastic Materials", *Proceedings of the First International Conference on Fracture,* (1965), Vol. 2, p 1139 See also Ph.D. Thesis, California Institute of Technology, Pasadena, California, 1963.
18. Knauss, W. G., "Stable and Unstable Crack Growth in Viscoelastic Media", *Trans. Soc. Rheology,* **13** (3), 291 (1969).
19. Mueller, H. K., and Knauss, W. G., "Crack Propagation in a Linearly Viscoelastic Strip", *J. Appl. Mech.,* **38,** Series E (No. 2), 483 (1971).
20. Mueller, H. K., and Knauss, W. G., "The Fracture Energy and Some Mechanical Properties of a Polyurethane Elastomer", *Trans. Soc. Rheology,* **15** (2), 217 (1971).
21. Knauss, W. G., "Delayed Failure – the Griffith Problem for Linearly Viscoelastic Materials", *Intern. J. Fracture Mech.,* **6** (1), 7 (1970). See also, *Fracture 1969,* Proceedings of the International Conference on Fracture, Brighton (1969), p 894.
22. Kostrov, B. V., and Nikitin, L. V., "Some General Problems of Mechanics of Brittle Fracture", *Archiwum Mechaniki Stosowanej,* **22** (6), English version p. 749 (1970).
23. Marshall, G. P., Culver, L. E., and Williams, J. G., "Crack and Craze Propagation in Polymers: A Fracture Mechanics Approach. I. Crack Growth in Polymethyl Methacrylate in Air", in *Plastics and Polymers,* The Plastics Institute Transactions and Journal, Headington Hill Hall, Oxford (1969), p 75.
24. Marshall, G. P., Culver, L. E., and Williams, J. G., *The Growth of Cracks and Crazes in Polystyrene: A Fracture Mechanics Approach,* Imperial College of Science and Technology, Mechanical Engineering Department, London (1971).
25. Prandtl, L., "Ein Gedankenmodell für den Zerreissvorgang spröder Körper", *ZAMM,* **13,** 129 (1933). See also Ludwig Prandtl, Gesammelte Abhandlungen, Erster Teil, Springer (1961).
26. Williams, M. L., "The Fracture of Viscoelastic Material", in *Fracture of Solids,* Drucker and Gilman (Eds). Interscience Publishers (1963), p 157.
27. Williams, M. L., "The Kinetic Energy Contribution to Fracture Propagation in a Linearly Viscoelastic Material", *Intern. J. Fracture Mech.,* **4** (1), 69 (1968).
28. Williams, M. L., "Initiation and Growth of Viscoelastic Fracture", *ibid.,* **1** (4), 292 (1965).
29. Wnuk, M. P., and Knauss, W. G., "Delayed Fracture in Viscoelastic-plastic Solids", *Intern. J. Solids Structures,* **6,** 995 (1970).
30. Wnuk, M. P., "Energy Criterion for Initiation and Spread of Fracture in Viscoelastic Solids", *South Dakota State University, Engineering Experiment Station Bulletin,* No. 7 (April, 1968).

31. Wnuk, M. P., "Effects of Time and Plasticity on Fracture", *Brit. J. Appl. Phys.*, Series 2, **2**, 1245 (1969).
32. Knauss, W. G., "The Mechanics of Polymer Fracture", *Appl. Mech. Reviews* (January, 1973).
33. Jones, W. J., North American Rockwell Corp., Rocketdyne Division, McGregor, Texas, personal communication.
34. Williams, M. L., Landel, R. F., and Ferry, J. D., "The Temperature Dependence of Relaxation Mechanisms in Amorphous Polymers and Other Glassforming Liquids", *J. Am. Chem. Soc.*, **77**, 3701-3707 (1955).
35. Kambour, R. P., "Mechanism of Fracture in Glassy Polymers. I. Fracture Surfaces in Polymethyl methacrylate", *J. Polymer Sci.*, Part A, **3**, 1713 (1965).
36. Kambour, R. P., "Mechanism of Fracture in Glassy Polymers, II. Survey of Crazing Response During Crack Propagation in Several Polymers", *J. Polymer Sci.*, Part A-2, **4**, 17 (1966).
37. Kambour, R. P., "Mechanism of Fracture in Glassy Polymers, III. Direct Observation of the Craze Ahead of the Propagating Crack in Poly(methyl methacrylate) and Polystyrene", *J. Polymer Sci.*, Part A-2, **4**, 349 (1966).
38. Kaelble, D. H., *Physical Chemistry of Adhesion*, Interscience Publishers, John Wiley and Sons, New York (1971).
39. McClintock, F. A., *Plasticity Aspects of Fracture, in Fracture, an Advanced Treatise*, Liebowitz, H. (Ed.), Academic Press, New York (1971), Vol 3, p 47.
40. Broek, D., *A Study on Ductile Fracture*, National Lucht-en Ruimtevaartlaboratorium, NLR TR 71021 U, The Netherlands.
41. Dugdale, D. S., "Yielding of Steel Sheets Containing Slits", *Jour. Mech. Phys. Solids*, **8**, 100 (1960).
42. Barenblatt, G. I., "The Mathematical Theory of Equilibrium Cracks in Brittle Fracture", in *Advances in Applied Mechanics*, Academic Press (1962), Vol. 7, p 55.
43. England, A. H., *Complex Variable Methods in Elasticity*, Wiley-Interscience, New York (1971).
44. Rice, J. R., *Mathematical Analysis in the Mechanics of Fracture*, Liebowitz, H. (Ed.), Academic Press, New York (1971), Vol 2, p 192.
45. Muskhelishvili, N. I., *Some Basic Problems of the Mathematical Theory of Elasticity*, J.R.M. Radok Transl.; Noordhoff Ltd., Groningen (1963).
46. Lake, G. J., and Thomas, A. G., "The Strength of Highly Elastic Materials", *Proc. Roy. Soc., London, Series A, Math. and Phys. Sci.*, **300** (1460), 108 (1967).
47. Bueche, F., and Halpin, J. C., "Molecular Theory for the Tensile Strength of Gum Elastomers", *J. Appl. Phys.*, **35** (1), 36 (1964).

DISCUSSION on Paper by W. G. Knauss

KANNINEN: From your presentation, it seemed as if you obtained a viscoelastic solution for a finite width domain directly from the linear elasticity solution for a partly loaded crack in an infinite domain as given in Muskhelishvili's book. Could you clarify this?

KNAUSS: I am dealing here with the problem of a viscoelastic planar solid under moving boundary tractions due to the unloading of the stresses on the surfaces of the enlarging crack. Linear viscoelasticity theory was used – employing the Kolosoff-Muskhelishvili complex variable theory for the elastic part of the calculations – to obtain the time-dependent stress components and deformations in a domain about the crack tip which is very small compared to the width of the sheet.

Because of the limitations on time, the details were not delineated, nor are they given in the written account because of the editorial constraint on its length.

THOMAS: Would you enlarge on the physical model you assume for the processes going on at the tip of the crack? In particular, does the dimension of 5000 A you mention have any relation to an effective diameter of the tip.

KNAUSS: You will recall that I professed ignorance of the detailed material behavior as it passes from the apparent continuum ahead of the crack tip through the fracture zone, since, in general, it eludes direct quantitative observation and measurement. It is only for those polymers in which relatively large voids are formed in a large crack-tip domain that such an observation becomes feasible for a moving crack. For the particular material used in this work, this disintegration region is so small – less than 5000 A – that direct observation is practically impossible. The mechanical response of the disintegrating material has therefore been replaced by an approximate distribution of "holding forces" shown as a bilinear distribution of tractions at the tip.

Regarding your second point, I think we need to agree first on the meaning of "an *effective* crack-tip radius" and I am afraid that an exploration of how the present model might relate to such a potential geometric crack-tip description is beyond the time remaining for discussion.

KOBAYASHI: Were the crack-opening displacement values mentioned measured experimentally?

KNAUSS: No. In view of the small size of the disintegration zone ($<$5000 A), it seems rather futile to attempt to do that, especially for a moving crack.

THE ROLE OF FAILURE CRITERIA IN THE FRACTURE ANALYSIS OF FIBER/MATRIX COMPOSITES

W. Knappe and W. Schneider

Deutsches Kunststoff-Institut
Darmstadt, Germany

ABSTRACT

In fiber/matrix composites (fmc), three types of failure occur:

1. Cohesive failure of the fiber
2. Cohesive failure of the matrix
3. Adhesive failure of the fiber/matrix interface.

Therefore, it seems unrealistic to apply only one failure criterion to fmc. Fiber failure takes place only when the stress in the direction of the fibers reaches a critical value, since the stress perpendicular to the fiber may be neglected. Fiber failure will lead mostly to total fracture of the composite. In many cases, before total fracture is achieved, partial fracture by cohesive failure of the matrix or by adhesive failure of the fiber/matrix interface occurs in the form of cracks between the fibers (interfiber failure). Hereby, the stress-strain-relation of the fmc is markedly changed.

By calculating the micromechanical stresses in the matrix and applying the criteria for cohesive-adhesive failure, one is able to predict cracking and its influence on stress-strain-behavior and total fracture. Some experimental methods for evaluation of fracture criteria for glass-reinforced plastics are described and some results are given.

1 INTRODUCTION

In the design with fiber/matrix composites (fmc), optimal mechanical properties of the composite in relation to the amount of material should be achieved by concentration and geometrical arrangements of the fibers. In the case of stiffness calculations, the optimizing problem can be solved with the continuum theory (see, for example, Refs. 1, 2, and 3). Optimization of strength is a much more difficult problem, since different kinds of failure occur. Thus we have to differentiate between the following processes[4]:

1. Cohesive failure of the fibers
2. Cohesive failure of the matrix
3. Adhesive failure of the interface between fiber and matrix.

In multi-ply fmc, total failure is caused by fiber failure (ff), mode 1. Before fiber failure occurs, there is mostly partial failure by cracking in the matrix, mode 2, or in the interface, mode 3. Modes 2 and 3 we denote interfiber failure (iff).

Since cracks can shorten the useful lifetime of constructions, it is sometimes desirable to optimize a fmc for interfiber failure (iff optimum). The optimum in strength for fiber failure alone (ff optimum) is given by the so-called network theory. In the latter case, it is assumed that the load will be carried only by the fibers, which will fail under uniaxial stress.

2 LAYERWISE FRACTURE ANALYSIS

Most fmc contain fibers lying in different directions. Despite the fact that there is often intermeshing of threads, as for example in fabrics and filament wound structures, a reasonable method of analysis seems to be to regard the composite as a system of so-called unidirectional (ud) layers, in which all fibers lie in the same (parallel) direction. Figure 1 gives a sketch of such a real laminate (a), which is dissolved into a system of unidirectional layers (b). We assume that all layers are perfectly bonded and therefore will have the same deformations as the laminate. Multi-ply fmc structures are mostly loaded by a plane state of stress, i.e., the single ud plies have a plane state of stress too (Fig. 2). Failure will occur in the following way[5]. Because of the stresses (σ_x, σ_y, τ_{xy}, Fig. 2), cracking in one unidirectional layer will begin. Thereby, the stiffness of this unidirectional layer decreases, except for Young's modulus in direction of the

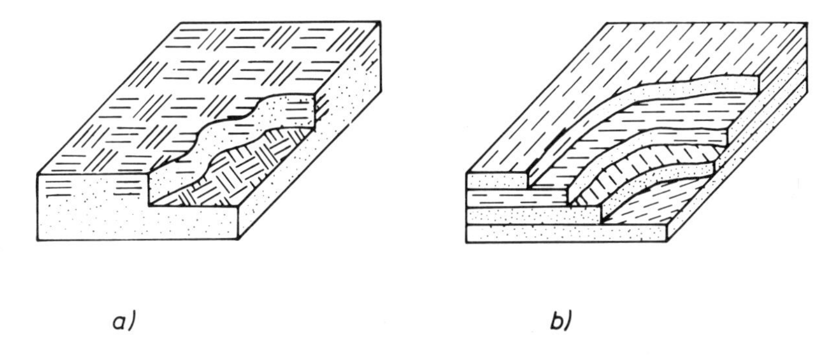

Fig. 1. Real fiber/matrix composite (a); idealized fiber/matrix composite (b) dissolved in unidirectional layers.

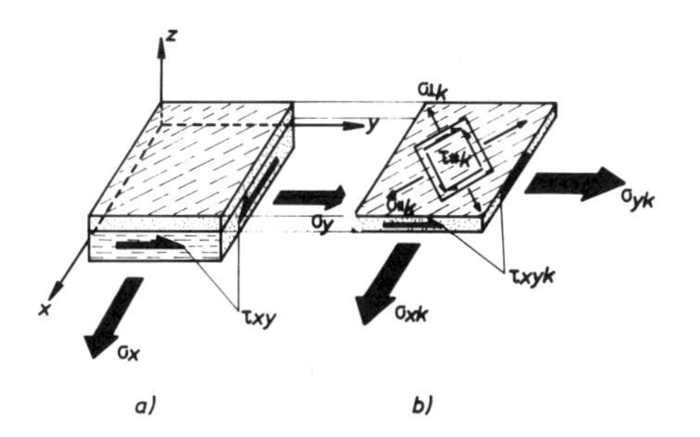

Fig. 2. Plane state of stress in a multi-ply composite (a) and in an unidirectional layer (b); k = index of the layer.

unbroken fibers ($E_{\|}$) which remains unaffected. Also, it should be kept in mind that cracking will not be suddenly accomplished at the critical stress level but will proceed with rising stress until all elastic constants of the layer except $E_{\|}$ are zero. By cracking in one ply, an altered stress distribution in the layers will be induced. When the stress of the whole composite is further increased, a second layer will crack, and so on until the first fiber failure mostly leads to total failure of the composite.[5]

For calculations, one needs the stresses in the layers as a function of the stress of the whole composite and fracture criteria for the above-discussed three modes of failure. By means of continuum theory, the constants of elasticity of the different layers and, from this, of the whole composite can be calculated.[1] Thus, for a given state of stress, we may calculate the three strain components of the whole composite (ϵ_x, ϵ_y, γ_{xy}), which are equal to the strain components in every layer (ϵ_{xk}, ϵ_{yk}, γ_{xyk}, k = index of the layer). These components can be easily transformed to the natural coordinate system of every layer given by the directions parallel and perpendicular to the fibers ($\epsilon_{\|k}$, $\epsilon_{\perp k}$, $\gamma_{\#k}$). $\epsilon_{\|}$ is the strain parallel to the fibers, $\epsilon_{\perp}$ is the strain perpendicular to the fibers, and $\gamma_{\#}$ is the shear deformation parallel and perpendicular to the fibers. With the strain components and the constants of elasticity of every layer, the stress components $\sigma_{\|k}$, $\sigma_{\perp k}$, $\tau_{\#k}$ (Fig. 2) can be calculated, and from fracture criteria of the unidirectional layer in the form

$$F\,(\sigma_{\|}, \sigma_{\perp}, \tau_{\#}) \leqslant 1 \quad , \tag{1}$$

the onset of failure may be predicted.

3 FIBER FAILURE

In most cases, the fibers have a much greater strength than the matrix. Therefore, $\sigma_{\perp}$ and $\tau_{\#}$ will be always much smaller than the fiber strength $\sigma_{\|ff}$. This means that the influence of $\sigma_{\perp}$ and $\tau_{\#}$ on fiber failure can be neglected, and we may use the criterion[4]

$$\sigma_{\|}/\sigma_{\|ff} \leqslant 1 \tag{2}$$

for fiber failure.

4 FAILURE CRITERION FOR INTERFIBER FAILURE

Failure criteria for ud layers can be illustrated by a failure surface in the $\sigma_{\|}$, $\sigma_{\perp}$, $\tau_{\#}$ space (see Fig. 3). Since two processes are involved in interfiber failure, we have two failure surfaces which may intersect, and we may have either cohesive failure of the matrix or adhesive failure of the interface, corresponding to the stress combination. This problem will be discussed later.

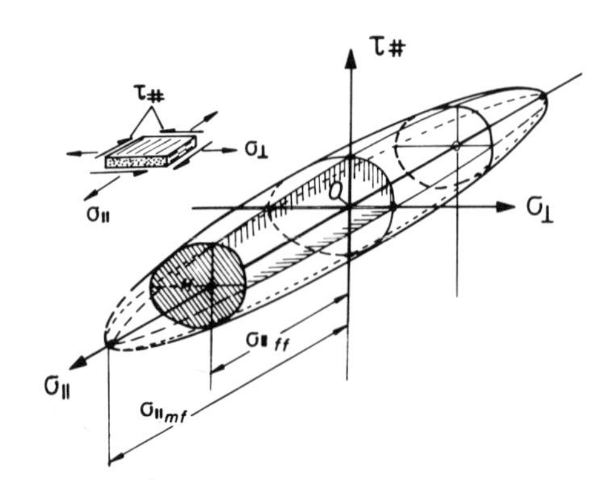

Fig. 3. Failure surface of the unidirectional layer.

Puck and Schneider[4] assumed that (iff) will occur by cohesive failure, and found a failure criterion which compares the three stresses $\sigma_{\|}$, $\sigma_{\perp}$, $\tau_{\#}$

with three calculated cohesive strengths ($\sigma_{\parallel mf}$, $\sigma_{\perp mf}$, $\tau_{\# mf}$). In later work[5], this theoretical failure criterion was transformed into a semi-empirical criterion of the following form:

$$(\sigma_{\parallel}/\sigma_{\parallel mf})^2 + (\sigma_{\perp}/\sigma_{\perp f})^2 + (\tau_{\#}/\tau_{\# f})^2 \leqslant 1 \quad . \tag{3}$$

$\sigma_{\perp f}$, the tensile or compressive strength perpendicular to the fibers and $\tau_{\# f}$, the shear strength, are experimentally achieved data. $\sigma_{\parallel mf}$ is a fictive cohesive strength, which would be reached if the fibers could endure the same elongation to rupture as the matrix. The fracture surface has the shape shown in Fig. 3. Thereby, one must take into account that the values of $\sigma_{\parallel mf}$ and $\sigma_{\perp f}$ are different for tension and compression.

Since in the direction of the fibers the strength is limited by $\sigma_{\parallel ff}$ (see Eq. 2), the ellipsoid given by Eq. 3 will be cut off by two planes perpendicular to the $\sigma_{\parallel}$ axis corresponding to the different values of $\sigma_{\parallel ff}$ in tension and compression. This failure surface seems to be more realistic than a failure criterion, which does not distinguish between interfiber and fiber failure.

Equation 3 has proved a reasonable base for strength calculations for filament-wound tubes of glass-reinforced plastics.[6,7] $\sigma_{\perp f}$ and $\tau_{\# f}$ were taken from measurements with specimens in the form of filament wound tubes with all fibers lying in the circumferential direction (see Fig. 4).

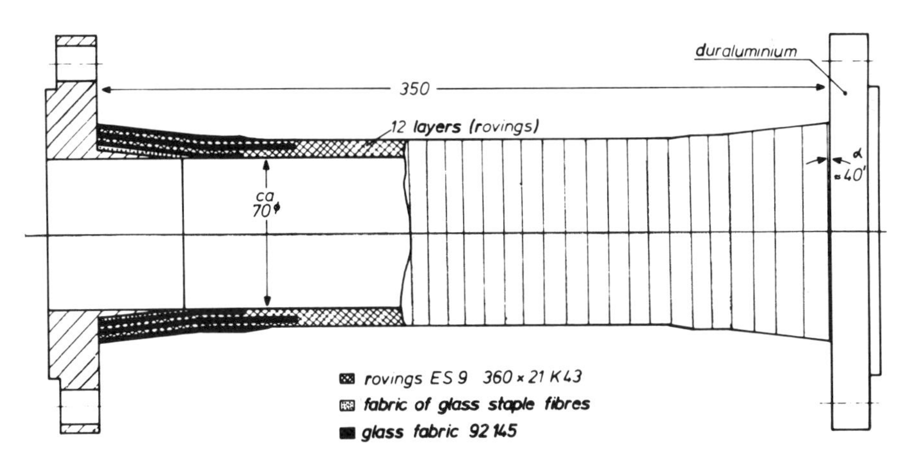

Fig. 4. Specimen for determination of failure curves of unidirectional layers.

Since the matrix (epoxy or unsaturated polyester resins) is viscoelastic, the relations between $\epsilon_{\perp}$ and $\sigma_{\perp}$, and especially between $\gamma_{\#}$ and $\tau_{\#}$, are nonlinear. One can take account of these nonlinearities by carrying out the calculations for the layerwise fracture analysis with strain-dependent coefficients of elasticity instead of strain-independent constants.

Computer programs are available, and the results show that a better approximation to the measured stress-strain curves of filament-wound tubes under internal pressure is reached. Figure 5 gives an example from the work of Förster[7], which shows that the predicted onset of cracking (interfiber failure limit) is in reasonable agreement with the measured values found by the onset of leakage (weeping). Also, the strong deviations from linearity in stress-strain behavior will be predicted by the layerwise fracture analysis.

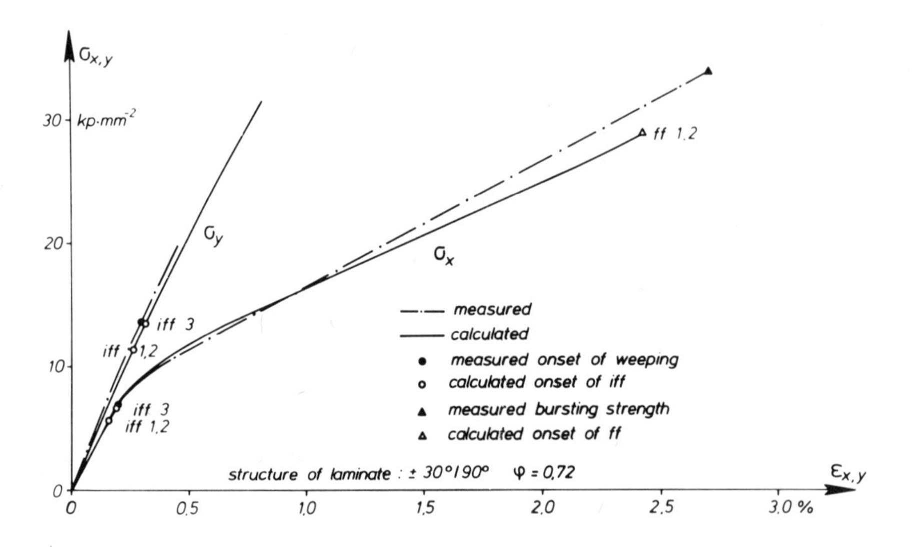

Fig. 5. Layerwise fracture analysis of a filament-wound tube.[7] σ_x = axial stress, σ_y = circumferential stress; three layers; angles with the axis: $\pm 30^\circ/90^\circ$; relative thickness of the layers: $t_1' = t_2' = 0.143$, $t_3' = 0.597$, t_m' (layer of resin at the outer surface) = 0.117; glass content by volume = 0.72.

For a given state of stress, an optimum for interfiber failure will be reached when cracking occurs in all layers at the same time. Analogously, the optimum for fiber failure is given by simultaneous failure of all fibers in the different layers.[8]

To provide the validity of Eq. 3, tests under combined torsional and axial stress were carried out with the specimens shown on Fig. 4. With $\sigma_\parallel = 0$, Eq. 3 will reduce to

$$(\sigma_\perp/\sigma_{\perp f})^2 + (\tau_\#/\tau_{\# f})^2 \leqslant 1 \quad . \tag{4}$$

Figure 6, taken from the work of W. Schneider[9], shows some results with an epoxy resin. Attempts were also made to investigate the influence of time, for it is believed that failure criteria are strongly dependent on time and temperature. In long-time experiments under combined stress, for

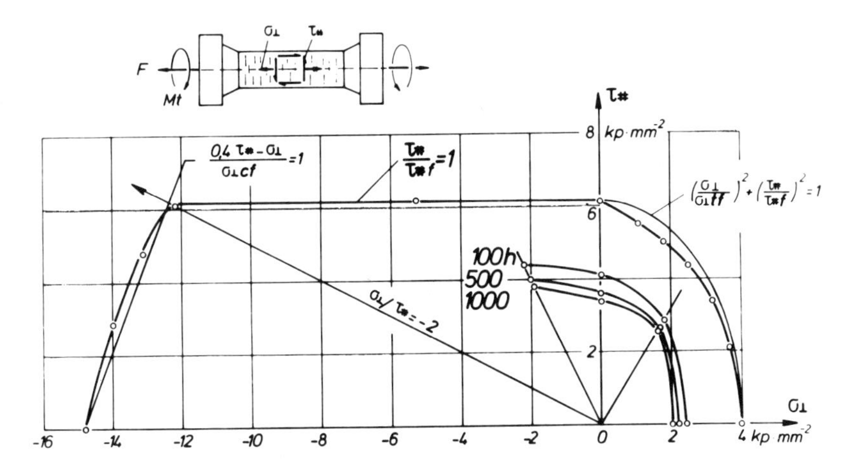

Fig. 6. Failure curves of the unidirectional layer.[9]

100-, 500-, and 1000-hour failure curves shown in Fig. 6 were determined, which are similar to the curve for short-time experiments. Under combined tensile and shear stress, the agreement with Eq. 4 is good enough in calculations for engineering purposes. Great deviations from Eq. 3 prevail for $\tau_{\#}/\sigma_{\perp} < 0$. Up to values of $\sigma_{\perp}/\sigma_{\perp cf} \approx 0.8$ ($\sigma_{\perp cf}$ = compressive strength perpendicular to the fibers), the endurable shear stress $\tau_{\#end.}$ will not be influenced by $\sigma_{\perp}$, so that the criterion $\tau_{\#}/\tau_{\#f} \leqslant 1$ is valid. At a compressive stress $\sigma_{\perp} > 0.8\,\sigma_{\perp cf}$, a steep decrease in $\tau_{\#end.}$ is observed. It follows from Fig. 6 that Eq. 3 is valid only in the range of $\sigma_{\perp}$ tensile stress. Good results were obtained by Förster[6,7] since he used tensile stresses or very low compressive stresses $\sigma_{\perp}$ in his calculations. From Fig. 6 and from similar measurements on other resins, it can be seen that $\tau_{\#f}$ is always greater than $\sigma_{\perp tf}$ ($\sigma_{\perp tf}$ = tensile strength perpendicular to the fiber). $\sigma_{\perp tf}$ is always smaller, more or less, than σ_{tf} of the resin, since the fibers have a higher Young's modulus than the matrix, and stress magnification in the matrix will be achieved (see Fig. 7). $\tau_{\#f}$ has approximately the same value as the shear strength of the matrix.

Since for other resins and equal glass concentrations with the same fibers we get markedly different values of $\tau_{\#f}$ and $\sigma_{\perp tf}$, it may be concluded that the fracture criterion of the unidirectional layer is also strongly influenced by the mechanical properties of the resin. Therefore, it should be of principal interest to derive the fracture criterion for the unidirectional layer from the fracture criterion of the matrix. This problem will be considered in a later publication.

5 MICROSTRESSES IN A UNIDIRECTIONAL LAYER UNDER PLANE STATE OF STRESS

In deriving the fracture criterion of the unidirectional layer from the criterion for cohesive failure of the matrix or adhesive failure of the fiber/matrix interface, it should be kept in mind that the stresses $\sigma_{\parallel}$, $\sigma_{\perp}$, and $\tau_{\#}$ (macrostresses) are different from the so-called microstresses acting in the matrix, the fiber, or the interface, which are strongly dependent on the coordinates in a microscopic scale (see Fig. 7).

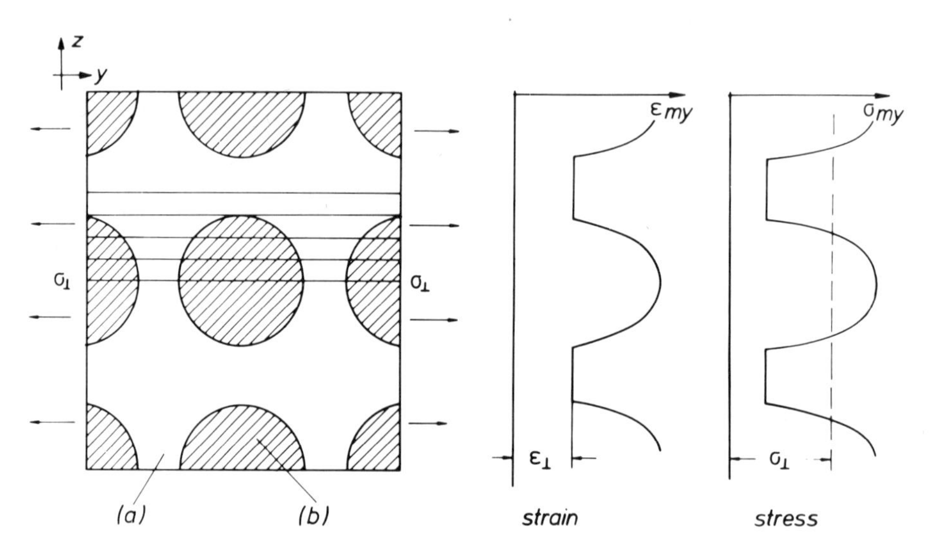

Fig. 7. Sliced sample of the unidirectional layer under $\sigma_{\perp}$ load: (a) matrix, (b) fibers; σ_{my} = matrix microstress, ϵ_{my} = matrix microstrain.

The experimental determination of the criteria for cohesive and adhesive failure is discussed below. If these criteria are known and also the field of microstresses as a function of $\sigma_{\parallel}$, $\sigma_{\perp}$ and $\tau_{\#}$, the onset of failure in an unidirectional layer with rising load can be predicted. To a first approximation, the unidirectional layer may be regarded as a regular array of fibers in a matrix. Figure 7, taken from the work of Puck[10], shows the cross section of such a specimen. The chosen array is a square one. At higher glass contents, a hexagonal array should be more probable. By a mixture of these two kinds of orders, one might obtain a good approximation to reality.

The whole field of microstresses is discussed by Puck and Schneider.[4] The calculation of microstresses will be simplified if one cuts the specimen into slices parallel to the plane of the layer, so that no forces will be carried over the cuts. Therefore, the slices will only have a planar state of stress in the matrix region, given by σ_{mx}, σ_{my}, and τ_{mxy}. It is assumed that all slices have the same macroscopic strain ($\epsilon_{\parallel}$, $\epsilon_{\perp}$, $\gamma_{\#}$). Fibers, applied in fmc, have, for the most part, a much greater stiffness than the matrix. From this it follows that, for instance, the stress (σ_{my}) and the strain (ϵ_{my}) in the matrix are higher than the macroscopic stress ($\sigma_{\perp}$) and strain ($\epsilon_{\perp}$) (stress and strain magnification), see Fig. 7. The highest matrix stresses and strains are achieved in the slice with the highest glass content (by volume). Therefore, fracture will begin in this region. In general, the microstresses in the matrix calculated with the sliced specimen will depend on the macrostresses by following equations[4]:

$$\begin{aligned} \sigma_{mx} &= a_{\parallel x}\sigma_{\parallel} + a_{\perp x}\sigma_{\perp} \\ \sigma_{my} &= a_{\parallel y}\sigma_{\parallel} + a_{\perp y}\sigma_{\perp} \\ \tau_{mxy} &= a_{\# xy}\tau_{\#} \end{aligned} \tag{5}$$

If the Poisson's ratio is approximately the same for the fibers and matrix, $a_{\parallel y}$ may be put zero.

Now we consider the interface between fiber and matrix at the point of highest microstress. If we neglect the differences in Poisson's ratio, we get only two stress components:

$$\begin{aligned} \sigma_{A} &= a_{\perp y}\sigma_{\perp} \\ \tau_{A} &= a_{\# xy}\tau_{\#} \end{aligned} \tag{6}$$

The coefficients a in Eqs. 5 and 6 are dependent on the geometrical arrangement and the concentration of the fibers and on the elastic properties (E, ν) of the fibers and matrix. If the stress-strain behavior of the matrix is nonlinear, as in the case of glass-reinforced plastics, the coefficients a depend also on the stress-strain relationship. With help of Eqs. 5 and 6, we should be able to evaluate the failure criterion for the unidirectional layer when the failure criteria for cohesive failure of the matrix and for adhesive failure of the interface are known.

6 FAILURE CRITERION OF THE MATRIX

Tubes of epoxy and unsaturated polyester resins were made by centrifugal casting and tested under combined torsional and axial load, and also under internal pressure. Some preliminary results from the work of W. Schneider[11] are shown in Fig. 8. Under biaxial tensile stress, none of the classical fracture criteria is fulfilled (see, for instance, Refs. 12 and 13).

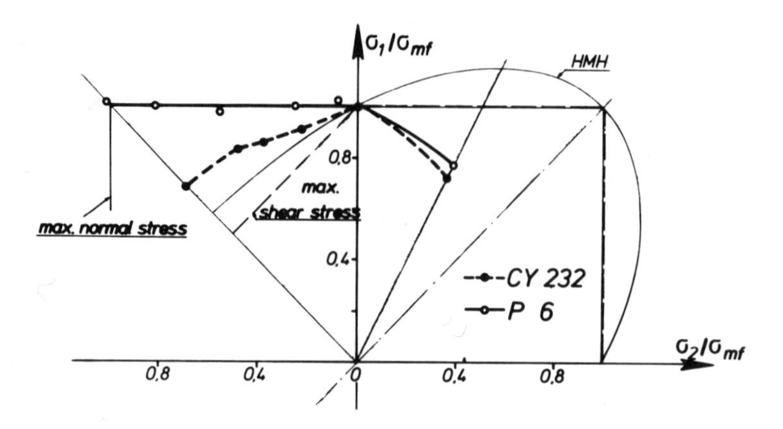

Fig. 8. Failure curves of resins plotted in the principal stress diagram. Tensile strength (σ_{mf}): 8.3 kp/mm^2 for Araldit CY 232, 4.4 kp/mm^2 for Palatal P 6.

Here it cannot be excluded that crazes or microcracks have a strength-decreasing effect, which would be more efficient under biaxial tensile stress than under uniaxial tensile stress. It should be pointed out that only tubes free of visual defects were tested. The tubes had a greater thickness at the ends, so that end effects could be excluded. Under combined tensile and torsional stress, the polyester resin follows the criterion of maximum normal stress, while the epoxy resin approximates the HMH criterion (Fig. 8). Final conclusions are not yet possible.

7 FAILURE CRITERION OF THE INTERFACE BETWEEN FIBER AND MATRIX

Since a coupling agent will often be used, the following types of failure can principally occur in the interface: adhesive failure between fiber and coupling agent, cohesive failure of the coupling-agent layer, and adhesive failure between coupling agent and matrix. It seems doubtful that all three types of failure exist in reality. Therefore, we shall try to describe the failure of the interface with an experimentally verified single criterion.

Experiments on a macroscopic scale are problematic, since the surface of a glass specimen may much differ from the surface of a glass fiber. Despite this objection, we tried to carry out some experiments.[14] The apparatus shown in Fig. 9 is somewhat similar to that used for determination of the failure criterion of the unidirectional layer under long-time load.[9] The specimen (a) is made of a cylindrical bar, cast from the resin. The bar is cut into two halves, which are bonded to a plate (b) of A-glass by the same resin. By deadweights, an axial and a torsional load (F_t, F_{to})

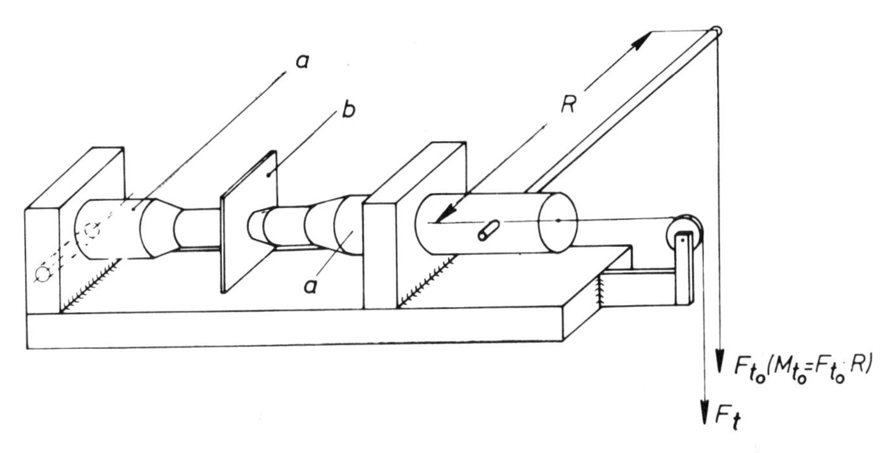

Fig. 9. Apparatus for investigating adhesive failure: (a) resin specimen, (b) glass plate.

is transferred to the specimen. In short-time tests, one load was kept constant, while the other load was raised continuously until failure occurred. Some preliminary results are given in Fig. 10.

Roughening of the surface by sandblasting and etching with hydrogen fluoride gave the highest strength values (curve a in Fig. 10). Without roughening, curve b resulted. In all cases, a constant ratio of adhesive shear strength and tensile strength $\tau_{\mathrm{Af}}/\sigma_{\mathrm{Af}} \approx 3$ was found. Thus, it seems

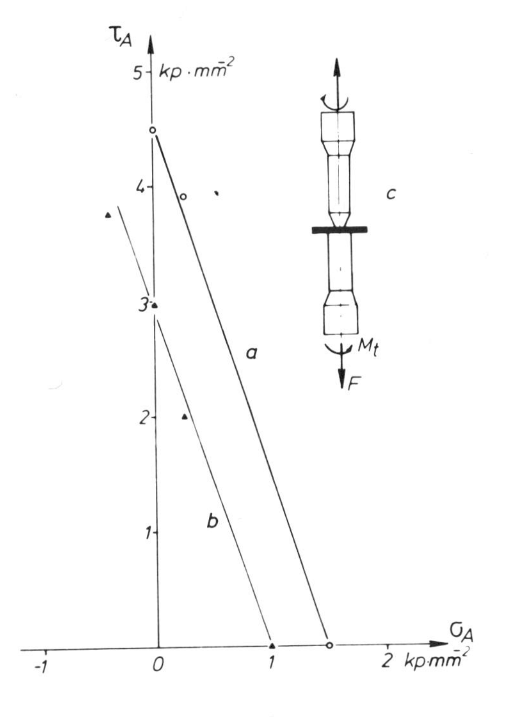

Fig. 10. Adhesive failure curves of the interface between A-glass and epoxy resin (Araldit CY 232); (a) sandblasted, 3-min treatment in HF (10 percent), followed by treatment in a 1 percent solution of A 1100; (b) 5-min treatment in HF (20 percent), followed by treatment in a 1 percent solution of A 1100; (c) specimen.

that Coulomb's criterion

$$\tau_A/\tau_{Af} + K\sigma_A/\sigma_{Af} \leqslant 1 \tag{7}$$

(where K is a constant) or a similar relation will be valid. If we assume that the ratio $\tau_{Af}/\sigma_{Af} \approx 3$ will hold independent of the surface treatment and the special character of the glass, we can draw some conclusions concerning our failure curves for the unidirectional layer of glass-reinforced plastics (see Fig. 11).

If we calculate with the values of curve (b) in Fig. 10 and with Eq. 6, we get the adhesive failure curve (a) in Fig. 11. The strength values of this curve are much too low. From this we may conclude that the adhesive strength of epoxy resin on roughened plates of A-glass is much lower than the adhesive strength of the resin on E-glass fibers.

If we assume that Eq. 7 with $\tau_{Af}/\sigma_{Af} = 3$ will hold and adhesive failure under shear occurs, then curve (b) in Fig. 11 results, which is so far from the experimental values (open circles) that adhesive shear failure may be excluded.

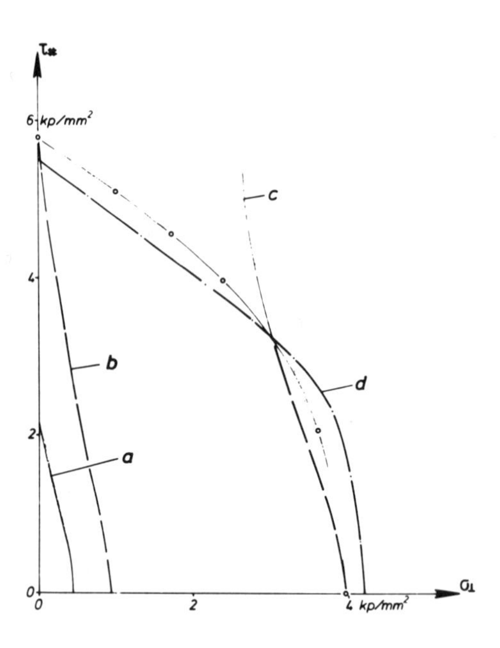

*Fig. 11. Computed failure curves for the unidirectional layer from E-glass and epoxy resin (Araldit CY 232); glass content by volume = 0.65 (o denotes experimental data); (a) calculated from adhesive-failure curve **b** in Fig. 10; (b) based on the assumption of adhesive failure under $\tau_{\#}$ load; (c) based on the assumption of adhesive failure under $\sigma_{\perp}$ load; (d) based on assumption of cohesive failure.*

If we assume that adhesive failure will occur under tensile stress, then curve (c) results, which has approximately the same slope as curve (b). Curve (c) will yield a high value for adhesive failure under shear which is much higher than the shear strength $\tau_{\#mf}$ for cohesive failure of the matrix, curve (d). This curve was calculated by Eq. 5 and from the failure curve of Fig. 8. Thus, it may be concluded that adhesive failure in shear is improbable, while tensile failure might be either cohesive or adhesive. Further measurements are needed to confirm this conclusion.

REFERENCES

1. Puck, A., *Kunststoffe,* **57**, 284 (1967).
2. Puck, A., *Kunststoffe,* **57**, 573 (1967).
3. Puck, A., *Kunststoffe,* **57**, 965 (1967).
4. Puck, A., and Schneider, W., *Plastics and Polymers,* p 33 (February, 1969).
5. Puck, A., *Kunststoffe,* **59**, 780 (1969).
6. Förster, R., and Knappe, W., *Kunststoffe,* **61**, 583 (1971).
7. Förster, R., *Kunststoffe,* **62**, 181 (1972).
8. Förster, R., *Kunststoffe,* **62**, 57 (1972).
9. Knappe, W., and Schneider, W., *Kunststoffe,* **62**, 864 (1972).
10. Puck, A., *Kunststoffe,* **55**, 913 (1965).
11. Schneider, W., unpublished results.
12. Ward, I. M., *Molecular Order-Molecular Motion: Their Response to Macroscopic Stresses,* H. H. Kausch (Ed.), Interscience Publ., New York (1971), p 195.
13. Tschoegl, N. W., *ibid.,* p 239.
14. Schneider, W., Paper presented at Third International Conference on Fracture, München, Germany, 1973.

DISCUSSION on Paper by W. Knappe

EIRICH: A frequent cause of composite failure is microcracks formed as a result of thermal stresses (differences in coefficients of expansion in adjacent layers) during the formation (crazing), or cooling, of the composite structure. In glass fiber-, or graphite-, epoxy structures, e.g., the epoxy contracts owing to crosslinking during the cure, and, further, contracts more than the fibers during cooling. The result is adhesive or cohesive cracks which may form with such speed that they cause audible cracking sounds (e.g., on mold opening).

KNAPPE: Under simplifying assumptions (complete adhesion between fiber and matrix, regular array of the fibers in the matrix), thermal stresses in fiber/matrix composites may be calculated if the coefficients of elasticity and the thermal-expansion coefficients of the matrix and the fibers are known. Calculations for the unidirectional layer and multi-ply composites of glass fibers and epoxy resin show that the thermal stress coefficients $\Delta\sigma_{\perp}/\Delta T$, $\Delta\sigma_{\parallel}/\Delta T$ and $\tau_{\#}/\Delta T$ at room temperature will normally not exceed 4 kp cm^{-2} grad^{-1} [see W. Schneider, *Kunststoffe,* **61**, 273 (1971)]. Temperature differences of about 100 degrees may therefore lead to failure. More complicated is the case of internal stress caused by irreversible shrinkage of the matrix during cure. Principally, it seems possible to calculate these stresses by the same model as before, if shrinkage of the matrix and the viscoelastic properties of the matrix as a function of shrinkage are known.

In our calculations we neglected internal stress. We used cold-setting resins and tried to keep internal stress low by annealing at 50 C.

THERMAL AND ENVIRONMENTAL EFFECTS IN CRAZE GROWTH AND FRACTURE

J. G. Williams and G. P. Marshall

Department of Mechanical Engineering
Imperial College of Science and Technology
London, England

ABSTRACT

A model is described for the growth of crazes in the presence of liquids and with changes in temperature. The controlling factor is the availability of the liquid at the craze tip, and changes in craze speed may be predicted from variations in other material properties. Observations of craze failure show that there are variations in craze structure with time and that gases are evolved under a number of conditions.

1 INTRODUCTION

Crazing is the precursor of all fractures in polymers. The presence of some crack, flaw, or inhomogeneity in a stressed polymer gives a region of high strain concentration and hydrostatic tension, resulting in the formation of voids. This process is common to all materials, but in polymers the molecular orientation in the ligaments between the voids gives sufficient work hardening to stabilize the porous structure so that it increases in extent rather than proceeding to failure directly as in most materials. Failure will eventually occur, either within or around the craze, but the existence of the intermediate stage of a stable, porous, plastic zone is of prime importance in an understanding of polymer fracture.

This paper sets out to describe the growth of these crazes and their subsequent failure in some detail. The presence of softening agents significantly affects both processes, since the high area-to-volume ratio of the crazed material gives a much larger effect than one would expect from bulk behavior. Earlier work has shown that the availability of the agent at the tip of the growing craze is the governing factor, and the consequences of this concept are explored over a wider range of variables than hitherto, with a view to testing its practical usefulness.

2 MODEL FOR CRAZE GROWTH IN ENVIRONMENTS

In a previous paper[1], the authors reported on a study of craze kinetics in PMMA tested in methanol. Single-edge-notch (SEN) specimens were tested under constant load, and the propagation of the crazes which formed at the crack tips was studied for a wide range of loading conditions with a variety of notch lengths to assess the influence of crack size. Photograph of a typical craze are shown in Figs. 1a and 1b. It was found that two types of growth history could be produced, depending on the applied load and notch length. In the first type – Fig. 2 (curve 2) – the craze would decelerate and eventually arrest as shown, whereas in the second type – Fig. 2 (curve 1) – the craze would decelerate initially and then finally propagate at a constant speed until failure occurred by crack propagation through the craze. The failure process occurred only in the latter part of any given test, the original crack remaining stationary for most of the craze-propagation history.

The parameter which governed the type of growth behavior and the rate of craze propagation was found to be the stress-intensity factor K_o – calculated using the load and initial crack length – since neither the gross nor the net section stress alone provided any degree of correlation when the crack length was varied. Two critical values of K_o were noted: that which governed the initiation of the craze (K_M) and that which determined the transition from the arrest to constant speed behavior (K_N). Examination of the craze-front geometries in the various tests showed that the crazes which arrested had a shape such as shown in Fig. 1(c), with the craze front having the shape of the original crack. In the constant-speed tests, the craze initially grew inwards from the specimen surface –

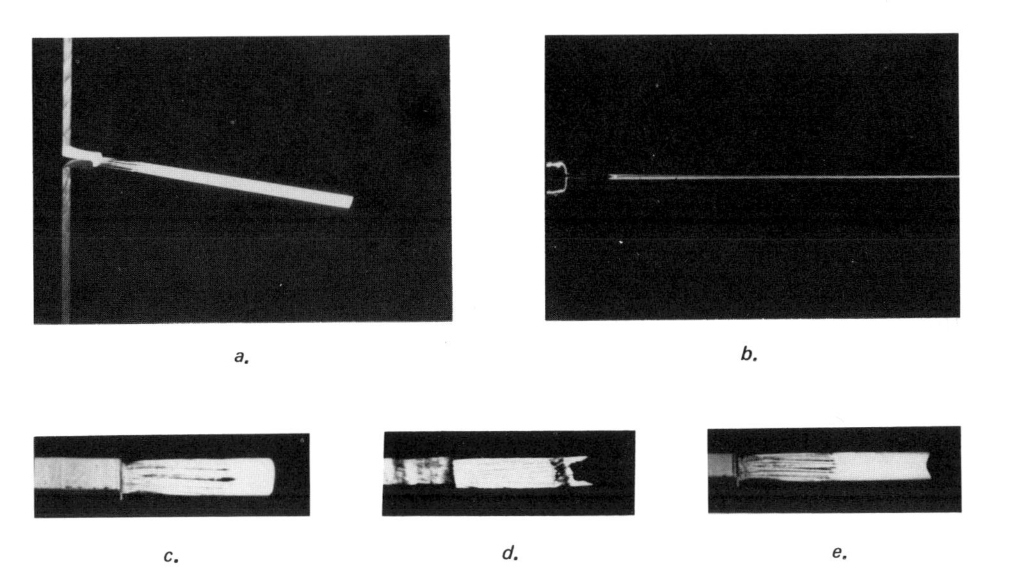

Fig. 1. Crazes in PMMA grown in methanol: (a) oblique view of craze; (b) side view of craze; (c) end flow alone; (d) onset of side flow; (e) side flow established.

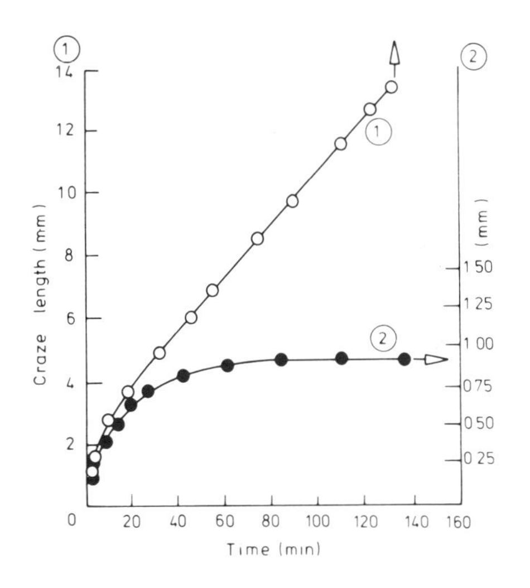

Fig. 2. Craze growth histories for PMMA in methanol at 20 C.

Fig. 1(d) – eventually achieving the shape shown in Fig. 1(e). It was inferred, and subsequently confirmed, that the direction of environmental attack was responsible for these transitions, the arresting crazes growing under the influence of environmental flow down the end of the crack alone and the constant speed crazes growing from the side flow through the surfaces of the specimen. Since the bulk absorption of the environment at the highly stressed crack and craze tip is expected to be very rapid, it was argued that the controlling factor affecting craze growth would be the availability of the environment at the craze tip, and that this would be governed by the rate of flow of the fluid through the existing softened craze. Accordingly, a model based on fluid motion in a craze was developed to explain the two types of growth observed and the nature of the K_o dependency. The expression for craze velocity was derived as:

$$\frac{dx}{dt} = \frac{\phi}{12\mu} \cdot \ell_o \cdot \delta \cdot \frac{dP}{dx} \quad , \tag{1}$$

where

ℓ_o is a function of the spacing between voids

δ is the crack-opening displacement for the unsoftened craze

$\frac{dP}{dx}$ is the pressure gradient in the direction of propagation

μ is the viscosity of the fluid.

The parameter ϕ is determined by the void content of the craze, V, such that:

$$\phi = \frac{1 - \sqrt{1-V}}{1 - V} \quad . \tag{2}$$

Because the fluid flows into a newly formed void where the pressure is effectively zero, the pressure gradient is determined by the applied pressure, in this case atmospheric $\overline{P}$, and the craze length, x, through which the fluid must flow, so that:

$$\frac{dP}{dx} = \frac{\overline{P}}{x} \quad . \tag{3}$$

The displacement, δ is determined from the stress-intensity factor for the particular specimen geometry used (in this case $K_o = \sigma\sqrt{\pi a}$ - for the infinite plate case), and δ is given by:

$$\delta = \frac{K_o^2}{\sigma_y E} \quad , \tag{4}$$

where σ_y and E are the yield stress and modulus, respectively, of the bulk unsoftened material.

Substitution of Eqs. 2, 3, and 4 into Eq. 1 and integrating gives:

$$x = \left\{ \frac{\phi \ell_o \overline{P}}{6\sigma_y E\mu} \right\}^{1/2} K_o t^{1/2} \quad . \tag{5}$$

This equation is particularly applicable to the case where environmental flow is down the end of the craze (not through the specimen surfaces), and the experimental results were replotted on an x vs $t^{1/2}$ basis for evaluation of the slope as a function of K_o. The results are shown in Fig. 3, a typical $x/t^{1/2}$ curve being inserted as an illustration and these confirm Eq. 5. The general agreement is considered sufficiently good to extend the model to cover the case of constant-speed growth. For the case of flow through the surfaces of the specimen, the pressure gradient is constantly maintained throughout the growth because of the proximity of

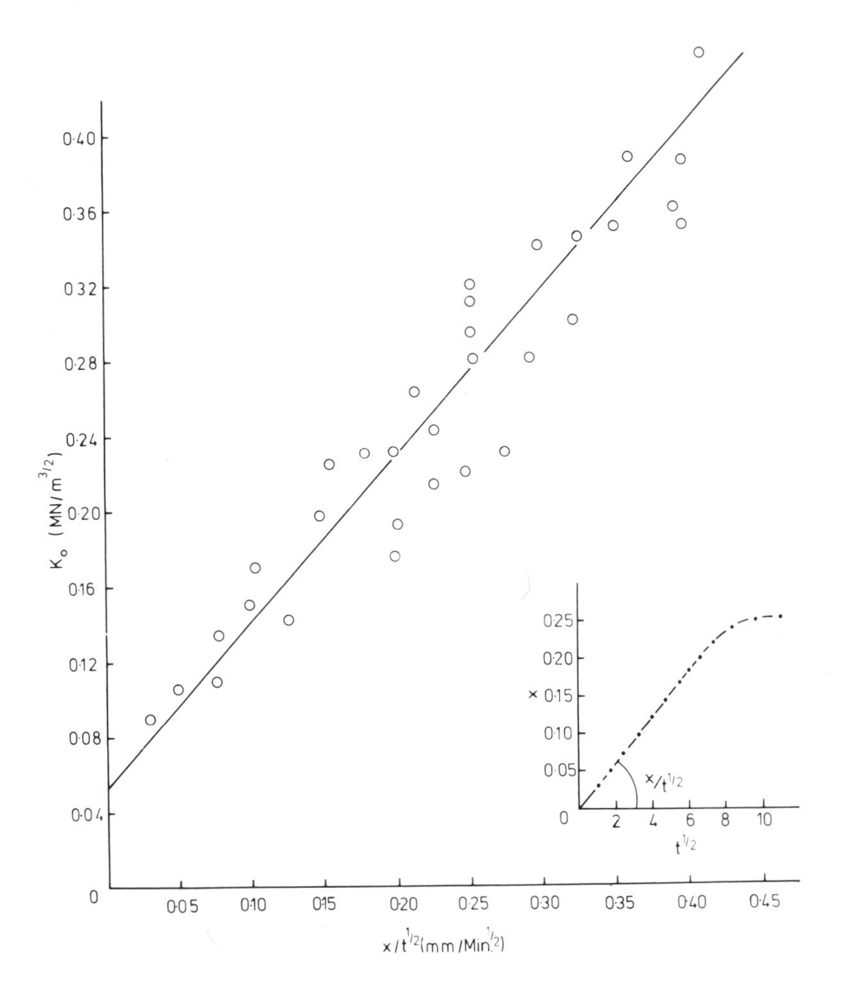

Fig. 3. End flow craze growth characteristics at 20 C.

the fluid supply, so that:

$$\frac{dP}{dx} = \frac{\overline{P}}{L} ,$$

where L is the implied length over which the pressure drop occurs. L depends on the ratio of the velocity of fluid into the specimen thickness and along the craze and also on the specimen thickness, so that:

$$L \alpha \frac{\dot{x}}{\dot{y}} \cdot H = \Delta H \quad , \qquad (6)$$

where Δ is a constant and H is the specimen thickness.

The modified version of Eq. 2 then becomes:

$$\frac{dx}{dt} = \left\{ \frac{\phi \ell_o \overline{P}}{6\sigma_y E\mu\Delta H} \right\} K_o^2 \quad . \qquad (7)$$

Having taken into account the craze initiation condition (K_M) and the end flow/side flow transition value K_N, Eq. 7 was used to predict the form of relationship expected between K_o and the final craze velocity observed experimentally. The results are shown in Fig. 4, and it can be seen that the analysis gives a good prediction of the shape of the curves for three different specimen thicknesses.

On the basis of these results it was concluded in the original paper that the fluid-flow model was basically correct and gave equations which could be used to predict the growth of a craze in an environment. However, the evidence at the time was limited to results in methanol alone and only one type of specimen geometry had been employed. To provide further support for the model, two further test series have been conducted varying firstly the fluid viscosity and secondly the specimen geometry.

(a) *Tests with ethylene glycol/methanol mixtures*

One obvious method of checking the validity of the flow equations in relation to craze growth is to change the environment such that the viscosity is significantly different. To do this, quantities of ethylene glycol were mixed with methanol and tests conducted as before. The ethylene glycol served essentially to change the viscosity alone since it is a poor crazing agent for PMMA. As before, with pure methanol, both types of craze-growth behavior were observed, the methanol obviously still acting as the controlling crazing agent. The results for end-flow only tests (i.e., craze-arrest condition) are plotted on a K_o vs $x/t^{1/2}$ basis in Fig. 5 for three mixtures – as indicated. The lines drawn through the data points represent the slopes which would be expected if only the viscosity of the

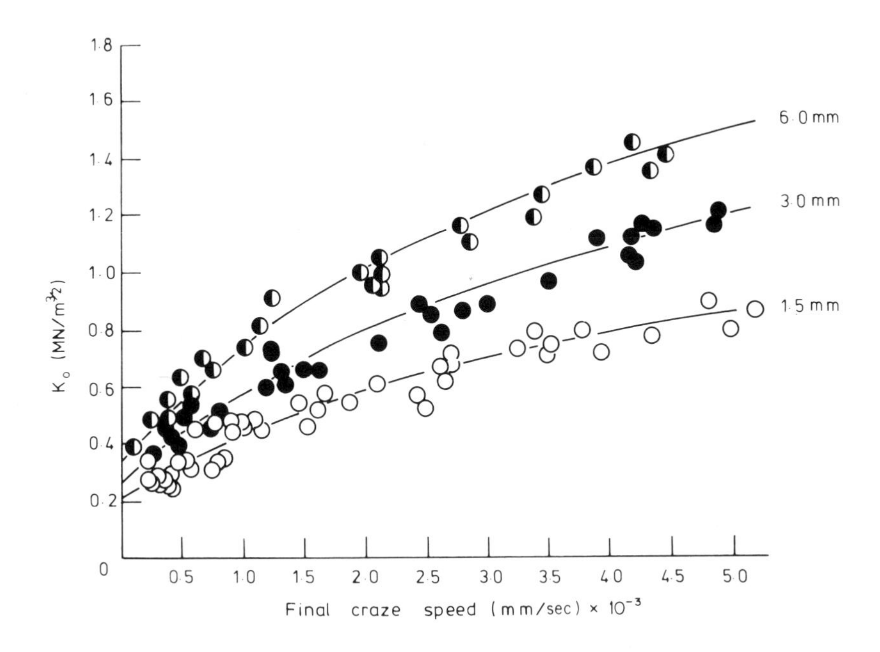

Fig. 4. Final craze speed as a function of original stress-intensity factor.

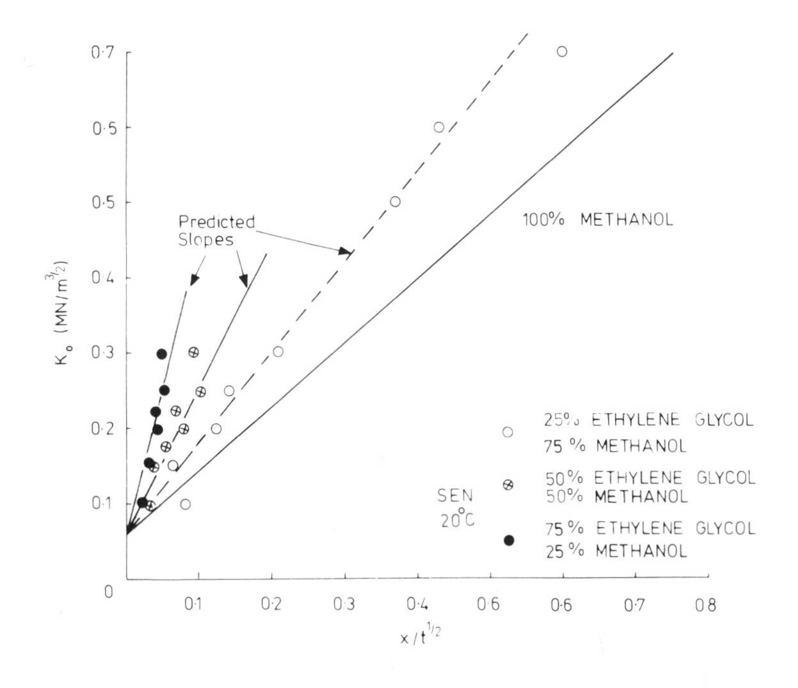

Fig. 5. Effect of varying fluid viscosity using ethylene glycol — methanol mixtures.

environment were changed. It can be seen that the agreement with experimental data is good, which supports the validity of the flow model. Sufficient results from constant-speed tests are not yet available for a full analysis, but on the basis of preliminary tests, it would appear that the viscosity change is again the controlling factor in reducing the craze speeds, at any given loading condition.

(b) *Tests using surface-notched specimens*

Perhaps the main limitation of the evidence for the model as originally proposed was its limitation to only one specimen geometry. The single-edge-notch (SEN) specimen has many advantages in experimental work, but is often not representative of practical cracking and crazing. In practice, crazes are initiated at the tips of small surface scratches and, accordingly, to broaden the range of applicability of the flow model, tests were conducted using specimens with semielliptical notches inserted in the face of the specimen by means of a curved scalpel blade. Since this geometry has had only limited usage in fracture testing, the K solution[2] was cross-checked by testing a number of specimens in air and measuring the fracture toughness K_{Ic}. The result, K_{Ic} = 1.70 $MN/m^{3/2}$, agreed almost exactly with previous results using SEN specimens[3], and the solution was considered valid.

For tests in methanol, two types of craze growth were obtained as before. The craze-propagation rates into the specimen thickness followed an $x/t^{1/2}$ form, whereas the growth along the specimen surface was at constant speed. The two types of craze geometry are shown in Fig. 6. In Fig. 6(a), the craze boundary is proceeding everywhere on an $x/t^{1/2}$ basis since critical conditions for side (or in this case "surface") flow had not been established. Once K_N for this geometry had been exceeded, the side (surface) flow pattern predominates, giving a nonuniform craze front – as shown in Fig. 6(b). Results for the $x/t^{1/2}$ type of flow have been analyzed

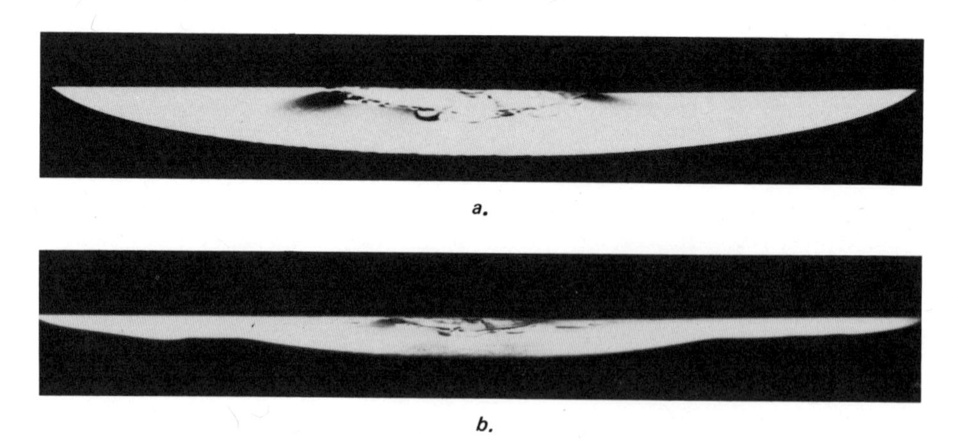

Fig. 6. Surface craze configurations: (a) end flow alone; (b) end + side flow.

and are plotted in Fig. 7. The straight line on the graph is that obtained for the SEN specimens, and it can be seen that the agreement is excellent – further supporting the flow model.

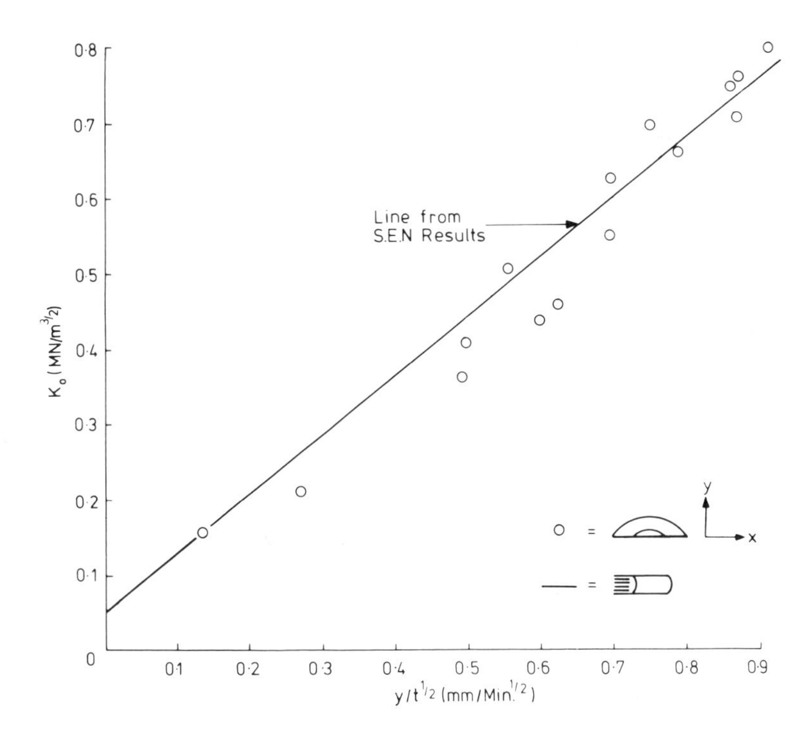

Fig. 7. Penetration rates of surface craze as a function of stress-intensity factor.

3 CRAZE GROWTH UNDER CYCLIC LOADS

The analysis of the effects of cyclic loading on PMMA in methanol[4] serves two useful purposes. In the first instance it provides another loading system against which the model for craze growth can be extended and further verified, and secondly it gives an opportunity for obtaining some information on corrosion fatigue in plastics since little work has been done on this subject hitherto.

As a preliminary to a test series using the conventional type of ramp load input, a few tests were carried out using a square-wave load variation on the constant-load machines. This type of loading is much easier to describe since the model ought to apply directly and it provides a simple description of cyclic effects. Accordingly, a number of specimens (SEN) were tested at various K_o levels, the load being changed six or seven times during any given test. An example of a typical craze-growth history is shown as Fig. 8. It can be seen that there is a virtually instantaneous response in growth rate when the K_o level is changed. Again no crack growth was observed until the very final moments of specimen life. The speeds obtained at each K_o level were noted and found to be commensurate with results obtained when tests were run under constant conditions, thereby indicating that the craze structure and nature are basically unchanged. The explanation of the changes in velocity with K_o changes is clearly that the void area is still determined by the current K_o value and that the observed speed changes are merely responses to changes in the void area. Changes in K_o produce different craze-tip deflections and, hence, different void areas.

The second, and main, test series involved continuous loading and unloading in a ramp cycle using an Instron Universal Testing Machine. At the "standard" frequency of 10^{-1} Hz, crazes were produced and the craze length/time curves were macroscopically similar to those shown in Fig. 2. However, at this frequency it is difficult to obtain readings during any given part of the cycle consistently, and tests at much lower frequencies were employed to observe the true form of craze response. A test result obtained at 5 x 10^{-4} Hz is shown in Fig. 9. It can be seen that there is a more complex variation in craze-growth pattern than that indicated in Fig. 2, since below K_N there is no growth and above K_N the craze speed increases in response to the increase in void area produced by the increas-

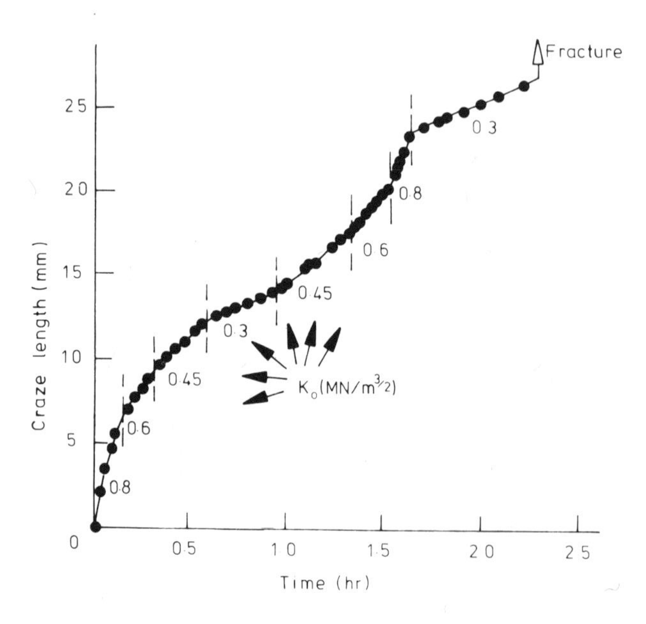

Fig. 8. Craze growth histories for varying loads.

ing deflections down the craze. The craze-front geometries reflect these changes as indicated in Fig. 9. The average response, which is observed to be linear growth at higher frequencies (10^{-1} Hz), is shown as a broken line.

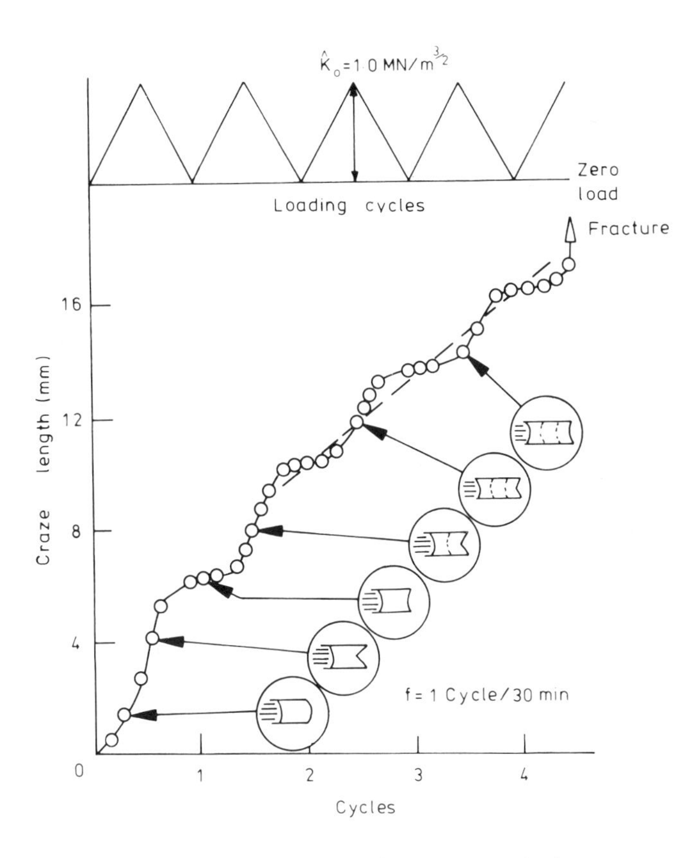

Fig. 9. Craze growth histories for slow cyclic loads.

To accommodate this varying load behavior, the constant-load model was modified to include both the varying stress and the limit introduced by the K_N transition. To simplify the situation, an effective mean stress-intensity factor ($\bar{K}_o$) which would be expected to give rise to a mean craze speed (the broken line of Fig. 9) was evaluated so that results obtained at higher frequencies of loading could be compared directly with each other and with constant-load data. The two expressions for $\bar{K}_o$ have been derived as[4]:

$$\text{For} \quad K_o \min (\check{K}_o) = 0 \quad , \qquad \bar{K}_o = \frac{\hat{K}_o}{\sqrt{3}} \left\{ 1 + 2 \left[\frac{K_N}{\hat{K}_o} \right]^3 \right\}^{1/2} \quad ;$$

For $\check{K}_o > K_N$, $\bar{K}_o = \frac{1}{\sqrt{3}} \left\{ \check{K}_o^2 + \hat{K}_o^2 + \hat{K}_o \check{K}_o \right\}^{1/2}$;

where $\hat{K}_o$ is the maximum level of K_o achieved during the load cycle.

A graph of "average" final craze speed vs $\bar{K}_o$ obtained at f = 10^{-1} Hz is shown in Fig. 10 for two specimen thicknesses and using both types of minimum loading conditions. The lines representing the model prediction for constant-loading results are shown, and it is clear that the results are consistent. The agreement between the two sets of data at this frequency amplifies the previous comments made about the effects of cyclic loading being to induce equivalent responses in void areas, thereby implying that up to this frequency, the hysteresis losses in the craze are small.

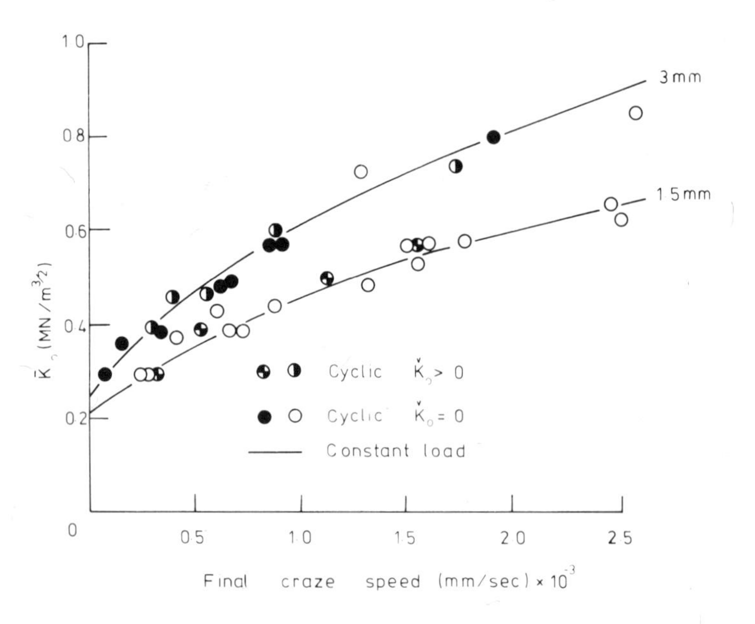

Fig. 10. Stress-intensity factor versus final craze speed for a loading frequency of 10^{-1} Hz.

At higher frequencies (~5 x 10^{-1} Hz), the agreement between the model predictions and the experimental data was poor since hysteresis effects start to influence the results. At these higher frequencies, the craze velocities were much greater than expected, thereby indicating that the craze does not open and close completely during each cycle as had been implied in the analysis, but remains open and tends to the constant-load condition. Further work will be needed before a complete analytical solution can be formulated to incorporate this frequency effect. In passing, it is pertinent to note that at very high frequencies, the nature of the growth pattern changed yet again. At 30 Hz, crazes were grown only for short periods (~15 min) before slow crack propagation started within the craze. Then, instead of producing rapid failure as in tests at lower frequencies and constant load, the crack advanced slowly across the specimen section until there was only a few microns of craze area ahead of the tip. This crack/craze system then propagated to total failure. In the authors' experience, this is the only occasion on which cracking per se has been observed with PMMA in a methanol environment. During the course of the crack growth in this manner, it was also observed that large quantities of gas bubbles were evolved at the crack tip and ejected in a continuous stream. Some of this gas was collected and analyzed, and apart from the expected methanol by-products, it was found that there was a substantial quantity of hydrogen present. Whether this hydrogen has been evolved as a result of free radical formation caused by chain scission or whether the whole gas evolution process occurs solely by cavitation remains unsolved at present.

4 EFFECT OF TEMPERATURE ON CRAZE PROPAGATION

In view of the practical problems which are encountered with environmental crazing, it is surprising to find that there has been virtually no work done on the combined effects of temperature and liquid environment on the crazing process. The two references[5,6] available do not deal directly with either craze propagation or the mechanics of fracture. A number of tests have therefore been conducted using SEN specimens of PMMA tested in methanol from -90 C to +40 C to explore the problem. The tests were all at constant load in environmental cabinets in which the methanol was heated by water-filled copper coils and cooled by means of evaporating liquid-nitrogen gas blown around the specimens by a fan. Since cooling below +20 C and heating to +40 C produced different effects, these will be described separately in the next two sections.

(a) *Craze Growth: -90 C to +20 C*

The most obvious effect of lowering the temperature is to make the craze-propagation rates decrease at any given loading condition. This follows the pattern of conventional toughness measurements in air, since at low temperatures the modulus and yield stress both increase with decreasing temperature. The effect is amply illustrated in Fig. 11 which shows a craze length vs time plot for a test which was established at -40 C until constant-speed conditions had been achieved; then the temperature was gradually allowed to increase (as shown) and the craze propagation continuously measured. It can be seen that as the temperature rises, the

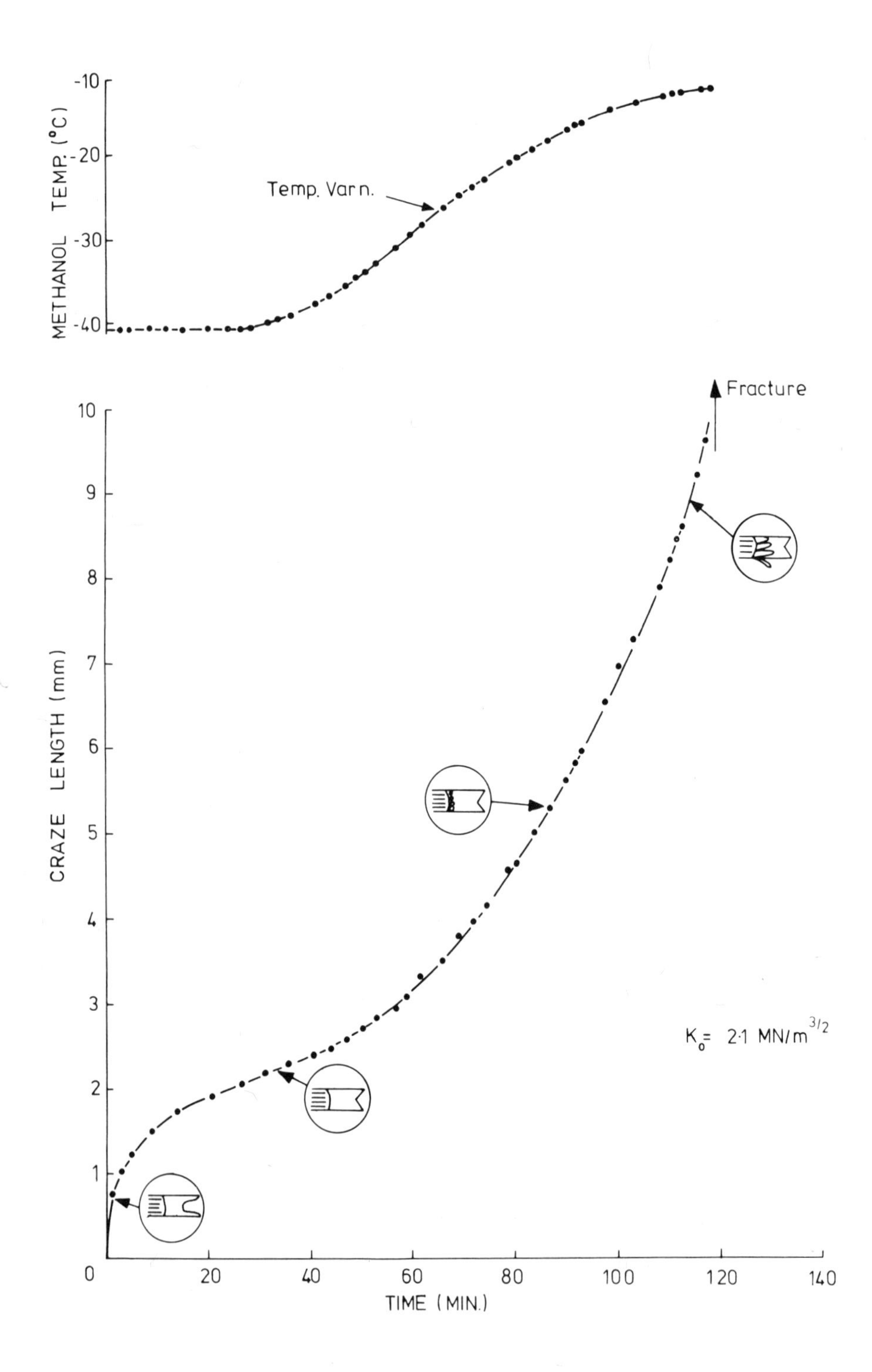

Fig. 11. Craze growth for a rising temperature.

craze-growth rate follows suit, reflecting the modulus, yield stress, and viscosity changes. To test whether these changes were sufficient to explain the results (via Eqs. 5 and 7), a series of tests was conducted at -40 C, -20 C, 0 C, and 10 C, with the specimens being used to evaluate both craze-arrest and constant-speed conditions. A number of tests were also conducted at -50 C, -60 C and -90 C to observe very low temperature effects. This latter series of tests produced surprising results in that it was found that only a very narrow range of final craze speeds could be obtained even for large changes in K_o. The results at many K_o levels are so close as to suggest that there is a definite limiting maximum craze speed which can be achieved below -50 C and that this speed is independent of temperature. In this temperature region, PMMA is transformed from a viscoelastic to an essentially elastic brittle material and, consequently, temperature has only a minor effect. This may account for the lack of sensitivity to temperature, but the independence of the speed on K_o requires further investigation since the mechanism implied in the model apparently ceases to operate in this region. The constant-speed results for -40 C to +20 C are summarized in Fig. 12 which shows the

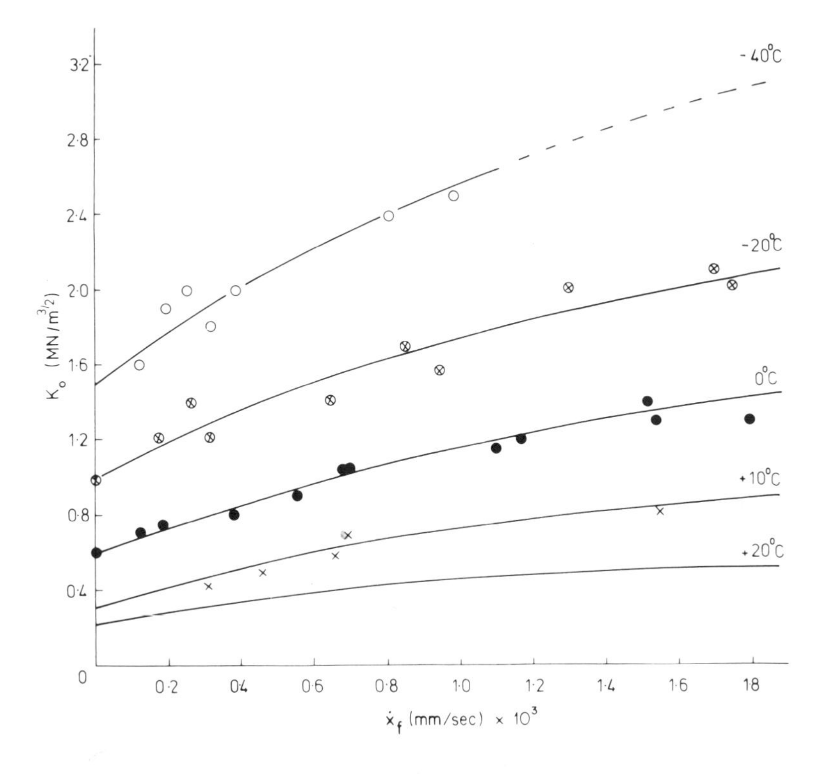

Fig. 12. Stress-intensity factor versus final craze speed at various temperatures.

plots of K_o vs final craze speed at each temperature. The effect of temperature is clearly indicated by the curves. The results of the craze-arrest tests were again plotted on $x/t^{1/2}$ and K_o vs $x/t^{1/2}$ bases as at 20 C, and the slopes of the K_o vs $x/t^{1/2}$ graphs are given in Table I. The $x/t^{1/2}$ data were good at each of the temperatures considered, the slopes being easier to identify precisely at the lower temperatures since the linear $x/t^{1/2}$ relationship was valid over a much longer period of craze growth. Also included in Table I are the slopes predicted from the model by substitution of corrected modulus, yield stress, and viscosity data at each temperature, with other parameters constant and the observed variations in K_M and K_N.

Table I. Variations of Flow Parameters with Temperature

Temp, C	Slope K_o vs $x/t^{1/2}$ ($MN/m^{3/2}/mm/min^{1/2}$) Expt.	Slope K_o vs $x/t^{1/2}$ ($MN/m^{3/2}/mm/min^{1/2}$) Calc.	K_M ($MN/m^{3/2}$)	K_N ($MN/m^{3/2}$)	L (mm) Expt.
+20	0.89	0.89	0.05	0.21	1.25
0	1.30	1.27	0.18	0.60	3.96
-20	1.80	1.73	0.27	1.00	4.75
-40	2.70	2.34	0.36	1.50	5.30

The agreement between the predicted and experimental values of the slopes of the K_o vs $x/t^{1/2}$ graphs is sufficiently good to justify the use of the model in predicting end-flow results. From Eq. 5 it follows necessarily that if the slopes are predicted in this way, then the void spacing parameter, ℓ_o, and the average void content, ϕ, have both remained unchanged. However, when a similar procedure was used to predict the side-flow results in Fig. 12, the predicted speeds were much greater than those observed experimentally. The implication is that the implied pressure-drop length L varies with temperature. By accounting for changes in σ_y, E, μ, K_M, and K_N and fitting the results via Eq. 7, the values of L at each temperature were calculated and these are given in Table I. There is a very pronounced variation in L with temperature, and the variation of both K_N and K_M (they turn out to be proportional) is much larger than one would expect from variations in yield stress and modulus which govern δ. K_N describes the initiation of flow through the surface and L is a measure of the pressure drop through the surface, so the fact that their temperature dependence is linked is not too surprising. It seems most likely that the methanol flows into the craze after first permeating a surface skin and that the temperature dependence observed is that for bulk diffusion in the surface skin. Both K_N and K_M are measures of the initiation of flow and, since they are through the surface and at the crack tip respectively, it would be expected that $K_N > K_M$ and that they would remain in proportion since the stress-concentration effect of the crack is the differentiating factor. The dependence of flow initiation on K is probably a reflection of the necessity of establishing the skin at some displacement (less for the crack tip so that $K_M < K_N$) in which the diffusion subsequently occurs. Below these displacements, the voiding process does not occur near the plane-stress surface regions.

(b) *Craze Growth: 20 C to 40 C*

The only previous work at temperatures above 20 C had indicated contradictory findings, since in Ref. 5 it was implied that crazes grow with greater facility, whereas in Ref. 6 it was stated that a plateau condition is achieved and crazes grow, or to be more precise, initiate at the same level as at 20 C. In the present tests, it was found that a somewhat different pattern emerges. At temperatures of 20 C to 30 C, the craze-growth rates increased quite markedly at any given loading, but at temperatures of 30 C to 40 C, there was a drastic change in that crazes would arrest at loadings where constant-speed behavior would be expected at 20 C. The effect is clearly illustrated in Fig. 13, which shows results of a test which was started at 15 C and a constant speed established, at which point the temperature was gradually increased. The marked decrease in the craze-growth rate at 30 C is obvious. The graph also includes diagrams of the craze-front geometry at various times during the test, and it can be seen from the long prongs which grow down the specimen surfaces as the craze slows down that there is some irregularity in the surface regions. It is possible that the temperature gradient across the section produces part of the observed effect, but it is thought that the increase in bulk absorption of methanol into the surface skin produces the major effect. Close observation of the region surrounding the craze showed that the refractive index had changed and that the change was commensurate with a large increase in the amount of methanol soaking into the bulk material. Other changes also manifest as the craze slowed down at 40 C were that all crazes changed color, becoming slightly orange or brown after a few minutes, thereby indicating an increase in void content. In some cases, the presumed increase in void content was made obvious by the appearance of small gas bubbles within thc craze as the material extended in a ductile manner prior to fracture. Failure would eventually occur by coalescence of these bubbles, as described later.

The apparent increase in bulk absorption rate of methanol into the skin could have a number of effects on the nature of craze propagation; in particular, compressive stresses would necessarily be generated in the soaked regions and these could have a deleterious effect on the side-flow

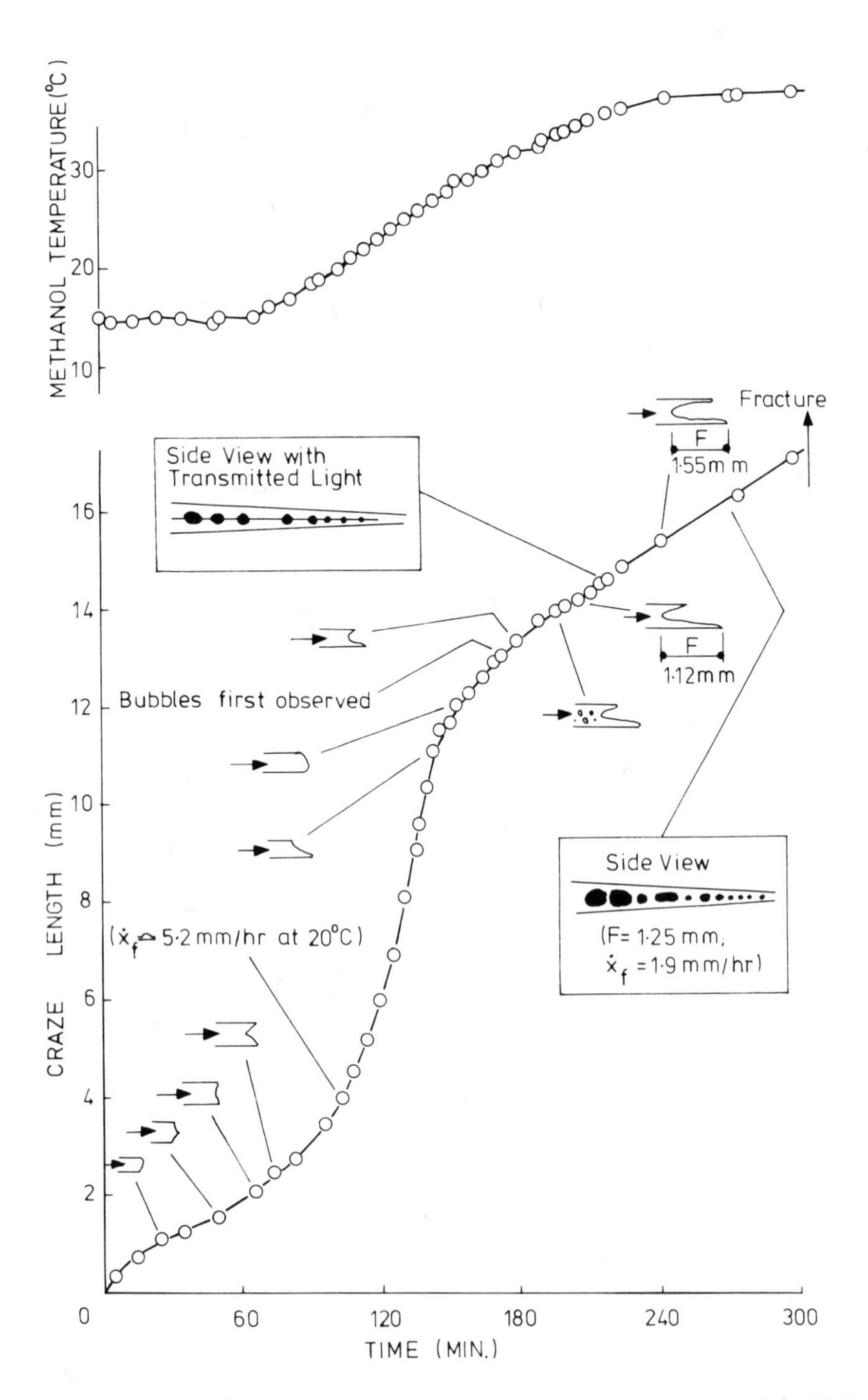

Fig. 13. Stress-intensity factor versus final craze speed with temperature increasing from 20 C to 40 C.

process. Accordingly, in order to separate out the effects of bulk absorption on craze propagation, a number of tests were conducted at 20 C using presoaking where there was already a body of work against which results could be compared.

5 EFFECT OF PRESOAKING ON CRAZE PROPAGATION

In all the previous tests involving PMMA in methanol, the specimens had been tested immediately following immersion in the liquid environment. In the present tests, the specimens were left to soak for varying times before the test started. Close examination of a specimen after soaking revealed that a distinct diffusion boundary could be observed – as shown in Fig. 14 – indicating that there had been a considerable uptake of methanol. In the tests conducted at room temperature, it was found that a number of features observed at temperatures >30 C were reproduced. Crazes which had been left unloaded and soaked for a few hours would not start growing immediately after reload; instead they would extend in a very ductile manner – giving rise to large-scale bubble formation – as shown in Fig. 15. Also, at the instant of failure, the craze would pull apart by ductile tearing through the holes – releasing gas from the bubbles as the crack front progressed. Very large side prongs were also observed in propagating crazes. The propagation rates and slopes of the $x/t^{1/2}$ curves in craze-arrest tests on specimens soaked for 12 hours were virtually unchanged from those normally expected at room temperature, which is not surprising since this type of growth is due to flow down the

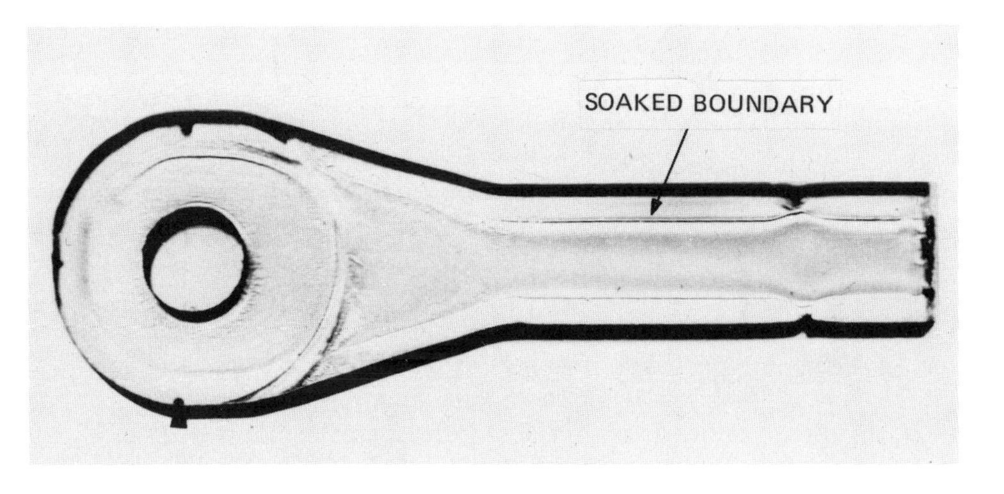

Fig. 14. Tension specimen soaked in methanol at 20 C.

Fig. 15. Bubbles forming in a craze in a presoaked specimen.

length of the craze and is independent of surface conditions to a large extent. The side-flow, constant-speed results were affected in all cases, however, the craze velocity being much slower than that obtained in unsoaked specimens. A graph of K_o vs $\dot{x}_f$ for specimens soaked for 12 hours is shown in Fig. 16 to illustrate the point.

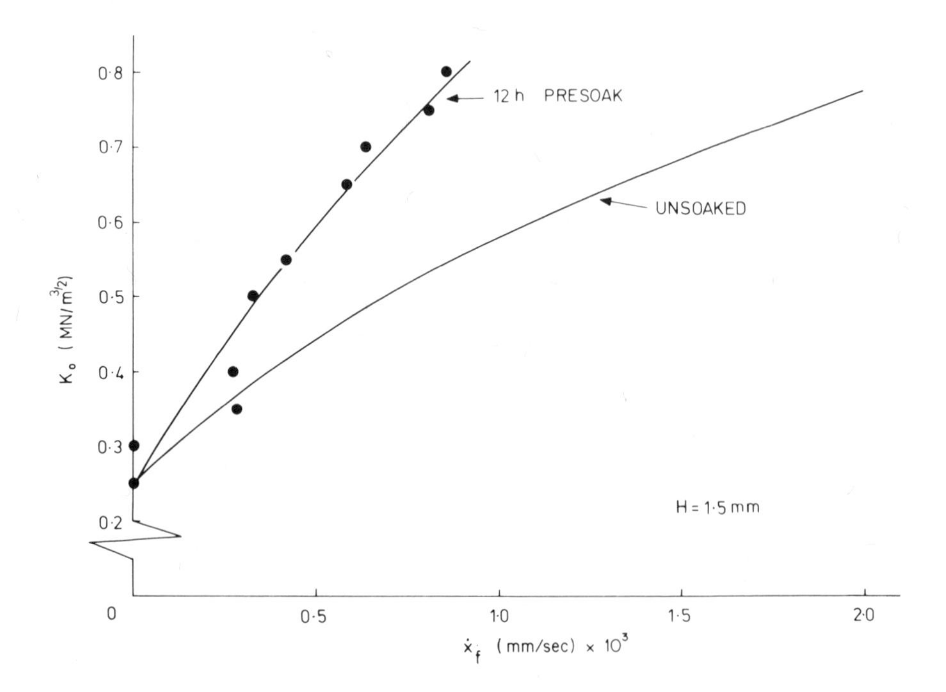

Fig. 16. Stress-intensity factor versus final craze speed for specimens after 12 hours presoaking.

As discussed for the low-temperature tests, it seems likely that the observations reflect the behavior of flow through a surface skin, which will be more pronounced for the constant-speed tests. Clearly, long presoaking times and unloaded soaks establish thicker skins through which the methanol must flow, and the higher temperatures simply produce these effects during the course of tests because the effects occur more quickly. More detailed quantitative evidence should elucidate the mechanisms further.

6 FRACTURE PROCESSES IN CRAZES

In the previous sections, craze kinetics have been described in detail and little mention has been made of the nature of the failure process in crazes, although this is probably the more acute practical problem. It has been observed experimentally that fracture within a "wet" craze can occur in a number of different ways, depending on the loading (constant or cyclic) and temperature. The fracture of the long crazes grown at room temperature under constant load has been investigated[7], and a mention of the observed processes is pertinent at this stage.

The construction of an analysis for predicting the onset of failure in one of the long crazes grown at room temperature presents a considerable challenge, since the crazes often extend from the crack tip almost totally across the specimen width. Because of this size effect, the conventional plastic-zone analyses of classical fracture mechanics are of little use in themselves since they bear no relation to the true situation. However, the large size of the craze does have a beneficial effect for the experimentalist since the growth of cracks can be observed with relative ease by means of high-powered microscopes. By viewing a growing craze from the side during the latter stages of many of the tests, it was seen that in practically all cases there was a considerable period (typically 10 minutes) of slow crack propagation before eventual fast fracture. Measurements of the crack length versus time characteristics showed that for the constant-loading conditions, the growth rate accelerated continuously until total fracture occurred.

Two forms of crack motion within the craze were observed. In the first and most common case, the slowly propagating crack grew through the center of the craze – as shown in Fig. 17(a), whereas in the second case, the crack was seen to oscillate between the craze/matrix interfaces – as shown in Fig. 17(b). In both cases, the crack grew only in the central region of the craze, a plane stress skin a few microns thick being left intact on the surfaces – breaking only when the crack had grown by a millimeter or so. The existence of this skin provides support for the bulk

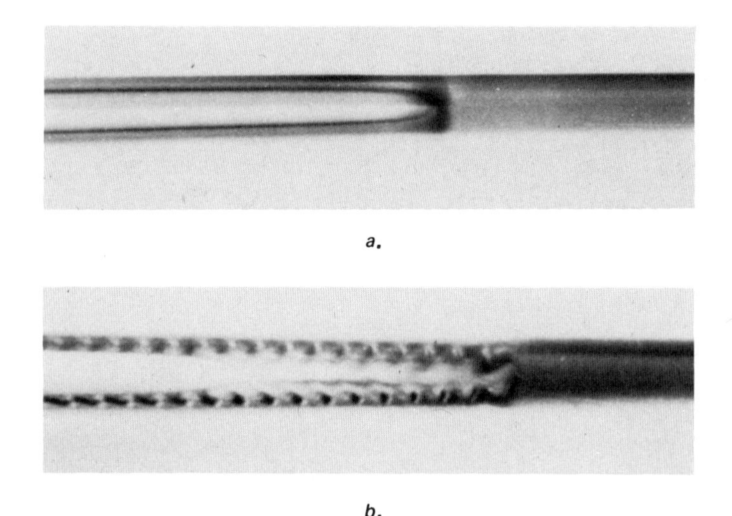

Fig. 17. Crack growth through a craze: (a) smooth fracture; (b) oscillatory fracture.

diffusion arguments discussed previously. Typical examples of the markings left on the fracture surfaces are shown in Fig. 19, together with schematic representations of the surfaces for both the smooth and oscillatory crack propagation. In the region immediately ahead of the crack front (A), one of two distinctly different surfaces was always observed. Either the surface would be completely mirror-smooth corresponding to fracture through the center of the craze – or it would be patterned with a series of "stripe" markings traversing the section, corresponding to the "hills" and "valleys" produced by the oscillatory fracture. The reasons for the onset of fracture in an oscillatory manner are, as yet, not fully understood, but it would appear that this type of failure occurs when there is a small fault or "bump" in the craze so that the crack is required to change direction in order to surmount the fault.[7]

The markings in the central and end regions of the broken craze which have been illustrated schematically in Fig. 18 are shown in more detail in the stereoscan micrographs (Fig. 19). In both cases, the opposite fracture surfaces mate exactly, the relief features on one surface corresponding to depressions on the other surface. The "line" markings of region B are analogous to the river patterns observed previously on slow crack propagation surfaces in PMMA in air by Berry[8] and are thought to be caused by crack propagation on slightly different levels across the specimen thickness. It is possible that there is a transformation to this form of crack propagation because of an inhomogeneous craze structure in this region, since this type of line marking has been observed within the structure of dried, unbroken crazes: e.g., see Fig. 19(b), which also shows

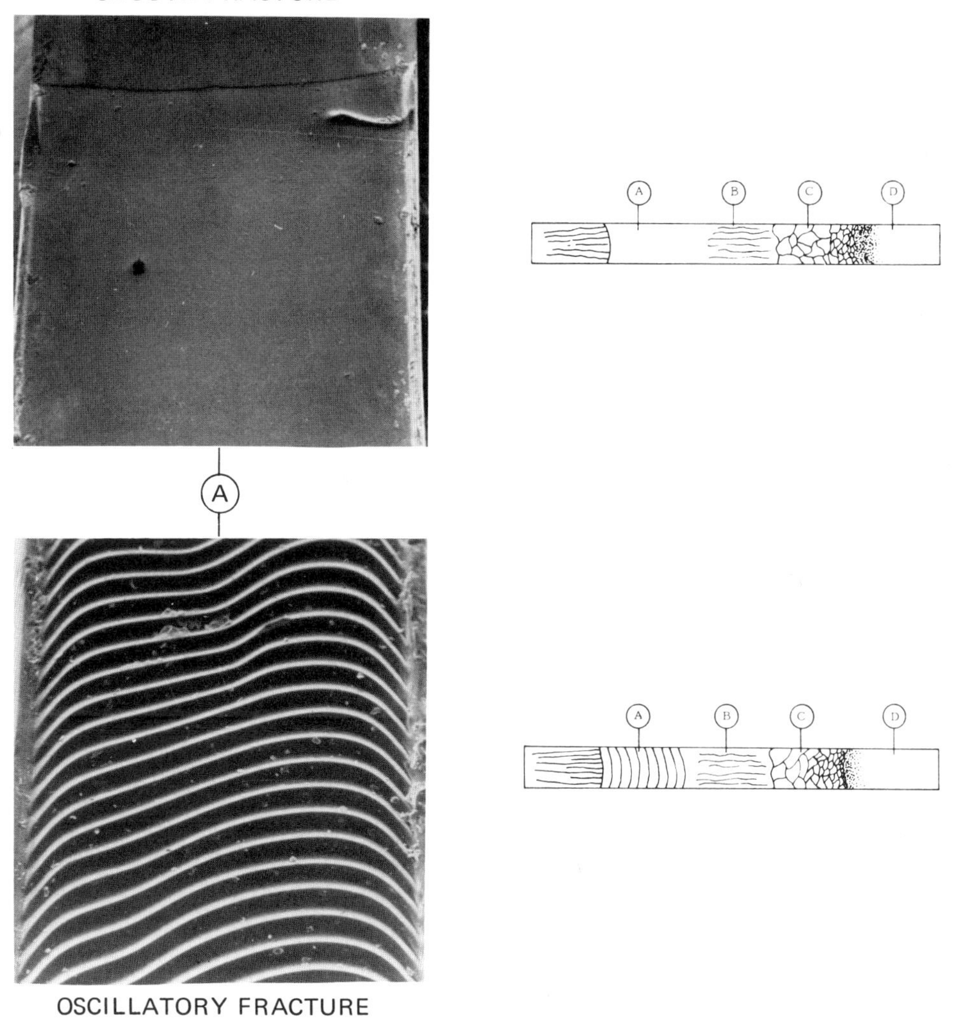

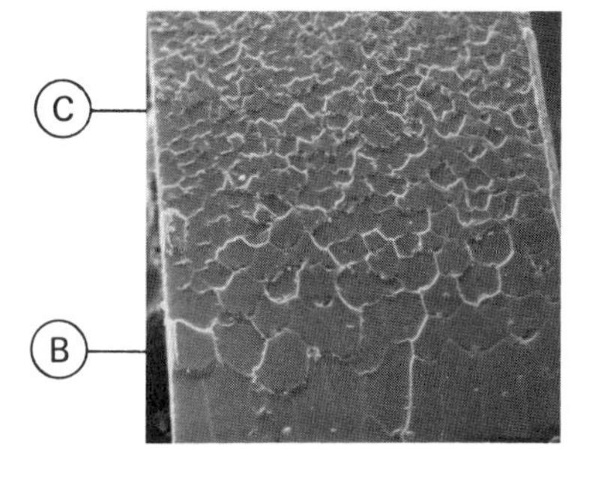

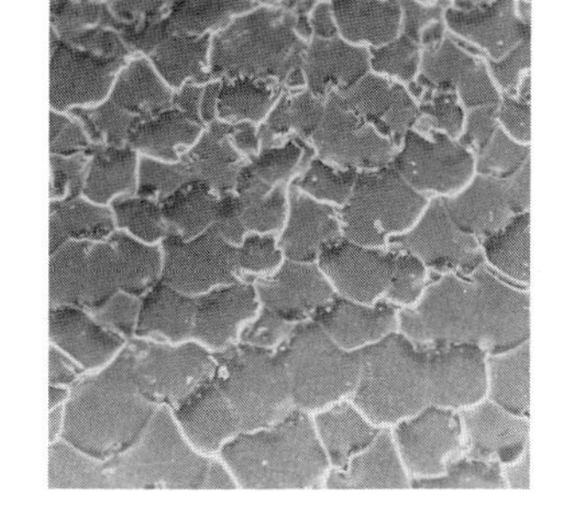

REGIONS B AND C (X27-1/2) REGION C (X55)

Fig. 18. Surface features on fractured, soaked crazes.

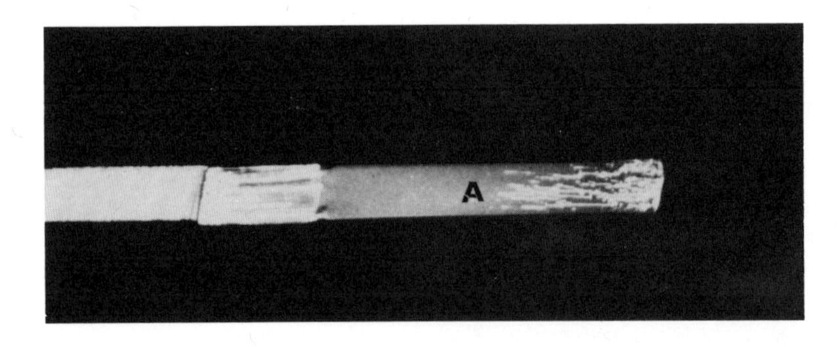

a.

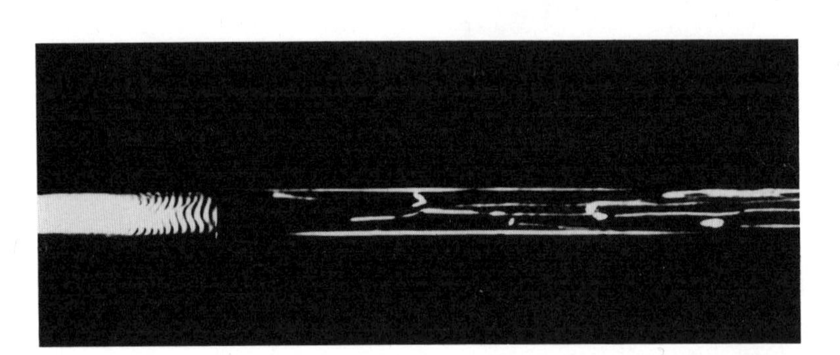

b.

Fig. 19. Structure observed in unbroken crazes: (a) markings at slow craze speeds; (b) residual markings in "dried" craze.

the onset of oscillatory fracture. Similarly, the "island" structure of region C may also be explained by the propagation of a crack through a preexisting type of structure, such as that shown in the stereoscans of Fig. 18, since the grainy features observed on the surfaces of *slowly* propagating "thin" crazes are similar in nature, e.g., see Fig. 19(a).

It would therefore appear that the fracture surface gives a true reflection of the type of structure existing within a craze at the time of fracture. Region C is continually converted into a more ordered structure – similar to B – as the craze extends and is oriented even further, eventually becoming completely homogeneous and giving rise to the smooth surface on fracture.

The failure process at low temperatures does not appear to differ significantly from that at 20 C, the fracture occurring predominantly through the center of the craze. At high temperatures and with the soaked specimens, however, there was a significant difference in that the presence of the bubbles in the crazes caused the slow growth to proceed in a more irregular fashion. When fracturing the ordinary craze material, the crack would propagate either through the craze center or would oscillate as before. A step-by-step description of an observed craze failure taken from a test history at 40 C is presented in Fig. 20 to illustrate the sequence of events in slow crack propagation.

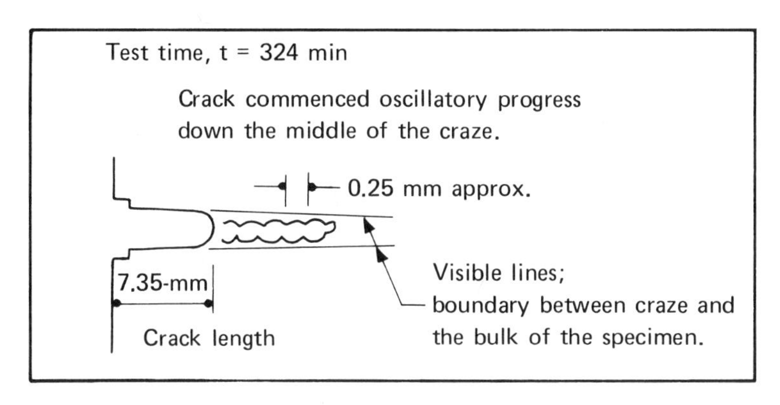

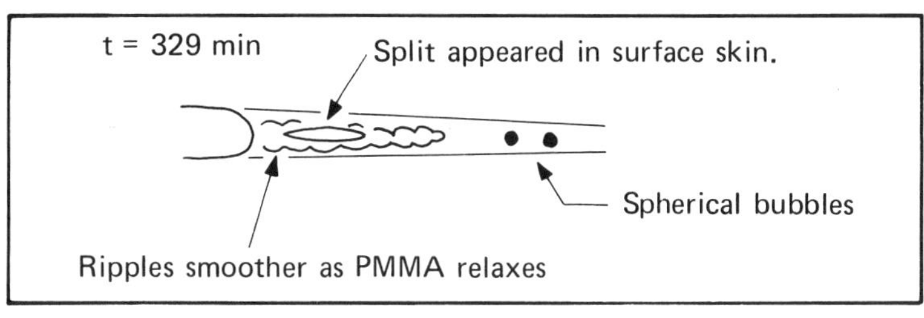

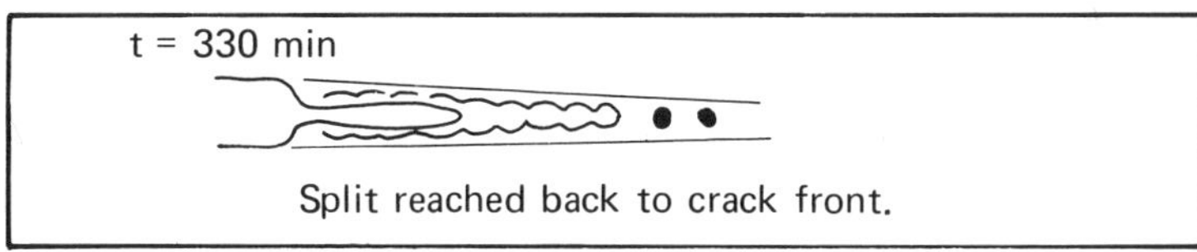

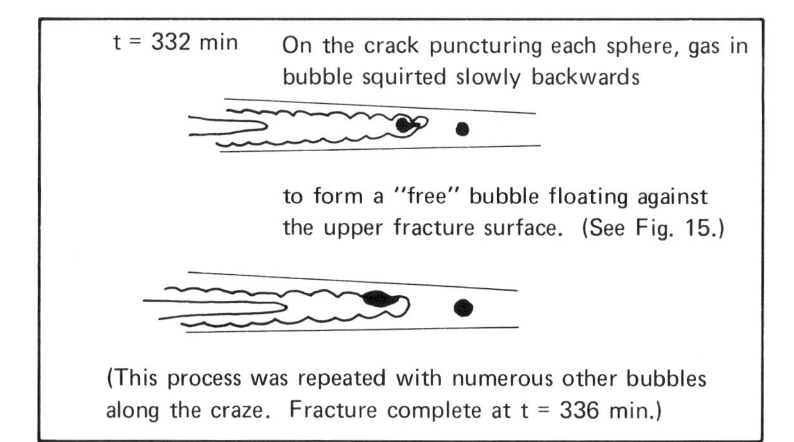

Fig. 20. Craze-failure sequence at 40 C.

7 CONCLUSIONS

In order to understand craze and crack growth in the practically important situations of changes in loads, different environments and varying temperatures, it is necessary to describe the processes involved in both craze and crack growth under these conditions. Earlier work had provided a basis for the study of craze kinetics, and this paper describes extensions to that work designed to give the original model more practical utility. In this it has been successful, in that load and geometry changes, and viscosity variations both by liquid changes and temperature changes, have been successfully described. The study of temperature effects and bulk absorption have thrown more light on the mechanisms involved, particularly in surface craze growth, and has also shown clearly that gas evolution occurs in several situations and may significantly affect the final craze failure. Earlier work had concentrated on describing the mechanics of failure in soaked crazes and attention here is confined to describing many interesting surface features. In particular, it should be noted that there is evidence for a change in craze structure with time so that young crazes tend to be inhomogeneous and become more uniform with time. The mechanisms involved could be of considerable significance.

ACKNOWLEDGMENTS

The authors wish to express their thanks to Mr. I. D. Graham, Mr. C. Webber and Mr. M. Parvin, some of whose experimental results appear in this paper, and to Mr. G. Weidmann for permission to use Fig. 14.

REFERENCES

1. Marshall, G. P., Culver, L. E., and Williams, J. G., *Proc. Roy. Soc. (London)*, **A319**, 165-187 (1970).
2. Paris, P. C., and Sih, G. C., *ASTM STP*, No. 381 (1965).
3. Marshall, G. P., and Williams, J. G., in press, *J. Mater. Sci.*
4. Marshall, G. P., and Williams, J. G., in press, *J. Appl. Polymer Sci.*
5. Sternstein, S. S., and Ongchin, L., *A.C.S. Polymer Prep.* (September, 1969).
6. Andrews, E. H., and Bevan, L., *Polymer*, **13** (July, 1972).
7. Graham, I. D., Marshall, G. P., and Williams, J. G., Proceedings of the International Conference on Dynamic Crack Propagation, Lehigh University, July 10-12, 1972, to be published.
8. Berry, J. P., *J. Appl. Phys.*, **33**, 1741 (1962).

DISCUSSION on Paper by J. G. Williams and G. P. Marshall

ANDREWS: The threshold conditions (critical surface work) for craze propagation is, we believe, controlled by cavitation in a solvent-swollen zone at the tip of the craze. We have measurements to suggest that this zone contains the equilibrium solvent concentration at the particular temperature of test. This suggests that processes in the uncrazed zone ahead of the craze are important as well as solvent flow through the craze itself, as proposed by Dr. Williams [see *Polymer,* 13, 337 (1972)].

WILLIAMS: The parameter K_M in the flow model, which is the initiation condition, is that to which Professor Andrews' results refer. His explanation is in reasonable accord with our observations of variations of initiation conditions.

EIRICH: Though your results seem to exclude this, I wonder whether it is really permissible to rule out surface effects from your model. The wetting and flow of liquids over free surfaces, or into porous media, is governed by a dimensionless number: $\frac{\gamma t}{\eta d}$, where γ is the (interfacial) surface free energy in erg/cm^2, η the coefficient of viscosity, d is a characteristic (pore) dimension, and t a time factor. Written as: $1/t = \gamma/\eta d$, the right side establishes a rate which, for small dimensions, can become quite large. Thus, the rate of liquid advance into open cracks and crazes may play an as yet to be determined role in your swelling rates.

WILLIAMS: This is a very interesting point and raises the whole question of the driving force involved in this phenomenon. The assumption that the resistance to flow is viscous seems reasonable, but the forces producing it are less certain. We have put them in the analysis as a pressure in the newly formed voids, but it is expected that at least two other mechanisms will be involved. One is vapor-pressure effects in the voids, which will tend to decrease the effective pressure, and the other is the surface tension or capillary effects. A simple calculation of the type you mention will give an equivalent pressure for this, and reasonable figures indicate that it would be around 5 percent of the atmospheric effect. It has, therefore, not been included in the analysis, but more detailed information is required to resolve the point definitely.

AGENDA DISCUSSION: FRACTURE

M. L. Williams *

College of Engineering
University of Utah
Salt Lake City, Utah

A. R. Rosenfield **

Metal Science Group
Battelle, Columbus Laboratories
Columbus, Ohio

1 INTRODUCTION

It is important to realize that fracture of design components is characteristically approached in two ways by the analyst.[1] Before recognition of the importance of inherent flaws in the material, he relied upon one of several average stress or strain criteria, e.g., maximum tensile stress, maximum principal strain, maximum octahedral stress, or others, depend-

*Chairman.
**Secretary.

ing usually on experimental evidence and experience. While the existence of microscopic or atomistic "holes" in material was recognized, it was generally assumed that their presence was of no design consequence, and, as long as material production control techniques were sufficiently reliable to produce microvoids of a small mean size with a low dispersion around the mean size, the use of an average stress or strain criterion was justified.

2 FRACTURE-CRITERION SELECTION

Hence, thinking of the material as a continuum, as in the case of a normal tensile specimen for example, the maximum tensile stresses measured in successive specimens of the same material were likely to be quite consistent. Actually, as discussed earlier by Menges[2], there is always some reasonably uniform distribution of small flaws present, whose size is related to the starting material and method of material fabrication. He found as such potential flaws, filler surfaces, boundaries of spherulites, and boundaries of other phases. Another simple example is a polymer which is mixed rapidly and contains finely dispersed air bubbles. Even with degassing, some distribution of flaws will exist on some dimensional scale. The average tensile strength therefore reflects their presence, and the dispersion of strength data about the norm describes the uniformity of the flaw distribution. Because most standard materials are made under reasonably strict quality-control conditions, it is not surprising to find that some sort of consistent (average) stress or stress-functional criterion can be used to predict failure.

Under more complicated conditions, such as the multiaxial stressing of a turbine disk, it is frequently customary to assume that the failure criterion is based on the octahedral shear stress (τ_{oct}) containing all three principal stresses and defined as

$$\tau_{oct} = K\sqrt{(\sigma_1 - \sigma_2)^2 + (\sigma_2 - \sigma_3)^2 + (\sigma_3 - \sigma_1)^2} \quad . \tag{1}$$

Assuming this criterion applies, one predicts failure whenever a combination of principal stresses at any point in the part exceeds τ_{oct}. And how is τ_{oct} determined? If Eq. 1 is a *universal* failure criterion, it must also apply to the failure of a simple uniaxial tensile specimen having stresses $\sigma_1 = \sigma_{tens}$, and $\sigma_2 = \sigma_3 = 0$. Thus, substituting into Eq. 1 one finds that

$$\tau_{oct} = K\sqrt{2\sigma_{tens}^2} \quad , \tag{2}$$

so that upon solving for the desired constant K and resubstituting into Eq. 1, one finds that failure is expected under a multiaxial principal stress combination whenever at some point in the body

$$\sqrt{(\sigma_1 - \sigma_2)^2 + (\sigma_2 - \sigma_3)^2 + (\sigma_3 - \sigma_1)^2} \geqslant \sqrt{2}\,\sigma_{tens} \quad . \tag{3}$$

In the more general average stress criterion case denoted as Region I in Fig. 1, the failure criterion based upon average principal stresses would have the form

$$F(\sigma_1, \sigma_2, \sigma_3) \geqslant \sigma_{F_{cr}} \quad , \tag{4}$$

where $F(\sigma_i)$ is some function of the principal stresses at a point in the material.

On the other hand, there are conditions in which discrete flaws substantially larger than the uniform size distribution normally present can exist in the material. Such inherent flaws may arise from localized corrosive attack, improper fabrication, cyclic loading, or accidental surface nicks or cuts. Because they are discrete, usually relatively sharp, and larger than the surrounding voids, they can induce additional stress concentrations and provide loci of cohesive-fracture initiation. Particularly if these inherent flaws are cracklike, ordinary elastic stress-concentration factors are essen-

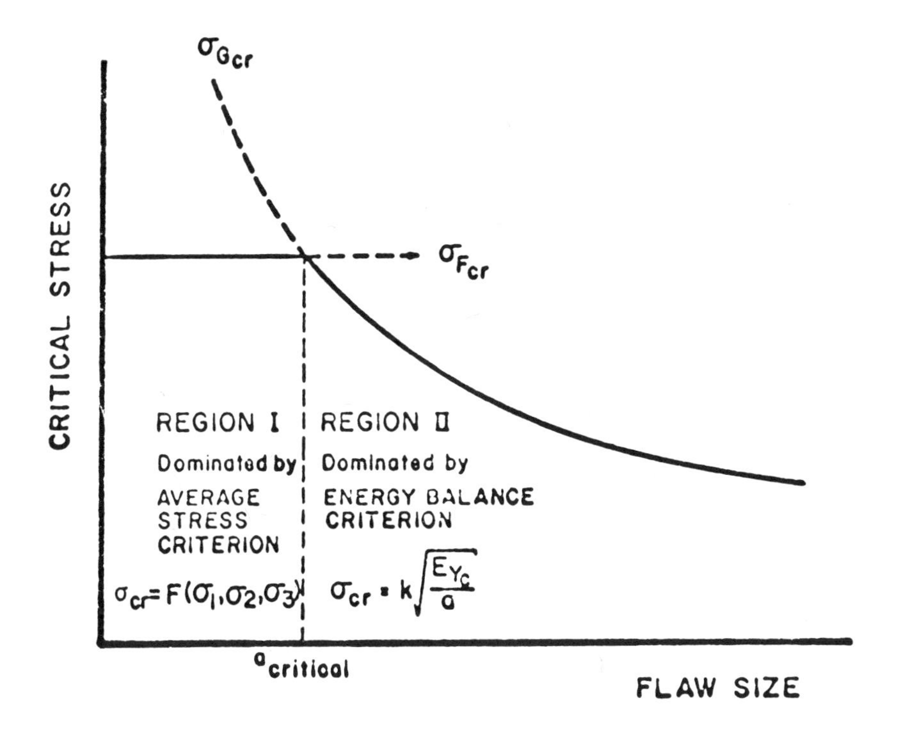

Fig. 1. Dominant fracture regions depending upon inherent flaw size.

tially useless because theoretical results predict an infinite concentration factor multiplying the average stress in the vicinity. Thus, the local stress value will exceed the finite allowable stress experimentally measured for the base material containing only the reasonably uniform distribution of inherent voids. The strength degradation in such a situation is illustrated by Region II of Fig. 1.

In a similar manner, the difficulty with many adhesive joints is that they also can possess very high stress concentrations at corners or along bond lines, and usually contain substantially larger than average internal flaws, frequently as the result of absorbing water or poor wetting of the interfaces. In any event, the flaw distribution becomes denser and/or of larger size than the average size for which an average tensile strength would be appropriate. Thus, the maximum permissible allowable stress for adhesive fracture is likewise decreased, phenomenologically similar to Region II behavior in cases of cohesive fracture.

Griffith[3] provided the first estimate of degradation as a function of the flaw size by considering the problem of a small, through, line crack in a thin sheet of brittle material. While theoretically the stress at the crack tips is (mathematically) infinite for an elastic body, thus giving rise to an infinite local stress at even small applied loadings – a degree of concentration for which Eq. 4 is useless – Griffith avoided this problem by considering changes in strain energy in the sheet, which, as integration of the stress, remained finite. He proposed that cohesive fracture would commence at a critical applied stress, σ_{cr}, when the incremental loss of strain energy of deformation with increasing fracture area just equaled the work required to create new fracture surface. Hence, in his case, with the elastic strain energy of deformation (U) due to the presence of the crack of length 2a being $U = \pi a^2 \sigma_{cr}^2/E$, one would have fracture whenever

$$\frac{\partial}{\partial a}\left[\pi a^2 \sigma_{cr}^2/E\right] \geqslant \frac{\partial}{\partial a}\left[4a\gamma_c\right] \quad , \tag{5}$$

where E is Young's modulus, and γ_c is the specific cohesive fracture energy density (in-lb/in.^{2}, or erg/cm^2).That is to say, the strain energy of deformation lost as the crack extended was converted into the work to create the new fracture area. From Eq. 5, the finite critical applied stress is determined (see Region II in Fig. 1) as

$$\sigma_{cr} > \sqrt{\frac{2}{\pi}\,\frac{E\gamma_c}{a}} \tag{6}$$

In general, the dissipative processes which occur around the crack tip cause the fracture energy to exceed the surface free energy. Provided these processes are confined to a small region near the tip, the functional dependence of fracture stress on crack size of Eq. 6 is retained.

The combination of the two criteria, one flaw insensitive (Region I) and the other dependent upon flaw size (Region II), thus permits the designer to select a maximum allowable design stress providing he knows,* or determines by tests in the laboratory on precracked thin sheet tensile specimens with known crack size a, the critical crack size, a_{cr} (Fig. 1). This critical crack size is deduced by the intersection of normal, nominally unflawed, tensile data $\left(\sigma_{F_{cr}}\right)$ and initially precracked sheet data following the Griffith curve data $\left(\sigma_{G_{cr}}\right)$. Once it is recognized that Eqs. 4 and 6 are not competing failure criteria, but instead are complementary, it is possible to approach the design against failure in a more direct manner.

3 THERMODYNAMICAL FORMULATION OF FAILURE CRITERIA

For materials whose properties are time, rate, or temperature dependent, it is not clear that a direct application of Griffith's energy balance will be satisfactory because a time-dependent fracture criterion would seem to be the more normal result for materials whose properties change with time, as might the loading itself.

As an outgrowth of Schapery's dissertation[4] on applications of irreversible thermodynamics in solids, a more general energy principle was developed for fracture. Based upon the first law of thermodynamics, the statement of energy conservation was expressed in terms of the rates of work input to the system being equated to the work absorbed by the system. In this way, any time dependency was implicitly incorporated[5], and the conditions for an energy balance in a continuum for which the equation of state (the relation between stress and strain) was known could be deduced.

In accordance with the law of conservation of energy, the thermodynamic requirement in a deformable solid requires that the rate of inputting work, $\dot{I}$, plus the rate of adding heat, $\dot{Q}$, must equal the rate of increase of the kinetic energy, $\dot{K}$, plus internal energy, $\dot{U}$, plus the rate of work absorbed in generating new fracture surface, $\dot{\Gamma}$. For isothermal systems, internal energy is composed of two parts, the intrinsic internal strain energy of deformation and the internal temperature change.

$$\dot{I} + \dot{Q} = \dot{K} + \dot{U} + \dot{\Gamma} \quad . \tag{7}$$

*The technically important problem of measuring the inherent flaw size in a part, preferably by some nondestructive test (NDT) method as ultrasonic wave reflection, X-ray, etc., will not be covered in this paper. Obviously, if the inherent flaw size is unknown, a priori, the analyst does not know whether to choose Region I or Region II criteria.

From the general definitions,

$$\dot{I} = \int_{\sigma} \overset{\upsilon}{T}_i \, \dot{u}_i \, d\sigma + \int_{\tau} F_i \, \dot{u}_i \, d\tau \tag{8}$$

$$\dot{Q} = \int_{\sigma} \dot{q}_i \, \upsilon_i \, d\sigma \tag{9}$$

$$\dot{K} = \frac{d}{dt} \int_{\tau} \frac{\rho^*}{2} \, \dot{u}_i \, \dot{u}_i \, d\tau \tag{10}$$

$$\dot{U} = \frac{d}{dt} \int_{\tau} \int_0^t \frac{\partial W}{\partial t} \, dt \, d\tau \tag{11}$$

$$\dot{\Gamma} = \frac{}{dt} \int_0^t \gamma(\xi) \frac{\partial \sigma}{\partial \xi} \, d\xi \quad , \tag{12}$$

in which the surface tractions are

$$\overset{\upsilon}{T}_i = \sigma_{ij} \cos(x_j, \upsilon) \quad , \tag{13}$$

and υ is the outward normal to the surface, ρ^* is the mass density, $\gamma(t)$ is the time-temperature dependent specific fracture energy, σ and τ denote the surface and volume of the region, respectively, and u_i and F_i are the components of the displacement and body force vectors, respectively. The internal strain-energy density rate, $\partial W/\partial t$, in terms of the heat flux vectors and the stress and strain tensors (σ_{ij} and e_{ij}), respectively, is

$$\frac{\partial W}{\partial t} = \sigma_{ij} \, \dot{e}_{ij} + \dot{q}_{i,i} \quad . \tag{14}$$

Presuming therefore that the specific fracture energy, body forces, and heat fluxes are known, one then proceeds to solve the equilibrium and compatibility equations with boundary conditions appropriate to a given physical problem. When this solution is substituted into Eq. 7 via Eqs. 8 through 14, there results the desired fracture criterion which will be of the form

$$\dot{a}(t) \, L \left\{ t, a(t), c_{ij}(t), \gamma(t), f(t) \right\} = 0 \quad , \tag{15}$$

in which the elementary solution $\dot{a}(t) = 0$ implies that the crack velocity is zero, or the crack length $a(t)$ is constant, i.e., the stationary, viscoelastic stress-strain solution. The second solution however, $L\{\ \} = 0$, is, in general, a nonlinear, integro-differential equation for the crack history $a(t)$ in terms of the constitutive law, c_{ij}, the specific fracture energy $\gamma(t)$, and the loading history, f(t).

Cherepanov[6] specialized the three-dimensional formulation to the case of plane stress, a thin flat sheet containing a crack of length a(t). Omitting the intervening details which are unessential to the philosophical purpose here, his conclusion can be expressed in the form

$$\frac{da}{dt}\left[\int_{-\pi}^{\pi}\left\{\left[\int_{0}^{t}\sigma_{ij}\,\dot{e}_{ij}\,dt + K_{*} - H\right]\cos\theta + \overset{\nu}{T}_{i}\,\frac{\partial u_i}{\partial a}\right\} Rd\theta - 2\gamma(t)\right] = 0. \quad (16)$$

Here σ_{ij} and e_{ij} are components of the stress and (small) strain tensor, respectively, K_* is identified with the kinetic energy and H with the body forces, R $(a/R < 1)$ is essentially the size of the specimen, and u_i are the components of displacement. Specifically, with ρ^* the mass density,

$$K_* = (\rho^*/2)\,\dot{u}_i\,\dot{u}_i \quad (17)$$

and H is defined implicitly through

$$\iint F_i\,(\partial u_i/\partial a)\,\Big|_{t,\ \text{constant}} dxdy \equiv -\iint(\partial H/\partial x)\,dxdy\ \Big|_{\substack{x = R\cos\theta \\ y = R\sin\theta}} . \quad (18)$$

In principle then, substitution of a known solution for $u_i(x_k, t)$ into the fundamental criterion (Eq. 7) will lead to a governing integro-differential equation for the crack position as a function of time.

It is interesting to note that for slow crack growth and in the absence of heat flux and body forces, the nonstationary part of Eq. 16 reduces in the special case of (time-independent) elasto-plastic deformation and time-independent fracture toughness, i.e., $2\gamma(t) \neq f(t) \equiv J$, to

$$R\int_{-\pi}^{\pi}\left[\left(\frac{1}{2}\right)\sigma_{ij}\,e_{ij}\cos\theta - \overset{\nu}{T}_i(\partial u_i/\partial x)\right]d\theta = J\ , \quad (19)$$

which is the form now being popularized by Rice[7] and others, as an outgrowth of the work of Eshelby[8], for calculating the path-independent fracture toughness of metal plates containing cracks and incorporating local plastic deformation at the crack tip before instability.

This brief summary of certain features of the physical separation mode of failure is intended to describe the present status of fracture analysis from the point of view of continuum mechanics. The essential feature is that there are two complementary criteria depending upon the distribution of actual microflaws in the material. In the important design case of an existing predominant flaw or crack, a general thermodynamic criterion for incipient fracture and subsequent growth can be formulated for, in principle, a crack of known shape in a three-dimensional solid of arbitrary material properties. These relatively recent developments permit a rational engineering investigation of potential fracture, freed from assumptions of purely elastic deformations, and specifically include consideration of elasto-plastic and viscoelastic mechanical-property behavior. The advent of high-speed computing capability makes possible consideration of even the most difficult problems now that the essential mathematical formulation has been derived.

4 MATERIAL CHARACTERIZATION

In the foregoing discussion, the material characteristics have entered in both the evaluation of the internal strain energy of deformation (Eq. 13) and of the fracture energy (Eq. 12), insofar as a continuous medium is assumed. If the constitutive (stress/strain/strain rate) law is known, then in principle the fracture characteristics can be deduced from Eq. 15. On the other hand, it is clear that the molecular structure of the material determines its mechanical behavior. Thus, a direct association between molecular structure and mechanical behavior would be desirable – particularly from the point of view of one's being able to eventually "design" a material, a priori, to do a particular job. As discussed by Barenblatt[9], the task is facilitated by dividing the test piece into two regions: a small strain region away from the crack tip and a large strain region in the vicinity of the crack tip.

4.1 Small Strain Region

As a way of showing one method by which the characteristics of continuum mechanics and polymer chemistry can be associated, consider the following figures which indicate what Dr. Kelley and the Chairman[10] have been popularizing in terms of an Interaction Matrix. The first (Fig. 2) is a typical relaxation modulus shown in terms of its log-log plot. They characterize the modulus by a modified power law representation containing five parameters: the glassy and rubbery moduli (E_g and E_e, respectively), the center and slope of the viscoelastic transition region (τ_o and n), plus the glass transition temperature, T_g. Can we relate these particular characteristics to the polymer structure? Table I shows their attempt at such an association, and lists, along with the mechanical (column) descriptors, ten structure (row) descriptors. The intersection of the appropriate rows and columns indicate the interaction. At present,

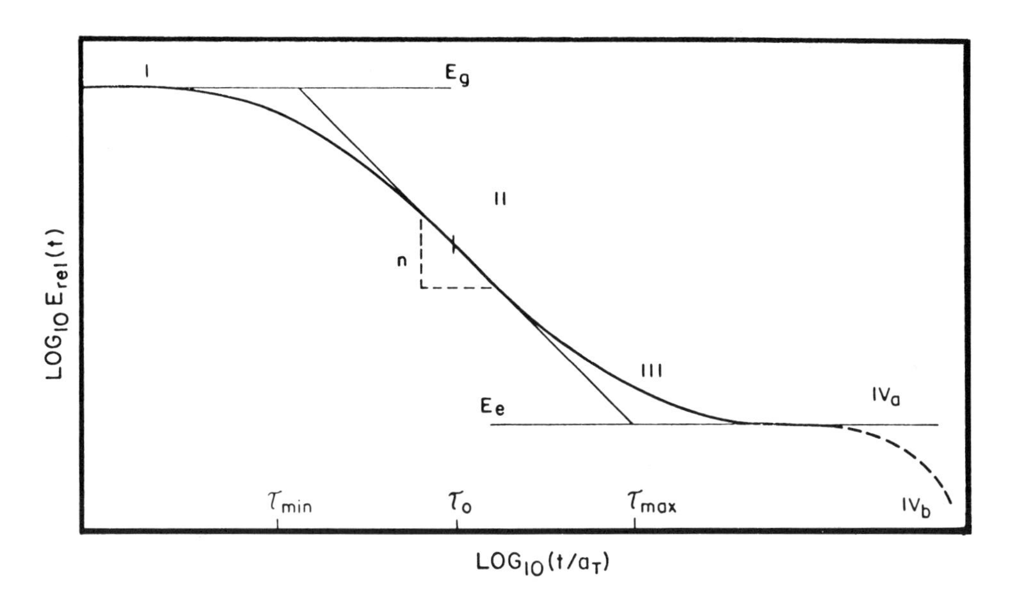

Fig. 2. Tensile relaxation modulus as a function of reduced times with power-law parameters shown.[10]

Table I. Interaction Matrix

Molecular Characteristics	Symbol	Modified Power-Law Parameters				
		E_g	E_e	τ_o	n	T_g
Cross-link density	ν_e	N	S	N	M	N(1)
Chain stiffness	N_s	N	N	U	M(3)	S
Monomeric friction coefficient	ζ_o	U	N	S	U	S
Solubility parameter	δ_p	M	N	U	U	S
Molecular weight	M	N	S(2)	N(3)	N(3)	S(4)
Heterogeneity index	M_w/M_n	N	N	M	N	M(5)
Molecular weight between entanglements	M_e	N	S(6)	N	N	N
Degree of crystallinity	Λ	N	S	S	S	N(7)
Volume fraction of filler	ϕ	N	S	M(8)	S	M
Volume fraction of plasticizer	V_p	N	S	S(9)	N	S

U = Unknown, N = negligible, M = moderate, S = strong; (1) except at very high values of ν_e; (2) effect of entanglements; (3) at high molecular weights; (4) at low molecular weights only; (5) chain end effects from short-chain fractions; (6) at high molecular weights producing a plateau or pseudo-equilibrium modulus; (7) except at very high Λ; (8) through the (WLF) time-temperature shift factor; (9) through ζ_o.[10]

most of the interrelations are unresolved and it was necessary to introduce a qualitative association (U, unknown; N, negligible; M, moderate; S, strong) based upon the considered collective judgment of representative workers in the field.

A quantitative interaction formula for the intersection of a row and column would be the desired association. For example, rubber elasticity theory tells us that the rubbery modulus, E_e, and cross-link density, υ, are theoretically related by $E_e = 3\upsilon kT$, in which k is the Boltzmann constant. One other semiempirical relation was given earlier[10] relating the chain stiffness, N_s, and slope of the relaxation modulus through the transition region, n (Fig. 3).

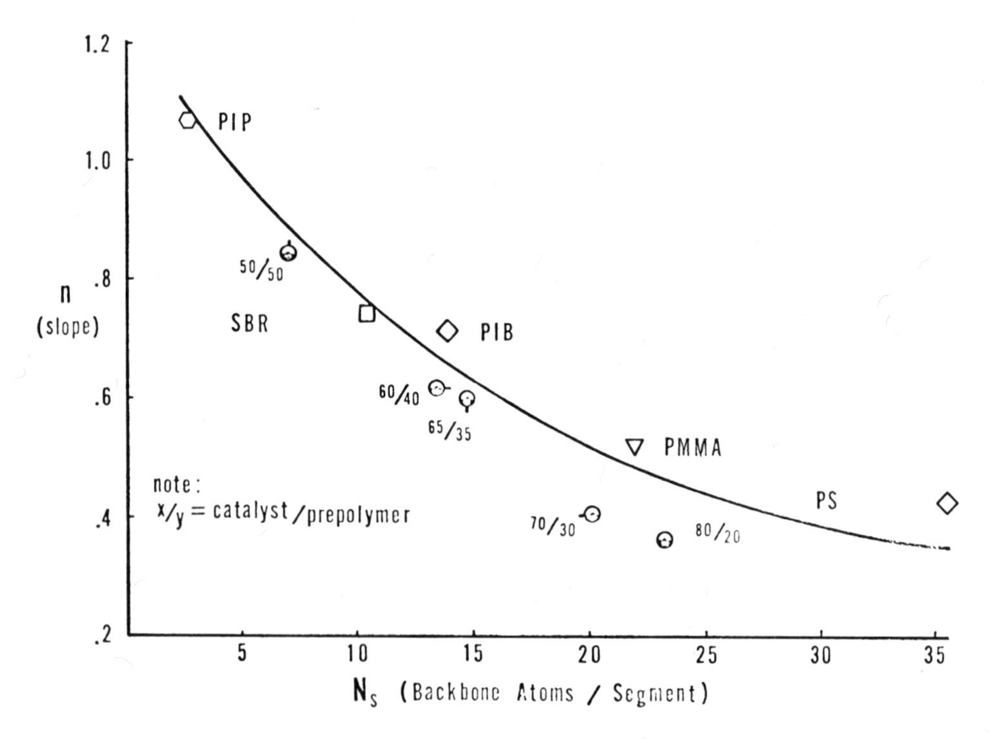

Fig. 3. Polymer backbone stiffness as a function of the slope of the relaxation curve. Half-filled circles – polyurethane of similar cross-link density, but varying catalyst-to-prepolymer ratio; hexagon – polyisoprene (PIP); square – styrene butadient rubber (SBR); triangle – polyisobutylene (PIB); inverted triangle – polymethyl methacrylate (PMMA); diamond – polystyrene (PS).[10]

When the discussion was opened, a few specific comments were made on the philosophy of the Interaction Matrix approach. In general, there were no challenges to the basic concept. Indeed, as one of its originators, Eirich stated that from the engineering standpoint, the parameters presented seemed to have successfully withstood the test of time. A few additions and modifications were suggested, however, by Vincent who felt that, in addition to specific inclusion of a free-volume parameter, molecular orientation (both that built into the structure during fabrication and that developed during deformation) should be included. Landel then noted that certain models based on molecular character, such as he then described, are not particularly good for the viscoelastic transition region. In fact, there is no good molecular picture of the transition region either in dilute solutions or in bulk materials. Landel also recommended that the kinetics of crystallization under strain and the rate of chemical attack should be added. Swartzl felt that the data, even the small amount presently existing, needed updating and certainly expansion. For example, he believes that the values of the double logarithmic slope of the modulus versus time curve are wrong, specifically polystyrene and polyurethane should be switched. Finally, Shen conjectured that the inclusion of supramolecular structure considerations, i.e., that polymer chains are not as random as one might think (nodules in rubber), might be revealing as well as tying in with the effect of molecular orientation mentioned by Vincent. However, Ferry cautioned that it may be quite difficult to include explicitly the characteristic features of molecular branching.

4.2 Large Strain Region

The connection between polymer chemistry and continuum mechanics in the large strain region requires a more specific knowledge of the dissipative mechanisms at the crack tip. Examples are stress-induced crystallization of natural rubber[11] and crazing of polystyrene[12]. Such mechanisms not only raise the fracture energy, γ_c, above the free surface value, γ_f, but also cause blunting of an initially sharp crack. In this way, the conceptual difficulty of infinite local stresses is removed. Furthermore, by adoption of simple strip-yielding models, the problem of evaluating the fracture energy is transformed to one of determining a critical stretch (or crack opening displacement) of the material directly at the crack tip (c.f., Knauss[13]).

In principle, an interaction matrix could also be developed for the fracture energy. A start has been made.[14] Although the form of the log γ_c versus log t curve may very well be similar to that shown in Fig. 2 (see Fig. 4 for results on a typical polybutadiene[14,15] for small strains), differences are expected in detail. It would be surprising if the large- and small-strain behaviors did have identifical rate dependence, and present indications appear to be that they do not.[14,16] Additional factors which can affect the two regions differently are multiaxiality of stress[17,18] and local corrosive attack[11,19].

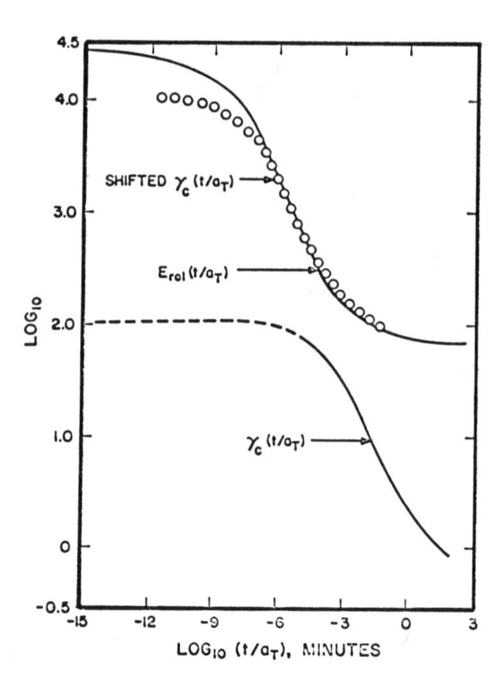

Fig. 4. Comparison of relaxation modulus and cohesive fracture energy[27]*; o o o o – γ_c (t/a_T) shifted 3.5 decades left and 2.0 decades up.*[14]

5 ENGINEERING PROBLEMS

Up to this point there seemed to be a balance established. The Introduction and Comments on the Interaction Matrix emphasized the means of analyzing the supply of energy to the crack-tip region, while the research papers emphasized the ways in which energy is absorbed within this region. There seemed to be bodies of opinion that one or the other of these aspects of the total problem could be safely ignored. However, as shown clearly by the thermodynamic power balance (Eq. 7), both aspects must be taken into account.

5.1 Fracture-Toughness Evaluation

Hull pointed out that reported values of fracture energy of PMMA vary by a factor of 6, while reported fracture energies in polystyrene have almost two orders of magnitude variation. Part of this variation can be ascribed to different test methods, including the method of introducing the crack. Another part may be due to a specimen thickness effect analogous to the plane-strain/plane-stress transition in high-strength metals.[20] Hull pointed out that PMMA always exhibits Dugdale strip-yielding zones[21] consisting of a thin zone of craze in the plane of the crack, whereas the plastic zone in polystyrene may be either a Dugdale zone or a much wider zone containing a large number of small crazes[22]. He claimed that the difference was due to differences in the molecular-weight distribution from one producer to another and that the changes in craze distribution in the zone accounted for the toughness difference. These differences cannot be related to the yield stress. J. G. Williams noted that the PMMAs tested by various authors are not very different, and much of the scatter is due to improper testing procedures, particularly the highest values. Polystyrene is somewhat different in that there are material differences that do show up in fracture experiments. This point has also been made by Rabinowitz and Beardmore[23] who point out that the transition in fracture stress as a function of temperature (Fig. 5) is closer to the ambient temperature in polystyrene, and changes in material properties can shift it sufficiently to cause apparent large variations in room-temperature fracture behavior. In PMMA, on the other hand, the transition is somewhat below ambient. As a related issue, the Chairman notes that the qualitative behaviors illustrated in Figs. 1 and 5 are not necessarily independent. At, say, similar experimental strain rates in the two materials, the temperature-reduced strain rate, $\dot{\epsilon}a_T$, in which a_T is the WLF factor, incorporates temperature change through a_T, which

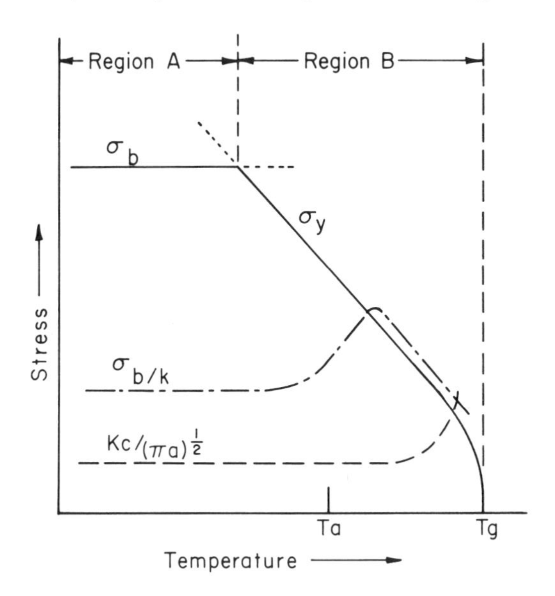

Fig. 5. Schematic temperature dependence of fracture stress of glassy polymers. Key: ——— smooth tensile bar, —— - —— notched tensile test, — — — precracked fracture specimen; T_g = glass transition temperature, T_a = ambient temperature, k = stress concentration of notch.

itself contains the material-property influence through the a_T dependence upon the glass temperature, T_g.

Vincent discussed the problem of excessive yielding accompanying fracture with the aid of Fig. 5. He pointed out that the lines representing yield and fracture of smooth bars were stress dominated. The line for sharp cracks is energy dominated. He also pointed out that a smooth bar test is too gentle for practical applications, while he claimed that a precracked test is also unsuitable. He advocated using notch specimens with a root radius on the order of 1/4 mm, which best models defects in practical situations. Unfortunately, practical failures often occur in the region where large plasticity occurs at the tip of the notch (T_a in Fig. 5). In this region, analyses involving *simple* small-scale yielding, and *simple* energy balances are not applicable. We are dealing with the knee of the curve of impact energy versus temperature. The material at the tip of the notch goes through the whole stress-strain curve of the polymer. The chairman pointed out that another way of looking at this problem which might still not be too complicated is to consider (statistically) whether enough holes form to degrade the stress.

5.2 Environmental Effects

Liquids. This part of the session began with a presentation by Andrews, who described previously published work.[24] The general sense was that there was a difference between his and J. G. Williams' work. Andrews was concerned only with the crazing phenomenon itself, and strongly suggested that the critical processes for craze propagation or arrest depend upon providing sufficient surface energy to grow new voids at the tip of the craze after the material is solvated by the solvent to its equilibrium swelling condition. The solvent reaches the material at the tip quickly by stress-assisted sorption. Cavitation then occurs when sufficient external constraint is reached. J. G. Williams, on the other hand, assumes that cavitation had already taken place, and that flow in the craze is the controlling factor, this being the difference in the two models. There is a parameter, K_m, in J. G. Williams' theory which could be construed to contain the initiation or arrest condition with which Andrews is concerned. Variations of K_m with temperature and water content appear consistent with Andrews' results.

J. G. Williams also interjected strong support for devoting part of the session to environmental effects, since this is one of the two major causes of practical failures in plastics, the other major cause being impact loading. Andrews pointed out the futility of performing engineering tests of craze-growth resistance in environment at a single temperature. These processes are very temperature sensitive and a 5 C change can easily change the relative rankings of behavior of two different solvents. In a reply to a question from Landel, Alfrey said that orientation had an important effect on the tendency to craze.

Vincent pointed out that the solubility parameter was an insufficient criterion for craze initiation. His experiments show that the possibility of hydrogen bonding between the polymer and the liquid environment needs to be taken into account.[25] Andrews replied that the hydrogen bonding was varying systematically among the liquids in his tests; the solubility parameter does not enter into his theory. What does enter is the depression of T_g by swelling which is affected by the total interaction between the solvent and the polymer. Also entering is the interfacial energy between the solvent and the swollen polymer.

Since both the solubility parameter and interfacial energy are measures of intermolecular forces, the smooth dependence is not surprising. Eirich pointed out that the times for diffusion of liquids into polymers can be quite short provided one considers thin (surface) layers. Using: $D = x^2/2t$, and taking even a very small value for D, e.g., 0.5×10^{-5}, a diffusive penetration of 1000 A may occur in 10^{-5} sec. This is quick enough for a practically instantaneous softening of polymer adjacent to the crack tip, or to crazes, thus permitting rupture initiation or cavitation. J. G. Williams replied that this rapid take up of the liquid into the polymer is assumed in his model since the fluid flow is taken as the slower, controlling variable.

Gases. Thomas pointed out that the presence of oxygen can change the rupture energy of rubber by a factor of 2 and wondered whether this effect of oxygen on bond strength had been investigated for other polymers. Hull pointed out that Brown[26] had conducted tensile tests to study the effect of gaseous atmosphere and crazing at temperatures as low as that of liquid helium, and he found significant effects on mechanical properties and, in particular, on crazing. Eirich said that gases can absorb on carbon-carbon bonds and reduce the bond energy. If we believe the chain scission mechanism, we do not need to wait for oxidation to occur, the mere absorption of gases on tertiary carbons in the hydrocarbon plays a role in reducing the bond strength for scission.

General Comments. Shen pointed out some particular problems of environmental effects. One was that tests on a freshly produced polymer may not be relevant after it has been exposed to service conditions for some time. Secondly, toxicity and biomedical effects should be included. Alfrey mentioned that vacuum compression molding improved the stress-crack resistance of a number of thermoplastics as shown by Charles Rogers at Case Western Reserve University. When Rogers used a vacuum and excluded oxygen, it was not necessary to add an antioxidant to polypropylene. Part of this evacuation effect seemed to be over and above the avoidance of oxygen and oxidation during high-temperature fabrication. Landel then commented that crazing or whitening due to environment had an effect in two-phase systems, not only on the matrix but also on the interface. In particulate-filled elastomers immersed in fluid and stretched, the fluid attacks the interface and the specimen yields, but the properties of the matrix are not particularly affected.

Other environmental effects mentioned by the Chairman were the insertion of a plastic tube in the human body and its interaction with natural tissue giving rise to adhesion problems, which is amenable to continuum mechanics analysis but requires a substantial knowledge of interfacial biochemistry. Also, he noted that similar problems arise in the filling of teeth with plastics. From the standpoint of continuum mechanics stress analysis, adhesive and cohesive fracture are formally similar.[27] It merely depends upon which value of specific fracture energy [cohesive (γ_c) or adhesive (γ_a)] is used in the results. In passing, however, the important comment of Knappe earlier is that frequently the crack trajectory is not known a priori, and hence the analyst is left in a quandry as to whether a cohesive or interfacial energy value is appropriate.

Professor Eirich concluded this part of the session with pertinent remarks which are best quoted: "I wish to mention two additional factors that will affect craze formation, and which might help to explain some of the more unusual observations. First, environmental gases (including water vapor) adsorb very rapidly on freshly formed surfaces, reduce the surface energy that must be provided, and thus facilitate additional surface formation. Moreover, if the adsorbed gas is oxygen, the adsorption energy contains a chemical term due to a reaction of oxygen with double bonds and tertiary hydrogens. In addition, resulting free radicals can start a chain reaction of polymer degradation. Second, it is well known in metallurgy that very small amounts of surface active agents can influence specimen strength greatly. The mechanism consists in a strong local bonding of the agent to various surface imperfections and dislocations, which thereby can become locally sessile. Similar effects should occur in polymers, especially on highly drawn sections or lamellae (as in crazes). Thus, minute impurities, or atmospheric agents (vapors), may have a large effect on the fate of crazes. This mechanism has to my knowledge not been considered for polymers."

5.3 Constrained Shrinkage

Alfrey then discussed the subject of treating delayed fracture of thin films on rigid substrates, specifically the problem of a thin latex film cast on glass. When the water is evaporated to about 70 percent solid, a gel is formed. As the remaining 30 percent of water evaporates, shrinkage is constrained by the rigid substrate, and biaxial tensile stresses are developed parallel to the surface, which can cause cracking. There are many possible characteristic patterns. He discussed the one shown in Fig. 6 which has a checkered appearance similar to a dried desert floor after a flash flood.

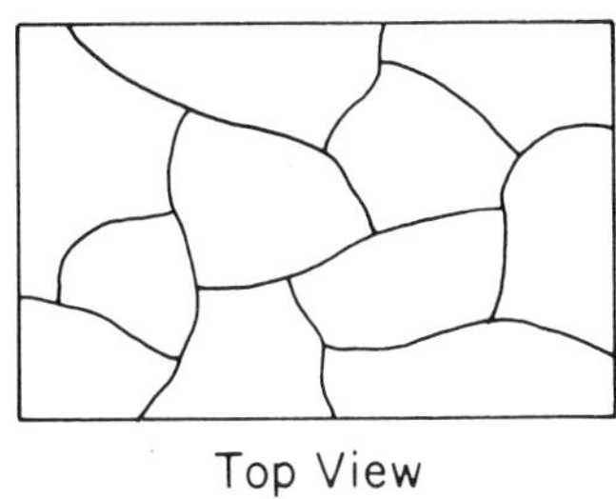

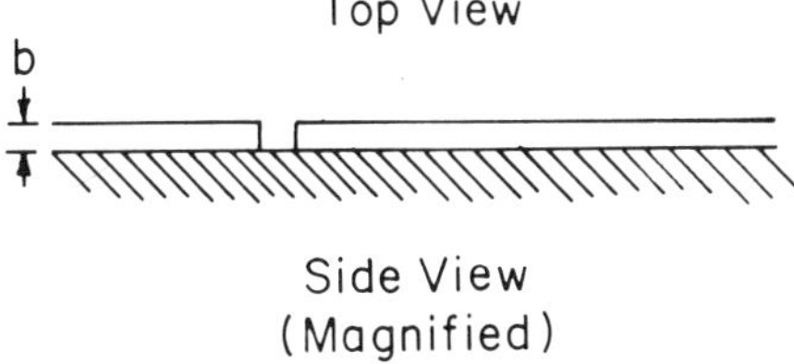

Fig. 6. Film drying on rigid substrate (typical pattern).

Consider a biaxially stressed plastic sheet of shear modulus G and thickness b separated from the substrate by a thin intervening layer of viscosity η and thickness h. One-dimensional behavior is found near the edge, and the elastic stress is given by the equilibrium equation:

$$b d\sigma/dx = \tau \tag{20}$$

and the viscous layer obeys

$$\tau = (\eta/h)\, \partial u/\partial t \quad , \tag{21}$$

where τ is the shear stress and u is the lateral displacement. Combining Eqs. 20 and 21 we get

$$4\,Gb \frac{\partial^2 u}{\partial x^2} = \frac{\eta}{h} \frac{\partial u}{\partial t} \quad , \tag{22}$$

which is the well-known diffusion equation. Therefore,

$$V\,(\text{edge}) = K\sqrt{t} \quad , \tag{23}$$

where V is the velocity of the edge and K contains material properties, dimensions, and the degree of stretch of the elastic membrane. This analysis actually is not particularly good for the edge because the degree of drying varies with position in the edge region. However, if a cut is made across the film after biaxial stress is set up, the separation of the cut is proportional to the square root of time. If we make a short cut in the film to initiate a growing crack, the tip advances with constant velocity while the separation of the faces varies as $t^{1/2}$. If there were no restraint from the substrate, there would be a fast-growing crack. Therefore, the crack tip would be expected to be a long narrow parabola. The motion of the crack faces relieves the tensile stress perpendicular to the crack; therefore, it becomes surrounded by a narrow zone in which the tensile stress parallel to the crack becomes much higher than that normal to it. If we start a new crack away from the original one, when the second crack enters this zone it curves to form a right angle with the original crack. The tip of the original crack is sharper and further advanced than one would expect of a parabola because two-dimensional effects arise at the tip. As a practical matter, this analysis is important to the drying of paint, for example. This cracking pattern occurs in the desert floor after a flash flood and in old statuary, and is used for decorative patterns in ceramics. Halpin mentioned biaxial stress fields at surface flaws and pointed out that Alfrey's analysis is the first practical analysis for combining polymer mechanics, physical chemistry, and environmental action at the surface as the crack penetrates the bulk. The problem is to formalize this model to the case of plane strain. This analysis may provide the basis for a practical screening test for environmental effects on polymers.

5.4 Crack Branching

Mayer initiated a discussion of crack branching, asking, in particular Is there a correlation between crack branching and any dynamic toughness parameter? What are the roles of velocity, strain-hardening behavior, flaws, etc? The chair answered that if too much energy is put into the propagation of a single crack, it has no option but to branch. There are calculations concerning the angle of the fork.[28] If the branched crack runs further it will branch again. Kobayashi pointed out that there is no specified dynamic stress intensity for branching in Homolite-100, but the static stress intensity provided a usable criterion.[29] Congleton and co-workers find similar results for PMMA.[30] However, for glass and tool steel, their experiments show that the nominal static stress intensity for branching can depend to some extent on the velocity at branching.[31] Kobayashi also pointed out that both velocity and dynamic stress intensity dropped before branching in his experiments. For one crack, which branched three times, branching always occurred at a critical static stress intensity. Andrews pointed out that one need not have a high velocity to produce branching. One only needs to decouple the crack from its stress field. The stressed crack overtakes its stress field and therefore it can't keep on axis. At low velocity, all one needs is large mechanical hysteresis, e.g., crystallization of natural rubber. The high velocity, in effect, is the same case. It is possible for the crack to travel so fast that it overtakes its stress distribution. Knauss then pointed out that the energy argument is a bit too simplistic. The crack runs only when the stress intensity at the tip is fairly high. There is then a relatively large area under high stress. The probability of generating flaws away from the tip is also large. Fractures can occur ahead of the crack and off-axis.[30,32-34] At sufficiently high velocity and/or stress intensity, branching is finally successful. Mayer than asked if the occurrence of a branched crack connotes higher toughness in polymers, and if impurities sometimes serve to increase toughness by promoting branching (as has been observed occasionally in steels). Eirich cited a case where this was true. He pointed out that off-line spawning of cracks is the reason for the high toughness of high-impact polystyrene, ABS rubber, etc. Local inhomogeneity off the crack plane causes crack branching. Alfrey returned to the phenomenon of slow crack advance in thin films on rigid substrates, and observed that bifurcation is sometimes encountered there. This results in (approximately) 120-degree crack junctions, in contrast to the 90-degree junctions discussed earlier.

5.5 Design Against Fracture

Menges provided the following comment relating fracture and engineering design: "We have worked out in the last few years a new and very simple calculation method for designing against failure. It is based on the existence of a strain limit, ϵ_m, for the production of voids, which limit also corresponds to the end of the linear viscoelastic range, and is effectively the first irreversible alteration in a strained part – the onset of fracture. This strain limit was found to be independent of rate of strain, of multiaxiality of stresses, of geometry, of homogeneity or inhomogeneity of stresses, and, finally whether or not the stresses were static, dynamic, or discontinuous. We therefore found that there exists only one failure criterion for all calculation of tensile stress failure. The range where our approach seems valid extends up to the glass or crystalline transition temperature."

What might be done in a practical calculation? First, calculate the principal stresses σ_1, σ_2, σ_3 in the normal way. Then transform the stresses into strains at the time range corresponding to the strain rate and temperature of the applied loading. This transformation can be made using an isochronal stress-strain diagram on which results for various temperatures are plotted (Fig. 7). The transformed strains become

$$\epsilon_{\sigma_i} = \sigma_i/E(t), \quad i = 1, 2, 3 \quad . \tag{24}$$

Finally, since we presume we are in the linear viscoelastic range, we

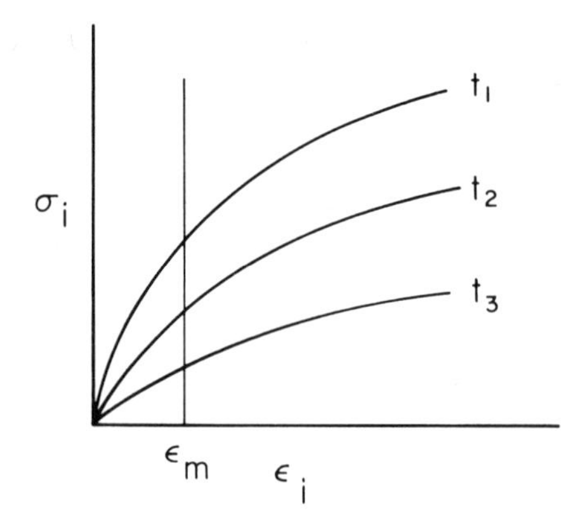

Fig. 7. Isochronal stress-strain diagram (t = time, $t_s > t_2 > t_1$).

calculate the strain in the part under load with Hooke's Law,

$$\epsilon_i = \epsilon_{\sigma_i} - \mu\epsilon_{\sigma_j} - \mu\epsilon_{\sigma_k} \quad , \tag{25}$$

and it follows now that we load our material only so that

$$\epsilon_1, \epsilon_2, \epsilon_3 < \epsilon_m \quad . \tag{26}$$

Since ϵ_m, the strain where voids are first found in the material, is essentially lower than the breaking strain under most practical loads, it is not necessary to incorporate a safety factor as is normally done in engineering calculations. We have found that this procedure is also useful under these corrosive atmospheres which do not swell or dissolve the material.[29]

This design procedure and failure criterion seems to be essentially the adoption of a maximum principal strain theory based upon an isochronal time. As an engineering approach, it falls under analysis methods for Region I (Fig. 1) and can be very useful. The Chairman notes, however, that its use must be verified for the specific material under consideration. In cases of nonproportional loading histories in multiaxial stress states and for nonlinear material behavior for example, its applicability in one engineering situation, namely, analyzing solid propellant (viscoelastic) rocket motor grains, has been found to be of limited value.

6 CONCLUSION

The discussion concluded by summarizing the main points enumerated above with the intent of setting the stage for the Final Agenda Discussion. As to important omissions, the Chairman pointed out that the problems of polymer processing, including fracture and characterization, had not been treated in a coordinated way during any of sessions, and suggested that the participants consider this subject for inclusion and discussion in the later concluding session. The chairman pointed out that he felt there was considerable help which could be obtained from continuum mechanics in this important engineering field, but that it likewise contained a large amount of polymer chemistry. The obvious conclusion, therefore, was that an Interaction Matrix approach to this interdisciplinary effort might be rewarding.

REFERENCES

1. Williams, M. L., "Remarks on the Mathematical Criterion for Fracture", Proceedings of Shell Symposium, California Institute of Technology, June, 1972. (See also UTEC DO 72-037, University of Utah, July, 1972).
2. Menges, G., paper included in this Materials Science Colloquia Proceedings.
3. Griffith, A. A., *Proceedings First International Congress on Applied Mechanics,* Delft (1924), p 55.
4. Schapery, R. A., "Irreversible Thermodynamics and Variational Principles with Applications to Viscoelasticity", Dissertation, California Institute of Technology, 1962.
5. Williams, M. L., *Intern. J. Fracture Mech.,* **1**, 292 (1965).
6. Cherepanov, G. P., *J. Structures of Solids,* **PPM 31** 3, 476-488 (1967).
7. Rice, J. R., *J. Appl. Mech.,* **35**, 379 (1968).
8. Eshelby, J. D., *Solid State Physics,* **3**, 79 (1956). See also *Inelastic Behavior of Solids,* M. F. Kanninen et al. (Eds.), McGraw-Hill, New York (1970), p 77.
9. Barenblatt, G. I., paper included in this Materials Science Colloquia Proceedings.
10. Kelley, F. N., and Williams, M. L., *Rubber Chem. Technol.,* **42**, 3 (1969). See also same authors, "The Relation Between Engineering Stress Analysis and Molecular Parameters in Polymeric Materials", *Proceedings of Fifth International Congress of the Society of Rheology,* Vol. 3, Kyoto, Japan, October 5-19, 1968.
11. Thomas, A. G., paper included in this Materials Science Colloquia Proceedings.
12. Hull, D., ibid.
13. Knauss, W., ibid.
14. Kelley, F. N., and Williams, M. L., in *Polymer Networks: Structural and Mechanical Properties,* A. J. Chompff (Ed.), Plenum Press (1971), pp 193-218.
15. Bennett, S. J., Anderson, G. P., and Williams, M. L., *J. Appl. Polymer Sci.,* **14**, 735 (1970).
16. Rosenfield, A. R., and Kanninen, M. F., *J. Macromol. Sci.,* to be published.
17. Rybicki, E., and Kanninen, M. F., paper included in this Materials Science Colloquia Proceedings.
18. Radcliffe, S. V., ibid.
19. Williams, J. G., ibid.
20. T. L. Smith, *J. Polymer Sci.,* Part C, **32** (*Polymer Symposia*), 269 (1971).
21. Brinson, H. F., *Exp. Mech.,* **10**, 72 (1970).
22. Bevis, M., and Hull, D., *J. Mater. Sci.,* **5**, 938 (1970).
23. Rabinowitz, S., and Beardmore, P., *Critical Reviews in Macromol. Sci.,* **1**, 1 (1972).
24. Andrews, E. H., and Bevan, L., *Polymer,* **13**, 337 (1972). See also the paper included in this Materials Science Colloquia Proceedings.
25. Vincent, P. I., and Raha, S., *Polymer,* **13**, 283-287 (June, 1972).
26. Parrish, M., and Brown, N., *Nature Physical Sciences,* **237** (77), 122-123 (June 19, 1972).
27. Williams, M. L., *J. Adhesion,* **4**, pp 307-332 (1972).
28. Kalthoff, J. F., "On the Propagation Direction of Bifurcated Cracks", International Conference on Dynamic Crack Propagation, Lehigh University, 1972, to be published.
29. Kobayashi, A. S., paper included in this Materials Science Colloquia Proceedings.

30. Anthony, S. R., Chubb, J. P., and Congleton, J., *Phil. Mag.*, **22,** 1201 (1970).
31. Anthony, S. R., and Congleton, J., *Metal Sci. J.*, **2,** 158 (1968).
32. Dvorak, G. J., *Eng. Fracture Mech.*, **3,** 351 (1971).
33. Dvorak, G. J., "Statistical Criteria for Microcrack Propagation in BCC Polycrystals", International Conference on Dynamic Crack Propagation, Lehigh University, 1972, to be published.
34. Hoagland, R. G., Rosenfield, A. R., Hahn, G. T., *Met. Trans.*, **3,** 123 (1972).

Part Six

CRITICAL ISSUES

CONCLUDING AGENDA DISCUSSION: CRITICAL ISSUES

J. C. Halpin*

Air Force Materials Laboratory
Wright-Patterson Air Force Base

J. A. Hassell**

Battelle-Columbus

1 INTRODUCTION

Historically, a colloquium of this nature documents a turning point in the science of polymeric solids. This was illustrated in the introductory lectures of H. F. Mark and T. Alfrey by delineating the traditional interests of the material scientists in achieving an understanding of the concept of a micromolecule (and the conditions of its formation) so as to enable our society to produce new and unique synthetic materials. This goal was projected and achieved within the past 60 years. Achievement in this area has been truly breathtaking in both its scope and rapidity. In

*Chairman.
**Secretary.

fact, as Dr. Hansen has pointed out in the banquet address, the polymer chemist has probably explored all the major domains of chemical bonding systems (organic and inorganic) suitable for polymer forming processes. While it may be correctly argued that the chemist continues to possess the capacity to produce new compounds, the realized or anticipated properties of these compounds do not show high promise of social or commercial utility solely on the basis of their general availability. In this regard, a substance (be it polymeric, ceramic, or metallic) becomes a useful material only when it is employed to make useful things.

Our problem today is not how to make a better system but how to better exploit the mechanical potential that the macromolecule offers via a mechanical-chemical manipulation of an existing polymeric system. The polymer scientist is attempting to formulate a technical attitude toward the upgrading of physical properties paralleling the advances made in metallurgy during the past three decades. It is for this reason that many members of this colloquium (Anthony, Gleiter, Hull, Li, Pechhold, Radcliffe, and Rosenfield) represent current competence in solid (defect) state physics and the metallurgical arts.

If the value of a chemical substance is gaged not by its intrinsic quality but by what it can be turned into, the value of applied mechanics, a companion theme of this colloquium, must also be judged in this perspective. The term "applied mechanics" suggests the study of Newton's Laws as related to the real world. A majority of the simpler problems in the mechanics areas have been solved, and we have now begun to consider more difficult questions of the type outlined by R. S. Rivlin, I. Barenblatt, and W. G. Knauss. A major effort involving classical solution methods would aid in solving urgent technical problems in engineering, structures, and material sciences. While we do possess useful first-order models for different classes of material behavior, recognizing that materials like people are different, as emphasized by Menges, our understanding of engineering structures and their relation to the characteristics of a material system (geometric scaling and material substitution) are not at all well developed. The material scientists believe in "molecular engineering" – the modification of a material at the primary level to adjust the mechanical properties so as to permit an engineering application. But it is a sad fact that the engineer, when questioned as to what it is that he wants, cannot identify with any degree of assurance what properties are required. This situation results because *each new material capability when matched against a usage requirement requires a structure of unique geometric structural form.* It is for this reason that most attempts at direct materials substitution in an existing structure have failed, the problem being that structures are optimized to match the capabilities of the then existing material technology and are formalized in design rules, safety factors, and codes. These design rules, safety factors and codes *constitute the impediment* to the introduction of new materials technology into our society.

In all fairness to the mechanics and design community, it must be acknowledged that the concept of "molecular engineering" is not as general a tool as we would wish. Materials are chemical systems with finite capabilities. *We need a better appreciation,* particularly in the polymer area, *of where we are relative to the best situation that can be achieved.*

Suppose for a moment that we possessed the capacity to define and produce a given part with given properties at specified points and in specified directions. In principal, the materials scientist and the solid mechanist would select both the material and the fabrication process to achieve the clearly stated objective at a price we can afford to pay. But the *basic knowledge to achieve this objective is so far from complete that the forming processes are mainly art forms at which some do well and some do poorly* when asked to produce new shapes from new materials. An appropriate process can produce desired surface and subsurface properties through microstructural orientation. However, an inappropriate process can and does produce severe flaws, inappropriate residual orientation patterns, etc., which results in a part with inadequate integrity prior to the application of environment and load. There is no question but that the *lack of product reliability has undermined the public's support of new high-technology enterprises.* Clearly then, the desire of our society to harness the technological capabilities that we, the technical community, have created constitutes a volatile critical issue in the changing of research philosophies.

During the course of this meeting, a strong theme developed around the conceptual model of the dislocation and its applicability to polymeric systems. In fact, strong, sometimes heated, exchanges occurred between the attendees regarding the viability of metallurgical concepts in polymer science. Unfortunately, these discussions were often obtuse in nature. In the discussion that follows, an attempt has been made to summarize objectively the relative positions of both groups in the section denoted as The Polymeric Model. In the succeeding section the results of the two models are used to define the limiting properties of polymeric solids. The issue here is whether strength and stiffness properties are dominated by long-range molecular orientation or by the impediment of short-range dislocation motion. Next, a summary of T. Alfrey's discussions is presented, emphasizing that fabrication paths to the projections of the models are achievable. In addition fabrication technology in polymers generally implies a change in the mechanical properties (stiffness, strength, impact resistance, etc.) as well as shape changes classically associated with forming procedures.

2 THE POLYMER MODEL

In the crystalline regions or phases of polymeric solids, the existence of dislocations (a line defect) and/or disclinations (a point defect)* is obvious unless one makes the difficult assumption that the region is perfect. There is, in fact, a body of experimental evidence (discussed by Li, Gleiter, Pechhold, and E. W. Fischer) supporting the thesis that imperfections do exist in polymeric crystals. A more radical position is the corollary statement that an amorphous solid or fluid (liquid or gaseous) behaves similarly to a crystalline solid possessing a large, statistically distributed number of defects which interact with one another (Gillman[1], Li[2], Pechhold[3]).

Dislocations and disclinations are classically defined as a deviation from the local order of a solid. For polymeric solids this implies the following:

Dislocation: slip processes are produced by a small or partial chain displacement. As a consequence, the motions of dislocations can provide a means of molecular translation but not configurational rearrangement.

Disclinations: local conformational variations in the polymer chain can produce a disturbance in the local structure at the molecular scale:

1.0 Chain twisting in one or two adjacent chains produces a twist disclination (the Reneker[4] and Kedzie[5] defects).

2.0 Chain kinking produces a wedge disclination. Kinking may be envisioned in a chain of, say, all gauche bond orientations, interrupted by a trans orientation.

3.0 Simultaneous kinking of many chains along a common plane produces kink band formation (common planes produce kink block structures).

Correspondingly, the polymer melt is envisioned[3] to be a random array of kink structures (predominantly gauche with some torsional defects) cooperatively arranged in the local area (Pechhold's Meander Model). Such a model implies that the thermodynamic restrictions on polymer configuration in the melt are different from those experienced in dilute solution at the Θ temperature.[6] The proponents of these defect-structure models suggest that they are more powerful tools because they

*Classically, in an elastic continuum, the disclination is a line defect (Nabarro[3a]). As shown below it is also a point defect which can exist in a polymer chain.

are two dimensional in character, whereas a random-coil model is essentially one dimensional in character.

Classically, polymer theory envisions the polymer-configuration problem as a statistical mechanics operation[6], in the Boltzmann sense, with its antecedence in the kinetic theory of gases. Configurational statistics are augmented with a cell or free-volume theory of a liquid to describe the viscoelastic[7-8] and rubberlike qualities. Ferry, Landel, and Halpin showed that in this theory, the polymer molecule exhibits the same configurational characteristics when studied in isolation (the dilute solution) or in the condensed liquid (fluid, viscous, or glassy) state. Thermodynamic restrictions administered in cooperation with nucleation and growth theory are envisioned to define the idealized crystalline and/or semicrystalline state, wherein there is a high degree of mechanical contrast between the amorphous and crystalline phases. Eyring's absolute rate theory is artfully exploited in both approaches (defect and classical). Similarly, both approaches rationalize energy-loss mechanisms in terms of viscously dampened vibrating string or membrane analogies.

These two conflicting but parallel viewpoints are not really unique, but constitute a contemporary reformulation of a historical duality in our understanding of the liquid state. Does one treat the liquid as an imperfect solid or as a nonideal gas? While it is often stated that a liquid is intermediate in its properties between a crystal and a gas, in fact most values for any specific physical property of a material in the liquid state are approximately equal to that of either the crystal or the gas. Intermediateness is then a statistical implication that a liquid has some of the properties of both crystal and gas.

It is not surprising then to see why the kinetic theory of rubberlike elasticity and viscoelasticity has been historically successful. This theoretical base is a hybrid model wherein the configurational statistics contain attributes of the gaseous state and the free-volume flow models (activated, cooperative rearrangement processes) contain attributes of the crystalline state. Although the kinetic theory of macromolecular response describes a multiplicity of attributes for macromolecular systems, it is also incomplete.[3] The work of Robertson[9] on the plastic deformation of glassy polymers illustrates both the robust quality of the classical theory and the difficulties in developing specific estimates for the local molecular processes. If a liberal attitude is adopted by the polymer physicist, the *intramolecular* and *intermolecular* energy changes for some classes of deformation can be estimated or approximated *by an analogy* with dislocation theory. For example, when a polymer undergoes plastic flow, the molecular chain segments undergo a configuration change, i.e., from a coiled or kinked conformation to an extended conformation. This process may take place by kinking and torsional motions which would be modeled as a wedge disclination loop. After all, the motion of a molecular segment in a liquid is formally analogous to the motion of a vacancy or hole in a crystal lattice.

An adoption of an analogy, as a theoretical device, and its correlation with laboratory data do not mean that dislocations exist in amorphous bodies in the same sense that they exist in a crystalline solid. Nor does the empirical correlation of polymer properties with the empirical laws developed in metals add much weight to the defect solid case, as most of these relations do not possess the property of uniqueness to specific molecular events. The production of shear (Lüders) bands as discussed by Li for "amorphous" materials is not a direct proof of the physical existence of dislocation structures in a body undergoing nonhomogeneous deformation. Arguments at the continuum mechanics level are sufficient to show that shear-band formation is controlled by and dependent upon the local state of stress independent of a specific molecular argument. Band formation does require a commitment to long-range plastic deformation. The physical question is – is it possible to have plasticity without dislocations within a noncrystalline body?

In any event, during the next decade theoreticians will relentlessly explore this area of bulk response and seek to demonstrate its value. The contested area will be the description of polymeric glasses; the reward, a potential means to control the rate sensitivity of polymeric glasses. Many manufacturing steps are dependent upon and frustrated by the rate-dependent plastic response of the polymer under rapid deformation and drawing conditions. (Can we make a polymeric substance with the properties of lead? Such a material would be ideal for cold-forming processes.) The techniques and tools developed in solid-state physics can be expected to modify and complement the existing kinetic theory, but they do not possess the range and capacity to displace the kinetic theory of macromolecular response.

Paralleling the continued exploitation of the solid-state-physics techniques will be the adoption of the theoretical models developed in the mechanics of composite materials[10] over the past decade. The mechanics of multiphase media in material science will be greatly strengthened through the use of modern micromechanics. Material morphological form, at a level intermediate to either the continuum or the molecular level, exerts a strong influence on its mechanical properties. This level of organization is rather well characterized or developed in the theoretical sense, but the misunderstanding of the importance of the morphological form has led to some erroneously contrived models of structural response[11] in polymer mechanics. Even in the metallurgical area the substructure of the fabricated items is becoming of increasing concern. For example, the metallurgist has produced high-yield-strength metals with inadequate fracture toughness. With only a small loss in toughness, modification of phase substructure by thermal-mechanical processing techniques is producing technologically acceptable high strength – high toughness metals.

3 LIMITING PROPERTIES IN POLYMERIC SYSTEMS

The discussion regarding the character of a polymeric material leads naturally to the question: are we achieving the maximum response from the macromolecular systems? The issue may be summarized in Figs. 1 and 2, wherein we define a lower bound on macromolecular response as the isotropic random coil with a Young's modulus of the order of 0.3 to 0.6 x 10^6 psi. In orthodox polymer science, the upper bound is envisioned to be a colinear array of molecules, highly anisotropic in character with an axial stiffness of the order 30 to 40 x 10^6 psi (for a polymethylene chain) in the principal axis direction. Transverse Young's moduli and strength are quite low, typically of the order of the glassy properties, and shear moduli within plane and out of plane are also of the order of 10^6 psi. Theoretical tensile strength in the chain direction of polymethylene will be of the order of 1 to 4 x 10^6 psi at room temperature. Table I lists the limiting engineering stiffnesses for different chemical structures. Polymer structures exhibiting low chain stiffness are also helical in conformation (Table I) and deform by bond rotation instead of bond stretching and bending.

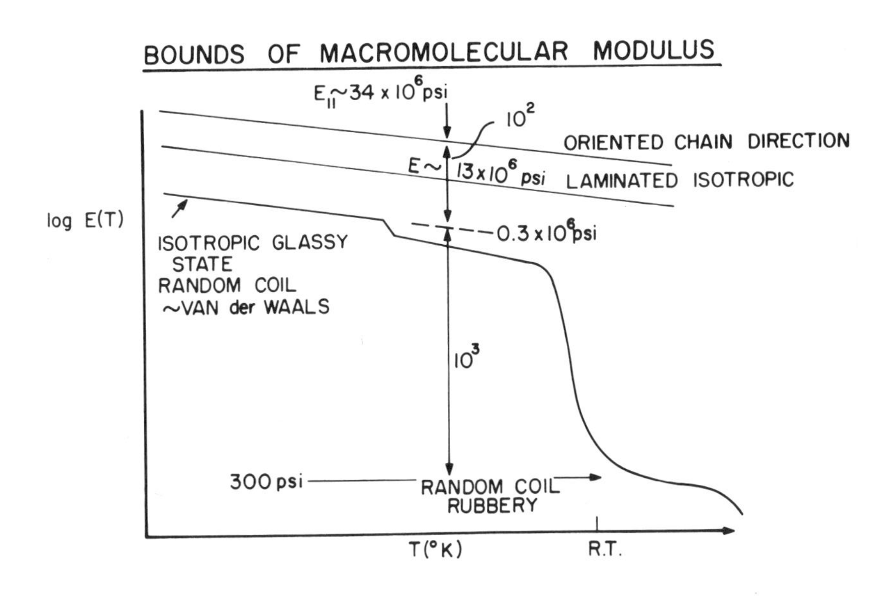

Fig. 1. Schematic of three limiting states of molecular conformation.

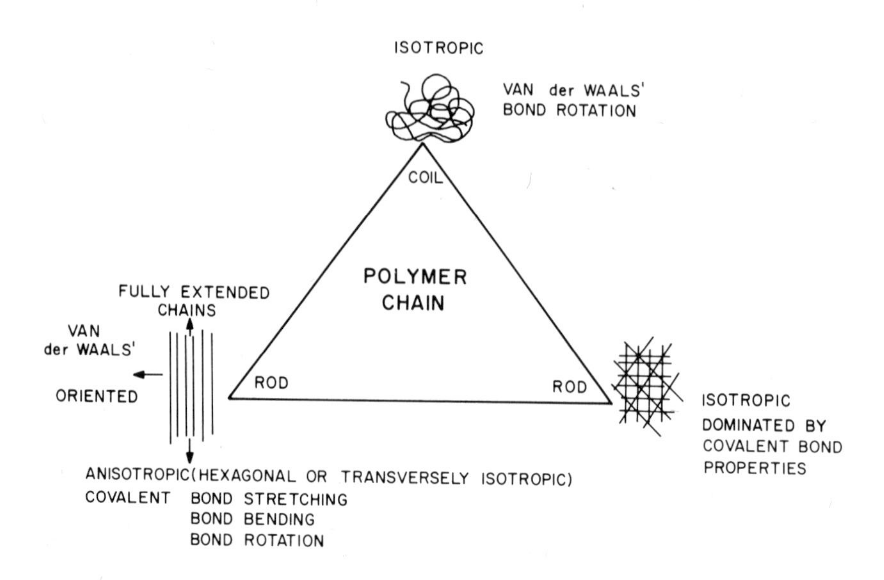

Fig. 2. Schematic of transition behavior of the three states of Fig. 1 for an amorphous solid.

Table I. Properties of Oriented Polymeric Chain Structures and Related Engineering Materials

Material	E_{11}, $\times 10^{-6}$ psi	E_{22}, $\times 10^{-6}$ psi	$\bar{E}_{iso}$, $\times 10^{-6}$ psi	Tensile Strength, ksi	Density, lb/in.^{3}
Steels and iron (cast and alloy)	–	–	28 - 30	13 - 300	0.25 - 0.29
Titanium	–	–	19	60 - 240	0.16
Poly(vinyl alcohol)	36.2	1.54	14.4	~ 100	0.05
Polyethylene	34	0.7	13.6	~ 100	0.05
Boron/Epoxy	30	2.7	11.89	83	0.07
Aluminum	–	–	10	22 - 90	0.10
Polytetrafluoroethylene	22.2	–	–	–	0.05
HTS-graphite/epoxy	21	1.7	10	62	0.05
Cellulose I	18.5	–	7.6	50	0.05
E-glass/epoxy	5.6	1.2	2.85	47	0.06
Polypropylene	6.0	0.42	2.5	–	0.05
Poly(ethylene oxide)	1.42	0.56	0.88	–	0.05

While extended-chain molecular structures are partially exploited in fibrous products, most structural usage of materials is in the bulk where uniformly extended molecular structures (synthetic polymers) have not been employed to date. For argument's sake, consider a material wherein the anisotropic material is layered and oriented at equal angular intervals of magnitude π/n where $n \geqslant 3$. Such a material is called "quasi-isotropic", as it will be isotropic in the plane of the layer[10-13] for all of the mechanical properties. Some quasi-isotropic properties are also listed in Table I for both anisotropic polymers and the advanced composites. It should be apparent that the fully oriented polymer chain structure can produce a nonmetallic material comparable to either a metal or an advanced composite material. For a density, ρ, of 0.05 pound per cubic inch, a polymer with a Young's modulus of 6×10^6 psi and a tensile strength of 60,000 psi is equal, on a specific weight basis, with high-performance steels.

These estimates are technically viable and illustrate the fact that *polymeric solids have not reached their theoretical maximum capacity in the isotropic bulk because they are highly coiled.* Vincent in the discussion session cited similar calculations that he had made wherein he observed a deviation of 165 from the theoretical estimates and isotropic (random molecular coil) glassy data. Both Vincent and Alfrey concurred that the deviation was too large for preexisting flaw corrections and must reflect the difference between the mechanics of an oriented chain versus a coiled molecular chain. Theoretically, the proper order of magnitude for the glassy moduli for a Gaussian coiled structure would be the apparent engineering stiffness, E_{22}, transverse to the oriented chain direction. The results shown in Table I and the preliminary calculations of de Boer[14] of Van der Waals' potential functions[15] suggest:

$$1.5 \times 10^6 > \bar{E}_{glassy} > 0.5 \times 10^6 \text{ psi}$$

at absolute zero. Most polymers exhibit glassy moduli at ambient conditions of the order of 0.1 to 0.5×10^6 psi. If one assumes a shear flow model, similar to Frenkel[16], for the theoretical yield strength

$$\frac{E}{20} > \sigma_y > \frac{E}{13}$$

we are looking at values of

$$25 \text{ to } 45 \text{ ksi} > \sigma_y > 5 \text{ to } 7 \text{ ksi}$$

for the range of tensile yield stresses. Thus, the order of magnitude of the experimental yield properties is consistent with the theoretical order of magnitude estimates for molecular flow processes dominated by the Van der Waals potential field controlling the configuration response of the randomly coiled macromolecule. It is not uncommon in metallic solids to find

$$\sigma_y \sim \frac{\bar{E}}{10^2 - 10^4} ,$$

which was a motivating force behind the early developments in dislocation theory by Orowan, Taylor, and Polanyi. *Generally, polymeric solids, both amorphous and semicrystalline, are within a factor of 2 of the Frenkel expectation.* This apparent correlation cannot be construed as evidence that defects do not exist in the polymeric bulk, because defective crystal structures have been experimentally demonstrated in the crystalline phases of macromolecular solids. A correlation with the Frenkel expectation may be only fortuitous or it may imply that the dynamics of slip processes play only a secondary role (the equivalent to the Peierls-Nabarro function stress is very high) in many polymeric bodies.

The projected upper bound in properties for polymeric materials is technically sound and achievable. Recently, two high-modulus fibers have been announced by Du Pont (PRD-45) and Monsanto (X-500). These materials are aromatic polyamides systems exhibiting significant engineering properties in the draw direction (Fig. 3). Employing the composite analogy outlined by Halpin and Kardos[12] for semicrystalline systems, a theoretical strength-modulus relationship was computed – the solid line in Fig. 3. Theoretical estimates of the assumed fibrillar crystallites were made by the techniques outlined in Reference 14. The fibrils were treated as slightly misaligned short fibers (finite length/diameter ratio) in a rubberlike matrix at two volumetric fractions of crystallinity content employing contemporary composite mechanics.[10-13] Notwithstanding the obvious comment that the striking correlation may again be fortuitous, it should be readily apparent that the polymer-science area has produced and commercialized the equivalent to the defect-free "whiskers" of the metallic area. Our problem now is how to exploit this latent potential, for if these polymer structures could be converted to the isotropic form illustrated in Figs. 1 and 2, we would possess polymeric solids comparable to any existing structural material.

In principle, the majority of the existing commercialized polymeric systems could be upgraded by an order of magnitude if an extended-chain quasi-isotropic solid could be achieved. Polyethylene provides a typical illustration of the difficulty in achieving this objective. There is no known fabrication technique for polyethylene to completely eliminate the chain folding crystal in favor of the fibrillar crystal. Consequently, current research efforts are typically producing (Table II) stiffness properties in

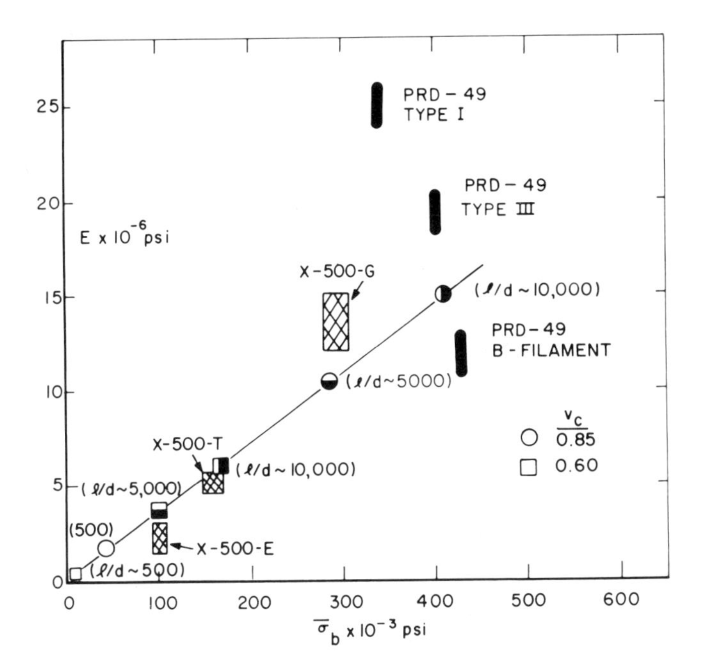

Fig. 3. Comparison of the theoretical projections with the experimental properties for highly oriented semicrystalline material from planar polymer chains (ℓ/d is the aspect ratio of the fibrillar crystals, and v_c is the volume fraction of crystalline content).

the 3 to 10 x 10^6-psi (orientation direction) range. While the stiffness in initial isotropic state is on the order of 0.1 to 0.3 x 10^6 psi, *the goal* of 30 to 35 x 10^6 psi *is still an order of magnitude away.* A strong research effort is expected in better defining the goals and acceptable paths to these goals during the next decade. Controlled orientation of a smaller order of magnitude will be used to upgrade existing thermoplastic molding materials. *The basic researcher must address the issue of what happens to transition temperatures and transition processes observed in isotropic systems (Fig. 2) as the material approaches the idealized state of total orientation.*

In conjunction with the previous discussion, Mark cited the existence of linear polyethylene samples that illustrate the high mechanical properties. Specifically, Dr. Wang (see Table II) of Bell Telephone Laboratories put droplets of molten polyethylene on the cold rolls of a rapidly moving calender. The films formed exhibited a modulus of 6 million psi and strengths of 250,000 to 300,000 psi. Analysis of the structure showed long thin crystalline units, which were highly oriented and in which all the folds had straightened out. He annealed the sample for 21 days at 121 C

Table II. Mechanical Forming[17]

Worker	Year	Method	$\bar{E}_{exp}$, x 10^{-6} psi	$\bar{\sigma}_6$, x 10^{-3} psi
		Material – Polyethylene ($\bar{E}_{Theo}$ ~ 35 x 10^6 psi)		
Wang (Bell Labs)	1971-72	Shearing the bulk melt	6	250 - 300
Pennings, et al.	1971-72	Shearing dilute solution	3.8	140
Takayanagi, et al.	1971	Hydrostatic extrusion	4.4	–
Weeks and Porter	1972	Hydrostatic extrusion	10	–
		Material: Polypropylene (E_{Theo} ~ 6 x 10^6 psi)		
Williams	1972	Hydrostatic extrusion	2.4	–

without observing any opaque appearance due to rearrangement into spherulitic structures. This shows that if one goes to extremes in processing, one can materialize the structures which approach the theoretical limits.

Andrews reconfirmed the influence of processing by citing his work where he investigated the morphology of natural rubber crystallized under different conditions. He changed the morphology without significantly changing the degree of crystallinity. He found that the lamella structure, produced by straining the material as it crystallizes, orients at right angles to the strained direction. The summary of the morphological effects of strain are: at zero strain, spherulitic structures are produced; at 50 percent, sheaths are produced; at 100 percent and up, row nucleated structures are produced in the strain direction; at 200 percent, more densely nucleated structures are produced; and finally, at 300 percent, filamented structures running in the strain direction (which he called gamma filaments) are produced which are essentially similar to synthetic fibers. As the morphology is changed, related Andrews, large changes are produced in the mechanical properties without necessarily changing the amount of crystallinity, but merely its arrangement in space.

Eirich pointed out that the review by Halpin emphasized impressively the widely applicable, but not widely understood, principle that composite, or multiphase, structure contributes greatly to any material's strength, crack resistance, and, therefore, toughness. He emphasized, apart from artificial composites, the importance of microcrystallinity. In this context I would like to mention other origins of multi(micro-)phase formation based on phase separation of incompatible components. Among these are the tactoid formations of separating elongated (chain) molecules and the island formations within ionomers and block, or segmented, polymers. Applying suitably imposed strains during cooling or gelling, one can enforce isotropic or anisotropic precipitate particle distributions. This may be much simpler and more effective than conventional processing methods of drawing, rolling, molding, etc.

D. Langbein noted that a parallel can be drawn between paramagnetism and polymer morphological changes. For example, in polymers, the strain is analogous to the magnetic field, and the orientation is analogous to magnetism. Langbein feels that he can compare the two theories point by point and use the type of mathematical modeling developed in paramagnetism to describe the morphological changes in the polymer field.

"I think it to be extremely important to emphasize the relations between different disciplines. We just saw that we may learn a lot from a comparison of polymers and lamellar systems. Another possibility of attack, which up to now has obtained little or even no attention, is ferromagnetism. There exists a phenomenological theory based upon susceptibility and a microscopic theory based on the temperature dependent orientation of spins, the Ising model. Let me point out a number of parallels and analogies: the state of lowest energy is the completely oriented ferromagnetic state, the state of highest energy is the alternating anti-ferromagnetic state. With increasing temperature there is a gradual transition to an intermediate state. Let me list a number of parallels and analogies between the theories of polymers and ferromagnetics." (This listing follows.)

Polymers	Ferromagnetics
Stress	Magnetic field
Deformation	Magnetization
Oriented regions	Whip regions
Phase boundaries	Phase boundaries
Linear / Nonlinear response	Linear / Nonlinear response
Frequency dependent compliance / modulus	Frequency-dependent susceptibility
Salination	Salination
Hysteresis	Hysteresis
Yield	Yield
Dissipation	Dissipation
Relaxation (with increasing temperature)	Relaxation (with increasing temperature)
Orientation models	Ising model
Oriented state	Ferromagnetic state
Meander state	Antiferromagnetic state
Transition temperature	Curie temperature
Viscous region	Paramagnetism

Alfrey then delineated the effects of processing on orientation and the changes that can occur in physical properties owing to the processing conditions. These comments formed the basis for the next section of the Discussion.

4 FABRICATION TECHNOLOGY

The mechanical performance of a polymeric article or part depends upon two general aspects of structure:

- The molecular structure governed by the conditions of chemical synthesis
- The spatial organization of polymer molecules, and phases, in the fabricated item, governed by the conditions of physical manipulation which accompanies or follows the chemical synthesis.

The sequel to Fig. 1 regarding the formation of controlled oriented and/or layered states can be illustrated by the schematic[18] of Fig. 4. Process [1] represents the usual melting or crystallization of polymers

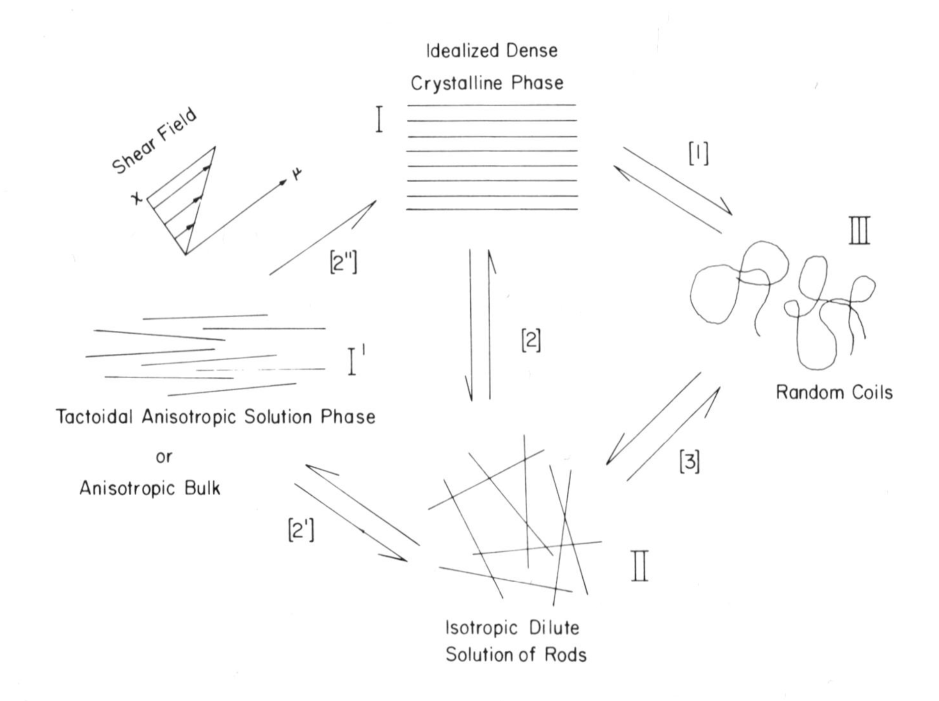

Fig. 4. Schematic of molecular phenomena required to achieve a controlled fabrication technology.

with a configurational change occurring during the phase transformation. A diluent may or may not be present in the amorphous state III, while state I represents the pure crystalline phase. The sequel to regular melting of a bulk polymer is the formation of a pure ordered phase, I′, from a dilute solution of anisotropic molecules. For example, a "helix-coil" transition does occur in dilute solutions of several macromolecules of biological interest and a few synthetic systems of technological interest in appropriate solvent media. This transition, identified as [3], produces an isotropic dilute solution of asymmetric rods of helices, II. Such a collection of molecules, wherein the individual species are uncorrelated and randomly arranged relative to one another, can exist as independent entities in a sufficiently dilute solution. However, rodlike molecules of high axial ratio cannot be arranged at random at high densities because of space requirements. Accordingly, a dilute tactoidal phase I′ is formed from the dilute isotropic phase by process [2′] through a small increase in the polymer concentration. If the material in state I′ is coagulated in a shear field [2″], a solid is achieved wherein the bulk state exhibits anisotropic properties of a great magnitude. These properties may be further optimized by traditional thermo-mechanical processing steps. It is important to note that states I, II and III are analogous to the configurational states presented in Fig. 1 in the definition of the mechanical bounds on polymer properties. In fact, the process route just outlined has recently been employed to achieve the outstanding properties of the polyamides of Fig. 3. These processing concepts will become topics of increasing importance in that states of high molecular orientations, employing purely mechanical stretching are not generally practical. Large mechanical energy input usually produces molecular fracture before ultimate elongation and orientation are achieved.

Comparable results on mechanical forming techniques are outlined in Table II. Solid-phase hydrostatic extrusion techniques[17] produce oriented rods of roughly 1/4 to 1/3 of the theoretical limit for tensile modulus. Hydrostatic extrusion is a process whereby a solid cylindrical billet of material is forced through a die by hydrostatic pressure transmitted to it via a surrounding liquid. This processing step is analogous to path [1] of Fig. 4. In metals technology, hydrostatic extrusion is employed to produce large cross-sectional items of controlled, complicated shapes. In polymer technology, such processes would both shape and produce physical property improvement.

The conceptual weakness of an extrusion process of the type outlined herein is that the formed object is dominated by a high state of uniaxial orientation. High axial orientation means low transverse properties as illustrated in Fig. 1 and Table I. Cleerman[19-22], working with Alfrey, has employed a cylindrical mold in which the inner and outer surfaces are counterrotated as the polymer melt is injected. This counterrotation produces helical flow lines which vary in pitch and eventually change sign with depth through the annulus thickness. This effect produces a layered material with a change in orientation through the thickness of the object. The product resembles plywood or a laminated composite in its substructure with enhanced biaxial strength, craze resistance, impact resistance, etc.; and is the first illustration of the successful production of the class of materials shown in the right-hand corner of Fig. 1. Cleerman applied these techniques to conventional thermoplastics systems but could not obtain a high state of local molecular orientation. It is interesting to speculate that Cleerman and Alfrey's techniques, when applied to polymeric solids in state II or I′ of Fig. 4, might produce the ultimate in polymeric solids.

In this example, multiple layers of the same material, possessing different orientations, were produced to obtain a mechanical advantage. Schrenk and Alfrey[23] have also produced multilayer (several hundred layers) laminates of different polymers. The key to such multilayer structures is control of the fluid kinematics of multiphased melts. While a two-, three-, or maybe a four-layer structure can be effected by "dies-within-a-die" technique, a large number of layers requires novel means. In the technique explored by Schrenk and Alfrey, different phases[23] were continuously introduced into adjacent sectors of an annular channel die. If one or both cylindrical surfaces are rotated, each sector is transformed into a continuous spiral path, Fig. 5, producing alternating layers[23] of controlled thickness. Controlled layer thickness of 600 A have been prepared in this manner. Surface-tension considerations limit the minimum layer thickness at 100 A; below this thickness, the layers break up into droplets. Such materials may be subsequently biaxially oriented and cold drawn to produce a shaped product of varying orientation, modulus, and fracture toughness (the alternating phases may be glassy and rubbery), permeability, and optical properties through the thickness of the as formed product. Coextrusion of multiphase melts without the rotation will produce extended fiber or rodlike phase geometry of one phase in a continuous second phase. If a glass and an elastomer is used, this material would, in fact, be comparable to a unidirectional composite in which the longitudinal response is dominated by the glassy fibers and the transverse response is dominated by the elastomeric phase. Such a material would be a polymer analogy to eutectic composites in metallurgy.

Beneficial orientation can be introduced into a billet of a bulk polymer to facilitate both a deep-drawing process and the types of orientation patterns deemed desirable for the finished product. For example, the *Cleerman-Alfrey techniques may be used to put in one set of orientations in a thermosetting system, which after a subsequent forming step (say parison blowing) will produce a favorable mechanical state unrealizable by*

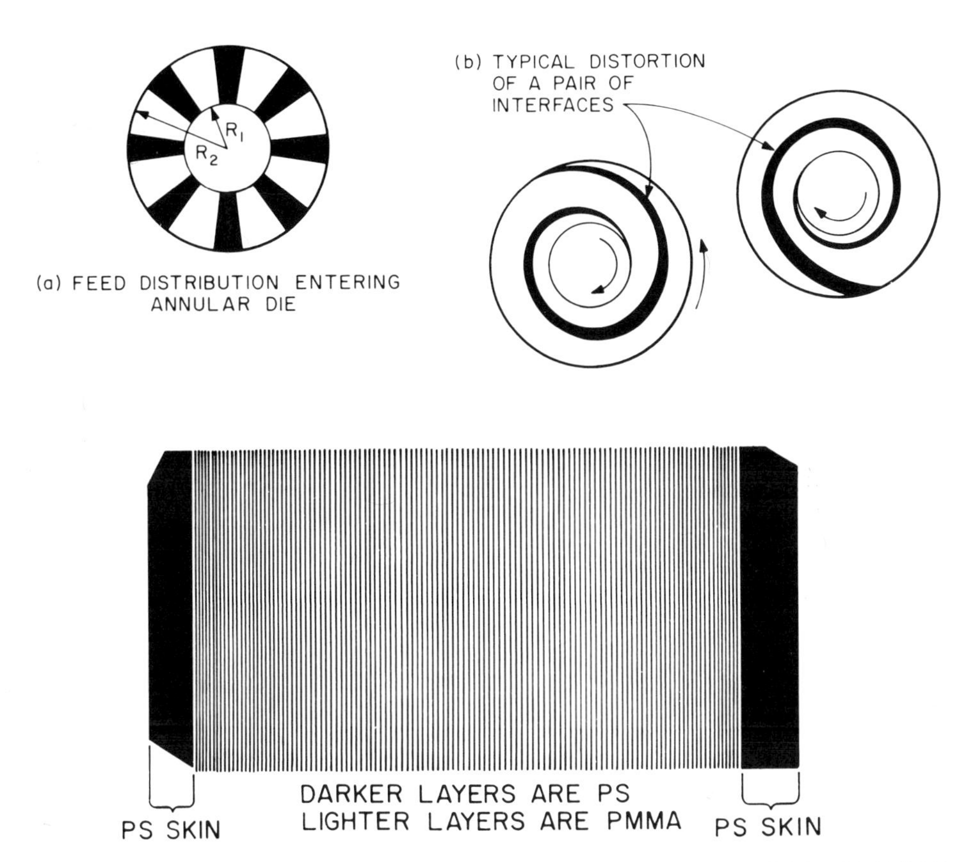

Fig. 5. Schematic of a multilayered melt extrusion.

the single application of processing step 2. As a second example, the results of Li et al.[24] show that the draw ratio in a deep-drawing fabrication step is limited by ratio of the drawing stress to the tensile strength or yield stress. To form a sheet of brittle glassy material into a deep seamless article the material is cross rolled, producing a biaxial orientation. This orientation decreases the apparent compressive yield stress in the out-of-plane direction while increasing the in-plane tensile strength, thus facilitating[24] the forming process. Both cold and hot forging techniques[24] have been explored, in a preliminary sense. Fabrication processes of this type are explicable with analytical models from metal-forging experiences, but they do not offer the potential for the precise control of either multiaxial orientations or phase geometries.

However, *the manufacturing techniques of "cold working" via deep drawing, hydrostatic extrusion, and forging are of serious interest to polymer technology, as virtually all traditional processes employed*[23] to fabricate articles from thermoplastics *require* the *polymer to be molten* and, subsequently, cooled. Cold working (means deforming a solid below either its melting and/or glass transition point) provides the ability to fabricate thick parts without the time limitations imposed by slow heat transfer as well as the forming of very high molecular weight materials, whose extremely high melt viscosity makes them difficult to process by ordinary melt techniques. Both melt and cold-working techniques are dominated by the physics of path 1 of Fig. 4. "Non-meltable" polymeric systems (excluding prepolymer and vulcanization technology) must be manipulated by the technology paths [3], [2], [2′], and [2″] outlined in Fig. 4. "Nonmeltable" means a system which undergoes thermal decomposition before noticeable flow is possible (cellulose is a typical example, as are many of the new thermal resistant polymers).

5 FRACTURE CONTROL, RELIABILITY, AND CONSUMERISM

The discussion within this section is limited to load-bearing applications of polymeric solids, which is not the largest use of polymers, but is the important usage in the motivation for fracture analysis. It is for this reason that the general problem in engineering structures has been summarized. This summary is followed by the open-discussion comments.

Current engineering structural systems are plagued with problems of overcost, behind-schedule development with unacceptable in-service structural maintenance, repair costs, and downtimes, and in some cases with marginal structural safety. As a result, the availability, that is the time an item is not undergoing maintenance, and cost of ownership (life-cycle costs) are now being placed on an equal footing with the cost of acquisition. The large number of product-liability suits arising in the courts against manufacturers of almost every type of consumer product bear witness to similar concerns in the general world economy.

Plaintiffs are accusing the material suppliers and the engineering design and manufacturing community of negligence, fraud, improper design, inadequate materials selection, implied warranty, etc. *Legal implications of these suits are that every person's duty is to exercise care to guard against any injury or major economic loss which may follow as a reasonable, probable, or foreseeable consequence of the use of his product or technology*. In this context it is important to recognize that a philosophy of design and prototype evaluation based on the prevention of failure or a quantified maintenance and repair program is more sound and workable than the stereotyped application of empiricisms, codes, specifications, and factors of safety now commonly used. *Irrational codes and safety factors not only constitute an impediment to product safety but also seriously frustrate attempts to introduce new materials into society.* It is for these reasons that the mechanics of fracture and reliability analysis are topics currently receiving serious attention in both the basic research laboratory and the engineering design firms.

The *products* that we use *generally contain preexisting flaws or defects* which are either *inherent in the materials or introduced during* a *fabrication* process. Relatively large flaws may be detected by quality control and inspection procedures which are largely deficient in nonmetallic materials and corrected by repair or maintenance procedures. Unfortunately, *undetectable or undetected* small surface *cracks* or *embedded flaws do grow to critical sizes.* In order to predict such failures, one must know the conditions under which subcritical flaw growth can occur, as well as either the actual initial flaw size or maximum possible initial flaw size in the structure when it is placed into service, plus the critical flaw size (Fig. 6). In order to obtain confidence that the structure will not fail in service, it is necessary to show that the largest possible initial flaw in the structure cannot grow to critical size during the required life span or that flaws may be prevented from attaining a critical size by detection and repair procedures.

Fracture-based design demands that the acceptable static loads be commensurate with the preexisting flaw size or sizes, the expected use and misuse of the item, and the expectation (growth law) that while the flaw grows in size during the lifetime of the item, the critical event of local fracture shall be prevented or controlled in some manner. *In effect, kinetic based fracture mechanics is a life model* in which static conditions are no longer divorced from lifetime safety and maintenance considerations.

Unfortunately, this model is constrained by an additional set of engineering uncertainties:

1. Uncertainties in the design and manufacturing process (for example, distribution of flaw sizes)
2. Uncertainties of the stresses and environment which a component must withstand in operation
3. Uncertainties in the description of a material system.

For example, if we accept the case illustrated in Fig. 6, the projected life of an object is the precision by which we know the distribution of flaw sizes and our ability to truncate that distribution at a defined maximum size (all other variables assumed to be known). With this example as motivation, let us define the term reliability as follows: the *reliability* of a component is its conditional probability that the item shall not fail to perform its function within specified performance limits at a given age for the time intended and under the operating stress and environmental conditions encountered. The reliability of an item is determined by a fracture-mechanics data base, which is generally of a probabilistic and time-dependent character, and must be a design parameter.

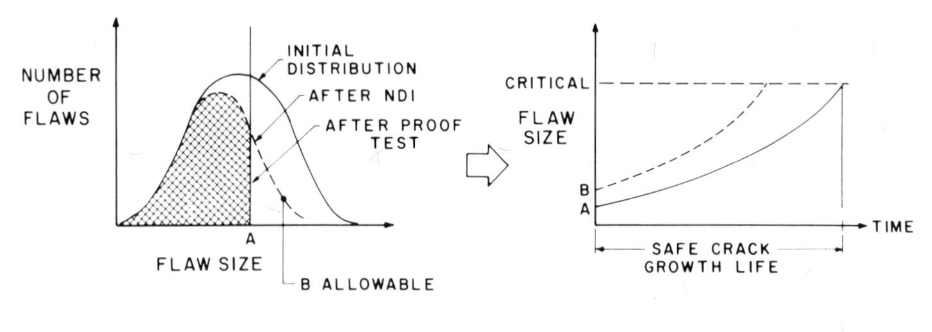

Fig. 6. Schematic of an engineering use of fracture mechanics.

During the next decade, serious effort will be expended in the development of comprehensive models of materials behavior[26] which complements the life-fracture mechanics-reliability models alluded to herein. Such models are necessary if advanced materials technologies are to be used to their fullest advantage in future structural systems. We fully expect that the polymer mechanics community will play a strong leading role in this area. The elastomer fracture problem constitutes the simplest fracture model in that it constitutes the simplest limiting form of a life model – time-dependent brittle fracture. In fact, the kinetic models of fracture developed by the polymer mechanics community are ahead of the general practice in the metallics area.

A topic of particular importance will be the development of an accelerated testing theory and technology, life estimation, within the framework of kinetic fracture theory. This technology will employ the now conventional flaw development and instability conditions augmented with the statistical attributes of the fracture properties to define the joint interaction between physical flaw development and the chemical changes that a body undergoes during its lifetime. In this model development, linear and factorized nonlinear viscoelastic theory, modified by reduced variables, will constitute a powerful tool. A critical issue, in the format of this problem, is whether the rate effects observed during idealized one-parameter tests, i.e., $t/a_{Temp.}$, $t/a_{Humidity}$, $t/a_{Chemical}$, can be summed as a product

$$\frac{t}{a_T \cdot a_H \cdot a_C}$$

and can be integrated through a time-varying environmental history;

$$t' = \int_0^t \frac{dt}{a_T\,[T(x,t)] \cdot a_H\,[H(x,t)] \cdot a_C\,[C(x,t)]} \quad .$$

Complementing this approach shall be the exploitation of physicochemical sensors of the type reviewed herein by Becht and Kaush. These sensors will identify the chemical paths leading to specific environmentally induced structural breakdowns, as well as provide diagnostic tools for the definition of suitable countermeasures.

Control of fracture toughness and impact resistance will be explored through a combination of controlled supermolecular morphological structure, orientation, and multiphased geometry via mechanical fabrication methods.

6 DISCUSSION

The general discussion was initiated with a brief review by Kobayaski of three problems. First, the onset of rapid fracture; second, the fatigue problem of predicting failure under fatigue conditions; and third, once the crack has started, how is it to be stopped? He then commented on the difficulty of formal calculations and that the presentation of the formal theory oftentimes involved extensive numerical calculations as well as an expensive data base. The discussions proceeded as follows:

Thomas related that in practice with rubber and many other polymer cases, we are not concerned with constant stress creep fracture but with repeated or cyclic stresses, and therefore with fatigue failure. Especially is this true in the case of tires, where cracks all grow in the sidewall, resulting in changes in the physical properties of the tire and consequently leading to fracture. His main emphasis was that in the polymer field in general, there is not much attention paid to cyclic loading. He pointed out also that this was of great importance, and in general, one cannot predict the failure under cyclic loading from that under a time-dependent dead loading.

Rosenfield, in reply to the comment that extensive computer time is necessary to develop useful fracture information related that the fracture behavior can be measured in the complex geometry, using just simple specimen compliance. He pointed out that the K_{Ic} fracture parameter may not be a very useful engineering property in the future, either for metals or in the polymer field, which will be impinging on the metal industry in mass production, the reason being that there are cheaper materials that can replace the highly brittle materials in which K_{Ic} can be measured. He suggested that some of the problems in dealing with polymers is to reduce the brittle transition in the polymer the same way as has been done in metal. Hassell commented that specified polymers usually have a brittle ductile transition which is defined by the structure of the polymer itself, and if you want to lower the brittle-ductile transition, you either blend it, copolymerize, or go to a different polymer.

The characterization of materials using fracture mechanics, related Andrews, could save a lot of money as opposed to using fracture mechanics for engineering design. He pointed out that there are only a few data available on fatigue and that large amounts of money will be required to obtain sufficient data. He feels that the approach he has taken along with Thomas on polyethylene and rubber, respectively, can save a lot of time and effort in using fracture mechanics to develop engineering design information on fatigue. He pointed out that there are two approaches to fracture. The metallurgist's approach using linear elasticity and the Rivlin-Thomas approach where the problem is not treated in terms of the near field, but in terms of the far field. He sees consequently a need of a theory that will bridge both the near- and the far-field elastic and plastic behavior.

M. Williams stated that there was no question that fracture mechanics does work, and that the fatigue problem is also an important problem that needs to be attacked. Fatigue can be estimated and predicted for plastic, elastic, viscoplastic, and viscoelastic materials. However, the problem needs to be attacked by considering the thermo-mechanical couplings, and how much temperature is introduced into the stresses and the influence this has on the ultimate behavior. He pointed out that one cannot get anywhere in the analysis of fracture mechanics until nondestructive testing of polymer materials has been perfected.

Kanninen underscored the need to better understand the material property called fracture toughness and how to use it in the polymer area. He related that progress is being made along this line in the metals area and suggests that this information could be of help also in the polymer area.

The two-dimensional fracture-mechanics system is not a completely solved field, stated Hulbert. It can give only far-field and not near-field information, and three-dimensional fracture mechanics needs much more than just computer development before it can be solved.

Menges emphasized the need to know more about shock resistance, corrosion resistance, stress-corrosion cracking, and the influence of orientation on crazes and cracks in practical parts and the need for a simple concept to calculate the fracture for practical purposes.

From the physical viewpoint, related Knauss, the understanding of time-dependent fracture in highly deformed materials will be indispensable in describing the constitutive behavior of these materials. This is one area where fracture mechanics can play a significant role.

The toughest of polymers is very much a question of the relaxation spectrum, related Retting. The task is to reduce the spectrum to shorter times and lower temperatures. Fracture mechanics is concerned with cracks in polymers. However, people are not interested in cracks in materials, but in materials which do not have cracks, and in ways of preventing initiation and propagation of cracks.

The deformation of polymers involves a competition of deformation and fracture because the fracture of polymer specimens starts very soon after the beginning of the deformation.

Seen from a molecules standpoint this means that only the so-called kinetic fracture theories (Buche-Halpin, Knauss, Kausch, J. G. Williams, etc.) can really describe the failure of polymers.

Seen from a phenomenological standpoint, one has to consider all the parameters which are influencing the mentioned competition and to try to estimate the direction and intensity of their efficiency.

One of the most important parameters influencing the fracture behavior of the viscoelastic polymers is the time which is available for a certain deformation. If this time is long enough to allow a suitable molecular relaxation mechanism to become freely movable, the growth of dangerous cracks may be prevented by stress relaxation, chain gliding, etc., and the rupture elongation may obtain very high values. Therefore, in view of this consideration, it seems reasonable to plot the fracture elongation (ϵ_b) as a function of the time measured from the beginning of the deformation up to the rupture of the specimen, and to compare this plot with the relaxation spectrum of the polymer in question. As we and some other authors discovered, the time positions of brittle-tough transitions and corresponding relaxation maxima agree very well in many cases unless other influences (like big cracks, e.g.) are covering the efficiency of the relaxation mechanism.

A proof of this conception would be the finding that the brittle-tough transitions were shifted by a temperature shift by the same amount and in the same direction as the relaxation maximum. We have found this, indeed, in many cases.

Another confirmation of this mentioned conception follows from the fact that if the relaxation maxima are shifted by other influences (e.g., orientation), the brittle-tough transitions are shifted too.

Andrews replied to Retting in that his comment on fracture is true for simple tensile failure, but in fatigue the cracks are there by way of intrinsic flaws. He agreed on the importance of the impact behavior because impact testing does not produce uniform stress fields. Andrews is developing a shock-tube method where the stress fields are more uniform.

J. Williams stated that the true value of G_{Ic} can be determined from simple impact tests.

7 CONCLUSIONS

Reflections upon this interdisciplinary meeting may be summarized as follows:

Structural-property concepts demand two developments:

(a) A molecular basis for the nonlinear constitutive description of a material

(b) The recognition of morphological form in the physics and mechanical models of polymeric solids. Modern analytical techniques developing out of the mechanics of composites area offer this potential.

Theoretical projections based upon the anisotropic character of the polymer chain, supported by recent fiber-technology developments, indicate that an *order of magnitude improvement* in bulk polymer properties *is* both *technically possible and achievable* by controlled fabrication.

Processing steps based upon the physical-chemistry properties of anisotropic solutions (similar to the helix-coil transition in biological polymers) offers a recognizable fabrication path toward this goal.

Controlled states of multiaxial orientation are desirable and achievable. Much effort is needed in this area, particularly in the techniques needed to describe the orientation patterns after several fabrication steps.

Reliability concepts, life-cycle models and scientific accelerated testing theories must be developed to support the engineering commitment to polymeric solids in structures.

The products of continuum mechanics are applicable to polymer science in that they are independent of any specific molecular model.

The proponents of the mechanics of dislocations and solid-state physics did present some evidence of potential capabilities but failed to present a concrete case that their point of view would provide a path to enhance the properties of bulk polymers. Workers in this area have failed to demonstrate that their model provides a unique description of physical reality.

The academic-engineering community has not recognized a fundamental problem in "molecular engineering": *each new material capability when matched against a usage requirement demands a structure of unique structural form.* Design concepts must be developed in parallel with the science of materials.

The traditional metal-forming techniques of "cold working" via deep drawing, hydrostatic extrusion, and forging are of significant interest to polymer technology. A mastery of viscoelastic-plastic analysis plus the results of Item 1 will be required to harness this potential.

Classical fracture mechanics in a kinetic format is immediately applicable to polymeric solids:

(a) Materials selection and characterization based upon fracture mechanics is practical. However, some work is needed to standardize characterization techniques and specimen dimensions.

(b) The real usefulness of fracture mechanics in design is an open question. Weakness in both analytical concepts and inspection techniques constitute limiting problems. Far-field solutions are generally sufficient for design purposes.

(c) Structural usage of polymers will require extensive fatigue analysis.

(d) Nonlinear constitutive relationships need to be simplified if they are to be used in fracture analysis.

Interdisciplinary working groups are often diffuse and frustrating in character because of:

(a) The resistance of an established group to even consider new ideas and concepts

(b) The tendency of the proponents of new concepts from other areas of scientific endeavor to enter a new field without spending the time and energy necessary to master the disciplines of the field they seek to influence.

The antagonism produced on both groups by these human failings frustrates the true intent of such a gathering – the communication of positions and the critical formulation of positions. Such an atmosphere oftentimes produces a superficial evaluation of issues. Nonetheless, such undertakings are needed and should be supported by the scientific-technical management community.

REFERENCES

1. Gilman, J. J., "A Unified View of Flow Mechanics in Materials", in *Physics of Strength and Plasticity,* A. S. Argon (Ed.), MIT Press (1969).

2. Li, J.C.M., in this volume.
3. Pechhold, W., (a) this volume and (b) in *Molecular Order – Molecular Motion: Their Response to Macroscopic Stress,* H. H. Kausch (Ed.), Polymer Symposium No. 32, Wiley-Interscience (1971).
4. Reneker, D. H., "Point Dislocations in Crystals in High Polymer Molecules", *J. Polymer Sci.,* **59,** 539 (1962).
5. Kedzie, R. W., paper presented at American Physical Society Meeting, Baltimore, March (1962).
6. Flory, P. J., *Statistical Mechanics of Chain Molecules,* Wiley-Interscience (1969).
7. Bueche, F., *Physical Properties of Polymers,* Wiley-Interscience (1962).
8. Ferry, J. D., *Viscoelastic Properties of Polymers,* John Wiley & Sons (1970).
9. Robertson, R. E., "An Equation for the Yield Stress of a Glassy Polymer", in *Polymer Modification of Rubbers and Plastics,* H. Keskkala (Ed.), Applied Polymer Symposium No. 7, Wiley-Interscience (1968).
10. Ashton, J. E., Halpin, J. C., and Petit, P. H., *Primer on Composite Materials: Analysis,* Technomic, Stamford, Conn. (1969).
11. Halpin, J. C., and Nicolais, L., "Materiali Compositi's Relazioni tra Proprieta e Struttura", *Ing. Chem. Italiano,* **7,** 173 (1971).
12. Halpin, J. C., and Kardos, J. L., "Moduli of Crystalline Polymers Employing Composite Theory", *J. Appl. Phys.,* **43,** 2235 (1972).
13. Jerina, K. L., and Halpin, J. C., "Strength of Molded Discontinuous Fiber Composites", AFML-TR-72-148.
14. De Boer, J. H., "The Influence of Van der Waal's Forces and Primary Bonds in Binding Energy Strength and Orientation, with Special Reference to Some Artificial Resins, *Trans. Faraday Soc.,* **32,** 19 (1936).
15. MacMillan, N. H., "The Theoretical Strength of Solids", *J. Mater. Sci.,* **7,** 239 (1972).
16. Frenkel, J., *Z. Physik,* **37,** 572 (1926).
17. Imada, K., Vamamoto, T., Shigematsu, K., and Takayanagi, M., "Crystal Orientation and Some Properties of Solid-State Extrudate of Linear Polyethylene", *J. Mater. Sci.,* **6,** 537 (1971); Pennings, A. J., Schonteten, C.J.H., and Kiel, A. M., "Hydrodynamically Induced Crystallization of Polymers from Solution: V. Tensile Properties of Fibrillar Polyethylene Crystals", *J. Polymer Sci.,* Part 6, No. 38, 167 (1972); Williams, T., "Hydrostatically-Extruded Polypropylene", *J. Mater. Sci.,* **8,** 59 (1973); Yang of Bell Laboratories, results were communicated by H. Mark.
18. Halpin, J. C., Jerina, K. L., and Johnson, T. A., "Characterization of Composites for the Purpose of Reliability Evaluation", ASTM STP 521, Philadelphia, Pa. (1973), p. 3; Halpin, J. C., and Polley, H. W., "Observations on the Fracture of Viscoelastic Bodies", *J. Comp. Mat.,* **1,** 64 (1967).
19. Flory, P. J., *J. Polymer Sci.,* **49,** 105 (1961).
20. Cleerman, K. J., Karam, H. J., and Williams, J. L., *Modern Plastics,* **30(a),** 119 (1953).
21. Cleerman, K. J., "Injection Molding of Shapes of Rotational Symmetry with Multiaxial Orientation", *SPE Journal,* **23,** 43-47 (1967); **25,** 55 (1969).
22. Cleerman, K. J., Schrenk, W. J., and Thomas, L. S., "Bottle Blowing Using Multiaxially Oriented Molded Parisons", *SPE Journal,* **24,** 27-31 (1968).
23. Schrenk, W. J., and Alfrey, T., "Some Physical Properties of Multilayered Films", *Polymer Eng. Sci.,* **9,** 393-399 (1969); Alfrey, T., Gurnee, E. F., and Schrenk, W. J., "Physical Optics of Iridescent Multilayered Films", *Ibid.,* **9,** 400-404 (1969); Schrenk, W. J., and Alfrey, T., "Co-Extrusion and Blown Multilayer Film", presented at the ACS Symposium on Coextruded Plastic Films, Fibers and Composites, April 9-14, 1972; Radford, J. A., Alfrey, T., and Schrenk, W. J., "Reflectivity of Iridescent Coextruded Multilayered Plastic Films", *Polymer Eng. Sci.,* **13,** 216-221 (1973).
24. Li, H. L., Koch, P. J., Prevorsek, D. C., and Oswald, H. J., "Factors Affecting the Depth of Draw in a Cold Forming Operation", *J. Macromol. Sci.-Phys.,* **B4(3),** 687 (1970).
25. Wissbrum, K. F., "Force Requirements in Forging of Crystalline Polymers", *Polymer Eng. Sci.,* **11,** 28 (1971).
26. McKelvey, J. M., *Polymer Processing,* John Wiley & Sons, N. Y. (1962), Chapter 12.4.

AUTHOR INDEX

SUBJECT INDEX

SUBJECT INDEX

SUBJECT INDEX

UBJECT INDEX